Quantum Field Theory and Manifold Invariants

Ω IAS/PARK CITY
MATHEMATICS SERIES
Volume 28

Quantum Field Theory and Manifold Invariants

Daniel S. Freed
Sergei Gukov
Ciprian Manolescu
Constantin Teleman
Ulrike Tillmann
Editors

American Mathematical Society
Institute for Advanced Study

Ian Morrison, Series Editor
Daniel S. Freed, Sergei Gukov, Ciprian Manolescu, Constantin Teleman
and Ulrike Tillmann, Volume Editors.

IAS/Park City Mathematics Institute runs mathematics education programs that bring together high school mathematics teachers, researchers in mathematics and mathematics education, undergraduate mathematics faculty, graduate students, and undergraduates to participate in distinct but overlapping programs of research and education. This volume contains the lecture notes from the Graduate Summer School program

2020 *Mathematics Subject Classification.* Primary 57K16, 57K18, 57K41, 57R56, 81Txx.

Library of Congress Cataloging-in-Publication Data

Names: Freed, Daniel S., editor. | Gukov, Sergei, 1977– editor. | Manolescu, Ciprian, 1978– editor. | Teleman, Constantin, 1968– editor. | Tillmann, Ulrike, 1962- editor.
Title: Quantum field theory and manifold invariants / Daniel S. Freed, Sergei Gukov, Ciprian Manolescu, Constantin Teleman, Ulrike Tillmann, editors.
Description: Providence : American Mathematical Society, 2021. | Series: IAS/Park City mathematics series, 1079-5634 ; Volume 28 | Includes bibliographical references.
Identifiers: LCCN 2021021514 | ISBN 9781470461232 (paperback) | (ebook)
Subjects: LCSH: Quantum field theory–Mathematics. | Invariant manifolds. | AMS: Manifolds and cell complexes – Differential topology – Topological quantum field theories (aspects of differential topology). | Quantum theory – Quantum field theory; related classical field theories.
Classification: LCC QC174.45 .Q3725 2021 | DDC 530.14/3–dc23
LC record available at https://lccn.loc.gov/2021021514

∞ The paper used in this book is acid-free and falls within the guidelines
established to ensure permanence and durability.
Visit the AMS home page at https://www.ams.org/

10 9 8 7 6 5 4 3 2 1 26 25 24 23 22 21

Contents

Preface

The IAS/Park City Mathematics Institute (PCMI) was founded in 1991 as part of the Regional Geometry Institute initiative of the National Science Foundation. In mid-1993 the program found an institutional home at the Institute for Advanced Study (IAS) in Princeton, New Jersey.

The IAS/Park City Mathematics Institute encourages both research and education in mathematics and fosters interaction between the two. The three-week summer institute offers programs for researchers and postdoctoral scholars, graduate students, undergraduate students, high school students, undergraduate faculty, K-12 teachers, and international teachers and education researchers. The Teacher Leadership Program also includes weekend workshops and other activities during the academic year.

One of PCMI's main goals is to make all of the participants aware of the full range of activities that occur in research, mathematics training and mathematics education: the intention is to involve professional mathematicians in education and to bring current concepts in mathematics to the attention of educators. To that end, late afternoons during the summer institute are devoted to seminars and discussions of common interest to all participants, meant to encourage interaction among the various groups. Many deal with current issues in education: others treat mathematical topics at a level which encourages broad participation.

Each year the Research Program and Graduate Summer School focuses on a different mathematical area, chosen to represent some major thread of current mathematical interest. Activities in the Undergraduate Summer School and Undergraduate Faculty Program are also linked to this topic, the better to encourage interaction between participants at all levels. Lecture notes from the Graduate Summer School are published each year in this series. The prior volumes are:

- Volume 1: *Geometry and Quantum Field Theory* (1991)
- Volume 2: *Nonlinear Partial Differential Equations in Differential Geometry* (1992)
- Volume 3: *Complex Algebraic Geometry* (1993)
- Volume 4: *Gauge Theory and the Topology of Four-Manifolds* (1994)
- Volume 5: *Hyperbolic Equations and Frequency Interactions* (1995)
- Volume 6: *Probability Theory and Applications* (1996)
- Volume 7: *Symplectic Geometry and Topology* (1997)
- Volume 8: *Representation Theory of Lie Groups* (1998)
- Volume 9: *Arithmetic Algebraic Geometry* (1999)
- Volume 10: *Computational Complexity Theory* (2000)

- Volume 11: *Quantum Field Theory, Supersymmetry, and Enumerative Geometry* (2001)
- Volume 12: *Automorphic Forms and their Applications* (2002)
- Volume 13: *Geometric Combinatorics* (2004)
- Volume 14: *Mathematical Biology* (2005)
- Volume 15: *Low Dimensional Topology* (2006)
- Volume 16: *Statistical Mechanics* (2007)
- Volume 17: *Analytical and Algebraic Geometry* (2008)
- Volume 18: *Arithmetic of L-functions* (2009)
- Volume 19: *Mathematics in Image Processing* (2010)
- Volume 20: *Moduli Spaces of Riemann Surfaces* (2011)
- Volume 21: *Geometric Group Theory* (2012)
- Volume 22: *Geometric Analysis* (2013)
- Volume 23: *Mathematics and Materials* (2014)
- Volume 24: *Geometry of Moduli Spaces and Representation Theory* (2015)
- Volume 25: *The Mathematics of Data* (2016)
- Volume 26: *Random Matrices* (2017)
- Volume 27: *Harmonic Analysis* (2018)

The American Mathematical Society publishes material from the Undergraduate Summer School in their Student Mathematical Library and from the Teacher Leadership Program in the series IAS/PCMI—The Teacher Program.

After more than 25 years, PCMI retains its intellectual vitality and continues to draw a remarkable group of participants each year from across the entire spectrum of mathematics, from Fields Medalists to elementary school teachers.

Rafe Mazzeo
PCMI Director
March 2021

IAS/Park City Mathematics Series
https://doi.org/10.1090/pcms/028/00873

Introduction

Dan Freed, Sergei Gukov, Ciprian Manolescu, Constantin Teleman, and Ulrike Tillmann

1. Background

The common theme of the lecture series in this volume is the notion of a *topological quantum field theory*. This concept has its origin in physics, where *(classical) field theories* describe the dynamics (evolution in time) of fields on a manifold M. In general, a field can be a section of any given sheaf over M. In practice, most common are scalar fields (maps $M \to \mathbb{R}$), vector fields (sections of the tangent bundle $TM \to M$), and gauge fields (connections in a vector bundle, which appear for example in Maxwell's theory of electromagnetism).

A *quantum field theory* (QFT) is a more complicated physical theory, which typically combines:

- a classical field theory;
- quantum mechanics (including dependence on a parameter $\hbar$, such that the limit $\hbar \to 0$ recovers the classical theory);
- special relativity (including Lorentz invariance).

Quantum field theory is the basis of the Standard Model of particle physics, which has been successfully used to explain three of the four fundamental forces of nature: electromagnetism, the weak interactions, and the strong interactions. The fourth force, gravity, can be understood (at large scales) using Einstein's theory of general relativity. It remains an open problem to formulate a unified theory that combines relativity with the other three forces. The search for unification has led physicists to propose new theories, such as string theory. In the process they studied various quantum field theories that can serve as toy models for some aspects of the unified theory.

One offshoot of the study of QFT's has been the discovery of interesting *topological* quantum field theories (TQFT's), which have captured the attention of mathematicians. Typically, the classical or quantum field theories that appear in physics depend essentially on the choice of a Riemannian metric or Lorentzian metric on the underlying manifold. TQFTs depend only on the topology. On the other hand, dropping the requirement that the manifold is Lorentzian makes the theory interesting in a different way. Indeed, this allows for arbitrary topological models for space-time, i.e., cobordisms between the initial and the final manifold.

2. Definition

Let us sketch the mathematical definition of a TQFT. We emphasize that, while TQFT's arose from the study of QFT's, there is nothing quantum in the definition, so mathematicians sometimes call them simply *topological field theories* (TFT's).

A *symmetric monoidal category* $\mathfrak{C}$ is a category equipped with a notion of tensor product $\otimes$ and isomorphisms

$$A \otimes B \xrightarrow{\cong} B \otimes A,$$

satisfying certain properties which extend the symmetry to a symmetric group action permitting the factors in a product to be of any length. Examples of symmetric monoidal categories include $\mathrm{Vect}_{\mathbf{k}}$ (the category of vector spaces over a field $\mathbf{k}$), R-mod (modules over a ring R), and the cobordism category Cob_{d+1}. The objects of Cob_{d+1} are d-dimensional closed smooth manifolds, the morphisms are smooth $(d+1)$-dimensional cobordisms, and the tensor product is the disjoint union. Then, a $\mathfrak{C}$*-valued TQFT* is a symmetric monoidal functor

$$Z : \mathrm{Cob}_{d+1} \to \mathfrak{C}.$$

Concretely, in a TQFT (say, with values in $\mathrm{Vect}_{\mathbf{k}}$) we assign a vector space $Z(Y)$ to each closed d-manifold, and a homomorphism

$$Z(W) : Z(Y_0) \to Z(Y_1)$$

to each $(d+1)$-dimensional cobordism W between manifolds Y_0 and Y_1. These are required to satisfy some properties, for example functoriality under gluing along a common boundary:

$$Z(W_0 \cup W_1) = Z(W_1) \circ Z(W_0).$$

For the empty set we have $Z(\emptyset) = \mathbf{k}$, and therefore to a closed $(d+1)$-dimensional manifold X we assign an invariant

$$Z(X) \in \mathrm{End}(\mathbf{k}) \cong \mathbf{k}.$$

This definition of a TQFT can be tweaked in many ways, by considering manifolds and cobordisms equipped with various structures, such as: orientations, basepoints and paths, spin structures, spin^c structures, embeddings in some other fixed manifolds. For example, a TQFT for links in $\mathbb{R}^3$ would assign a vector space $Z(L)$ to each link $L \subset \mathbb{R}^3$, and a map $Z(\Sigma) : Z(L_0) \to Z(L_1)$ for each smoothly embedded cobordism $\Sigma \subset \mathbb{R}^3 \times [0,1]$ from L_0 to L_1.

3. Examples

Here are a few examples of TQFT's:

(1) $(1+1)$*-dimensional TQFT's* with values in $\mathrm{Vect}_{\mathbf{k}}$. These consist of a vector space $A = Z(S^1)$ and four operations

$$\mu : A \otimes A \to A, \quad \Delta : A \to A \otimes A, \quad \eta : \mathbf{k} \to A, \quad \varepsilon : A \to \mathbf{k}$$

corresponding to the four cobordisms shown at the top of the next page.

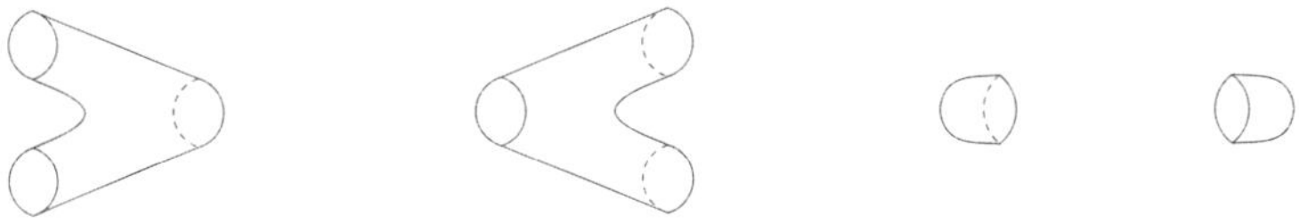

The four operations are called the multiplication, comultiplication, unit, and counit. The TQFT structure imposes certain properties on them, which turns A into a *Frobenius algebra*. Conversely, any Frobenius algebra produces a $(1+1)$-dimensional TQFT.

(2) *Khovanov homology* is a TQFT for links in $\mathbb{R}^3$. This is built by starting from the Frobenius algebra

$$A = \mathbf{k}[x]/(x^2) \cong H^*(S^2;\mathbf{k})$$

with the usual polynomial multiplication μ, the unit 1, and the comultiplication and counit given by

$$\Delta(1) = 1 \otimes x + x \otimes 1,\ \Delta(x) = x \otimes x,\ \varepsilon(1) = 0,\ \varepsilon(x) = 1.$$

The Frobenius algebra A is what is assigned to the unknot. For a general link, Khovanov described a procedure to construct a chain complex by taking a cube of resolutions for a planar diagram, and using the operations μ and Δ to define the chain maps. The homology of this complex is an interesting link invariant, whose Euler characteristic is the celebrated Jones polynomial. Khovanov homology and its generalizations ($\mathfrak{sl}(n)$ Khovanov-Rozansky homologies, HOMFLY-PT homology, colored homologies) are the focus of the Jacob Rasmussen's lectures.

(3) *Invertible TQFT's* are those where the objects and the morphisms are invertible under the tensor product operation. For example, in dimension $1+1$, they correspond to Frobenius algebras such that $A \cong \mathbf{k}$, $\eta = 1$, $\mu : \mathbf{k} \to \mathbf{k} \otimes \mathbf{k} \cong \mathbf{k}$ is given by a unit $\mu \in \mathbf{k}^*$, and Δ by μ^{-1}. In higher dimensions, general TQFT's are hard to classify, whereas it is easier to get a handle on invertible TQFT's, using the tools of homotopy theory. Invertible TQFT's are related to classifying spaces of diffeomorphism groups, and thus to the classification of manifold bundles. These topics are discussed in the lectures of Søren Galatius.

More interesting examples of TQFT's come from actual quantum field theories studied in physics. There are two major kinds of TQFT's that arise from physics, sometimes called *Schwarz-type* and *Witten-type*.

4. TQFT's from path integrals

Schwarz-type TQFT's are obtained via path integrals, i.e., integrals over an (infinite-dimensional) space of paths or connections. It is hard to make mathematical sense of such integrals, but they are fundamental in QFT. The typical example of a Schwarz-type TQFT is Chern-Simons theory, in dimension $2+1$. To a closed 3-manifold X, this theory associates the path integral

$$Z(X) = \int_{\mathcal{A}} e^{2\pi i k\, CS(A)} dA.$$

Here, $\mathcal{A}$ is the space of gauge equivalence classes of connections in the (trivial) SU(2)-bundle over X, and CS is the Chern-Simons functional

$$CS(A) = \frac{1}{4\pi^2} \int_X tr\big(A \wedge dA + \frac{2}{3} A \wedge A \wedge A\big) \in \mathbb{R}/\mathbb{Z}.$$

This definition of $Z(X)$ is due to Witten. Later, Reshetikhin and Turaev gave a rigorous mathematical definition of $Z(X)$, using representation theory and surgery presentations of 3-manifolds in terms of links in $\mathbb{R}^3$. The Witten-Reshetikhin-Turaev (WRT) invariants $Z(X)$ are the main object of study in a branch of mathematics called quantum topology. They admit an extension for links $L \subset X$, where the path integral incorporates Wilson loops. For example, for a link $L \subset S^3$, the resulting invariant is the Jones polynomial.

Recent developments in Chern-Simons theory are the focus of the lectures of Pavel Putrov.

5. TQFT's from supersymmetry

Witten-type TQFT's are obtained from some QFT's by a process called "topological twisting." We will not describe the process in any detail here, but let us mention a few key terms that are appear in this context.

For a QFT to admit a topological twist, one needed ingredient is *supersymmetry* (SUSY). This means that the theory contains fields of two types:

- bosons (those whose spin is an integer), e.g. scalar fields or vector fields. In particle physics, the photon is the standard example of a boson;
- fermions (with half-integer spin), e.g. spinors. In particle physics, electrons, neutrons and protons are all fermions.

Furthermore, the QFT should admit a symmetry that interchanges the bosons and the fermions; this is called a supersymmetry. A QFT may have several supersymmetries. A *supercharge* is an odd element of the Lie algebra of SUSY transformations.

When physicists talk about a 3d $\mathcal{N} = 2$ or 4d $\mathcal{N} = 4$ theory (for example), the number d refers to the dimension of the underlying spacetime, and $\mathcal{N}$ captures the number of independent supercharges. More precisely, the number of independent supercharges is $k\mathcal{N}$, where $k = k(d)$ depends only on d, being the dimension of the minimal super-Poincaré algebra in dimension d. For example, we have $k = 2$ in dimension $d = 3$, and $k = 4$ for $d = 4$. Other cases mentioned below are $d = 2$ and $d = 6$; for those, supersymmetries come in two types, so one talks about $\mathcal{N}$ being a pair of numbers, such as $(2, 2)$.

Another concept familiar to physicists, but sometimes mysterious to mathematicians, is that of *BPS states*, named after Bogomolnyi, Prasad, and Sommerfeld. These are the states in a supersymmetric QFT that are annihilated by a supercharge. Mathematically, they are described as solutions of a first order PDE, and they minimize some energy functional. As a typical example, in Yang-Mills theory, the Yang-Mills equations are of order two:

$$d_A^* F_A = 0,$$

whereas the anti-self-dual (ASD) equations

$$\star F_A = -F_A$$

are of order one and minimize the energy $\int_M |F_A|^2 dvol$. The solutions to the ASD equations are BPS states, as are many other interesting mathematical objects: Seiberg-Witten monopoles, J-holomorphic curves, etc.

It is worth mentioning here that the limit as $\hbar \to 0$ of a quantum system is a classical system, described by solutions to differential equations. In a TQFT of Witten-type, since there is no metric dependence, there is also no dependence on $\hbar$. That is why such theories are formulated in terms of differential equations.

Supersymmetry and BPS states are discussed further in Andrew Neitzke's lectures. In particular, these lectures include a detailed description of $\mathcal{N} = (2,2)$ supersymmetry in dimension $d = 2$.

In the next subsections we mention various other supersymmetric theories that are of interest to mathematicians.

Pure $SU(2)$ super Yang-Mills theory This is a 4d $\mathcal{N} = 2$ theory, which admits a topological twist giving rise to a Witten-type TQFT. At high energy (i.e., small distance scales), the topological theory is called *Donaldson theory*, and is described by the ASD equations. At low energy (high distance scales), it gives rise to *Seiberg-Witten theory*, described by the Seiberg-Witten equations.

The Donaldson and Seiberg-Witten theories are fundamental tools in low dimensional topology. They give rise to numerical invariants of smooth 4-manifolds, which are able to detect exotic smooth structures (pairs of manifolds homeomorphic, but not diffeomorphic to each other). Since they are two aspects of the same theory, the Seiberg-Witten and Donaldson invariants are closely related, by a formula called Witten's Conjecture (proved in many cases by Feehan and Leness).

The Seiberg-Witten and ASD equations have similar properties, such as invariance under an infinite-dimensional gauge group. The study of these and other gauge-invariant equations forms the object of *mathematical gauge theory*, which is discussed (with a particular focus on Seiberg-Witten theory) in the lectures of Andriy Haydys. In dimension 3, the Donaldson and Seiberg-Witten TQFT's associate to a 3-manifold Abelian groups called *instanton Floer homology* and *monopole Floer homology*, respectively. These have many topological applications, to questions about surgery, cobordisms, contact structures, foliations, etc. Unlike for the 4-dimensional invariants, the relation between instanton and monopole Floer homology is still a mystery. Instanton Floer homology is the topic of the lectures of Tomasz Mrowka.

Seiberg-Witten theory has a symplectic-geometric counterpart developed by Ozsváth and Szabó. To a three-manifold Y, they associated an invariant called *Heegaard Floer homology*, defined by decomposing Y into two handlebodies (glued along a surface Σ), and then using J-holomorphic curves on a symmetric product of Σ. It is now known that the Heegaard Floer and monopole Floer homology are isomorphic. Ozsváth and Szabó also associated numerical invariants to four-manifolds, which are conjecturally the same as the Seiberg-Witten invariants. There is also a version of Heegaard Floer homology for knots in 3-manifolds, called *knot Floer homology*. Heegaard Floer theory is the topic of Jennifer Hom's lectures.

More 4d $\mathcal{N} = 2$ theories In addition to pure super Yang-Mills theory, there is a whole family of 4d $\mathcal{N} = 2$ theories, called *of class S*. These are parametrized by triples $(\mathfrak{g}, C, D)$ where $\mathfrak{g}$ is a Lie algebra of ADE type, C is a Riemann surface with punctures, and D is certain data at the punctures. For example, pure $SU(2)$ super Yang-Mills theory corresponds to $\mathfrak{g} = \mathfrak{su}(2)$, the planar surface $C = \mathbb{P}^1 \setminus \{0, \infty\}$, and some data D.

Theories of class S are all derived from a superconformal 6d $\mathcal{N} = (2,0)$ theory, called "theory $\mathcal{X}$" or "fivebrane theory". This involves considering theory $\mathcal{X}$ on $M \times C$ ("compactifying on C"), where M is a 4-manifold and C is the punctured Riemann surface above.

If we write a theory of class S on a product $M = Y \times S^1$, we get a 3d theory that can be described in terms of maps $Y \to \mathfrak{M}_{\text{Higgs}}(C)$, where $\mathfrak{M}_{\text{Higgs}}(C)$ is the *Hitchin moduli space* of Higgs bundles on C. The Hitchin moduli space has a hyperkähler structure, and appears in various guises in different branches of mathematics. If C is a closed Riemann surface and $\mathfrak{g} = \mathfrak{su}(2)$, then the Hitchin moduli space is homeomorphic to the $SL(2, \mathbb{C})$ character variety of C:

$$\mathfrak{M}_{\text{Higgs}}(C) \cong \{\pi_1(C) \to SL(2, \mathbb{C})\}/\text{ conjugation}$$

The mathematics around the Hitchin moduli space is discussed in the lectures of by Laura Schaposnik.

The count of BPS states in the 3d theory (coming from the 4d theory class of S) gives information about the hyperkähler metric on $\mathfrak{M}_{\text{Higgs}}(C)$. A discussion of this appears in Andy Neitzke's lectures.

4d $\mathcal{N} = 4$ super Yang-Mills theory This is obtained from the 6d $\mathcal{N} = (2,0)$ theory $\mathcal{X}$ by writing it on $M^4 \times T^2$, i.e., compactifying it on a torus T^2. The 4d $\mathcal{N} = 4$ super Yang-Mills theory admits several topological twists, giving rise to some interesting gauge-invariant PDE such as the Kapustin-Witten and Vafa-Witten equations. These have recently been studied by mathematicians, with an eye toward new topological applications (beyond those coming from Donaldson, Seiberg-Witten, or Heegaard Floer theory). A specific proposal in this direction was made by Witten: interpret the coefficients of the Jones polynomial (of knots in $\mathbb{R}^3$) as counts of solutions to the Kapustin-Witten equations on $\mathbb{R}^3 \times (0, \infty)$,

IAS/Park City Mathematics Series
Volume 28, Pages 1–76
https://doi.org/10.1090/pcms/028/00874

Introduction to Gauge Theory

A. Haydys

Abstract. I describe the basics of a gauge-theoretic approach to construction of invariants of manifolds. The main example considered here is the Seiberg–Witten gauge theory. However, I tried to present the material in a form, which is suitable for other gauge-theoretic invariants too.

Contents

2020 *Mathematics Subject Classification.* Primary 53C05, 53C07, 58E15; Secondary 57R57, 58B15, 58J05.
Key words and phrases. gauge theory, Fredholm map, Seiberg–Witten theory.

1. Introduction

Gauge theory by now is a vast subject with many connections in geometry, analysis, and physics. In these notes I focus on gauge theory as it is used in the construction of manifolds invariants, other uses of gauge theory remain beyond the scope of these notes.

The basic scheme for constructing invariants of manifolds via gauge theory is quite simple. Let me describe some concrete details. Thus, let $\mathcal{G}$ be a Lie group acting on a manifold $\mathcal{C}$. The common convention, which I will follow, is that $\mathcal{G}$ acts *on the right*, although this is clearly nonessential. The quotient $\mathcal{B} := \mathcal{C}/\mathcal{G}$ may fail to be a nice space[1] in a few ways, for example due to the presence of points with nontrivial stabilizers. Denote by $\mathcal{C}_{irr} \subset \mathcal{C}$ the subspace consisting of all points with the trivial stabilizer. Then $\mathcal{G}$ acts on $\mathcal{C}_{irr}$ and the quotient $\mathcal{B}_{irr} := \mathcal{C}_{irr}/\mathcal{G}$ is better behaved. Let me assume that $\mathcal{B}_{irr}$ is in fact a manifold.

Pick a $\mathcal{G}$–representation V and a smooth $\mathcal{G}$–equivariant map $F\colon \mathcal{C} \to V$, where the equivariancy means the following:

$$F(a \cdot g) = g^{-1} \cdot F(a).$$

Here V is thought of as a left $\mathcal{G}$–module.

More to the point, assuming that $0 \in V$ is a regular value of F, we obtain a submanifold $\mathcal{M}_{irr} := F^{-1}(0) \cap \mathcal{C}_{irr}/\mathcal{G} \subset \mathcal{B}_{irr}$. Assume further $d := \dim \mathcal{M}_{irr} < \infty$. The space $\mathcal{M}_{irr}$ will be referred to as the 'moduli space', although the terminology may seem odd at this moment. If $\mathcal{M}_{irr}$ is compact and oriented, it has the fundamental class $[\mathcal{M}_{irr}] \in H_d(\mathcal{B}_{irr}; \mathbb{Z})$. In particular, for any cohomology class $\eta \in H^d(\mathcal{B}_{irr}; \mathbb{Z})$ we obtain an integer

$$\langle [\mathcal{M}_{irr}], \eta \rangle = \int_{\mathcal{M}_{irr}} \eta,$$

where η is thought of as a closed form of degree d on $\mathcal{B}_{irr}$. This is the 'invariant' we are interested in.

One way to construct cohomology classes on $\mathcal{B}_{irr}$ is as follows. Assume there is a normal subgroup $\mathcal{G}_0 \subset \mathcal{G}$ so that $G := \mathcal{G}/\mathcal{G}_0$ is a Lie group. Then the 'framed moduli space' $\hat{\mathcal{M}}_{irr} := F^{-1}(0) \cap \mathcal{C}_{irr}/\mathcal{G}_0$ is equipped with an action of $\mathcal{G}/\mathcal{G}_0 = G$ such that $\hat{\mathcal{M}}_{irr}/G = \mathcal{M}_{irr}$. In other words, $\hat{\mathcal{M}}_{irr}$ can be viewed as a principal

[1] Ideally, one wishes the quotient to be a smooth manifold.

G bundle, whose characteristic classes yield the cohomology classes we are after. More details on this are provided in Section 3.

While this scheme is clearly very general, the details in each particular case may differ to some extent. The aim of these notes is to explain this basic scheme in some detail rather than variations enforced by a concrete setup. As an illustration, I consider the Seiberg–Witten theory in dimension four, where this scheme works particularly well. Due to the tight timeframe of the lectures, this will be essentially the only gauge–theoretic problem considered in these notes.

At present a number of monographs in gauge theory is available, for example [6, 7, 9, 16, 20, 21] just to name a few. All of them are however pretty much specialized to a concrete setting or problem while many features are common to virtually any theory, which is based on counting solutions of non-linear elliptic PDEs, even not necesserily of gauge–theoretic origin. While the presentation here does not differ substantially from the treatment of [7, 21], I tried to keep separate general principles from peculiar features of concrete problems. Also, I hopefully streamlined somewhat some arguments, for example the proof of the compactness for the Seiberg–Witten moduli space. Although I did not cover any other examples of gauge–theoretic invariants except the Seiberg–Witten invariant, the reader will be hopefully well prepared to read advanced texts on other enumerative invariants such as [7] on his own.

The reader may ask why should we actually care about gauge theories beyond the Seiberg–Witten one, since most of the results obtained by gauge–theoretic means can be reproduced with the help of the Seiberg–Witten gauge theory. One answer is that the Seiberg–Witten theory does not generalize to higher dimensions, whereas for example the anti–self–duality equations do generalize and are currently a topic of intense research. Even in low dimensions physics suggests that gauge theories based on non-abelian groups such as $SL(2,\mathbb{C})$ or more traditional $SU(2)$ may provide insights that the Seiberg–Witten theory is not capable of. For example, the Kapustin–Witten equations play a central role in Witten's approach to the construction of the Jones polynomial.

These notes are organized as follows. In introductory Sections 2–4, I describe those properties of the 'points' in $\mathcal{C}$ that are important for the intended applications. Sections 5 and 6 are devoted to analytic tools used in the studies of gauge–theoretic moduli spaces. This can be viewed as studies of properties of F from a (somewhat) abstract point of view. The concepts and theorems considered in this part are not really specific to gauge theory, the same toolbox is used in virtually any enumerative problem based on elliptic PDEs. The last section is devoted to the Seiberg–Witten theory.

I tried to keep these notes self-contained where this did not require long and technical detours. I hope this will make the notes available to early graduate students. Part of such course is necessarily certain tools from PDEs. I tried to state clearly the facts that play a role in the main part of these notes, however

I did not intend to give a detailed description or complete proofs of those. The reader may wish to consult a more specialized literature on this topic, for example relevant chapters of [8, 24].

2. Bundles and connections

2.1. Vector bundles

Basic notions. Roughly speaking, a vector bundle is a family of vector spaces parametrized by points of a manifold (or, more generally, of a topological space).

More formally, the notion of a vector bundle is defined as follows.

Definition 2.1.1. Choose a non-negative integer k. A real smooth vector bundle of rank k is a triple (π, E, M) such that the following holds:

(i) E and M are smooth manifolds, $\pi\colon E \to M$ is a smooth submersion (the differential is surjective at each point);

(ii) For each $m \in M$ *the fiber* $E_m := \pi^{-1}(m)$ has the structure of a vector space and $\dim E_m = k$;

(iii) For each $m \in M$ there is a neighborhood $U \ni m$ and a smooth map ψ_U such that the following diagram

$$\begin{array}{ccc} \pi^{-1}(U) & \xrightarrow{\psi_U} & U \times \mathbb{R}^k \\ & {\scriptstyle\pi}\searrow \quad \swarrow{\scriptstyle pr_1} & \\ & U & \end{array}$$

commutes. Moreover, ψ_U is a fiberwise linear isomorphism.

The following terminology is commonly used: E is the total space, M is the base, π is the projection, and ψ_U is the local trivialization (over U).

Example 2.1.2.

(a) The product bundle: $M \times \mathbb{R}^k$;

(b) The tangent bundle TM of any smooth manifold M.

Let E and F be two vector bundles over a common base M. A *homomorphism* between E and F is a smooth map $\varphi\colon E \to F$ such that the diagram

$$\begin{array}{ccc} E & \xrightarrow{\varphi} & F \\ & {\scriptstyle\pi_E}\searrow \quad \swarrow{\scriptstyle\pi_F} & \\ & M & \end{array}$$

commutes and φ is a fiberwise linear map.

Two bundles E and F are said to be isomorphic, if there is a homomorphism φ, which is fiberwise an isomorphism.

A bundle E is said to be *trivial*, if E is isomorphic to the product bundle.

Operations on vector bundles. Let E and F be two vector bundles over a common base M. Then we can construct new bundles E^*, $\Lambda^p E$, $E \oplus F$, $E \otimes F$, and $\mathrm{Hom}(E, F)$ as follows:

(∗) $(E^*)_m = (E_m)^*$;
(Λ) $(\Lambda^p E)_m = \Lambda^p(E_m)$;
(⊕) $(E \oplus F)_m := E_m \oplus F_m$;
(⊗) $(E \otimes F)_m := E_m \otimes F_m$;
(Hom) $\mathrm{Hom}(E, F)_m := \mathrm{Hom}(E_m, F_m)$.

If $f\colon M' \to M$ is a smooth map, we can define the *pull-back* of $E \to M$ via

$$(f^*E)_{m'} := E_{f(m')}.$$

For example, if M' is an open subset of M and ι is the inclusion, then $E|_{M'} := \iota^* E$ is just the restriction of E to M'.

The reader should check that the families of vector spaces defined above satisfy the properties required by Definition 2.1.1.

Exercise 2.1.3. Prove that $E^* \otimes F$ is isomorphic to $\mathrm{Hom}(E, F)$.

Exercise 2.1.4. Prove that the tangent bundle of the 2-sphere is non-trivial. (Hint: Apply the hairy ball theorem).

Sections.

Definition 2.1.5. A smooth map $s\colon M \to E$ is called a section, if $\pi \circ s = \mathrm{id}_M$.

In other words, a section assigns to each point $m \in M$ a vector $s(m) \in E_m$ such that $s(m)$ depends smoothly on m.

Sections of the tangent bundle TM are called vector fields. Sections of $\Lambda^p T^*M$ are called differential p-forms.

Example 2.1.6. If f is a smooth function on M, then the differential df is a 1-form on M. More generally, given functions $f_1, \dots, f_p$ we can also construct a p-form $\omega := df_1 \wedge \dots \wedge df_p$.

Exercise 2.1.7. Let $E \to M$ be a vector bundle of rank k and $U \subset M$ be an open subset. Prove that E is trivial over U if and only if there are k sections $e = (e_1, \dots, e_k)$, $e_j \in \Gamma(U; E)$, such that $e(m)$ is a basis of E_m for each $m \in U$. More precisely, given e show that ψ_U can be constructed according to the formula

$$\psi_U^{-1}\colon U \times \mathbb{R}^k \longrightarrow E|_U, \qquad (m, x) \mapsto e(m) \cdot x.$$

In fact this establishes a one-to-one correspondence between k-tuples of pointwise linearly independent sections and local trivializations of E.

We denote by $\Gamma(E) = \Gamma(M; E)$ the space of all smooth sections of E. Clearly, $\Gamma(E)$ is a vector space, where the addition and multiplication with a scalar are defined pointwise. In fact, $\Gamma(E)$ is a $C^\infty(M)$-module.

For a local trivialization e over U (cf. Exercise 2.1.7) a section s, we can write

$$s(m) = \sum_{j=1}^{k} \sigma_j(m) e_j(m)$$

for some functions $\sigma_j\colon U \to \mathbb{R}$. Thus, locally any section of a vector bundle can be thought of as a map $\sigma\colon U \to \mathbb{R}^k$.

It is important to notice that σ depends on the choice of a local trivialization. Indeed, if e' is another local trivialization of E over U', then there is a map

$$g\colon U \cap U' \longrightarrow GL_n(\mathbb{R}) \quad \text{such that} \quad e = e' \cdot g. \tag{2.1.8}$$

If $\sigma'\colon U' \to \mathbb{R}^k$ is a local representation of s with respect to e', we have

$$s = e'\sigma' = eg^{-1}\sigma' = e\sigma \qquad \Longrightarrow \qquad \sigma' = g\sigma.$$

Covariant derivatives. The reader surely knows from the basic analysis course that the notion of the derivative is very useful. It is natural to ask whether there is a way to differentiate sections of bundles too.

To answer this question, recall the definition of the derivative of a function $f\colon M \to \mathbb{R}$. Namely, choose a smooth curve $\gamma\colon (-\varepsilon, \varepsilon) \to M$ and let $m := \gamma(0)$ and $v := \dot\gamma(0)$ in T_mM. Then

$$df(v) = \lim_{t\to 0} \frac{f(\gamma(t)) - f(m)}{t}. \tag{2.1.9}$$

Trying to replace f by a section s of a vector bundle, we immediately run into a problem, namely the difference $s(\gamma(t)) - s(m)$ is ill-defined in general since these two vectors may lie in different vector spaces.

Hence, instead of trying to mimic (2.1.9) we will define the derivatives of sections axiomatically, namely asking that the most basic property of the derivative—the Leibniz rule—holds.

Definition 2.1.10. Let $E \to M$ be a vector bundle. *A covariant derivative* is an $\mathbb{R}$-linear map

$$\nabla\colon \Gamma(E) \to \Gamma(T^*M \otimes E)$$

such that

$$\nabla(fs) = df \otimes s + f\nabla s \tag{2.1.11}$$

holds for all $f \in C^\infty(M)$ and all $s \in \Gamma(E)$.

Example 2.1.12. Let $M \subset \mathbb{R}^N$ be an embedded submanifold. Then the tangent bundle TM is naturally a subbundle of the product bundle $\underline{\mathbb{R}}^N := M \times \mathbb{R}^N$. In particular, any section s of TM can be regarded as a map $M \to \mathbb{R}^N$. With this at hand we can define a connection on TM as follows

$$\nabla s := \mathrm{pr}(ds),$$

where pr is the orthogonal projection onto TM. A straightforward computation shows that this satisfies the Leibniz rule, i.e., ∇ is a connection indeed.

Theorem 2.1.13. *For any vector bundle $E \to M$ the space of all connections $\mathcal{A}(E)$ is an affine space modelled on $\Omega^1(\mathrm{End}\, E) = \Gamma\big(T^*M \otimes \mathrm{End}(E)\big)$.*

To be somewhat more concrete, the above theorem consists of the following statements:

(a) $\mathcal{A}(E)$ is non-empty.

(b) For any two connections ∇ and $\hat{\nabla}$ the difference $\nabla - \hat{\nabla}$ is a 1-form with values in $\mathrm{End}(E)$;

(c) For any $\nabla \in \mathcal{A}(E)$ and any $a \in \Omega^1(\mathrm{End}\, E)$ the following

$$(\nabla + a)s := \nabla s + as$$

is a connection.

For the proof of Theorem 2.1.13 we need the following elementary lemma, whose proof is left as an exercise.

Lemma 2.1.14. *Let* $A\colon \Gamma(E) \to \Omega^p(F)$ *be an* $\mathbb{R}$*-linear map, which is also* $C^\infty(M)$*-linear, that is,*

$$A(fs) = fA(s) \qquad \forall f \in C^\infty(M) \quad \textit{and} \quad \forall s \in \Gamma(E).$$

Then there exists $a \in \Omega^p\big(\mathrm{Hom}(E,F)\big)$ *such that* $A(s) = a \cdot s$. □

Proof of Theorem 2.1.13. Notice first that $\mathcal{A}(E)$ is convex, i.e., for any $\nabla, \hat{\nabla} \in \mathcal{A}(E)$ and any $t \in [0,1]$ the following $t\nabla + (1-t)\hat{\nabla}$ is also a connection.

If ψ_U is a local trivialization of E over U, then we can define a connection ∇_U on $E|_U$ by declaring

$$\nabla_U s := \psi_U^{-1}\, d(\psi_U(s)).$$

Using a partition of unity and the convexity property, a collection of these local covariant derivatives can be sewed into a global covariant derivative just like in the proof of the existence of Riemannian metrics on manifolds, cf. [2, Thm. 3.3.7]. This proves (a).

By (2.1.11), the difference $\nabla - \hat{\nabla}$ is $C^\infty(M)$–linear. Hence, (b) follows by Lemma 2.1.14. The remaining step, namely (c), is straightforward. This finishes the proof of this theorem. □

While Theorem 2.1.13 answers the question of the existence of connections, the reader may wish to have a more direct way to put his hands on a connection. One way to do this is as follows.

Let e be a local trivialization. Since e is a pointwise base we can write

$$\nabla e = e \cdot A, \tag{2.1.15}$$

where

$$A = A(\nabla, e)$$

is a $k \times k$-matrix, whose entries are 1-forms defined on U. A is called the connection matrix of ∇ with respect to e.

If σ is a local representation of a section s, then

$$\nabla s = \nabla(e\sigma) = \nabla(e)\sigma + e \otimes d\sigma = e\big(A\sigma + d\sigma\big).$$

Hence, it is common to say that locally

$$\nabla = d + A,$$

which means that $d\sigma + A\sigma$ is a local representation of ∇s. In particular, ∇ is uniquely determined by its connection matrix over U and any $A \in \Omega^1(U; \mathfrak{gl}_k(\mathbb{R}))$ appears as a connection matrix of some connection (cf. Theorem 2.1.13).

Just like σ, the connection matrix also depends on the choice of e. If $e' = eg$, we have

$$\nabla e' = \nabla(eg) = (\nabla e)g + e \otimes dg = e(Ag + dg) = e'(g^{-1}Ag + g^{-1}dg).$$

Hence, the connection matrix A' of ∇ with respect to e' can be expressed as:

$$A' = g^{-1}Ag + g^{-1}dg. \tag{2.1.16}$$

The curvature. While Definition 2.1.10 yields a way to differentiate sections, some properties very well known from multivariable analysis are *not* preserved. One of the most important cases is that covariant derivatives with respect to two variables do not need to commute. The failure of the commutativity of partial covariant derivatives is closely related to the notion of curvature, which is described next.

Denote $\Omega^p(E) := \Gamma(\Lambda^p T^*M \otimes E)$. We can extend the covariant derivative ∇ to a map $d_\nabla\colon \Omega^p(E) \to \Omega^{p+1}(E)$ as follows. If $s \in \Gamma(E)$ and $\omega \in \Omega^p(M)$ we declare

$$d_\nabla(\omega \otimes s) := d\omega \otimes s + (-1)^p \omega \wedge \nabla s, \tag{2.1.17}$$

which yields a unique $\mathbb{R}$-linear map d_∇ defined on all of $\Omega^p(E)$. This satisfies the following variant of the Leibniz rule:

$$d_\nabla(\alpha \wedge \omega) = d\alpha \wedge \omega + (-1)^q \alpha \wedge d_\nabla \omega, \quad \text{for all } \alpha \in \Omega^q(M) \text{ and } \omega \in \Omega^p(M; E).$$

Thus, we obtain a sequence

$$0 \to \Omega^0(E) \xrightarrow{d_\nabla = \nabla} \Omega^1(E) \xrightarrow{d_\nabla} \cdots \to \Omega^n(E) \to 0,$$

where $n = \dim M$. However, unlike in the case of the de Rham differential, the above does not need to be a complex.

Proposition 2.1.18. *There is a 2-form* F_∇ *with values in* End E *such that*

$$d_\nabla \circ d_\nabla = F_\nabla. \tag{2.1.19}$$

This equality means that for any $\omega \in \Omega^p(E)$ we have $d_\nabla(d_\nabla(\omega)) = F_\nabla \wedge \omega$, where the right hand side of the letter equality is a combination of the wedge-product and the natural contraction $\mathrm{End}(E) \otimes E \to E$.

Proof of Proposition 2.1.18. We prove the proposition for $p = 0$ first. Applying the Leibniz rule twice, we see that $d_\nabla \circ \nabla\colon \Omega^0(E) \to \Omega^2(E)$ is $C^\infty(M)$–linear. Hence, by Lemma 2.1.14 we obtain a 2-form F_∇ such that

$$d_\nabla(\nabla s) = F_\nabla s$$

holds for any $s \in \Omega^0(E)$.

It remains to consider the case $p > 0$. For any $\eta \in \Omega^p(M)$ and $s \in \Gamma(E)$ we have

$$d_\nabla(d_\nabla(\eta \otimes s)) = d_\nabla(d\eta \otimes s + (-1)^p \eta \wedge \nabla s)$$

$$= 0 + (-1)^{p+1} d\eta \wedge \nabla s + (-1)^p d\eta \wedge \nabla s + (-1)^{2p} \eta \wedge d_\nabla(\nabla s)$$
$$= \eta \wedge F_\nabla s$$
$$= F_\nabla \wedge (\eta \otimes s).$$

Here the last equality holds because F_∇ is a differential form of even degree. □

Definition 2.1.20. The 2-form F_∇ defined by (2.1.19) is called the curvature form of ∇.

Our next aim is to clarify somewhat the meaning of the curvature form. If v is a tangent vector at some point $m \in M$, we call

$$\nabla_v s := \iota_v \nabla s$$

the covariant derivative of s in the direction of v.

Choose local coordinates $(x_1, \dots, x_n)$ and denote by $\partial_i = \frac{\partial}{\partial x_i}$ the tangent vector of the curve

$$\gamma(t) = (x_1^0, \dots, x_{i-1}^0, t, x_{i+1}^0, \dots, x_n^0).$$

We may call

$$\nabla_i s := \nabla_{\partial_i} s$$

"partial covariant derivatives". With these notations at hand we can write

$$\nabla s = \sum_{i=1}^n dx_i \otimes \nabla_i s,$$

which is just a form of the familiar expression for the differential of a function: $df = \sum \frac{\partial f}{\partial x_i} dx_i$.

Furthermore, we have

$$\begin{aligned} d_\nabla(\nabla s) &= -\sum_{i=1}^n dx_i \wedge \nabla(\nabla_i s) \\ &= -\sum_{i,j=1}^n dx_i \wedge dx_j \otimes \nabla_j(\nabla_i s) \\ &= \sum_{i,j=1}^n dx_i \wedge dx_j \otimes \nabla_i(\nabla_j s). \end{aligned}$$

From this we conclude

$$F_\nabla(\partial_i, \partial_j) = \nabla_i(\nabla_j s) - \nabla_j(\nabla_i s),$$

i.e., the curvature form measures the failure of partial covariant derivatives to commute as mentioned at the beginning of this section.

Sometimes it is useful to have an expression of the curvature form in a local frame. Thus, pick a local frame e of E and let A be the connection form of ∇ with respect to e. If σ is a local representation of some $s \in \Gamma(E)$, then we have

$$d_\nabla(\nabla(e\sigma)) = d_\nabla(e(d\sigma + A\sigma)) = \nabla e \wedge (d\sigma + A\sigma) + e(dA\,\sigma - A \wedge d\sigma)$$

$$= e\Big(A\wedge d\sigma + A\wedge A\,\sigma + dA\,\sigma - A\wedge d\sigma\Big)$$
$$= e\big(dA + A\wedge A\big)\,\sigma.$$

Hence, we conclude that locally

$$F_\nabla = dA + A\wedge A. \tag{2.1.21}$$

In particular, the curvature form is a first order non-linear operator in terms of the connection form.

Remark 2.1.22. It turns out to be useful to think of A as a 1–form with values in the Lie algebra $\mathfrak{gl}_k(\mathbb{R}) = \mathrm{End}(\mathbb{R}^k)$. From this perspective it is more suitable to write (2.1.21) in the following form

$$F_\nabla = dA + \frac{1}{2}[A\wedge A], \tag{2.1.23}$$

where the last term is a combination of the wedge product and the Lie brackets.

If e' is another local frame such that $e = e'\cdot g$, then (2.1.16) one can show that

$$F'_\nabla = dA' + A'\wedge A' = g^{-1}F_\nabla\, g.$$

The reader is strongly encouraged to check the details of this computation.

Local expression (2.1.21) implies immediately the following.

Proposition 2.1.24. *If $a \in \Omega^1(\mathrm{End}\,E)$, then the curvature forms of ∇ and $\nabla + a$ are related by the equality*

$$F_{\nabla+a} = F_\nabla + d_\nabla a + a\wedge a,$$

where d_∇ is defined on Page 8. □

The gauge group. Pick a vector bundle E and consider the set

$$\mathcal{G} = \mathcal{G}(E) := \Big\{g \in \Gamma\big(\mathrm{End}(E)\big) \mid \forall m \in M\ g(m) \in GL(E_m)\Big\},$$

which is endowed with the C^∞–topology. The elements of $\mathcal{G}$ are called *gauge transformations*.

The pointwise multiplication defines a group structure on $\mathcal{G}$. Thus, $\mathcal{G}$ is a topological group, which is called *the gauge group*.

If E is endowed with an extra structure, for example an orientation or a scalar product, we also require that gauge transformations respect this structure.

If ∇ is a connection on E and $g \in \mathcal{G}$, we can define another connection by

$$\nabla^g s := g^{-1}\nabla(gs).$$

This yields a right action of $\mathcal{G}$ on $\mathcal{A}(E)$.

Definition 2.1.25. Two connections are called gauge equivalent, if there is a gauge transformation that transforms one of the connections into the other one.

Pick a local trivialization e of E over an open subset U. Let $A = A(e,\nabla)$ in $\Omega^1(U;\mathfrak{gl}_k(\mathbb{R}))$ be a local representation of ∇, i.e., $\nabla e = e\cdot A$. Let us compute the local representation of ∇^g. Notice first, that using e we may think of g as a map $U \to GL_k(\mathbb{R})$, i.e., $g(e) = e \cdot g$. Then

$$\begin{aligned} \nabla^g e &= g^{-1}\big(\nabla(e\cdot g)\big) = g^{-1}(e\cdot Ag + e\cdot dg) = g^{-1}(e)\cdot(Ag+dg) \\ &= e\cdot\big(g^{-1}Ag + g^{-1}dg\big). \end{aligned}$$

We conclude

$$A(\nabla^g, e) = g^{-1}Ag + g^{-1}dg = A(\nabla,\ e\cdot g),$$

where the last equality follows by (2.1.16). Thus, (2.1.16) expresses both the change of the connection matrix under the change of local trivializations and the action of the gauge group.

2.2. Principal bundles The computations we met in the previous sections on the dependence of our objects on the choice of a local frame become formidable quite soon. The notion of a principal bundle is useful in dealing with this and turns out to have other advantages as we will see below. The idea is to consider all possible frames at once rather than choosing local trivializations when needed.

The frame bundle and the structure group. Let G be a Lie group.

Definition 2.2.1. A principal bundle with structure group G is a triple (P, M, π), where

(i) P and M are smooth manifolds and $\pi\colon P \to M$ is a surjective submersion;
(ii) G acts on P on the right such that $\pi(p\cdot g) = \pi(p)$ for all $p \in P$ and all $g \in G$;
(iii) G acts freely and transitively on each fiber $\pi^{-1}(m)$;
(iv) For each $m \in M$ there is a neighborhood $U \ni m$ and a map ψ_U such that the diagram

$$\begin{array}{ccc} \pi^{-1}(U) & \xrightarrow{\ \psi_U\ } & U\times G \\ & {\scriptstyle\pi}\searrow \quad \swarrow{\scriptstyle pr_1} & \\ & U & \end{array}$$

commutes. Moreover, ψ_U is G–equivariant $\psi_U(p\cdot g) = \psi_U(p)\cdot g$, where G acts on $U\times G$ by the multiplication on the right on the second factor.

A fundamental example of a principal bundle is the frame bundle of a vector bundle $E \to M$. We take a moment to describe the construction of this bundle in some detail.

Thus, for a fixed $m \in M$ let $\mathrm{Fr}(E_m)$ denote the set of all bases of E_m. The group $GL_k(\mathbb{R})$ acts freely and transitively on $\mathrm{Fr}(E_m)$ so that we can in fact identify $\mathrm{Fr}(E_m)$ with $GL_k(\mathbb{R})$ even though in a non-canonical way. In any case, $\mathrm{Fr}(E_m)$ can be viewed as an open subset of $\mathbb{R}^{k^2}$.

Consider

$$\mathrm{Fr}(E) := \bigsqcup_{m\in M} \mathrm{Fr}(E_m).$$

Clearly, there is a well-defined projection $\pi\colon \operatorname{Fr}(E) \to M$ determined uniquely by the property: $\pi(p) = m$ if and only if $p \in \operatorname{Fr}(E_m)$.

We introduce a smooth structure on $\operatorname{Fr}(E)$ as follows. Pick a chart U on M. By shrinking U if necessary we can assume that there is a local frame e of E defined on U. Then we have the bijective map

$$\Psi_U\colon U \times GL_k(\mathbb{R}) \longrightarrow \pi^{-1}(U), \qquad (m, h) \mapsto e(m) \cdot h. \tag{2.2.2}$$

Using this we can think of $\pi^{-1}(U)$ as an open subset of an Euclidean space so that we can declare $\pi^{-1}(U)$ to be a chart on $\operatorname{Fr}(E)$ with an obvious choice of coordinates.

Let U' be another chart on M such that there is a local frame e' defined on U'. A straightforward computation yields

$$\Psi_{U'}^{-1} \circ \Psi_U(m, h) = \big(m,\ g(m)h\big),$$

where g is defined by (2.1.8). This implies that the transition maps between $\pi^{-1}(U)$ and $\pi^{-1}(U')$ are smooth, i.e., we have constructed a smooth atlas on $\operatorname{Fr}(U)$. The rest of the properties required in the definition of the principal bundle are clear from the construction.

Exercise 2.2.3. Show that local triviality of a principal bundle, i.e., Property (iv) of Definition 2.2.1, is equivalent to the existence of local sections. More precisely, if P admits a trivialization over U, then there is a section of $P|_U$ and conversely, if $P|_U$ admits a section, then $P|_U$ is also trivializable over U. In particular, show that the frame bundle of TS^2 does not admit any global sections.

Often vector bundles come equipped with an extra structure, for example orientations of each fiber and/or scalar product on each fiber. In the language of principal bundles this corresponds to the notion of a G-structure.

Definition 2.2.4. Let G be a Lie subgroup of $GL_k(\mathbb{R})$. A G-structure on E is a G–subbundle P of the frame bundle. If so, G is called the structure group of E.

To illustrate this notion, let us consider the following example. Assume E is an Euclidean vector bundle, which means that each fiber E_m is equipped with an Eucliden scalar product $\langle \cdot, \cdot \rangle_m$, which depends smoothly on m. Here the dependence is said to be smooth if for any two smooth sections s_1 and s_2 the function $\langle s_1, s_2 \rangle$ is also smooth.

It is natural to consider the subset

$$O(E) := \big\{ e \in \operatorname{Fr}(E) \mid e \text{ is orthonormal} \big\}.$$

The restriction of π yields a surjective map $O(E) \to M$, which is still denoted by π. If e is any local frame of E over an open subset U, the Gram-Schmidt orthogonalization process shows that there is also a smooth pointwise orthonormal frame e_O defined on U. Just like in the case of $\operatorname{Fr}(E)$ we can cover $O(E)$ by open subsets $\pi^{-1}(U)$ such that

$$\Psi_U\colon U \times O(k) \to \pi^{-1}(U), \qquad (m, h) \mapsto e_O(m) \cdot h$$

is a bijection. While $O(k)$ is not an open subset of an Euclidean space, it is a manifold, and therefore we can cover $O(k)$—hence, also $\pi^{-1}(U)$—by a collection of charts. The same argument as in the case of the frame bundle shows that the transition functions are smooth so that $O(E)$ is a principal $O(k)$–bundle.

We see that an Euclidean structure on E determines an $O(k)$–structure on E. Conversely, an $O(k)$-structure $P \subset Fr(E)$ determines an Euclidean structure on E. Indeed, pick any $p \in P_m$ and any two vectors $v_1, v_2 \in E_m$. Since p is a basis of E_m, we can write $v_i = p \cdot x_i$, where $x_i \in \mathbb{R}^k$. We define a scalar product on E_m by

$$\langle v_1, v_2 \rangle = x_1^t x_2.$$

It is straightforward to check that this does not depend on the choice of p. Moreover, the scalar product defined in this way depends smoothly on m.

To summarize, an Euclidean structure on a vector bundle is equivalent to an $O(k)$–structure.

Exercise 2.2.5. A fiberweise volume form is by definition a nowhere vanishing section of $\Lambda^k E^*$, where $k = \operatorname{rk} E$. Show that there is a one-to-one correspondence between fiberwise volume forms and $SL_k(\mathbb{R})$–structures.

Let V be a complex vector space of complex dimension k. One can view V as a real vector space of dimension $2k$ equipped with an endomorphism I in $\operatorname{End}_{\mathbb{R}}(V)$, $Iv = iv$ so that $I^2 = -\operatorname{id}$. Conversely, given a real vector space equipped with an endomorphism $I \in \operatorname{End}_{\mathbb{R}}(V)$ such that $I^2 = -\operatorname{id}$ we can regard V as a complex vector space, where $i \cdot v := Iv$. In this case $\dim_{\mathbb{R}} V$ is necessarily even. The map I is called a complex structure.

With the above understood, a complex vector bundle is just a real vector bundle equipped with $I \in \Gamma(\operatorname{End} E)$ such that $I^2 = -\operatorname{id}$. In particular, each fiber E_m is endowed with a complex structure $I(m)$. Thus, a complex vecor bundle is essentially a family of complex vector spaces parametrized by points of the base.

Exercise 2.2.6.

(i) Show that a complex vector bundle can be defined as a locally trivial family of complex vector spaces akin to Definition 2.1.1;

(ii) Show that there is a one-to-one correspondence between complex structures on a real vector bundle E or rank $2k$ and $GL_k(\mathbb{C})$–structures. Here $GL_k(\mathbb{C})$ is viewed as a subgroup of $GL_{2k}(\mathbb{R})$

$$GL_k(\mathbb{C}) = \{A \in GL_{2k}(\mathbb{R}) \mid A \circ I_{st} = I_{st} \circ A\},$$

where I_{st} is the standard complex structure on $\mathbb{R}^{2k}$:

$$I_{st}(x_1, y_1, \dots, x_k, y_k) := (-y_1, x_1, \dots, -y_k, x_k).$$

Exercise 2.2.7. A Hermitian structure on a complex vector bundle is a smooth family of Hermitian scalar products on each fiber. Prove that the following holds:

(i) Any complex vector bundle admits a Hermitian structure;

(ii) There is a one-to-one correspondence between Hermitian structures and $U(k)$–structures.

The associated vector bundle. Let $\pi\colon P \to M$ be a principal G–bundle. For any representation $\rho\colon G \to GL(V)$, where V is a vector space, we can construct a vector bundle over M as follows.

Define the right G–action on the product bundle $P \times V$ via

$$(p, v) \cdot g = \big(p \cdot g,\ \rho(g^{-1})v\big).$$

Clearly, this action is free and properly discontinuous so that the quotient

$$P \times_\rho V := \big(P \times V\big)/G \tag{2.2.8}$$

is a smooth manifold. The map π yields a well-defined projection

$$P \times_\rho V \to M, \qquad [p, v] \mapsto \pi(p)$$

which we still denote by the same letter π. Its fibers are isomorphic to V, hence these have a canonical structure of vector spaces. Moreover, given a local section $s \in \Gamma(U; P)$ we can construct a local trivialization of E via the map

$$\psi_U^{-1}\colon U \times V \to \pi_E^{-1}(U), \qquad (m, v) \mapsto [s(m), v].$$

Hence, if P is locally trivial, so is E.

Definition 2.2.9. The vector bundle $E = E(P, \rho, V) := P \times_\rho V$ is called the vector bundle associated with (P, ρ), or simply *the associated bundle.*

Example 2.2.10. Let $P = \mathrm{Fr}(E)$, $V = \mathbb{R}^k$, and $\rho = \mathrm{id}$ be the tautological representation of $GL_k(\mathbb{R})$. Then the map

$$\mathrm{Fr}(E) \times \mathbb{R}^k \longrightarrow E, \qquad (e, x) \mapsto \sum_{i=1}^{n} x_i e_i$$

induces an isomorphism $E\big(\mathrm{Fr}(E), \mathrm{id}\big) \cong E$.

This example shows that the construction of an associated bundle allows one to recover a vector bundle from its frame bundle. However, by varying the representation ρ we can obtain other bundles as well. The following example illustrates this.

Example 2.2.11. Consider $P = \mathrm{Fr}(E)$ again, but this time we take $V = \Lambda^p(\mathbb{R}^k)^*$ and ρ the natural representation of $GL_k(\mathbb{R})$ on $\Lambda^p(\mathbb{R}^k)^*$, i.e.,

$$\rho\colon GL_k(\mathbb{R}) \times \Lambda^p(\mathbb{R}^k)^* \to \Lambda^p(\mathbb{R}^k)^*, \qquad (g, \alpha) \mapsto \alpha\big(g^{-1}\cdot, \ldots, g^{-1}\cdot\big). \tag{2.2.12}$$

We have the map

$$\mathrm{Fr}(E) \times \Lambda^p(\mathbb{R}^k)^* \to \Lambda^p E^*, \qquad (e, \alpha) \mapsto \alpha(e^{-1}\cdot, \ldots, e^{-1}\cdot)$$

where we think of a frame e as an isomorphism $\mathbb{R}^k \to E_{\pi(e)}$. This induces an isomorphism of vector bundles

$$\mathrm{Fr}(E) \times_\rho \Lambda^p(\mathbb{R}^k)^* \cong \Lambda^p E^*.$$

Thus, $\Lambda^p E^*$ can be recovered as the vector bundle associated with $(\mathrm{Fr}(E), \rho)$, where ρ is given by (2.2.12).

Exercise 2.2.13. Let V be the space of $k \times k$-matrices $M_k(\mathbb{R}) \cong \mathrm{End}(\mathbb{R}^k)$ viewed as a $GL_k(\mathbb{R})$–representation as follows:

$$\rho\colon (g, A) \mapsto gAg^{-1}.$$

Show that $\mathrm{Fr}(E) \times_\rho \mathrm{End}(\mathbb{R}^k)$ is isomorphic to $\mathrm{End}(E)$.

Denote

$$C^\infty(P; V)^G := \left\{ \hat{s}\colon P \to V \mid \hat{s}(p \cdot g) = \rho(g^{-1})\hat{s}(p) \quad \forall p \in P \text{ and } g \in G \right\}.'$$

Pick any $\hat{s} \in C^\infty(P; V)^G$ and denote by $\varpi\colon P \times V \to P \times_\rho V$ the natural projection. Consider the map

$$\varpi \circ (\mathrm{id}, \hat{s})\colon P \longrightarrow P \times_\rho V, \qquad p \mapsto [p,\ \hat{s}(p)].$$

The letter map is G-invariant, where G acts trivially on the target space. Hence, there is a unique map $s\colon M \to P \times_\rho V$ such that

$$s \circ \pi = \varpi \circ (\mathrm{id}, \hat{s}). \tag{2.2.14}$$

In other words, s is defined by requiring that the diagram below commutes.

$$\begin{array}{ccc} P & \xrightarrow{(\mathrm{id}, \hat{s})} & P \times V \\ {\scriptstyle\pi}\big\downarrow & & \big\downarrow{\scriptstyle\varpi} \\ M & \xrightarrow[s]{} & P \times_\rho V \end{array}$$

Proposition 2.2.15. *The map*

$$C^\infty(P; V)^G \to \Gamma(P \times_\rho V), \qquad \hat{s} \mapsto s,$$

where s is defined by (2.2.14), *is a bijection.*

Proof. We only need to construct the inverse map. Thus, let s be a section of $P \times_\rho V$. For any $p \in P$ there is a unique $\hat{s}(p) \in V$ such that

$$s\big(\pi(p)\big) = [p,\ \hat{s}(p)].$$

It is straightforward to check that $\hat{s}$ is equivariant. □

It will be useful below to have a description of differential forms with values in an associated bundle in the spirit of Proposition 2.2.15. Before stating the claim, we need a few notions.

For $g \in G$ denote $R_g\colon P \to P$, $R_g(p) = p \cdot g$. The infinitesimal action of the Lie algebra $\mathfrak{g}$ is given by the vector field

$$K_\xi(p) := \frac{d}{dt}\Big|_{t=0}\big(p \cdot \exp(t\xi)\big), \qquad \xi \in \mathfrak{g}.$$

Since G acts freely on the fibers, we have

$$\mathcal{V}_p := \ker \pi_*|_p = \{K_\xi(p) \mid \xi \in \mathfrak{g}\} \cong \mathfrak{g}.$$

$\mathcal{V}_p$ is called *the vertical subspace.*

Definition 2.2.16.

- A q–form ω with values in some G-representation V is said to be G-*equivariant*, if $R_g^*\,\omega = \rho(g^{-1})\omega$;
- A q–form ω with values in some G-representation V is said to be *basic*, if $\omega(v_1,\dots,v_q) = 0$ whenever one of the arguments belongs to the vertical subspace.

Denote by $\Omega^q_{bas}(P;V)^G$ the space of all basic and G-equivariant q-forms on P with values in V.

Proposition 2.2.17. *For any* $a \in \Omega^q(M; P\times_\rho V)$ *the pull-back* $\pi^* a$ *can be viewed as an element of* $\Omega^q_{bas}(P;V)^G$. *Moreover, the map* $a \mapsto \pi^* a$ *establishes a bijective correspondence between* $\Omega^q(M; P\times_\rho V)$ *and* $\Omega^q_{bas}(P;V)^G$.

Proof. For any tangent vectors $\hat v_1,\dots,\hat v_q$ to P the equality

$$a(\pi_*\hat v_1,\dots,\pi_*\hat v_q) = [p,\ \hat a_p(\hat v_1,\dots,\hat v_q)]$$

determines uniquely $\hat a \in \Omega^q(P;V)$. Since by definition the vertical space is $\ker \pi_*$, the fact that $\pi^* a$ is basic is clear. The equivariancy of $\hat a$ follows from a straighforward computation.

Conversely, let $\hat a$ be given. Pick any $m \in M$ and any $p \in \pi^{-1}(m)$. Pick also any $v_1,\dots,v_q \in T_mM$ and choose $\hat v_1,\dots,\hat v_q \in T_pP$ such that $\pi_*(\hat v_j) = v_j$ for all $j \in \{1,\dots,q\}$. These lifts do exist because π_* is surjective, however these need not be unique. With this at hand, we can define a by the equality

$$a_m(v_1,\dots,v_q) = [p,\ \hat a_p(\hat v_1,\dots,\hat v_q)].$$

Since $\hat a$ is basic and equivariant, a does not depend on the choices involved. □

Connections on principal bundles. The Lie algebra $\mathfrak{g}$ can be viewed as a G-representation, where the action is the adjoint one: $\mathrm{ad}\colon G \to GL(\mathfrak{g})$. For example, if G is a subgroup of $GL_k(\mathbb{R})$, then

$$\mathrm{ad}_g\,\xi = g\,\xi\,g^{-1}.$$

Definition 2.2.18. *A connection form,* or simply *a connection,* on a principal G-bundle P is a G–equivariant 1–form a with values in the Lie algebra $\mathfrak{g}$ such that

$$a(K_\xi) = \xi \qquad \forall \xi \in \mathfrak{g}.$$

Denote

$$\mathrm{ad}\,P = P\times_{\mathrm{ad}}\mathfrak{g}.$$

Theorem 2.2.19. *For any principal bundle the space of all connections* $\mathcal{A}(P)$ *is an affine space modelled on* $\Omega^1(\mathrm{ad}\,P)$.

This theorem can be proved in the same manner as Theorem 2.1.13. Instead of going through the details, we describe the only essential modification of the argument used in the proof of Theorem 2.1.13. Namely, given $a,\ a' \in \mathcal{A}(P)$ the difference $b := a - a'$ is basic and G-invariant so that by applying Proposition 2.2.17 one can think of b as a 1–form on M with values in $\mathrm{ad}\,P$.

Notice that Theorem 2.2.19 states in particular, that $\mathcal{A}(P)$ is non-empty.

Example 2.2.20. $U(1)$ acts freely on

$$S^{2n+1} := \{z \in \mathbb{C}^{n+1} \mid |z_0|^2 + \cdots + |z_n|^2 = 1\}$$

in the diagonal manner. This represents S^{2n+1} as the total space of the $U(1)$-bundle

$$\pi\colon S^{2n+1} \to \mathbb{C}P^n, \qquad z \mapsto [z], \tag{2.2.21}$$

which is sometimes called the *Hopf fibration*.

The infinitesimal action of $U(1)$ on S^{2n+1} is given by the vector field

$$v(z) = (iz_0, \dots, iz_n).$$

There is a unique connection $a \in \Omega^1(S^{2n+1}; \mathbb{R}i)$ such that $\ker a = v^\perp$. Explicitly,

$$a_z(u) = \langle v(z), u\rangle\, i, \qquad \text{where } u \in T_z S^{2n+1}.$$

Since the action of $U(1)$ preserves the Riemannian metric of the sphere, a is an invariant 1-form. It remains to notice that for abelian groups the notions of equivariant and invariant forms coincide. Thus, a is a connection 1-form on (2.2.21).

Exercise 2.2.22. Prove that the associated bundle $S^{2n+1} \times_{U(1)} \mathbb{C}$ is (canonically) isomorphic to *the tautological line bundle:*

$$\mathcal{O}(-1) := \Big\{ \big([z], w\big) \in \mathbb{C}P^n \times \mathbb{C}^{n+1} \mid w \in [z] \cup \{0\} \Big\}.$$

The definitions of a connection on a vector and principal bundles differ significantly and the reader may wonder what is the relation between these two notions. The following result yields an answer to this question.

First notice that by differentiating $\rho\colon G \to GL(V)$ we obtain an action of $\mathfrak{g}$ on V, i.e., a Lie algebra homomorphism

$$\rho_*|_{\mathbb{1}}\colon \mathfrak{g} \to \mathrm{End}(V). \tag{2.2.23}$$

This way one can think of a connection a as a 1-form with values in $\mathrm{End}(V)$ provided a representation V is given.

Theorem 2.2.24.

(i) *Let E be a vector bundle. Any connection ∇ on E determines a unique connection a on the frame bundle such that*

$$e^* a = A(\nabla, e).$$

Here we think of a local frame e as a local section of $\mathrm{Fr}(E)$ and $A(\nabla, e)$ is the local connection 1–form of ∇ with respect to e.

(ii) *Let P be a principal G–bundle. Any connection a on P induces a unique connection ∇ on any associated vector bundle $P \times_\rho V$ such that*

$$\pi^*\nabla s = d\hat{s} + a \cdot \hat{s}. \tag{2.2.25}$$

Here the right hand side is a basic and G-invariant 1–form on P and the equality is understood in the sense of Proposition 2.2.17.

Proof. The proof consists of a number of the following steps.

Step 1. *Denote*

$$\hat{R}\colon P \times G \longrightarrow P, \qquad \hat{R}(p,g) = p \cdot g.$$

Then the differential $\hat{R}_*$ *at* (p,g) *satisfies:*

$$\hat{R}_*(v,w) = (R_g)_* v + K_{(L_{g^{-1}})_* w}(p \cdot g),$$

where $L_{g^{-1}}\colon G \to G$, $h \mapsto g^{-1}h$, $v \in T_pP$, *and* $w \in T_gG$.

The proof of this step is a simple computation, which is left as an exercise.

Step 2. *Let* e *be a local frame of* E *over* U. *For any 1-form* $A \in \Omega^1(U; \mathfrak{gl}_k(\mathbb{R}))$ *there is a unique connection* a *on* $\mathrm{Fr}(E)|_U$ *such that*

$$e^* a = A.$$

Since e is a section of $\mathrm{Fr}(E)|_U$, we have $\pi_* \circ e_* = \mathrm{id}_U$ and therefore e_* is injective. Hence, by dimensional reasons, we have $T_{e(m)}P = \mathcal{V}_{e(m)} \oplus \operatorname{Im} e(m)_*$. Therefore, we can define

$$a_{e(m)}(K_\xi + e_* v) = \xi + A(v).$$

This extends to a unique G-invariant 1-form on $\mathrm{Fr}(E)|_U$. The equality

$$(R_g)_* K_\xi = K_{\mathrm{ad}_{g^{-1}}\xi},$$

which can be checked by a straightforward computation, implies that a is a connection.

Step 3. *We prove* (i).

Choose two local frames e and e', which we may assume to be defined on the same open set U (restrict to the intersection of the corresponding domains if necessary). The equalities

$$a_{e(m)}(e_* \cdot) = A_m(\cdot) \qquad \text{and} \qquad a'_{e'(m)}(e'_* \cdot) = A'_m(\cdot)$$

determine unique $\mathrm{GL}_k(\mathbb{R})$–invariant 1-forms a and a' on $\mathrm{Fr}(E)|_U$ by Step 2.

We have

$$\begin{aligned} e'^* a(\cdot) &= a_{e'}(e'_* \cdot) = a_{e\cdot g}\Big(\big(\hat{R} \circ (e,g)\big)_* \cdot \Big) \\ &= a_{e\cdot g}\Big((R_g)_* \circ e_* \cdot \Big) + a_{e\cdot g}\big(K_{g^{-1}dg}(e \cdot g)\big) \\ &= g^{-1} a_e(e_* \cdot) g + g^{-1} dg = A'(\cdot). \end{aligned}$$

Here the second equality follows by Step 1 and the last one by (2.1.16). Thus, we conclude that $a = a'$ on the intersection of the domains of local frames e and e'. Thus, a is globally well-defined.

Step 4. *We prove* (ii).

By the equivariancy of $\hat{s}\colon P \to V$ we have

$$(d\hat{s} + a \cdot \hat{s})(K_\xi) = -\xi \cdot \hat{s} + a(K_\xi) \cdot \hat{s} = 0,$$

i.e., $d\hat{s} + a \cdot \hat{s}$ is a basic 1-form, which is G-invariant as well. Hence, there is a 1-form ∇s on M such that (2.2.25) holds. Moreover, for any G-invariant function f on P we have the equality

$$d(f \cdot \hat{s}) + a \cdot (f \cdot \hat{s}) = df \otimes \hat{s} + f\big(d\hat{s} + a \cdot \hat{s}\big),$$

which shows that ∇ is a connection. This finishes the proof of this theorem. □

One immediate corollary of this theorem is that a connection on E induces a connection on E^*, $\mathrm{End}(E)$ and so on. Indeed, these bundles can be interpreted as associated bundles for $P = \mathrm{Fr}(E)$ and the corresponding $GL_k(\mathbb{R})$–representations. A more direct characterization of the induced connections is given in the following.

Exercise 2.2.26. Let ∇ be a connection on E. Show that the following holds.

(i) There is a unique connection on E^*, still denoted by ∇, such that

$$d\langle \alpha, s\rangle = \langle \nabla\alpha, s\rangle + \langle \alpha, \nabla s\rangle \qquad \forall \alpha \in \Gamma(E^*) \quad \text{and} \quad \forall s \in \Gamma(E),$$

where $\langle \cdot, \cdot \rangle$ denotes the natural pairing $E^* \otimes E \to \mathbb{R}$. This connection coincides with the induced one.

(ii) There is a unique connection on $\mathrm{End}(E)$, still denoted by ∇, such that

$$\nabla\big(\varphi(s)\big) = (\nabla\varphi)(s) + \varphi(\nabla s) \qquad \forall \varphi \in \Gamma(\mathrm{End}(E)) \quad \text{and} \quad \forall s \in \Gamma(E),$$

This connection coincides with the induced one.

The curvature of a connection on a principal bundle. Local expression (2.1.23) of the curvature form suggests the following construction. Let a be a connection on a principal G–bundle P. The 2-form $da + \frac{1}{2}[a \wedge a]$ with values in $\mathfrak{g}$ is clearly G-invariant. This form is also basic as the following computation shows:

$$\begin{aligned}
\imath_{K_\xi}\big(da + \frac{1}{2}[a \wedge a]\big) &= \mathcal{L}_{K_\xi} a - d\big(\imath_{K_\xi} a\big) + \frac{1}{2}\big[a(K_\xi), a(\cdot)\big] - \frac{1}{2}\big[a(\cdot), a(K_\xi)\big] \\
&= -[\xi, a] + 0 + \frac{1}{2}[\xi, a] + \frac{1}{2}[\xi, a] \\
&= 0.
\end{aligned}$$

Here $\mathcal{L}_K$ denotes the Lie derivative with respect to a vector field K and the first equality uses Cartan's magic formula

$$\mathcal{L}_K = d\imath_K + \imath_K d.$$

Hence, by Proposition 2.2.17 we obtain that there is some $F_a \in \Omega^2(M; \mathrm{ad}\ P)$ such that

$$\pi^* F_a = da + \frac{1}{2}[a \wedge a]. \tag{2.2.27}$$

Definition 2.2.28. The 2–form F_a defined by (2.2.27) is called *the curvature form* of a.

Example 2.2.29. Let us compute the curvature of the connection a that was constructed in Example 2.2.20 in the simplest case $n = 1$. On the 3-sphere we have the following vector fields

$$v_1 := (-x_1, x_0, -x_3, x_2), \quad v_2 := (-x_2, x_3, x_0, -x_1), \quad \text{and} \quad v_3 := (-x_3, -x_2, x_1, x_0),$$

which at each point yield an orthonormal oriented basis of the tangent space to the sphere. It is worthwhile to notice that v_1 coincides with the vector field v from Example 2.2.20.

By the definition of the connection form a, we have

$$a = \big(-x_1\, dx_0 + x_0\, dx_1 - x_3\, dx_2 + x_2\, dx_3\big)\, i.$$

Hence,

$$\pi^* F_a = da = 2\big(dx_0 \wedge dx_1 + dx_2 \wedge dx_3\big)\, i.$$

In particular, we have

$$F_a(\pi_* v_2, \pi_* v_3) = \pi^* F_a(v_2, v_3) = 2\,i.$$

It can be shown that the quotient metric on $S^3/U(1)$ yields the round metric on the sphere of radius $1/2$. Explicitly, the corresponding isometry is given by

$$(z_0, z_1) \mapsto \big(z_0 \bar{z}_1, \tfrac{1}{2}(|z_0|^2 - |z_1|^2)\big).$$

Here we think of S^3 as a subset of $\mathbb{C}^2$ and, likewise, of S^2 as a subset of $\mathbb{C} \times \mathbb{R}$.

Since $(\pi_* v_2, \pi_* v_3)$ is an oriented orthonormal basis of the tangent space of $S^2_{1/2}$, we conclude that $F_a = 2\, \mathrm{vol}_{S^2_{1/2}}\, i$, where $\mathrm{vol}_{S^2_{1/2}}$ is the volume form of the standard round metric on $S^2_{1/2}$.

Notice that

$$\int_{S^2} F_a = 2\, \mathrm{Vol}(S^2_{1/2})\, i = 2\pi\, i.$$

The proof of the following proposition is left as an excercise.

Proposition 2.2.30. *Let a be a connection on a principal bundle P and let V be a G–representation. Using (2.2.23) we can think of F_a as a 2-form with values in $\mathrm{End}(P \times_\rho V)$. With these identifications in mind, the curvature of the induced connection ∇ on $P \times_\rho V$ equals F_a.* □

Given a local trivialization of P, i.e., a section σ over an open subset $U \subset M$, we say that $A := \sigma^* a \in \Omega^1(U; \mathfrak{g})$ is a local representation of a with respect to σ. Then over U we have

$$F_a = \sigma^* \pi^* F_a = \sigma^* \left(da + \frac{1}{2}[a \wedge a]\right) = dA + \frac{1}{2}[A \wedge A].$$

One of the most basic properties of the curvature form is the so called Bianchi identity.

Proposition 2.2.31 (Bianchi identity)**.** *Let a be a connection on P. Then the curvature form F_a satisfies*

$$d_{\nabla^a} F_a = 0,$$

where ∇^a is the connection on $\mathrm{ad}\, P$ induced by a.

Proof. Let A be a local representation of a as above. We have

$$\begin{aligned} d_{\nabla^a} F_a &= d\Big(dA + \frac{1}{2}[A \wedge A]\Big) + \Big[A \wedge \big(dA + \frac{1}{2}[A \wedge A]\big)\Big] \\ &= \frac{1}{2}[dA \wedge A] - \frac{1}{2}[A \wedge dA] + [A \wedge dA] + \frac{1}{2}\big[A \wedge [A \wedge A]\big] \\ &= 0. \end{aligned}$$

Here the first three summands sum up to zero because $[\omega, \eta] = -[\eta, \omega]$ for any $\omega \in \Omega^2(U; \mathfrak{g})$ and $\eta \in \Omega^1(U; \mathfrak{g})$; The vanishing of the last summand follows from the Jacobi identity. □

Remark 2.2.32. Thinking of $\hat{F}_A := \pi^* F_A = dA + \frac{1}{2}[A \wedge A]$ also as a curvature form of A, the computation in the proof of Proposition 2.2.31 yields the following equivalent form of the Bianchi identity:

$$d\hat{F}_A = [\hat{F}_A \wedge A]. \tag{2.2.33}$$

Indeed, this can be obtained by the computation: $d\hat{F}_A = [dA \wedge A] = [\hat{F}_A \wedge A]$.

Given a bundle $P \to M$ and a map $f \colon N \to M$ we can construct the pull-back bundle $f^*P \to N$ just like in the case of vector bundles. Informally speaking, we have the equality of fibers $(f^*P)_n = P_{f(n)}$ and this defines f^*P. More formally, define

$$f^*P := \big\{(p, n) \in P \times N \mid f(n) = \pi(p)\big\}.$$

It is easy to see that f^*P is a submanifold of $P \times N$. Moreover, we have a projection $\varpi \colon f^*P \to N$, which is just the restriction of the natural projection $(p, n) \mapsto n$. The structure group G clearly acts on P so that the action on the fibers $\varpi^{-1}(n) = P_{f(n)}$ is transitive. Moreover, we also have a natural G-equivariant map $\hat{f} \colon f^*P \to P$, which covers f, i.e., the diagram below commutes.

$$\begin{array}{ccc} f^*P & \xrightarrow{\ \hat{f}\ } & P \\ {\scriptstyle\varpi}\downarrow & & \downarrow{\scriptstyle\pi} \\ N & \xrightarrow[\ f\ ]{} & M \end{array}$$

For future use we note the following statement, whose proof is left to the reader as an exercise.

Proposition 2.2.34. *Let a be a connection on P. Then $f^*a := \hat{f}^*a$ is a connection on f^*P and $F_{f^*a} = f^*F_a$.* □

The gauge group. For a principal bundle $P \to M$, the automorphism group

$$\mathcal{G}(P) := \big\{\psi \colon P \to P \mid \pi \circ \psi = \pi,\ \psi(pg) = \psi(p)g \quad \forall p \in P,\ g \in G\big\}$$

is called the gauge group.

Since $\psi(p)$ lies in the same fiber as p and G acts transitively on the fibers, we can write

$$\psi(p) = p\,\hat{f}(p), \tag{2.2.35}$$

where $\hat{f}\colon P \to G$ is some map. The equivariancy of ψ is then equivalent to

$$\hat{f}(pg) = g^{-1}\hat{f}(p)g. \tag{2.2.36}$$

In other words, $\hat{f}$ can be identified with some section f of the bundle

$$\mathrm{Ad}\,P := P \times_G G,$$

where G acts on itself by conjugation.

Conversely, given $f \in \Gamma(\mathrm{Ad}\,P)$, we can construct an automorphism ψ of P via (2.2.35). In other words we have a natural bijective map

$$\mathcal{G}(P) \to \Gamma(\mathrm{Ad}\,P).$$

Exercise 2.2.37. Fix some non-negative integer k. Prove that $C^k(M;\,\mathrm{Ad}\,P)$ is a Banach Lie group. Moreover, show that the Lie algebra of this group is $C^k(M;\,\mathrm{ad}\,P)$.

Example 2.2.38. Assume G is abelian. Then (2.2.36) just means that $\hat{f}$ is an *invariant* map so that the corresponding section f can be identified with a map $M \to G$. This shows that *for an abelian group* G we have

$$\mathcal{G}(P) \cong C^\infty(M;\,G).$$

Just like in the case of vector bundles, the gauge group acts naturally on the space of all connections, i.e., we have a map

$$\mathcal{A}(P) \times \mathcal{G}(P) \to \mathcal{A}(P), \qquad (a, \psi) \mapsto a \cdot \psi := \psi^* a. \tag{2.2.39}$$

Exercise 2.2.40. Let $E := P \times_{G,\rho} V$ be an associated vector bundle. Show that the following holds:

(i) The representation ρ induces a natural homomorphism of gauge groups $\gamma\colon \mathcal{G}(P) \to \mathcal{G}(E)$;
(ii) The corresponding infinitesimal map $d\rho_e\colon \mathfrak{g} \to \mathrm{End}\,V$ induces a Lie algebra homomorphism $C^k(M;\,\mathrm{ad}\,P) \to C^k(M;\,\mathrm{End}\,E)$;
(iii) $\nabla^{a\cdot g} = \gamma(g)^{-1}\nabla^a\gamma(g)$. Here $a \in \mathcal{A}(P)$, $g \in \Gamma(\mathrm{Ad}\,P)$, and ∇^a denotes the connection on E induced by a.

The relation of (iii) is written usually simply as

$$\nabla^{a\cdot g} = g^{-1}\nabla^a g,$$

with the identification (i) being understood.

Exercise 2.2.41. Show that the infinitesimal action of the gauge group is given by

$$\mathcal{A}(P) \times \Gamma(\mathrm{ad}\,P) \to \Omega^1(\mathrm{ad}\,P), \quad (a, \xi) \mapsto -d_a\xi.$$

2.3. The Levi–Civita connection As we have seen above, any vector bundle of positive rank over a manifold of positive dimension admits a large family of connections. Moreover, in general there is also no preferred connections. The situation is somewhat different in the case of the tangent bundle of a manifold, where it turns out that a choice of a preferred connection does exist under certain circumstances.

Thus, pick a connection ∇ on $TM \to M$. For any vector fields v, w on M define *the torsion* of ∇ by

$$T(v,w) := \nabla_v w - \nabla_w v - [v,w].$$

Clearly, T is antisymmetric. It is also easy to check that T is tensorial in both v and w, i.e., $T(f_1 v, f_2 w) = f_1 f_2 T(v,w)$ for any $f_1, f_2 \in C^\infty(M)$. Hence, applying Lemma 2.1.14 we obtain that $T \in \Omega^2(M; TM)$.

Definition 2.3.1. A connection ∇ on the tangent bundle TM is called *torsion-free,* if $T \equiv 0$, i.e., if the following

$$\nabla_v w - \nabla_w v = [v,w] \tag{2.3.2}$$

holds for any vector fields v and w.

Exercise 2.3.3. Let ∇ be a connection on the tangent bundle. Slightly abusing notations, denote also by ∇ the induced connection on T^*M. Consider the map

$$\Omega^1(M) \xrightarrow{\nabla} \Gamma(T^*M \otimes T^*M) \xrightarrow{\mathrm{Alt}} \Omega^2(M), \tag{2.3.4}$$

where the last map is the natural projection (alternation) $T^*M \otimes T^*M \to \Lambda^2 T^*M$. Show that ∇ is torsion-free if and only if (2.3.4) coincides with the de Rham differential.

Theorem 2.3.5. *For any Riemannian manifold (M, g) there is a unique metric torsion-free connection ∇.* □

The proof of this classical result, which is sometimes called the fundamental theorem of Riemannian geometry, can be found for instance in [23, Prop. 2.1]. The unique connection described in the above theorem is called *the Levi-Civita connection.* It is uniquely characterized by (2.3.2) together with

$$d\big(g(v,w)\big) = g(\nabla v, w) + g(v, \nabla w).$$

The curvature form R of the Levi-Civita connection can be viewed as a section of $\Lambda^2 T^*M \otimes \mathrm{End}(TM)$. This can be decomposed into various components. For example, in terms of a local orthonormal frame e_i we can define a quadratic form

$$\mathrm{Ric}_g(v,w) := -\sum_i g\big(R(e_i,v)e_i, w\big),$$

which is called *the Ricci curvature.* The trace of the Ricci curvature, i.e.,

$$s_g := \sum_i \mathrm{Ric}(e_i, e_i)$$

is called *the scalar curvature*. These are important characteristics of the metric g. An interested reader may wish to consult [3, 14, 23] for more information on these matters. For us, the importance of the scalar curvature is explained by its appearance in the Weitzenböck formula below, see Corollary 4.4.6.

2.4. Classification of U(1) and SU(2) bundles It is more convenient in this section to work in a topological category rather than the smooth one. The reader will have no difficulties to adopt the corresponding notions to this setting.

Definition 2.4.1. Let G be a compact Lie group. A topological space E equipped with an action of G is said to be *a classifying bundle* for G, if E is contractible and the G–action is free.

Denoting $B := E/G$, we obtain a natural projection $E \to B$ so that E could be thought of as a principal[2] G–bundle over B. If E exists, it is easy to see that E is unique up to a homotopy equivalence.

One can prove that for a compact Lie group a classifying space always exist. An interested reader may wish to consult [13, Sect. 1.2]. Also, we take the following result as granted.

Theorem 2.4.2 ([13, Thm. 1.1.1 and Rem. 2]). *Let $P \to M$ be a (topological) principal G–bundle over a manifold M. Then there exists a continuous map $f\colon M \to B$ such that P is isomorphic to f^*E. In fact f is unique up to a homotopy so that the map*

$$f \mapsto f^*E$$

yields a bijective correspondence between the set of isomorphism classes of principal G-bundles over M and the set $[M;\, B]$ of homotopy classes of maps $M \to B$. □

In some sense, the above theorem yields a classification of principal—hence, also vector—bundles. In praxis it is not always easy to describe the set $[M;\, B]$ though. One way to deal with this problem is via the so called characteristic classes, which I describe next.

Definition 2.4.3. Let $f\colon M \to B$ be a continuous map. Pick any $c \in H^\bullet(B; R)$, where R is a ring. Then $f^*c \in H^\bullet(M; R)$ is called a characteristic class of f^*E.

It is worth pointing out that characteristic classes depend on the isomorphism class of the bundle only.

Usually, characteristic classes are easier to deal with than the set of homotopy classes of maps. The most common choices for the ring R are $\mathbb{Z}$, $\mathbb{Z}/n\mathbb{Z}$, $\mathbb{R}$, and $\mathbb{C}$.

In some cases the classifying bundle can be constructed fairly explicitly. I restrict myself to the following two cases, namely $G = U(1)$ and $G = SU(2)$, which are most commonly used in gauge theoretic problems.

Complex line bundles. It follows from Exercise 2.2.7 that the classification problems for complex line bundles and principal $U(1)$-bundles are equivalent. Even though what follows below can be described in terms of vector bundles only, the language of $U(1)$–bundles has certain advantages and will be mainly used below.

Consider the following commutative diagram

$$\begin{array}{ccccccccc} S^3 & \longrightarrow & S^5 & \longrightarrow & \dots & \longrightarrow & S^{2n+1} & \longrightarrow & \dots \\ \downarrow & & \downarrow & & & & \downarrow & & \\ \mathbb{CP}^1 & \longrightarrow & \mathbb{CP}^2 & \longrightarrow & \dots & \longrightarrow & \mathbb{CP}^n & \longrightarrow & \dots \end{array}$$

[2]The notations E and B are traditional in this context so I will keep to this tradition. The reader should not be confused by the fact that here E denotes a principal rather than vector bunlde.

where the horizontal arrows are natural inclusions. For example, the inclusion of the spheres is given by

$$\mathbb{C}^n \supset S^{2n-1} \ni \quad (z_0, \ldots, z_{n-1}) \mapsto (z_0, \ldots, z_{n-1}, 0) \quad \in S^{2n+1} \subset \mathbb{C}^{n+1},$$

which is a U(1)–equivariant map.

The direct limit construction yields a CW-complex S^∞ equipped with a free U(1)-action. S^∞ can be shown to be contractible. Furthermore, we have also a CW-complex $\mathbb{C}P^\infty$ so that the natural quotient map

$$S^\infty \to \mathbb{C}P^\infty$$

is the classifying bundle for the group U(1).

Example 2.4.4 (Classification of line bundles on 2–manifolds). Let M be an oriented two–manifold. A continuous map $f\colon M \to \mathbb{C}P^\infty$ is homotopic to a map, which takes values in the 2-skeleton $\mathbb{C}P^1 \subset \mathbb{C}P^\infty$ so that we have the equality $[M; \mathbb{C}P^\infty] = [M; \mathbb{C}P^1]$. Topologically, $\mathbb{C}P^1$ is just the 2-sphere so that we have a well–defined degree-map

$$[M; S^2] \to \mathbb{Z}, \qquad [f] \mapsto \deg f,$$

which is in fact a bijection. Thus, a complex line bundle L on an oriented two–manifold is classified by an integer d, which is called the degree of L.

It is easy to see that the cohomology ring $H^\bullet(\mathbb{C}P^\infty; \mathbb{Z})$ is generated by a single element $a \in H^2(\mathbb{C}P^\infty; \mathbb{Z})$. We can fix the choice of the generator by requiring

$$\langle a, [\mathbb{C}P^1] \rangle = 1, \tag{2.4.5}$$

where $\langle \cdot, \cdot \rangle$ is the pairing between the homology and cohomology groups.

With this understood, to any principal U(1)-bundle $P \to M$ (equivalently, to any complex line bundle $L \to M$) we can associate a cohomology class as follows. If $f\colon M \to \mathbb{C}P^\infty$ is a map such that P is the pull-back of the bundle $S^\infty \to \mathbb{C}P^\infty$, then

$$c_1(P) := -f^*a \in H^2(M; \mathbb{Z}). \tag{2.4.6}$$

The minus sign in this definition is a convention.

Definition 2.4.7. The class $c_1(P)$ defined by (2.4.6) is called *the first Chern class* of P.

If $L \to M$ is a complex line bundle, we can choose a Hermitian scalar product, or, in other words, we can choose a U(1)-structure $P \subset \mathrm{Fr}(L)$. Then

$$c_1(L) := c_1(P)$$

is also called the first Chern class of L.

Exercise 2.4.8. Check that the first Chern class of L does not depend on the choice of the Hermitian scalar product on L.

Theorem 2.4.9. *The first Chern class of the complex line bundle has the following properties:*

(i) $c_1(\underline{\mathbb{C}}) = 0$, *where* $\underline{\mathbb{C}}$ *is the product bundle;*

(ii) $c_1(L_1 \otimes L_2) = c_1(L_1) + c_1(L_2)$ *for all line bundles* L_1 *and* L_2 *over the same base* M*;*

(iii) $c_1(L^\vee) = -c_1(L)$, *where* $L^\vee := \operatorname{Hom}(L; \mathbb{C})$ *is the dual line bundle;*

(iv) $c_1(f^*L) = f^*c_1(L)$ *for all line bundles* $L \to M$ *and all (continuous) maps* $f \colon N \to M$. □

Quaternionic line bundles. Let $\mathbb{H}$ denote the $\mathbb{R}$-algebra of quaternions. One can think of the (compact) symplectic group

$$\mathrm{Sp}(1) := \{ q \in \mathbb{H} \mid |q|^2 = q\bar{q} = 1 \}$$

as a quaternionic analogue of $U(1)$. It is easy to see that $\mathrm{Sp}(1)$ is isomorphic to $SU(2)$. Indeed, it is easy to write down an isomorphism explicitly:

$$\mathrm{Sp}(1) \to SU(2), \qquad q = z + wj \mapsto \begin{pmatrix} z & w \\ -\bar{w} & \bar{z} \end{pmatrix}.$$

Just like $U(1)$ acts freely on S^{2n+1}, $\mathrm{Sp}(1)$ acts freely in a diagonal manner on $S^{4n+3} = \{(h_0, \cdots, h_n) \in \mathbb{H}^{n+1} \mid |h_0|^2 + \cdots + |h_n|^2 = 1\}$ in so that we have a natural principal $\mathrm{Sp}(1)$-bundle:

$$S^{4n+3} \to \mathbb{HP}^n.$$

The commutative diagram

$$\begin{array}{ccccccccc} S^7 & \longrightarrow & S^{11} & \longrightarrow & \cdots & \longrightarrow & S^{4n+3} & \longrightarrow & \cdots \\ \downarrow & & \downarrow & & & & \downarrow & & \\ \mathbb{HP}^1 & \longrightarrow & \mathbb{HP}^2 & \longrightarrow & \cdots & \longrightarrow & \mathbb{HP}^n & \longrightarrow & \cdots \end{array}$$

leads to the construction of the classifying bundle for $\mathrm{Sp}(1)$:

$$S^\infty \to \mathbb{HP}^\infty.$$

Notice that $\mathbb{HP}^\infty$ is (an infinite-dimensional) CW-complex, which has exactly one cell of dimension $4n$. This implies in particular that the cohomology ring $H^\bullet(\mathbb{HP}^\infty; \mathbb{Z})$ is generated by a single generator b of degree 4. The choice of b can be fixed for example by requiring

$$\langle b, [\mathbb{HP}^1] \rangle = 1.$$

Proposition 2.4.10. *Let* M *be a manifold of dimension at most* 3*. Any* $\mathrm{Sp}(1)$*-bundle over* M *is trivial.*

Proof. Given a principal $\mathrm{Sp}(1)$-bundle P, by Theorem 2.4.2 we can find a map $f \colon M \to \mathbb{HP}^\infty$ such that f^*S^∞ is isomorphic to P. Further, since f is homotopic to a map f_1 that maps into the 3-skeleton of $\mathbb{HP}^\infty$, which is a point, P is trivial. □

Let M be a manifold of arbitrary dimension. To any map $f \colon M \to \mathbb{HP}^\infty$ we can associate a class $f^*b \in H^4(M; \mathbb{Z})$, where b is a generator of $H^\bullet(\mathbb{HP}^\infty; \mathbb{Z})$ as above.

Definition 2.4.11. Let $P \to M$ be a principal $Sp(1)$–bundle. If $f\colon M \to \mathbb{HP}^\infty$ is a map such that $f^*S^\infty \cong P$, then the *the second Chern class* of P is defined by

$$c_2(P) := -f^*b \in H^4(M; \mathbb{Z})$$

Remark 2.4.12. The terminology may seem to be somewhat odd at this point. The reason is that for a principal $U(r)$–bundle P one can define r characteristic classes $c_1(P), c_2(P), \dots, c_r(P)$, where $c_j(P) \in H^{2j}(M; \mathbb{Z})$. In the particular cases described above, this yields the constructions of the first and the second Chern classes, see Subsection 3.1 below for more details.

3. The Chern–Weil theory

3.1. The Chern–Weil theory In this section we could equally well work with both $\mathbb{R}$ and $\mathbb{C}$ as ground fields. I opt for $\mathbb{C}$ mainly for the sake of definiteness. The modifications needed for the case of $\mathbb{R}$ as a ground field are straightforward.

Let $p\colon \mathfrak{g} \to \mathbb{C}$ be an ad–invariant homogeneous polynomial of degree d. This means the following:

- Given a basis $\xi_1, \dots, \xi_n$ of $\mathfrak{g}$, the expression $p(x_1\xi_1 + \dots + x_n\xi_n)$ is a polynomial of degree d in $x_1, \dots, x_n$;
- $p(\mathrm{ad}_g\, \xi) = p(\xi)$ for all $g \in G$ and all $\xi \in \mathfrak{g}$;
- $p(\lambda\xi) = \lambda^d p(\xi)$ for all $\lambda \in \mathbb{R}$ and all $\xi \in \mathfrak{g}$.

Example 3.1.1.

(a) For $\mathfrak{g} = \mathfrak{u}(r)$ the map $p_d(\xi) = i \operatorname{tr} \xi^d$ is an ad-invariant polynomial of degree d.

(b) Choose $\mathfrak{g} = \mathfrak{u}(r)$ again and define polynomials $c_1, \dots, c_r$ of degrees $1, \dots, r$ respectively by the equality

$$\det\Big(\lambda\mathbb{1} + \frac{i}{2\pi}\xi\Big) = \lambda^r + c_1(\xi)\lambda^{r-1} + \dots + c_r(\xi).$$

For example, $c_r(\xi) = \frac{i^r}{(2\pi)^r} \det \xi$ and $c_1(\xi) = \frac{i}{2\pi} \operatorname{tr} \xi$. Notice also that the equality

$$\overline{\det\Big(\lambda\mathbb{1} + \frac{i}{2\pi}\xi\Big)} = \det\Big(\bar\lambda\mathbb{1} + \frac{i}{2\pi}\xi\Big)$$

implies that each c_j takes values in $\mathbb{R}$.

Let $P \to M$ be a principal G–bundle equipped with a connection $a \in \Omega^1(P; \mathfrak{g})$. Think of the curvature form $\pi^*F_a = da + \frac{1}{2}[a \wedge a]$ as a matrix, whose entries are 2-forms on P. Since forms of even degrees commute, the expression $p(\pi^*F_a)$ makes sense as an $\mathbb{R}$-valued differential form of degree at most $2d$ on P. Since each entry of π^*F_a is basic, so is $p(\pi^*F_a)$. Moreover, the ad–invariancy of p implies that $p(\pi^*F_a)$ is G–invariant. By Proposition 2.2.17 applied in the case of the trivial G–representation we obtain that there is a form $p(F_a)$ on M of degree $2d$ such that

$$\pi^*p(F_a) = p(\pi^*F_a).$$

Lemma 3.1.2. *The following holds:*

(i) $p(F_a)$ *is closed;*

(ii) *The de Rham cohomology class of* $p(F_a)$ *does not depend on the choice of connection* a.

Proof. The proof consists of the following steps.

Step 1. *We prove* (i).

Pick any $\xi, \xi_1, \dots, \xi_d \in \mathfrak{g}$. Thinking of $\mathfrak{g}$ as a matrix Lie algebra, I write temporarily $\mathrm{ad}_{e^{t\xi}}\xi_j = e^{t\xi}\xi_j e^{-t\xi}$. Slightly abusing notations, let $p\colon \mathrm{Sym}^d(\mathfrak{g}) \to \mathbb{C}$ be the d-multilinear function whose restriction to the diagonal yields the original polynomial p. Then, differentiating the equality

$$p\big(e^{t\xi}\xi_1 e^{-t\xi}, \dots, e^{t\xi}\xi_d e^{-t\xi}\big) = p(\xi_1, \dots, \xi_d)$$

with respect to t, yields

(3.1.3)
$$p\big([\xi, \xi_1], \xi_2, \dots, \xi_d\big) + p\big(\xi_1, [\xi, \xi_2], \dots, \xi_d\big) + \dots + p\big(\xi_1, \xi_2, \dots, [\xi, \xi_d]\big) = 0.$$

Denoting by $\hat F_A := \pi^* F_A$, the last equation implies that

$$p\big([\hat F_A \wedge A], \hat F_A, \dots, \hat F_A\big) + p\big(\hat F_A, [\hat F_A \wedge A], \dots, \hat F_A\big) + \dots + p\big(\hat F_A, \hat F_A, \dots, [\hat F_A \wedge A]\big)$$

equals 0. Hence,

$$\begin{aligned} d\big(p(\hat F_A)\big) &= p(d\hat F_A, \hat F_A, \dots, \hat F_A) + p(\hat F_A, d\hat F_A, \dots, \hat F_A) + \dots + p(\hat F_A, \hat F_A, \dots, d\hat F_A) \\ &= p\big([\hat F_A \wedge A], \hat F_A, \dots, \hat F_A\big) + p\big(\hat F_A, [\hat F_A \wedge A], \dots, \hat F_A\big) + \dots \\ &\qquad\qquad + p\big(\hat F_A, \hat F_A, \dots, [\hat F_A \wedge A]\big) \\ &= 0. \end{aligned}$$

Here the second equality uses (2.2.33). This finishes the proof of (i).

Step 2. *Let* $I = [0, 1]$ *be the interval and* $\iota_0, \iota_1\colon M \to M \times I$ *be the natural inclusions correspodnding to the endpoints of the interval. There exist linear maps* $Q\colon \Omega^k(M \times I) \to \Omega^{k-1}(M)$ *such that for any* $\omega \in \Omega^k(M \times I)$ *we have*

$$\iota_1^*\omega - \iota_0^*\omega = d\,Q\omega - Q\,d\omega.$$

The argument goes just like in the proof of the Poincaré lemma, see for example [4, Prop. 4.1.1]. I omit the details.

Step 3. *We prove* (ii).

Pick any two connection A_0 and A_1 and think of $A_t := (1-t)A_0 + tA_1$ as a connection on $\varpi^*P \to M \times I$, where $\varpi\colon M \times I \to M$ is the natural projection. Then

$$p(F_{A_1}) - p(F_{A_0}) = \iota_1^* p(F_{A_t}) - \iota_0^* p(F_{A_t}) = d\,Qp(F_{A_t})$$

by the previous step. This proves (ii). □

The Chern classes. Let c_j be the polynomial of degree j from Example 3.1.1.

Let $P \to M$ be a principal $U(r)$–bundle with a connection A. By Lemma 3.1.2, $c_j(F_A)$ is closed, real valued, and the de Rham cohomology class of $c_j(F_A)$ does not depend on the choice of the connection.

Definition 3.1.4. The class $c_j(P) := [c_j(F_A)] \in H^{2j}_{dR}(M;\mathbb{R})$ is said to be the jth Chern class of P and

$$c(P) := 1 + c_1(P) + \cdots + c_r(P) \in H^\bullet(M;\mathbb{R})$$

is called the total Chern class of P.

Remark 3.1.5. The above definition yields Chern classes as elements of the de Rham cohomology groups only. In fact, one can show that they lie in the image of $H^\bullet(M;\mathbb{Z}) \to H^\bullet_{dR}(M;\mathbb{R})$. I will discuss this briefly in the case of the first two Chern classes below. Also, at this point we have two a priori unrelated definitions of the first two Chern classes. We will see below that in fact they agree.

Remark 3.1.6. Let E be a complex vector bundle of rank r. Choosing a fiberwise Hermitian structure on E, we obtain a principal $U(r)$–bundle Fr_U so that we can define $c_j(E) := c_j(Fr_U)$. It is easy to show that this does not depend on the choice of the Hermitian structure.

Theorem 3.1.7. *The Chern classes satisfy the following properties*

(i) $c_0(E) = 1$ *for any vector bundle* E*;*

(ii) $c(f^*E) = f^*c(E)$ *for all vector bundles* $E \to M$ *and all maps* $f\colon N \to M$*;*

(iii) $c(E_1 \oplus E_2) = c(E_1) \cup c(E_2)$*;*

(iv) $c\big(\mathcal{O}(-1)\big) = 1 - a$*, where* $\mathcal{O}(-1)$ *is the tautological line bundle over* $\mathbb{P}^1$ *and* a *is the generator of the cohomology group of* $\mathbb{P}^1$ *such that* (2.4.5) *holds.*

The first property above is just the definition, the remaining properties are called naturality, Whitney sum formula, and normalization respectively.

Proof. The naturality follows immediately from Proposition 2.2.34. The normalization is equivalent to $\frac{i}{2\pi}\int_{\mathbb{P}^1} F_a = -1$, where a is a unitary connection on the tautological line bundle. This was established in Example 2.2.29.

Thus, we only need to prove the Whitney sum formula. If ∇_1 and ∇_2 are unitary connections on E_1 and E_2 respectively, then the curvature of the corresponding connection on the Whitney sum is a block diagonal matrix. More precisely, this means that the curvature is a 2-form with values in $\mathrm{End}(E_1) \oplus \mathrm{End}(E_2)$. If A and B are any square matrices, we have

$$\det\begin{pmatrix} A & 0 \\ 0 & B \end{pmatrix} = \det A \det B \quad \Longrightarrow \quad c(F_{\nabla_1 \oplus \nabla_2}) = \det\begin{pmatrix} \mathbb{1} + \frac{i}{2\pi}F_{\nabla_1} & 0 \\ 0 & \mathbb{1} + \frac{i}{2\pi}F_{\nabla_2} \end{pmatrix}$$
$$= c(F_{\nabla_1}) \wedge c(F_{\nabla_2}).$$

The latter equality clearly implies the Whitney sum formula. □

Exercise 3.1.8. Let E be a vector bundle. Prove that the following holds:

(a) The Chern classes depend on the isomorphism class of E only;
(b) $c_j(E^\vee) = (-1)^j c_j(E)$ for all j;
(c) If E is trivial, then $c(E) = 1$;
(d) If $E \cong E_1 \oplus \underline{\mathbb{C}}^k$, then $c_j(E) = 0$ for $j > \operatorname{rk} E - k$.

Exercise 3.1.9. Show that the tangent bundle of S^2 is non-trivial.

Theorem 3.1.10. *Let L be a complex line bundle. Then the first Chern class in the sense of Definition 3.1.4 coincides with the image in $H^2_{dR}(M;\mathbb{R})$ of the first Chern class in the sense of Definition 2.4.7.*

Sketch of proof. Let $L \to M$ be a complex line bundle. It is not too hard to show that there is $N < \infty$ and a smooth map $f\colon M \to \mathbb{P}^N$ such that $f^*\mathcal{O}(-1)$ is isomorphic to L, where $\mathcal{O}(-1)$ is the tautological bundle of $\mathbb{P}^N$. Notice that $H^2_{dR}(\mathbb{P}^N;\mathbb{R})$ is one dimensional and generated by the class Poincaré dual to $[\mathbb{P}^1]$, where $\iota\colon \mathbb{P}^1 \subset \mathbb{P}^N$ is a standard embedding.

Pick a Hermitian structure on $\mathcal{O}(-1)$ and a Hermitian connection ∇. Then $\iota^*\mathcal{O}(-1)$ is the tautological bundle of $\mathbb{P}^1$, so that $\frac{i}{2\pi}\iota^*F_\nabla$ represents the first Chern class of $\mathcal{O}_{\mathbb{P}^1}(-1)$. Hence, by Example 2.2.29 we have

$$\left\langle [\tfrac{i}{2\pi}\iota^*F_\nabla],\, [\mathbb{P}^1] \right\rangle = -1$$

Hence, $c_1(\mathcal{O}(-1)) = -a$ so that $c_1(L) = c_1(f^*L) = -f^*a$. □

Remark 3.1.11. One can prove arguing along similar lines that in the case of $SU(2)$–bundles the two definitions of the second Chern class agree on the level of the de Rham cohomology groups. I leave the details to the readers. Moreover, one can also show that the infinite Grassmannian $Gr_k(\mathbb{C}^\infty)$ is a classifying space for the group $U(k)$. Thus one could also define the Chern classes as pull-backs of certain classes on $Gr_k(\mathbb{C}^\infty)$.

Remark 3.1.12. One corollary of Definition 2.4.7 is as follows. Let $L \to M$ be a Hermitian line bundle and $h\colon \Sigma \to M$ a smooth map, where Σ is a compact oriented two–manifold. Then we have

$$\frac{i}{2\pi}\int_M h^*F_\nabla \in \mathbb{Z},$$

where ∇ is any Hermitian line bundle. This property may be quite surprising if one's starting point is Definition 3.1.4.

In particular, if M is itself a compact oriented two-dimensional manifold (and h is the identity map), then

$$\frac{i}{2\pi}\int_M F_A$$

is an integer, which coincides with the degree of L, cf. Example 2.4.4.

Remark 3.1.13. A straightforward computation yields that for any matrix $\xi \in \mathfrak{su}(2)$ we have $\operatorname{tr}\xi^2 = -2\det\xi$. Hence, for an SU(2)–bundle P we have

$$c_2(P) = \frac{1}{8\pi^2}\big[\operatorname{tr}(F_A \wedge F_A)\big] \in H^4_{dR}(M;\mathbb{R}).$$

In particular, if M is a closed oriented four-manifold, the integration yields an isomorphism $H^4_{dR}(M;\mathbb{R}) \cong \mathbb{R}$. In fact, just as in the case of line bundles above, we have

$$c_2(P) = \frac{1}{8\pi^2}\int_M \operatorname{tr}(F_A \wedge F_A) \in \mathbb{Z}. \tag{3.1.14}$$

3.2. The Chern–Simons functional In this section I will restrict myself to dimension three and $G = SU(2)$. Thus, let M be a three manifold equipped with an $SU(2)$–bundle $P \to M$. Notice that P is trivial by Proposition 2.4.10.

As a matter of fact, any closed oriented three-manifold is a boundary of a compact oriented four-manifold, say $\partial X = M$. Assume there is an extension of P to X, i.e., a bundle P_X such that $P_X|_M = P$. In this case any connection A on P can be extended to a connection A_X on P_X so that we can form the integral

$$\frac{1}{8\pi^2}\int_X \operatorname{tr}\big(F_{A_X} \wedge F_{A_X}\big).$$

If we take any other extension (X', P'_X, A'_X), we can glue X and X' along their common boundary to form a four-manifold without boundary. Strictly speaking, when performing the gluing we have to change the orientation of X' so that $\partial X'$ is equipped with the orientation opposite to that of M to have the resulting manifold oriented. This together with (3.1.14) yields, that the difference

$$\frac{1}{8\pi^2}\int_X \operatorname{tr}\big(F_{A_X} \wedge F_{A_X}\big) - \frac{1}{8\pi^2}\int_{X'} \operatorname{tr}\big(F_{A'_X} \wedge F_{A'_X}\big)$$

is an integer. Hence,

$$\vartheta(A) := \frac{1}{8\pi^2}\int_X \operatorname{tr}\big(F_{A_X} \wedge F_{A_X}\big) \tag{3.2.1}$$

is well-defined as a function with values in $\mathbb{R}/\mathbb{Z}$. This is called *the Chern–Simons functional.*

While this definition makes transparent the relation of the Chern–Simons functional with the Chern–Weil theory, it is possible to compute the value of the Chern–Simons functional directly without extending A to a four-manifold. In fact, choosing a trivialization of P we can think of A as a 1-form on M with values in $\mathfrak{su}(2)$. Then

$$\vartheta(A) = \frac{1}{8\pi^2}\int_M \operatorname{tr}\big(A \wedge dA + \frac{2}{3}A \wedge A \wedge A\big).$$

Notice that this expression does not yield an $\mathbb{R}$–valued function. The reason is that by changing the trivialization of P the value of ϑ changes by an integer so that we obtain again a map to the circle.

Exercise 3.2.2. Denote $CS(A) := \operatorname{tr}\big(A \wedge dA + \frac{2}{3}A \wedge A \wedge A\big)$.

(a) For $A \in \Omega^1(X; \mathfrak{su}(2))$, where X is a four-manifold, prove the equality

$$d\, CS(A) = \operatorname{tr}(F_A \wedge F_A). \tag{3.2.3}$$

Notice the following: In the special case $X = M \times \mathbb{R}$ denote by A_t the pull-back of A to $M \times \{t\}$. Then (3.2.3) clearly implies

$$\vartheta(A_t) - \vartheta(A_{t_0}) = \frac{1}{8\pi^2} \int_{M \times [t_0, t]} \operatorname{tr}(F_A \wedge F_A).$$

(b) Prove that the two definition of the Chern–Simons functional agree.

(c) Let g be a gauge transformation, and view g as a map $M \to SU(2) \cong S^3$. For $A \in \Omega^1(M)$, prove that

$$CS(A \cdot g) - CS(A) - \frac{1}{24\pi^2} \operatorname{tr}(g^{-1}dg \wedge g^{-1}dg \wedge g^{-1}dg)$$

is an exact 3-form.

(d) Using (c), prove the equality

$$\vartheta(A \cdot g) = \vartheta(A) + \deg g.$$

Conclude from this that the values of the Chern–Simons functional with respect to two different trivializations differ by an integer.

Let us compute the differential of ϑ. For $a \in \Omega^1(M; \mathfrak{su}(2))$ we have

$$d\vartheta_A(a) = \frac{1}{8\pi^2} \int_M \operatorname{tr}(a \wedge dA + A \wedge da + 2a \wedge A \wedge A) = \frac{1}{4\pi^2} \int_M \operatorname{tr}(F_A \wedge a).$$

In the second equality the integration by parts is used. Hence, we conclude the following.

Proposition 3.2.4. *The critical points of the Chern–Simons functional are flat connections, i.e., connections A such that* $F_A = 0$. □

3.3. The modui space of flat connections Even though in Section 3.2, I opted to work with three-manifolds and $G = SU(2)$, the notion of a flat connection clearly makes sense for any background manifold and any structure group. Thus, we do not need to impose these restrictions in this section.

Let $P \to M$ be a principal G–bundle. Denote by $\mathcal{A}^\flat(P)$ the space of all flat connections on P. The gauge group $\mathcal{G}(P)$ acts on $\mathcal{A}^\flat(P)$ so that we can form *the moduli space of flat connections:*

$$\mathcal{M}^\flat(P) := \mathcal{A}^\flat(P)/\mathcal{G}(P).$$

Such moduli spaces are typical objects in gauge theory that we will meet many times below. The main questions we are interested in are the following: Is $\mathcal{M}^\flat(P)$ compact? Is $\mathcal{M}^\flat(P)$ a manifold?

An important point to notice is that $\mathcal{M}^\flat$ represents the space of all solutions of a non-linear PDE modulo an equivalence relation, so that in essence the question is to describe topological properties of the space of all solutions of a non-linear PDE. In general, this may be a hard question, however, in this particular case we will see below that this can be done with a little technology involved.

However, why could one be potentially interested in spaces like $\mathcal{M}^\flat$? The two main reasons are as follows: First, sometimes $\mathcal{M}^\flat$ encodes a subtle information about the background manifold M (as well as the bundle P) and thus can be used for instance as a tool in studies of the topology of M; Secondly, moduli spaces come often equipped with an extra structure, which may be of interest on its own. In these notes I will mainly emphasize the first point, while the second one will be only briefly mentioned below.

Parallel transport and holonomy. Let $E \to M$ be a vector bundle equipped with a connection ∇. For any (smooth) curve $\gamma\colon [0,1] \to M$, $\gamma^*\nabla$ is a connection on γ^*E. A section $s \in \Gamma(\gamma^*E)$ is said to be *parallel along* γ, if $(\gamma^*\nabla)(s) = 0$.

Remark 3.3.1. If γ is a simple embedded curve, then s can be thought of as a section of E defined along the image of γ.

Since any bundle over an interval is trivial, we can trivialize $\gamma^*E \cong \mathbb{R}^k \times [0,1]$ so that $\gamma^*\nabla$ can be written as $\frac{d}{dt} + B(t)\,dt$, where $B\colon [0,1] \to M_k(\mathbb{R})$ is a map with values in the space of $k \times k$-matrices. Thinking of s as a map $[0,1] \to \mathbb{R}^k$, we obtain that s is parallel along γ if and only if s is a solution of the equations:

$$\dot{s} + A(t)s(t) = 0.$$

The main theorem of ordinary differential equations states that this equation has a unique solution defined on the whole interval $[0,1]$ for any initial value s_0.

Definition 3.3.2. If $s \in \Gamma(\gamma^*E)$ is parallel along γ, then $s(1) \in E_{\gamma(1)}$ is called the parallel transport of $s(0) = s_0 \in E_{\gamma(0)}$ with respect to ∇.

The above consideration shows in fact that for any connection ∇ and any curve γ we have a linear isomorphism

$$PT_\gamma\colon E_{\gamma(0)} \to E_{\gamma(1)},$$

which is called *the parallel transport* along γ.

In a special case, namely when γ is a loop, the parallel transport yields an isomorphism of a fiber. If we concatenate two loops, the parallel transport is the composition of the parallel transports corresponding to the initial loops. Hence, for a fixed connection the set of all parallel transports along loops based at some fixed point is in fact a group.

Definition 3.3.3. Pick a point $m \in M$. The group

$$\mathrm{Hol}_m(\nabla) := \big\{\, PT_\gamma \in GL(E_m) \mid \gamma \text{ is a loop based at } m \,\big\}$$

is called the holonomy group of ∇ based at m.

Choosing a basis of E_m, we can think of $\mathrm{Hol}_m(\nabla)$ as a subgroup of $GL_k(\mathbb{R})$. A standard argument shows that if m and m' lie in the same connected component, then the holonomy groups $\mathrm{Hol}_m(\nabla)$ and $\mathrm{Hol}_{m'}(\nabla)$ are conjugate, i.e., there is $A \in GL_k(\mathbb{R})$ such that $\mathrm{Hol}_{m'}(\nabla) = A\,\mathrm{Hol}_m(\nabla)A^{-1}$. With this understood, we can drop the basepoint from the notation. Even though $\mathrm{Hol}(\nabla)$ is defined up to a conjugacy only, it is still commonly referred to as a subgroup of $GL_k(\mathbb{R})$.

Exercise 3.3.4. Show that the following holds:

- ∇ is Euclidean $\Longrightarrow$ PT_γ is orthogonal $\Longrightarrow$ $\mathrm{Hol}(\nabla) \subset O(k)$.
- ∇ is complex $\Longrightarrow$ PT_γ is complex linear $\Longrightarrow$ $\mathrm{Hol}(\nabla) \subset GL_{k/2}(\mathbb{C})$.
- ∇ is complex Hermitian $\Longrightarrow$ PT_γ is unitary $\Longrightarrow$ $\mathrm{Hol}(\nabla) \subset U(k/2)$.

Remark 3.3.5. The concept of the parallel transport also makes sense for connections on principal bundles. The construction does not differ substantially from the case of vector bundles. The details are left to the readers.

The monodromy representation of a flat connection. Let ∇ be a flat connection on a vector bundle E of rank k. We can view the parallel transport as a map

$$\gamma \mapsto \mathrm{Hol}(\nabla; \gamma),$$

where γ is a loop based at some fixed point m. If A is flat, this map depends on the homotopy class of γ only [15, II.9] so that effectively we obtain a representation of the fundamental group: $\rho_A \colon \pi_1(M) \to GL_k(\mathbb{R})$.

Exercise 3.3.6. Show that a gauge-equivalent connection yields a conjugate representation.

Conversely, from the representation $\rho\colon \pi_1(M) \to GL_k(\mathbb{R})$ we obtain the bundle

$$E := \widetilde{M} \times_{\pi_1(M),\rho} \mathbb{R}^k.$$

Here $\widetilde{M}$ is the universal cover of M and $\pi_1(M)$ acts on $\widetilde{M}$ by the deck transformations. This means that $\widetilde{M}$ can be viewed as a principal $\pi_1(M)$–bundle so that E is the associated bundle corresponding to the representation ρ.

This bundle is equipped with a natural flat connection. Indeed, interpreting a section s of E as a $\pi_1(M)$–equivariant map $\hat{s}\colon \widetilde{M} \to \mathbb{R}^k$, we can define ∇s via

$$\pi^* \nabla s = d\hat{s},$$

cf. (2.2.25).

These constructions establish a bijective correspondence between the space of all flat connections $\mathcal{M}^\flat$ and the representation variety

$$\mathcal{R}(M; GL_k(\mathbb{R})) := \big\{\rho\colon \pi_1(M) \to GL_k(\mathbb{R}) \text{ is a group homomorphism}\big\}/\mathrm{Conj},$$

where two representations are considered to be equivalent if they are conjugate.

Remark 3.3.7. We could equally well consider flat connections on Euclidean or Hermitian vector bundles. This requires only cosmetic changes and the outcome is the representation space $\mathcal{R}(M;\, O(k))$ and $\mathcal{R}(M;\, U(k))$ respectively. Even more generally, the constructions above can be modified to the case of principal G–bundles so that the space of flat G–connections correspond to $\mathcal{R}(M;\, G)$. I leave the details to the readers.

Since the fundamental group of a manifold is finitely presented, we can choose a finite number of generators of $\pi_1(M)$, say $\gamma_1, \dots, \gamma_N$. Then any representation ρ is uniquely specified by the images of the generators $g_i = \rho(\gamma_i) \in G$, which satisfy a finite number of relations. This shows the inclusion

$$\mathcal{R}(M; G) \subset G^N/G,$$

where G acts by the adjoint action on each factor. This implies in particular the following.

Proposition 3.3.8. *Let* M *be a manifold. If* G *is a compact Lie group, then the space* $\mathcal{M}^\flat$ *of all flat* G*-connections is compact.* □

Example 3.3.9. For $M = \mathbb{T}^n$, we have clearly

$$\mathcal{R}(\mathbb{T}^n; U(1)) = \mathrm{Hom}(\mathbb{T}^n, U(1)) = U(1)^n \cong \mathbb{T}^n.$$

Example 3.3.10. Let Σ be a compact Riemann suface of genus γ without boundary. It is well-known that the fundamental group of Σ has the following representation

$$\pi_1(\Sigma) \cong \Big\langle a_1, \dots, a_\gamma, b_1, \dots, b_\gamma \mid \prod_i [a_i, b_i] = 1 \Big\rangle.$$

Hence,

$$\mathcal{R}(\Sigma, G) = \Big\{ A_1, \dots, A_\gamma, B_1, \dots, B_\gamma \in G \mid \prod_i [A_i, B_i] = 1 \Big\} / G.$$

For instance, for $G = SL(n; \mathbb{C})$ the representation variety has a rich geometric structure, which is being actively studied, see for example [5, 11, 22] and references therein.

4. Dirac operators

4.1. Spin groups and Clifford algebras In this subsection I recall briefly the notions of Clifford algebra and spin group focusing on low dimensions. More details can be found for instance in [17].

Since $\pi_1(SO(n)) \cong \mathbb{Z}/2\mathbb{Z}$ for any $n \geqslant 3$, there is a simply connected Lie group denoted by $\mathrm{Spin}(n)$ together with a homomorphism $\mathrm{Spin}(n) \to SO(n)$, which is a double covering. This characterizes $\mathrm{Spin}(n)$ up to an isomorphism. The spin groups can be constructed explicitly with the help of Clifford algebras, however in low dimensions this can be done more directly with the help of quaternions.

Since $Sp(1) \cong SU(2)$ is diffeomorphic to the 3–sphere, this is a connected and simply connected Lie group. Identify $\mathrm{Im}\,\mathbb{H} = \{\bar{h} = -h\}$ with $\mathbb{R}^3$ and consider the homomorphism

$$\alpha\colon Sp(1) \to SO(3), \qquad q \mapsto A_q, \tag{4.1.1}$$

where $A_q h = qh\bar{q}$. It is easy to check that the corresponding Lie-algebra homomorphism is in fact an isomorphism. Since $SO(3)$ is connected, α is surjective. Moreover, $\ker\alpha = \{\pm 1\}$. Hence, (4.1.1) is a non-trivial double covering, i.e., $\mathrm{Spin}(3) \cong Sp(1)$.

To construct the group $\mathrm{Spin}(4)$, recall first that the Hodge operator $*$ yields the splitting $\Lambda^2(\mathbb{R}^4)^* = \Lambda^2_+(\mathbb{R}^4)^* \oplus \Lambda^2_-(\mathbb{R}^4)^*$, where $\Lambda^2_\pm(\mathbb{R}^4)^* = \{\omega \mid *\omega = \pm\omega\}$. Since $\mathfrak{so}(4) \cong \Lambda^2(\mathbb{R}^4)^* = \Lambda^2_+(\mathbb{R}^4)^* \oplus \Lambda^2_-(\mathbb{R}^4)^* = \mathfrak{so}(3) \oplus \mathfrak{so}(3)$, the adjoint representation yields a homomorphism $\mathrm{SO}(4) \to \mathrm{SO}(3) \times \mathrm{SO}(3)$.

Identify $\mathbb{R}^4$ with $\mathbb{H}$ and consider the homomorphism[3]

$$\beta\colon \mathrm{Sp}_+(1) \times \mathrm{Sp}_-(1) \to \mathrm{SO}(4), \qquad (q_+, q_-) \mapsto A_{q_+, q_-},$$

where $A_{q_+,q_-}h = q_+ h \bar{q}_-$. An explicit computation shows that the composition $\mathrm{Sp}_+(1) \times \mathrm{Sp}_-(1) \to \mathrm{SO}(4) \to \mathrm{SO}(3) \times \mathrm{SO}(3)$ is given by $(q_+, q_-) \mapsto (A_{q_+}, A_{q_-})$. Hence, the Lie algebra homomorphism corresponding to β is an isomorphism and $\ker\beta$ is contained in $\{(\pm1, \pm1)\}$. The equality $\ker\beta = \{\pm(1,1)\} \cong \mathbb{Z}/2\mathbb{Z}$ is readily checked,. Hence, $\mathrm{Sp}_+(1) \times \mathrm{Sp}_-(1) \cong \mathrm{Spin}(4)$.

Let U be an Euclidean vector space. Then the Clifford algebra $Cl(U)$ is the tensor algebra $TU = \mathbb{R} \oplus U \oplus U \otimes U \oplus \ldots$ modulo the ideal generated by elements $u \otimes u + |u|^2 \cdot 1$. In other words, $Cl(U)$ is generated by elements of U subject to the relations $u \cdot u = -|u|^2$. For instance, $Cl(\mathbb{R}^1) \cong \mathbb{R}[x]/(x^2+1) \cong \mathbb{C}$. The algebra $Cl(\mathbb{R}^2)$ is generated by $1, e_1, e_2$ subject to the relations $e_1^2 = -1 = e_2^2$ and $e_1 \cdot e_2 = -e_2 \cdot e_1$, which follows from $(e_1 + e_2)^2 = -2$. In other words, $Cl(\mathbb{R}^2) \cong \mathbb{H}$. In general, $Cl(\mathbb{R}^n)$ is generated by $1, e_1, \ldots, e_n$ subject to the relations $e_i^2 = -1$ and $e_i \cdot e_j = -e_j \cdot e_i$ for $i \neq j$.

It can be shown that the subgroup of $Cl(\mathbb{R}^n)$ generated by elements of the form $v_1 \cdot v_2 \cdot \ldots \cdot v_{2k}$ is isomorphic to $\mathrm{Spin}(n)$, where each $v_j \in \mathbb{R}^n$ has the unit norm. In particular, this yields $\mathrm{Spin}(n) \subset Cl(\mathbb{R}^n)$.

It is convenient to have some examples of modules over Clifford algebras. Such module is given by a vector space V together with a map

$$U \otimes V \to V, \qquad u \otimes v \mapsto u \cdot v,$$

which satisfies $u \cdot (u \cdot v) = -|u|^2 v$ for all $u \in U$ and $v \in V$. An example of a $Cl(U)$–module is $V = \Lambda U^*$, where the $Cl(U)$–module structure is given by the map

$$u \otimes \varphi \mapsto \imath_u \varphi - \langle u, \cdot \rangle \wedge \varphi. \tag{4.1.2}$$

Let V be a quaternionic vector space. Then the quaternionic multiplication gives rise to the map $\mathrm{Im}\,\mathbb{H} \otimes V \to V$ by sending $h \otimes v \mapsto h \cdot v$ and this satisfies $h \cdot (h \cdot v) = -h\bar{h}v = -|h|^2 v$. Thus, any quaternionic vector space is a $Cl(\mathbb{R}^3)$–module. In particular, the fundamental representation $\$ \cong \mathbb{H}$ of $\mathrm{Sp}(1) \cong \mathrm{Spin}(3)$ with the action given by the left multiplication is a $Cl(\mathbb{R}^3)$–module.

The multiplication on the right by $\bar{\imath}$ endows $\$$ with the structure of a complex $\mathrm{Sp}(1)$–representation, which is in fact also Hermitian (this is just another manifestation of the isomorphism $\mathrm{Sp}(1) \cong \mathrm{SU}(2)$). It is then an elementary exercise in

[3] We adopt the common convention $\mathrm{Sp}_\pm(1) = \mathrm{Sp}(1)$. The significance of the subscripts "$\pm$" will be clear below.

the representation theory to show the isomorphisms

$$\operatorname{Im}\mathbb{H}\otimes\mathbb{C}\cong \operatorname{End}_0(\mathbb{S}), \tag{4.1.3}$$

where the left hand side is viewed as an Sp(1)–representation via the homomorphism α and $\operatorname{End}_0(\mathbb{S})$ denotes the subspace of traceless endomorphisms. Moreover, the real subspace $\operatorname{Im}\mathbb{H}$ can be identified with the subspace of traceless Hermitian endomorphisms.

Furthermore, for any quaternionic vector space V the space $V\oplus V$ is a $Cl(\mathbb{R}^4)$–module. Indeed, the $Cl(\mathbb{R}^4)$–module structure is induced by the map

$$\mathbb{H}\otimes_{\mathbb{R}}(V\oplus V)\to V\oplus V,\quad h\otimes(v_1,v_2)\mapsto(hv_2,-\bar{h}v_1)=\begin{pmatrix}0 & h\\ -\bar{h} & 0\end{pmatrix}\begin{pmatrix}v_1\\ v_2\end{pmatrix}. \tag{4.1.4}$$

In particular, the $Sp_+(1)\times Sp_-(1)$–representation $\mathbb{S}^+\oplus\mathbb{S}^-$ is a $Cl(\mathbb{R}^4)$–module. Here, as the notation suggests, $\mathbb{S}^\pm$ is the fundamental representation of $Sp_\pm(1)$.

Just like in the case of dimension three, we have an isomorphism of Spin(4)–representations

$$\mathbb{H}\otimes\mathbb{C}\cong \operatorname{Hom}(\mathbb{S}^+;\mathbb{S}^-),$$

where the left hand side is viewed as a Spin(4)–representation via the homomorphism β.

4.2. Dirac operators Let M be a Riemannian oriented manifold of dimension n. Denote by Fr_{SO} the principal $SO(n)$ bundle of orthonormal oriented frames. The tautological action of $SO(n)$ on $\mathbb{R}^n$ extends to an action on $Cl(\mathbb{R}^n)$ so that we can construct the associated bundle

$$Cl(M):=Fr_{SO}\times_{SO(n)}Cl(\mathbb{R}^n).$$

This can view this as the bundle with fiber $Cl(T_mM)\cong Cl(T^*_mM)$ at the point $m\in M$. Notice that the Levi–Civita connection yields a connection on $Cl(M)$. This is denoted by the same symbol ∇^{LC}.

Let $E\to M$ be a bundle of $Cl(M)$–modules, i.e., there is a morphism of vector bundles

$$Cl\colon TM\otimes E\to E,\qquad (v,e)\mapsto v\cdot e,$$

such that $v\cdot(v\cdot e)=-|v|^2e$. Then E is called a Dirac bundle if it is equipped with an Euclidean scalar product and a connection ∇ such that the following conditions hold:

- ∇ is Euclidean;
- $\langle v\cdot e_1, v\cdot e_2\rangle=|v|^2\langle e_1,e_2\rangle$ for any $v\in T_mM$ and $e_1,e_2\in E_m$;
- $\nabla(\varphi\cdot s)=(\nabla^{LC}\varphi)\cdot s+\varphi\cdot\nabla s$ for any $\varphi\in\Gamma(Cl(M))$ and $s\in\Gamma(E)$.

Definition 4.2.1. If E is a Dirac bundle, the operator

$$\mathcal{D}\colon\Gamma(E)\xrightarrow{\nabla}\Gamma(T^*M\otimes E)\xrightarrow{Cl}\Gamma(E)$$

is called the Dirac operator of E.

In other words, if $e_1, \dots, e_n$ is a local orthonormal oriented frame of TM, then

$$\mathcal{D}\, s = \sum_{i=1}^{n} e_i \cdot \nabla_{e_i} s$$

Example 4.2.2. The bundle $\Lambda T^*M = \oplus_{k=0}^{n} \Lambda^k T^*M$ has a natural structure of a Dirac bundle, where the Clifford multiplication is given by (4.1.2). The corresponding Dirac operator equals $d + d^*$ [17, Thm 5.12], where d^* is the formal adjoint of d, see Section 5.3 below for more details.

Exercise 4.2.3. Pick two sections $s_1, s_2 \in \Gamma(E)$ and define a vector field v by the following condition:

$$\langle v, w \rangle = -\langle s_1, w \cdot s_2 \rangle \qquad \text{for all vector fields } w.$$

(a) For any $m \in M$ prove the equality

$$\langle \mathcal{D}\, s_1, s_2 \rangle_m = (\operatorname{div} v)_m + \langle s_1, \mathcal{D}\, s_2 \rangle_m,$$

where $\operatorname{div} v$ is the divergence of v.

(b) Show that the Dirac operator on a closed manifold is formally self-adjoint, that is,

$$\int_M \langle \mathcal{D}\, s_1,\, s_2 \rangle = \int_M \langle s_1,\, \mathcal{D}\, s_2 \rangle.$$

4.3. Spin and Spin^c structures Let $Fr_{SO} \to M$ be the principal bundle of orthonormal oriented frames of M.

Definition 4.3.1. M is said to be spinnable, if there is a principal $Spin(n)$ bundle P equipped with a $Spin(n)$–equivariant map $\tau\colon P \to Fr_{SO}$, which covers the identity map on M and is a fiberwise double covering. Here $Spin(n)$ acts on Fr_{SO} via the homomorphism $Spin(n) \to SO(n)$.

A choice of a bundle P as above is called a spin structure. A manifold equipped with a spin structure is called a spin manifold.

For a given M a spin structure may or may not exist. If M is spinnable, there may be many non-equivalent spin structures. The questions on existence and classification of spin structures may be completely answered in terms of the Stiefel–Whitney classes [17]. However, I will not go into the details here. For the remaining part of this section I assume throughout that M is spin.

Let $\omega \in \Omega^1(Fr_{SO}; \mathfrak{so}(n))$ be the connection 1-form of the Levi–Civita connection. Since the homomorphism $Spin(n) \to SO(n)$ is a local diffeomorphism. Hence we have an isomorphism of Lie algebras $\mathfrak{spin}(n) \cong \mathfrak{so}(n)$ and the pullback $\tau^*\omega \in \Omega^1(P; \mathfrak{spin}(n))$ is a connection on P. Slightly abusing terminology, this is still called the Levi–Civita connection.

For any $n \geqslant 3$ there is a unique complex representation $\rho\colon Spin(n) \to End(\$)$ distinguished by the property that it extends to a complex irreducible representation of $Cl(\mathbb{R}^n)$. Notice that this means neither that $\$$ is a unique $Spin(n)$–representation, nor that $\$$ is an irreducible $Spin(n)$–representation. For example, for $n = 3$ this representation coincides with the fundamental representation of $Sp(1) = Spin(3)$. For $n = 4$ we have $\$ = \$^+ \oplus \$^-$, see (4.1.4).

If M is spin, we can construct *the spinor bundle*

$$\$:= P \times_{Spin(n),\rho} \$.$$

Here, as it is quite common, we use the same notation both for the representation and the associated vector bundle.

Since ρ extends to a representation of $Cl(\mathbb{R}^n)$, the spinor bundle is in fact a bundle of $Cl(M)$–modules. Hence, the construction of Subsection 4.2 yields the spin Dirac operator

$$\not{D}\colon \Gamma(\$) \to \Gamma(\$).$$

Remark 4.3.2. Recall that when $\dim M = 4$, the spinor bundle splits: $\$ = \$^+ \oplus \$^-$. By (4.1.4), the Clifford multiplication with 1–forms changes the chirality, i.e., for any $\omega \in \Omega^1(M)$ we have $\omega\cdot\colon \$^\pm \to \$^\mp$. Hence,

$$\not{D} = \begin{pmatrix} 0 & \not{D}^- \\ \not{D}^+ & 0 \end{pmatrix}, \qquad \text{where } \not{D}^\pm\colon \Gamma(\$^\pm) \to \Gamma(\$^\mp).$$

The following two variations of this construction are frequently used. First, let P be a principal G–bundle equipped with a connection A and let $\tau\colon G \to U(n)$ be a unitary representation of G so that we have the associated bundle $E = P \times_{G,\tau} \mathbb{C}^n$, which is Hermitian. Then the twisted spinor bundle $\$ \otimes E$ is also a Dirac bundle so that we have a twisted Dirac operator

$$\not{D}_A\colon \Gamma(\$ \otimes E) \to \Gamma(\$ \otimes E).$$

Example 4.3.3. Let us assume that $\dim M = 3$ for the sake of definiteness. Choose $E = \$$, which is equipped with the Levi–Civita connection. Then we have

$$\$ \otimes \$ = \mathrm{Sym}^2(\$) \oplus \Lambda^2\$ \cong T^*_{\mathbb{C}}M \oplus \mathbb{C},$$

cf. (4.1.3). Hence, the twisted spinors can be identified with the complexification of odd forms. Of course, the Hodge $*$-operator yields and isomorphism between odd and even forms so that we can identify the twisted spinors with even forms too. We already have seen above a Dirac operator acting on forms, namely

$$d + d^*\colon \Omega^{odd}(M) \to \Omega^{even}(M),$$

cf. Example 4.2.2. One can show that the complexification of $d + d^*$ coincides with the twisted Dirac operator on $\Gamma(\$ \otimes \$)$.

Exercise 4.3.4. The Clifford multiplication combined with the map $\mathrm{ad}\,P \to \mathrm{End}(E)$ yields the 'twisted' Clifford multiplication

$$T^*M \otimes \mathrm{ad}\,P \to \mathrm{End}(\$) \otimes \mathrm{End}(E) \cong \mathrm{End}(\$ \otimes E).$$

Show that for $a \in \Omega^1(\mathrm{ad}\,P)$ the following holds:

$$\not{D}_{A+a}\psi = \not{D}_A\psi + a \cdot \psi.$$

Let me explain the second variation, which in essence is not really much different from the first one, however the details are somewhat involved.

Thus, by the construction of $\mathrm{Spin}(n)$ we have the exact sequence

$$\{1\} \to \{\pm 1\} \to \mathrm{Spin}(n) \to \mathrm{SO}(n) \to \{1\}.$$

Moreover, one can show that the kernel $\{\pm 1\}$ lies in the center of $\mathrm{Spin}(n)$. This is clear anyway in our main cases of interest, namely for $n = 3$ and $n = 4$. Denote

$$\mathrm{Spin}^c(n) := \mathrm{Spin}(n) \times \mathrm{U}(1)/\pm 1,$$

where $\{\pm 1\}$ is embedded diagonally. Notice that both $\mathrm{Spin}(n)$ and $\mathrm{U}(1)$ are subgroups of $\mathrm{Spin}^c(n)$ and $\mathrm{U}(1)$ lies in fact in the center of $\mathrm{Spin}^c(n)$.

Furthermore, we have the following exact sequences:

$$\{1\} \to \mathrm{U}(1) \to \mathrm{Spin}^c(n) \xrightarrow{\rho_0} \mathrm{Spin}(n)/\pm 1 = \mathrm{SO}(n) \to \{1\},$$

$$\{1\} \to \mathrm{Spin}(n) \to \mathrm{Spin}^c(n) \xrightarrow{\rho_{\det}} \mathrm{U}(1)/\pm 1 \cong \mathrm{U}(1) \to \{1\}.$$

These in turn give rise to the exact sequence

$$\{1\} \to \{\pm 1\} \to \mathrm{Spin}^c(n) \xrightarrow{(\rho_0, \rho_{\det})} \mathrm{SO}(n) \times \mathrm{U}(1) \to \{1\}, \tag{4.3.5}$$

which shows that $\mathrm{Spin}^c(n)$ is a double covering of $\mathrm{SO}(n) \times \mathrm{U}(1)$.

Example 4.3.6.

(a) For $n = 3$ we have $\mathrm{Spin}^c(3) = \mathrm{SU}(2) \times \mathrm{U}(1)/\pm 1 \cong \mathrm{U}(2)$. In particular, $\rho_{\det}(A) = \det A$ and the sequence (4.3.5) has the following form

$$\{1\} \to \{\pm \mathbb{1}\} \to \mathrm{U}(2) \to \mathrm{SO}(3) \times \mathrm{U}(1) \to \{1\},$$

where the homomorphism $\mathrm{U}(2) \to \mathrm{SO}(3) = \mathrm{PU}(2)$ is the natural projection.

(b) For $n = 4$ we have

$$\begin{aligned}\mathrm{Spin}^c(4) &= \Big(\big(\mathrm{SU}(2) \times \mathrm{SU}(2)\big) \times \mathrm{U}(1)\Big)/\pm 1 \\ &= \big\{(A_+, A_-) \in \mathrm{U}(2) \times \mathrm{U}(2) \mid \det A_+ = \det A_-\big\}.\end{aligned}$$

In particular, $\rho_{\det}(A_+, A_-) = \det A_+ = \det A_-$.

Notice that in this case we also have the homomorphisms

$$\rho_\pm \colon \mathrm{Spin}^c(4) \to \mathrm{U}(2), \qquad \rho_\pm(A_+, A_-) = A_\pm.$$

Definition 4.3.7. A spin^c structure on M is a principal $\mathrm{Spin}^c(n)$–bundle $P \to M$ equipped with a $\mathrm{Spin}^c(n)$–equivariant map $P \to \mathrm{Fr}_{\mathrm{SO}}$ which induces an isomorphism $P/\mathrm{U}(1) \cong \mathrm{Fr}_{\mathrm{SO}}$.

Just like in the case of spin structures, spin^c structures may or may not exist. If a spin^c structure exists, it is rarely unique. It is also clear that if M is spin, then M is also spin^c.

It turns out that four–manifolds are somewhat special as the following result shows.

Proposition 4.3.8. *Any closed oriented four–manifold admits a spin^c structure.* □

A proof of this claim can be found for example in [21, Lem. 3.1.2]. Notice however, that there are closed four–manifolds that are not spin. In dimension three the situation is different: Any closed oriented three–manifold is spin [10, Rem. 1.4.27], and hence also spin^c.

With these preliminaries at hand, let M be a manifold equipped with a spin^c structure P. Define *the determinant line bundle*

$$L_{\det} := P \times_{\rho_{\det}} \mathbb{C}.$$

This is clearly a Hermitian line bundle and its $U(1)$–structure is

$$P_{\det} := P/\operatorname{Spin}(n).$$

By (4.3.5), we have a double cover map

$$\tau\colon P \to P/\{\pm 1\} = \mathrm{Fr}_{SO} \times_M P_{\det}.$$

Hence, if A is a connection on $P_{\det}$, then $\tau^*(\omega + A) \in \Omega^1\big(P; \mathfrak{spin}^c(n)\big)$, where ω is the connection 1-form of the Levi-Civita connection. Notice also that we have used the isomorphism $\mathfrak{spin}^c(n) \cong \mathfrak{so}(n) \oplus \mathfrak{u}(1)$. Thus, the choice of a unitary connection on the determinant line bundle together with the Levi–Civita connection determines a connection on the spin^c bundle.

Let $\$$ be the distinguished representation of $\operatorname{Spin}(n)$. This can be clearly extended to a $\operatorname{Spin}^c(n)$–representation, which is still denoted by $\$$: $[g,z]\cdot s = z\rho(g)\, s$. This in turn yields the spin^c spinor bundle

$$\$:= P \times_{\operatorname{Spin}^c(n)} \$,$$

which is a Dirac bundle. Hence, in turn, we obtain the spin^c Dirac operator $\not{D}_A\colon \Gamma(\$) \to \Gamma(\$)$.

Example 4.3.9. Assume M is spin and pick a spin structure P_{Spin}. Pick also a principal $U(1)$–bundle P_0. Then

$$P := P_{\mathrm{Spin}} \times_M P_0/\pm 1$$

is a principal $\big(\operatorname{Spin}(n) \times U(1)\big)/\pm 1 = \operatorname{Spin}^c(n)$ bundle. In this case the spin^c spinor bundle is just the twisted spinor bundle $\$ \otimes L_0$, where $\$$ is the pure spinor bundle and $L_0 := P_0 \times_{U(1)} \mathbb{C}$ is the associated complex line bundle. Likewise, the spin^c Dirac operator is just the twisted Dirac operator.

Remark 4.3.10. In some sense, any spin^c spinor bundle can be thought of as the twisted spinor bundle just like in the example above. The problem is that in general neither $\$$ nor L_0 is globally well-defined, however their product $\$ \otimes L_0$ is well-defined. The notion of a spin^c structure is just a convenient way to make this 'definition' precise.

Exercise 4.3.11. For $a \in \Omega^1(M; \mathbb{R}i)$ prove that

$$\not{D}_{A+a}\psi = \not{D}_A\psi + \frac{1}{2}\, a \cdot \psi. \tag{4.3.12}$$

The coefficient $\frac{1}{2}$ in (4.3.12) can be explained as follows: Think of the spin^c spinor bundle as $\$ \otimes L_0$ just like in Remark 4.3.10. Then the determinant line bundle is $\Lambda^2(\$ \otimes L_0) = L_0^2$ so that $A_0 \in \mathcal{A}(L_0)$ induces some connection A on L_0^2. Then $A_0 + a_0$ corresponds to $A + 2\,a_0$.

On the classification of spin^c structures. Let P be a spin^c structure with the spinor bundle $\$$. If L is any Hermitian line bundle, then $\$ \otimes L$ is a spin^c spinor bundle corresponding to a spin^c structure P_L. This defines an action of $H^2(M; \mathbb{Z})$ on the set $\mathcal{S} = \mathcal{S}(M)$ of all spin^c structures on M. This action can be shown to be free and transitive, hence $\mathcal{S}(M)$ can be identified with $H^2(M; \mathbb{Z})$, however, such an identification is not canonical.

For example, let M be spin and let $\$$ be the pure spinor bundle. The spin structure of M can be also viewed as a distinguished spin^c structure so that $\mathcal{S}(M)$ has a preferred point, namely, the origin. This choice fixes an isomorphism $\mathcal{S}(M) \cong H^2(M; \mathbb{Z})$.

4.4. The Weitzenböck formula For a function $u\colon \mathbb{R}^4 \to \mathbb{H}$ the operators

$$\begin{aligned} \not{D}^+(u) &= \frac{\partial u}{\partial x_0} + i\frac{\partial u}{\partial x_1} + j\frac{\partial u}{\partial x_2} + k\frac{\partial u}{\partial x_3}, \\ \not{D}^-(u) &= -\frac{\partial u}{\partial x_0} + i\frac{\partial u}{\partial x_1} + j\frac{\partial u}{\partial x_2} + k\frac{\partial u}{\partial x_3} \end{aligned}$$

can be thought of as four-dimensional analogues of the $\bar\partial = \frac{1}{2}(\frac{\partial}{\partial x} + i\frac{\partial}{\partial y})$ and $\partial = \frac{1}{2}(\frac{\partial}{\partial x} - i\frac{\partial}{\partial y})$ operators for complex–valued functions of one complex variable $z = x + yi$ respectively. In fact, tracing through the construction, it is easy to see that for the flat four–manifold $\mathbb{R}^4$ the Dirac operator can be written as follows

$$\not{D} = \begin{pmatrix} 0 & \not{D}^- \\ \not{D}^+ & 0 \end{pmatrix},$$

cf. (4.1.4). Here I use the fact, that the spinor bundle of $\mathbb{R}^4$ is (canonically) the product bundle. A straightforward computation yields

$$\not{D}^2 = \begin{pmatrix} \not{D}^-\not{D}^+ & 0 \\ 0 & \not{D}^+\not{D}^- \end{pmatrix} = \Delta, \tag{4.4.1}$$

where $\Delta = -\sum_{i=0}^{3} \frac{\partial^2}{\partial x_i^2}$ is the Laplacian. This is usually phrased as "the Dirac operator is the square root of the Laplacian".

Remark 4.4.2. The Dirac operator on $\mathbb{R}^2$ has the form

$$\begin{pmatrix} 0 & 2\frac{\partial}{\partial z} \\ -2\frac{\partial}{\partial \bar z} & 0 \end{pmatrix} : C^\infty(\mathbb{R}^2; \mathbb{C}^2) \to C^\infty(\mathbb{R}^2; \mathbb{C}^2)$$

and squares to the Laplacian just like in dimension four. Notice, however, that the case of dimension two requires a special treatment because $\pi_1(SO(2)) \cong \mathbb{Z}$, which was one of the reasons I assumed $n \geqslant 3$ in this section.

On $\mathbb{R}^3$ the spinor bundle can be identified with the product bundle $\underline{\mathbb{H}}$ so that the corresponding Dirac operator is

$$\not{D}(u) = i\frac{\partial}{\partial x_1} + j\frac{\partial}{\partial x_2} + k\frac{\partial}{\partial x_3},$$

which also squares to the Laplacian.

On general Riemannian manifolds the relation $\not{D}^2 = \Delta$ still holds up to a zero order operator. This is known as the Weitzenböck formula, which I describe next.

Thus, given an Euclidean vector bundle E with a connection ∇ *the connection Laplacian* is defined by

$$\nabla^*\nabla \colon \Gamma(E) \xrightarrow{\nabla} \Omega^1(E) \xrightarrow{\nabla^* = -*d_\nabla *} \Gamma(E). \tag{4.4.3}$$

Exercise 4.4.4.

(a) Let $(e_1, \dots, e_n)$ be a local orthonormal frame of TM over an open subset of M. Show that the connection Laplacian can be expressed as follows

$$\nabla^*\nabla s = -\sum_{i=1}^n \left(\nabla_{e_i}\nabla_{e_i} s - \nabla_{\nabla_{e_i} e_i} s\right),$$

where ∇e_i means the Levi–Civita connection applied to e_i.

(b) Prove that the connection Laplacian is formally self-adjoint.

(c) In the case M is closed, prove the identity $\langle \nabla^*\nabla s,\, s\rangle_{L^2} = \|\nabla s\|^2_{L^2}$.

Assume that E is a Dirac bundle and let $R \in \Omega^2(\mathrm{End}(E))$ be the curvature 2-form of the corresponding connection ∇. Using the Clifford multiplication, we obtain a homomorphism $\Lambda^2T^*M \otimes \mathrm{End}(E) \to \mathrm{End}(E)$, which maps R to an endomorphism $\mathfrak{R}$. In a local orthonormal frame (e_i) of TM this can be expressed as follows

$$\mathfrak{R}(s) := \frac{1}{2}\sum_{i,j=1}^n e_i \cdot e_j \cdot R_{e_i, e_j}(s).$$

Theorem 4.4.5 (Weitzenböck formula)**.** *Let $\not{D}$ be the Dirac operator for the Dirac bundle E. Then the following holds:*

$$\not{D}^2 = \nabla^*\nabla + \mathfrak{R}.$$

Proof. Pick a point $m \in M$ and a local orthonormal frame (e_i) such that $\nabla_{e_i} e_j$ vanishes at m. Then at m we have the following:

$$\begin{aligned}
\not{D}^2 s &= \sum_{i=1}^n e_i \cdot \nabla_{e_i}\Big(\sum_{j=1}^n e_j \cdot \nabla_{e_j} s\Big) = \sum_{i,j=1}^n e_i \cdot e_j \cdot \nabla_{e_i}\nabla_{e_j} s \\
&= \sum_{i=1}^n e_i \cdot e_i \cdot \nabla_{e_i}\nabla_{e_i} s + \frac{1}{2}\sum_{i\neq j} e_i \cdot e_j \cdot (\nabla_{e_i}\nabla_{e_j} - \nabla_{e_j}\nabla_{e_i}) s \\
&= \nabla^*\nabla s + \mathfrak{R}s.
\end{aligned}$$

□

In the case of the spin^c spinor bundle, a straightforward computation yields the following corollary, whose proof is left as an exercise.

Corollary 4.4.6. *Let* A *be a Hermitian connection on the determinant line bundle with the curvature form* F_A*. Then the spin*^{c} *Dirac operator satisfies:*

$$\not{D}^2\psi = \nabla^*\nabla\psi + \frac{1}{4}s_g\,\psi + \frac{1}{2}F_A\cdot\psi,$$

where s_g *is the scalar curvature of the background metric* g. □

In particular, in the case M is spin and $\not{S}$ is the pure spinor bundle, we have

$$\not{D}^2\psi = \nabla^*\nabla\psi + \frac{1}{4}s_g\,\psi.$$

This implies, for example, that for a metric with positive scalar curvature the space of harmonic spinors $\ker\not{D}$ is trivial.

Recall that in dimension four, the spinor bundle splits: $\not{S} = \not{S}^+ \oplus \not{S}^-$. With respect to this splitting the Dirac operator takes the form

$$\not{D} = \begin{pmatrix} 0 & \not{D}^- \\ \not{D}^+ & 0 \end{pmatrix},$$

which we already have seen above in the case of the flat space. By Corollary 4.4.6 for $\psi \in \Gamma(\not{S}^+)$ we obtain

$$\not{D}^-\not{D}^+\psi = (\not{D}^+)^*\not{D}^+\psi = \nabla^*\nabla\psi + \frac{1}{4}s_g\,\psi + \frac{1}{2}F_A^+\cdot\psi.$$

Here we used the fact that anti-self-dual 2-forms act trivially on $\not{S}^+$.

5. Linear elliptic operators

5.1. Sobolev spaces Consider the following classical boundary value problem in the theory of partial differential equations: Let $\Omega \subset \mathbb{R}^n$ be a bounded domain with a smooth boundary $\partial\Omega$. Does there exist a function $u \in C^2(\Omega)\cap C^0(\bar{\Omega})$ such that

$$\Delta u = 0 \quad \text{in } \Omega \qquad \text{and} \qquad u|_{\partial\Omega} = \varphi, \tag{5.1.1}$$

where φ is a given function on $\partial\Omega$?

Consider *the energy functional*

$$E(u) := \frac{1}{2}\int_\Omega |\nabla u(x)|^2\,dx,$$

where dx denotes the standard volume form on $\mathbb{R}^n$. Assume there exists an absolute minimum of E, i.e., a function $u \in C^2(\Omega)\cap C^0(\bar{\Omega})$ such that

$$E(u) = \inf\{E(v) \mid v \in C^2(\Omega)\cap C^0(\bar{\Omega}),\ v|_{\partial\Omega} = \varphi\}.$$

A straightforward computation using integration by parts yields

$$0 = \frac{d}{dt}\Big|_{t=0}E(u+tw) = \int_\Omega w\,\Delta u\,dx$$

for all w such that $w|_{\partial\Omega} = 0$. This implies that u is harmonic in Ω. Thus, we can find a solution of (5.1.1) if we can show that the energy functional attains its minimum.

Hence, a strategy for proving the existence of solutions of (5.1.1) may be the following: Pick a sequence u_k such that $E(u_k)$ converges to the infimum of the energy functional and prove that u_k converges to a limit u possibly after extracting a subsequence. This strategy does work indeed, however, it requires certain technology known as the theory of Sobolev spaces, which turned out to be very useful in gauge theory as well. What follows below is a crash course in Sobolev spaces. For details, interested readers may consult more specialized literature.

First recall that the space $L^2(\mathbb{R}^n)$ of square integrable functions equipped with a scalar product

$$\langle u, v\rangle_{L^2} := \int_{\mathbb{R}^n} uv\, dx$$

is a Hilbert space, i.e., it is complete with respect to the norm $\|u\|_{L^2} := \sqrt{\langle u, u\rangle}$. This space can be viewed as a completion of the space $C_0^\infty(\mathbb{R}^n)$ of all smooth functions with compact support with respect to the norm $\|\cdot\|_{L^2}$.

This definition admits a number of variations, which will be of use below. First, we can pick any $p > 1$ and put

$$\|u\|_{L^p} := \left(\int_{\mathbb{R}^n} |u(x)|^p\, dx\right)^{\frac{1}{p}},$$

where $u \in C_0^\infty(\mathbb{R}^n)$. The completion of $C_0^\infty(\mathbb{R}^n)$ with respect to $\|\cdot\|_{L^p}$ is then the Banach space $L^p(\mathbb{R}^n)$.

Secondly, we can also consider

$$\|u\|_{W^{k,p}} := \Big(\sum_{i=0}^{k} \|\nabla^i u\|_{L^p}\Big)^{\frac{1}{p}}.$$

The completion of $C_0^\infty(\mathbb{R}^n)$ with respect to this norm is denoted by $W^{k,p}(\mathbb{R}^n)$. These spaces are called *Sobolev spaces*.

Remark 5.1.2. Many different and inconsistent notations are in use for Sobolev spaces. For example, sometimes L^s_r may mean $W^{r,s}$ and sometimes $W^{s,r}$. Of all notations being used for Sobolev spaces, the symbol $W^{k,p}$ seems to be most consistently used, hence I opted for this one.

Third, we can replace the domain $\mathbb{R}^n$ by any open subset of $\mathbb{R}^n$ or, even more generally, by a Riemannian manifold M. The corresponding spaces will be denoted by $W^{k,p}(M)$. Here ∇u should be understood as the differential of u, that is $\nabla u \in \Gamma(T^*M)$. Fixing a connection on T^*M, for example the Levi–Civita one, ∇u can be differentiated again. This clarifies the meaning of symbols such as $\nabla^2 u \in \Gamma(T^*M \otimes T^*M)$ and so on up to the dependence of the space $W^{k,p}(M)$ on the choice of connection.

This naturally leads us to one more variation of the definitions above. Pick a vector bundle $E \to M$ and connections $\nabla^E \in \mathcal{A}(E), \nabla^M \in \mathcal{A}(T^*M)$. This yields a connection on $T^*M \otimes E$ so that the higher derivatives $\nabla^i(\nabla s)$ are well defined for any $s \in \Gamma(E)$. This yields the Sobolev spaces $W^{k,p}(M;E)$ in the same manner as above. While the norm $\|\cdot\|_{W^{k,p}}$ does depend on ∇^E and ∇^M, different choices yield equivalent norms so that the resulting topology is independent of the choices made.

With this understood, for any $p > 1$ we have the sequence of inclusions

$$L^p(M;E) = W^{0,p}(M;E) \supset W^{1,p}(M;E) \supset W^{2,p}(M;E) \supset \dots$$

Relations between all these spaces is given by the following theorem, which is of fundamental importance in the theory of PDEs.

Theorem 5.1.3. *Let M be a compact manifold.*

(i) *If $s \in W^{k,p}(M;E)$, then $s \in W^{m,q}(M;E)$ provided*

$$k - \frac{n}{p} \geqslant m - \frac{n}{q} \quad \textit{and} \quad k \geqslant m,$$

where $n = \dim M$, and provided there is a constant C independent of s such that $\|s\|_{W^{m,s}} \leqslant C\|s\|_{W^{k,p}}$. In other words, the natural embedding

$$j\colon W^{k,p}(M;E) \subset W^{m,q}(M;E)$$

is continuous.

(ii) *j is a compact operator provided*

$$k - \frac{n}{p} > m - \frac{n}{q} \quad \textit{and} \quad k > m. \tag{5.1.4}$$

This means that any sequence bounded in $W^{k,p}$ has a subsequence, which converges in $W^{m,q}$ provided (5.1.4) holds.

(iii) *We have a natural continuous embedding*

$$W^{k,p}(M;E) \subset C^r(M;E)$$

provided $k - \frac{n}{p} > r$. In particular, if $s \in W^{k,p}(M;E)$ for some fixed p and for all $k \geqslant 0$, then $s \in C^\infty(M;E)$.

(iv) (a) *In the case $kp > n$ the space $W^{k,p}(M;\mathbb{R})$ is an algebra.*

(b) *In the case $kp < n$, we have a bounded map*

$$W^{k_1,p_1} \otimes W^{k_2,p_2} \to W^{k,p}, \qquad \textit{provided} \;\; k_1 - \frac{n}{p_1} + k_2 - \frac{n}{p_2} \geqslant k - \frac{n}{p}. \qquad \square$$

Remark 5.1.5. Although the proof of this theorem goes beyond the goals of these notes, it may be instructive to see "the spirit of the proof" in a special case. Thus, let $M = S^1$ and $u \in C^\infty(S^1;\mathbb{R})$. Denote $\bar u := \frac{1}{2\pi}\int u(\theta)\,d\theta$ and $u_0(\theta) := u(\theta) - \bar u$. By the mean value theorem, there is $\theta_0 \in S^1$ such that $u_0(\theta_0) = 0$. Hence, for any $\theta \in S^1$ we have

$$|u_0(\theta)| = \left|\int_{\theta_0}^{\theta} u_0'(\varphi)\,d\varphi\right| \leqslant \sqrt{\int_{\theta_0}^{\theta} |u_0'(\varphi)|^2\,d\varphi}\,\sqrt{\int_{\theta_0}^{\theta} 1^2\,d\varphi} \leqslant \sqrt{2\pi}\,\|u_0\|_{W^{1,2}}, \tag{5.1.6}$$

where the first inequality follows by the the Cauchy–Schwarz inequality. This yields the estimate $\|u\|_{C^0} \leqslant C\|u\|_{W^{1,2}}$, which in turn shows that there is a continuous embedding $W^{1,2}(S^1) \subset C^0(S^1)$.

Furthermore, by tracing through (5.1.6) it is easy to see that

$$|u(\theta_1) - u(\theta_2)| \leqslant \sqrt{2\pi}\, \|u\|_{W^{1,2}}\, \mathrm{dist}(\theta_1, \theta_2)^{\frac{1}{2}}.$$

Hence, if u_k is any sequence bounded in $W^{1,2}(S^1)$, then this sequence consists of uniformly bounded and equicontinuous functions on S^1. By the Arzela–Ascoli theorem, this sequence has a convergent subsequence in $C^0(S^1)$, thus proving the compactness of the embedding $W^{1,2}(S^1) \subset L^p(S^1)$ for any p.

5.2. Elliptic operators A map $L\colon C^\infty(\Omega; \mathbb{R}^r) \to C^\infty(\Omega; \mathbb{R}^s)$, where $\Omega \subset \mathbb{R}^n$ is an open subset, is said to be a linear differential operator of order ℓ if L can be expressed in the form

$$Lf = \sum_{|\alpha| \leqslant \ell} A_\alpha(x) \frac{\partial^{|\alpha|}}{\partial x^\alpha} f, \tag{5.2.1}$$

where $\alpha = (\alpha_1, \ldots, \alpha_n)$, $\alpha_i \in \mathbb{Z}_{\geqslant 0}$, is a multi-index, $|\alpha| := \sum \alpha_i$, and A_α lies in $C^\infty\big(\Omega; \mathrm{Hom}(\mathbb{R}^r; \mathbb{R}^s)\big)$. For example, in the case of operator of order $\ell = 2$ acting on functions of two variables, we have

$$Lf = A_{20}\frac{\partial^2}{\partial x_1^2} + A_{11}\frac{\partial^2}{\partial x_1 \partial x_2} + A_{02}\frac{\partial^2}{\partial x_2^2} + A_{10}\frac{\partial}{\partial x_1} + A_{01}\frac{\partial}{\partial x_2} + A_{00},$$

where A_{ij} are smooth functions on Ω.

It is intuitively clear that the highest order terms determine some essential properties of L. Thus, we say that

$$\sigma_L(\xi) := \sum_{|\alpha| = \ell} A_\alpha(x)\xi^\alpha = \sum_{|\alpha|=\ell} A_\alpha(x)\xi_1^{\alpha_1} \cdot \ldots \cdot \xi_n^{\alpha_n} \qquad \text{where } \xi \in (\mathbb{R}^n)^* \cong \mathbb{R}^n$$

is *the principal symbol* of L, or simply *the symbol*. It can be conveniently viewed as a map $\Omega \times \mathbb{R}^n \to \mathrm{Hom}(\mathbb{R}^r; \mathbb{R}^s)$, which is polynomial in the ξ-variable.

Definition 5.2.2. A linear differential operator L is called *elliptic*, if for all $x \in \Omega$ and all $\xi \in \mathbb{R}^n$, $\xi \neq 0$, we have: $\sigma_L(\xi)$ is an invertible homomorphism.

Notice in particular that we must require $r = s$ to have an elliptic operator.

Example 5.2.3. For the Laplacian $\Delta = -\sum \frac{\partial^2}{\partial x_i^2}$ on the flat space $\mathbb{R}^n$ acting on functions we have

$$\sigma_\Delta(\xi) = -\sum_{i=1}^n \xi_i^2 = -|\xi|^2.$$

Hence, Δ is an elliptic operator. This is in fact a prototypical example of an elliptic operator.

The concepts above make sense for a more general setting of vector bundles. To spell some details, let E and F be vector bundles over a manifold M of rank r and s respectively. We say that a map $L\colon \Gamma(E) \to \Gamma(F)$ is a linear differential operator of order ℓ, if for any choice of local coordinates on $\Omega \subset M$ and any trivializations of $E|_\Omega$ and $F|_\Omega$ the map L can be represented as in (5.2.1), where the coefficients A_α are allowed to depend on the choices made.

Let $\pi\colon T^*M \to M$ be the natural projection. A straightforward computation shows that the symbol makes sense as a section of $\mathrm{Hom}(\pi^*E; \pi^*F)$. Then L is said to be elliptic if the symbol is a pointwise invertible homomorphism away from the zero section of T^*M.

Example 5.2.4. Let M be an oriented Riemannian manifold. Consider the Laplace–Beltrami operator acting on the space of functions on M:

$$\Delta f = - * d * df,$$

where $*\colon \Lambda^k T^*M \to \Lambda^{n-k} T^*M$ is the Hodge operator.

Choose local coordinates $(x_1, \dots, x_n)$ and write $g = g_{ij}\, dx_i \otimes dx_j$, $|g| = \det(g_{ij})$. Denoting by (g^{ij}) the inverse matrix, a straightforward computation yields the local form of the Laplace–Beltrami operator:

$$\Delta f = -|g|^{-\frac{1}{2}} \sum_{i,j=1}^{n} \frac{\partial}{dx_i}\left(|g|^{\frac{1}{2}} g^{ij} \frac{\partial f}{\partial x_j}\right)$$

From this it is easy to compute the symbol. Indeed, if $\xi = \sum \xi_i\, dx_i$, then

$$\sigma_\Delta(\xi) = -|g|^{-\frac{1}{2}} \sum_{i,j=1}^{n} \xi_i |g|^{\frac{1}{2}} g^{ij} \xi_j = -|\xi|^2,$$

where $|\cdot|$ denotes the norm on T^*X. In particular, the Laplace–Beltrami operator is elliptic.

Example 5.2.5. The Dirac operator $\mathcal{D}\colon \Gamma(E) \to \Gamma(E)$ corresponding to any Dirac bundle E is elliptic. Indeed, it is easy to see that the principal symbol $\sigma_\xi(D)$ is just the Clifford multiplication with ξ. This is invertible with the inverse given by the Clifford multiplication with $|\xi|^{-2}\xi$.

In dimension four (and in fact in any even dimension) the chiral Dirac operators $\not{D}^{\pm}$ are also elliptic.

Any linear elliptic operator $L\colon C^\infty(M; E) \to C^\infty(M; F)$ of order ℓ extends as a bounded linear map

$$L\colon W^{k+\ell,p}(M; E) \to W^{k,p}(M; F) \tag{5.2.6}$$

for any $k \geqslant 0$ and any $p > 1$.

Theorem 5.2.7 (Elliptic estimate)**.** *Let M be a compact manifold. For any linear elliptic operator L there is a constant $C > 0$ with the following property: If $Ls \in W^{k,p}(M; F)$, then $s \in W^{k+\ell,p}(M; E)$ and*

$$\|s\|_{W^{k+\ell,p}} \leqslant C\big(\|Ls\|_{W^{k,p}} + \|s\|_{L^p}\big).$$

Here C depends on k and p but not on s. □

Assume that both E and F are equipped with an Euclidean structure. An operator $L^*\colon C^\infty(M; F) \to C^\infty(M; E)$ is said to be formal adjoint of L if

$$\langle Ls, t\rangle_{L^2} = \langle s, L^* t\rangle_{L^2} \tag{5.2.8}$$

holds for any $s \in C^\infty(M; E)$ and any $t \in C^\infty(M; F)$. One can show that L^* exists and is a linear differential operator of order ℓ. Moreover, L is elliptic if and only if L^* is elliptic.

One of the most important results in the theory of elliptic differential operators is the following.

Theorem 5.2.9 (Fredholm alternative). *For* L *elliptic,* M *compact, and* $t \in C^\infty(M; F)$, *the equation*

$$Ls = t$$

has a smooth solution if and only if t *is* L^2*-orthogonal to* $\ker L^*$. □

Remark 5.2.10. A corollary of Theorem 5.2.9 can be formulated as follows. Under the hypotheses of Theorem 5.2.9 one and only one of the following statements holds:

(i) The homogeneous equation $L^*s = 0$ has a non-trivial solution.
(ii) The inhomogeneous equation $Ls = t$ has a unique solution for any smooth t.

This form of Theorem 5.2.9 is widely used in the theory of PDEs.

Definition 5.2.11. A bounded linear map $B\colon X \to Y$ between two Banach spaces is called Fredholm, if the following conditions hold:

(a) $\dim \ker B < \infty$;
(b) $\operatorname{Im} B$ is a closed subspace of Y;
(c) $\operatorname{coker} B := Y/\operatorname{Im} B < \infty$.

If B is Fredholm, the integer

$$\operatorname{index} B := \dim \ker B - \dim \operatorname{coker} B$$

is called the index of B.

Remark 5.2.12. One can show that (b) follows from the other two conditions. Nevertheless, it will be useful to know that $\operatorname{Im} B$ is closed, even if one does not necessarily need to check this.

For example, any linear map between finite dimensional spaces is Fredholm and its index equals $\dim X - \dim Y$. In fact, Fredholm operators resemble linear maps between finite dimensional vector spaces well known from the basic course of linear algebra and this largely explains the importance of Fredholm operators for us.

Exercise 5.2.13. For a Fredholm operator $B_0\colon X \to Y$ choose X_0 and Y_0 to be a closed complement to the kernel and the cokernel of B respectively (in the latter case this means that the natural map $Y_0 \to Y \to Y/\operatorname{Im} B_0$ is an isomorphism). With respect to the decompositions

$$X = X_0 + \ker B_0, \qquad Y = \operatorname{Im} B_0 + Y_0$$

any linear operator $B\colon X \to Y$ can be written as

$$B = \begin{pmatrix} a & b \\ c & d \end{pmatrix}.$$

For example,

$$B_0 = \begin{pmatrix} a_0 & 0 \\ 0 & 0 \end{pmatrix}$$

with a_0 being just the restriction of B_0 to X_0.

(a) Show that if $\|B\|$ is sufficiently small, then $a_0 + a$ is invertible, where $\|\cdot\|$ means the operator norm in the space of bounded linear maps $X \to Y$. Furthermore, define

$$C := \begin{pmatrix} \mathbb{1} & -(a_0 + a)^{-1} b \\ 0 & \mathbb{1} \end{pmatrix} \quad \text{and} \quad D = \begin{pmatrix} \mathbb{1} & 0 \\ -c(a_0 + a)^{-1} & \mathbb{1} \end{pmatrix}.$$

and check the equality

$$D\,(B_0 + B)\,C = \operatorname{diag}(a_0 + a,\ f),$$

where $f = -c\,(a_0 + a)^{-1}\,b + d$.

(b) Using (a), prove that

$$\ker(B_0 + B) = \ker f \quad \text{and} \quad \operatorname{coker}(B_0 + B) \cong \operatorname{coker} f.$$

(c) Show that for any linear Fredholm operator B_0 there is an $\varepsilon > 0$ such that any bounded operator B with

$$\|B - B_0\| < \varepsilon$$

is also Fredholm. Prove also that $\operatorname{index} B = \operatorname{index} B_0$.

Theorem 5.2.14. *For any elliptic operator L of order $\ell > 0$ on a compact manifold M, (5.2.6) is a Fredholm operator. Moreover, $\ker L$ consists of smooth sections only.*

Sketch of proof. If $s \in W^{\ell,p}$ is in the kernel of L, then $s \in W^{k,p}$ for any $k \geqslant 0$ by Theorem 5.2.7. Hence, s is smooth by Theorem 5.1.3.

Let $s_j \in W^{\ell,p}$ be any sequence such that $Ls_j = 0$ and $\|s_j\|_{W^{\ell,p}} \leqslant 1$. Then the sequence $\|s_j\|_{W^{\ell+1,p}}$ is bounded by Theorem 5.2.7. Hence, by Theorem 5.1.3 after passing to a subsequence if necessary, s_j converges to some limit s_∞ in $W^{\ell,p}$. Since $L\colon W^{\ell,p} \to L^p$ is bounded, the limit $s_\infty \in \ker L$. In other words, the unit ball in $\ker L$ is compact with respect to the $W^{\ell,p}$ norm and hence $\ker L$ is finite dimensional.

Next, let us assume that $p = 2$, which simplifies the discussion somewhat. Denote $V := \left(\operatorname{Im} L\colon W^{k+\ell,2} \to W^{k,2}\right)^\perp$, where the orthogonal complement is understood in the sense of L^2–scalar product. One can show that all sections lying in V are in fact in $W^{\ell,p}$. Then (5.2.8) implies that $V = \ker L^*$. This in turn yields that V is finite dimensional (and consists of smooth sections only). □

5.3. Elliptic complexes Fix a manifold M and a sequence of vector bundles $E_1, \dots, E_k$, which I assume to be finite for the simplicity of exposition. Let

$$0 \to \Gamma(E_1) \xrightarrow{L_1} \Gamma(E_2) \xrightarrow{L_2} \dots \xrightarrow{L_{k-1}} \Gamma(E_k) \to 0 \tag{5.3.1}$$

be a sequence of differential operators such that $L_j \circ L_{j-1} = 0$ for every integer j in $[1, k-1]$, i.e., (5.3.1) is a complex. In particular, we can define the corresponding cohomology groups

$$H^j(E) := \ker L_j / \operatorname{im} L_{j-1},$$

where $j \in \{1, 2, \dots, k\}$.

Associated to (5.3.1) is the sequence of principal symbols:

$$0 \to \pi^* E_1 \xrightarrow{\sigma_{L_1}} \pi^* E_2 \xrightarrow{\sigma_{L_2}} \pi^* E_3 \xrightarrow{\sigma_{L_3}} \dots \xrightarrow{\sigma_{L_{k-1}}} \pi^* E_k \to 0.$$

If this is exact on the complement of the zero section, (5.3.1) is called an elliptic complex. For example, a very short complex $0 \to \Gamma(E) \xrightarrow{L} \Gamma(F) \to 0$ is elliptic if and only if L is elliptic.

With these preliminaries at hand, we can construct the associated Laplacians:

$$\Delta = \Delta_j \colon \Gamma(E_j) \to \Gamma(E_j), \qquad \Delta_j = L_j^* L_j + L_{j-1} L_{j-1}^*.$$

Exercise 5.3.2. Prove that each Δ_j is an elliptic operator provided the initial complex is elliptic.

In the sequel, I will drop the index of the differentials L_i to simplify the notations so that all maps in (5.3.1) are denoted by the same symbol L. I assume also that all differentials L are of order 1, since only this case will show up in these notes, but operators of other orders could be considered as well. Also, I assume that all bundles E_i are equipped with an Euclidean structure.

Denote

$$\mathcal{H}_j(E) := \big\{ s \in \Gamma(E_j) \mid \Delta s = 0 \big\}.$$

Elements of $\mathcal{H}_j(E)$ are called harmonic sections.

Theorem 5.3.3. *For an elliptic complex on a compact manifold the following holds:*

(i) *Each $\mathcal{H}_j(E)$ is a finite dimensional vector space.*
(ii) $s \in \mathcal{H}_j(E)$ *if and only if* $Ls = 0$ *and* $L^* s = 0$.
(iii) *The natural homomorphism*

$$\mathcal{H}_j(E) \to H^j(E), \qquad s \mapsto [s]$$

is an isomorphism.

Proof. The first statement follows from the ellipticity of Δ. The second statement follows easily from the equality

$$\begin{aligned} \langle \Delta s, s \rangle_{L^2} &= \big\langle (LL^* + L^*L)s, s \big\rangle_{L^2} \\ &= \|L^* s\|_{L^2}^2 + \|Ls\|_{L^2}^2. \end{aligned}$$

It remains to prove the last claim. Thus pick any $s_0 \in \Gamma(E_j)$ such that $Ls_0 = 0$. We wish to show that the equation

$$(L + L^*)(s_0 + Lt) = 0$$

has a solution $t \in \Gamma(E_{j-1})$. Notice that this equation is equivalent to

$$L^* Lt = -L^* s_0. \tag{5.3.4}$$

Consider instead the equation $\Delta t = -L^* s_0$, whose right hand side is clearly L^2-orthogonal to $\ker \Delta^* = \ker \Delta$. Hence, by Theorem 5.2.9 there is a unique solution of $\Delta t = -L^* s_0$, which can be rewritten as

$$L(L^* t) + L^*(s_0 + Lt) = 0.$$

A moment's thought shows that $\operatorname{Im} L$ is L^2-orthogonal to $\operatorname{Im} L^*$. Hence, t is a solution of (5.3.4) in fact. This finishes the proof of the existence part.

The uniqueness is easy to show: If $s_1, s_2 \in \mathcal{H}_j(E)$ are such that $s_1 - s_2 = Lt$ for some $t \in \Gamma(E_{j-1})$, then

$$\|Lt\|^2_{L^2} = \langle L^* Lt, t\rangle_{L^2} = \langle L^*(s_1 - s_2), t\rangle_{L^2} = 0. \qquad \square$$

A refinement of the argument in the proof of the above theorem shows that in fact we have the following decomposition

$$\Gamma(E_j) = \operatorname{Im} L \oplus \mathcal{H}_j(E) \oplus \operatorname{Im} L^*, \tag{5.3.5}$$

which is L^2–orthogonal. Details can be found for example in [24].

Exercise 5.3.6. Show that the short complex

$$0 \to \Gamma(E_1) \xrightarrow{L_1} \Gamma(E_2) \xrightarrow{L_2} \Gamma(E_3) \to 0$$

is elliptic if and only if the operator

$$(L_2, L_1^*)\colon \Gamma(E_2) \to \Gamma(E_3 \oplus E_1)$$

is elliptic.

A gauge-theoretic interpretation. The cohomology group $H^j(E)$ is an example of a linear gauge–theoretic moduli space, cf. Section 1. Let me spell some details. The manifold[4] $\mathcal{B} = \Gamma(E_j)$ carries an action of the additive group $\mathcal{G} := \Gamma(E_{j-1})$ by translations: $(s, t) \mapsto s + Lt$. If $\Gamma(E_{j+1})$ is viewed as a trivial $\mathcal{G}$–representation, $L\colon \Gamma(E_j) \to \Gamma(E_{j+1})$ is a $\mathcal{G}$-equivariant map, which in this particular case just means that L is $\mathcal{G}$-invariant. The action of the gauge group on $L^{-1}(0)$ is not free in general, however $\ker L$ acts trivially so that the corresponding 'moduli space' $L^{-1}(0)/\mathcal{G} = H^j(E)$ is a finite dimensional manifold (a vector space, in fact) provided the complex is elliptic and the base manifold is compact. This explains our interest in the theory of elliptic operators.

Notice, however, that because of the linear setting we can not expect that the 'moduli space' $H^j(E)$ will be compact. The reason is that $H^j(E)$ inherits the action

[4]One can consider $\Gamma(E_j)$ as a Banach manifold by taking a Sobolev completion. Since this is not really important at this point, I describe the setting in the smooth category.

of $\mathbb{R}_{>0}$ by dilations and if we take the quotient of $H^j \setminus \{0\}$ by this action the resulting space is compact. Thus, in this setting $\dim H^j(E) < \infty$ is a suitable replacement for the compactness of the moduli space.

One more important feature we have seen is the so called gauge fixing. Namely, we have shown that for each point $s \in L^{-1}(0)$ there is a unique representative $h(s)$ in the 'gauge-equivalence class of s' such that $h(s)$ is harmonic. Moreover, the map

$$L^{-1}(0) \to \mathcal{H}_j(E), \qquad s \mapsto h(s)$$

induces a diffeomorphism $L^{-1}(0)/\mathcal{G} \to \mathcal{H}^j(E)$.

Furthermore, an isomorphism class of a finite dimensional vector space is determined by a unique non-positive integer, namely its dimension. Hence, $b_j(E) := \dim H^j(E)$ is an 'invariant' of E. In many cases, these invariants capture a subtle information about the underlying manifold M.

The example of the de Rham complex in the following subsection will make these constructions more concrete.

The de Rham complex. Recall that for any manifold M of dimension n we have the de Rham complex

$$0 \to \Omega^0(M) \xrightarrow{d} \Omega^1(M) \xrightarrow{d} \Omega^2(M) \longrightarrow \ldots \longrightarrow \Omega^n(M) \to 0.$$

Exercise 5.3.7. Show that the de Rham complex is elliptic.

Any oriented Riemanian manifold M is equipped with a Hodge star operator $*\colon \Lambda^j T^*M \to \Lambda^{n-j}T^*M$. Using this, the formal adjoint of the exterior derivative can be expressed as

$$d^* = (-1)^{n(j-1)+1} * d\, *\colon \Omega^j(M) \to \Omega^{j-1}(M)$$

so that the Hodge–de Rham Laplacian is $\Delta = dd^* + d^*d$. If M is in addition compact, the space $\mathcal{H}_j$ of harmonic forms of degree j is naturally isomorphic to the jth de Rham cohomology group. Then $b_j(M) := \dim \mathcal{H}_j$ is the jth Betti number of M. This is a topological invariant of M even though in this approach a smooth structure has been used.

6. Fredholm maps

6.1. The Kuranishi model and the Sard–Smale theorem Let X and Y be Banach manifolds.

Definition 6.1.1. A map $F \in C^\infty(X; Y)$ is said to be Fredholm if the differential dF is a Fredholm linear map at each point.

Hence, for each $x \in X$ the index of d_xF is well defined. If X is connected, then index d_xF does not depend on x and this common value is denoted by index F.

Fredholm maps have a lot in common with smooth maps between finite dimensional manifolds. A manifestation of this is the following result.

Theorem 6.1.2 (Kuranishi model)**.** *Let X and Y be Banach spaces and $F\colon X \to Y$ a Fredholm map. Pick a point $p \in F^{-1}(0)$ and denote $X_0 = \ker d_pF$, $Y_0 := \operatorname{im} d_pF$. Furthermore, choose subspaces $X_1 \subset X$ and $Y_1 \subset Y$ such that*

$$X = X_0 \oplus X_1 \qquad \textit{and} \qquad Y = Y_0 \oplus Y_1.$$

Then there is a diffeomorphism φ of a neighborhood of the origin in X onto a neighbourhood of p such that $\varphi(0) = p$, a linear isomorphism $T\colon X_1 \to Y_0$, and a smooth map $f\colon X \to Y_1$ such that

$$F \circ \varphi(x_0, x_1) = Tx_1 + f(x_0, x_1)$$

for all $(x_0, x_1) \in X_0 \oplus X_1$ in the neighborhood of the origin.

In particular, if $f_0\colon X_0 \to Y_1$ denotes the restriction of f to X_0, then a neighborhood of p in $F^{-1}(0)$ is homeomorphic to a neighborhood of the origin in $f_0^{-1}(0)$. □

Corollary 6.1.3. *Assume the hypotheses of Theorem 6.1.2. Suppose also that 0 is a regular value of F, i.e., the d_pF is surjective for all $p \in F^{-1}(0)$. Then $F^{-1}(0)$ is a smooth manifold of dimension* index F.

Proof. Since $\operatorname{Im} d_pF = Y$, we necessarily have $Y_1 = \{0\}$ so that f_0 is a constant map. Hence, $F^{-1}(0)$ is diffeomorphic to a neighborhood of the origin in X_0, i.e., a manifold of dimension $\dim X_0 = \operatorname{index} F$. □

For a smooth map between finite dimensional manifolds, almost any value is regular by Sard's theorem [2, Thm. 9.5.4]. There is a generalization of this statement for an infinite dimensional setting due to Smale. This is commonly known as the Sard–Smale theorem.

Recall that a subset A of a topological space is said to be of *second category*, if A can be represented as a countable intersection of open dense subsets. If the underlying topological space is a Banach manifold, then a subset of second category is dense.

Theorem 6.1.4. *Let F be a smooth Fredholm map between paracompact Banach manifolds. Then the set of regular values of F is of second category, in particular dense.* □

Let $Z \subset Y$ be a smoothly embedded finite dimensional submanifold. A map F is said to be *transverse* to Z, if for any $z \in Z$ and any $x \in F^{-1}(Z)$ the following holds:

$$\operatorname{Im} d_xF + T_zZ = T_zY.$$

In particular, F is transverse to $Z = \{z\}$ is and only if z is a regular value of F. It is well-known that the notion of transversality is a useful generalization of the notion of regular value, see for instance [12].

Just as in the finite dimensional case we have the following result.

Theorem 6.1.5. *Let $Z \subset Y$ be a smoothly embedded finite dimensional submanifold. If F is transverse to Z, then $F^{-1}(Z)$ is a smooth submanifold of X and*

$$\dim F^{-1}(Z) = \operatorname{index} F + \dim Z.$$ □

6.2. The $\mathbb{Z}/2\mathbb{Z}$ degree Recall that a map $F\colon X \to Y$ is called *proper* if preimages of compact subsets are compact.

Let $F\colon X \to Y$ be a proper Fredholm map between (paracompact) Banach manifolds of index zero, where Y is connected. Then for any regular value $y \in Y$ the preimage $F^{-1}(y)$ is a compact manifold of dimension zero, hence a finite number of points. The number

$$\deg_2 F := \# f^{-1}(y) \mod 2$$

is called the $\mathbb{Z}/2\mathbb{Z}$ degree of F.

Theorem 6.2.1.

(i) $\deg_2 F$ *does not depend on the choice of the regular value* y*;*

(ii) *If* F_0 *and* F_1 *are homotopic within the class of proper Fredholm maps of vanishing index, then* $\deg_2 F_0 = \deg_2 F_1$.

Proof. The proof requires a number of steps.

Step 1. *If* F *is a proper Fredholm map, then the set of regular values of* F *is open and dense.*

First notice that the set of critical points $\mathrm{Crit}(F)$ is closed. Since any proper map is closed, the set of critical values $F(\mathrm{Crit}(F))$ is closed.

Step 2. *The function*

$$y \mapsto \# F^{-1}(y) \mod 2$$

is locally constant on the set of regular values.

Let y be a regular point of F, $F^{-1}(y) = \{x_1, \dots, x_k\}$. By the inverse function theorem, there is a neighborhood U of y and a neighborhood V_j of x_j such that $F\colon V_j \to U$ is a diffeomorphism. Then, for any $y' \in U$ we have $\# f^{-1}(y') = k$, hence the claim.

Step 3. *Let* F_0 *and* F_1 *be homotopic so that the homotopy is within the class of proper Fredholm maps. Then*

$$\# F_0^{-1}(y) = \# F_1^{-1}(y) \mod 2 \tag{6.2.2}$$

for any y*, which is a regular value for both* F_0 *and* F_1.

Let F_t, $t \in [0,1]$, be a homotopy. Let me assume first that, y is a regular value for $F\colon X \times [0,1] \to Y$. Then $F^{-1}(y)$ is a 1-dimensional manifold with boundary such that $\partial F^{-1}(y) = F_0^{-1}(y) \cup F_1^{-1}(y)$. Hence, (6.2.2) holds.

If y is not a regular value of F, then we can choose y_1 arbitrarily close to y such that y_1 is a regular value for any of F_0, F_1, F. The conclusion of this step follows by Step 2.

Step 4. *Let* x *be an arbitrary point in the unit ball of a Banach space* B*. Then there is a diffeomorphism* φ *of* B *such that the following holds:* $\varphi(0) = x$*,* φ *is the identity map on* *the complement of the Ball of radius* 2*, and* φ *is homotopic to the identity map relative to the complement of the ball of radius* 2*.*

This is really a finite dimensional statement. Indeed, choose decomposition $B = V \oplus V'$, where V is finite dimensional and $x \in V$. By [19, P. 22] there is a 1-parameter family $\psi_t\colon V \to V$ of diffeomorphisms such that $\psi_1(0) = x$, $\psi_0 = \mathrm{id}_V$, and ψ_t is identity outside the unit ball in V.

Choose a smooth function $\chi\colon \mathbb{R}_{\geqslant 0} \to \mathbb{R}_{\geqslant 0}$ such that $\chi(0) = 1$ and $\chi(t) = 0$ for $t \geqslant 1$. Then

$$\varphi_t(v, v') := \psi_{t\chi(|v'|)}(v) + v'$$

is a family of diffeomorphisms such that $\varphi_0 = \mathrm{id}$ and $\varphi_1(0,0) = \psi_1(0) = x$. Clearly, for any t we have $\|\varphi_t(v, v')\| \leqslant 2$ by the triangle inequality.

Let us check that φ_t is identity on the complement of the ball of radius 2. Thus, pick any (v, v') such that $\|v\| + \|v'\| \geqslant 2$. Then either $\|v\| \geqslant 1$ or $\|v'\| \geqslant 1$. The first case is clear and the second one we have

$$\varphi_t(v, v') = \psi_0(v) + v' = v + v'$$

since $\chi(\|v'\|) = 0$ by the construction.

Step 5. *If* Y *is a connected Banach manifold, then for any* $y_1, y_2 \in Y$ *there is a diffeomorphism* $\varphi \in \mathrm{Diff}_0(Y)$ *such that* $\varphi(y_1) = y_2$*, where* $\mathrm{Diff}_0(Y)$ *denotes the subgroup of all diffeomorphisms homotopic to the identity.*

Fix $y_2 \in Y$ and denote

$$C(y_2) := \{y_1 \in Y \mid \exists \varphi \in \mathrm{Diff}_0(Y) \text{ such that } \varphi(y_1) = y_2\}$$

By the previous step, $C(y_2)$ is open and non-empty, hence $C(y_2) = Y$.

Step 6. *We prove the claim of this theorem.*

Let F be as in the statement of this theorem. Pick any regular values y_1 and y_2 and $\varphi \in \mathrm{Diff}_0(Y)$ such that $\varphi(y_1) = y_2$. Then y_2 is a common regular value for F and $\varphi \circ F$. Since these maps are homotopic, by Step 3 we have

$$F^{-1}(y_2) = (\varphi \circ F)^{-1}(y_2) = F^{-1}(y_1) \mod 2.$$

This proves (i).

Furthermore, if $F\colon X \times [0,1] \to Y$ is a homotopy between F_0 and F_1, we can choose $y \in Y$ which is a regular value for any of the following maps: F_0, F_1, and F. This implies the second assertion. □

Corollary 6.2.3. *Let* $F\colon X \to Y$ *be a proper Fredholm map between (paracompact) Banach manifolds of index zero, where* Y *is connected. If* $\deg_2 F \neq 0$*, then* F *is surjective.* □

6.3. The parametric transversality Given a 'random' point y in the target, there is no reason to expect that the preimage $F^{-1}(y)$ will be a submanifold. Of course, if F is Fredholm the theorem of Sard–Smale implies that we can choose y' arbitrarily close to y so that $F^{-1}(y')$ is a manifold indeed. In practice, however, it is not always the case that one has a freedom to choose a point in the target. This is typically the case, which will be considered in some detail below, if X and Y are equipped with an action of a Lie group $\mathcal{G}$ and F is $\mathcal{G}$–equivariant. In this case it is natural to choose y as a fixed point of the $\mathcal{G}$–action on Y so that $F^{-1}(y)$ inherits a $\mathcal{G}$–action.

This choice, however, restricts severely possible choices of y so that the Sard–Smale theorem is not applicable in a straightforward manner.

One way to deal with this problem is as follows. Assume there is a connected Banach manifold W and a smooth Fredholm map $\mathcal{F}\colon X \times W \to Y$ with the following properties:

(A) For each $w \in W$ the map $\mathcal{F}_w\colon X \times \{w\} \to Y$ is Fredholm;

(B) There is a point $w_0 \in W$ such that $\mathcal{F}_{w_0} = F$;

(C) y is a regular value of $\mathcal{F}$;

Typically, it is not too hard to construct a map $\mathcal{F}$ satisfying these properties.

Since y is a regular value, $\mathcal{F}^{-1}(y) \subset X \times W$ is a smooth Banach submanifold. We have a natural projection $\pi\colon \mathcal{F}^{-1}(y) \to W$, which is just the restriction of the projection $X \times W \to W$.

Lemma 6.3.1.

(i) *π is a Fredholm map and* $\operatorname{index}\pi = \operatorname{index}F$;

(ii) *There is a subset $W_0 \subset W$ of second category such that y is a regular value for $\mathcal{F}_w$ for all $w \in W_0$.* □

Hence, by the Sard–Smale theorem, there is w arbitrarily close to w_0 such that

$$\pi^{-1}(w) = \mathcal{F}_w^{-1}(y)$$

is a smooth submanifold of dimension $\operatorname{index}F$.

Suppose that in addition to hypotheses (A)–(B), the following holds:

(D) Each map $\mathcal{F}_w$ is proper.

In what follows I will also assume that $\operatorname{index}F = 0$, since this is the setting where the degree of a Fredholm map was defined. However, this is by no means essential.

Thus, if (A)–(D) holds, $\mathcal{F}_w^{-1}(y) = \pi^{-1}(w)$ consists of finitely many points. Since W is connected by assumption, there is a path connecting w_0 and w so that $\mathcal{F}_w$ is homotopic to $\mathcal{F}_{w_0} = F$. Hence,

$$\deg_2 F = \#\mathcal{F}_w^{-1}(y) \mod 2.$$

In particular, $\#\mathcal{F}_w^{-1}(y) \mod 2$ does not depend on w.

Of course, this conclusion is pretty much straightforward thanks to the facts we have established in the preceding subsection. The point is that the number $\#\mathcal{F}_w^{-1}(y) \mod 2$ could be taken as a definition of the degree of F thus omitting the construction of Subsection 6.2. For instance, this is commonly used in the equivariant setting as discussed at the beginning of this section.

Let me describe this alternative construction of the degree of a proper Fredholm map in some detail. Thus, choose any $w_1, w_2 \in W_0$ so that y is a regular value for $\mathcal{F}_j := \mathcal{F}_{w_j}$. For a fixed $k \geqslant 0$ denote

$$\Gamma(w_1, w_2) := \Big\{\gamma \in C^k\big([1,2]; W\big) \mid \gamma(1) = w_1 \text{ and } \gamma(2) = w_2\Big\},$$

which is a Banach manifold. Consider the map

$$\mathbf{F}\colon X \times \Gamma(w_1, w_2) \times [1,2] \to Y, \qquad \mathbf{F}(x, \gamma, t) := \mathcal{F}(x, \gamma(t)).$$

Clearly, the point y is a regular value for $\mathbf{F}$ too so that $\mathbf{F}^{-1}(y)$ is a submanifold of $X \times \Gamma(w_1, w_2) \times [1,2]$. Applying the Sard–Smale theorem to the restriction of the projection $X \times \Gamma(w_1, w_2) \times [1,2] \to \Gamma(w_1, w_2)$ we obtain that for a generic $\gamma \in \Gamma(w_1, w_2)$ the subset

$$\mathcal{M}_\gamma := \big\{(x,t) \in X \times [1,2] \mid \mathbf{F}(x, \gamma, t) = y\big\}$$

is a smooth submanifold of dimension 1 and

$$\partial \mathcal{M}_\gamma = \mathcal{F}_1^{-1}(y) \cup \mathcal{F}_2^{-1}(y).$$

Hence,

$$\#\mathcal{F}_1^{-1}(y) = \#\mathcal{F}_2^{-1}(y) \mod 2$$

so that the number $\#\mathcal{F}_w^{-1}(y) \mod 2$ does not depend on $w \in W_0$. This common value, as we already know, is just $\deg_2(F)$.

6.4. The determinant line bundle If X and Y are closed *oriented* manifolds of the same finite dimension, for any map $F\colon X \to Y$ the degree can be extended to take values in $\mathbb{Z}$ rather than $\mathbb{Z}/2\mathbb{Z}$ [12]. This generalization requires orientations of the background manifolds and this tool is not readily available in infinite dimensions. In this section I will describe how to deal with this problem.

Let $\mathcal{P}$ be a topological space and $\{T_p \mid p \in \mathcal{P}\}$ be a continuous family of linear Fredholm maps. This means that the map

$$\mathcal{P} \to \operatorname{Fred}(X; Y), \qquad p \mapsto T_p$$

is continuous with respect to the operator norm, where X, Y are Banach spaces and $\operatorname{Fred}(X;Y)$ denotes the subspace of linear Fredholm maps.

Pick a point $p \in \mathcal{P}$. Since both the kernel and cokernel of T_p are finite dimensional, we can construct the (real) line

$$\det T_p := \Lambda^{\mathrm{top}} \ker T_p \otimes \Lambda^{\mathrm{top}} (\operatorname{coker} T_p)^*.$$

As p varies, we obtain a family of vector spaces of dimension one. It turns out that this family is actually a locally trivial vector bundle, which is somewhat non-obvious given that the dimensions of the kernel as well as cokernel may 'jump' as p varies.

To understand why this is the case, pick a point p_0 and a finite dimensional subspace $V \subset Y$ transverse to $\operatorname{Im} T_{p_0}$. The linear map

$$T_{p,V}\colon X \oplus V \to Y, \qquad T_{p,V}(x, v) = T_p\, x + v$$

is surjective for $p = p_0$ and, hence, also for all p sufficiently close to p_0. Therefore, we have the exact sequence

$$(6.4.1) \qquad 0 \to \ker T_p \to \ker T_{p,V} \to V \to V/\operatorname{Im} T_p \cap V \to 0$$

Notice that $\operatorname{coker} T_p = (\operatorname{Im} T_p + V)/\operatorname{Im} T_p = V/\operatorname{Im} T_p \cap V$.

Exercise 6.4.2. Let $0 \to U_0 \to U_1 \to U_2 \to 0$ be a short exact sequence of real vector spaces. Show that the 'inner product map'

$$\Lambda^{top}U_0 \otimes \Lambda^{top}U_1^* \to \Lambda^{top}(U_1/U_0)^* = \Lambda^{top}U_2^*$$

induces an isomorphism $\Lambda^{top}U_1 \cong \Lambda^{top}U_0 \otimes \Lambda^{top}U_2$.

More generally, show that if $0 \to U_0 \to U_1 \to U_2 \to \cdots \to U_k \to 0$ is any exact sequence of real vector spaces, there is a canonical isomorphism

$$\bigotimes \Lambda^{top}U_{even} \cong \bigotimes \Lambda^{top}U_{odd}.$$

Hence, by (6.4.1) we have a canonical isomorphism

$$\det T_p \cong \Lambda^{top} \ker T_{p,V} \otimes \Lambda^{top}V^*.$$

It can be shown not only that $\dim \ker T_{p,V}$ is constant in p near p_0 but also that $\ker T_{p,V}$ is trivial in a neighborhood W of p_0. Details can be found for instance in [18, Thm. A.2.2]. Thus, the family of real lines $\{\det T_p \mid p \in W\}$ admits a trivialization, i.e., $\det T$ is a vector bundle.

Definition 6.4.3. Two continuous families $\{T_{p,i} \mid p \in \mathcal{P}\}$, $i \in \{0, 1\}$ of Fredholm operators are said to be *homotopic*, if there is a continuous family of Fredholm operators $\{T_{p,t} \mid (p,t) \in \mathcal{P} \times [0,1]\}$, whose restriction to $\mathcal{P} \times \{0\}$ and $\mathcal{P} \times \{1\}$ yields the initial families.

Assume that $\{T_{p,0}\}$ is homotopic to $\{T_{p,1}\}$ and the determinant line bundle $\det T_{p,1}$ is trivial. Since $\mathcal{P} \times [0,1]$ is homotopy equivalent to $\mathcal{P}$, the bundle $\det T_{p,t}$ is also trivial. In particular, $\det T_{p,0}$ is trivial. More precisely, we have the following statement.

Proposition 6.4.4. *Let $\{T_{p,0}\}$ and $\{T_{p,1}\}$ be homotopic families of Fredholm operators such that $\det T_{p,1}$ is trivial. Then $\det T_{p,0}$ is also trivial. Moreover, a choice of a trivialization of $\det T_{p,1}$ and a homotopy $\{T_{p,t} \mid t \in [0,1]\}$ yields a trivialization of $\det T_{p,0}$. This is unique up to a multiplication with a positive function.* □

6.5. Orientations and the $\mathbb{Z}$–valued degree Let X, Y, and W be Banach manifolds. Let $F\colon X \times W \to Y$ be a smooth map such that the following holds:

(a) $F_w = F|_{X \times \{w\}}\colon X \to Y$ is a Fredholm map of index d for each $w \in W$;
(b) $y \in Y$ is a regular value of F;
(c) $F_w^{-1}(y)$ is compact for any $w \in W$;
(d) $\det d_x F_w$ is trivial over $X \times W$.

Furthermore, I assume that a trivialization of the determinant line bundle has been fixed.

Pick any $w \in W$ such that y is a regular value of F_w. Then for any $x \in F_w^{-1}(y)$ we have

$$\det d_x F_w = \Lambda^{\mathrm{top}} \ker d_x F_w = \Lambda^{\mathrm{top}} T_x F_w^{-1}(y)$$

so that the hypotheses above imply that $\mathcal{M}_w := F_w^{-1}(y)$ is an oriented d–manifold.

Furthermore, let $\gamma\colon [0,1] \to W$ be a path connecting w_0 and w_1 akin to the situation considered in Subsection 6.3. For generic γ the space

$$\mathcal{M}_\gamma := \big\{(x,t) \in X \times [0,1] \mid F_{\gamma(t)}(x) = y\big\}$$

is an oriented manifold of dimension $d+1$ such that

$$\partial \mathcal{M}_\gamma = \mathcal{M}_{w_1} \sqcup \overline{\mathcal{M}}_{w_0}, \tag{6.5.1}$$

where $\overline{\mathcal{M}}_{w_0}$ means that the orientation of $\mathcal{M}_{w_0}$ is reversed.

In the particular case $d = 0$, $\mathcal{M}_w$ is just a finite collection of points $\{m_1, \dots, m_k\}$ equipped with signs $\{\varepsilon_1, \dots, \varepsilon_k\}$ so that we can define

$$\deg F_w := \sum_{i=1}^{k} \varepsilon_k \in \mathbb{Z}. \tag{6.5.2}$$

In fact, for any two choices w_0 and w_1 as above, (6.5.1) implies that we have $\deg F_{w_0} = \deg F_{w_1}$ provided w_0 and w_1 are in the same connected component. Thus, $\deg F_w$ does not depend on w as long as W is connected.

The following result summarizes the considerations above.

Theorem 6.5.3. *Assume* $F\colon X \times W \to Y$ *satisfies Hypotheses* (a)–(d) *above. If* W *is connected, then* $\deg F_w$ *does not depend on* w. □

Clearly, the case $d = 0$ is not really special. Indeed, as we already know for any $d \geqslant 0$, $\mathcal{M}_w = F_w^{-1}(y)$ is a smooth oriented manifold for all w in a dense subset in W. Then (6.5.1) shows that for any two choices w_0 and w_1 in this subset, $\mathcal{M}_{w_0}$ and $\mathcal{M}_{w_1}$ are cobordant, i.e., the oriented cobordism class of $[\mathcal{M}_w]$ is well–defined and does not depend on w. One can take this oriented cobordism class as an invariant, however, in practice it may be hard to deal with. One way to extract a number out of this cobordism class is as follows.

Let $P \to X$ be a (principal) bundle. Assume there are characteristic classes $\alpha_1, \dots \alpha_k$ of P such that $\alpha := \alpha_1 \cup \dots \cup \alpha_k \in H^d(X;\mathbb{Z}) \subset H^d(X;\mathbb{R})$. Since the restriction of α to $\mathcal{M}_w$ can be represented by a closed d–form, say ω, Stokes' theorem implies that

$$\langle \alpha, [\mathcal{M}_w] \rangle = \int_{\mathcal{M}_w} \omega$$

does not depend on w. In fact, this is an integer, since $[\omega]$ represents an integral cohomology class.

6.6. An equivariant setup Let X be a Banach manifold equipped with an action of a Banach Lie group $\mathcal{G}$. For any $x \in X$ the infinitesimal action of $\mathcal{G}$ at x is given by the linear map

$$R_x\colon \mathrm{Lie}(\mathcal{G}) \to T_x X, \tag{6.6.1}$$

whose image is the tangent space to the orbit through x.

For the sake of simplicity of exposition let me assume that $\mathcal{G}$ acts freely on X. It will be also convenient to assume that X and $\mathcal{G}$ are Hilbert manifolds.

Definition 6.6.2. A Hilbert submanifold $S \subset X$ containing x is said to be a local slice of the $\mathcal{G}$-action at x, if the set $\mathcal{G}S := \{g \cdot s \mid g \in \mathcal{G},\ s \in S\}$ is open in X and the natural map

$$\mathcal{G} \times S \to \mathcal{G}S, \qquad (g, s) \mapsto g \cdot s$$

is a diffeomorphism.

Proposition 6.6.3. *Assume $\mathcal{G}$ acts freely on X. If the $\mathcal{G}$–action admits a slice at any point, then the quotient $X/\mathcal{G}$ is a manifold.*

The proof of this proposition is clear: S can be identified with a neighbourhood of the orbit $\mathcal{G} \cdot x$ in the quotient space $X/\mathcal{G}$.

Example 6.6.4. The group $U(1)$ acts on S^2 by rotations around the z-axis. If we remove the north and the south poles, this action is free. The submanifold

$$S := \{(x, 0, z) \mid x^2 + z^2 = 1, x > 0\} \cong (-1, 1)$$

is a global slice for the $U(1)$–action. In particular, the quotient is a manifold, which in this case is naturally diffeomorphic to an interval.

Let Y be a smooth manifold equipped with a $\mathcal{G}$–action and $F\colon X \times W \to Y$ be a smooth map such that each $F_w\colon X \to Y$ is $\mathcal{G}$–equivariant. Let $y \in Y$ be a fixed point of the $\mathcal{G}$–action, i.e., $g \cdot y = y$ for all $g \in \mathcal{G}$. For any $x \in F_w^{-1}(y)$ we can construct the sequence

$$0 \to \mathrm{Lie}(\mathcal{G}) \xrightarrow{R_x} T_x X \xrightarrow{d_x F_w} T_y Y \to 0, \tag{6.6.5}$$

which is in fact a complex. This follows immediately from the equivariancy of F_w (and the assumption that y is a fixed point). This is called *the deformation complex* at x.

The zero's cohomology group of the deformation complex is just the Lie algebra of the stabilizer of x. Our assumption implies that this is trivial.

The second cohomology group is just the cokernel of $d_x F_w$. This is trivial if and only if y is a regular value for F_w.

It is also easy to understand the meaning of the first cohomology group. Indeed, $\ker d_x F_w$ is the Zariski tangent space to $F_w^{-1}(y)$. Since y is fixed, $\mathcal{G}$ acts on $F_w^{-1}(y)$ so that the first cohomology group can be thought of as the tangent space to the 'moduli space'

$$\mathcal{M}_w := F_w^{-1}(y)/\mathcal{G}$$

at $\mathcal{G} \cdot x$.

Since T_xX is by assumption a Hilbert space, we have a linear map

$$D_x := (R_x^*,\ d_xF_w)\colon T_xX \to \mathrm{Lie}(\mathcal{G}) \oplus T_yY \tag{6.6.6}$$

whose kernel can be identified with the first cohomology group of the deformation complex.

Theorem 6.6.7. *Let y is a fixed point of the $\mathcal{G}$–action on Y. Assume that the following holds:*

(i) *$\mathcal{G}$ acts freely on X;*
(ii) *y is a regular value for F_w;*
(iii) *There is a local slice at each point $x \in F_w^{-1}(y)$;*
(iv) *D_x is a Fredholm linear map of index d.*

Then $\mathcal{M}_w$ is a smooth manifold of dimension d.

Proof. The statement is local, so we can restrict our attention to a neighborhood of a point $x \in F_w^{-1}(y)$. Let S be a slice at x so that T_xS and $\mathrm{Im}\, R_w$ are complementary subspaces in T_xX. Then y is still a regular value for $F_w|_S$ and

$$\ker d_xF_w|_S = \ker(R_x^*,\ d_xF_w)$$

so that $F_w^{-1} \cap S$ is a manifold of dimension $\dim\ker D_x = d$. □

By tracing through the discussion of Subsection 6.5 it is easy to see that the following theorem holds.

Theorem 6.6.8. *Assume that in addition to Hypotheses (i)–(iv) of Theorem 6.6.7 the determinant line bundle $\det D_x$ is trivialized and this trivialization is preserved by the action of $\mathcal{G}$. Then $\mathcal{M}_w$ is oriented. If in addition $\mathcal{M}_w$ is compact for any w, then the oriented bordism class of $\mathcal{M}_w$ does not depend on w.* □

7. The Seiberg–Witten gauge theory

7.1. The Seiberg–Witten equations Recall that in dimension 4 we have an isomorphism of $\mathrm{Spin}(4)$ representations

$$\mathbb{R}^4 \otimes \mathbb{C} \cong \mathrm{Hom}(\mathcal{S}^+;\mathcal{S}^-), \tag{7.1.1}$$

where $\mathrm{Spin}(4)$ acts on $\mathbb{R}^4$ via the homomorphism $\mathrm{Spin}(4) \to SO(4)$. (7.1.1) is still valid as an isomorphisms of $\mathrm{Spin}^c(4)$–representations. Somewhat more explicitly, the standard inclusion $\mathbb{R}^4 \to Cl(\mathbb{R}^4)$ yields a monomorphism

$$\mathbb{R}^4 \to \mathrm{Hom}(\mathcal{S}^+;\mathcal{S}^-), \qquad v \mapsto (\psi \mapsto v \cdot \psi).$$

Of course, we have also an inclusion $\mathbb{R}^4 \to \mathrm{Hom}(\mathcal{S}^-;\mathcal{S}^+)$. Hence, the Clifford multiplication with a 2–form yields a map $\Lambda^2\mathbb{R}^4 \to \mathrm{End}(\mathcal{S}^{\pm})$, whose kernel is $\Lambda^2_{\mp}\mathbb{R}^4$ as a straightforward computation shows. Moreover, the image of this map consists of skew-Hermitian endomorphisms so that we obtain an isomorphism

$$\Lambda^2_+\mathbb{R}^4 \to \mathfrak{su}(\mathcal{S}^+),$$

whose complexification yields $\Lambda^2_+\mathbb{R}^4 \otimes \mathbb{C} \cong \mathrm{End}_0(\mathcal{S})$, cf. (4.1.3).

Notice also that we have a quadratic map

$$\mu\colon \mathbb{S}^+ \to i\,\mathfrak{su}(\mathbb{S}^+) \subset \mathrm{End}_0(\mathbb{S}^+), \qquad \mu(\psi) = \psi\psi^* - \frac{1}{2}|\psi|^2,$$

where the expression $\mu(\psi)$ on the right hand side is defined by the following: $\mu(\psi)(\varphi) = \langle\varphi,\psi\rangle\psi - \frac{1}{2}|\psi|^2\,\varphi$. Somewhat more concretely, μ is just the map

$$\mathbb{C}^2 \to i\,\mathfrak{su}(2), \qquad \begin{pmatrix}\psi_1\\ \psi_2\end{pmatrix} \mapsto \frac{1}{2}\begin{pmatrix}|\psi_1|^2 - |\psi_2|^2 & 2\,\psi_1\bar\psi_2\\ 2\,\bar\psi_1\psi_2 & |\psi_2|^2 - |\psi_1|^2\end{pmatrix}.$$

Hence, we can think of $\mu(\psi)$ as a purely imaginary self–dual 2-form.

A more global version of these identifications is as follows. Pick an oriented Riemannian four–manifold M equipped with a spin^c structure. Denote by $\mathbb{S}^\pm$ the corresponding spinor bundles, see Subsection 4.3 for details. Then we have the isomorphisms of vector bundles

$$T^*_{\mathbb{C}}M \cong \mathrm{Hom}(\mathbb{S}^+;\mathbb{S}^-), \tag{7.1.2}$$

$$i\Lambda^2_+T^*M \cong i\,\mathfrak{su}(\mathbb{S}^+) \tag{7.1.3}$$

and a fiberwise quadratic map

$$\mu\colon \mathbb{S}^+ \to i\Lambda^2_+T^*M.$$

With this understood, *the Seiberg–Witten equations* for a pair $(\psi, A) \in \Gamma(\mathbb{S}^+) \times \mathcal{A}(P_{\det})$ are as follows:

$$\not{D}^+_A\psi = 0 \quad\text{and}\quad F^+_A = \mu(\psi). \tag{7.1.4}$$

It is also convenient to introduce *the Seiberg–Witten map* by

$$SW\colon \Gamma(\mathbb{S}^+)\times\mathcal{A}(P_{\det}) \to \Gamma(\mathbb{S}^-)\times\Omega^2_+(M;\mathbb{R}i), \qquad SW(\psi,A) = \Big(\not{D}^+_A\psi, F^+_A - \mu(\psi)\Big).$$

The gauge group action. Since the structure group of $P_{\det}$ is the abelian group $U(1)$, we have an identification $\mathcal{G} := \mathcal{G}(P_{\det}) \cong C^\infty(M; U(1))$. This acts on $\mathcal{A}(P_{\det})$ on the right by gauge transformations:

$$A\cdot g = A + 2\,g^{-1}dg. \tag{7.1.5}$$

We can extend this to the right action of $\mathcal{G}$ on *the configuration space* $\Gamma(\mathbb{S}^+)\times\mathcal{A}(P_{\det})$ as follows:

$$(\psi, A)\cdot g = (\bar g\psi,\ A\cdot g).$$

We let also $\mathcal{G}$ act on $\Gamma(\mathbb{S}^-)$ on the left in the obvious manner. Extending this action by the trivial one on $\Omega^2_+(M;\mathbb{R}i)$, we obtain a left action of $\mathcal{G}$ on $\Gamma(\mathbb{S}^-)\times\Omega^2_+(M;\mathbb{R}i)$.

Lemma 7.1.6. *The Seiberg–Witten map is $\mathcal{G}$-equivariant, i.e.,*

$$SW\big((\psi,A)\cdot g\big) = \bar g\cdot SW(\psi,A). \qquad \square$$

From this we obtain that $\mathcal{G}$ acts on the space of solutions of (7.1.4). The quotient

$$\mathcal{M}_{SW} = \big\{(\psi,A)\text{ is a solution of (7.1.4)}\big\}/\mathcal{G}$$

is called *the Seiberg–Witten moduli space*.

Notice that the action of $\mathcal{G}$ on the configuration space is *not* free. Indeed, if $g \in \mathcal{G}$ is the stabilizer of a point (ψ, A) in the configuration space, then (7.1.5) implies that g is constant. Hence,

$$\mathrm{Stab}(\psi, A) \neq \{1\} \iff \psi \equiv 0.$$

The point (ψ, A) with a non-vanishing spinor ψ are called *irreducible*, while points of the form $(0, A)$ are called *reducible*.

Denote also

$$\mathcal{M}^{irr}_{SW} := \big\{ (\psi, A) \text{ is an irreducible solution of (7.1.4) } \big\} / \mathcal{G}.$$

The deformation complex. By Subsection 6.6, we can associate to any solution (ψ, A) of the Seiberg–Witten equations the deformation complex[5]:

$$(7.1.7)\quad 0 \to \Omega^0(M; \mathbb{R}i) \xrightarrow{R_{(\psi,A)}} \Gamma(\mathbb{S}^+) \oplus \Omega^1(M; \mathbb{R}i) \xrightarrow{d_{(\psi,A)}SW} \Gamma(\mathbb{S}^-) \oplus \Omega^2_+(M; \mathbb{R}i) \to 0.$$

To explain, since $\mathcal{A}(P_{det})$ is an affine space modelled on $\Omega^1(M; \mathbb{R}i)$, the tangent space to the configuration space at any point can be naturally identified with the middle space of the complex. It is easy to compute the infinitesimal action of the gauge group:

$$R_{(\psi,A)}\,\xi = \big(-\xi\psi,\ 2\,d\xi\big), \qquad \xi \in \Omega^0(M; \mathbb{R}i).$$

Lemma 7.1.8. *We have*

$$d_{(\psi,A)}SW(\dot\psi, \dot a) = \Big(\mathfrak{D}^+_A\dot\psi + \frac{1}{2}\,\dot a \cdot \psi,\ d^+\dot a - 2\mu(\psi, \dot\psi)\Big),$$

where the second summand of the first component means the Clifford multiplication of $\dot a$ *and* ψ, $d^+\dot a$ *is the projection of* $d\dot a$ *onto the space of self–dual 2-forms, and* $\mu(\cdot,\cdot)$ *is the polarization of* μ. □

Proposition 7.1.9. *For any solution* (ψ, A) *of the Seiberg–Witten equations,* (7.1.7) *is an elliptic complex.*

The proof of this proposition hinges on the following result, which is of independent interest.

Proposition 7.1.10. *For any Riemannian oriented four–manifold* X*, the Atiyah complex*

$$0 \to \Omega^0(X) \xrightarrow{d} \Omega^1(X) \xrightarrow{d^+} \Omega^2_+(X) \to 0$$

is elliptic. □

One can prove this proposition either by computing the principal symbols in local coordinates or by noticing that $d^+ + d^*$ is a twisted Dirac operator. I leave the details to the reader.

Proof of Proposition 7.1.9. Modulo zero order terms, which are clearly immaterial for the statement of this proposition, (7.1.7) can be written as the direct sum of

[5]Strictly speaking, at this point the constructions of Subsection 6.6 are not applicable, but we will see below how to fix this.

the Atiyah complex and

$$0 \to 0 \to \Gamma(\mathcal{S}^+) \xrightarrow{\not{D}^+} \Gamma(\mathcal{S}^-) \to 0.$$

The claim follows from the ellipticity of both complexes. □

Sobolev completions. The smooth category we used to define the Seiberg–Witten map does not allow us to use the technique developed in Section 6. Hence, we will work with Sobolev completions of the spaces under consideration. This is described next.

Pick any $A_0 \in \mathcal{A}(P_{\det})$ so that we can identify $\mathcal{A}(P_{\det})$ with $\Gamma(T^*M \otimes \mathbb{R}i)$ even if not canonically so. For any fixed (k,p) the space

$$\mathcal{A}^{k,p}(P_{\det}) := A_0 + W^{k,p}(T^*M \otimes \mathbb{R}i).$$

is an affine Banach space, hence a Banach manifold. It is easy to see that the resulting structure is independent of the choice of A_0.

Just like in the case of the configuration space, it is also convenient to complete the gauge group. Namely, a map $M \to S^1$ is said to be of class $W^{k,p}$ if the composition $M \to S^1 \subset \mathbb{R}^2$ is in $W^{k,p}(M;\mathbb{R}^2)$. The subset $\mathcal{G}^{k,p}$ of all such maps is not a vector space, however this is a Banach manifold. Moreover, if $kp > 4$ the Sobolev multiplication theorem shows that $\mathcal{G}^{k,p}$ is closed under the pointwise multiplication, so that $\mathcal{G}^{k,p}$ is in fact a Banach Lie group. Its Lie algebra is given by

$$\mathrm{Lie}(\mathcal{G}^{k,p}) = W^{k,p}(M;\mathbb{R}i).$$

Proposition 7.1.11. *For any* k *and any* $p > 1$ *such that* $kp > 4 = \dim M$ *the Seiberg–Witten map extends as a smooth map*

$$SW\colon W^{k+1,p}(\mathcal{S}^+) \times \mathcal{A}^{k+1,p}(P_{\det}) \to W^{k,p}(\mathcal{S}^-) \times W^{k,p}(\Lambda^2_+ T^*M \otimes \mathbb{R}i).$$

The action of the gauge group extends to a smooth action of $\mathcal{G}^{k+2,p}$ *and SW is equivariant with respect to this action.*

Proof. Pick a smooth connection A_0 as a reference point. Then recalling (4.3.12) we obtain by Theorem 5.1.3 (iv)

$$\not{D}^+_{A_0+a}\psi = \not{D}^+_{A_0}\psi + \frac{1}{2}a\cdot\psi \in W^{k,p}(\mathcal{S}^-),$$
$$F^+_{A_0+a} - \mu(\psi) = F^+_{A_0} + d^+a - \mu(\psi) \in W^{k,p}(\Lambda^2_+ T^*M \otimes \mathbb{R}i),$$

where $a \in W^{k+1,p}(T^*M \otimes \mathbb{R}i)$.

The fact that the action of the gauge group extends follows from the Sobolev multiplication theorem and (7.1.5), which explains the choice $k+2$ when completing the gauge group. □

In what follows, any (k,p) with k sufficiently large would work. For the sake of definiteness, I will stick to $(k,p) = (5,2)$, which suffices for the arguments invoked below.

Compactness of the Seiberg–Witten moduli space. The most important property of the Seiberg–Witten moduli space is its compactness. In this section I explain why $\mathcal{M}_{SW}$ enjoys this property.

Before going into details, let me briefly explain the approach. Given any sequence (ψ_n, A_n) we need to show that there is a convergent subsequence. This would follow from the compactness of Sobolev embedding, if we could establish that the sequence $\|(\psi_n, A_n)\|_{W^{6,2}}$ is bounded from above by a constant independent of n. Here we have the freedom to change solutions by the gauge group action, since we want to extract a subsequence, which converges in the quotient space $\mathcal{C}/\mathcal{G}$. In fact, we will see below that (ψ_n, A_n) is bounded in $W^{k,2}$ for any $k \geqslant 0$ possibly after applying gauge transformations.

The formal proof requires a number of technical lemmas. The key property is as follows.

Lemma 7.1.12. *There is a constant* $C > 0$ *such that for any solution*

$$(\psi, A) \in \mathcal{C}^{5,2} := W^{5,2}(\mathcal{S}^+) \times \mathcal{A}^{5,2}(P_{\det})$$

of the Seiberg–Witten equations we have

$$\|\psi\|_{C^0} \leqslant C.$$

Proof. First notice that $\psi \in W^{5,2}$ implies that $\psi \in C^2$.

Let x_0 be the point of maximum of the function $|\psi|^2$. The pointwise equality

$$\Delta|\psi|^2 = 2\,\langle \nabla_A^* \nabla_A \psi, \psi\rangle - 2\,|\nabla_A \psi|^2$$

implies that

$$\langle \nabla_A^* \nabla_A \psi, \psi\rangle \geqslant |\nabla_A \psi|^2 \qquad \text{at } x_0.$$

Furthermore, notice also that we have the following pointwise equality:

$$\langle \mu(\psi)\psi, \psi\rangle = \Big\langle |\psi|^2\psi - \frac{1}{2}|\psi|^2\psi,\ \psi\Big\rangle = \frac{1}{2}\,|\psi|^4.$$

Combining this with the Weitzenböck formula, we obtain

$$0 = \langle \nabla_A^* \nabla_A \psi, \psi\rangle + \frac{1}{4} s_g |\psi|^2 + \frac{1}{2}\,|\psi|^4.$$

Hence, at x_0 the following holds:

$$\frac{1}{4} s_g(x_0)|\psi|^2(x_0) + \frac{1}{2}\,|\psi|^4(x_0) \leqslant -|\nabla_A \psi|^2(x_0) \leqslant 0.$$

If $|\psi(x_0)| = 0$, there is nothing to prove. If $|\psi|(x_0)| > 0$, then the above inequality yields

$$|\psi|^2(x_0) \leqslant -\frac{1}{2}\, s_g(x_0),$$

thus providing the required estimate. □

Remark 7.1.13. Notice that the proof of this lemma does not go through for the equations $\not{D}_A^+\psi = 0, F_A^+ = -\mu(\psi)$, which differ from (7.1.4) just by a sign.

Corollary 7.1.14. *For any* $p > 1$ *there is a non-negative constant* κ_p*, which depends on the background Riemannian metric* g *only, such that for any solution* $(\psi, A) \in \mathcal{C}^{5,2}$ *of the Seiberg–Witten equations the following estimate holds:* $\|\psi\|_{L^p} \leqslant \kappa_p$. □

Corollary 7.1.15. *For any solution* (ψ, A) *of the Seiberg–Witten equations we have the estimates*

$$\|F_A^+\|_{L^2} \leqslant C \qquad \text{and} \qquad \|F_A^-\|_{L^2} \leqslant C - 4\pi^2 c_1(L_{\det})^2, \tag{7.1.16}$$

where the constants depend on the background metric only.

Proof. The first inequality follows immediately from Corollary 7.1.14. To prove the second bound, recall that the first Chern class of $L_{\det}$ is represented by $\frac{i}{2\pi}F_A$ so that we have

$$\begin{aligned} c_1(L_{\det})^2 &= \frac{i^2}{(2\pi)^2}\int_M F_A \wedge F_A = -\frac{1}{4\pi^2}\int_M (F_A^+ \wedge F_A^+ + F_A^- \wedge F_A^-) \\ &= \frac{1}{4\pi^2}(\|F_A^+\|_{L^2}^2 - \|F_A^-\|_{L^2}^2), \end{aligned}$$

which yields the required bound. □

Remark 7.1.17. Of course, the right hand side of the second inequality of (7.1.16) is just a constant independent of the solution. However, this explicit form will be useful below.

The proof of the proposition below hinges on the following technical lemma, which follows essentially from the elliptic estimate.

Lemma 7.1.18 ([21, Lemma 5.3.1])**.** *Let* L *be any Hermitian line bundle over* M*. Fix a smooth reference connection* A_0*. For any* $k \geqslant 0$ *there are positive constants* C_1 *and* C_2 *with the following property: For any* $W^{k,2}$*–connection* A *on* L *there is a gauge transformation* $g \in \mathcal{G}^{k+1,2}$ *such that* $A \cdot g = A_0 + \alpha$*, where* $\alpha \in W^{k,2}(T^*M \otimes \mathbb{R}i)$ *satisfies*

$$d^*\alpha = 0 \qquad \text{and} \qquad \|\alpha\|_{W^{k,2}} \leqslant C_1\|F_A^+\|_{W^{k-1,2}} + C_2.$$

Moreover, the harmonic component α_h *of* α *can be assumed to be bounded in* L^2 *by a constant independent of* k. □

Proposition 7.1.19. *For each* $k \geqslant 0$ *there are positive constants* $C_k > 0$ *with the following property: For any solution* (ψ, A) *of the Seiberg–Witten equations there is a gauge transformation* $g \in \mathcal{G}^{k+1,2}$ *such that*

$$\|(\psi, \alpha)\|_{W^{k,2}} \leqslant C_k, \tag{7.1.20}$$

where $A \cdot g = A_0 + \alpha$.

Proof. The proof is given via the induction on k. However, a few first values of k require a special treatment.

For $k = 0$ we know already that $\|(\psi, F_A^+)\|_{L^2}$ is bounded by a constant independent of (ψ, A). Lemma 7.1.18 yields immediately the required estimate for α.

For $k = 1$, using the Seiberg–Witten equations we have

$$0 = \not{D}_A^+\psi = \not{D}_{A_0}^+\psi + \frac{1}{2}\alpha \cdot \psi,$$

where the second summand is bounded in L^2 by Lemma 7.1.12. Invoking the elliptic estimate for $\not{D}^+_{A_0}$, we obtain a bound for ψ in $W^{1,2}$. This proves (7.1.20) for $k = 1$.

Let us consider the case $k = 2$. Notice that by the Seiberg–Witten equations we have the pointwise estimate

$$|\nabla^{LC} F_A^+| = |\nabla^{LC} \mu(\psi)| \leqslant C|\nabla^{A_0}\psi||\psi| \leqslant C|\nabla^{A_0}\psi|,$$

where the first inequality follows from the fact that μ is quadratic and the second one follows by Lemma 7.1.12. This clearly implies a bound on the $W^{1,2}$-norm of F_A^+. This in turn, yields a bound on the $W^{2,2}$–norm of α.

Furthermore, the bound for $\|\alpha\|_{W^{2,2}}$ together with the Sobolev multiplication theorem yields that $\alpha \cdot \psi$ is bounded in $W^{1,2}$. By the same token as above, this yields the $W^{2,2}$–bound on ψ.

By now, the reader will have no difficulties in proving (7.1.20) for $k = 3$. I leave this as an exercise.

We are now prepared to make the induction step. Thus, assume that (7.1.20) has been established for some $k \geqslant 3$. Since $W^{k,2}$ is an algebra for $k \geqslant 3$, we have a bound on F_A^+ in $W^{k,2}$. Lemma 7.1.18 yields a $W^{k+1,2}$–bound on α.

By a similar argument, $\alpha \cdot \psi$ is bounded in $W^{k,2}$ so that the elliptic estimate yields a $W^{k+1,2}$–bound for ψ. This finishes the proof of this proposition. □

Corollary 7.1.21. *The Seiberg–Witten moduli space*

$$\mathcal{M} := \{(\psi, A) \in \mathcal{C}^{5,2} \mid SW(\psi, A) = 0\}/\mathcal{G}^{6,2}$$

is compact.

Proof. Any sequence of solutions (ψ_n, A_n) is gauge equivalent to a sequence $(\psi_n, A_0 + \alpha_n)$ such that (ψ_n, α_n) is bounded in $W^{6,2}$. By the Sobolev embedding theorem, a subsequence converges in $\mathcal{C}^{5,2}$ and the limit is a solution of the Seiberg–Witten equations. □

In fact, Corollary 7.1.21 can be substantially strengthened, as the following result shows.

Theorem 7.1.22. *For each solution $(\psi, A) \in \mathcal{C}^{5,2}$ of the Seiberg–Witten equations there is a gauge transformation $g \in \mathcal{G}^{6,2}$ such that $(\psi, A) \cdot g$ is smooth. Furthermore, $\mathcal{M}$ is homeomorphic to*

$$\mathcal{M}^\infty := \{(\psi, A) \in \Gamma(\mathcal{S}^+) \times \mathcal{A}(P_{\det}) \mid SW(\psi, A) = 0\}/C^\infty(M; U(1)).$$

This space is compact in the C^∞–topology.

Proof. Fix a solution $(\psi, A) \in \mathcal{C}^{5,2}$. Then we have $F_A = F_{A\cdot g} = F_{A_0} + d\alpha \in W^{k-1,2}$ for all k by Proposition 7.1.19. In particular, F_A is smooth.

Writing $A \cdot g = A_0 + \alpha$ as before, we obtain

$$(d + d^*)\,\alpha = d\alpha = F_A - F_{A_0} \in C^\infty.$$

By the elliptic regularity, α is smooth, i.e., A is smooth. Hence, ψ is also smooth as an element of the kernel of a smooth elliptic differential operator $\not{D}_A^+$.

Furthermore, let (ψ_n, A_n) be any sequence of smooth solutions. By Corollary 7.1.21, this contains a subsequence still denoted by (ψ_n, A_n), which converges in the $W^{5,2}$–topology after possibly applying a sequence of $\mathcal{G}^{6,2}$ gauge transformations and the limit lies also in $\mathcal{C}^{5,2}$.

Since (ψ_n, A_n) is bounded in $W^{7,2}$, a subsequence converges in $W^{6,2}$ after possibly applying a sequence of $\mathcal{G}^{7,2}$ gauge transformations. Repeating this process for each $k \geqslant 6$ and choosing the diagonal subsequence, we obtain the claim. □

Slices. In this section we wish to construct local slices for the gauge group action on the subspace of irreducible configurations

$$\mathcal{C}^{5,2}_{\text{irr}} := \{(\psi, A) \in \mathcal{C}^{5,2} \mid \psi \not\equiv 0\},$$

where $\mathcal{G}^{6,2}$ acts freely.

The tangent space to the slice at a point (ψ, A) must be transversal to $\operatorname{Im} R_{(\psi,A)}$, hence it is natural to consider the kernel of the formal adjoint operator

$$R^*_{(\psi,A)}(\dot\psi, \dot a) = 2\, d^* \dot a + i\, \operatorname{Re}\langle \psi, i\, \dot\psi \rangle.$$

Here $\langle \cdot, \cdot \rangle$ denotes the Hermitian scalar product on the spinor bundle.

Proposition 7.1.23. *For any irreducible configuration* (ψ, A) *the subspace*

$$(\psi, A) + \ker R^*_{(\psi,A)}$$

is a slice for the $\mathcal{G}^{6,2}$*–action on* $\mathcal{C}^{5,2}_{\text{irr}}$*. Here* $\ker R^*_{(\psi,A)}$ *means the kernel of the map* $R^*_{(\psi,A)} \colon W^{5,2} \to W^{4,2}$. □

A perturbation. In general, there is no reason to expect that the origin will be a regular value for the Seiberg–Witten map. However, a family of perturbations just like in Subsection 6.3 can be constructed by hand.

Proposition 7.1.24. *The origin is a regular value of the map*

$$\mathcal{SW} \colon \mathcal{C}^{5,2}_{\text{irr}} \times W^{4,2}(\Lambda^2_+ T^*M \otimes \mathbb{R}i) \to W^{4,2}(\mathbb{S}^-) \times W^{4,2}(\Lambda^2_+ T^*M \otimes \mathbb{R}i),$$

$$\mathcal{SW}(\psi, A, \eta) = \big(\not{D}_A^+ \psi,\ F_A^+ - \mu(\psi) - \eta\big).$$

Proof. The perturbation has been chosen so that

$$\operatorname{pr}_2 \frac{\partial \mathcal{SW}}{\partial \eta} \colon W^{4,2}(\Lambda^2_+ T^*M \otimes \mathbb{R}i) \to W^{4,2}(\Lambda^2_+ T^*M \otimes \mathbb{R}i)$$

is surjective. Hence, it is enough to show that the map

$$T := \operatorname{pr}_1 d_{(\psi,A)} \mathcal{SW}_\eta \colon T_{(\psi,A)} \mathcal{C}^{5,2} \to W^{4,2}(\mathbb{S}^-), \qquad (\dot\psi, \dot A) \mapsto \not{D}_A \dot\psi + \frac{1}{2} \dot A \cdot \psi$$

is also surjective for any solution (ψ, A) of the perturbed Seiberg–Witten equations.

Assume this is not the case. Then there is a non-zero vector

$$\varphi \in \operatorname{Im} T^\perp := \{\varphi \in W^{4,2} \mid \langle T(\dot\psi, \dot A), \varphi \rangle_{L^2} = 0 \quad \forall (\dot\psi, \dot A)\}.$$

In particular, we have $0 = \langle \not{D}_A^+ \dot\psi, \varphi\rangle_{L^2} = \langle \dot\psi, \not{D}_A^- \varphi\rangle_{L^2}$, which implies

$$\not{D}_A^- \varphi = 0. \tag{7.1.25}$$

Also,

$$0 = \langle T(0, \dot A), \varphi\rangle_{L^2} = \langle \dot A \cdot \psi, \varphi\rangle_{L^2} \tag{7.1.26}$$

for all $\dot A \in W^{5,2}(T^*M \otimes \mathbb{R}i)$.

Since $W^{5,2} \subset C^0$ in dimension four, ψ is continuous and by assumption does not vanish identically. Hence, there is a point $m \in M$ such that ψ does not vanish on a neighborhood U of m.

Furthermore, notice that the Clifford multiplication with a fixed non-zero vector $\psi_0 \in \$^+$ is surjective, i.e., the map

$$\mathbb{R}^4 \to \$^-, \qquad v \mapsto v \cdot \psi_0$$

(here $\$^\pm$ are thought of as $\mathrm{Spin}^c(4)$–representations) is surjective. This implies the following: if φ does not vanish on U, then there is some $\dot A$ supported in U such that $\langle \dot A \cdot \psi, \varphi\rangle_{L^2} > 0$. However, this contradicts (7.1.26) so that we conclude that φ must vanish on an open set. Then, by a theorem of Aronszajn [1], (7.1.25) implies that φ vanishes identically. This in turn proves the surjectivity of T and finishes the proof of this proposition. □

The deformation complex at a solution of the perturbed Seiberg–Witten equations

$$\not{D}_A^+ \psi = 0, \qquad F_A^+ = \mu(\psi) + \eta$$

is given again by (7.1.7) since the differentials of $SW = \mathcal{S}\mathcal{W}_0$ and $\mathcal{S}\mathcal{W}_\eta$ coincide. Hence, this complex is elliptic and therefore the operator

$$\left(d_{(\psi,A)}\mathcal{S}\mathcal{W}_\eta,\ R^*_{(\psi,A)}\right) : W^{5,2}(\$^+ \oplus T^*X \otimes \mathbb{R}i) \to W^{4,2}(\$^- \oplus \Lambda^2_+ T^*X \otimes \mathbb{R}i) \tag{7.1.27}$$

is also elliptic by Exercise 5.3.6. Thus, (7.1.27) is Fredholm so that by appealing to Theorem 6.6.7, we obtain the following result.

Corollary 7.1.28. *There is a subset* $H \subset W^{4,2}(\Lambda^2_+ T^*X \otimes \mathbb{R}i)$ *of the second category such that for any* $\eta \in H$ *the space*

$$\mathcal{M}_\eta^{irr} := \left\{(\psi, A) \in \mathcal{C}^{5,2} \mid \mathcal{S}\mathcal{W}(\psi, A, \eta) = 0, \quad \psi \not\equiv 0\right\}/\mathcal{G}^{6,2}$$

is a smooth manifold of dimension

$$d = \frac{1}{4}\Big(c_1(L_{\det})^2 - 2\chi(M) - 3\,\mathrm{sign}(M)\Big), \tag{7.1.29}$$

where $\chi(M)$ *and* $\mathrm{sign}(M) := b_2^+ - b_2^-$ *are the Euler characteristic and the signature of* M *respectively.*

Proof. We only need to prove the formula for the dimension of the moduli space. To this end the following fact will be useful: The index is a locally constant function on the space of all Fredholm operators. Equivalently, if $\{T_t \mid t \in [0,1]\}$ is a one-parameter family of Fredholm operators, then $\operatorname{index} T_0 = \operatorname{index} T_1$.

With this understood, consider the operator

$$\begin{aligned}(7.1.30)\qquad & D_{(\psi,A)} \\ & := \left(d_{(\psi,A)} SW_\eta,\ R^*_{(\psi,A)}\right) : T_{(\psi,A)}\mathcal{C}^{5,2} \to W^{4,2}\left(\mathcal{S}^- \oplus \Lambda^2_+ T^*M \otimes \mathbb{R}i \oplus \underline{\mathbb{R}i}\right).\end{aligned}$$

Writing

$$(7.1.31)\qquad D_{(\psi,A)} = \begin{pmatrix} \not{D}_A^+ & \\ & d^+ + d^* \end{pmatrix} + B = D_0 + B,$$

where B is a zero order operator, we see that $D_{(\psi,A)}$ is homotopic through Fredholm operators to D_0. Since D_0 decouples, we have

$$\operatorname{index} D_0 = \operatorname{index} \not{D}_A^+ + \operatorname{index}(d^+ + d^*).$$

The last index is easily computed: $\operatorname{index}(d^+ + d^*) = b_1 - b_0 - b_2^+ = -\frac{1}{2}(\chi + \operatorname{sign})$. The index of $\not{D}_A^+$ can be computed by applying the Atiyah–Singer index theorem:

$$\operatorname{index}_{\mathbb{C}} \not{D}_A^+ = \frac{1}{8}\left(c_1(L_{\det})^2 - \operatorname{sign}\right).$$

This yields the result. □

Remark 7.1.32. Strictly speaking, at this point we should redo the analysis of the compactness of the Seiberg–Witten moduli space for the perturbed equations. However, this requires cosmetic changes only. The reader should have no difficulties to check that this is indeed the case.

Reducible solutions. While Corollary 7.1.28 yields a smooth moduli space, by removing some points we lost an essential property, namely compactness. The following lemma demonstrates how to deal with this problem.

Proposition 7.1.33. *If $b_2^+ \geqslant 1$, there is an affine subspace $Q \subset W^{4,2}(\Lambda^2_+ T^*M \otimes \mathbb{R}i)$ of codimension b_2^+ with the following property: If $\eta \notin Q$, then there are no reducible solutions of the Seiberg–Witten equations.*

Proof. Pick a reference connection A_0 on $L_{\det}$ and denote $Q := F^+_{A_0} + \operatorname{Im} d^+$, which is a subspace of codimension b_2^+. Clearly, if $\eta \notin Q$, then there are no reducible solutions. □

Corollary 7.1.34. *Assume $b_2^+ \geqslant 1$. Then for a generic η the perturbed Seiberg–Witten moduli space $\mathcal{M}_\eta$ contains no reducible solutions.* □

Orientability of the Seiberg–Witten moduli space. As we have already seen in the proof of Corollary 7.1.28, $D_{(\psi,A)}$ defined by (7.1.30) is homotopic through elliptic operators to D_0, which is given by (7.1.31). This can be also used to orient the Seiberg–Witten moduli space by appealing to Proposition 6.4.4.

Indeed, to orient the Seiberg–Witten moduli space it suffices to check that $\det D_0$ is trivial and pick a trivialization of this bundle. Since D_0 splits, we have

$$\det D_0 \cong \det \not{D}_A^+ \otimes \det(d^+ + d^*).$$

Notice that $\not{D}_A^+$ is a complex linear map. In particular, both $\operatorname{coker} D_A^+ \cong \ker D_A^-$ and $\ker D_A^+$ are complex linear subspaces, hence oriented. This implies that the real determinant bundle $\det \not{D}_A^+$ is trivial.

Furthermore, we have

$$\ker(d^+ + d^*) = H^1_{dR}(M; \mathbb{R}\, i), \qquad \operatorname{coker}(d^+ + d^*) = H^0(M; \mathbb{R}\, i) \oplus H^2_+(M; \mathbb{R}\, i).$$

Hence, by picking an orientation of these cohomology groups, a trivialization of $\det(d^+ + d^*)$ is fixed. This in turn yields a trivialization of $\det D_0$, hence also of $\det D_{(\psi,A)}$. Thus, the Seiberg–Witten moduli space is orientable and a choice of orientations of the cohomology groups $H^0(M; \mathbb{R}\, i)$, $H^1(M; \mathbb{R}\, i)$ and $H^2_+(M; \mathbb{R}\, i)$ yields an orientation of the Seiberg–Witten moduli space.

Remark 7.1.35. Strictly speaking, the orientation of the Seiberg–Witten moduli space we constructed in this section is *not* canonical. First, there are at least two commonly used non-equivalent conventions how to orient complex linear spaces. Namely, if V is a complex linear space and $(v_1, \dots, v_k)$ is a complex basis of V, the underlying real vector space $V_{\mathbb{R}}$ can be oriented by saying that

$$(v_1, \dots, v_k, i v_1, \dots, i v_k) \qquad \text{or} \qquad (v_1, i v_1, \dots, v_k, i v_k)$$

determines an orientation of $V_{\mathbb{R}}$. This yields the opposite orientations in the case $k = \dim_{\mathbb{C}} V$ is even.

Secondly, orientations of $H^0(M; \mathbb{R}\, i)$, $H^1(M; \mathbb{R}\, i)$ and $H^2_+(M; \mathbb{R}\, i)$ is again a choice. Notice however, that these vector spaces depend on the topological structure of M only.

7.2. The Seiberg–Witten invariant Combining results obtained in the preceding sections we arrive at our main result.

Theorem 7.2.1. *Assume $b_2^+(M) \geqslant 2$. For any spin^c-structure $\sigma \in \mathcal{S}(M)$ and any generic η the perturbed Seiberg–Witten moduli space $\mathcal{M}_\eta$ is a smooth compact oriented manifold of dimension d, which is given by* (7.1.29). *Moreover, if η_0 and η_1 are any two generic perturbations, then $\mathcal{M}_{\eta_0}$ and $\mathcal{M}_{\eta_1}$ are oriented-bordant.* □

The only thing, which perhaps needs explanation, is the hypothesis $b_2^+(M) \geqslant 2$. The point is that this ensures that a generic bordism between $\mathcal{M}_{\eta_0}$ and $\mathcal{M}_{\eta_1}$ does not contain reducible solutions and therefore is smooth.

Remark 7.2.2. Notice that $\mathcal{M}_\eta$ depends on the choice of σ but this dependence is suppressed in the notations.

With this understood, we can proceed to the definition of the Seiberg–Witten invariant. Pick a point $m_0 \in M$ and consider *the based gauge group* defined by

$$\mathcal{G}_0 := \{ g \in W^{6,2}(M; U(1)) \mid g(m_0) = 1 , \}$$

which fits into the exact sequence

$$\{1\} \to \mathcal{G}_0 \to \mathcal{G} \xrightarrow{ev_{m_0}} U(1) \to \{1\},$$

where ev_{m_0} is the evaluation at m_0. Then the quotient

$$\hat{\mathcal{M}}_\eta := \{(\psi, A) \in \mathcal{C}^{5,2} \mid SW_\eta(\psi, A) = 0\}/\mathcal{G}_0$$

is called *the framed moduli space* and is equipped with a free action of $U(1) = \mathcal{G}/\mathcal{G}_0$. Clearly, the quotient $\hat{\mathcal{M}}_\eta/U(1)$ is just $\mathcal{M}_\eta$ so that $\hat{\mathcal{M}}_\eta$ is a principal $U(1)$–bundle over $\mathcal{M}_\eta$. Let μ be the first Chern class of this $U(1)$–bundle. Notice that this $\hat{\mathcal{M}}_\eta$ is just a restriction to $\mathcal{M}_\eta$ of the following bundle: $\mathcal{C}_{irr}/\mathcal{G}_0 \to \mathcal{C}_{irr}/\mathcal{G}$.

Definition 7.2.3. The Seiberg–Witten invariant of M is a function $sw\colon \mathcal{S}(M) \to \mathbb{Z}$ defined by

$$sw(\sigma) := \begin{cases} \langle [\mathcal{M}_\eta], \mu^{d/2} \rangle & \text{if d is even,} \\ 0 & \text{if d is odd.} \end{cases}$$

Theorem 7.2.4. *The Seiberg–Witten invariant vanishes for all but finitely many spin^c structures.*

Proof. If $sw(\sigma) \neq 0$, then $\mathcal{M}_\eta$ is a smooth manifold of dimension $d \geqslant 0$. Hence,

$$c_1(L_{det})^2 \geqslant 2\chi(M) + 3\,\mathrm{sign}(M).$$

By (7.1.16), we have[6] $\|F_A^-\|_{L^2}^2 \leqslant C$, where C depends on the background Riemannian metric and the topology of M only. This yields that $\|F_A\|_{L^2}^2$ is also bounded by a constant, which depends on the background Riemannian metric and the topology of M only. This in turn shows that

$$|c_1(L_{det})|^2 = |c_1(L_{det}) \cup PD(c_1(L_{det}))| = \frac{1}{4\pi^2}\|F_A\|_{L^2}^2 \leqslant C,$$

where PD stays for the Poincaré dual class. This proves the statement of this theorem. ☐

Sample application of the Seiberg–Witten invariant. The Seiberg–Witten invariant are most well–known for its applications to the topology of smooth four–manifolds. I restrict myself just to a list of some sample applications.

Theorem 7.2.5 (Witten). *If M with $b_2^+ \geqslant 2$ admits a metric of positive scalar curvature, then $sw_M \equiv 0$.* ☐

The proof of this theorem follows easily from the Weitzenböck formula.

Theorem 7.2.6 (Witten). *Let M_1 and M_2 be closed four-manifolds both with $b_2^+ \geqslant 1$. Then the Seiberg–Witten invariant of the connected sum $M_1 \# M_2$ vanishes.* ☐

These vanishing theorems are complemented by the following non-vanishing result, which can be also viewed as the statement about obstructions for the existence of symplectic strutures on closed smooth four-manifolds.

Theorem 7.2.7 (Taubes). *If M is symplectic and $b_2^+(M) \geqslant 2$, then $sw_M \not\equiv 0$.* ☐

[6] At this point we should have used an analogous a priory estimate for the perturbed Seiberg–Witten equations, cf. Remark 7.1.32. However, this does not change the argument.

The results below were originally obtained by other methods, however can be also obtained with the help of the Seiberg–Witten theory.

Theorem 7.2.8 (Donaldson). *There are (many) closed topological four-manifolds that do not admit a smooth structure.* □

Theorem 7.2.9 (Fintushel–Stern). *There are infinitely many closed four-manifolds, all homeomorphic, but pairwise non-diffeomorphic.* □

Put differently, the last two theorems say the following: On a given closed topological four-manifold there may or may not be a smooth structure and if a smooth structure exists, there may well be infinitely many such structures.

References

[1] N. Aronszajn, *A unique continuation theorem for solutions of elliptic partial differential equations or inequalities of second order*, J. Math. Pures Appl. (9) **36** (1957), 235–249. MR92067 ↑70

[2] D. Barden and C. Thomas, *An introduction to differential manifolds*, Imperial College Press, London, 2003, DOI 10.1142/p285. MR1992457 ↑7, 54

[3] A. L. Besse, *Einstein manifolds*, Classics in Mathematics, Springer-Verlag, Berlin, 2008. Reprint of the 1987 edition. MR2371700 ↑23

[4] R. Bott and L. W. Tu, *Differential forms in algebraic topology*, Graduate Texts in Mathematics, vol. 82, Springer-Verlag, New York-Berlin, 1982. MR658304 ↑28

[5] S. B. Bradlow, O. García-Prada, and P. B. Gothen, *What is... a Higgs bundle?*, Notices Amer. Math. Soc. **54** (2007), no. 8, 980–981. MR2343296 ↑35

[6] S. K. Donaldson, *Floer homology groups in Yang-Mills theory*, Cambridge Tracts in Mathematics, vol. 147, Cambridge University Press, Cambridge, 2002. With the assistance of M. Furuta and D. Kotschick, DOI 10.1017/CBO9780511543098. MR1883043 ↑3

[7] S. K. Donaldson and P. B. Kronheimer, *The geometry of four-manifolds*, Oxford Mathematical Monographs, The Clarendon Press, Oxford University Press, New York, 1990. Oxford Science Publications. MR1079726 ↑3

[8] L. C. Evans, *Partial differential equations*, 2nd ed., Graduate Studies in Mathematics, vol. 19, American Mathematical Society, Providence, RI, 2010, DOI 10.1090/gsm/019. MR2597943 ↑4

[9] D. S. Freed and K. K. Uhlenbeck, *Instantons and four-manifolds*, 2nd ed., Mathematical Sciences Research Institute Publications, vol. 1, Springer-Verlag, New York, 1991, DOI 10.1007/978-1-4613-9703-8. MR1081321 ↑3

[10] R. E. Gompf and A. I. Stipsicz, *4-manifolds and Kirby calculus*, Graduate Studies in Mathematics, vol. 20, American Mathematical Society, Providence, RI, 1999, DOI 10.1090/gsm/020. MR1707327 ↑41

[11] P. B. Gothen, *Representations of surface groups and Higgs bundles*, Moduli spaces, London Math. Soc. Lecture Note Ser., vol. 411, Cambridge Univ. Press, Cambridge, 2014, pp. 151–178. MR3221295 ↑35

[12] V. Guillemin and A. Pollack, *Differential topology*, AMS Chelsea Publishing, Providence, RI, 2010. Reprint of the 1974 original, DOI 10.1090/chel/370. MR2680546 ↑54, 58

[13] V. W. Guillemin and S. Sternberg, *Supersymmetry and equivariant de Rham theory*, Mathematics Past and Present, Springer-Verlag, Berlin, 1999. With an appendix containing two reprints by Henri Cartan [MR0042426 (13,107e); MR0042427 (13,107f)], DOI 10.1007/978-3-662-03992-2. MR1689252 ↑24

[14] D. D. Joyce, *Riemannian holonomy groups and calibrated geometry*, Oxford Graduate Texts in Mathematics, vol. 12, Oxford University Press, Oxford, 2007. MR2292510 ↑23

[15] S. Kobayashi and K. Nomizu, *Foundations of differential geometry. Vol. I*, Wiley Classics Library, John Wiley & Sons, Inc., New York, 1996. Reprint of the 1963 original; A Wiley-Interscience Publication. MR1393940 ↑34

[16] P. Kronheimer and T. Mrowka, *Monopoles and three-manifolds*, New Mathematical Monographs, vol. 10, Cambridge University Press, Cambridge, 2007, DOI 10.1017/CBO9780511543111. MR2388043 ↑3

[17] H. B. Lawson Jr. and M.-L. Michelsohn, *Spin geometry*, Princeton Mathematical Series, vol. 38, Princeton University Press, Princeton, NJ, 1989. MR1031992 ↑35, 38

[18] D. McDuff and D. Salamon, J*-holomorphic curves and symplectic topology*, 2nd ed., American Mathematical Society Colloquium Publications, vol. 52, American Mathematical Society, Providence, RI, 2012. MR2954391 ↑59

[19] J. W. Milnor, *Topology from the differentiable viewpoint*, Princeton Landmarks in Mathematics, Princeton University Press, Princeton, NJ, 1997. Based on notes by David W. Weaver; Revised reprint of the 1965 original. MR1487640 ↑56

[20] J. D. Moore, *Lectures on Seiberg-Witten invariants*, Lecture Notes in Mathematics, vol. 1629, Springer-Verlag, Berlin, 1996, DOI 10.1007/BFb0092948. MR1439049 ↑3

[21] J. W. Morgan, *The Seiberg-Witten equations and applications to the topology of smooth four-manifolds*, Mathematical Notes, vol. 44, Princeton University Press, Princeton, NJ, 1996. MR1367507 ↑3, 41, 67

[22] S. Rayan, *Aspects of the topology and combinatorics of Higgs bundle moduli spaces*, SIGMA Symmetry Integrability Geom. Methods Appl. **14** (2018), Paper No. 129, 18, DOI 10.3842/SIGMA.2018.129. MR3884746 ↑35

[23] S. Salamon, *Riemannian geometry and holonomy groups*, Pitman Research Notes in Mathematics Series, vol. 201, Longman Scientific & Technical, Harlow; copublished in the United States with John Wiley & Sons, Inc., New York, 1989. MR1004008 ↑23

[24] R. O. Wells Jr., *Differential analysis on complex manifolds*, 2nd ed., Graduate Texts in Mathematics, vol. 65, Springer-Verlag, New York-Berlin, 1980. MR608414 ↑4, 52

Institute of Mathematics, University of Freiburg, Ernst-Zermelo-Str. 1, D-79104 Freiburg im Br., Germany

Email address: `andriy.haydys@math.uni-freiburg.de`

IAS/Park City Mathematics Series
Volume 28, Pages 77–170
https://doi.org/10.1090/pcms/028/00875

Knots, Polynomials, and Categorification

Jacob Rasmussen

Abstract. These lectures give an introduction to knot polynomials and their categorifications. Topics covered include the Jones and Alexander polynomials, Khovanov homology of links and tangles, HOMFLY-PT homology, and the Λ^k-colored HOMFLY-PT polynomial.

Contents

2020 *Mathematics Subject Classification.* Primary 57M27; Secondary 57M25.
Key words and phrases. Knot polynomial, knot homology, Khovanov homology, Alexander polynomial, Jones polynomial, HOMFLY-PT.

1. Prelude: Knots and the Jones polynomial

1.1. Knots

Definition 1.1.1. An *oriented knot* in $\mathbb{R}^3$ is a smooth embedding $K : S^1 \hookrightarrow \mathbb{R}^3$. We say that two knots $K_0, K_1 : S^1 \hookrightarrow \mathbb{R}^3$ are *isotopic* if there is a smooth map $\Phi : S^1 \times [0,1] \to \mathbb{R}^3$ such that $K_t := \Phi|_{S^1 \times t}$ is a knot for each t.

We declare two knots to be equivalent if they are isotopic, and leave it to the reader to check that isotopy is indeed an equivalence relation. Since any two orientation preserving homeomorphisms of S^1 are isotopic, the equivalence class of knot is determined by its oriented image in $\mathbb{R}^3$.

To draw pictures of a knot, we consider its image under a linear projection $\pi : \mathbb{R}^3 \to \mathbb{R}^2$. We say that the image is *generic* if 1) $\pi \circ K : S^1 \to \mathbb{R}^2$ is an immersion, and 2) $\pi \circ K$ is injective except at transverse double points, which we refer to as *crossings*. By a small isotopy, we may arrange that the image of K under any given projection π is generic.

Figure 1.1.2. A crossing

Let us identify the image of the projection π with the xy plane in $\mathbb{R}^3$. If π is generic, the isotopy class of K is determined by the image $\pi \circ K$ together with the relative z-coordinate of the two strands near each double point. The strand with the larger z-coordinate is an *overcrossing*, while the strand with the smaller z-coordinate is an *undercrossing*.

We record this information by briefly picking up our pen near each undercrossing, as shown in the figure. The resulting picture is a *planar diagram* of the knot K. Here are some examples:

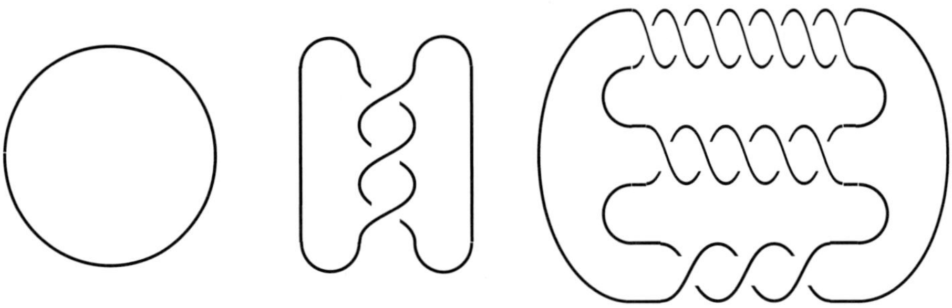

Unknot O Trefoil T Pretzel knot $P(-3,5,7)$

Figure 1.1.3. Examples of Knots

We declare two planar diagrams to be equivalent if there is an orientation preserving homeomorphism of the plane which carries one to the other. For example, any crossingless planar diagram is equivalent to the standard diagram of the unknot shown above. (This seemingly obvious fact is far from trivial — it is the 2-dimensional Schoenflies theorem.)

Note that any given knot will have many inequivalent planar diagrams. For example, the first three diagrams in Figure 1.1.4 all represent the unknot.

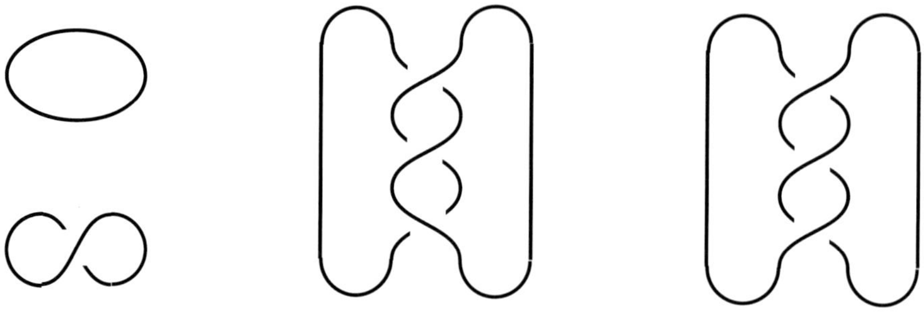

Figure 1.1.4. One of these knots is not like the others.

Definition 1.1.5. Planar diagrams D_1 and D_2 are related by a *Reidemeister move* if there are embeddings $\varphi_1, \varphi_2 : B^2 \hookrightarrow \mathbb{R}^2$ such that

1) $D_1 \cap (\mathbb{R}^2 - \varphi_1(B^2)) = D_2 \cap (\mathbb{R}^2 - \varphi_2(B^2))$
2) $\varphi_1^{-1}(D_1)$ and $\varphi_2^{-1}(D_2)$ are one of the three pairs of diagrams shown in Figure 1.1.6.

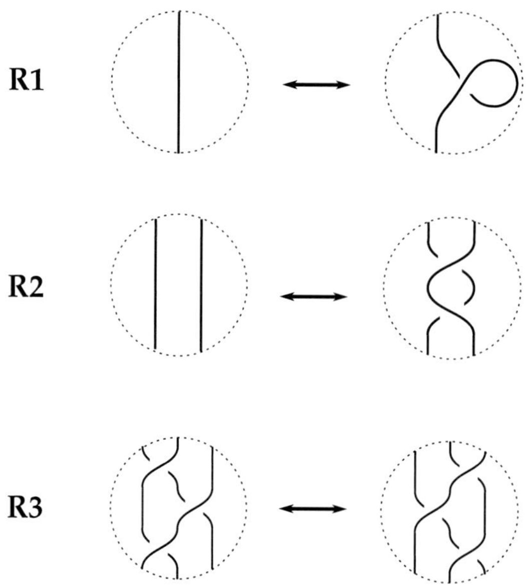

Figure 1.1.6. Reidemeister moves

The reader should convince themselves that diagrams related by a Reidemeister move are indeed isotopic. For example, R1 corresponds to giving a strand of the diagram a half twist; R2 to sliding one strand of the diagram over another, and R3 to sliding a crossing under a strand. As a handy mnemonic, note that the number of the move is equal to the number of distinct strands involved in it.

Theorem 1.1.7 (Reidemeister). *If two planar diagrams represent the same knot, they are related by a sequence of Reidemeister moves.*

1.2. Generalizations One way to generalize Definition 1.1.1 is to change the domain of the embedding.

Definition 1.2.1. We define an *oriented* n*-component link* in $\mathbb{R}^3$ to be a smooth embedding $\amalg^n S^1 \to \mathbb{R}^3$. Two such links L_0 and L_1 are isotopic if there is a smooth map $\Phi : (\amalg^n S^1) \times [0,1] \to \mathbb{R}^3$ such that $L_t := \Phi_{(\amalg^n S^1)\times t}$ is a link for all $t \in [0,1]$.

A few examples are shown in Figure 1.2.2.

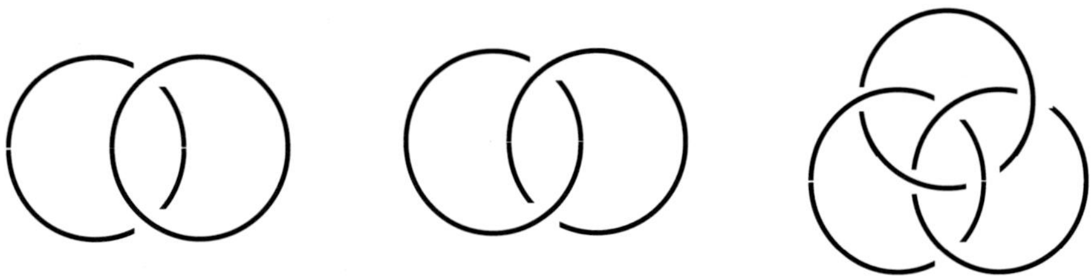

Unlink Hopf link Borromean rings

Figure 1.2.2. Examples of Links

Links have planar diagrams just like knots do, and satisfy an analog of Theorem 1.1.7. We can also vary the target of the embedding. If Y is a smooth 3-manifold, an oriented knot in Y is a smooth embedding $K : S^1 \hookrightarrow Y$. Viewing S^3 as the one-point compactification of $\mathbb{R}^3$, we see that every knot in $\mathbb{R}^3$ gives rise to a knot in S^3. Conversely, if K is a knot in S^3 and p is a point, it's easy to see that K is isotopic to a knot which does not contain p. Hence every knot in S^3 arises in this way. Similarly, if two knots are isotopic in S^3, they are isotopic in $\mathbb{R}^3$. (The trace of the isotopy is 2-dimensional, so the isotopy can itself be isotoped to avoid p.) In summary, there is a natural bijection between the set of knots in $\mathbb{R}^3$ and the set of knots in S^3.

1.3. New knots from old

Definition 1.3.1. If $K \hookrightarrow \mathbb{R}^3$ is a knot, its *mirror knot* $\overline{K}$ is $\rho(K)$, where $\rho : \mathbb{R}^3 \to \mathbb{R}^3$ is an orientation reversing homeomorphism.

Note that any two such ρ are isotopic, so it doesn't matter which one we pick. Taking r to be reflection in the xy-plane, we see that if D is a diagram of K, the diagram $\overline{D}$ obtained by switching all overcrossings for undercrossings is a diagram of $\overline{K}$.

Definition 1.3.2. Choose orientation preserving embeddings $i_1, i_2 : \mathbb{R}^3 \to \mathbb{R}^3$ whose images are disjoint balls in $\mathbb{R}^3$. If L_1 and L_2 are oriented links in $\mathbb{R}^3$, their *disjoint union* is the union of $i_1(L_1)$ and $i_2(L_2)$.

All orientation preserving embeddings $\mathbb{R}^3 \to \mathbb{R}^3$ are isotopic, so again, the exact choice of the maps i_1 and i_2 is irrelevant.

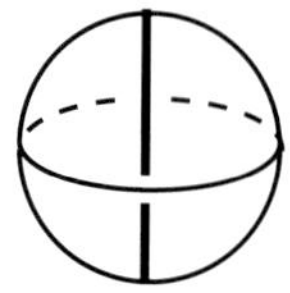

Finally, we describe the connected sum operation. If K_1, K_2 are knots in S^3, choose balls $B_1, B_2 \subset S^3$ such that $B_i \cap K_i$ is a single unknotted arc as shown in the figure to the right. Let $Y_i = S^3 \setminus \text{int}(B_i)$. The orientation on $K_i \cap B_i$ induces an orientation on $K_i \cap \partial B_i$; one of the two points is positively oriented, the other negatively. Let $\varphi : \partial B_1 \to \partial B_2$ be reflection in the vertical direction of the figure (so $\varphi(K_1 \cap B_1) = K_2 \cap B_2$, but with the orientation reversed.)

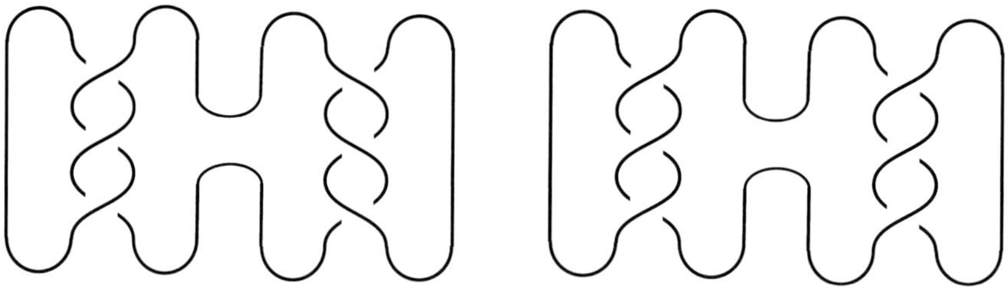

Square Knot $T\#\overline{T}$ Granny Knot $T\#T$

Figure 1.3.3. Connected sums

Definition 1.3.4. The connected sum $K_1 \# K_2$ is the knot obtained by taking

$$(K_1 \cap Y_1) \cup_\varphi (K_2 \cap Y_2) \subset Y_1 \cup_\varphi Y_2 = S^3$$

and smoothing.

Again, this definition does not depend on the exact choice of the balls B_1 and B_2. Note that it does depend on the orientation on K_1 and K_2 (just as with connected sum of manifolds). If we want to take the connected sum of two links, we must specify the components we are summing along.

1.4. The Jones polynomial It would be nice to have some way of telling when two diagrams represent different links. For example, we'd like to be able to say that the rightmost diagram in Figure 1.1.4 is not the unknot. Our experience with shoelaces and electrical cables suggests that this is the case, but how can we prove it?

One approach is suggested by Theorem 1.1.7. Let $\mathcal{D}$ be the set of all link diagrams up to isotopy, let $\mathcal{K}$ the set of all links in $\mathbb{R}^3$, and let X be some other set. If we can define a function $I : \mathcal{D} \to X$ such that $I(D) = I(D')$ whenever D and D' are related by a Reidemeister move, I will descend to a function $I : \mathcal{K} \to X$. In other words, I is a *link invariant*.

We can define a link invariant with the help of the *Kauffman bracket*, which is a function $\langle \cdot \rangle : \mathcal{D} \to \mathbb{Z}[A^{\pm 1}, B]$ satisfying the local relations

1) $\langle \times \rangle = A^{-1} \langle \asymp \rangle + A \langle)(\rangle$
2) $\langle \bigcirc \rangle = B\langle \ \rangle$

together with the normalization $\langle \emptyset \rangle = 1$, where $\emptyset$ denotes the empty diagram.

Relation 1) should be interpreted in the following way: if we have any three planar diagrams which agree outside of a disk $B^2 \subset \mathbb{R}^2$ and whose intersections with B^2 are the three diagrams shown, then their brackets satisfy the relation. For example, applying this relation to the upper crossing in our diagram of the Hopf link gives

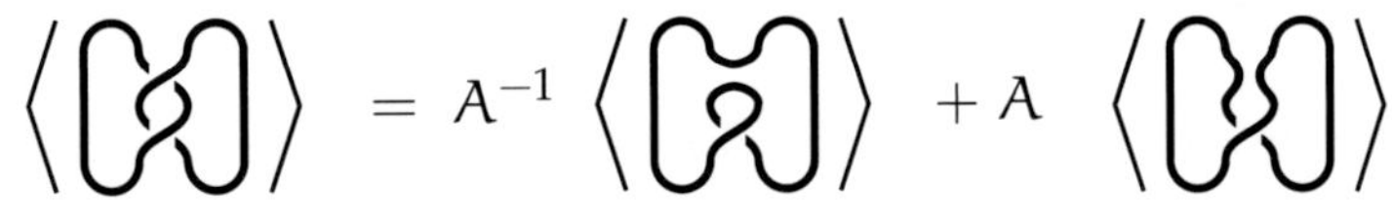

Applying the same relation to the lower crossing, we get

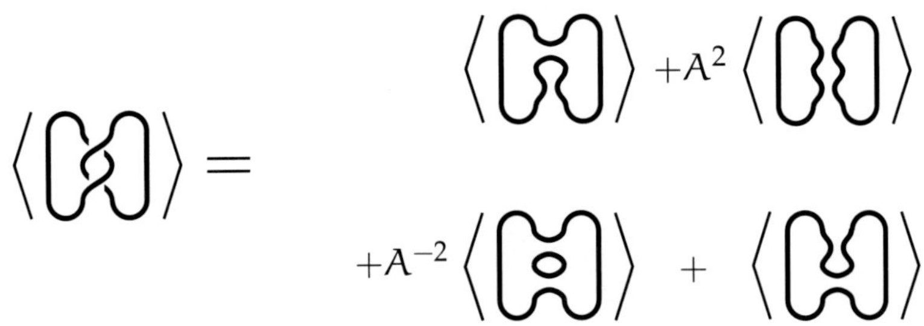

Applying the second relation, which allows us to remove a crossingless circle, we get

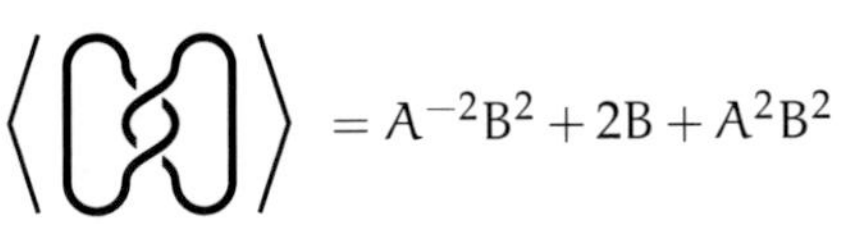

The procedure above can be applied to any planar diagram D. By turning our heads, we can arrange that any unoriented crossing in D looks like the one on the left-hand side of relation 1). Hence by applying relation 1) repeatedly, we can reduce $\langle D \rangle$ to a sum of terms involving the brackets of planar diagrams with no crossings, which can then be simplified using relation 2).

We can formalize this a bit by saying that each crossing has two *resolutions*, which we call the *0 resolution* and *1 resolution*, as shown in the figure to the right. If D has n crossings, there will be 2^n ways to resolve all n, and these 2^n diagrams are in bijection with the vertices of the n-dimensional cube $[0,1]^n$. If $D_{\mathbf{v}}$ denotes the crossingless planar diagram at vertex $\mathbf{v}$, then

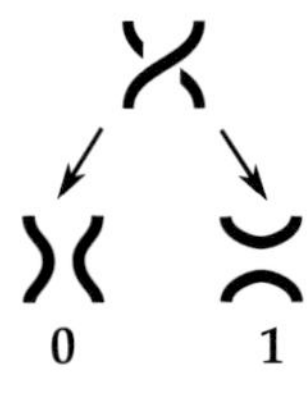

(1.4.1) $$\langle D \rangle = \sum_{\mathbf{v}} A^{n-2|\mathbf{v}|} B^{|D_{\mathbf{v}}|}$$

where $|\mathbf{v}|$ denotes the sum of the coefficients of $\mathbf{v}$, and $|D_{\mathbf{v}}|$ denotes the number of components of $D_{\mathbf{v}}$. To summarize, we have the following

Lemma 1.4.2. *There is a unique function* $\langle \cdot \rangle : \mathcal{D} \to \mathbb{Z}[A^{\pm 1}, B]$ *satisfying the local relations*

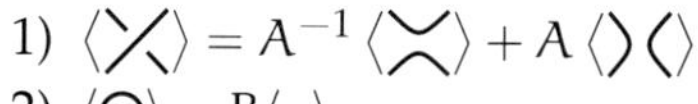

2) $\langle \bigcirc \rangle = B\langle\ \rangle$

3) $\langle \emptyset \rangle = 1.$

Proof. We've shown above that relations 1)-3) imply that $\langle D \rangle$ is given by equation (1.4.1). Conversely, if D is any planar diagram, we may define $\langle D \rangle$ by equation (1.4.1). It is then easy to see that D satisfies relations 1) and 2). □

Next, we consider how the bracket changes under Reidemeister moves. Skipping over R1 for the moment, we consider the effect of the R2 move. We compute

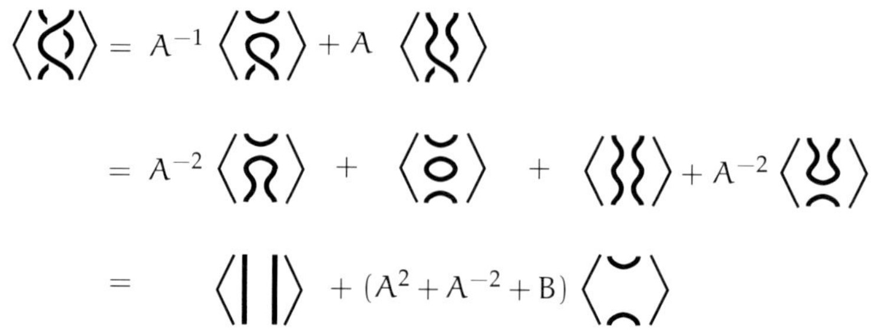

and see that in order for the bracket to be invariant under R2, we must set

$$\boxed{B = -A^{-2} - A^{2}.}$$

Although this is sufficient to ensure invariance under the R2 move, it's not very promising. We still have two Reidemeister moves to go, and have already eliminated one of our two variables. Nevertheless, we press on and consider the R3 move.

At first, it seems like we'll have to resolve three crossings, resulting in eight different diagrams and a very messy calculation. In fact, since we've already proved that the bracket is invariant under R2, it's enough to resolve one:

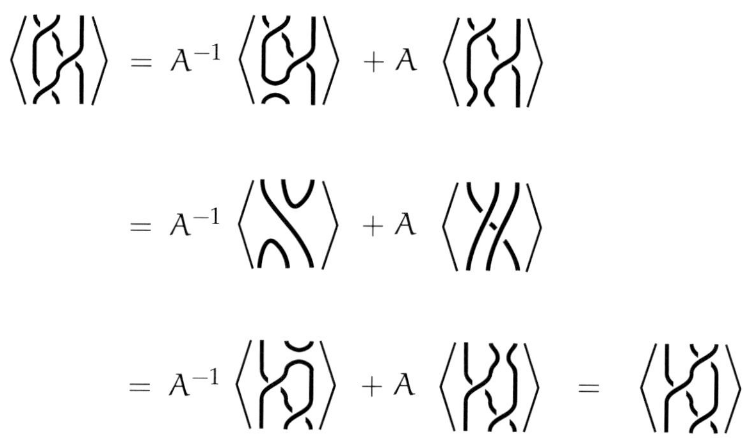

It follows that the bracket is invariant under the R3 move without any further specialization. It remains only to consider the R1 move. We compute

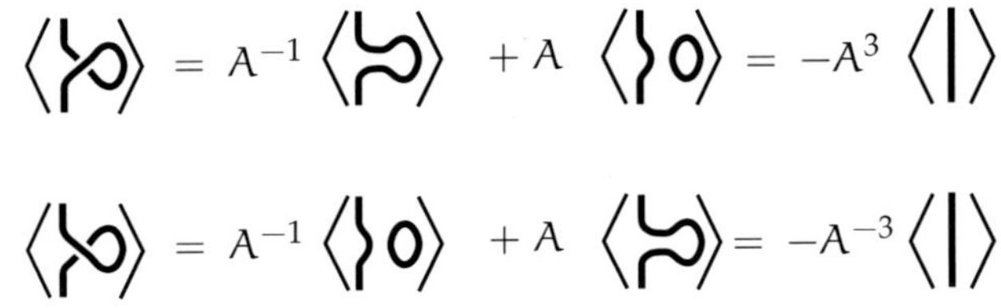

Disappointingly, the we find that the bracket is **not** invariant under the R1 move, and hence **not** a link invariant. But it's very close — close enough, in fact, that there is a fix for our problem.

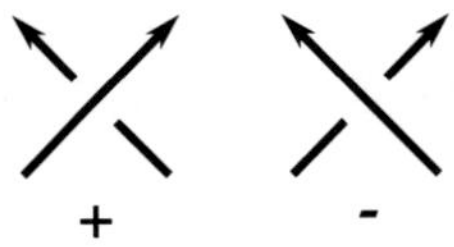

Figure 1.4.3. Positive and negative crossings

The Fix: So far, we have been thinking about unoriented link diagrams. If we pay attention to the orientations, it is no longer true that every crossing looks like the one in Figure 1.1.2. Instead there are two possible types, which we refer to as *positive* and *negative* crossing. These are shown in the figure above. If D is a planar diagram, we write $n_\pm(D)$ for the number of positive/ negative crossings of D, and define the *writhe* of D to be

$$w(D) = n_+(D) - n_-(D).$$

Lemma 1.4.4. *The writhe is invariant under Reidemeister moves 2 and 3.*

Proof. The two additional crossings in the right-hand diagram for R2 will always have opposite signs, no matter how we orient the strands. Hence the net change to the writhe is 0. For the R3 move, consider the figure below:

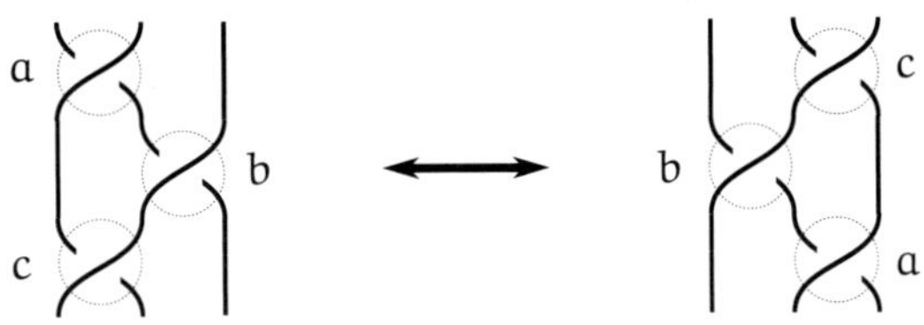

The crossings labelled a will have the same sign on both sides of the figure, regardless of the orientation of the strands. Similarly for the crossings b and c, so the writhe will be the same. □

On the other hand, an R1 move will either increase or decrease the writhe by 1. We can use this to counteract the change in the Kauffman bracket under an R1 move.

Definition 1.4.5. If D is an oriented link diagram, we define its *unreduced Jones polynomial* to be

$$\overline{V}(D) = (-A^3)^{-w(D)}\langle D\rangle.$$

Theorem 1.4.6. *The unreduced Jones polynomial is an invariant of oriented links.*

Proof. We have already seen that both the writhe and the bracket are invariant under R2 and R3, so $\overline{V}(D)$ is invariant under R2 and R3. If D and D′ are the left and right-hand figures in the diagram of the R1 move in Figure 1.1.6, then $w(D') = w(D) + 1$ and $\langle D' \rangle = -A^3 \langle D' \rangle$. It follows that $\overline{V}(D') = \overline{V}(D)$. □

Example 1.4.7. With this definition, we have $V(\emptyset) = 1$, where $\emptyset$ denotes the empty link, and $\overline{V}(\bigcirc) = -A^{-2} - A^2$. More generally, $\overline{V}(\bigcirc^n) = (-A^{-2} - A^2)^n$, where $\bigcirc^n$ denotes the n-component unlink. If H_+ is the positively oriented Hopf link (shown to the right), we have already computed that

$$\langle H_+ \rangle = B(A^{-2}B + 2 + A^2 B) = (-A^{-2} - A^2)(-A^{-4} - A^4) = A^{-6} + A^{-2} + A^2 + A^6,$$

so

$$\overline{V}(H_+) = A^{-12} + A^{-8} + A^{-4} + 1.$$

We deduce that the Hopf link is not the unlink.

Normalizations If D is a nonempty link diagram, then every complete resolution D_v will have at least one component. It follows that every term of the sum in equation (1.4.1) is divisible by $B = \overline{V}(\bigcirc)$, so $\overline{V}(D)$ is divisible by $\overline{V}(\bigcirc)$.

Definition 1.4.8. A nonempty link L has a *reduced Jones polynomial* $V(L)$ given by

$$V(L) = \overline{V}(L)/\overline{V}(\bigcirc).$$

Example 1.4.9. Let T be the positive trefoil knot, as shown in Figure 1.1.3. We compute the bracket as in the figure above.

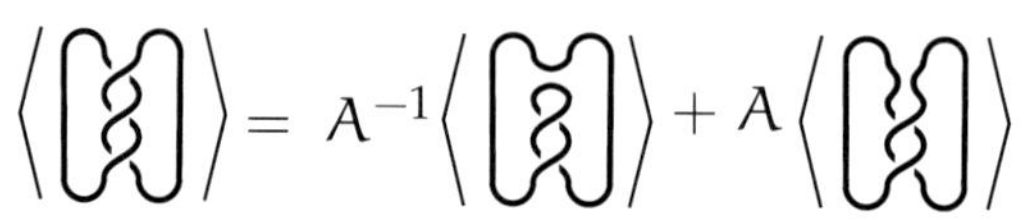

$$= A^{-1}(-A^{-3})^2 B + AB(-A^{-4} - A^4)$$
$$= (A^{-7} - A^{-3} - A^5)B$$

Since $w(D) = 3$, we get $V(T) = -A^{-16} + A^{-12} + A^{-4}$. We conclude that the trefoil is not the unknot.

Although the variable A is most natural for the Kauffman bracket, when we work with the Jones polynomial, we will generally use the variable

$$\boxed{q = -A^{-2}.}$$

We have $\overline{V}(\bigcirc^n) = (q + q^{-1})^n$, $V(H_+) = q + q^5$, and $V(T) = q^2 + q^6 - q^8$ in this normalization. When we want to emphasize the variable being used, we put the link in the subscript, writing $V_L(q)$ for what we previously denoted by $V(L)$.

As the examples above suggest, $V_L(q) \in \mathbb{Z}[q^{\pm 1}]$ for any link L, although *a priori* we might expect to see terms involving $(-q)^{1/2}$. One way to prove this is via the following

Proposition 1.4.10. *The Jones polynomial satisfies the* oriented skein relation

$q^2 V$ () $- q^{-2} V$ () $= (q - q^{-1}) V$ ().

As usual, this is to be interpreted as a local relation. For example, applying the skein relation to a crossing in our standard diagram for the trefoil gives

$$q^2 V(O) - q^{-2} V(T) = (q - q^{-1}) V(H_+).$$

1.5. Connections and Further Reading The Kauffman bracket was introduced in a classic paper by Kauffman [37], which is well worth reading. This paper also contains an important early application of the Jones polynomial which we have not discussed: it gives a lower bound on the crossing number of a link.

Theorem 1.5.1. *Let* $M(L)$ *and* $m(L)$ *be the maximum and minimum degrees of the Laurent polynomial* $V_L(q)$*. If* D *is any planar diagram representing* L*, then*

$$2c(D) \geqslant M(L) - m(L).$$

If D *is reduced and alternating then equality holds. In contrast, if* D *is non-alternating, then the inequality is strict.*

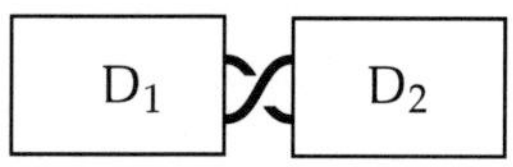

Here a diagram D is *reduced* if it does not contain a crossing of the form shown in the figure below. D is *alternating* if crossings alternate between over and under as we traverse any component. For example, the diagrams of the trefoil and Borromean rings in Figures 1.1.3 and 1.2.2 are both alternating, while the pretzel knot $P(-3, 5, 7)$ is not.

Theorem 1.5.1 resolved one of the oldest open problems in knot theory, due to Peter Guthrie Tait:

Corollary 1.5.2 (First Tait Conjecture). *Let* D *be a reduced alternating diagram. If* D' *is any other diagram representing the same link, then* $c(D) \leqslant c(D')$.

In particular, a reduced alternating diagram with > 0 crossings is always knotted.

Our strategy of checking invariance under the Reidemeister moves is an effective way of proving that some quantity is a link invariant, but it is a terrible way of finding such invariants to start with. The definition of the Jones polynomial we have given is completely elementary, but remained undiscovered for 100 years after mathematicians first started thinking about knots. Jones arrived at hs original definition [36] by thinking about something entirely different — representations of von Neumann algebras. We'll return to the representation-theoretic approach in lecture 4, when we talk about the HOMFLY-PT polynomial.

After Jones's discovery, Witten realized that the Jones polynomial should fit into a much broader theory of invariants of 3-manifolds defined using Chern-Simons theory. His paper [83] launched an entire industry devoted to the study of these *quantum invariants*, both in physics and mathematics. We'll discuss these more general invariants in lecture 6, but for now it is worth knowing that Witten's approach assigns a polynomial invariant of knots to each complex Lie algebra $\mathfrak{g}$ equipped with a representation V; the Jones polynomial corresponds to the vector representation of $\mathfrak{sl}_2$.

Exercises

1. If K is a knot in S^3, show that $H_1(S^3 - K) = \mathbb{Z}$ and that $H_*(S^3 - K) = 0$ for $* > 1$. More generally, if $L \subset S^3$ is an n-component link, show that $H_1(S^3 - L) = \mathbb{Z}^n$.
2. Suppose $K \subset \mathbb{R}^3$, and let $f : K \to \mathbb{R}$ be projection on the z-coordinate. Show that if f has a single local maximum on K, then K is unknotted. Deduce that any knot diagram may be unknotted by a sequence of crossing changes, in which we replace  by  while leaving the rest of the diagram unchanged.
3. Suppose D is an oriented 2-component link diagram, and let $n^*_{\pm}(D)$ be the number of positive/negative crossings in which the two strands belong to *different* components of L. Show that the *linking number* $\mathrm{lk}(D)$ defined by setting $\mathrm{lk}(D) = \frac{1}{2}[n^*_+(D) - n^*_-(D)]$ is invariant under all three Reidemeister moves, and hence an invariant of L.
4. Suppose that L is as in the previous exercise, and that L_1 and L_2 are its components. If $S \subset \mathbb{R}^3$ is an oriented embedded surface with $\partial S = L_1$, show that $\mathrm{lk}(L)$ is the intersection number $L_2 \cdot S$
5. Show that $V_{\overline{L}}(q) = V_L(q^{-1})$. Deduce that T and $\overline{T}$ are not equivalent.
6. Show that $V(K_1 \# K_2) = V(K_1)V(K_2)$. Deduce that the square knot $T\#\overline{T}$ and the granny knot $T\#T$ are not equivalent.
7. Use the Kauffman bracket to compute the Jones polynomial of the $(2, n)$ torus knots and of the figure-eight knot, which are shown below. (For something harder, but still doable by hand, try computing $V(P(-3, 5, 7))$.)
8. Prove the skein relation of Proposition 1.4.10. Deduce that $\overline{V}_L(1) = 2^n$, where n is the number of components of L.

(n crossings)
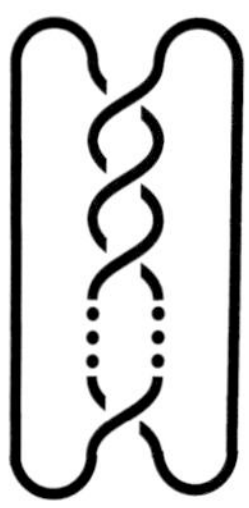
T(2, n)

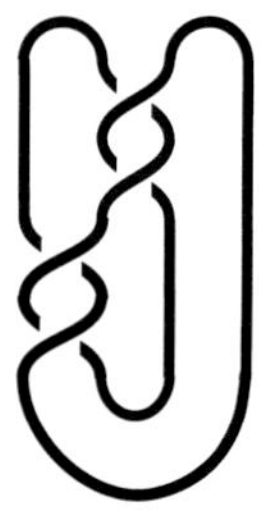
Figure-eight

2. The Alexander Polynomial

In this lecture, we'll discuss another polynomial invariant of knots, known as the Alexander polynomial. This polynomial is much older than the Jones polynomial, and there's a good reason it was discovered first. It can be derived from a standard invariant of algebraic topology — the fundamental group.

2.1. The knot group If K is a knot in S^3, we let $M_K = S^3 - K$ be its complement.

Lemma 2.1.1. *If K_0 and K_1 are isotopic knots in S^3, then their complements M_{K_0} and M_{K_1} are orientation preserving homeomorphic.*

Proof. (Sketch) If $\Psi : S^1 \times [0,1] \to S^3$ is the isotopy, then we can define a time-dependent vector field $\mathbf{v}_t$ on the image $\Psi(S^1, t)$ by

$$\mathbf{v}_t|_{\Psi(\theta,t)} = d\Psi\left(\frac{\partial}{\partial t}|_{(\theta,t)}\right).$$

We can extend $\mathbf{v}_t$ to a time-dependent vector field $\mathbf{v}_t$ on all of S^3. This is best done by thinking of $\mathbf{v}_t$ as a vector field on the embedded cylinder

$$\{(\Psi(\theta,t),t) \mid (\theta,t) \in S^1 \times [0,1]\} \subset S^3 \times [0,1].$$

If $\psi_t : S^3 \to S^3$ is the flow defined by $\mathbf{v}_t$, then $\psi_1 : S^3 \to S^3$ is an orientation-preserving diffeomorphism satisfying $\psi_1(K_0) = K_1$. □

As a consequence, any topological invariant of the knot complement M_K is an invariant of K. The most obvious one to try is the ordinary homology $H_*(M_K)$, but this turns out not to be so useful as an invariant: $H_1(M_K) \simeq \mathbb{Z}$ for every knot K, and $H_*(M_K) = 0$ for $* > 1$. (Lecture 0, Exercise 1). In contrast, $\pi_1(M_K)$ turns out to be a much better invariant. We sketch a method for computing it from a diagram D of K.

Definition 2.1.2. If D is an oriented planar diagram, an arc of D is a segment of D between two undercrossings. The *Wirtinger presentation* associated to D has generators a_i, where $i \in \{\text{arcs of } D\}$ and relations w_c, where $c \in \{\text{crossings of } D\}$. Here we take our basepoint $*$ to lie above the plane of the paper, and a_i to be a loop which runs down from the basepoint, around arc i, and then back up to the basepoint, as shown in the figure on the left below. The orientation of the loop a_i and the arc i should should satisfy the right-hand rule as shown. The relations w_c associated to positive and negative crossings are shown in the two figures on the right.

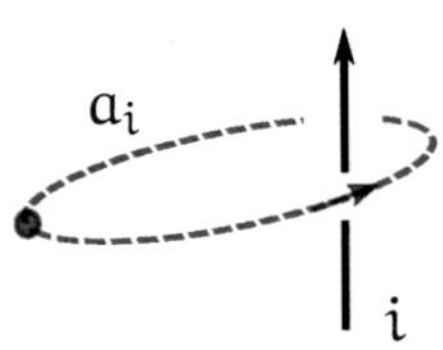

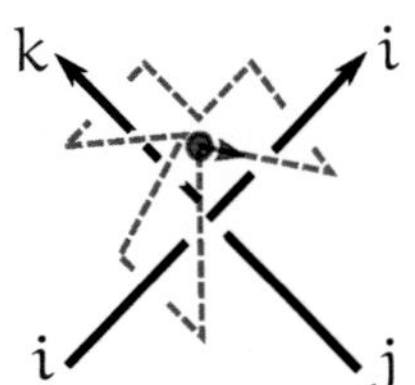

$a_j = a_i a_k a_i^{-1}$

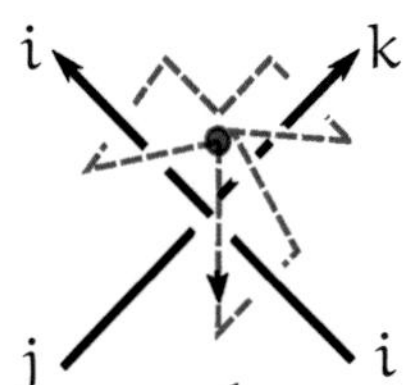

$a_j = a_i^{-1} a_k a_i$

The reader should take a moment to convince her/himself that these relations hold, as illustrated in the figure. Of course, we should really prove that the a_i generate $\pi_1(M_K)$, and that there are no more relations. One approach to this (sketched in the exercises) is to find a handle decomposition of $S^3 - \nu(K)$.

Example 2.1.3. If we use the standard diagram of the trefoil shown to the right, the Wirtinger presentation is

$$\pi_1(M_T) = \langle a_1, a_2, a_3 \mid a_1 = a_3^{-1}a_2a_3, a_3 = a_2^{-1}a_1a_2, a_2 = a_1^{-1}a_3a_1\rangle.$$

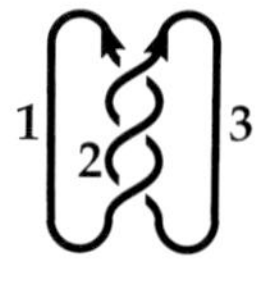

which we rewrite as

$$\pi_1(M_T) = \langle a, b, c \mid a = c^{-1}bc, c = b^{-1}ab, b = a^{-1}ca\rangle.$$

We can eliminate c using the second relation; when we do so, both new relations we get are equivalent to the single relation $abab^{-1}a^{-1}b^{-1} = 1$. Hence

$$\pi_1(M_T) = \langle a, b \mid abab^{-1}a^{-1}b^{-1} = 1\rangle.$$

Remark 2.1.4. The simplest topological space with this fundamental group is the finite cell complex consisting of a single 0-cell, two 1-cells a and b, and a single 2-cell attached along the curve $abab^{-1}a^{-1}b^{-1}$. In fact, $S^3 - \nu(T)$ has a handle decomposition consisting of a single 0-handle e_0, two 1-handles a and b, and a 2-handle f_2 attached to the one-skeleton along the word $abab^{-1}a^{-1}b^{-1}$. The proof is sketched in the exercises.

If K and $\overline{K}$ are mirror knots, then $S^3 - K$ is (orientation-reversing) homeomorphic to $S^3 - \overline{K}$. Hence $\pi_1(M_K) \simeq \pi_1(M_{\overline{K}})$, and the fundamental group does not distinguish between mirrors. This seemingly trivial example can be leveraged to give less trivial ones by using connected sums. For example, the complements of $T\#T$ and $T\#\overline{T}$ are not homeomorphic, but they have isomorphic fundamental groups.

However, this is essentially the only way in which things can go wrong. If K is a prime knot in S^3 (that is, it cannot be decomposed as a nontrivial connected sum) and $\pi_1(S^3 - K) \simeq \pi_1(S^3 - K')$, then it can be shown that either $K' = K$ or $K' = \overline{K}$ [26,82]. This is a deep result that combines Waldhausen's work on Haken manifolds [23] with Gordon and Luecke's theorem [26] that knots are determined by their complements.

2.2. The infinite cyclic cover The knot group is a very powerful invariant, but it can be hard to tell whether two finitely presented groups are isomorphic. What we'd like is some invariant of a group which can be derived from a presentation, but is easy to compute. The most obvious such invariant is the abelianization. For example, if we abelianize the presentation of $G = \pi_1(M_T)$ given above, we see that

$$G^{ab} = [a, b \mid a + b + a - b - a - b = 0] = [a, b \mid a = b] \simeq \mathbb{Z}$$

where we've used the square brackets to indicate we're writing a presentation of an abelian group. Of course this is no surprise; the abelianization of $\pi_1(M_K)$ is $H_1(M_K) \simeq \mathbb{Z}$, as you computed in the exercises for section 1.

This may seem like a dead end, but in fact it leads somewhere interesting. Let

$$|\cdot| : \pi_1(M_K) \to H_1(M_K) \simeq \mathbb{Z} = \langle t \rangle$$

be the abelianization map. Corresponding to the diagram

$$\begin{array}{ccc} \ker|\cdot| & \longrightarrow & \pi_1(M_K) \\ & & \big\downarrow |\cdot| \\ & & H_1(M_K) \end{array} \qquad \text{there is a regular covering map} \qquad \begin{array}{c} \overline{M}_K \\ p \big\downarrow \\ M_K \end{array}$$

where $\pi_1(\overline{M}_K) = \ker|\cdot|$.

Definition 2.2.1. The covering space $\overline{M}_K$ is the *infinite cyclic cover* of M_K.

The group of deck transformations is $H_1(M_K) = \mathbb{Z}$, which we take to be generated by $\varphi : \overline{M}_K \to \overline{M}_K$.

Example 2.2.2. Consider the handle decomposition of $S^3 - \nu(T)$ described in Remark 2.1.4. The preimages of the handles under p give a handle decomposition of $\overline{M}_K$. We build this decomposition up one dimension at a time.

Dimension 0: The preimage of the 0-handle e_0 is an infinite disjoint union of 0-handles $t^k\widetilde{e}_0$, $k \in \mathbb{Z}$. The deck transformation φ acts by $\varphi(t^k\widetilde{e}_0) = t^{k+1}\widetilde{e}_0$.

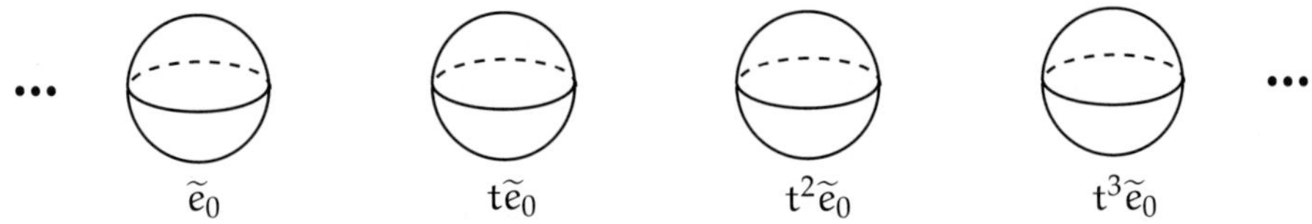

Dimension 1: Let $\widetilde{a}$ be the lift of a whose foot lies on $\widetilde{e}_0$. The head of $\widetilde{a}$ will lie on $|a|\widetilde{e}_0 = t\widetilde{e}_0$. All the other lifts of $\widetilde{a}$ will be orbits of this lift under the action of the deck group. The lifts of b can be described similarly.

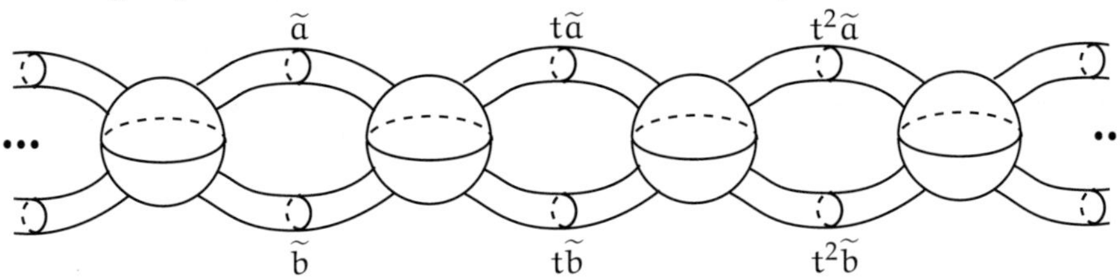

Dimension 2: The attaching circle of the 2-handle f_2 traces out a path given by the word $abab^{-1}a^{-1}b^{-1}$, one of whose lifts is shown in the figure below. The attaching circle of the 2-handle $\widetilde{f}_2$ runs along this lift. All the other lifts (not shown) are orbits of this one under the action of the deck group.

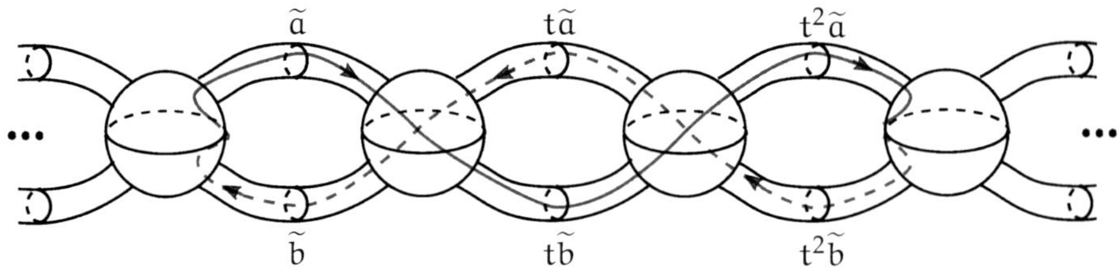

2.3. The Alexander polynomial Now we consider the group $H_1(\overline{M}_K;\mathbb{Q})$. The deck group acts on this group via $t \cdot x = \varphi_*(x)$. Extending linearly, we get an action of the rational group ring $R = \mathbb{Q}[\mathbb{Z}] = \mathbb{Q}[t^{\pm 1}]$ on $H_1(\overline{M}_K;\mathbb{Q})$. In other words, $H_1(\overline{M}_K;\mathbb{Q})$ is an R-module. By the structure theorem for finitely generated modules over a PID, we know that $H_1(\overline{M}_K;\mathbb{Q}) \simeq R^k \oplus R/(p_1) \oplus \dots R/(p_n)$ for some $p_1, \dots p_n \in R$.

Lemma 2.3.1. $H_1(\overline{M}_K;\mathbb{Q})$ *is a torsion module over* R; i.e. $k = 0$.

Proof. The chain complex $C_*(\overline{M}_K;\mathbb{Q})$ is free over R. If we define $N_1 = R/(t-1)$, then we have

$$C_*(\overline{M}_K;\mathbb{Q}) \otimes_R N_1 \simeq C_*(S^3 - K;\mathbb{Q}).$$

Since R is a PID, we can apply the universal coefficient theorem to see that

$$H_1(M_K) = H_1(\overline{M}_K;\mathbb{Q}) \otimes_R N_1 \bigoplus \mathrm{Tor}_R(H_0(\overline{M}_K;\mathbb{Q}), N_1).$$

$\overline{M}_K$ is connected, so $H_0(\overline{M}_K;\mathbb{Q}) \simeq \mathbb{Q}$, and if $x \in H_0(\overline{M}_K;\mathbb{Q})$, $\varphi_*(x) = x$. We conclude that $H_0(\overline{M}_K;\mathbb{Q}) \simeq N_1$ as an R-module, so $\mathrm{Tor}_R(H_0(\overline{M}_K;\mathbb{Q}), N_1) \simeq N_1$ has dimension 1 over $\mathbb{Q}$. On the other hand, we know that $H_1(M_K;\mathbb{Q}) \simeq \mathbb{Q}$, so we must have $H_1(\overline{M}_K;\mathbb{Q}) \otimes_R N_1 = 0$. It follows that $H_1(\overline{M}_K;\mathbb{Q})$ is torsion. □

If N is a finitely-generated torsion module over a PID R, then we can write $N = R/(\alpha_1) \oplus \dots R/(\alpha_n)$.Although this decomposition is not unique, the *order* $\operatorname{ord} N := \alpha_1 \dots \alpha_n$ is well defined up to multiplication by units in R. To indicate this ambiguity, we write $\operatorname{ord} N \sim p(t)$, rather than $\operatorname{ord} N = p(t)$.

If $R = \mathbb{Z}$, so N is a finitely generated abelian group, this definition reduces to the usual notion of the order of a group. Many well-known properties of the order of a group extend to the general situation. Two which we will use are

1) The order is multiplicative. If $N_1 \subset N_2$, we define $[N_2 : N_1] \sim \operatorname{ord} N_2/N_1$. Then if $N_1 \subset N_2 \subset N_2$, we have $[N_3 : N_1] \sim [N_3 : N_2][N_2 : N_1]$.
2) A matrix $A \in M_{n\times n}(R)$ defines a map $A : R^n \to R^n$ for which we have $\operatorname{ord}\operatorname{coker} A \sim [R^n : \operatorname{im} A] \sim \det A$.

Definition 2.3.2. The *Alexander polynomial* of a knot K in S^3 is defined to be $\Delta_K(t) \sim \operatorname{ord} H_1(\overline{M}_K;\mathbb{Q})$.

A priori, $\Delta_K(t)$ is well-defined up to multiplication by units in R; *i.e.* up to multiplication by ct^k, where $c \in \mathbb{Q}$. In fact, as we describe in the next section, this ambiguity can be eliminated to give a well-defined polynomial $\Delta_K(t) \in \mathbb{Z}[t^{\pm 1}]$.

We remark that in contrast to the Jones polynomial, which we defined using a diagram of K, and thus requires the fact that $K \subset S^3$, the definition of $\Delta_K(t)$ only depended on $\pi_1(S^3 - K)$. Hence it extends without change to knots in a homology sphere, and (with a little more work) to knots in an arbitrary 3-manifold Y.

Example 2.3.3. From the handle decomposition in Example 2.2.2, we see that $C_*^{cell}(\overline{M}_T;\mathbb{Q})$ has the form:

$$\begin{array}{ccccc} C_2^{cell}(\overline{M}_T) & \xrightarrow{\quad d_2 \quad} & C_1^{cell}(\overline{M}_T) & \xrightarrow{\quad d_1 \quad} & C_0^{cell}(\overline{M}_T) \\ \| & \begin{bmatrix} 1+t^2-t \\ t-t^2-1 \end{bmatrix} & \| & \begin{bmatrix} t-1 & t-1 \end{bmatrix} & \| \\ \langle \widetilde{f}_2 \rangle & \longrightarrow & \langle \widetilde{a}, \widetilde{b} \rangle & \longrightarrow & \langle \widetilde{e}_0 \rangle \end{array}$$

We see that $\ker d_1 = \langle \widetilde{a} - \widetilde{b} \rangle$, while $\operatorname{im} d_2 = \langle (1 - t + t^2)(\widetilde{a} - \widetilde{b}) \rangle$. Hence $H_1(\overline{M}_T) = R/(t^2 - t + 1)$, and $\Delta(T) \sim t^2 - t + 1$.

2.4. Fox calculus The procedure in Examples 2.2.2 and 2.3.3 can be applied to any group presentation. To be specific, suppose

$$G = \langle a_1, \dots a_m \,|\, w_1, \dots w_n \rangle$$

is a finitely generated group, and let F be the free group generated by the a_i. We can build a 2-dimensional cell complex X with $\pi_1(X) \simeq G$ by starting with a 0-cell e_0, attaching one 1-cell for each generator a_i, and then attaching one 2-cell along the loop corresponding to each word w_i.

Definition 2.4.1. The *Fox derivative* (or *free derivative*) $d_i : F \to \mathbb{Z}[F]$ is the unique map satisfying the following properties:

1) $d_i a_j = \delta_{ij}$
2) $d_i(ww') = d_i w + [w] d_i w'$

The first property says that d_i behaves like partial derivative with respect to a_i, but the "Leibniz rule" is a bit funny.

In order for the definition to make sense, we must check that there is a unique map d_i satisfying these two properties. If $1 \in F$ is the identity element, then applying the Leibniz rule to the relation $1 \cdot 1 = 1$ gives $d_i(1) = 2d_i(1)$, so $d_i(1) = 0$. Next, by applying d_i to the relation $1 = a_j^{-1} a_j$, we see that

$$d_i a_j^{-1} = -[a_j^{-1}] d_i a_j = -[a_j^{-1}] \delta_{ij}.$$

Since every $w \in F$ can be uniquely expressed as a reduced word in the a_i's and a_i^{-1}'s, it follows that $d_i w$ is determined by property 2). For example,

$$d_a(ca^{-1}b^{-1}a^2ca^{-1}) = -[ca^{-1}] + [ca^{-1}b^{-1}] + [ca^{-1}b^{-1}a] - [ca^{-1}b^{-1}a^2ca^{-1}].$$

We leave it to the reader to check that for any two reduced words w, w' this definition gives $d_i(ww') = d_i w + [w] d_i w'$.

The geometric significance of the Fox derivative is as follows. Let G and X be as above, and suppose $\varphi : G \to G'$ is a surjective homomorphism. Let $\widetilde{X}$ be the corresponding regular cover with deck group G'. The attaching circle w_j lifts up to a closed loop $\widetilde{w}_j$ in the 1-skeleton $\widetilde{X}^{(1)}$.

The cellular chain complex $C_*^{cell}(\widetilde{X})$ is a free module over the group ring $\mathbb{Z}[G']$, generated by lifts of the cells of X. We fix a lift $\widetilde{e}_0$ of the 0-cell, and choose lifts $\widetilde{a}_i$ of the 1-cells by requiring that $\widetilde{a_i}$ has one foot on $\widetilde{e}_0$ and points away from it. Similarly, we choose $\widetilde{w}_j$ so that it is based at $\widetilde{e}_0$.

Lemma 2.4.2. *With these choices we have,*

$$[\widetilde{w}_j] = \sum_{i=1}^{n} \psi(d_i w_j) \cdot [\widetilde{a}_i] \in H_1^{cell}(\widetilde{X}^{(1)}),$$

where $\psi : \mathbb{Z}[F] \to \mathbb{Z}[G']$ *is induced by the composition of* φ *and the projection* $F \to G$.

Proof. Each appearance of a_i in w_j will lift to $g\widetilde{a}_i$ for some $g \in G'$ and hence will contribute $g \cdot [\widetilde{a}_i]$ to $[\widetilde{w}_j]$. To determine g, write $w_j = wa_iw'$, where we've broken out the appearance of a_i we are interested in. Then $g = \psi(w)$, and $g \cdot [\widetilde{a}_i]$ is exactly the term of d_iw_j corresponding to this appearance of a_i. Similarly, an appearance of a_i^{-1} of the form $w = wa_i^{-1}w'$ will contribute $-\psi(wa_i^{-1}) \cdot [\widetilde{a}_i]$ to $[\widetilde{w}_j]$. □

The complex $C_*^{cell}(\widetilde{X})$ is a free chain complex over the ring $R = \mathbb{Z}[G']$. Using the lemma, we see that it has the form

$$R^n \xrightarrow{A} R^m \xrightarrow{B} R$$

where

$$A = \psi \begin{bmatrix} d_1w_1 & d_1w_2 & \cdots & d_1w_n \\ d_2w_1 & d_2w_2 & \cdots & d_2w_n \\ \vdots & & \ddots & \vdots \\ d_mw_1 & d_mw_2 & \cdots & d_mw_n \end{bmatrix} \quad \text{and } B = \begin{bmatrix} \psi(a_1)-1 & \cdots & \psi(a_m)-1 \end{bmatrix}.$$

The matrix A is called the *Alexander matrix*.

We now specialize to the case where $G' = \mathbb{Z}$. If our presentation of G comes from a handle decomposition of the knot complement M_K, then $n = m-1$, and $\widetilde{X}$ is homotopy equivalent to $\overline{M}_K$. In this case, we let A_i be Alexander matrix with its ith row deleted, so A_i is a square matrix.

Proposition 2.4.3. $\det A_i \sim \left(\dfrac{\psi(a_i)-1}{t-1}\right)\Delta_K(t)$.

Proof. Let $b_i = \psi(a_i) - 1$ be the ith entry of B, and write $b = \gcd\{b_1, \dots b_{n+1}\}$. Since $\psi(a_i)$'s generate $\mathbb{Z} = \langle t \rangle$, $b = t - 1$. If $\pi_i : R^{n+1} \to R^n$ is the projection which forgets the ith coordinate, then

$$[R^n : \pi_i(\ker B)] \sim \frac{b_i}{b} \sim \frac{\psi(a_i)-1}{t-1}.$$

By definition, $\Delta_K(t) \sim \operatorname{ord} H_1(\widetilde{X}) \sim [\ker B : \operatorname{im} A]$. If $b_i \neq 0$, then π_i maps $\ker B$ injectively to R^n, so $[\ker B : \operatorname{im} A] = [\pi_i(\ker B) : \pi_i(\operatorname{im} A)]$. We see that

$$\begin{aligned}\det A_i &\sim [R^n : \pi_i(\operatorname{im} A)] \\ &\sim [\pi_i(\ker B) : \pi_i(\operatorname{im} A)][R^n : \pi_i(\ker B)] \\ &\sim \Delta_K(t)\frac{\psi(a_i) - 1}{t-1}.\end{aligned}$$

If $b_i = 0$, let B_i be the matrix obtained by deleting the ith entry from B. We know that $d^2 = 0$ in the cellular chain complex, so $B_i A_i = BA = 0$, which implies that the columns of A_i are linearly dependent. Hence $\det A_i = 0$. □

Example 2.4.4. We use the presentation $\pi_1(M_T) = \langle a, b, c \mid ac^{-1}b^{-1}c, cb^{-1}a^{-1}b\rangle$ from Example 2.1.3, and find that the Alexander matrix is

$$A = \begin{bmatrix} 1 & -t^{-1} \\ -t^{-1} & -1+t^{-1} \\ -1+t^{-1} & 1 \end{bmatrix}.$$

$\psi(a) = \psi(b) = \psi(c) = t$, so we expect $\det A_i \sim 1 - t + t^2$ for each 2×2 minor A_i, as is indeed the case.

The proposition implies that $\Delta_K(t) \sim \gcd \det A_i$. More generally, suppose that $\langle a_1, \ldots, a_m \mid w_1, \ldots, w_n\rangle$ is a presentation of a group G and $\psi : G \to \mathbb{Z}$ is a surjective homomorphism. Then we can form an Alexander matrix A_ψ exactly as we did above.

Theorem 2.4.5. *Suppose $\psi : G \to \mathbb{Z}$ is as above. Then*

$$\Delta_\psi(G) := \gcd \{\det A'_\psi \mid A'_\psi \text{ is a } m-1 \times m-1 \text{ minor of } A_\psi\} \in \mathbb{Z}[\mathbb{Z}] = \mathbb{Z}[t^{\pm 1}]$$

is an invariant of the pair (G, ψ) which is well-defined up to multiplication by units in $\mathbb{Z}[t^{\pm 1}]$.

Sketch of proof: see [17]. A Tietze move is one of the following operations on a group presentation:

1) Duplicating a relation.
2) Multiplying one relation by another.
3) Conjugating a relation by a generator.
4) Adding a new generator a' together with the relation $a' = 1$.

Any two presentations of G are related by a sequence of Tietze moves, and one checks directly that the gcd does not change under each of the moves. □

The theorem implies that $\Delta_K(t)$ is both well-defined up to multiplication by $\pm t^k$ $(k \in \mathbb{Z})$ and an invariant of $\pi_1(M_K)$.

2.5. Fibred knots We say that a 3-manifold Y *fibres over* S^1 if there is a submersion $f : Y \to S^1$. If this is the case, all of the fibres $f^{-1}(p)$ are diffeomorphic to some surface Σ, there is a diffeomorphism $\varphi : \Sigma \to \Sigma$ called the *monodromy*, and $Y \simeq \Sigma \times [0,1]/\sim$, where $(x, 1) \sim (\varphi(x), 0)$.

Suppose Y is connected, and let θ be a generator of $H^1(S^1)$. We leave it as an exercise to show that the fibre Σ is connected if only if $f^*(\theta)$ is a primitive element of $H^1(Y)$. If $f^*(\theta)$ is divisible by n, then there is a lift $\widetilde{f}$ as shown

$$\begin{array}{ccc} & & S^1 \\ & \overset{\widetilde{f}}{\nearrow} & \big\downarrow{\scriptstyle z \mapsto z^n} \\ Y & \xrightarrow{\ f\ } & S^1 \end{array}$$

and $\widetilde{f}$ is a submersion with connected fibre. From now on, we will only consider fibrations of this form.

We say that a knot K is *fibred* if M_K fibres over S^1.

Example 2.5.1. Identify S^3 with the set $\{(z, w) \in \mathbb{C}^2 \mid |z|^2 + |w|^2 = 1\}$. If p and q are relatively prime, the (p, q) *torus knot* is the set

$$T(p, q) = \{(z, w) \in S^3 \mid z^p = w^q\}.$$

The map $f : M_{T(p,q)} \to S^1$ given by

$$f(z, w) = \frac{z^p - w^q}{|z^p - w^q|}$$

is a submersion, and $T(p, q)$ is a fibred knot.

If K is fibred, we can write $M_K \simeq \Sigma \times [0,1]/\sim$, where $(\varphi(x), 1) \sim (x, 0)$. Consider the map $\Phi : \Sigma \times \mathbb{R} \to \Sigma \times \mathbb{R}$ given by $\Phi(x, t) = (\varphi(x), t + 1)$. Φ generates a free, properly discontinous action of $\mathbb{Z}$ on $\Sigma \times \mathbb{R}$. The set $\Sigma \times [0, 1]$ is a fundamental domain for the action of Φ, so $(\Sigma \times \mathbb{R})/\langle\Phi\rangle \simeq \Sigma \times [0,1]/\sim \ \simeq M_K$. The quotient map $p : \Sigma \times \mathbb{R} \to M_K$ is a covering map with deck group $\mathbb{Z}$. The corresponding homomorphism $\pi_1(M_K) \to \mathbb{Z}$ must be the abelianization map, since since any such map factors through $H_1(M_K) \simeq \mathbb{Z}$. Hence p is the infinite cyclic cover. To summarize, we have proved

Proposition 2.5.2. *If M_K fibres over S^1 with monodromy $\varphi : \Sigma \to \Sigma$, then $\overline{M}_K \simeq \Sigma \times \mathbb{R}$. The action of the deck group is generated by the map $(x, t) \to (\varphi(x), t + 1)$.*

Corollary 2.5.3. *If M_K fibres over S^1 with monodromy $\varphi : \Sigma \to \Sigma$, then*

$$\Delta_K(t) \sim \det(tI - \varphi_*)$$

where $\varphi_* : H_1(\Sigma) \to H_1(\Sigma)$ is the homomorphism induced by the monodromy.

Proof. For the isomorphism $H_1(\overline{M}_K) \simeq H_1(\Sigma)$, the map $\Phi_* : H_1(\overline{M}_K) \to H_1(\overline{M}_K)$ is given by $\varphi_* : H_1(\Sigma) \to H_1(\Sigma)$. Hence if $e_1, \dots e_{2g}$ is a basis for $H_1(\Sigma)$ over $\mathbb{Z}$, the $\mathbb{Z}[t^{\pm 1}]$-module $H_1(\overline{M}_K)$ will be generated by the e_i, with relations $te_i = \varphi_*(e_i)$. In other words, $H_1(\overline{M}_K)$ has a square presentation matrix of the form $tI - \varphi_*$. □

Corollary 2.5.4. *If K is a fibred knot, then $\Delta_K(t)$ is a monic polynomial of degree $2g$, where g is the genus of the fibre.*

The condition of the corollary provides an effective (but not perfect) obstruction to a knot being fibred: a knot of 10 or fewer crossings, is fibred if and only if its Alexander polynomial is monic. More generally, alternating knots are fibred if and only if they have monic Alexander polynomial [57].

2.6. The Seifert genus

Definition 2.6.1. Let K be a knot in S^3. A *Seifert surface* for K is an embedded, orientable, connected surface $\Sigma \hookrightarrow S^3$ with $\partial\Sigma = K$.

Equivalently, we may think of a Seifert surface as being a connected orientable surface Σ with one boundary component that is properly embedded in $S^3 - \nu(K)$ and whose homology class generates $H_2(S^3 - \nu(K), \partial(S^3 - \nu(K))$. (There are many ways to check the latter condition. Perhaps the easiest is to check that $\partial\Sigma$ represents a nonzero class in $H_1(\partial(S^3 - \nu(K))$.)

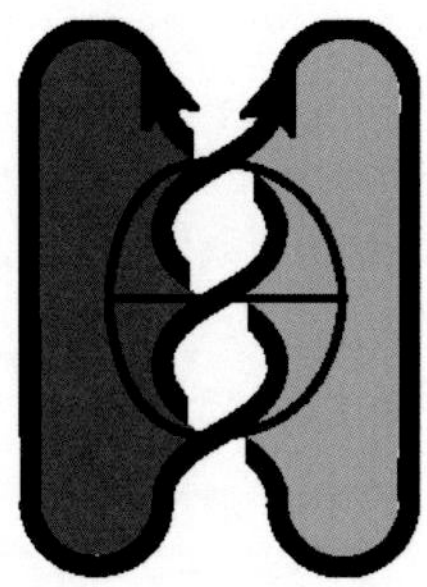

The figure above exhibits a Seifert surface for the trefoil knot T. Note that the surface (being orientable) has two sides, light and dark grey. It is perhaps easiest to visualize the surface as consisting of two large disks of cloth (0-handles) one light side up, the other dark, to which we have attached 3 small strips of cloth (1-handles), one for each crossing. We give each strip half a twist before gluing its ends to the light and dark disks. The surface deformation retracts to the graph shown in black, which has Euler characteristic 2-3 = -1. Hence the surface is homeomorphic to a once-punctured torus.

Lemma 2.6.2. *Any* $K \subset S^3$ *has a Seifert surface.*

We give two proofs, one constructive, and the other more abstract. Both are worth knowing.

Proof 1. (Seifert's algorithm)

Step 1: Orient K, and replace each crossing with its *oriented resolution*, as shown in the figure to the right. The resulting diagram is a collection of oriented circles, known as *Seifert circles*. Each circle C in the diagram has a *nesting height* $h(C)$ which is equal to the number of other circles one must cross to get from C to the point at infinity in the diagram.

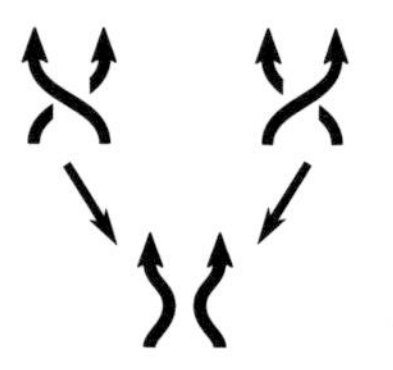

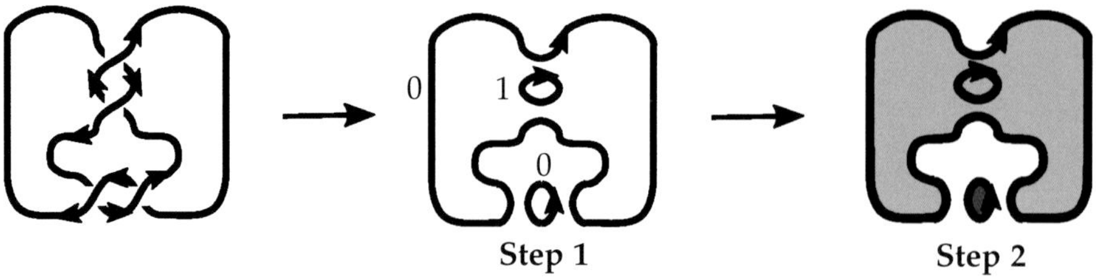

Figure 2.6.3. Seifert's algorithm applied to the figure-8 knot. The numbers in the middle figure indicate the nesting heights of the three Seifert circles. On the right, there are two disks with the light side up: a large one lying in the plane of the paper, and a small one above it.

Step 2: Fill in each circle C with a disk parallel to the xy plane at height $z = h(C)$. If C is oriented clockwise, the dark side of the disk faces up; if it is counterclockwise, the light side. Suppose c is a crossing of our original diagram D. Resolve every crossing other than c and consider the component of the resulting diagram which contains c. It must be in one of the two configurations below.

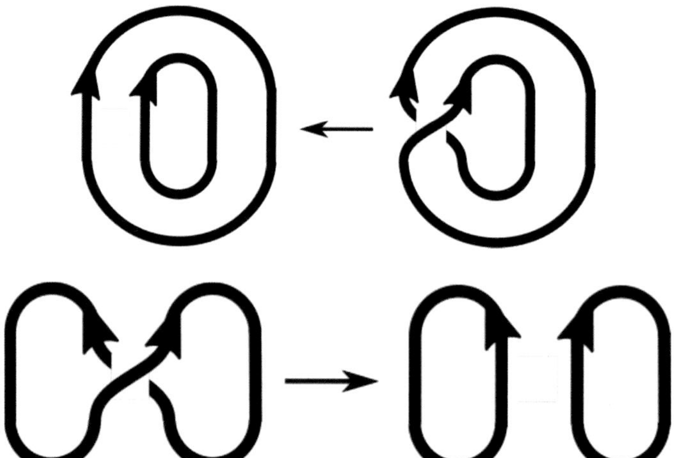

When we resolve c, we get either a pair of nested disks with the same color facing up, or a pair of unnested disks with opposite colors.

Step 3: Attach a 1-handle for each crossing, as shown in the figure on the right. For visualization purposes, it's helpful to think of the one-handle as a band of cloth with light and dark sides. We give the handle a half twist, and attach it to the disks on either side of the crossing. In each of the two cases considered above, we can arrange that both feet of the one-handle are compatible with the colors of the disks they attach to. Hence the result is an oriented surface. □

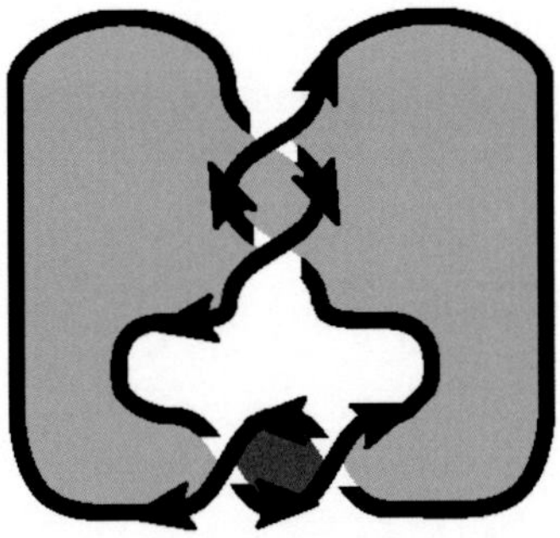

Figure 2.6.4. Seifert surface for the figure-8 knot. The surface in the top half of the figure consists of a large disk (light side up) lying in the plane of the paper together with a twisted band coming out of the paper towards the viewer.

Note that the resulting surface will have $\chi = n - c$, where n is the number of Seifert circles, and c is the number of crossings.

Proof 2. Let $\alpha \in \Omega^1(S^3 - \nu(K))$ be a closed 1-form generating $H^1(S^3 - \nu(K); \mathbb{R})$, and let β be generator of $H_1(S^3 - \nu(K), \mathbb{Z}) \simeq \mathbb{Z}$. After scaling α, we may assume $\langle[\alpha], \beta\rangle = 1$. Fix a basepoint $* \in S^3 - \nu(K)$. There is a well-defined smooth map $f : S^3 - K \to S^1$ given by $f(x) = \int_{\gamma_x} \alpha \mod 1$, where γ_x is any path from $*$ to x, and $f^*(d\theta) = [\alpha]$. If $p \in S^1$ is a regular value of f, then $f^{-1}(p)$ is a submanifold of $S^3 - \nu(K)$ which is Poincare dual to $[\alpha]$. Hence $f^{-1}(p)$ is a Seifert surface for K. $\square$

If Σ is a Seifert surface for K, we can find a Seifert surface of higher genus by taking the connected sum of Σ with a closed surface embedded in S^3. On the other hand, it may not be possible to reduce the genus of Σ. For example, we have

Lemma 2.6.5. *If $K \subset S^3$ bounds an embedded disk, then K is unknotted.*

Proof. Suppose $\Phi : D^2 \to S^3$ is an embedded disk with $\Phi(S^1) = K$. For $t \in (0, 1]$, we define $K_t : S^1 \to S^3$ by $K(\theta) = \Phi(t\theta)$, so Φ is an isotopy between K_t and K_1. Let $T \subset \mathbb{R}^3$ be the tangent plane to $\Phi(D^2)$ at $\Phi(0)$, and let $\pi : \mathbb{R}^3 \to T$ be orthogonal projection. If D^2_ε is the disk of radius ε, the map $\pi \circ \Phi : D^2_\varepsilon \to T$ will be an embedding for small ε. It follows that K_ε is isotopic to the unknot. $\square$

Definition 2.6.6. If K is a knot in S^3, its *Seifert genus* is

$$g(K) = \min\{g(\Sigma) \mid \Sigma \text{ is a Seifert surface for } K\}.$$

For example, the trefoil and figure-8 knot both have Seifert surfaces of genus 1. Neither is the unknot, so both knots have genus 1.

One of the most important properties of the Alexander polynomial is the fact that it gives a lower bound on the Seifert genus.

Theorem 2.6.7. *(Seifert)* $2g(K) \geqslant \deg \Delta_K(t)$.

Proof. Suppose $\Sigma \subset M_K$ is a Seifert surface with tubular neighborhood $\nu(\Sigma)$, and let $Y = M_K - \nu(\Sigma)$. Y and Σ are orientable, so $\nu(\Sigma) = \Sigma \times [-1,1]$. We have maps $\iota^{\pm} : \Sigma \to \partial Y$ given by the inclusions of $\Sigma \times \{\pm 1\}$, and $M_K \simeq Y/\sim$, where $\iota^-(x) \sim \iota^+(x)$.

Lemma 2.6.8. $H_1(Y) \simeq \mathbb{Z}^{2g}$, *where* $g = g(\Sigma)$.

Proof. We have

$$\begin{aligned} &H^*(S^3,\Sigma)\\ \simeq &H^*(S^3,\nu(\Sigma))\\ \simeq &H^*(Y,\partial Y) \quad \text{by excision}\\ \simeq &H_{3-*}(Y) \quad \text{by Poincaré duality.}\end{aligned}$$

The long exact sequence of the pair (S^3,Σ) gives

$$0 = H^1(S^3) \to H^1(\Sigma) \to H^2(S^3,\Sigma) \to H^2(S^3) = 0$$

So $H_1(Y) \simeq H^2(S^3,\Sigma) \simeq H^1(\Sigma) \simeq \mathbb{Z}^{2g}$. □

Returning now to the proof of Theorem 2.6.7, let $Z = (Y \times \mathbb{Z})/\sim$, where $(\iota^+(x),t) \sim (\iota^-(x),t+1)$. The map $p : Z \to M_K$ given by $p(y,t) = y$ is a covering map with deck group $\mathbb{Z}$ (generated by the map $(x,t) \mapsto (x,t+1)$), so $Z \simeq \overline{M}_K$. Finally, we have $H_1(Z) = H_1(Y) \otimes \mathbb{Z}[t^{\pm 1}]/\sim$, where the relations $\sim$ are given by $t \cdot \iota_*^-(e) = \iota_*^+(e)$ for $e \in H_1(\Sigma)$. In other words, $H_1(Z) \simeq \operatorname{coker} B$, where $B : R^{2g} \to R^{2g}$ is given by $B(e) = t\iota_*^-(e) - \iota_*^+(e)$. The entries of the $2g \times 2g$ matrix B are linear polynomials in t, so $\operatorname{ord} H_1(Z) = \det A$ is a polynomial in t with degree $\leqslant 2g$. □

Corollary 2.6.9. *If* K *is a fibred knot with fibre* Σ, *then* $g(K) = g(\Sigma)$.

Proof. If Σ is a fibre surface for M_K, $[\Sigma]$ generates $H_2(M_K,\partial M_K)$ (exercise). Hence Σ is a Seifert surface for K. By Corollary 2.5.4, $2g(\Sigma) = \deg \Delta_K(t) \leqslant 2g(K)$, so $g(\Sigma) = g(K)$. □

As with Corollary 2.5.4, the bound of Theorem 2.6.7 is sharp for many knots, including all alternating knots [16, 56] and all knots of 10 crossings or fewer. However there are many nontrivial knots K with $\Delta_K(t) = 1$. (Perhaps the simplest example is the pretzel knot $P(-3,5,7)$ from Figure 1.1.3.) Any such knot has $g(K) \geqslant 1$, so the inequality of Theorem 2.6.7 is not sharp.

2.7. The Seifert Matrix The maps $\iota_*^{\pm}$ can be described much more concretely. Let $\alpha : H_1(Y) \to H^1(\Sigma)$ be the isomorphism of Lemma 2.6.8.

Lemma 2.7.1. *If* $y \in H_1(\Sigma)$, *then* $\langle \alpha(x), y\rangle = \operatorname{lk}(x, \iota_*(y))$, *where* $\iota : \Sigma \to S^3$ *is the inclusion.*

Here $\operatorname{lk}(a,b)$ is the linking number of the two-component link $a \amalg b$, as discussed in the exercises to section 1.

Proof. The Poincaré duality isomorphism $PD : H_1(Y) \to H^2(Y, \partial Y)$ satisfies

$$\langle PD(x), S\rangle = x \cdot S$$

for $S \in H_2(Y, \partial Y)$ and the pair of boundary maps $\delta : H^1(\Sigma) \to H^2(S^3, \Sigma)$ and $\partial : H_2(S^3, \Sigma) \to H_1(\Sigma)$ are dual to each other, so

$$\langle \alpha(x), y\rangle = \langle \delta^{-1}(PD(x)), y\rangle = \langle PD(x), \partial^{-1}(y)\rangle = x \cdot \partial^{-1}(y)$$

The class $\partial^{-1}(y)$ is represented by an embedded surface bounding $\iota_*(y)$, so we have $x \cdot \partial^{-1}(y) = \mathrm{lk}(x, \iota_*(y))$. □

Choose embedded oriented curves $x_1, \ldots, x_{2g} \subset \Sigma$ which form a basis of $H_1(\Sigma)$, and let $x^1, \ldots x^{2g}$ be the dual basis of $H^1(\Sigma)$. We define $\alpha^\pm : H_1(\Sigma) \to H^1(\Sigma)$ by $\alpha^\pm(x) = \alpha(\iota_*^\pm(x))$.

The lemma says that with respect to the bases $\{x_1, \ldots, x_{2g}\}$ and $\{x^1, \ldots, x^{2g}\}$, the map α^+ is represented by the $2g \times 2g$ matrix A^+ with entries

$$a_{ij}^+ = \mathrm{lk}(\iota_*^+(x_j), \iota_*(x_i)),$$

while α^- is represented by the $2g \times 2g$ matrix A^- with entries

$$a_{ij}^- = \mathrm{lk}(\iota_*^-(x_j), \iota_*(x_i)).$$

Equivalently, using the bases $\{x_1, \ldots, x_{2g}\}$ for $H_1(\Sigma)$ and $\{\alpha^{-1}(x^1), \ldots, \alpha^{-1}(x^{2g})\}$ for $H_1(Y)$, the maps $\iota_*^\pm : H_1(\Sigma) \to H_1(Y)$ are given by the matrices $A^\pm$. Comparing with end of the proof of Theorem 2.6.7, we see that

$$\Delta_K(t) \sim \det B = \det(tA^- - A^+).$$

The link $\iota^-(x_j) \cup \iota(x_i)$ is isotopic to $\iota(x_j) \cup \iota^+(x_i)$, so $a_{ij}^- = a_{ji}^+$, or equivalently, $A^- = (A^+)^T$.

Definition 2.7.2. If Σ is a Seifert surface for K and $x_1, \ldots, x_{2g} \subset \Sigma$ is a basis of $H_1(\Sigma)$, the associated *Seifert matrix* $A := A^+$ is given by

$$A = [a_{ij}] = [\mathrm{lk}(\iota_*^+(x_j), \iota_*(x_i))].$$

To sum up, we have proved

Proposition 2.7.3. $\Delta_K(t) \sim \det(A - tA^T)$, *where* A *is a Seifert matrix of* K.

The Seifert matrix lets us pick out a canonical representative for the Alexander polynomial as an element of $\mathbb{Z}[t^{\pm 1}]$.

Definition 2.7.4. The *symmetrized Alexander polynomial* of a knot $K \subset S^3$ is defined to be

$$\Delta_K(t) = \det(t^{-1/2}A - t^{1/2}A^T),$$

where A is a Seifert matrix for K.

Proposition 2.7.5. *The symmetrized Alexander polynomial satisfies the following properties:*

1) *(Integrality)* $\Delta_K(t) \in \mathbb{Z}[t^{\pm 1}]$.

2) *(Symmetry)* $\Delta_K(t^{-1}) = \Delta_K(t)$.
3) *(Normalization)* $\Delta_K(1) = 1$.

Proof of Proposition 2.7.5. For item 1), note that A is a $2g \times 2g$ matrix, so

$$\det(t^{-1/2}A - t^{1/2}A^T) = t^{-g}\det(A - tA^T).$$

For 2) note that the expression $\det(t^{-1/2}A - t^{1/2}A^T)$ is invariant under the involution of $\mathbb{Z}[t^{\pm 1/2}]$ which sends $t^{1/2}$ to $-t^{-1/2}$. The restriction of this involution to $\mathbb{Z}[t^{\pm 1}]$ sends t to t^{-1}.

Finally, for property 3), note that $\Delta_K(1) = \det(A - A^T) = \det(C)$, where

$$c_{ij} = \mathrm{lk}(\iota_*^+(x_j), \iota_*(x_i)) - \mathrm{lk}(\iota_*^-(x_j), \iota_*(x_i)) = X_j \cdot x_i,$$

where X_j is the annulus $x_j \times [-1,1] \subset \Sigma \times [-1,1]$. On the other hand, it is easy to see that $X_j \cdot x_i = x_j \cdot x_i$, where the intersection number on the left is in S^3, and the one on the right is taken in Σ. In other words C is the matrix representing the intersection form on Σ with respect to the basis $x_1, \ldots x_{2g}$. Hence $\det C = 1$. □

Remark 2.7.6. In order for Definition 2.7.4 to make sense, we must check that it does not depend on the choice of Seifert matrix A. This follows from Proposition 2.7.5. Indeed, if we are given are two Seifert matrices A_1 and A_2 for K, let $p_i(t) = \det(t^{-1/2}A_i - t^{1/2}A_i^T)$. Then $p_1(t)$ and $p_2(t)$ are both normalized and symmetric, and $p_1(t) \sim \Delta_K(t) \sim p_2(t)$. It is easy to see that this implies $p_1 = p_2$.

2.8. Links If $L \subset S^3$ is a link, we have $H_1(S^3 - L) = \mathbb{Z}^{|L|}$, so there are many surjective homomorphisms $\varphi : \pi_1(S^3 - L) \to \mathbb{Z}$. Each such homomorphism gives rise to an infinite cyclic cover X_φ, and we can define a corresponding Alexander polynomial $\Delta_\varphi(L)$ to be the order of $H_1(X_\varphi; \mathbb{Q})$ as a module over $R = \mathbb{Q}[t^{\pm 1}]$. Note that $H_1(X_\varphi; \mathbb{Q})$ need not be a torsion module over R; if it is not, its order $\Delta_\varphi(L)$ is defined to be 0. (With this definition, we still have $\operatorname{ord} M = \det A$ if A is a $n \times n$ presentation matrix for M.)

If L is an oriented link with n components, the oriented meridians $m_1, \ldots, m_n$ form a basis for $H_1(S^3 - L)$.

Definition 2.8.1. If L is an oriented link, we define $\Delta(L) = \Delta_\varphi(L)$ where the homomorphism $\varphi : \pi_1(S^3 - L) \to \mathbb{R}$ is given by $\varphi(m_i) = 1$ for all i. Note that if L^r denotes L with the orientation on each component reversed, the homomorphisms corresponding to L and L^r have the same kernel, so $\Delta(L) = \Delta(L^r)$.

Similarly, if L is an oriented link, we define a Seifert surface for L to be an embedded oriented surface whose oriented boundary is L. With this definition, the results of section 2.6 carry over more or less unchanged. In particular, if Σ is a Seifert surface for L, the degree of the (unsymmetrized) Alexander polynomial $\Delta(L)$ gives a lower bound on the rank of $H_1(\Sigma)$.

Note that changing the orientation of some (but not all) components of L will usually have a drastic effect on the Alexander polynomial and Seifert genus; see Figure 2.8.2 for an example.

$\Delta(L) = t^{3/2} - t^{1/2} + t^{-1/2} - t^{-3/2}$ $\qquad$ $\Delta(L) = 2t^{1/2} - 2t^{-1/2}$

Figure 2.8.2. Minimal genus Seifert surfaces for two different orientations on the torus link $T(2,4)$. The corresponding Alexander polynomials are shown.

Finally, we remark that if L is an oriented link, and Σ is a Seifert surface for L, then the definition of the Seifert matrix given in Definition 2.7.2 still makes sense. As before, we define the symmetrized Alexander polynomial $\Delta_L(t)$ to be $\det(t^{-1/2}A - t^{1/2}A^T)$, where A is the Seifert matrix.

We can use the Seifert matrix to prove that the Alexander polynomial satisfies a skein relation similar to the oriented skein relation satisfied by the Jones polynomial.

Proposition 2.8.3. *(Skein Relation)* Δ () $- \Delta$ () $= (t^{1/2} - t^{-1/2})$ () .

Proof. Let $D_\pm$ and D_0 be the three planar diagrams represented by 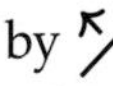, , and , and let Σ_0 be the Seifert surface for  obtained by applying Seifert's algorithm to D_0. The surfaces $\Sigma_\pm$ that we get by applying Seifert's algorithm to $D_\pm$ are the result of attaching one more 1-handle to Σ_0.

If we choose a basis $x_1, \dots, x_n$ for $H_1(\Sigma_0)$, we get bases for $H_1(\Sigma_\pm)$ by appending a single curve $x_\pm$, which runs from a point p in Σ_0, over the new 1-handle, and back to p. Clearly, these satisfy $\mathrm{lk}(i_*^+(x_+), i_*(x_i)) = \mathrm{lk}(i_*^+(x_-), i_*(x_i))$. Similarly, the diagrams for the two-component links $i_*^+(x_+) \amalg i_*(x_+)$ and $i_*^+(x_-) \amalg i_*(x_-)$ will be the same except in a neighborhood of the new crossing, where the first diagram will have a positive crossing and the second diagram will have a negative one.

It follows that if A_0 is a Seifert matrix for Σ_0, the Seifert matrix $A_\pm$ for $\Sigma_\pm$ will have the form

$$A_+ = \begin{bmatrix} A_0 & \mathbf{x} \\ \mathbf{y} & k \end{bmatrix} \qquad A_- = \begin{bmatrix} A_0 & \mathbf{x} \\ \mathbf{y} & k-1 \end{bmatrix}.$$

If $B_0 = t^{-1/2}A_0 - t^{1/2}A_0^T$ and similarly for $B_\pm$, we have

$$B_+ = \begin{bmatrix} B_0 & \mathbf{z} \\ \mathbf{w} & (t^{-1/2} - t^{1/2})k \end{bmatrix} \qquad B_- = \begin{bmatrix} B_0 & \mathbf{z} \\ \mathbf{w} & (t^{-1/2} - t^{1/2})(k-1) \end{bmatrix}.$$

Expanding both determinants along the lower row, we see that

$$\det B_+ - \det B_- = (t^{-1/2} - t^{1/2}) \det B_0$$

which gives the skein relation stated above. $\square$

2.9. Connections and Further Reading The Alexander polynomial of a knot is a special case of a more general invariant of 3-manifolds the multi-variable Alexander polynomial defined as follows. If Y is a 3-manifold with toroidal boundary, we can consider the covering space $\widetilde{Y}$ corresponding to the surjective map $\pi_1(Y) \to H_1(Y) \to H_1(Y)/\mathrm{Tors}$ where $\mathrm{Tors} \subset H_1(Y)$ is the torsion subgroup. If $H_1(Y)$ has rank k, $H_1(\widetilde{Y})$ is a module over $\mathbb{Z}[\mathbb{Z}^k]$, which is the ring of Laurent polynomials in k variables. This ring is not a PID, but it is a UFD, so we can use a definition analogous to that of Theorem 2.4.5 to get an invariant $\Delta(Y) \in \mathbb{Z}[H_1(Y)/\mathrm{Tors}]$. In the case where $Y = S^3 - \nu(L)$, the single-variable Alexander polynomial can be obtained from $\Delta(Y)$ by specialization and multiplication by an appropriate factor. There are useful analogs of Theorem 2.6.7 and Corollary 2.5.4 which relate the multivariable Alexander polynomial to the Thurston norm. McMullen's paper [52] is a good reference.

Closely related to the multivariable Alexander polynomial is the notion of the Alexander polynomial as a *torsion*, which was developed in a beautiful paper by Milnor [54]. A key property of the torsion is that it satisfies a product formula, which relates the torsion of two manifolds glued together along a torus to the torsion of the individual pieces. This is especially useful for understanding the effect of Dehn filling and satellite operations on the Alexander polynomial. Turaev's book [79] provides a nice introduction to this subject.

Exercises

1. Find a presentation of $\pi_1(M_K)$, where K is the figure-8 knot. Show that it can be reduced to a 2-generator, 1-relator presentation. Compute the Alexander polynomial using Fox calculus and using the skein relation.
2. Show that $S^3 - T(p,q)$ can be written as the union of two solid tori glued together along an annulus. Deduce that $\pi_1(M_{T(p,q)}) = \langle a, b \mid a^p b^q = 1\rangle$. Compute $\Delta(T(p,q))$. What is $g(T(p,q))$?
3. If L is a split link (*i.e* a disjoint union of two links), show that $\Delta(L) = 0$ by a) showing that $H_1(\widetilde{M_L})$ is not a torsion module and b) using the skein relation.
4. If K is a knot in S^3, use the fact that $H_1(M_K) = \mathbb{Z}$ to show that $\Delta_K(1) = 1$.
5. Use the skein relation to show that $\Delta_K(-1) = V_K(-1)$. The quantity $|\Delta_K(-1)|$ appears in many contexts, and is known as the *determinant* of K.
6. Let L be the $(2, 2n)$ torus link, oriented so that it is the boundary of an embedded annulus, as in the right-hand link in Figure 2.8.2. Compute $\Delta(L)$ both by using the skein relation and by finding a Seifert matrix.
7. Let $P(p,q,r)$ be the (p,q,r) pretzel knot, as in Figure 1.1.3, where p, q and r are all odd. Show that K bounds a genus 1 Seifert surface. Find the associated Seifert matrix and compute $\Delta(K)$. Use this to show that $\Delta(P(-3,5,7)) = 1$. Compare this with Exercise 7, section 1), which shows that $P(-3,5,7)$ is not the unknot.

8. Suppose that H_g is a 3-dimensional handlebody obtained by starting with a 0-handle and attaching g 1-handles $D^1 \times D^2$ in an oriented fashion. Consider the compressing disks $0 \times D^2$ inside the one-handles, and let $\alpha_1, \dots, \alpha_g \subset \Sigma_g$ be their boundaries. If β is a curve in Σ_g, let w be its image in $\pi_1(H_g) = \langle a_1, \dots, a_g \rangle$. Explain why there is a bijection between $\beta \cap \alpha_i$ and occurrences of $a_i^{\pm 1}$ in the word w. Now suppose that $\beta_1, \dots, \beta_g$ are g distinct curves in Σ_g and consider the set of unordered g-tuples of points $\{p_1, \dots, p_g\}$, where $p_i \in \alpha_i \cap \beta_{\sigma(i)}$ for some permutation $\sigma \in S_g$.

   Show that the elements of this set are in bijection with monomials in $\det d_i w_j$, where we expand everything out without cancelling any monomials with opposite signs.
9. We say K is a g-bridge knot if K has a diagram with k maxima for the z-coordinate. If this is the case, show that $S^3 - \nu(K)$ can be decomposed as a handlebody of genus K with $k-1$ 2-handles attached. (Hint: we can assume the maxima all occur at $z = 1$ and the minima at $z = -1$. Consider the intersection of K with the half-spaces $z \geqslant 0$ and $z \leqslant 0$.) Can you draw the resulting handle decomposition for the complement of the trefoil?

3. Khovanov Homology

In this lecture, we transition from studying polynomial invariants of knots to their categorifications. Khovanov homology is an invariant of oriented knots and links in S^3. It is a homological generalization of the Jones polynomial in the following sense: if $L \subset S^3$ is an oriented link, its Khovanov homology is a bigraded abelian group $Kh^{i,j}(L)$, whose graded Euler characteristic

$$\chi(Kh(L)) := \sum_{i,j} (-1)^i q^j \dim Kh^{i,j}(L) = \overline{V}(L).$$

To define $Kh(L)$, we first represent L by a planar diagram D. To such a diagram, Khovanov assigns a bigraded chain complex $CKh(D)$; $Kh(L)$ is its homology. Khovanov showed that if D_1 and D_2 are two planar diagrams representing the same link L, $CKh(D_1)$ and $CKh(D_2)$ are chain homotopy equivalent, hence have the same homology.

The definition of $CKh(D)$ is neatly encapsulated in the following

Slogan ([3]). $CKh(D)$ is obtained by applying a certain $1+1$ dimensional TQFT $\mathcal{A}$ to the cube of resolutions of D.

whose meaning we now explain.

3.1. Cube of resolutions If c is a crossing of the diagram D, we may turn the paper (or our head) until the crossing appears as shown in Figure 3.1.1. As illustrated in the figure, the crossing c can be resolved in two ways, which we call the 0 and 1-resolutions.

Figure 3.1.1. The 0 and 1 resolutions of a crossing

Vertices: If D has n crossings, there are 2^n ways to resolve all n of them, which (after ordering the crossings) are naturally in bijection with the vertices of the cube $[0,1]^n$. If v is a vertex of the cube, we write D_v for the planar diagram of the corresponding resolution. D_v has no crossings, so it is a collection of embedded circles in the plane. We will (mostly) ignore the embedding, and view D_v as a 1-dimensional manifold.

The figure below illustrates this process for the Hopf link.

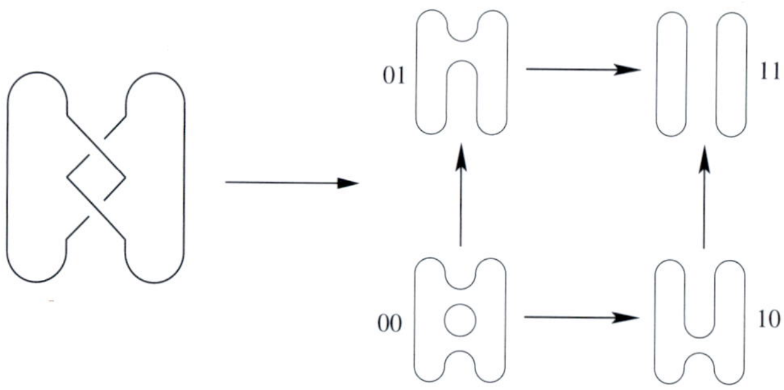

Figure 3.1.2. The cube of resolutions of the Hopf link

Edges: Along each edge e of the cube, one coordinate varies from 0 to 1, while all the other coordinates are fixed. We orient e to point from the vertex v_0 where the variable coordinate is 0 to the vertex v_1 where the variable component is 1, as shown in Figure 3.1.2, and write $e : v_0 \to v_1$.

To each edge $e : v_0 \to v_1$, we assign a surface S_e with $\partial S_e = D_{v_0} \cup D_{v_1}$. The diagrams D_{v_0} and D_{v_1} are identical away from a neighborhood of a single crossing c, and we define S_e to be the product $D_{v_0} \times I$ away from this neighborhood. Inside this neighborhood, S_e is given by the saddle cobordism shown in Figure 3.1.3.

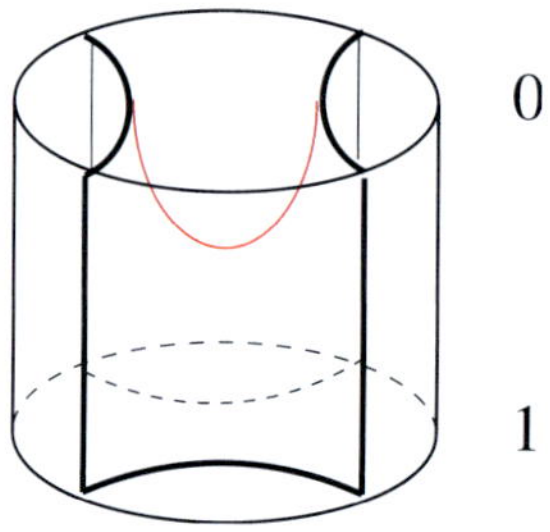

Figure 3.1.3. Cobordism from D_{v_0} to D_{v_1}

Equivalently, S_e is obtained from $D_{v_0} \times I$ by attaching a 1-handle whose core is the red curve shown in the figure.

3.2. The Cobordism Category If Y_1 and Y_2 are compact oriented n-manifolds, we define a *cobordism* W from Y_1 to Y_2 to be a compact oriented $n+1$-manifold with $\partial W = -Y_1 \coprod Y_2$. We write $W : Y_1 \to Y_2$. Two cobordisms $W, W' : Y_1 \to Y_2$ are *equivalent* if there is a homeomorphism $f : W \to W'$ whose restriction to ∂W is the identity. If $W_1 : Y_1 \to Y_2$ and $W_2 : Y_2 \to Y_3$ are cobordisms, their *composition* $W_2 \circ W_1 := W_1 \cup_{Y_2} W_2$ is a cobordism from $Y_1 \to Y_3$, as shown in Figure 3.2.1.

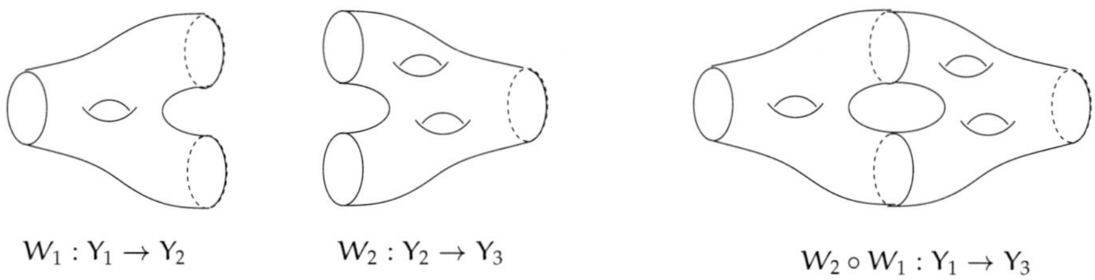

Figure 3.2.1. Composition of Cobordisms

We define the $n+1$ *dimensional cobordism category* to be the category whose objects are compact oriented $n-$manifolds and whose morphisms are equivalence classes of cobordisms between them. If Y is an object in this category, the identity morphism 1_Y is the product $Y \times [0,1]$.

We would like to view the vertices of the cube of resolutions as being decorated by objects of the $1+1$ dimensional cobordism category, and its edges as being decorated by morphisms. To do this, we must orient the objects involved. First note that any circle embedded in $\mathbb{R}^2$ has a standard (counterclockwise) orientation.

Definition 3.2.2. Let D_v be a set of embedded circles in $\mathbb{R}^2$. The *canonical orientation* on D_v is defined by giving the ith circle C_i $(-1)^{n_i}$ times the standard orientation, where n_i is the number of circles separating C_i from infinity in $\mathbb{R}^2$.

This process is illustrated in the figure below.

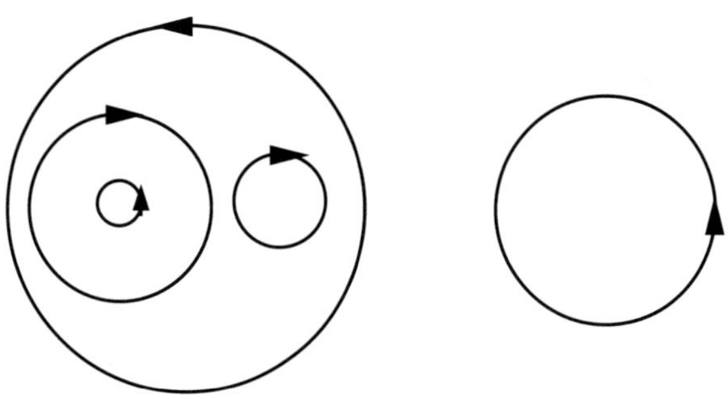

Figure 3.2.3. The canonical orientation on circles embedded in $\mathbb{R}^2$.

Exercise 3.2.4. Let $e : v_0 \to v_1$ be an edge in the cube of resolutions. If we give D_{v_0} and D_{v_1} their canonical orientations, show that S_e is an oriented cobordism from D_{v_0} to D_{v_1}.

Remark 3.2.5. Since S_e is given by a single one-handle attachment and is orientable, it is the disjoint union of a single pair of pants with some cylinders.

Exercise 3.2.6. A knot $K \subset T^2 \times I$ can be represented by a planar diagram in T^2, and we can form the cube of resolutions for K as above. Show by example that the surfaces S_e may be nonorientable in this case.

Faces: Each 2-dimensional face of the cube of resolutions corresponds to a square of morphisms in the cobordism category. This square commutes, since 1-handles can be added in any order without changing the homeomorphism type of the resulting surface. The situation is summarized in the first two columns of the table below.

vertex v	$\Longrightarrow$	1-manifold D_v	$\Longrightarrow$	group $\mathcal{A}(D_v)$
edge $e : v_0 \to v_1$	$\Longrightarrow$	cobordism $S_e : D_{v_0} \to D_{v_1}$	$\Longrightarrow$	linear map $\mathcal{A}(S_e) : \mathcal{A}(D_{v_0}) \to \mathcal{A}(D_{v_1})$
2 dim'l faces		commute		commute

3.3. Applying a TQFT A $1+1$ dimensional topological quantum field theory (or TQFT, for short) is a monoidal functor

$$\mathcal{A} : \begin{Bmatrix} \text{1-manifolds} \\ \text{cobordisms} \end{Bmatrix} \to \begin{Bmatrix} \text{abelian groups} \\ \text{linear maps} \end{Bmatrix}$$

Saying that that $\mathcal{A}$ is *monoidal* means that it behaves well under disjoint unions: if Y and Y' are 1-manifolds, then $\mathcal{A}(Y \coprod Y') = \mathcal{A}(Y) \otimes \mathcal{A}(Y)$. Similarly if $W : Y_1 \to Y_2$, $W' : Y_1' \to Y_2'$ are morphisms, then $\mathcal{A}(W \coprod W') = \mathcal{A}(W) \otimes \mathcal{A}(W')$.

Applying $\mathcal{A}$ to the cube of resolutions, we get a cube whose vertices are labeled by abelian groups $\mathcal{A}(D_v)$, and whose edges are labeled by linear maps $\mathcal{A}(S_e)$. Since $\mathcal{A}$ is a functor, the two dimensional faces in this cube still commute. The situation is summarized in the last column of the table above.

We can now define the Khovanov complex. As a group

$$CKh(D) = \bigoplus_v \mathcal{A}(D_v)$$

where the sum runs over all vertices of the cube $[0,1]^n$. For $x \in \mathcal{A}(D_v)$, the differential on $CKh(D)$ is given by

$$dx = \sum_{e : v \to v'} (-1)^{\sigma(e)} \mathcal{A}(S_e)(x)$$

where σ is a map from the set of edges to $\{0,1\}$, which we insert in order to make $d^2 = 0$. Indeed, we have

Lemma 3.3.1. *If σ is chosen so that each two dimensional face of the cube has an odd number of edges with $\sigma(e) = 1$, then $d^2 = 0$.*

Proof. If v'' is a vertex of the cube obtained by changing two 0 coordinates of v to 1's, then the component of $d^2(x)$ which lies in the summand in $\mathcal{A}(v'')$ is

$$(-1)^{\sigma(e_1)+\sigma(e_2)}\mathcal{A}(S_{e_2}) \circ \mathcal{A}(S_{e_1})(x) + (-1)^{\sigma(e_3)+\sigma(e_4)}\mathcal{A}(S_{e_3}) \circ \mathcal{A}(S_{e_4})(x)$$

where e_1, e_2, e_3 and e_4 are the edges of the two dimensional face containing v and v'', labeled clockwise starting from v. Since two-dimensional faces of the cube commute,

$$\mathcal{A}(S_{e_2}) \circ \mathcal{A}(S_{e_1}) = \mathcal{A}(S_{e_3}) \circ \mathcal{A}(S_{e_4})$$

Thus to ensure that $d^2 = 0$, it suffices to choose σ so that each two-dimensional face of the cube has an odd number of edges with $\sigma(e) = 1$. □

The following exercise shows that it is possible to choose such a σ, and that any two such choices give rise to isomorphic chain complexes.

Exercise 3.3.2. Let $C^* = C^*_{cell}([0,1]^n; \mathbb{Z}/2)$ be the cellular cochain complex of the cube with $\mathbb{Z}/2$ coefficients.

(1) Viewing σ as an element of C^1, show that the condition of the lemma is satisified if and only if $d\sigma = \tau$, where $\tau \in C^2$ is the cochain which assigns 1 to each two-dimensional face of the cube. Deduce that it is possible to choose σ satisfying the conditions of the lemma.
(2) Suppose $\sigma' \in C^1$ also satisfies $d\sigma' = \tau$, and let $CKh(D)$ and $CKh'(D)$ be the Khovanov complexes defined using σ and σ'. Show that $\sigma - \sigma' = d\rho$ for some $\rho \in C^0$, and use ρ to define an isomorphism between $CKh(D)$ and $CKh'(D)$.

3.4. The TQFT $\mathcal{A}$ Up until this point, the construction we have described works for any TQFT, but the resulting homology depends on the planar diagram D, rather than its underlying link L. To get a chain complex whose homology is a link invariant, we will use a particular TQFT $\mathcal{A}$ for which

$$\mathcal{A}(S^1) = \langle \mathbf{1}, \mathbf{x} \rangle =: V$$

is a free abelian group of rank two. Since $\mathcal{A}$ is a monoidal functor, we must have

$$\mathcal{A}(\coprod^n S^1) = V^{\otimes n}.$$

This completely specifies the functor $\mathcal{A}$ at the level of objects.

If D_v is a closed 1-manifold, we define a *state* of D_v to be a labeling of each component of D_v by either **1** or **x**. The $\mathbb{Z}$-module $\mathcal{A}(D_v)$ has a basis consisting of states of D_v.

More generally, if D is a planar diagram, we define a *state* of D to be a choice of a complete resolution of D, together with a state of the complete resolution. $CKh(D)$ has a basis consisting of states of D.

Example 3.4.1. If D is the diagram of the Hopf link shown in Figure 3.1.2, $\mathrm{CKh}(D)$ has rank 12. A set of generators for the complex is shown in Figure 3.4.2.

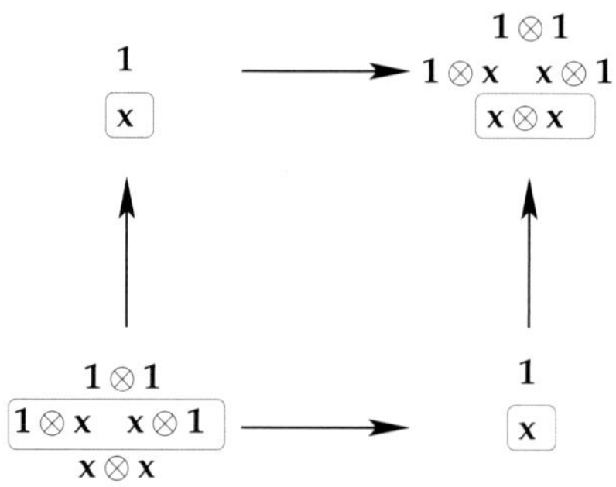

Figure 3.4.2. $\mathrm{CKh}(D)$, where D is the diagram of the Hopf link shown in Figure 3.1.2. The generators in the boxes all have the same q-grading.

Morse theory tells us that any cobordism is the composition of handle attachments. It follows that any morphism in the cobordism category can be built up as a composition of the *elementary cobordisms* shown in Figure 3.4.3 and their disjoint unions with cylinders

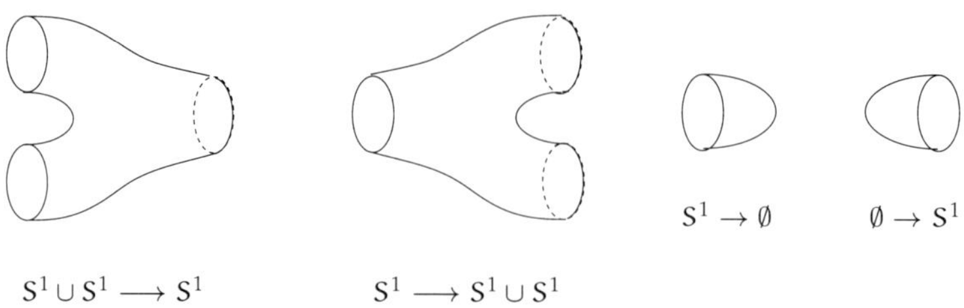

Figure 3.4.3. Elementary cobordisms: merge, split, death, and birth

The functor $\mathcal{A}$ is monoidal, so to understand how it acts on morphisms, it is enough to describe it for the cobordisms shown in the figure. The split and merge morphisms give rise to maps

$$\begin{aligned} m : V \otimes V &\to V \\ \mathbf{1} \otimes \mathbf{1} &\mapsto \mathbf{1} \\ \mathbf{1} \otimes \mathbf{x}, \mathbf{x} \otimes \mathbf{1} &\mapsto \mathbf{x} \\ \mathbf{x} \otimes \mathbf{x} &\mapsto 0 \end{aligned} \qquad \begin{aligned} \Delta : V &\to V \otimes V \\ \mathbf{1} &\mapsto \mathbf{1} \otimes \mathbf{x} + \mathbf{x} \otimes \mathbf{1} \\ \mathbf{x} &\mapsto \mathbf{x} \otimes \mathbf{x} \end{aligned}$$

while the birth and death morphisms give maps

$$\begin{aligned} \eta : \mathbb{Z} &\to V \\ 1 &\mapsto \mathbf{1} \end{aligned} \qquad \begin{aligned} \varepsilon : V &\to \mathbb{Z} \\ \mathbf{1} &\mapsto 0 \\ \mathbf{x} &\mapsto 1. \end{aligned}$$

We have now given a complete definition of the complex $CKh(D)$. Given enough patience and/or computer memory, it is straightforward to compute the complex and its homology for any planar diagram D.

Exercise 3.4.4. Compute the homology of the complex shown in Figure 3.4.2.

Exercise 3.4.5. Show that $CKh(D)$ and $CKh(\overline{D})$ are dual chain complexes.

Exercise 3.4.6. Suppose that $\mathcal{A}$ is a $1+1$ dimensional TQFT, and let $A = \mathcal{A}(S^1)$. A can be equipped with maps $m : A \otimes A \to A$, $\Delta : A \to A \otimes A$, $1 : \mathbb{Z} \to A$, and $\varepsilon : A \to \mathbb{Z}$ as above. Let $\iota : A \otimes A \to A \otimes A$ be the involution given by $\iota(a \otimes b) = b \otimes a$. Use the fact that $\mathcal{A}$ is a TQFT to show that

(1) $m(a \otimes b) = m(b \otimes a)$ and $m(m(a \otimes b) \otimes c) = m(a \otimes m(b \otimes c))$
(2) $\iota \circ \Delta = \Delta$ and $(1_A \otimes \Delta) \circ \Delta = (\Delta \otimes 1_A) \circ \Delta$
(3) $m(\mathbf{1} \otimes a) = a$ and $(\varepsilon \otimes 1_A) \circ \Delta = 1_A$
(4) $\Delta \circ m = (m \otimes 1_A) \circ (1_A \otimes \Delta)$.

A set A equipped with maps m (multiplication), Δ (comultiplication), 1 (unit) and ε (counit) satisfying the properties above is called a *Frobenius algebra*. The exercise above shows that any $1+1$ dimensional TQFT determines a Frobenius algebra. Conversely, it can be shown [46] that any Frobenius algebra determines such a TQFT.

At this point, it is quite natural to ask whether it is possible to replace the particular TQFT $\mathcal{A}$ used above with some other TQFT $\mathcal{A}'$. As the following exercise shows, the possible choices of $\mathcal{A}$ are strongly restricted by the condition that the homology should be a link invariant. (In fact, we will see in section 3.8 that the TQFT used here is essentially the only one whose homology gives an interesting link invariant.)

Exercise 3.4.7. Suppose $\mathcal{A}'$ is a $1+1$ dimensional TQFT for which $\mathcal{A}'(S^1)$ is a free abelian group of rank n. If D is a planar diagram, let $CKh_{\mathcal{A}'}$ be the result of applying the construction of section 3.3 with $\mathcal{A}'$ in place of $\mathcal{A}$. Let D be the zero-crossing diagram of the unknot, and let D' be a one-crossing diagram of the unknot. Show that if $CKh_{\mathcal{A}'}(D)$ and $CKh_{\mathcal{A}'}(D')$ have isomorphic homology, then $n = 2$.

3.5. Gradings $CKh(D)$ can be equipped with a natural bigrading:

$$CKh(D) = \bigoplus_{i,j} CKh^{i,j}(D)$$

such that $d : CKh^{i,j} \to CKh^{i+1,j}$. The first grading is the *(co)homological* grading on $CKh(D)$. If v is a vertex of $[0,1]^n$, we write $|v|$ for the sum of the coefficients of v. An element of $\mathcal{A}(D_v)$ has (co)homological grading $|v|$. It is clear from the definition that d raises the cohomological grading by 1. To define the second grading, which we call the q-grading, we first define a grading $\widetilde{q}$ on V by

$$\widetilde{q}(\mathbf{1}) = 1 \qquad \widetilde{q}(\mathbf{x}) = -1$$

and extend it to $V^{\otimes n}$ by the relation $\widetilde{q}(a \otimes b) = \widetilde{q}(a) + \widetilde{q}(b)$. With respect to this grading, $\mathcal{A}$ is a *graded TQFT*, in the sense that

$$\widetilde{q}(\mathcal{A}(S)(x)) = \chi(S) + \widetilde{q}(x)$$

for any cobordism S. The reader can easily check this for the maps induced by the four elementary cobordisms above (note that any element of the ground ring $\mathbb{Z}$ has $\widetilde{q}$-grading 0). Any cobordism S is a composition of elementary cobordisms, so the general case follows from the fact that $\chi(S_1 \circ S_2) = \chi(S_1) + \chi(S_2)$.

Since each cobordism S_e is the union of a pair of pants with some cylinders, we see that if $x \in \mathcal{A}(D_v)$, then $\widetilde{q}(dx) = \widetilde{q}(x) - 1$. For $x \in \mathcal{A}(D_v)$, we define

$$q(x) = \widetilde{q}(x) + |v|$$

so that $q(dx) = q(x)$.

Remark 3.5.1. As the above discussion shows, $\widetilde{q}$ is also a natural choice of homological grading on $CKh(D)$; it appears as the homological grading in Seidel and Smith's *symplectic Khovanov homology* [73].

Since d preserves the q grading, $CKh(D)$ decomposes as a direct sum of chain complexes:

$$CKh(D) = \bigoplus_j CKh^{*,j}(D).$$

We can combine the Euler characteristics of the summands together to form a generating function

$$\chi(Kh(D)) := \sum_{i,j} (-1)^i q^j \operatorname{rk} CKh^{i,j}(D)$$

which is called the *graded Euler characteristic* of $Kh(D)$.

If we further define the the *bigraded Poincaré polynomial* of $Kh(D)$ to be

$$P(Kh(D)) := \sum_{i,j} t^i q^j \operatorname{rk} Kh^{i,j}(D),$$

then the usual argument shows that $\chi(Kh(D)) = P(Kh(D))|_{t=-1}$.

3.6. Invariance Let D be a planar diagram of L. To pin down the exact normalization of the Jones polynomial $V(L)$, we needed to fix an orientation on L and shift $\langle D \rangle$ by some power of q depending on the number of positive and negative crossings in D. The situation for Kh is similar, but now we need to shift both the homological and q-gradings. Our notation for this is as follows. For $n, m \in \mathbb{Z}$, we define $t^m q^n CKh(D)$ to be the bigraded chain complex whose i, jth group is $CKh^{i-m,j-n}(D)$. The notation is chosen so that

$$P(t^m q^m Kh(D)) = t^m q^n P(Kh(D)).$$

If o is an orientation on our diagram D, we let $n_\pm(D, o)$ be the number of positive/negative crossings in D, and define

$$CKh(D, o) = t^{-n_-(D,o)} q^{n_+(D,o) - 2n_-(D,o)} CKh(D).$$

Exercise 3.6.1. Let D_o be the diagram obtained by taking the oriented resolution at each crossing of D. Show that in the shifted complex $\mathrm{CKh}(D, o)$, the summand $\mathcal{A}(D_o)$ is in homological grading 0.

The first main result of Khovanov's original paper is

Theorem 3.6.2 ([39]). *If* (D, o) *and* (D', o') *are oriented diagrams related by a Reidemeister move, then* $\mathrm{CKh}(D, o)$ *is chain homotopy equivalent to* $\mathrm{CKh}(D', o')$.

The proof of this theorem will be given in the next lecture. It justifies

Definition 3.6.3. If L is an oriented link in S^3 represented by a planar diagram D with orientation o, then $\mathrm{Kh}(L) := \mathrm{Kh}(D, o)$.

The second main result of [39] is that the graded Euler characteristic of $\mathrm{Kh}(L)$ is given by the Jones polynomial:

Theorem 3.6.4. *For any oriented link* L *as above ,*

$$\chi(\mathrm{Kh}(L)) = \overline{V}(L)\,.$$

Proof. Up to overall shifts, this follows easily from equation 1.4.1, which expresses $\langle D \rangle$ as a sum over vertices in the cube of resolutions. We have

$$\begin{aligned} \chi(\mathrm{CKh}(D)) &\sim \sum_v (-q)^{|v|}\, \mathrm{qdim}\, \mathcal{A}(D_v) \\ &\sim \sum_v (-q)^{|v|} (q + q^{-1})^{|D_v|} \\ &\sim \sum_v A^{n-2|v|} B^{|D_v|} = \langle D \rangle \end{aligned}$$

where $q = -A^{-2}$ and $B = -A^2 - A^{-2} = q + q^{-1}$. We leave the reader to check that the overall shifts match up correctly. □

3.7. Functoriality One important property of the Khovanov homology is the fact that it is projectively functorial: cobordisms of links induce maps on Kh which are well-defined up to an overall sign.

Definition 3.7.1. Suppose that L_0 and L_1 are oriented links in $\mathbb{R}^3$. A *cobordism* from L_0 to L_1 is an smooth properly embedded oriented surface $\Sigma \subset \mathbb{R}^3 \times I$ such that $\partial\Sigma = -L_0 \times \{0\} \cup L_1 \times \{1\}$.

Two cobordisms Σ_0, Σ_1 from L_0 to L_1 are equivalent if there is a smooth isotopy $\Phi : \Sigma \times I \to \mathbb{R}^3 \times I$ such that the restriction of Φ to each $\Sigma \times t$ is a cobordism from L_0 to L_1, $\Phi|_{\Sigma\times 0} = \Sigma_0$, $\Phi|_{\Sigma\times 1} = \Sigma_1$, and for each $x \in \partial\Sigma$, $\Phi|_{x\times I}$ is a constant map. (In other words, the isotopy fixes both L_0 and L_1).

As we did with abstract cobordisms, we can form a category whose objects are links in $\mathbb{R}^3$ and whose morphisms are equivalence classes of cobordisms between them. Following [39], we would like to define a functor from this category to the category of abelian groups and linear maps, which on objects is given by Kh.

The basic philosophy behind this definition is to divide the given cobordism into a sequence of elementary cobordisms. We then define the map induced by each elementary cobordism at the chain level, as a map on CKh.

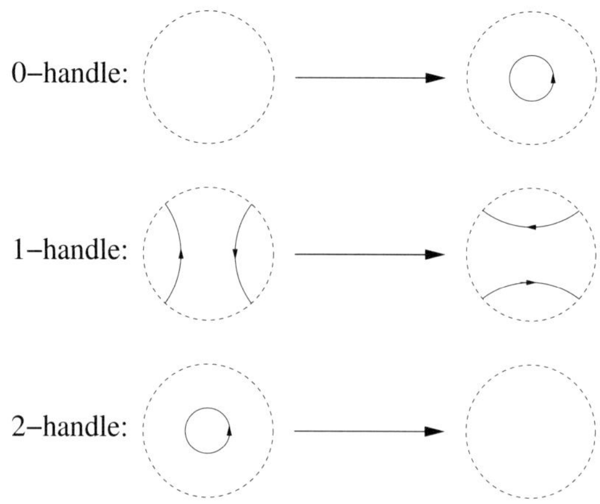

Figure 3.7.2. The three Morse moves.

Definition 3.7.3. A *movie* is a finite subset $S = \{t_1, \ldots, t_n\} \subset (0, 1)$, together with a planar diagram D_t for each $t \in I - S$, such that for $t \in (t_i, t_{i+1})$, D_t is an isotopy of planar diagrams, and for small ε, the diagrams $D_{t_i - \varepsilon}, D_{t_i + \varepsilon}$ are related by either a Reidemeister move or one of the *Morse moves* shown in Figure 3.7.2.

Let $\Sigma : L_0 \to L_1$ be a cobordism. After a small isotopy of Σ, we may arrange that $L_t := \Sigma \cap (\mathbb{R}^3 \times \{t\})$ is a link for all but finitely many values of t, and that the associated planar diagrams D_t form a movie, which is said to represent Σ. Just as any two planar diagrams representing a given link are related by Reidemeister moves, it can be shown that any two movies representing the same cobordism are related by a finite sequence of elementary moves, known as *movie moves* [12].

To each Reidemeister move R_i ($i = 1, 2, 3$) relating diagrams D_0 and D_1, we assign a chain map $\Phi_{R_i} : \mathrm{CKh}(D_0) \to \mathrm{CKh}(D_1)$. Φ_{R_i} is the chain homotopy equivalence provided by theorem 3.6.2. These maps will be described explicitly in the next lecture.

Similarly, for each Morse move M_i ($i = 0, 1, 2$) relating diagrams D_0 and D_1, we assign a chain map $\Phi_{M_i} : \mathrm{CKh}(D_0) \to \mathrm{CKh}(D_1)$. Φ_{M_i} is given by the corresponding map in the TQFT $\mathcal{A}$. To be precise, the crossings of D_0 can naturally be identified with those of D_1, so the vertices of the cube of resolutions for D_0 and D_1 can also be identified. For each vertex v, the Morse move gives a natural cobordism (corresponding to the addition of a single handle) $S_v : (D_0)_v \to (D_1)_v$. For $x \in \mathcal{A}((D_0)_v)$, we define

$$\Phi_{M_i}(x) = \mathcal{A}(S_v)(x).$$

We leave it as an exercise to the reader to check that Φ_{M_i} is a chain map.

To a movie $\{D_t\}$, we want to assign a chain map

$$\Phi_{D_t} : CKh(D_0) \to CKh(D_1).$$

As t varies in the interval (t_i, t_{i+1}), the diagrams D_t change only by isotopy, so the chain complexes $CKh(D_t)$ are all isomorphic. The diagrams $D_{t_i-\varepsilon}$ and $D_{t_i+\varepsilon}$ are related by either a Reidemeister or a Morse move, and we then let $\Phi_{t_i} : CKh(D_{t_i-\varepsilon}) \to CKh(D_{t_i+\varepsilon})$ be the map defined above. Finally, we let Φ_{D_t} be their composition:

$$\Phi_{D_t} := \Phi_{t_n} \circ \Phi_{t_{n-1}} \circ \cdots \circ \Phi_{t_1}.$$

To sum up, we have assigned a chain map $\Phi_{D_t} : CKh(D_0) \to CKh(D_1)$ to each movie D_t, which induces a map on homology $\Phi_{D_t} : Kh(D_0) \to Kh(D_1)$. These maps were defined in Khovanov's first paper [39]. The question of whether these maps are functorial (that is, invariant under the movie moves) was first investigated by Jacobsson [35].

Theorem 3.7.4 ([35]). *The map Φ_{D_t} is projectively functorial; that is, if D_t and D'_t are two movies representing equivalent cobordisms, then Φ_{D_t} is chain homotopic to $\pm\Phi_{D'_t}$.*

The sign ambiguity is necessary: Jacobsson found explicit pairs of movies representing the same cobordism for which the induced maps have opposite sign. Subsequently, Clark, Morrison and Walker [15] and Blanchet [10] showed that the sign ambiguity can be eliminated, but at the cost of redefining the Khovanov homology.

Theorem 3.7.5. [10, 15] *There is a functor $\mathcal{F}$ from the category of oriented links and cobordisms between them to the category of abelian groups such that*

(1) *A planar diagram D for L yields an isomorphism $\iota_D : Kh(D) \to \mathcal{F}(L)$.*
(2) *If D_t is a movie representing a cobordism Σ, then $\Phi_{D_t} = \pm\iota_{D_1}^{-1} \circ \mathcal{F}(\Sigma) \circ \iota_{D_0}$.*

3.8. Deformations To define $CKh(D)$, we applied the TQFT $\mathcal{A}$ to the cube of resolutions of D. If we apply a different TQFT $\mathcal{A}'$, the result will still be a chain complex, but its homology need not be invariant under the Reidemeister moves. The structure of a (1+1) dimensional TQFT is more or less determined by the multiplicative ring of its underlying Frobenius algebra. For Khovanov homology, this is $A = \mathbb{Z}[X]/X^2$. We have already seen (Exercise 3.4.7) that in order to get a knot invariant $A' = \mathcal{A}'(S^1)$ must have rank 2, so a natural guess is to consider a Frobenius algebra with multiplicative ring $A' = \mathbb{Z}[X]/(X^2 - aX - b)$ for $a, b \in \mathbb{Z}$. (In other words, $m(\mathbf{x} \otimes \mathbf{x}) = a\mathbf{x} + b$.) We can find a Frobenius algebra of this form; the comultiplication is given by

$$\Delta(\mathbf{1}) = \mathbf{1} \otimes \mathbf{x} + \mathbf{x} \otimes \mathbf{1} - a\mathbf{1} \otimes \mathbf{1} \quad \text{and} \quad \Delta(\mathbf{x}) = \mathbf{x} \otimes \mathbf{x} + b\mathbf{1} \otimes \mathbf{1}.$$

It can be shown that any such $\mathcal{A}'$ determines a knot invariant $Kh_{\mathcal{A}'}(L)$.

Semi-simple TQFT's: The structure of $Kh_{\mathcal{A}'}(L)$ was largely determined by Lee and Turner [50, 80]. We begin with the observation that if $X^2 - aX - b$ has distinct roots, $\mathcal{A}'$ can be written in a particularly simple form.

Definition 3.8.1. Suppose $\mathcal{A}'$ is a $1+1$ dimensional TQFT defined over a field F, and let A be the corresponding Frobenius algebra. We say $\mathcal{A}'$ is *semi-simple* if $A' = \bigoplus_{i=1}^{n} A_i$, where each A_i is a 1-dimensional Frobenius algebra.

The existence of a unit and co-unit imply that if $\mathbf{v}_i$ is a nonzero element of A_i, then $m(\mathbf{v}_i \otimes \mathbf{v}_i) = k_i \mathbf{v}_i$ and $\Delta(\mathbf{v}_i) = c_i \mathbf{v}_i \otimes \mathbf{v}_i$, where $c_i \neq 0 \neq k_i$.

Lemma 3.8.2. *Suppose $\mathcal{A}'$ is a $1+1$ dimensional TQFT whose Frobenius algebra has multiplicative ring $A' = F[X]/(p(X))$, where $p(X)$ splits over F with distinct roots. Then $\mathcal{A}'$ is semi-simple.*

Proof. Let $n = \dim A' = \deg p$. We view multiplication by X as a linear operator $X : A' \to A'$. This operator satisfies the relation $p(X) = 0$, so p divides the characteristic polynomial of X. But $n = \deg p$, so p must be the characteristic polynomial. Since p splits over F, we have distinct eigenvalues $\lambda_1, \ldots, \lambda_n \in F$. It follows that X is diagonalizable. Let $\{\mathbf{v}_1, \ldots, \mathbf{v}_n\}$ be a basis of eigenvectors with $X(\mathbf{v}_i) = \lambda_i \mathbf{v}_i$.

Let A_i be the subspace of A' spanned by $\mathbf{v}_i$. We will show that $A' \simeq \oplus A_i$ (as Frobenius algebras). First consider the multiplication. We have

$$X(m(\mathbf{v}_i \otimes \mathbf{v}_j)) = m(m(X \otimes \mathbf{v}_i) \otimes \mathbf{v}_j) = \lambda_i m(\mathbf{v}_i \otimes \mathbf{v}_j),$$

and

$$X(m(\mathbf{v}_i \otimes \mathbf{v}_j)) = X(m(\mathbf{v}_j \otimes \mathbf{v}_i)) = \lambda_j m(\mathbf{v}_j \otimes \mathbf{v}_i) = \lambda_j m(\mathbf{v}_i \otimes \mathbf{v}_j).$$

Since $\lambda i \neq \lambda_j$, we must have $m(\mathbf{v}_i \otimes \mathbf{v}_j) = 0$ for $i \neq j$. Similarly, we must have $X(m(\mathbf{v}_i \otimes \mathbf{v}_i)) = \lambda_i m(\mathbf{v}_i \otimes \mathbf{v}_i)$, which implies $m(\mathbf{v}_i \otimes \mathbf{v}_i) \in A_i$.

Next, the comultiplication: on $A' \otimes A'$ we have two linear operators X_l and X_r which act by multiplication by X on the left and right-hand tensor factors. $\Delta(\mathbf{v}_i)$ is an eigenvector for both X_l and X_r; the eigenvalue in both cases is λ_i. It follows that $\Delta(\mathbf{v}_i) = c_i \mathbf{v}_i \otimes \mathbf{v}_i$. □

In contrast, the TQFT $\mathcal{A}$ used to define Khovanov homology is not semi-simple; the Jordan normal form of the operator X is given by the upper triangular matrix

$$\begin{pmatrix} 0 & 1 \\ 0 & 0 \end{pmatrix}.$$

Canonical Generators: Let D be a diagram representing L. If $\mathcal{A}'$ is semi-simple, Lee describes an explicit set of generators for $Kh_{\mathcal{A}'}(D)$. The elements of this set are in bijection with orientations on D.

Given an orientation o of D, the oriented resolution of D with respect to o is an oriented collection of circles in the plane. Let $h(C)$ be the nesting height (as in the proof of Lemma 2.6) of such a circle, and let

$$p(C) = \begin{cases} h(C) & \text{if } C \text{ is oriented counterclockwise} \\ h(C)+1 & \text{if } C \text{ is oriented clockwise} \end{cases}$$

Definition 3.8.3. Suppose $\mathcal{A}'$ is semi-simple, with simple summands generated by $\mathbf{v}_1$ and $\mathbf{v}_2$. If o is an orientation of D, the *canonical generator* $\mathfrak{s}_o \in \mathrm{CKh}_{\mathcal{A}'}(D)$ is state obtained by taking the oriented resolution of D with respect to o and labeling each circle C by $\mathbf{v}_1$ if $p(C)$ is odd, and $\mathbf{v}_2$ if $p(C)$ is even.

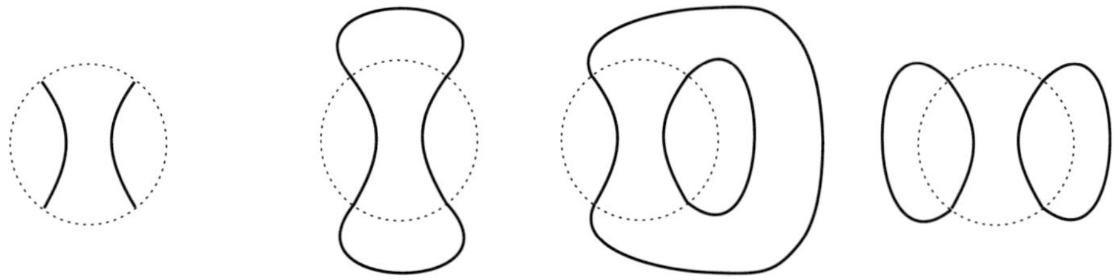

Figure 3.8.4. On the left, two strands in a disk. On the right are some possible closures.

Lemma 3.8.5. *Consider an embedded disk which intersects the diagram for $\mathfrak{s}_o$ in two strands, as shown on the left of Figure 3.8.4. If the orientations on the two strands are parallel, they have different labels; if the orientations are opposite, the labels are the same.*

Proof. Consider the circles containing the two strands. Up to a rotation of the diagram, they must be in one of the three configurations shown on the right-hand side of the figure. If there is one circle in the closure, the two orientations must point in opposite directions and have the same label. If there are two nested circles C_1 and C_2, then $h(C_1) \equiv h(C_2) + 1 \mod 2$. In this situation, the orientations are either parallel, and then $p(C_1) \not\equiv p(C_2)$, or opposite, and then $p(C_1) \equiv p(C_2)$. Either way, the statement holds. The case where C_1 and C_2 are not nested is similar. □

Corollary 3.8.6. *$\mathfrak{s}_o$ is closed in* $\mathrm{CKh}_{\mathcal{A}'}(D)$.

Proof. Let v_o be the vertex of the cube of resolutions associated with the oriented resolution of L, and let $e : v_o \to v'$ be an edge. The two strands near the crossing have parallel orientations. By the lemma, they have opposite labels. Thus they belong to different components of D_{v_o}, and the cobordism S_e is a merge. Since $m(\mathbf{v}_1 \otimes \mathbf{v}_2) = 0$, we have $\mathcal{A}'(S_e)(\mathfrak{s}_o) = 0$. □

Theorem 3.8.7. ([50, 80]) *If $\mathcal{A}'$ is a semisimple TQFT defined over F, then $\mathrm{Kh}_{\mathcal{A}'}(L; F)$ is freely generated by the classes $\mathfrak{s}_o$, where o runs over the set of orientations on L. In particular, $\mathrm{Kh}_{\mathcal{A}'}(L; F)$ has dimension $2^{|L|}$.*

Proof. By induction on the number of crossings in D. When D has no crossings, $\mathrm{Kh}_{\mathcal{A}'}(D) = \mathcal{A}'(D)$ has dimension $2^{|D|}$ and is generated by states in which we label each component of D by either $\mathbf{v}_1$ or $\mathbf{v}_2$. These are the canonical generators.

In general, chose a crossing c of D and let D_0 and D_1 be the diagrams obtained by resolving c. There is a short exact sequence

$$0 \to \mathrm{CKh}_{\mathcal{A}'}(D_1) \to \mathrm{CKh}_{\mathcal{A}'}(D) \to \mathrm{CKh}_{\mathcal{A}'}(D_0) \to 0$$

which comes from splitting the cube of resolutions for D into two halves, according to the way that c is resolved. We will compute the boundary map $\partial : \mathrm{Kh}_{\mathcal{A}'}(D_0) \to \mathrm{Kh}_{\mathcal{A}'}(D_1)$ in the associated long exact sequence.

By induction, $\mathrm{Kh}_{\mathcal{A}'}(D_0)$ and $\mathrm{Kh}_{\mathcal{A}'}(D_1)$ are generated by canonical generators. Let o be an orientation on D_0, and let v_o be the vertex of the cube of resolutions for D_0 given by taking the oriented resolution with respect to o. The canonical generator $\mathfrak{s}_o \in \mathrm{CKh}(D_0)$ is supported at v_o. In the cube of resolutions for D, there is a unique edge e which points from v_o to a vertex v'_o in the cube of resolutions for D_1; namely, the edge corresponding to changing the resolution at c from the 0 resolution to the 1 resolution. It follows that $\partial[\mathfrak{s}_o] = [\mathcal{A}'(S_e)(\mathfrak{s}_o)]$.

At this point, the argument splits into three cases, depending on how the ends at the crossing c are joined up in L. The three possibilities are shown in Figure 3.8.8 below.

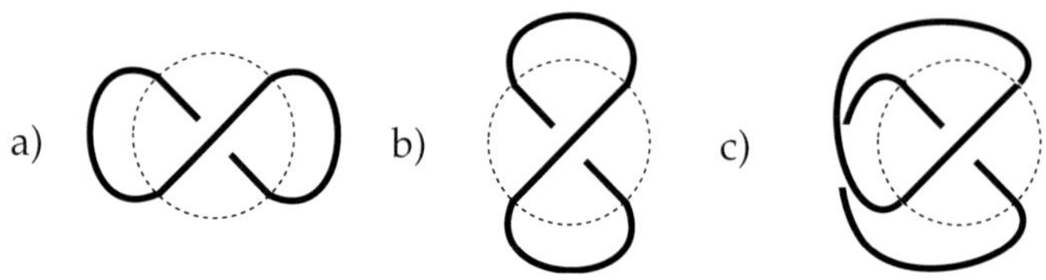

Figure 3.8.8. The three possible ways of joining up the ends of a crossing.

In case a), $|D| = |D_1| = |D_0| - 1$, so there are twice as many orientations on D_0 as on D or D_1. We can divide the orientations on D_0 into two types: those in which the orientation on the two strands near c is parallel, and those in which it is opposite. As shown in the figure below, orientations of the first type are in bijection with orientations on D, and orientations of the second type are in bijection with orientations on D_1.

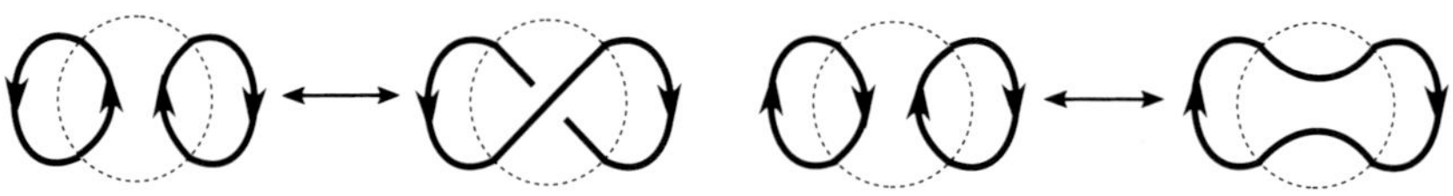

If o is of the first type, Lemma 3.8.5 implies that the labels on the two strands are different. It follows that S_e is a merge, and $\mathcal{A}'(S_e)(\mathfrak{s}_o) = 0$. If o is of the second type, the labels are the same. In this case, $\mathcal{A}'(S_e)(\mathfrak{s}_o) = c\mathfrak{s}_{o_1}$, regardless of whether S_e is a merge or a split. Here o_1 is the orientation on D_1 which corresponds to o, and $c \neq 0$. We conclude that $\partial : \mathrm{Kh}_{\mathcal{A}'}(D_0) \to \mathrm{Kh}_{\mathcal{A}'}(D_1)$ is surjective, and $\mathrm{Kh}_{\mathcal{A}'}(D) = \ker \partial$ is freely generated by the classes $[\mathfrak{s}_o]$, where o runs over orientations of D.

Cases b) and c) are very similar. In case b), D_1 has twice as many orientations as D or D_0, and the boundary map is a surjection, while in case c), D has twice as many orientations as D_0 and D_1, and the boundary map vanishes. We leave the details for the reader. $\square$

In order to fix the homological grading on $Kh(L)$, we needed to fix an orientation o on L. The vertex D_o of the cube of resolutions is in homological grading 0. If K is a knot either orientation gives the same homological grading, and both canonical generators are in homological grading 0. More generally, the two generators $\mathfrak{s}_o, \mathfrak{s}_{-o}$ will be in homological grading 0; the grading of all the other generators is determined by the pairwise linking numbers of the components of L [50].

Lee's spectral sequence: The existence of the q-grading on Khovanov homology depended on the fact that $\mathcal{A}$ is a graded TQFT: $\widetilde{q}(\mathcal{A}(S)(x)) = \widetilde{q}(x) + \chi(S)$. A quick look at the formulas for m' and Δ' shows that this does not hold for $\mathcal{A}'$; however $\mathcal{A}'$ is *filtered*, in the sense that the $\widetilde{q}$-grading of every term that appears in $\mathcal{A}(S)(x)$ is at least $\widetilde{q}(x) + \chi(S)$. The associated graded TQFT (obtained by ignoring all terms that strictly raise the q-grading) is $\mathcal{A}$. In turn, $CKh_{\mathcal{A}'}(D)$ is a filtered chain complex. The associated graded chain complex (obtained by ignoring all terms in the differential that strictly raise the q-grading) is $CKh(D)$. We deduce from these observations that

Proposition 3.8.9. *There is a spectral sequence with* E_1 *term* $Kh(L)$ *which abuts to* $Kh_{\mathcal{A}'}(L)$. *The filtration grading in this spectral sequence is given by the* q*-grading on* $Kh(L)$.

We should think of the combination of Theorem 3.8.7 and Proposition 3.8.9 and as categorifying the fact that $V_L(1) = 2^{|L|}$. This is the first incidence of a general principle: identities where we specialize a variable in a polynomial typically categorify to give a spectral sequence. In this case (and in many others) it can be shown that the E_i term of the sequence is an invariant of L for all $i > 0$. When this is the case we get a bonus invariant: the filtration grading of the surviving generators in the spectral sequence.

For concreteness, we now specialize to the situation originally considered by Lee, in which $F = \mathbb{Q}$ and $A' = \mathbb{Q}[X]/(X^2 - 1)$. In this case, it can be shown [66] that the filtration grading of the two generators in the spectral sequence for $CKh_{\mathcal{A}'}(K; \mathbb{Q})$ are of the form s_{min}, s_{max}, where $s_{max} - s_{min} = 2$

Definition 3.8.10. If K is a knot in S^3, $s(K) = s_{min} + 1 = s_{max} - 1$.

For example if $Kh(\bigcirc) = \mathcal{A}(S^1)$ has dimension 2, so the spectral sequence has already converged at the E_1 term. In this case $s_{min} = -1, s_{max} = 1$, and $s(\bigcirc) = 0$.

Maps induced by cobordism: If $\Sigma : L_0 \to L_1$ is a cobordism, we can define a map $\Phi'_\Sigma : Kh_{\mathcal{A}'}(L_0) \to Kh_{\mathcal{A}'}(L_1)$ exactly as we did for Khovanov homology, but using $\mathcal{A}'$ in place of $\mathcal{A}$. Φ'_Σ is a filtered map of degree $\chi(\Sigma)$.

If $\widehat{o}$ is an orientation on Σ, let $-\widehat{o}_0$ and $\widehat{o}_1$ be the induced orientations on L_0 and L_1. (With this convention, if $\Sigma = L \times [0,1]$, $\widehat{o}_0$ and $\widehat{o}_1$ are the same orientation on L.) Let $\mathcal{O}(\Sigma)$ be the set of all orientations on Σ.

Theorem 3.8.11 ([66]). *If* o *is an orientation on* L_0 *and* $\Sigma : L_0 \to L_1$ *is a cobordism, then*

$$\Phi'_\Sigma(\mathfrak{s}_o) = \sum_{\{\widehat{o} \in \mathcal{O}(\Sigma) \mid \widehat{o}_0 = o\}} c_{\widehat{o}} \mathfrak{s}_{\widehat{o}_1}$$

where $c_{\widehat{o}} \neq 0$ *for all orientations* $\widehat{o} \in \mathcal{O}(\Sigma)$.

Sketch of Proof. First, note that

$$\mathcal{O}(\Sigma \circ \Sigma') = \{(\widehat{o}, \widehat{o}') \in \mathcal{O}(\Sigma) \times \mathcal{O}(\Sigma') \mid \widehat{o}_0 = \widehat{o}'_1\}.$$

If the statement holds for Σ and Σ', then

$$\begin{aligned}
\Phi'_{\Sigma \circ \Sigma'}(\mathfrak{s}_o) &= \Phi'_\Sigma \circ \Phi'_{\Sigma'}(\mathfrak{s}_o) \\
&= \sum_{\{\widehat{o} \in \mathcal{O}(\Sigma) \mid \widehat{o}_0 = \widehat{o}'_1\}} \sum_{\{\widehat{o}' \in \mathcal{O}(\Sigma') \mid \widehat{o}'_0 = o\}} c_{\widehat{o}} c_{\widehat{o}'} \mathfrak{s}_{\widehat{o}_1} \\
&= \sum_{\{\widehat{o}'' \in \mathcal{O}(\Sigma \circ \Sigma') \mid \widehat{o}''_0 = o\}} c_{\widehat{o}''} \mathfrak{s}_{\widehat{o}''_1},
\end{aligned}$$

so it holds for $\Sigma \circ \Sigma'$ as well. Hence it is enough to check that the statement holds for elementary cobordisms. The case where Σ is cobordism corresponding to a Reidemeister move is checked in [66].

We check the statement holds for Morse moves. Let o be an orientation on L_0. First, suppose Σ is a 0-handle attachment, and let C be the new circle created in L_0. There are two orientations on Σ compatible with o, corresponding to the two possible orientations on C. Now $\Phi'_\Sigma(\mathfrak{s}_o)$ is the state in which all the labels on the old circles remain the same and C is labeled by 1. Since $1 = c_1\mathbf{v}_1 + c_2\mathbf{v}_2$ with both $c_i \neq 0$, the statement holds.

Next, suppose Σ is obtained by attaching a 1-handle. The 1-handle attachment can either be compatible with o or incompatible. If it is compatible, there is a unique orientation $\widehat{o} \in \mathcal{O}(\Sigma)$ with $\widehat{o}_0 = o_0$. The two strands on which the attachment takes place have opposite orientations in $\mathfrak{s}_o$, hence the same label, so we have $\Phi'(\Sigma)(\mathfrak{s}_o) = c\mathfrak{s}_{\widehat{o}_1}$ with $c \neq 0$ regardless of whether the cobordism is a split or a merge.

If the orientation on the 1-handle is incompatible with o, then there is no $\widehat{o} \in \mathcal{O}(\Sigma)$ with $\widehat{o}_0 = o$. The two strands on which the attachment takes place have the same orientation in $\mathfrak{s}_0$ hence opposite labels. It follows that the cobordism is a merge and $\Phi'(\Sigma)(\mathfrak{s}_o) = 0$. Thus the statement holds in this case.

Finally, we consider the case where Σ is a 2-handle attachment. In this case, there is a unique $\widehat{o} \in \mathcal{O}(\Sigma)$ with $\widehat{o}_0 = o$. Since $\varepsilon'(\mathbf{v}_1) \neq 0 \neq \varepsilon'(\mathbf{v}_2)$, it is easy to see that $\Phi'_\Sigma(\mathfrak{s}_o) = c\mathfrak{s}_{\widehat{o}_1}$ with $c \neq 0$, and the statement holds in this case as well. □

Corollary 3.8.12. *If* K_0 *and* K_1 *are knots and* $\Sigma : K_0 \to K_1$ *is a connected cobordism, then* $\Phi'_\Sigma : Kh_{\mathcal{A}'}(K_0; \mathbb{Q}) \to Kh_{\mathcal{A}'}(K_1; \mathbb{Q})$ *is an isomorphism.*

Corollary 3.8.13. *If* $\Sigma : K_0 \to K_1$ *is a connected cobordism, then* $s(K_1) \geqslant s(K_0) + \chi(\Sigma)$.

Proof. If D is a diagram of K we have

$$s_{max}(K) = \max\{q(x) \mid [x] \neq 0 \in Kh'(D)\}.$$

Let D_0 be a diagram of K_0, and choose an $x \in CKh'(D_0)$ which realizes the maximum. Then $[\Phi_\Sigma(x)] \neq 0 \in Kh'(K_1)$, so

$$s_{max}(K_1) \geqslant q(\Phi_\Sigma(x)) \geqslant q(x) + \chi(\Sigma) = s_{max}(K_0) + \chi(\Sigma). \qquad \square$$

The slice genus:

Definition 3.8.14. If K is a knot in S^3, the *smooth slice genus* of K is

$$g_*(K) = \min\{g(\Sigma) \mid \Sigma \hookrightarrow B^4 \text{ is a smoothly embedded, orientable, and } \partial\Sigma = K\}$$

If Σ is a surface as in the definition, removing a small ball around a point in Σ gives a cobordism $\widetilde{\Sigma} : K \to \bigcirc$ with $\chi(\widetilde{\Sigma}) = -2g(\Sigma)$. Applying Corollary 3.8.13, we get $s(K) \leqslant 2g(\Sigma) + g(\bigcirc) = 2g(\Sigma)$.

Exercise 3.8.15. Using exercise 3.4.5, show that $s(\overline{K}) = -s(\overline{K})$.

Since $g_*(K) = g_*(\overline{K})$, we deduce

Corollary 3.8.16. $|s(K)| \leqslant 2g_*(K)$.

Unlike the Seifert genus, the slice genus can vanish for a nontrivial knot. If $g_*(K) = 0$, we say that K is *slice.*

Exercise 3.8.17. Show that $g_*(K\#\overline{K}) = 0$ for any knot K.

A surface $\Sigma \subset B^4$ is *locally flat* if every point of $\Sigma \subset B^4$ has an open neighborhood U such that the pair $(U, \Sigma \cap U)$ is homeomorphic to $(\mathbb{R}^4, \mathbb{R}^2)$. The *topological slice genus* is defined to be

$$g_*^{top}(K) = \min\{g(\Sigma) \mid \Sigma \subset B^4 \text{ is locally flat, orientable, and } \partial\Sigma = K\}.$$

A smooth surface is locally flat, but it turns out that the converse is false.

Theorem 3.8.18 (Freedman). *If* $\Delta(K) = 1$, *then* $g_*^{top}(K) = 0$.

Let $P(-3,5,7)$ be the $(-3,5,7)$ pretzel knot from Figure 1.1.3. As we saw in execise 7, section 2, $\Delta(P(-3,5,7)) = 1$, so this knot is topologically slice. In contrast, direct computation shows that $s(P(-3,5,7)) = 2$ and hence that $g_*(P(-3,5,7)) = g(P(-3,5,7)) = 1$. Hence s can be used to detect the difference between smooth and locally flat embeddings. In combination with a theorem of Freedman and Quinn on smoothing topological 4-manifolds with boundary, this can be used to construct a manifold homeomorphic, but not diffeomorphic, to the standard $\mathbb{R}^4$. For details, see [25].

3.9. Connections and Further Reading There is an extensive literature on Khovanov homology — far more than we can survey here. We mention only a few highlights. First, its module structure. As observed by Khovanov [42], Kh(L) is a module over the ring $\mathbb{Z}[X_1, \ldots, X_n]/(X_1^2, \ldots X_n^2)$, where there is one variable for each component of X. When $L = K$ is a knot, this gives reduced homology groups $Kh_r(K)$ satisfying $\chi(Kh_r(K)) = V(K)$ (the normalized Jones polynomial).

Second, its behavior for the simplest knots. If K is an alternating knot, $Kh(K)$ is *thin*: $Kh_r^{i,j}(K) = 0$ unless $2i - j = \sigma(K)$. It follows that $Kh_r(K)$ (and also $Kh(K)$) is determined by the Jones polynomial and signature of K. This was conjectured (in the unreduced case) by Bar-Natan [3], and proved by Lee [50], who created her epynomous spectral sequence in the process.

Third, its relations with Floer homology. The most immediate comparison is with knot Floer homology $\widehat{HFK}(K)$, which categorifies the Alexander polynomial and is discussed in Hom's lectures. The invariant $s(K)$ was defined in analogy with the Ozsváth-Szabó τ-invariant and gives a similar bound on the slice genus.

In another direction, there are many invariants which satisfy unoriented skein exact sequences similar to that satisfied by Khovanov homology. If we use $\mathbb{Z}/2$ coefficients, such theories are often the target of spectral sequences whose E_1 term is $Kh(K; \mathbb{Z}/2)$. The first example is in a beautiful paper of Ozsváth and Szabó [61], who constructed a spectral sequence from $Kh(K; \mathbb{Z}/2)$ to $\widehat{HF}(\Sigma(-K); \mathbb{Z}/2)$ — the Heegaard Floer homology of the branched double cover of K. Another is due to Kronheimer and Mrowka [48], who constructed a spectral sequence from $Kh(K; \mathbb{Z}/2)$ to their instanton knot homology $KHI(K; \mathbb{Z}/2)$. As a consequence, they proved that Khovanov homology detects the unknot.

Finally, the symplectic Khovanov homology of Seidel and Smith [74] is defined geometrically using Lagrangian Floer homology. It is isomorphic to Khovanov homology, as proved by Abouzaid and Smith [2].

Two more recent developments worth knowing are the Batson-Seed spectral sequence [6], which relates the Khovanov homology of a link to that of its components, and Grigsby-Licata-Wehrli's $\mathfrak{sl}_2$ action on sutured annular Khovanov homology [30].

4. Khovanov Homology for Tangles

In this lecture, we will extend the definition of Khovanov homology to the 2-category of tangles and cobordisms between them. There are many ways to do this, all of them more or less equivalent [5, 41, 45, 77]. We will follow Bar-Natan [5], whose construction is appealingly geometric. The ideas in this lecture will play a central role in the subsequent lectures, so it's worth spending time thinking about them.

4.1. Tangles For each positive integer n, we fix a set X_n of n points in $\mathbb{R}$.

Definition 4.1.1. An (n, m)-tangle is a proper smooth embedding

$$T : \left(\coprod_{i=1}^{a} I \right) \amalg \left(\coprod_{j=1}^{b} S^1 \right) \hookrightarrow \mathbb{R}^2 \times I$$

such that the set $\amalg_{i=1}^{a} \partial I$ is mapped bijectively onto $(X_n \times 0 \times 0) \amalg (X_m \times 0 \times 1)$. In particular, $a = (n+m)/2$.

We label the coordinates on $\mathbb{R}^2$ by (x, y), and the coordinate on $[0,1]$ by s. Two tangles T_0, T_1 are equivalent if there is an isotopy

$$\Phi : \left((\amalg_{i=1}^{a} I) \amalg (\amalg_{j=1}^{b} S^1) \right) \times I \to \mathbb{R}^2 \times I$$

such that $T_t := \Phi(\cdot, t)$ is a tangle for all $t \in [0,1]$. Note that this definition implies that the boundary of the tangle is fixed under the isotopy.

Definition 4.1.2. We let $\mathcal{T}_{n,m}$ be the set of (n,m) tangles, and let $\mathbf{T}_{n,m}$ be the set of (n,m) tangles up to isotopy.

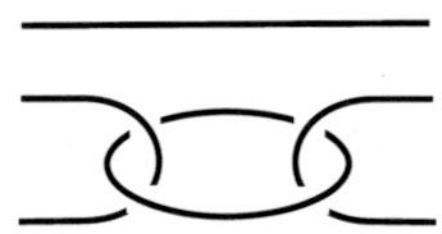

For example, $\mathbf{T}_{0,0}$ is the set of isotopy classes of links in $\mathbb{R}^2 \times I$, which can be naturally identified with the set of isotopy classes of links in $\mathbb{R}^3$. Just as with knots and links, a tangle may be represented by a planar diagrams which records the projection of the tangle to the sx plane, together with over and under-crossings. As an example, a planar diagram of a (3,3) tangle is shown in the figure to the right. As with links, it can be shown that two diagrams which represent the same tangle are related by a sequence of Reidemeister moves.

An important property of tangles is that they satisfy a partial composition rule: if $T \in \mathcal{T}_{n,m}$ and $T' \in \mathcal{T}_{m,l}$, then $TT' \in \mathcal{T}_{n,l}$ is the tangle obtained by stacking T and T' horizontally (in the s direction), as shown in the figure below.

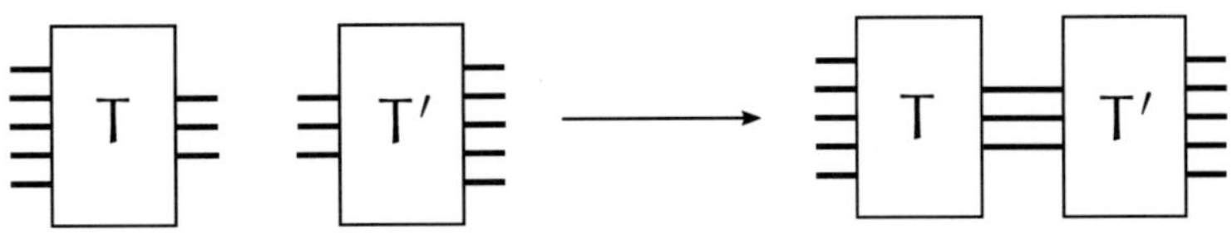

Figure 4.1.3. Horizontal composition of tangles.

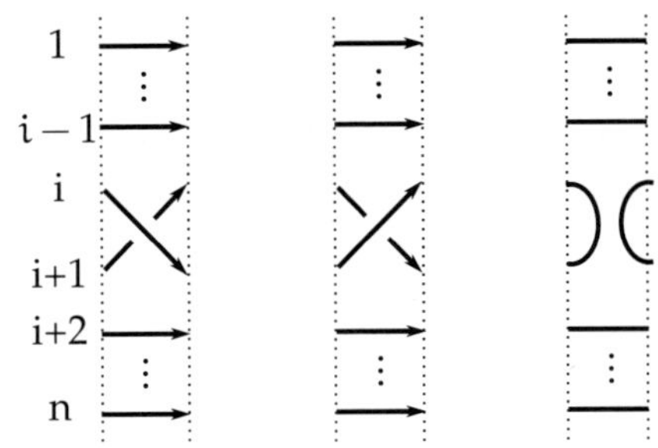

Figure 4.1.4. From left: the elementary braids σ_i σ_i^{-1} and the non-invertible tangle U_i.

Example 4.1.5. It is easy to see that $\mathbf{T}_{n,n}$ forms a monoid under composition. The *braid group on* n *strands* (written Br_n) is the submonoid of $\mathbf{T}_{n,n}$ generated by the *elementary braids* $\sigma_i^{\pm 1}$ shown in Figure 4.1.4. The reader should convince him/herself that Br_n is indeed a group. (Note that $\mathbf{T}_{n,n}$ is not a group; for example the tangle U_i shown in the figure is not invertible.)

It can be shown that Br_n has a presentation

$$\mathrm{Br}_n = \left\langle \sigma_1, \ldots \sigma_{n-1} \;\middle|\; \begin{matrix} \sigma_i\sigma_j = \sigma_j\sigma_i & |i-j| > 1 \\ \sigma_i\sigma_{i+1}\sigma_i = \sigma_{i+1}\sigma_i\sigma_{i+1} & \end{matrix} \right\rangle.$$

The first set of relations is known as *far-commutativity* while the second corresponds to the third Reidemeister move.

Definition 4.1.6. If T_0, T_1 are (n, m) tangles, a *cobordism* $\Sigma : T_0 \to T_1$ is a smooth properly embedded orientable surface $\Sigma \subset \mathbb{R}^2 \times I \times I$ such that

$$\partial\Sigma = T_0 \times 0 \cup T_1 \times 1 \cup (X_n \times 0 \times 0 \times I) \cup (X_m \times 0 \times 1 \times I).$$

We say cobordisms $\Sigma, \Sigma' : T_0 \to T_1$ are equivalent if there is a diffeomorphism $\varphi : \mathbb{R}^2 \times I \times I \to \mathbb{R}^2 \times I \times I$ such that φ restricts to the identity on $\partial(\mathbb{R}^2 \times I \times I)$ and $\varphi(\Sigma) = \Sigma'$.

We will denote the coordinates on $\mathbb{R}^2 \times I \times I$ by (x, y), s and t. Tangle cobordisms can be composed in two different ways. First suppose $\Sigma : T_0 \to T_1$ is a cobordism of (n, m) tangles, and $\Sigma : T_0' \to T_1'$ is a cobordism of (m, l) tangles. Then $\Sigma\Sigma' : T_0T_0' \to T_1T_1'$ is the cobordism obtained by stacking Σ and Σ' in the s direction. This is called *horizontal composition*. Second, if T_0, T_1, T_2 are all (m, n) tangles, $\Sigma_0 : T_0 \to T_1$, and $\Sigma_1 : T_1 \to T_2$, we get a cobordism $\Sigma_0 \circ \Sigma_1 : T_0 \to T_2$ by stacking in the t direction. This is called *vertical composition*. It is not hard to see that both of these notions are well-behaved with respect to equivalence of cobordisms. (Tangle cobordisms are hard to draw, but see Figures 4.2.4 and 4.2.5, which illustrate the analogous compositions for planar tangles.)

A good way to think about these various compositions is to say that there is a category whose objects are the sets X_n for $n \geqslant 0$ and for which $\mathrm{Mor}(X_n, X_m)$ is the set of (n, m) tangles. For each n and m, the set $\mathrm{Mor}(X_n, X_m)$ is itself the set of objects of a category whose morphisms are equivalence classes of cobordisms. A more sophisticated way of saying this is that we have a 2-*category* with objects X_n $(n \geqslant 0)$, 1- morphisms given by tangles, and 2-morphisms given by equivalence classes of cobordisms between tangles.

Exercise 4.1.7. Show that if T_0 and T_1 are isotopic (n, m) tangles, they are isomorphic as objects of the category $\mathrm{Mor}(X_n, X_m)$.

4.2. Planar Tangles

Definition 4.2.1. A *planar* (n, m) *tangle* is a proper smooth embedding

$$T : \left(\coprod_{i=1}^{a} I\right) \amalg \left(\coprod_{j=1}^{b} S^1\right) \hookrightarrow \mathbb{R} \times I$$

such that the set $\amalg_{i=1}^{a} \partial I$ is mapped bijectively onto $(X_n \times 0) \amalg (X_m \times 1)$.

We denote the set of all such tangles by $\mathcal{P}_{n,m}$. We say a planar tangle is *simple* if it has no closed components, and let $\mathbf{P}_{n,m}$ be the set of isotopy classes of simple planar (n, m) tangles. For example, $\mathbf{P}_{2,2}$ consists of the two tangles $\mathbf{1}$ and U shown in the figure below.

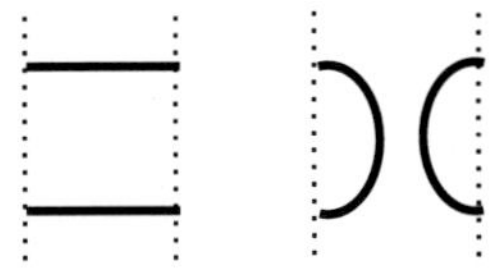

$\mathbf{1}$ $\quad$ U

Exercise 4.2.2. Show that $\mathbf{P}_{n,m}$ has $C_{(n+m)/2}$ elements, where $C_r = \dfrac{1}{r+1}\dbinom{2r}{r}$ is the r*th Catalan number*.

Definition 4.2.3. If T_0, T_1 are planar (n, m) tangles, a *cobordism* $\Sigma : T_0 \to T_1$ is a smooth properly embedded orientable surface $\Sigma \subset \mathbb{R} \times I \times I$ such that

$$\partial\Sigma = T_0 \times 0 \cup T_1 \times 1 \cup (X_n \times 0 \times I) \cup (X_m \times 1 \times I).$$

We say cobordisms $\Sigma, \Sigma' : T_0 \to T_1$ are equivalent if there is a diffeomorphism $\varphi : \mathbb{R} \times I \times I \to \mathbb{R} \times I \times I$ such that φ restricts to the identity on $\partial(\mathbb{R} \times I \times I)$ and $\varphi(\Sigma) = \Sigma'$.

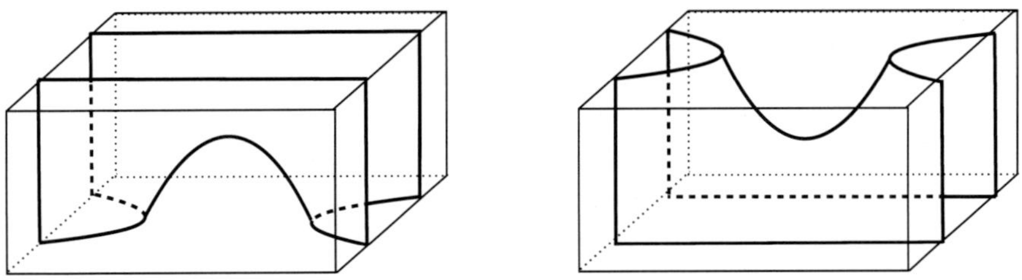

$S : U \to \mathbf{1}$ $\qquad$ $S' : \mathbf{1} \to U$

For example, the figure above shows the saddle cobordism $S : U \to \mathbf{1}$ and its reverse $S' : \mathbf{1} \to U$.

As before, if we are given $\Sigma : T_0 \to T_1$ a cobordism of planar (n, m) tangles and $\Sigma : T_0' \to T_1'$ a cobordism of planar (m, l) tangles, then we can form their horizontal composition $\Sigma\Sigma' : T_0T_0' \to T_1T_1'$ by stacking in the s direction, as illustrated in the figure below:

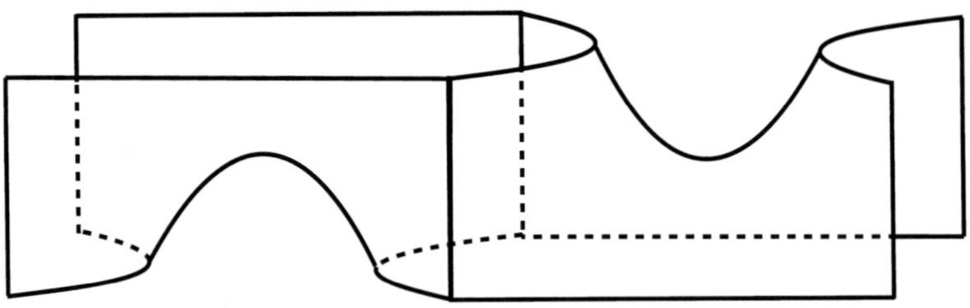

Figure 4.2.4. The horizontal composition $SS' : U\mathbf{1} \to \mathbf{1}U$.

If T_0, T_1, T_2 are all planar (m, n) tangles, $\Sigma_0 : T_0 \to T_1$, and $\Sigma_1 : T_1 \to T_2$, their vertical composition $\Sigma_0 \circ \Sigma_1 : T_0 \to T_2$ is obtained by stacking in the t direction:

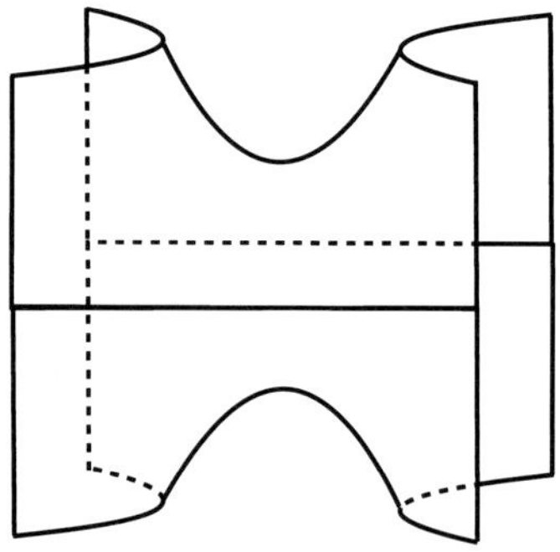

Figure 4.2.5. The vertical composition $S \circ S' : U \to U$.

4.3. The Kauffman Bracket Let $R = \mathbb{Z}[q^{\pm 1}]$, and define $V_{n,m}$ to be the free R-module generated by $\mathbf{P}_{n,m}$. Suppose D is a planar diagram representing an (n, m) tangle. If D has k crossings, there will be 2^k ways to resolve all k, and the set of possible resolutions is in bijection with the vertices of $[0, 1]^k$. The diagram D_v corresponding to the resolution at vertex v is a planar (n, m) tangle.

Definition 4.3.1. For D an (n, m) tangle diagram, its Kauffman bracket is

$$\langle D \rangle = \sum_v (-q)^{|v|}(q + q^{-1})^{|D_v|}\widetilde{D_v} \in V_{n,m}$$

where $|D_v|$ denotes the number of closed components of D_v, and $\widetilde{D_v}$ is the simple planar tangle obtained by erasing all closed components of D_v.

Exercise 4.3.2. Let R_i and R'_i be the (i, i) tangle diagrams corresponding to the ith Reidemeister move. Check that $\langle R_i \rangle \sim \langle R'_i \rangle$, where $p_1 \sim p_2$ means $p_1 = \pm q^k p_2$ for some $k \in \mathbb{Z}$.

Composition: We define $\cdot : V_{n,m} \times V_{m,l} \to V_{n,l}$ by setting $D \cdot D' = \langle DD' \rangle$ when D and D' are simple planar tangles, and extending bilinearly.

Proposition 4.3.3. *For* D *an* (n, m) *tangle diagram and* D' *an* (m, l) *tangle diagram,*

$$\langle DD' \rangle = \langle D \rangle \cdot \langle D' \rangle.$$

Proof. The cube of resolutions for DD' is the product of the cube of resolutions for D and the cube for D', so

$$\begin{aligned}\langle DD' \rangle &= \sum_{(v,v')} (-1)^{|v|+|v'|}(q + q^{-1})^{|D_v D'_{v'}|}[\widetilde{D_v D_{v'}}] \\ &= \sum_{(v,v')} (-1)^{|v|+|v'|}(q + q^{-1})^{|D_v|+|D'_{v'}|}\widetilde{D_v} \cdot \widetilde{D_{v'}} \\ &= \langle D \rangle \langle D' \rangle.\end{aligned}$$

□

Proposition 4.3.4. *If two planar diagrams* D *and* D' *represent the same tangle, then we have* $\langle D\rangle \sim \langle D'\rangle$.

Proof. We checked in the exercise above that the statement holds for the planar diagrams R_i and R_i' corresponding to the ith Reidemeister move. From this, it is easy to see that the statement holds if D_r and D_r' are (n,n) tangle diagrams obtained by adding some horizontal arcs to a pair of Reidemeister diagrams. Finally, we observe that if D and D' are any two tangles related by a Reidemeister move, then we can factor $D = D_0D_rD_1$ and $D' = D_0D_r'D_1$ where D_r and D_r' are as above. Then

$$\langle D\rangle = \langle D_0\rangle \cdot \langle D_r\rangle \cdot \langle D_1\rangle = \langle D_0\rangle \cdot (\pm q^k\langle D_r'\rangle) \cdot \langle D_1\rangle = \pm q^k\langle D'\rangle. \qquad \square$$

Note that the proof of invariance of the Jones polynomial that we gave in section 1.4 silently used the multiplicativity property of the bracket that we have made explicit here. In order to eliminate the factor of $\pm q^k$, we would need to consider framed tangles, but we will not do this here.

Definition 4.3.5. The multiplication above makes the R-module $V_{n,n}$ into an algebra, which is the *Temperly-Lieb algebra* TL_n.

TL_n is generated by the elementary planar tangles $U_1,\dots,U_n$ shown in Figure 4.1.4. It has a presentation

$$TL_n = \left\langle U_1,\dots U_{n-1} \;\middle|\; \begin{matrix} U_i^2 = (q+q^{-1})U_i \\ U_iU_j = U_jU_i \text{ for } |i-j|>1 \\ U_iU_jU_i = U_i \text{ for } |i-j|=1 \end{matrix} \right\rangle.$$

Exercise 4.3.6. Check that the relations above hold. Write the five simple planar $(3,3)$ tangles as products of the generators U_1, U_2, and compute the effects of multiplying each of these basis elements by U_1 and U_2.

Exercise 4.3.7. Show that there is a homomorphism $\psi_2 : Br_n \to TL_n$ given by

$$\psi_2(\sigma_i) = 1 - qU_i \qquad \psi_2(\sigma_i^{-1}) = 1 - q^{-1}U_i,$$

and that $\langle\sigma\rangle \sim \psi_2(\sigma)$. Compute $\psi_2(\sigma_1\sigma_2\sigma_3)$ and $\psi_2(\sigma_1^n)$.

4.4. Category theory The main goal of this lecture is to categorify the Kauffman bracket introduced in the previous section. First, we review some basic notions from category theory.

Gradings: A category $\mathcal{C}$ is *graded* if there is an invertible functor $q : \mathcal{C} \to \mathcal{C}$, called the *grading shift*. The graded categories we will consider all arise from the following construction.

A category $\mathcal{C}$ is *pre-graded* if there is a function $q : \amalg_{X,Y}\mathrm{Mor}(X,Y) \to \mathbb{Z}$ such that $q(\alpha\circ\beta) = q(\alpha)+q(\beta)$ whenever the composition makes sense. The category $\mathcal{T}_{n,m}$ is an important example. For $T, T' \in \mathcal{T}_{n,m}$ and $\Sigma : T \to T'$, we define $q(\Sigma) = \chi(\Sigma) - (m+n)/2$. If $T_0, T_1, T_2 \in \mathcal{T}_{n,m}$ and $\Sigma_0 : T_0 \to T_1$, $\Sigma_1 : T_1 \to T_2$, then

$$\chi(\Sigma_0\circ\Sigma_1) = \chi(\Sigma_0)+\chi(\Sigma_1)-\chi(T_1) = \chi(\Sigma_0)+\chi(\Sigma_1)-(n+m)/2$$

by additivity of the Euler characteristic. It follows that $q(\Sigma_0 \circ \Sigma_1) = q(\Sigma_0) + q(\Sigma_1)$.

If $\mathcal{C}$ is pre-graded, we construct a graded category $\mathcal{C}^{gr}$ as follows. Objects of $\mathcal{C}^{gr}$ are pairs (X, n) where X is an object of $\mathcal{C}$ and $n \in \mathbb{Z}$. and we then define

$$\mathrm{Mor}_{\mathcal{C}^{gr}}((X,n),(Y,m)) = \{\alpha \in \mathrm{Mor}_{\mathcal{C}}(X,Y) \mid q(\alpha) = n - m\}.$$

The grading shift functor is given on objects by $q((X,n)) = (X, n+1)$ and on morphisms by $q(\alpha) = \alpha$.

Exercise 4.4.1. Check that $\mathcal{C}^{gr}$ is a category and that q is a functor.

If $\mathcal{C}$ is a pre-graded cateory, we will denote the object (X, n) of $\mathcal{C}^{gr}$ by $q^n X$.

Additive Categories: The category $\mathcal{C}$ is *additive* if

1) For all $X, Y \in \mathrm{Obj}(\mathcal{C})$, $\mathrm{Mor}(X, Y)$ is a $\mathbb{Z}$-module;
2) composition of morphisms is bilinear; and
3) there is an object $X \oplus Y \in \mathrm{Obj}(\mathcal{C})$ satisfying the usual categorical definition of direct sum.

If $\mathcal{C}$ is an additive category and $X, Y \in \mathrm{Obj}(\mathcal{C})$, we write $\mathrm{Hom}(X, Y)$ in place of $\mathrm{Mor}(X, Y)$. Let $X_1, \ldots, X_r$ be a set of objects of $\mathcal{C}$. If every object of $\mathcal{C}$ is isomorphic to a finite direct sum of the C_i's, we say that $X_1, \ldots, X_r$ *generate* $\mathcal{C}$.

More generally, if $\mathcal{C}$ is graded, we say that $X_1, \ldots, X_r$ generate $\mathcal{C}$ if any object is isomorphic to a finite direct of shifted copies of the X_i and write

$$\mathrm{Hom}^q(X, Y) = \bigoplus_{i \in \mathbb{Z}} \mathrm{Hom}(X, q^i Y).$$

For any category $\mathcal{C}$, we have an additive category $\mathrm{Add}(\mathcal{C})$ with objects given by finite formal direct sums $\bigoplus_{j=1}^n X_n$, $X_j \in \mathrm{Obj}\,\mathcal{C}$, morphisms by matrices of formal linear combinations of morphisms in $\mathcal{C}$, and composition by multiplication of matrices. More precisely, if G_{ij} is the free $\mathbb{Z}$-module generated by $\mathrm{Mor}(X_j, Y_i)$, then we have

$$\mathrm{Hom}(\bigoplus_{j=1}^n X_j, \bigoplus_{i=1}^m Y_i) = \{m \times n \text{ matrices } [\alpha_{ij}] \mid \alpha_{ij} \in G_{ij}\}.$$

Field Coefficients: If $\mathcal{C}$ is an additive category and A is a $\mathbb{Z}$-module, we define the category $\mathcal{C} \otimes A$ to be the category whose objects are the objects of $\mathcal{C}$ and whose morphism spaces are given by $\mathrm{Hom}_{\mathcal{C}\otimes A}(X, Y) = \mathrm{Hom}_{\mathcal{C}}(X, Y) \otimes_{\mathbb{Z}} A$. A case of particular interest to us is when $A = \mathbb{F}$ is a field.

Quotient Categories: Suppose $\mathcal{C}$ is an additive category. An *ideal* $\mathcal{I}$ in $\mathcal{C}$ consists of a subgroup $\mathcal{I}_{XY} \subset \mathrm{Hom}(X, Y)$ for each X and Y in $\mathrm{Obj}(\mathcal{C})$ such that for every $\alpha \in \mathrm{Hom}(W, X)$, $\beta \in \mathcal{I}_{XY}$, and $\gamma \in \mathrm{Hom}(Y, Z)$, $\alpha \circ \beta \in \mathcal{I}_{WY}$ and $\beta \circ \gamma \in \mathcal{I}_{YZ}$ hold.

If $\mathcal{I}$ is an ideal in $\mathcal{C}$, the quotient category $\mathcal{C}/\mathcal{I}$ is defined to have the same objects as $\mathcal{C}$ and morphisms given by $\mathrm{Hom}_{\mathcal{C}/\mathcal{I}}(X, Y) = \mathrm{Hom}_{\mathcal{C}}(X, Y)/\mathcal{I}_{XY}$. Note that although the two categories have the same objects, two objects which are not isomorphic in $\mathcal{C}$ may well become isomorphic in $\mathcal{C}'$.

Example 4.4.2. Let $\mathcal{C}$ be the category of chain complexes over a ring R. $\mathcal{C}$ is an additive category. It contains an ideal $\mathcal{J}$ which consists of all chain maps f which are chain homotopic to the 0 map (*i.e.* $f = dh + hd$ for some h.) The quotient $\mathcal{C}/\mathcal{J}$ is the homotopy category of chain complexes over R.

Exercise 4.4.3. Let $\mathcal{P}$ be the category of planar tangles, and let $R = \mathbb{Z}[q^{\pm}1]$. Define **P** to be the quotient of $\mathrm{Add}(\mathcal{P}) \otimes R$ by the ideal generated by the local relation $\bigcirc = q + q^{-1}$. Show that $\mathrm{Hom}_{\mathbf{P}}(X_n, X_m) = V_{n,m}$, and that $\cdot : V_{n,m} \times V_{m,l} \to V_{n,l}$ is composition in **P**.

4.5. The Krull-Schmidt property

Definition 4.5.1. An additive graded category $\mathcal{C}$ is *positively graded* if $\mathcal{C}$ is generated by objects $X_1, X_2, \ldots X_r$ which satisfy

(1) $\mathrm{Hom}(X_i, q^n X_j) = 0$ for any $n < 0$ and all i, j.
(2) $\mathrm{Hom}(X_i, X_j) = 0$ for $i \neq j$.
(3) $\mathrm{Hom}(X_i, X_i) \simeq \mathbb{Z}$ is generated by id_{X_i}.

If $\mathcal{C}$ is positively graded, it has a decreasing sequence of ideals $\mathcal{J}_0 \supset \mathcal{J}_1 \supset \ldots$ where $\mathcal{J}_n$ is generated by morphisms in $\mathrm{Hom}(X_i, q^m X_j)$ for all i, j and every $m > n$. Note that $\mathrm{Hom}(Y, Y') \cap \mathcal{J}_n = 0$ for all but finitely many n, since Y and Y' can each be written as a finite direct sum of the X_i's.

The quotient $\mathcal{C}/\mathcal{J}_1$ is the *associated graded category*; it satisfies

$$\mathrm{Hom}_{\mathcal{C}/\mathcal{J}_1}(q^a X_i, q^b X_j) = \begin{cases} \mathrm{Hom}_{\mathcal{C}}(X_i, X_j) & a = b \\ 0 & a \neq b \end{cases}.$$

Proposition 4.5.2. *Suppose* Y *and* Y' *are objects of a positively graded category* $\mathcal{C}$. *If*

$$Y = \bigoplus_{j=1}^{N} q^{a_j} X_{f(j)} \simeq \bigoplus_{k=1}^{M} q^{b_k} X_{g(k)} = Y',$$

then $N = M$ *and there is a bijection* $\varphi : \{1, \ldots, N\} \to \{1, \ldots, N\}$ *such that* $b_{\varphi(j)} = a_j$ *and* $g(j) = f(\varphi(j))$.

In other words, Y can be uniquely decomposed into a direct sum of shifted copies of X_i. This is known as the *Krull-Schmidt property.*

Proof. First suppose all the a_j's and b_k's are 0. Then $\mathrm{Hom}(Y, Y')$ is a direct sum of algebras $M_{n_i \times m_i}(\mathbb{Z})$, where m_i is the size of the set $f^{-1}(i)$ and n_i is the size of $g^{-1}(i)$. For $\mathrm{Hom}(Y, Y')$ to contain an invertible element, we must have $n_i = m_i$ for all $i = 1, \ldots, r$.

For the general case, let $\overline{Y} = \bigoplus_{n \in \mathbb{Z}} q^n \overline{Y}_n$ and $\overline{Y}' = \bigoplus_{n \in \mathbb{Z}} q^n \overline{Y}'_n$ be the images of Y and Y' in $\mathcal{C}/\mathcal{J}_1$. Then $\mathrm{Hom}(\overline{Y}, \overline{Y}') = \bigoplus_{n \in \mathbb{Z}} \mathrm{Hom}(\overline{Y}_n, \overline{Y}'_n)$. Since $\overline{Y} \simeq \overline{Y}'$, we see that $\overline{Y}_n \simeq \overline{Y}'_n$ for all n. The lemma now follows from the the first case. □

Corollary 4.5.3. *If* $f \in \mathrm{Hom}(Y, Y')$ *and the associated graded morphism* $\overline{f} : \overline{Y} \to \overline{Y}'$ *is an isomorphism, then* f *is an isomorphism.*

Proof. Write $f = f_0 + f_1$, where f_0 consists of those components of f in summands of the form $\mathrm{Hom}(q^n X_i, q^n X_j)$ and f_1 consists of those components in summands $\mathrm{Hom}(q^n X_i, q^m X_j)$ where $n < m$. Then $f_0 = \overline{f}$ is an isomorphism and $f_1 \in \mathcal{J}_1$. Write $g = -f_0^{-1} f_1$, so $f = f_0(\mathrm{id}_Y - g)$. Then $f^{-1} = \left(\sum_{i=0}^{\infty} g^i\right) f_0^{-1}$. Note that $g^i \in \mathcal{J}_i$, so all but finitely many terms of the sum are 0. □

4.6. Chain complexes over a category

Definition 4.6.1. A chain complex over an additive category $\mathcal{C}$ is a sequence

$$\cdots \xrightarrow{d_{i+2}} C_{i+1} \xrightarrow{d_{i+1}} C_i \xrightarrow{d_i} C_{i-1} \xrightarrow{d_{i-1}} \cdots$$

where each $C_i \in \mathrm{Obj}(\mathcal{C})$, $d_i \in \mathrm{Hom}(C_i, C_{i-1})$, and $d_{i-1} \circ d_i = 0$ for all i.

The usual definitions of chain map, chain homotopy, and chain homotopy equivalence can be easily generalized to complexes over $\mathcal{C}$.

In general, it can be hard to tell whether two chain complexes over an arbitrary category $\mathcal{C}$ are homotopy equivalent. If $\mathcal{C}$ is not abelian, we cannot take the homology of a complex over it. However when $\mathcal{C}$ is positively graded, there is an effective way of addressing this question.

If (C, d) is a chain complex over a positively graded category $\mathcal{C}$, define the *associated graded complex* $(\overline{C}, \overline{d})$ to be the image of (C, d) in the associated graded category $\mathcal{C}/\mathcal{J}_1$.

Definition 4.6.2. Suppose that (C, d) is a chain complex over a positively graded category $\mathcal{C}$, and that $(\overline{C}, \overline{d})$ is the associated graded complex. We say C is *minimal* if $\overline{d} = 0$.

Proposition 4.6.3. *Suppose C and C' are finitely generated minimal chain complexes over a positively graded category $\mathcal{C}$. If C and C' are chain homotopy equivalent, they are isomorphic.*

Proof. Suppose $f : C \to C'$, $g : C' \to C$ are chain maps inducing a chain homotopy equivalence. Then $\overline{f} : \overline{C} \to \overline{C}'$, $\overline{g} : \overline{C}' \to \overline{C}$ induce a chain homotopy equivalence in the associated graded category. Since C and C' are minimal, $\overline{d} = \overline{d}' = 0$, and $\overline{f}$ and $\overline{g}$ are isomorphisms. Since C and C' are finitely generated, Corollary 4.5.3 implies that the map $f_i : C_i \to C'_i$ is an isomorphism for each i. Hence f is an isomorphism of chain complexes. □

Starting with an arbitrary complex C, we can try to find a minimal complex by repeatedly applying the following lemma, whose proof is left as an exercise to the reader.

Lemma 4.6.4 (Cancellation). *If $\iota \in \mathrm{Hom}(A, A)$ is an isomorphism, the complex*

$$\to C_{n+1} \xrightarrow{\begin{bmatrix} d_{n+1} \\ f \end{bmatrix}} C'_n \oplus A \xrightarrow{\begin{bmatrix} \alpha & \beta \\ \gamma & \iota \end{bmatrix}} C'_{n-1} \oplus A \xrightarrow{\begin{bmatrix} d_{n-1} & g \end{bmatrix}} C_{n-2} \to$$

is homotopy equivalent to the complex

$$\to C_{n+1} \xrightarrow{d_{n+1}} C'_n \xrightarrow{\alpha-\beta\iota^{-1}\gamma} C'_{n-1} \xrightarrow{d_{n-1}} C_{n-2} \to$$

We describe the operation of passing from the larger complex to the smaller one by saying that we *cancel* the two objects isomorphic to A.

Proposition 4.6.5. *Suppose $\mathcal{C}$ is a positively graded category with coefficients in a field $\mathbb{F}$. Then any finitely generated complex over $\mathcal{C}$ is homotopy equivalent to a minimal complex.*

Proof. By induction on the total number of summand in C. Since $\mathcal{C}$ has coefficients in a field, any nonzero component of d which belongs to a summand of the form $\mathrm{Hom}(q^n X_i, q^n X_i)$ is an isomorphism. If $\overline{d} \neq 0$, we can find such a component of d. Canceling produces a homotopy equivalent complex which has fewer summands. □

In practice, many complexes considered here can be reduced to minimal form by repeated application of Lemma 4.6.4 over $\mathbb{Z}$, with no need to pass to a field.

4.7. The Cube of Resolutions We now turn our attention to the problem of categorifying the Jones polynomial of a tangle.

Definition 4.7.1. Suppose that $D_0, D_1 \in \mathcal{P}_{n,m}$ are planar n-tangles. A *marked cobordism* $\mathcal{S} : D_0 \to D_1$ is a pair $\mathcal{S} = (S, P)$ where $S \hookrightarrow \mathbb{R} \times I \times I$ is a cobordism from D_0 to D_1, and $P \subset \mathrm{int}\, S$ is a finite set of *marked points*. Marked cobordisms $(S_1, P_1), (S_2, P_2) : D_0 \to D_1$ are said to be equivalent if there is a homeomorphism $\varphi : \mathbb{R} \times I \times I$ which takes (S_1, P_1) to (S_2, P_2) and is the identity on $\partial\mathbb{R} \times I \times I$.

Let $\mathcal{C}$ be the category whose objects are planar (n, m) tangles and whose morphisms are marked cobordisms between them. $\mathcal{C}$ is pre-graded, with

$$q(S, P) = \chi(S) - 2|P| - (m+n)/2.$$

We define $\mathbf{Cob}_{n,m} = \mathcal{C}^{gr}$. Just as for tangles in $\mathbb{R}^2 \times I$, there is a 2-category **Cob** whose 1-morphisms are planar tangles and whose 2-morphisms are marked cobordisms between them. In particular, we have horizontal and vertical compositions of marked tangles whose definitions are completely analogous to those for planar tangles.

Given a (m, n) tangle diagram D, we can form the cube of resolutions of D. We label each vertex v of the cube with its associated resolution $q^{|v|}D_v$, which is an object of $\mathbf{Cob}_{n,m}$. Similarly, we label each edge $e : v_0 \to v_1$ with the morphism $\mathcal{S}_e = (S_e, \emptyset) : q^{|v_0|}D_{v_0} \to q^{|v_1|}D_{v_1}$. Here S_e is the standard saddle cobordism in a neighborhood of the crossing, and the product cobordism outside of it. We have $q(\mathcal{S}_e) = \chi(S_e) - (n+m)/2 = -1 = |v_0| - |v_1|$, so $\mathcal{S}_e$ is a morphism in the graded category $\mathbf{Cob}_{n,m}$.

To build a cochain complex $C(D)$ from this cube, we pass to the additive category $\mathrm{Add}(\mathbf{Cob}_{m,n})$, in which we are allowed to take formal direct sums of objects and formal linear combination of morphisms.

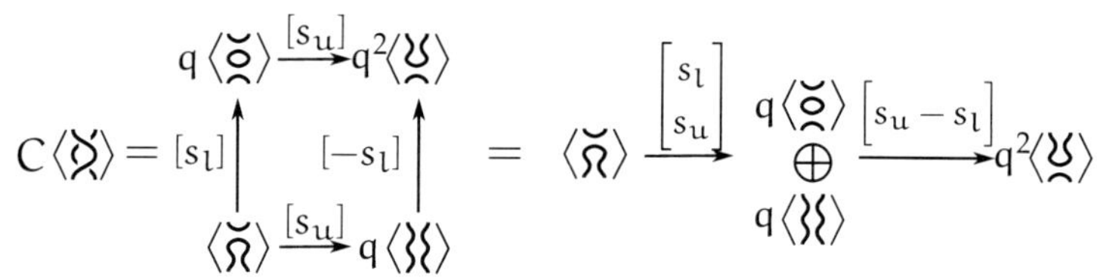

Figure 4.7.2. An example of a complex $C(D)$

Definition 4.7.3. At the level of objects, the complex $C(D)$ is given by

$$C^i(D) = \bigoplus_{|v|=i} q^i D_v,$$

The differential $d_i : C^i(D) \to C^{i+1}(D)$ will be a matrix whose columns are labeled by vertices with $|v| = i$ and whose rows are labeled by vertices with $|v| = i+1$. If $|v_0| = i$ and $e : v_0 \to v_1$, the corresponding entry in the matrix for d_i will be $(-1)^{\sigma(e)}\mathcal{S}_e$, where $\sigma(e)$ is a sign assignment as in Lemma 3.3.1; if there is no such edge, the entry will be the 0 morphism. The fact that $d^2 = 0$ follows exactly as in the closed case.

Example 4.7.4. Figure 4.7.2 shows the complex $C(D)$, where D is the planar diagram at the left of the figure. The morphisms s_u and s_l are handle attachments at the site of the upper and lower crossings.

Composition: Horizontal composition gives a functor $\mathbf{Cob}_{n,m} \times \mathbf{Cob}_{m,l} \to \mathbf{Cob}_{n,l}$. We extend this functor bilinearly over the ring $R = \mathbb{Z}[q^{\pm 1}]$ to get a functor $\mathcal{H} : \mathbf{Cob}_{n,m} \otimes_R \mathbf{Cob}_{m,l} \to \mathbf{Cob}_{n,l}$ which on objects is given by $\mathcal{H}(D \otimes D') = DD'$.

Proposition 4.7.5. $C(DD') = \mathcal{H}(C(D) \otimes C(D'))$.

Proof. At the level of objects, this follows just as in the proof of Proposition 4.3.3. For the differential note that an edge in the cube of resolutions for DD' is either of the form $e : v_0v_0' \to v_1v_0'$ where $e : v_0 \to v_1$ is an edge of the cube of resolutions for D, or of the form $e' : v_0v_0' \to v_0v_1'$, where $e' : v_0' \to v_1'$ is an edge in the cube of resolutions for D'. Edges of the first and second types contribute terms respectively of the form $d \otimes 1$, and $1 \otimes d'$ to the differential. □

4.8. Bar-Natan's category At this point, we have effectively generalized most of the steps in the definition of Khovanov homology to tangles, but the complex we have defined is not invariant (even up to homotopy equivalence) under the Reidemeister moves. It remains to find an appropriate generalization of the final step: applying the TQFT $\mathcal{A}$. It turns out that the right thing to do is to pass to a certain quotient of the category $\mathrm{Add}(\mathbf{Cob}_{n,m})$.

Definition 4.8.1. *Bar-Natan's category* $\mathcal{C}^{BN}_{n,m}$ is the quotient of $\mathrm{Add}(\mathbf{Cob}_{n,m})$ by the ideal generated by the following local relations. Suppose $\mathcal{S} = (S, P)$ is a morphism in $\mathbf{Cob}_{n,m}$.

- (Sphere relations) If a component of S is a sphere with one marked point, $\mathcal{S} = \mathcal{S}'$, where $\mathcal{S}'$ is the cobordism obtained by deleting this component. If some component of S is a sphere with any number of marked points other than one, $\mathcal{S}$=0.
- (Neck-cutting relation) If $A \subset S$ is an annulus, let $\mathcal{S}_r$ (resp. $\mathcal{S}_l$) be the diagram obtained by removing A replacing it with a pair of disks, and putting a marked point on the left-hand (resp. right-hand) disk. Then $\mathcal{S} = \mathcal{S}_r + \mathcal{S}_l$.

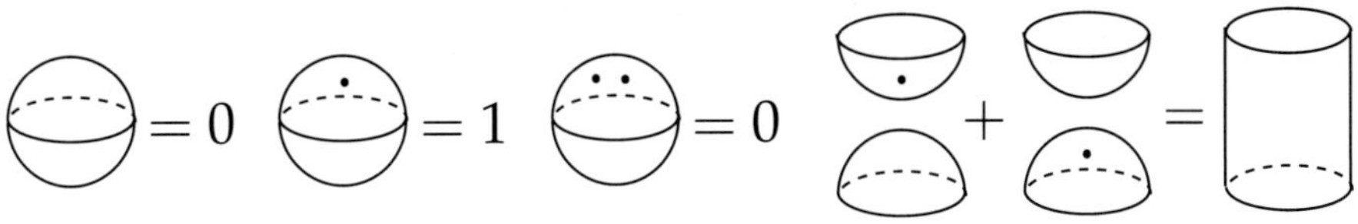

Figure 4.8.2. Relations in $\mathcal{C}^{BN}_{n,m}$

Exercise 4.8.3. Check that these relations respects the q-grading on morphisms.

Definition 4.8.4. Suppose $D \in \mathcal{D}_{n,m}$. We define the *Khovanov bracket* $\mathrm{CKh}(D)$ to be the image of $C(D)$ under the quotient functor $\mathrm{Add}(\mathbf{Cob}_{n,m}) \to \mathcal{C}^{BN}_{n,m}$.

Using the additional relations in $\mathcal{C}^{BN}_{n.m}$, we will prove that the homotopy type of $\mathrm{CKh}(D)$ is invariant under Reidmeister moves. As an example of how the relations are applied we prove

Lemma 4.8.5. *If any component of* S *contains two or more elements of* P*, then* $\mathcal{S} = 0$ *in* $\mathcal{C}^{BN}$.

Proof. We can find an embedded disk in S containing two marked points. Apply the neck cutting relation to the boundary of this disk. Both the resulting cobordisms contain a sphere with more than one marked point. □

At first glance, the relations in Definition 4.8.1 may seem somewhat mysterious. In fact, they are a natural extension of the TQFT $\mathcal{A}$ to the category of tangles. To see this, consider the category $\mathbf{Cob}_{0,0}$, whose objects are closed 1-manifolds embedded in $\mathbb{R} \times I$, and whose morphisms are marked cobordisms between them. Note that any marked cobordism can be written as a composition of ordinary cobordisms (without markings) and product cobordisms in which one of the product cylinders is marked with a single dot.

We define a functor $\overline{\mathcal{A}} : \mathbf{Cob}_{0,0} \to \mathbb{Z}$-mod. $\overline{\mathcal{A}}$ agrees with $\mathcal{A}$ on objects and unmarked morphisms. If $X_i : D \to D$ is a product cobordism marked with a single point on the ith cylinder,

$$\overline{\mathcal{A}}(X_i) : \mathcal{A}(D) \to \mathcal{A}(D)$$

is given by multiplication by $\mathbf{x}$ on the ith tensor factor of $\mathcal{A}(S^1)$.

Exercise 4.8.6. Check that $\overline{\mathcal{A}}$ descends to a functor $\mathcal{C}^{BN}_{0,0} \to \mathbb{Z}$-mod.

We now consider the structure of $\mathcal{C}^{BN}_{n,m}$.

Proposition 4.8.7 (Eliminating Closed Circles). *Suppose* $D \in \mathcal{P}_{n,m}$ *contains a small closed circle* c*, and let* $\widetilde{D}$ *be the same diagram, but with* c *removed. In* $\mathcal{C}^{BN}_{n,m}$, D *is isomorphic to* $q\widetilde{D} \oplus q^{-1}\widetilde{D}$.

The proposition categorifies the circle-removal relation $\langle D \rangle = (q+q^{-1})\langle \widetilde{D} \rangle$ for the Kauffman bracket.

Proof. We give the proof in the case where $\widetilde{D}$ is empty; the argument in the general case is the same. There are morphisms $\eta : D \to \widetilde{D}$ and $\varepsilon : \widetilde{D} \to D$ given by addition of a 2-handle and 0-handle, respectively. Define morphisms

$$\varphi : D \to q\widetilde{D} \oplus q^{-1}\widetilde{D} \quad \text{and} \quad \psi : q\widetilde{D} \oplus q^{-1}\widetilde{D} \to D$$

by

$$\varphi = \begin{pmatrix} \eta X \\ \eta \end{pmatrix} \qquad \psi = \begin{pmatrix} \varepsilon & X\varepsilon \end{pmatrix}$$

where $X : D \to D$ is the product cobordism with a single marked point. Then

$$\varphi\psi = \begin{pmatrix} \eta X \varepsilon & \eta X^2 \varepsilon \\ \eta\varepsilon & \eta X \varepsilon \end{pmatrix}$$

The cobordism $\eta\varepsilon$ is a closed sphere, so the sphere relations imply that the diagonal entries of this matrix are 1, and the off-diagonal entries are 0. Similarly

$$\psi\varphi = \varepsilon\eta X + X\varepsilon\eta = 1_D$$

by the neck-cutting relation. □

Corollary 4.8.8. $\mathcal{C}^{BN}_{n,m}$ *is generated (as an additive category) by simple planar tangles.*

Figure 4.8.9. Trading genus for marked points

Next, we consider the morphisms of $\mathcal{C}^{BN}_{n,m}$. Let $\mathcal{S} : D_0 \to D_1$ be a marked cobordism between simple planar tangles. We say that $\mathcal{S}$ is *simple* if every component of $\mathcal{S}$ is a disk with either 0 or 1 marked point.

Now suppose that $\mathcal{S} : D_0 \to D_1$ is an arbitrary marked cobordism. Applying the neck-cutting relation as shown in Figure 4.8.9, we see that we can reduce the genus of any component of $\mathcal{S}$ at the cost of adding marked points. Similarly, if any component of $\mathcal{S}$ has more than one boundary component, we can use the neck-cutting relation to separate these components. Finally, we can use the sphere relations to eliminate any closed components. What remains is a linear combination of simple cobordisms.

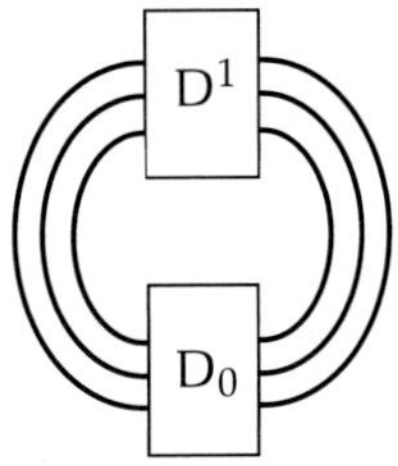

Figure 4.8.10. The closed diagram $\overline{D^1 D_0}$. The planar diagram D^1 is obtained by reflecting D_1 across a horizontal line.

If $\mathcal{S} = (S, P)$ is a simple marked cobordism, the components of S are determined up to isotopy by

$$\partial S = D_0 \times 0 \cup X_n \times 0 \times I \cup D_1 \times 1 \cup X_m \times 1 \times I \subset \partial(\mathbb{R} \times I \times I).$$

We can identify $\partial(\mathbb{R} \times I \times I)$ with $\mathbb{R}^2 - 0$. Under this identification, ∂S is identified with the diagram $\overline{D^1 D_0}$ shown in Figure 4.8.10. In particular, the number of disks is given by the number of components in $\overline{D^1 D_0}$, and each disk can have either 0 or 1 marked point. In summary, we have proved

Proposition 4.8.11. *If D_0 and D_1 are simple planar tangles, $\mathrm{Hom}^q(D_0, D_1)$ is a free $\mathbb{Z}$-module of rank 2^r, where r is the number of components of $\overline{D^1 D_0}$.*

The simple marked cobordism with maximal grading is $(S, \emptyset)$, for which we have $q(S, \emptyset) = r - (m+n)/2$. It follows that as a graded module, $\mathrm{Hom}^q(D_0, D_1)$ is naturally isomorphic to $q^{(m+n)/2}\mathcal{A}(\overline{D^1 D_0})$. In particular, $\mathrm{Hom}(D_0, q^i D_1) = 0$ for $i < r - (m+n)/2$. It is easy to see that $r \leqslant (m+n)/2$, with equality if and only if $D_0 = D_1$, and that $\mathrm{Hom}(D_0, D_0) \simeq \mathbb{Z}$ is generated by id_{D_0}. Hence we have proved

Corollary 4.8.12. *$\mathcal{C}^{BN}_{n,m}$ is positively graded.*

This corollary has two powerful consequences which make computations with $CKh(D)$ feasible. First, $\mathcal{C}^{BN}_{n,m}$ has the Krull-Schmidt property. Second, by Proposition 4.6.3, any two minimal representatives of $CKh(D)$ are isomorphic.

4.9. How to Compute Given a tangle diagram D, we can simplify $CKh(D)$ by the following steps. First, we apply the isomorphism of Proposition 4.8.7 to eliminate any objects of $CKh(D)$ which contain closed circles. The result is a chain complex whose objects are direct sums of simple tangles. Next, we use the relations in Definition 4.8.1 to express all morphisms in the resulting complex in terms of elementary morphisms, as in the proof of Proposition 4.8.11. Finally, we repeatedly apply the Cancellation Lemma to eliminate any components of d which correspond to the identity morphism.

As an example of this process, we prove that CKh is invariant under the first Reidemeister move. Let R_1 and R_1' be the two diagrams for the Reidemeister I move shown in Figure 1.1.6.

Proposition 4.9.1. $\mathrm{CKh}(R_1) \sim \mathrm{CKh}(R_1')$, *where* $\sim$ *indicates that the two complexes are homotopy equivalent up to an overall shift of the homological and* q*-gradings.*

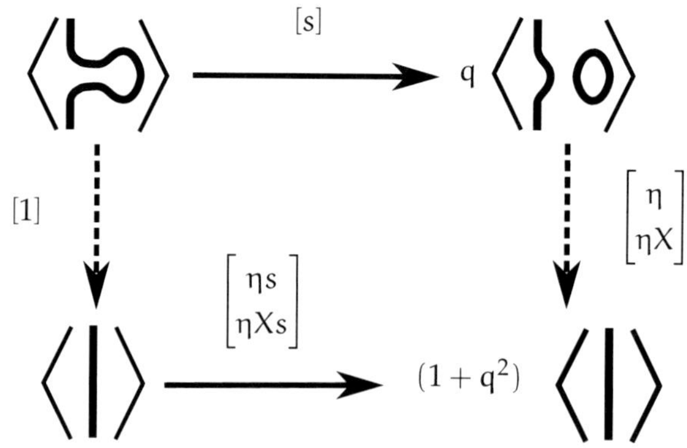

Proof. The complex $\mathrm{CKh}(R_1')$ is shown in the upper row of the figure. The map s is a 1-handle cobordism. The dashed arrows give an isomorphism between the complexes in the upper and lower rows. The morphism ηs is a 1-handle followed by a cancelling 2-handle, so it is isotopic to the identity. Hence we can cancel $\langle\,|\,\rangle$ on the left with $\langle\,|\,\rangle$ on the right, leaving only a summand of $q^2\langle\,|\,\rangle$ in homological grading 1. This is $\mathrm{CKh}(R_1)$ with the homological grading shifted by 1 and the q-grading shifted by q^2. □

Exercise 4.9.2. Show that CKh is invariant under the mirror image of this Reidemeister move.

Invariance under the Reidemeister II move is similar.

Proposition 4.9.3. $\mathrm{CKh}(R_2) \sim \mathrm{CKh}(R_2')$.

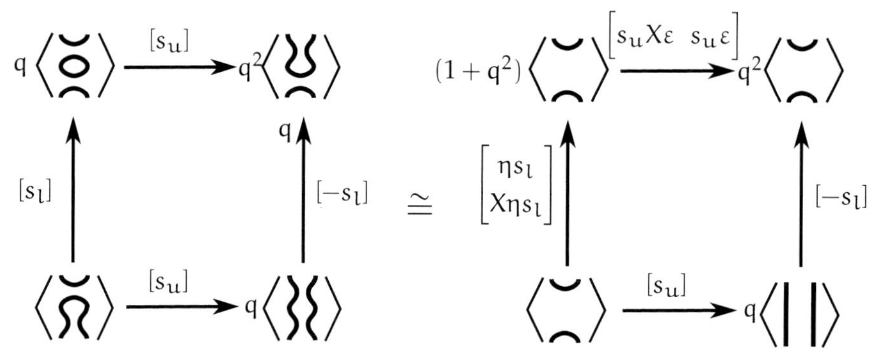

Proof. The complex $\mathrm{CKh}(R_2')$ is shown on the left in the figure above. The subscripts s_u and s_l indicate whether the 1-handle should be added to the upper or lower half of the figure. After eliminating the extra circle in the tangle at the upper left, we arrive at the complex on the right-hand side of the figure. The left-hand side of the complex is the same as the complex we considered for the Reidemeister 1 move, so we can cancel the two objects labeled $\langle\asymp\rangle$. Doing this does not affect the morphism on the upper arrow, so we can cancel the two objects labeled $q^2\langle \asymp \rangle$, leaving only $\rangle\langle$ in the lower-right. □

When we compute CKh of a tangle represented by a braid, it is often simpler to represent tangles by elements of TL_n, rather than drawing pictures. Under this correspondence, we have

$$\mathrm{CKh}(\sigma_1) = 1 \to qU_1 \quad \text{and} \quad \mathrm{CKh}(\sigma_1^{-1}) = U_1 \to q1.$$

Example 4.9.4. Let us compute CKh of a full twist on two strands. Using Proposition 4.7.5, we see that

$$\mathrm{CKh}(\sigma_1)^2 = \mathrm{CKh}(\sigma_1) \otimes \mathrm{CKh}(\sigma_1) = (1 \to qU_1) \otimes (1 \to qU_1)$$

Expanding this out gives

$$\begin{array}{ccc} qU_1 & \xrightarrow{s_u} & q^2U_1^2 \\ {\scriptstyle s_l}\uparrow & & \uparrow{\scriptstyle -s_l} \\ 1 & \xrightarrow{s_u} & qU_1 \end{array} \qquad\qquad \begin{array}{ccc} qU_1 & \xrightarrow{1\oplus X_u} & (q+q^3)U_1 \\ {\scriptstyle s_l}\uparrow & & \uparrow{\scriptstyle 1\oplus X_l} \\ 1 & \xrightarrow{s_u} & qU_1 \end{array}$$

where all the maps have already been computed in the proofs of Propositions 4.9.1 and 4.9.3. Cancelling the identity map in the top row, we obtain

$$\mathrm{CKh}(\sigma_1^2) \sim 1 \xrightarrow{s} qU_1 \xrightarrow{X_l - X_u} q^3U_1.$$

Since the saddle cobordism has only one component, $X_u s = X_l s$ and $d^2 = 0$ as it should.

We could do the same thing to compute $\mathrm{CKh}(\sigma_1^3)$, but the following lemma makes the computation a little easier:

Lemma 4.9.5. *If* C, C' *and* C_1 *are complexes and* $C \sim C'$, *then* $C \otimes C_1 \sim C' \otimes C_1$.

The proof is left to the reader. Writing

$$\mathrm{CKh}(\sigma_1^3) = \mathrm{CKh}(\sigma_1^2) \otimes \mathrm{CKh}(\sigma_1)$$

and applying the lemma, we see that $\mathrm{CKh}(\sigma_1^3)$ is equivalent to the complex below:

$$\begin{array}{ccccc} qU_1 & \longrightarrow & (q+q^3)U_1 & \longrightarrow & (q^3+q^5)U_1 \\ \uparrow & & \uparrow & & \uparrow \\ 1 & \longrightarrow & qU_1 & \longrightarrow & q^3U_1 \end{array}$$

The complex in the top row gives CKh(D), where D is U_1 with two Reidemeister one moves applied to it. By R1 invariance, we see that we can cancel all terms in the top row except q^5U_1. The resulting complex is

$$\mathrm{CKh}(\sigma_1^3) \sim 1 \xrightarrow{s} qB_1 \xrightarrow{X_l - X_u} q^3U_1 \xrightarrow{X_l + X_u} q^5U_1.$$

This technique was used to great effect by Bar-Natan and Green in their program [4], which made a revolutionary advance in ability to compute Khovanov homology.

Exercise 4.9.6. Show that in general $\mathrm{CKh}(\sigma_1^n) \sim$

$$1 \xrightarrow{s} qU_1 \xrightarrow{X_l - X_u} q^3U_1 \xrightarrow{X_l + X_u} q^5U_1 \xrightarrow{X_l - X_2} \dots q^{2n-3}U_1 \xrightarrow{X_l \pm X_u} q^{2n-1}U_1$$

where the boundary map alternates between $X_l - X_u$ and $X_l + X_u$. By considering one closure of the tangle σ_1^n, compute the Khovanov homology of the torus knot $T(2, 2k+1)$ and the torus link $T(2, 2k)$. Show that the complex obtained by taking the other closure of the tangle correctly computes $\mathrm{Kh}(\bigcirc)$.

We can now check the invariance of CKh under the third Reidemeister move.

Proposition 4.9.7. $\mathrm{CKh}(\sigma_1\sigma_2\sigma_1) = \mathrm{CKh}(\sigma_1\sigma_2\sigma_1)$.

Proof. We compute

$$\begin{aligned}\mathrm{CKh}(\sigma_1\sigma_2) &= (1 \to qU_1) \otimes (1 \to qU_2)\\ &= 1 \to q(U_1 + U_2) \to q^2U_1U_2\end{aligned}$$

where every matrix entry in each boundary map is a saddle cobordism. Thus $\mathrm{CKh}(\sigma_1\sigma_2\sigma_1)$ has the form shown below:

$$\begin{array}{ccccc} qU_1 & \longrightarrow & (q+q^3)U_1 + q^2U_2U_1 & \longrightarrow & q^3U_1 \\ \uparrow & & \uparrow & & \uparrow \\ 1 & \longrightarrow & q(U_1+U_2) & \longrightarrow & q^3U_1U_2 \end{array}$$

Consider the morphisms $F : qU_1 \to qU_1$ and $G : q^3U_1 \to q^3U_1$ appearing in the upper row. The morphism F has already been studied in our calculation of $\mathrm{CKh}(\sigma_1^2)$, where we saw it is the identity. G is given by the composition of the map $\varepsilon : q^3U_1 \to q^2D$, where D is the diagram to the right with the 1-handle cobordism indicated by the dotted line. This is the addition of a 0-handle followed by a cancelling 1-handle, so the composition is isotopic to the identity.

It follows that we can cancel all terms in the upper row except U_1U_2. We are left with a complex

$$1 \to q(U_1 + U_2) \to q^2(U_1U_2 + U_2U_1)$$

in which every possible map is given (up to sign) by a saddle. But this complex is invariant under the reflection which switches U_1 and U_2. Hence the same argument shows it is isomorphic to $\mathrm{CKh}(\sigma_2\sigma_1\sigma_2)$. □

Theorem 4.9.8. *If* D *and* D′ *two diagrams having the same tangle, then we have* $\mathrm{CKh}(D) \sim \mathrm{CKh}(D')$.

Proof. We've already checked the statement for the three pairs of Reidemeister diagrams. The general case follows exactly as in the proof of Proposition 4.3.4, using Proposition 4.7.5 and Lemma 4.9.5. □

4.10. Connections and Further Reading The Khovanov homology of tangles serves as a template for many subsequent developments in categorification. For the beginner, it is worth taking the time to understand it well; Bar-Natan's paper [5] is still a good reference. One important feature which we have not touched on in this section is maps induced by tangle cobordisms. The proof of (projective) functoriality is much easier in this setting [5,41] than if we restrict ourself to the context of link cobordisms.

In a different direction, Ozsváth and Szabó [62,63] have developed a knot Floer analog of the Khovanov complex of a tangle. The underlying categories are more complicated than the Bar-Natan category, and it seems there is still much to be learned by studying them.

Finally, we mention the recent work of Kotelskiy, Watson, and Zibrowius, [47], who represent the Khovanov complex of a tangle as an object in the Fukaya category of the 4-punctured sphere.

5. HOMFLY-PT Homology

The similarity between the skein relations for the Jones and Alexander polynomials is hard to miss, and it's natural to ask if they have a common generalization. In fact, this question is so natural that it was answered by at least eight people working in five different groups (Freyd-Yetter, Hoste-Lickorish, Morton, Ocneanu, and Pryztycki-Traczyk) not long after Jones announced the existence of his new polynomial.

Theorem 5.0.1 (HOMFLY-PT [24] [64]). *There's an invariant which assigns to an oriented link* $L \subset S^3$ *a rational function* $P_L(a, q) \in \mathbb{Z}(a, q)$ *which satisfies the skein relation*

aP () $- a^{-1}P$ () $= (q - q^{-1})P$ () .

The function P_L is known as the (normalized) HOMFLY-PT polynomial. It is completely determined by the skein relation plus the normalization $P(\bigcirc) = 1$. By applying the skein relation to the one-crossing diagram of the unknot, we find that

$$P(\bigcirc\,\bigcirc) = \frac{a - a^{-1}}{q - q^{-1}}$$

and more generally, that

$$P(\bigcirc^n) = \left(\frac{a - a^{-1}}{q - q^{-1}}\right)^{n-1}.$$

Notation: We write

$$\{n\} = \frac{aq^{-n} - a^{-1}q^n}{q - q^{-1}} \quad \text{and} \quad [n] = \frac{q^n - q^{-n}}{q - q^{-1}}.$$

so that $P(\bigcirc^n) = \{0\}^{n-1}$. Note that $\{n\}|_{a=q^k} = [k - n]$. The polynomial $[n]$ is known as *quantum* n.

As with the Jones polynomial, there's also an *unnormalized* version $\overline{P}_L$ which is defined by

$$\overline{P}(L) = P(L)\left(\frac{a-a^{-1}}{q-q^{-1}}\right).$$

Exercise 5.0.2. Use the skein relation to show that $(q-q^{-1})^{|L|}\overline{P}(L) \in \mathbb{Z}[a^{\pm 1}, q^{\pm 1}]$. In particular, if K is a knot, $P(K) \in \mathbb{Z}[a^{\pm 1}, q^{\pm 1}]$.

Our goal in this lecture is to discuss one way of defining P_L (following the method of Jones [36] and Ocneanu, and then outline its categorification, due to Khovanov and Rozansky.

5.1. Braid Closures Given $\sigma \in Br_n$, we can form the *braid closure* $\overline{\sigma}$, which is an oriented link in S^3, as illustrated in the figure above. By sliding σ "around the circle" as shown, we see that $\overline{\sigma\tau} = \overline{\tau\sigma}$; in other words, that the operation of taking the closure is invariant under conjugation in the braid group.

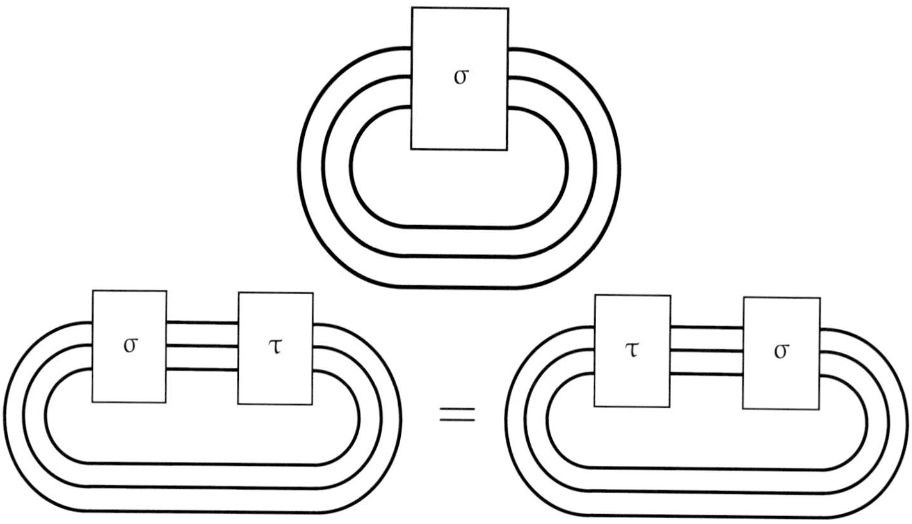

Figure 5.1.1. From the schematic diagram of closure of the braid σ above we see that $\overline{\sigma\tau} = \overline{\tau\sigma}$ below.

The significance of this operation stems from the following two theorems:

Theorem 5.1.2 (Alexander). *Every oriented link* $L \subset S^3$ *can be written as* $L = \overline{\sigma}$ *for some* $\sigma \in Br_n$.

Note that n depends on L, and that neither n nor σ are uniquely determined by L. This failure of uniqueness is described by Markov's theorem:

Theorem 5.1.3 (Markov). *If* $\sigma \in Br_n$, $\sigma' \in Br_{n'}$ *have isotopic closures, then* σ *and* σ' *are related by a sequence of the following moves:*

1) *(Conjugation) Replace* $\sigma \in Br_m$ *with* $\tau\sigma\tau^{-1}$, *where* $\tau \in Br_m$.
2) *(Stabilization) Replace* $\sigma \in Br_m$ *with* $\sigma\sigma_m^{\pm 1} \in Br_{m+1}$ *or vice-versa.*

Next, we recall the Temperly-Lieb algebra TL_n from the previous lecture. If $U \in TL_n$ is a basis element corresponding to a crossingless planar diagram, we can form its closure $\overline{U}$ just as we formed the closure of a braid. We define a map $Tr : TL_n \to \mathbb{Z}[q^{\pm 1}]$ by setting

$$Tr(U) = (q + q^{-1})^{|\overline{U}|}$$

if U is a crossingless planar tangle, and extending linearly to all of TL_n. This map is called the *Jones trace*. Since $\overline{UU'} = \overline{U'U}$, it satisfies the defining property of a trace: $Tr(UU') = Tr(U'U)$.

Using the homomorphism $\Psi_2 : Br_m \to TL_m$ defined in exercise 4.3.7, we see that the unnormalized Jones polynomial can be expressed as

$$\overline{V}(\overline{\sigma}) = Tr\,\Psi_2(\sigma).$$

5.2. The Hecke algebra

Definition 5.2.1. The Hecke algebra of type A_n is the algebra defined over the ring $R = \mathbb{Z}[q^{\pm 1}]$ by the presentation

$$H_n = \left\langle T_1, \dots T_{n-1} \;\middle|\; \begin{array}{c} T_i T_j = T_j T_i \quad |i-j| > 1 \\ T_i T_{i+1} T_i = T_{i+1} T_i T_{i+1} \\ T_i^2 = (q - q^{-1}) T_i + 1 \end{array} \right\rangle.$$

There are obvious homomorphisms,

$$Br_n \xrightarrow{\Psi} H_n \xrightarrow{\rho} \mathbb{Z}[S_n]$$

given by $\Psi(\sigma_i) = T_i$ and $\rho(T_i) = s_i$, where S_n is the nth symmetric group, which has a presentation in terms of elementary transpositions s_i for $i = 1, \dots, n-1$ as

$$S_n = \left\langle s_1, \dots s_{n-1} \;\middle|\; \begin{array}{c} s_i s_j = s_j s_i \quad |i-j| > 1 \\ s_i s_{i+1} s_i = s_{i+1} s_i s_{i+1} \\ s_i^2 = 1 \end{array} \right\rangle.$$

The presentation of H_n above was chosen to make this sequence of homomorphisms obvious. It is also useful to work with the alternate generators $B_i = q - T_i$, with respect to which the presentation has the form

$$H_n = \left\langle B_1, \dots B_{n-1} \;\middle|\; \begin{array}{c} B_i B_j = B_j B_i \quad |i-j| > 1 \\ B_i B_{i+1} B_i - B_i = B_{i+1} B_i B_{i+1} - B_{i+1} \\ B_i^2 = (q + q^{-1}) B_i \end{array} \right\rangle.$$

With this set of generators, it's clear there is a homomorphism $p_2 : H_n \to TL_n$ given by $p_2(B_i) = U_i$, and that the following diagram commutes:

$$\begin{array}{ccc} Br_n & & \\ \downarrow \Psi & \searrow^{\Psi_2} & \\ H_n & \xrightarrow{p_2} & TL_n. \end{array}$$

Here Ψ_2 is basically the map discussed in section 4.3, but with q and q^{-1} switched.

In analogy with the Jones trace, we have the *Jones-Ocneanu trace* on the Hecke algebra:

Theorem 5.2.2 (Oceanu). *n There is a unique family of* R*-linear maps*

$$\mathrm{Tr}_n : H_n \to \mathbb{Z}[a^{\pm 1}, q^{\pm 1}, (q-q^{-1})^{-1}]$$

satisfying the following properties

1) $\mathrm{Tr}_{n+1} BB' = \mathrm{Tr}_{n+1} B'B$
2) $\mathrm{Tr}_{n+1} \iota(B) = \{0\} \mathrm{Tr}_n(B)$
3) $\mathrm{Tr}_{n+1} \iota(B) B_n = \{1\} \mathrm{Tr}_n(B)$
4) $\mathrm{Tr}_0(1) = 1.$

where $\iota : H_n \to H_{n+1}$ *is the inclusion which sends* $B_i \mapsto B_i$.

We can use the Jones-Ocneanu trace to give a definition of the HOMFLY-PT polynomial.

Definition 5.2.3. If $\sigma \in \mathrm{Br}_n$, we define $\overline{P}(\overline{\sigma}) = a^{-w(\sigma)} \mathrm{Tr}_n \Psi(\sigma)$.

By Alexander's theorem, any link in S^3 can be represented as a braid closure. To see that P is well-defined, we use Markov's theorem, which says that it is enough to check that P is invariant under the two Markov moves.

Exercise 5.2.4. Use the properties of the Ocneanu trace to check that $\overline{P}$ is invariant under the Markov moves. Next, use the quadratic relations in H_n to show that if D is any diagram of a braid closure, then $P(D)$ satisfies the HOMFLY-PT skein relation.

5.3. Structure of H_n In this section, we give a proof of Theorem 5.2.2. Along the way, we must establish some facts about the structure of H_n.

Let $\mathcal{W}_n$ be the set of all words in the letters $1, 2, \dots n-1$. If $I = i_1 i_2 \cdots i_k \in \mathcal{W}_n$, we let $\ell(I) = k$ be its length. We write $s(I) = s_{i_1} s_{i_2} \cdots s_{i_k} \in S_n$ and similarly for $B(I) = B_{i_1} B_{i_2} \cdots B_{i_k} \in H_n$. If $j \leqslant i$, We write $I_{j,i} = j(j+1)(j+2)\cdots(i-1)$; note that $I_{i,i}$ is the empty word. It is easy to see that $s(I_{j,i})(i) = j$.

Definition 5.3.1. A *reduced word* is a word of the form $I_{j_n,n} I_{j_{n-1},n-1} \cdots I_{j_1,1}$, where $j_k \leqslant k$ for all $1 \leqslant k \leqslant n$.

Let $\mathcal{R}_n = \{I \in \mathcal{W}_n \mid I \text{ is reduced}\}$.

Lemma 5.3.2. *Every* $s \in S_n$ *has a unique expression as* $s = s(I)$, *where* $I \in \mathcal{R}_n$.

Proof. By induction on n. The result for $n = 1$ is obvious. Suppose the lemma holds for $n-1$, and let $s \in S_n$. Let $j = s(n)$. Then $s(I_{j,n})^{-1} s(n) = n$, so we can view $s(I_{j,n})^{-1} s$ as an element of S_{n-1}. By the induction hypothesis we can write $s(I_{j,n})^{-1} s = s(I')$, where I' is a reduced word in the letters $1, \dots n-2$. Then $s = s(I_{j,n} I')$, and $I_{j,n} I'$ is a reduced word in $n-1$ letters. For uniqueness, note that if $s = s(I)$, where I is a reduced word, we must have $j_k = s(k)$. □

This lemma has an easy graphical interpretation. Any $s \in S_n$ can be represented by a *string diagram*, which is a braid diagram in which we have forgotten the information about overcrossings and undercrossings, and the different strands are allowed to slide freely through each other. The string diagram associated to the elementary permutation s_i is shown in the figure below.

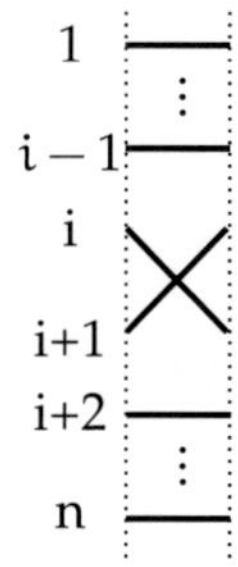

Given a string diagram for a permutation s, suppose the strand that ends at n starts at position i_n. We take this strand and pull it down and to the right, so that the initial part of the word associated to the new diagram reads $i_n i_{n+1} \dots n-1$. The rest of the diagram consists of a diagram on $n-1$ strings together with a single string at the bottom. Repeat this process gives the desired reduced word. For example, the figure below shows that we can write the permutation given in cycle notation by $(134)(2)$ as $s_1 s_2 s_3 s_1$.

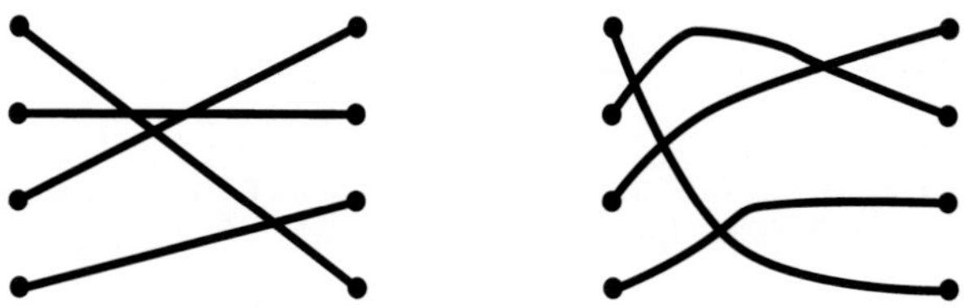

Next, we prove an analogous theorem for H_n. Let

$$\mathcal{B}_{n,k} = \text{span}\,\{B(I) \mid I \in \mathcal{R}_n \text{ and } \ell(I) \leqslant k\} \subset H_n.$$

Lemma 5.3.3. *If* $I \in \mathcal{W}_n$, $B(I) \in \mathcal{B}_{n,\ell(I)}$.

Proof. The proof is a double induction. First, we induct on n. The case $n = 1$ is trivial, so suppose the statement holds for $1, \dots, n-1$. We show that it holds for n by induction on $\ell(I)$. Again, the case where $\ell(I) = 0$ is trivial, so suppose the statement holds whenever $\ell(I) < k$. If $\ell(I) = k$, write $I = I' i_k$. By the induction hypothesis, $B(I') \in \mathcal{B}_{n,k-1}$, so it suffices to show that if I is a reduced word of length $k-1$, then $B(I)B_{i_k} \in \mathcal{B}_{n,k}$.

Now suppose that I is a reduced word of length $k-1$ and write $I = I_{i_n,n} I_1$, where $I_1 \in \mathcal{R}_{n-1}$. If $i_k < n-1$, then by the induction hypothesis

$$B(I_1)B_{i_k} = \sum_{J \in \mathcal{R}_{n-1}, \ell(J) \leqslant \ell(I_1 i_k)} c_J B(J).$$

Now if $J \in \mathcal{R}_{n-1}$, $I_{i_n,n}J \in \mathcal{R}_n$, and if $\ell(J) \leqslant \ell(I_1 i_k)$, then $\ell(I_{i_n,n}J) \leqslant \ell(Ii_k)$. Hence

$$B(I)B_{i_k} = \sum_J c_J B(I_{i_n,n}J) \in \mathcal{B}_{n,k}.$$

Thus it remains to consider the case where $i_k = n-1$. In this case, write $I = I_{i_n,n}I_{i_{n-1},n-1}I''$, where $I'' \in \mathcal{R}_{n-2}$. If I'' is nonempty, we can commute $i_k = n-1$ with the last letter of I'' to write $B(I)B_{i_k} = B(\tilde{I})B_j$ where $j < n-1$. By the inner induction hypothesis, $B(\tilde{I}) \in \mathcal{B}_{n,k-1}$, so we have reduced to the case of the previous paragraph.

We now suppose that the word I'' is empty. In this case $B(I)$ can be written in one of the following forms: a) $B(I) = B(I_{i_{n-1},n-1})$, b) $B(I) = B(I_{i_n,n})$, or c) $B(I) = B(I_0)B_{n-1}B_{n-2}$, where $I_0 \in \mathcal{W}_{n-2}$. In case a), $B(I)B_{n-1} = B(I_{i_{n-1},n})$, and in case b) $B(I)B_{n-1} = (q+q)^{-1}B(I)$. Finally, in case c)

$$B(I)B_{n-1} = B(I_0)B_{n-2}B_{n-1}B_{n-2} - B(I_0)B_{n-2} + B(I_0)B_{n-1}.$$

The first term on the right is in $\mathcal{B}_{n,k}$ by the argument in the second paragraph, and the other two terms are in $\mathcal{B}_{n,k}$ by the induction hypothesis. This proves the claim. □

Proposition 5.3.4. *The set* $\mathcal{B}_n = \{B(I) \mid I \in \mathcal{R}_n\}$ *is a basis for* H_n.

Proof. By Lemma 5.3.2, $\mathcal{B}_n$ contains exactly $|S_n| = n!$ elements, and it spans H_n by Lemma 5.3.3. On other hand, there is a surjective homomorphism from, H_n to $\mathbb{Z}[S_n]$, so any basis for H_n must contain at least $n!$ elements. □

Proof of Theorem 5.2.2. Suppose Tr_n has been defined; we will inductively construct Tr_{n+1}. We first define Tr_{n+1} on the elements of the basis $\mathcal{B}_{n+1}$; by linearity, this determines Tr_{n+1} on all of H_{n+1}. Now every element of $\mathcal{B}_{n+1}$ is either a) of the form $a \in H_n$ or b) of the form aB_nb, where $a, b \in H_n$. (For simplicity, we have dropped ι from the notation.) In order to satisfy properties 1)-3), we must have $\mathrm{Tr}_{n+1}\, a = \{0\}\,\mathrm{Tr}_n\, a$ and $\mathrm{Tr}_{n+1}\, aB_nb = \{1\}\,\mathrm{Tr}_n(ab)$. Hence Tr_{n+1} is uniquely determined.

We define Tr_{n+1} on $\mathcal{B}_{n+1}$ by these relations and extend linearly to H_{n+1}. With this definition, it is easy to see that $\mathrm{Tr}_{n+1}\, a = \{0\}\,\mathrm{Tr}_n\, a$ whenever $a \in H_n$, and $\mathrm{Tr}_{n+1}\, aB_nb = \{1\}\,\mathrm{Tr}_n\, ab$ whenever $a, b \in H_n$. It therefore remains to check that $\mathrm{Tr}_{n+1}\, xy = \mathrm{Tr}_{n+1}\, yx$ for $x, y \in H_{n+1}$. To do this, it is enough to check that $\mathrm{Tr}_{n+1}\, xB_i = \mathrm{Tr}_{n+1}\, B_ix$ for all $x \in \mathcal{B}_{n+1}$. For $i < n$, this follows from the induction hypothesis.

It remains to prove that $\mathrm{Tr}_{n+1}\, xB_n = \mathrm{Tr}_{n+1}\, B_nx$. If $x \in H_n$, this follows from the definition. Otherwise, we can write $x = aB_nb$, for $B, B' \in H_n$. We must check that $\mathrm{Tr}_{n+1}\, B_naB_nb = \mathrm{Tr}_{n+1}\, aB_nbB_n$. We consider four cases:

(1) $a, b \in H_{n-1}$.
(2) $a \in H_{n-1}$, $b = b_1B_{n-1}b_2$ with $b_1, b_2 \in H_{n-1}$.
(3) $a = a_1B_{n-1}a_2$ with $a_1, a_2 \in H_{n-1}$, $b \in H_{n-1}$.
(4) $a = a_1B_{n-1}a_2$, $b = b_1B_{n-1}b_2$ with $a_1, a_2, b_1, b_2 \in H_{n-1}$.

In case (1), a and b commute with B_n, so the result is obvious. In case (2), we compute

$$\mathrm{Tr}_{n+1} B_n a B_n b = [2]\{1\} \mathrm{Tr}_n ab = [2]\{1\}^2 \mathrm{Tr}_{n-1} ab_1b_2$$

while

$$\begin{aligned}
\mathrm{Tr}_{n+1} aB_n bB_n &= \mathrm{Tr}_{n+1} aB_n b_1 B_{n-1} b_2 B_n \\
&= \mathrm{Tr}_{n+1} ab_1 B_n B_{n-1} B_n b_2 \\
&= \mathrm{Tr}_{n+1} ab_1 (B_{n-1} B_n B_{n-1} - B_{n-1} + B_n) b_2 \\
&= ([2]\{1\}^2 - \{0\}\{1\} + \{1\}\{0\}) \mathrm{Tr}_{n-1} ab_1 b_2 \\
&= 2]\{1\}^2 \mathrm{Tr}_{n-1} ab_1 b_2
\end{aligned}$$

so the result holds. Case (3) is similar. Finally, for case 4, we compute

$$\begin{aligned}
\mathrm{Tr}_{n+1} aB_n bB_n &= \mathrm{Tr}_{n+1} ab_1 B_n B_{n-1} B_n b_2 \\
&= \mathrm{Tr}_{n+1} ab_1 (B_{n-1} B_n B_{n-1} - B_{n-1} + B_n) b_2 \\
&= ([2]\{1\} - 1) \mathrm{Tr}_n ab + \{1\}^2 \mathrm{Tr}_{n-1} a_1 a_2 b_1 b_2
\end{aligned}$$

A similar computation shows

$$\mathrm{Tr}_{n+1} B_n a B_n b = ([2]\{1\} - 1) \mathrm{Tr}_n ab + \{1\}^2 \mathrm{Tr}_{n-1} a_1 a_2 b_1 b_2$$

which concludes the proof. □

5.4. The cube of resolutions The definition of the HOMFLY-PT polynomial we have given in the preceding sections is modeled on the definition of the Jones polynomial via the Temperly-Lieb algebra. We can rephrase this definition to more closely resemble the Kauffman state model. From section 5.2, we see that

$$\Psi(\sigma_i) = q - B_i \qquad \Psi(\sigma_i^{-1}) = q^{-1} - B_i. \tag{5.4.1}$$

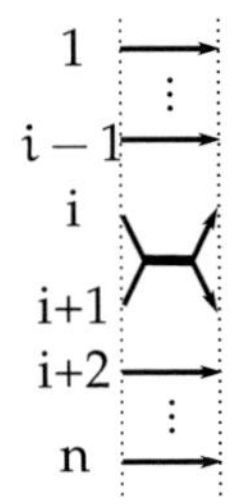

Figure 5.4.2. B_i is represented by a thick edge.

Graphically, we represent B_i by a diagram with a thick edge, as shown in Figure 5.4.2. The resulting diagrams are known as *MOY diagrams*, and were first introduced by Murakami, Ohtsuki, and Yamada.

Suppose that c is a crossing in a braid diagram D corresponding to an appearance of $\sigma_i^{\pm 1}$ in the braid word. We consider two ways to "resolve" c: the oriented resolution (corresponding to $1 \in H_n$ and the "thick edge" corresponding to B_i.

We call one of these the 0-resolution, and the other the 1-resolution. Which is which depends on the sign of the crossing, as shown in the figure below.

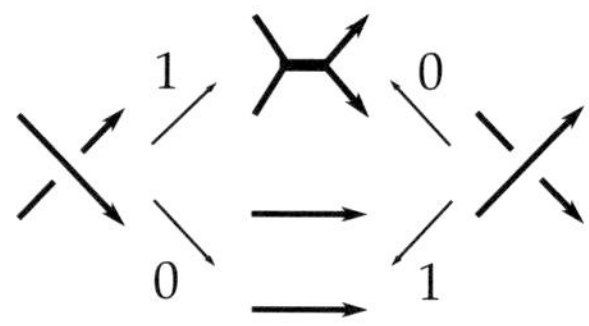

(**Warning:** if we replace the thick edge by the unoriented resolution we get a convention for 0 and 1 resolutions which is exactly opposite of the one we used in Khovanov homology.)

Just as in Khovanov homology, we can form the cube of resolutions of D, as illustrated below.

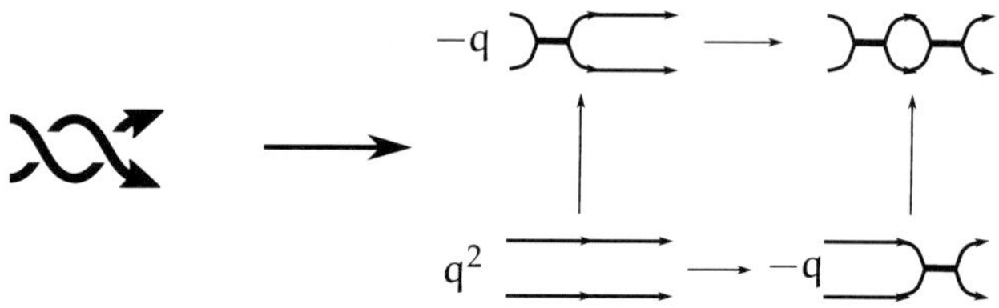

To each vertex v of the cube we associate a planar MOY diagram D_v, which we view as representing an element $D_v \in H_n$. Using equation 5.4.1, we see that

$$\Psi(D) = q^{n_+(D)} \sum_v (-q)^{-|v|} D_v.$$

If D is an n-strand MOY diagram, we can form its closure $\overline{D}$ exactly as we formed the closure of a braid. In analogy with the Kauffman bracket, we define $\langle \overline{D} \rangle = \mathrm{Tr}_n\, D$. We can evaluate the trace by applying the *MOY relations* shown in the next figure:

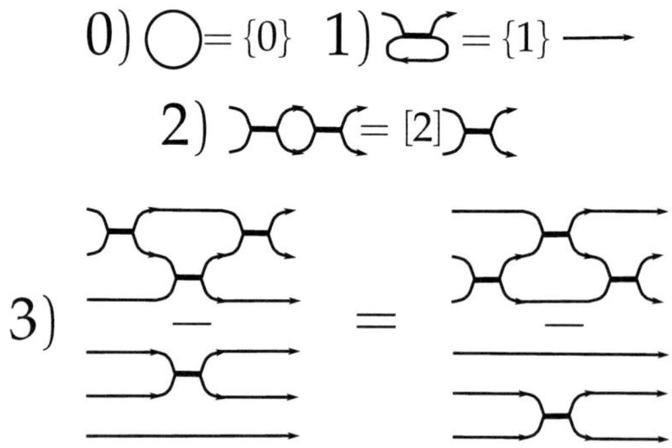

MOY relations 0) and 1) are properties 2) and 3) of the trace, while relations 2) and 3) are the defining relations of the Hecke algebra. For example, we have

$$\langle \overline{B_1^2} \rangle = [2]\langle \overline{B_1} \rangle = [2]\{1\}\langle \bigcirc \rangle = [2]\{1\}\{0\}.$$

Finally, comparing with Definition 5.2.3, we see that

$$P(\overline{D}) = a^{-w(D)} q^{n_+(D)} \sum_v (-q)^{-|v|} \langle D_v \rangle.$$

5.5. The Kazhdan-Lusztig basis We'll use the cube of resolutions in the previous section to categorify the HOMFLY-PT polynomial. Before we do so, we must discuss one more thing about the Hecke algebra.

The algebra H_2 has a (rather trivial) filtration by ideals: we have $B_1^2 = [2]B_1$, so $(B_1) \subset (1) = H_2$. More interestingly, a similar phenomenon holds for H_3. Define

$$B_w = B_1B_2B_1 - B_1 = B_2B_1B_2 - B_2.$$

If we use the basis

$$\{1, B_1, B_2, B_{12} := B_1B_2, B_{21} := B_2B_1, B_w\}$$

for the algebra H_3, we see that all the structure constants for multiplication have positive coefficients and there is an increasing sequence of ideals

$$(B_w) \subset (B_1, B_2) \subset (1).$$

Kazhdan and Lusztig [38] defined a similar basis for every H_n, and conjectured that the structure constants defining the multiplication in this basis were all positive. This conjecture was first proved by Beilinson-Bernstein and Brylinski-Kashiwara, and the basis is now known as the Kazhdan-Lusztig basis. We summarise some relevant facts about it in the following:

Theorem 5.5.1 ([7, 11, 38]). H_n *has a basis* $\{B_s \mid s \in S_n\}$ *such that*

1) $B_sB_t = \sum_u C^u_{st}B_u$, *where* $C^u_{st} \in \mathbb{N}[q^{\pm 1}]$.
2) $\{B_s \mid \ell(s) \leqslant k\}$ *is a basis for* $\mathcal{B}_{n,k}$.
3) *There is a natural map* $|\cdot| : S_n \to \Pi(n)$ *(the set of partitions of* n*) such that the set* $I_\lambda = \langle B_s \mid |s| \succeq \lambda\rangle$ *is an ideal in* H_n.

The partial order $\succeq$ on $\Pi(n)$ is the usual one: $\lambda \succeq \mu$ if the for all k the sum of the k largest elements of λ is $\geqslant$ the sum of the k largest elements of μ. The map $S_n \to \Pi_n$ is much used in the representation theory of S_n, *e.g.* in the Robinson-Schensted correspondence [8].

Exercise 5.5.2. Use the MOY rules to compute Tr_3 of each of the Kazhdan-Lusztig basis elements for H_3. You should be able to express the answers nicely in terms of $\{i\}$ for various values of i. Can you guess where the pattern you see in the answers comes from?

5.6. Soergel Bimodules The proof of Theorem 5.5.1 is one of the first appearances of categorification in representation theory. It proceeds by finding a graded monoidal category $\mathcal{C}_n$ which is generated by objects $\{\mathbb{B}_s \mid s \in S_n\}$ which satisfy

$$\mathbb{B}_s \otimes \mathbb{B}_t \simeq \bigoplus_u C^u_{s,t}\mathbb{B}_s.$$

Since this is the case, we must have $C^u_{s,t} \in \mathbb{N}[q^{\pm 1}]$. There are many different ways to descibe such a category, but the simplest is due to Soergel [75]. The category which he constructed is known as the category of Soergel bimodules, and is denoted by SBim_n. This category will play the same role in our definition of the HOMFLY-PT complex of a braid as Bar-Natan's category in the definition of the Khovanov complex of a tangle.

The objects of SBim_n are graded R_n-R_n bimodules over the graded $\mathbb{C}$-algebra $R_n = \mathbb{C}[X_1, \ldots, X_n]$. (Saying that $\mathbb{B}$ is a bimodule over a $\mathbb{C}$-algebra means that $\mathbb{B}$ is vector space over $\mathbb{C}$, and the $\mathbb{C}$ action on both left and right is given by scalar multiplication.) The grading on R_n is given by $q(X_i) = 2$ for all i, and the monoidal structure on SBim_n is given by tensor product of bimodules.

Since R_n is commutative, any bimodule over R_n can be thought of as a module over the larger polynomial ring $R_n^e := R_n \otimes_{\mathbb{C}} R_n \simeq \mathbb{C}[X_1, \ldots, X_n, Y_1, \ldots Y_n]$, where the X_i's act by multiplication by X_i on the left, and the Y_i's as multiplication by X_i on the right.

Example 5.6.1. R_n is a bimodule over itself, where the left action of R_n is given by multiplication on the left, and the right action is given by multiplication on the right. (Of course, R_n is commutative, so which side we multiply on doesn't actually matter.) As a module over R_n^e, $R_n = R_n^e/(X_1 - Y_1, \ldots X_n - Y_n)$.

The symmetric group S_n acts on R_n by permuting the X_i's. Let $R_n^{s_i} \subset R_n$ be the ring of elements which are invariant under the elementary permutation s_i. $R_n^{s_i}$ is itself a polynomial ring:

$$R_n^{s_i} \simeq \mathbb{C}[X_1, \ldots, X_{i-1}, e_1^i, e_2^i, X_{i+2}, \ldots, X_n].$$

where $e_1^i = X_i + X_{i+1}$ and $e_2^i = X_i X_{i+1}$ are the elementary symmetric functions in X_i and X_{i+1}. Viewed as a module over $R_n^{s_i}$, R_n is free of rank 2, with basis $\{1, X_i\}$. Hence $R_n \simeq (1+q^2)R_n^{s_i}$ as graded $R_n^{s_i}$-modules. As usual, the notation $q^2 R_n^{s_i}$ indicates a grading shift: $1 \in R_n^{s_i}$ has grading 0, but the same element in $q^2 R_n^{s_i}$ has grading 2.

Definition 5.6.2. The elementary Soergel bimodule $\mathbb{B}_i := \mathbb{B}_{s_i}$ is defined to be

$$\mathbb{B}_i = q^{-1} R_n \otimes_{R_n^{s_i}} R_n.$$

Viewed as a module over R_n^e, we have

$$\begin{aligned}
\mathbb{B}_i &= q^{-1} R_n^e/(e_1^i(X) - e_1^i(Y), e_2^i(X) - e_2^i(Y)) \\
&= q^{-1} R_n^e/((X_i + X_{i+1}) - (Y_i + Y_{i+1}), X_i X_{i+1} - Y_i Y_{i+1}) \\
&= q^{-1} R_n^e/(X_i + X_{i+1} - Y_i - Y_{i+1}, (Y_i - X_i)(Y_i - X_{i+1}))
\end{aligned}$$

The $\mathbb{B}_i$'s satisfy the relations given by the Hecke algebra.

Proposition 5.6.3. *$\mathbb{B}_i^2 \simeq (q + q^{-1})\mathbb{B}_i$ as R_n-R_n bimodules.*

Proof. We prove this for $\mathbb{B}_1 \in \mathrm{SBim}_2$; the general case is completely analogous. Write $R = R_2$, $R' = R_2^{s^1}$. We have

$$\begin{aligned}
\mathbb{B}_1 \otimes \mathbb{B}_1 &\simeq q^{-2}(R \otimes_{R'} R) \otimes_R (R \otimes_{R'} R) \\
&\simeq q^{-2} R \otimes_{R'} R \otimes_{R'} R \\
&\simeq q^{-2} R \otimes_{R'} (1 + q^2) R' \otimes_{R'} R
\end{aligned}$$

$$\simeq (q^{-1}+q)q^{-1}R \otimes_{R'} R \simeq (q^{-1}+q)\mathbb{B}_1 \qquad \square$$

This proof is algebraically slick, but can be unilluminating if you've never seen something like it before. A more hands-on approach is to work with modules over R_n^e. Suppose $\mathbb{B}, \mathbb{B}'$ are R_n-R_n bimodules, and that $\mathbb{B} \simeq \mathbb{C}[\mathbf{X},\mathbf{Y}]/(f_i(\mathbf{X},\mathbf{Y}))$, $\mathbb{B}' \simeq \mathbb{C}[\mathbf{Y},\mathbf{Z}]/(g_j(\mathbf{Y},\mathbf{Z}))$. Then

$$\mathbb{B} \otimes \mathbb{B}' \simeq \mathbb{C}[\mathbf{X},\mathbf{Y},\mathbf{Z}]/(f_i(\mathbf{X},\mathbf{Y}), g_j(\mathbf{Y},\mathbf{Z})),$$

which we view as a module over $\mathbb{C}[\mathbf{X},\mathbf{Z}] = R_n^e$.

Example 5.6.4. From this perspective, we have

$$\mathbb{B}_1^2 = q^{-2}\mathbb{C}[X_1,X_2,Y_1,Y_2,Z_1,Z_2]/(\mathcal{R})$$

viewed as a module over $\mathbb{C}[X_1,X_2,Z_1,Z_2]$, where the ideal of relations $\mathcal{R}$ is given by

$$\begin{aligned}\mathcal{R} &= (e_j(\mathbf{X}) - e_j(\mathbf{Y}), e_j(\mathbf{Y}) - e_j(\mathbf{Z}))(j=1,2)\\ &= (e_j(\mathbf{X}) - e_j(\mathbf{Y}), e_j(\mathbf{X}) - e_j(\mathbf{Z}))\end{aligned}$$

viewed as a module over $\mathbb{C}[X_1,X_2,Z_1,Z_2]$. The relations $X_1+X_2 = Y_1+Y_2$ and $X_1X_2 = Y_1Y_2$ can be rewritten as $Y_2 = X_1+X_2-Y_1$ and $Y_1^2 = (X_1+X_2)Y_1 - X_1X_2$. We use these relations to eliminate Y_2 and Y_1^2, so

$$\mathbb{B}_1^2 = q^{-2}(\mathbb{B} \oplus Y_1\mathbb{B}) = (q+q^{-1})q^{-1}\mathbb{B}$$

where

$$\begin{aligned}\mathbb{B} &= \mathbb{C}[\mathbf{X},\mathbf{Z}]/(e_j(\mathbf{X}) - e_j(\mathbf{Z}))\ (j=1,2)\\ &= q\mathbb{B}_1.\end{aligned}$$

The relation in Proposition 5.6.3 corresponds to the quadratic relation in the Hecke algebra. It is easy to see that the analog of the far-commutativity relation is satisified: $\mathbb{B}_i\mathbb{B}_j \simeq \mathbb{B}_j\mathbb{B}_i$ if $|i-j|>1$. To prove the analog of the braid relation, we define a bimodule $\mathbb{B}_w$ over R_3 by $\mathbb{B}_w = q^{-3}R_3 \otimes_{R_3^{S_3}} R_3$, where $R_3^{S_3}$ denotes the ring of invariants under the action of S_3 on R_3 by permuting the coordinates. Since $R_3^{S_3}$ is a polynomial ring in the elementary symmetric functions e_1, e_2, e_3, we have

$$\mathbb{B}_w = q^{-3}R_3^e/(e_j(\mathbf{X}) = e_j(\mathbf{Y})) \quad (j=1,2,3).$$

Proposition 5.6.5. $\mathbb{B}_1\mathbb{B}_2\mathbb{B}_1 \simeq \mathbb{B}_w \oplus \mathbb{B}_1$ *as* R_3-R_3 *bimodules.*

Proof. (Sketch) $\mathbb{B}_1\mathbb{B}_2\mathbb{B}_1 = R_3^e[a,b,c]/\mathcal{I}$, where $\mathcal{I}$ is the ideal generated by the relations

$$\begin{aligned}X_1+X_2 &= a+b & X_1X_2 &= ab\\ b+X_3 &= c+Y_3 & bX_3 &= cY_3\\ a+c &= Y_1+Y_2 & ac &= Y_1Y_2\end{aligned}$$

where the variables corresponding to different edges are shown in the figure below.

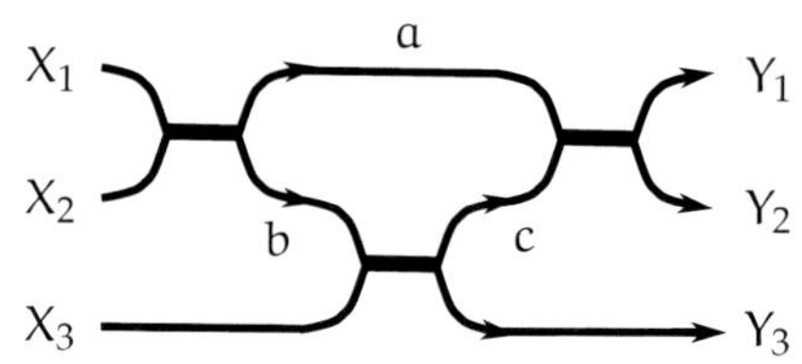

We can use the linear relations to eliminate a and c, and one of the quadratic relations to eliminate b^2. View $\mathbb{B}_1\mathbb{B}_2\mathbb{B}_1$ as a module over R_3^e, and let M_1 and M_2 be the submodules generated by 1 and $b - X_3$, respectively. Then as a module over R_3^e, $\mathbb{B}_1\mathbb{B}_2\mathbb{B}_1 \simeq M_1 \oplus M_2$. It is easy to check that the maps $R_3^e \to M_1$ and $R_3^e \to M_2$ given by $p \mapsto p$ and $p \mapsto (b - X_3)p$ factor through $\mathbb{B}_w$ and $\mathbb{B}_1$ respectively. Finally, we note that as modules over R_3, $\mathbb{B}_1\mathbb{B}_2\mathbb{B}_1$ is free of rank 8, $\mathbb{B}_w$ is free of rank 6, and $\mathbb{B}_1$ is free of rank 2. It follows that we must have $M_1 \simeq \mathbb{B}_w$ and $M_2 \simeq \mathbb{B}_1$. □

Definition 5.6.6. A *Bott-Samuelson bimodule* is a R_n-R_n bimodule obtained by taking tensor products of the $\mathbb{B}_i$'s. A *Soergel bimodule* is a direct summand of a Bott-Samuelson bimodule.

The category SBim_n is the subcategory of the category of R_n-R_n bimodules whose objects are Soergel bimodules. Its split Grothendieck group $K(\mathrm{SBim}_n)$ is the $\mathbb{Z}[q^{\pm 1}]$-module generated by objects of SBim_n, modulo relations of the form $[\mathbb{B} \oplus \mathbb{B}'] = [\mathbb{B}] + [\mathbb{B}']$ and $[q\mathbb{B}] = q[\mathbb{B}]$. Tensor product on SBim_n makes $K(\mathrm{SBim}_n)$ into a ring: $[\mathbb{B}][\mathbb{B}'] = [\mathbb{B} \otimes \mathbb{B}']$. Propositions 5.6.3 and 5.6.5 imply that there is a homomorphism $\rho : H_n \to K(\mathrm{SBim}_n)$ given by $\rho(B_i) = [\mathbb{B}_i]$. Soergel showed that this map is an isomorphism.

Theorem 5.6.7 (Soergel [75]). *There are indecomposable Soergel bimodules* $\mathbb{B}_s$ $(s \in S_n)$ *which generate* SBim_n *and satisfy* $\rho(B_s) = [\mathbb{B}_s]$.

Exercise 5.6.8. Let $\mathbb{B}_{w_n}$ be the $R_n - R_n$ bimodule defined by

$$\mathbb{B}_{w_n} = q^{-n(n-1)/2} R_n \otimes_{R_n^{S_n}} R_n.$$

Show that $\mathbb{B}_i\mathbb{B}_{w_n} = [2]\mathbb{B}_{w_n}$. What is $\mathbb{B}_{w_n}\mathbb{B}_{w_n}$?

Importantly for us, SBim_n satisfies the analog of Corollary 4.8.12.

Theorem 5.6.9 ([75]). SBim_n *is a positively graded category;* $\{\mathbb{B}_s \mid s \in S_n\}$ *is a set of positive generators.*

5.7. Hochschild homology and cohomology

Definition 5.7.1. Suppose that R is a $\mathbb{C}$-algebra. If M is an R-R bimodule, its *Hochschild homology* and cohomology are defined as $HH_*(M) = \mathrm{Tor}_*^{R^e}(R, M)$ and $HH^*(M) = \mathrm{Ext}^*_{R^e}(R, M)$.

Here Tor, as usual, denotes the derived tensor product:

$$\mathrm{Tor}_i^{R^e}(R, M) = H_i(C_R \otimes M) = H_i(R \otimes C_M)$$

where C_R and C_M are free resolutions of R and M as R^e modules. Similarly, Ext denotes derived Hom. If R is a graded ring and M is a graded bimodule, then $HH_*(M)$ and $HH^*(M)$ are graded as well.

A priori, $HH_*(M)$ should be a module over R^e, or equivalently, an R-R bimodule. However the left and right actions of R on R are the same, which implies that the same is true for the left and right actions of R on $HH_*(M)$ and $HH^*(M)$, so we can view them as R modules. Hence we have functors

$$HH_*, HH^* : \text{R-gMod-R} \to \text{R-gMod},$$

where R-gMod-R denotes the category of graded R-R bimodules. Note that HH_* and HH^* are both *covariant* functors of M.

When $R = R_n$ is a polynomial ring, the situation simplifies considerably, since R has a very simple resolution as an R^e-module. Namely, if $\mathcal{K}(X_i - Y_i)$ is the short complex $q^2R^e \xrightarrow{X_i - Y_i} R^e$, then the *Koszul complex*

$$C_{R_n} = \bigotimes_{i=1}^{n} \mathcal{K}(X_i - Y_i)$$

is a free resolution of R as a module over R^e.

Example 5.7.2. $HH_*(R_n) = H_*(C_{R_n} \otimes R_n)$. Since $X_i = Y_i$ in R_n,

$$C_{R_n} \otimes R_n \simeq (q^2R_n \xrightarrow{0} \underline{R_n})^{\otimes n}$$

where the underlined terms are in homological grading 0, and the differential lowers the homological grading by 1. The differential in this complex is trivial, so $HH_*(R_n) = R_n \otimes \Lambda^*(a_1, \ldots, a_n)$, where $q(a_i) = 2$ and the homological grading is given by the degree in the exterior algebra.

To compute the Hochschild cohomology, we tensor with the dual chain complex $C^*_{R_n}$. We have

$$C^*_{R_n} \otimes R_n \simeq (\underline{R_n} \xrightarrow{0} q^{-2}R_n)^{\otimes n}$$

where the underlined terms are in homological grading 0 and the differential raises grading by 1. Again, the cohomology is an exterior algebra tensored with R_n.

Example 5.7.3. Let $n = 2$. To compute $HH^*(B_1)$, we use a free resolution of $\mathbb{B}_1$, which is given by the Koszul complex

$$C(\mathbb{B}_1) = (q^2R_2^e \xrightarrow{Y_1+Y_2-X_1-X_2} R_2^e) \otimes (q^4R_2^e \xrightarrow{Y_1Y_2-X_1X_2} R_2^e).$$

Then $HH^*(\mathbb{B}_1) = \mathrm{Hom}(R_n, C(\mathbb{B}_1)) = R_n \otimes \Lambda^*(a_1, a_2)$, but now we have $q(a_1) = 2$, $q(a_2) = 4$.

Proposition 5.7.4. *If* M *and* N *are* $R_n - R_n$*-bimodules, then*

(1) $HH_*(M \otimes N) \simeq HH_*(N \otimes M)$
(2) $\mathrm{Hom}(N, M) \simeq HH^0(M \otimes N^{op})$ *where* N^{op} *denotes* N *with the right and left actions reversed.*

Proof. We prove the first part; the second follows from standard properties of Ext. Let C_M and C_N be free resolutions of M and N. We view them as being defined over polynomial rings with variables X, Y and X', Y', respectively. A free resolution of $M \otimes N$ is given by $C_M \otimes_{\mathbb{C}} C_N/(Y = X')$. Tensoring this with R gives $C_M \otimes_{\mathbb{C}} C_N/(Y = X', X = Y'))$. $HH_*(M \otimes N)$ is the homology of this complex. Similarly, a free resolution of $N \otimes M$ is $C_M \otimes_{\mathbb{C}} C_N/(Y' = X)$, so $HH_*(M \otimes N)$ is also the homology of $C_M \otimes_{\mathbb{C}} C_N/(Y = X', X = Y'))$. □

The first property in the proposition says that HH_* behaves like a trace. In fact, Khovanov showed that HH_* categorifies the Jones-Ocneanu trace. If $\mathbb{B} \in \mathrm{SBim}_n$ we define

$$P(\mathbb{B}) = (aq)^{-n} \sum (-a^2q^2)^i \,\mathrm{qdim}\, HH^i(\mathbb{B}).$$

Theorem 5.7.5 ([43]). *If* $\mathbb{B} \in \mathrm{SBim}_n$, $P(\mathbb{B}) = \mathrm{Tr}_n[\mathbb{B}]$, *where* $[\mathbb{B}]$ *denotes the image of* $\mathbb{B}$ *in* $K(\mathrm{SBim}_n) \simeq H_n$.

Proof. It is enough to check the equality when $\mathbb{B}$ is a Bott-Samuelson bimodule. Since the Jones-Ocneanu trace satisfies and is determined by the MOY rules, it suffices to check that P satisfies the MOY rules as well. Rules 2) and 3) are a direct consequence of Propositions 5.6.3 and 5.6.5.

For MOY 0), suppose that $\mathbb{B}$ is a Bott-Samuelson bimodule in SBim_n, and let $\iota(\mathbb{B})$ be its image in SBim_{n+1}. If $C_{\mathbb{B}}$ is a free resolution of $\mathbb{B}$ over R^e_n, then

$$C_{\mathbb{B}} \otimes \left(q^2 R^e_{n+1} \xrightarrow{Y_{n+1}-X_{n+1}} R^e_{n+1}\right)$$

will be a free resolution of $\iota(\mathbb{B}_n)$ over Tr_{n+1}. After taking the tensor product with R_{n+1}, this becomes

$$(C_{\mathbb{B}} \otimes_{R^e_n} R_n) \otimes_{\mathbb{C}} \left(q^2\mathbb{C}[X_{n+1}] \xrightarrow{0} \mathbb{C}[X_{n+1}]\right).$$

It follows that $HH^*(\iota(\mathbb{B})) = HH^*(\mathbb{B}) \otimes_{\mathbb{C}} \mathbb{C}[X_{n+1}] \otimes \Lambda^*(a_1)$ so, as desired,

$$P(\iota(\mathbb{B})) = a^{-1}q^{-1}P(\mathbb{B})\left(\frac{q^2 - a^2q^2}{1-q^2}\right) = \{0\}P(\mathbb{B}).$$

The proof for MOY 1) is similar: $\iota(\mathbb{B}) \otimes \mathbb{B}_n$ has a free resolution

$$C_{\mathbb{B}} \otimes \left(q^2 R^e_{n+1} \xrightarrow{Y_n - Y' + Y_{n+1} - X_{n+1} pt} R^e_{n+1}\right) \otimes \left(q^4 R^e_{n+1} \xrightarrow{(X_{n+1}-Y_{n+1})(X_{n+1}-Y')} R^e_{n+1}\right).$$

After tensoring with $\mathbb{R}_n$, this becomes

$$C_{\mathbb{B}} \otimes_{R^e_n} \left(q^2\mathbb{C}[Y'] \xrightarrow{Y_n - Y'} \mathbb{C}[Y']\right) \otimes_{\mathbb{C}} \left(q^4\mathbb{C}[X_{n+1}] \xrightarrow{0} \mathbb{C}[X_{n+1}]]\right).$$

As a chain complex over R_{n+1}, this is homotopy equivalent to the chain complex $C_{\mathbb{B}} \otimes (q^3 R_{n+1} \xrightarrow{0} q^{-1}R_{n+1})$, so

$$P(\iota(\mathbb{B}) \otimes \mathbb{B}_n) = (aq)^{-1}P(\mathbb{B})\frac{q^4 - a^2q^2}{1-q^2} = \{1\}P(\mathbb{B}).$$ □

Proposition 5.7.6 ([67]). *If* $\mathbb{B} \in \mathrm{SBim}_n$, *then* $HH_*(\mathbb{B})$ *is free over* R_n.

Together with Theorem 5.7.5, the proposition says that $HH_*(\mathbb{B})$ can be easily computed using the MOY rules.

Example 5.7.7. In $SBim_2$, $Hom(\mathbb{B}_1, \mathbb{B}_1) = HH_0(\mathbb{B}_1^2)$. Earlier, we computed that $Tr_2(B_1^2) = [2]\{0\}\{1\}$, so

$$\text{qdim}\, Hom(\mathbb{B}_1, \mathbb{B}_1) = \frac{1+q^2}{(1-q^2)^2}.$$

Since $\text{qdim}\, R_n = (1-q^2)^{-2}$, $Hom(\mathbb{B}_1, \mathbb{B}_1)$ is free of rank 2 over R_2. The identity map and the map "multiplication by Y_1" are a basis for $Hom(\mathbb{B}_1, \mathbb{B}_1)$ over R_2.

A similar computation shows that $Hom(\mathbb{1}, \mathbb{B}_1) \simeq qR_2 \simeq Hom(\mathbb{B}_1, \mathbb{1})$. We can describe the generators $S : \mathbb{1} \to \mathbb{B}_i$ and $S' : \mathbb{B}_i \to \mathbb{1}$ explicitly as follows. Identify $\mathbf{1} = R_2^e/\mathcal{I}_1$ and $\mathbb{B}_1 = R_2^e/\mathcal{I}_2$, where

$$\mathcal{I}_1 = (e_1, X_1 - Y_1) \quad \text{and} \quad \mathcal{I}_2 = (e_1, (Y_1 - X_1)(Y_2 - X_2))$$

with $e_1 = Y_1 + Y_2 - X_1 - X_2$. Clearly $\mathcal{I}_2 \subset \mathcal{I}_1$, and we define $S' : q\mathbb{B}_1 \to \mathbb{1}$ to be the quotient map. For the other direction, note that $(Y_1 - X_2)\mathcal{I}_1 \subset \mathcal{I}_2$, so there is a well-defined map $S : q\mathbb{1} \to \mathbb{B}_1$ given by $S(p) = (Y_1 - X_2)p$. Similarly, in $SBim_n$, we can define morphisms $S_i : q\mathbb{1} \to \mathbb{B}_i$ and $S_i' : q\mathbb{B}_i \to \mathbb{1}$ by acting on the variables with index i and $i+1$.

Exercise 5.7.8. Use the above method to compute $Hom(\mathbb{B}_s, \mathbb{B}_{s'})$, where $s, s' \in S_3$. Describe generators for the Hom-spaces, and verify that Theorem 5.6.9 holds in this case.

5.8. The Rouquier complex We can now describe the HOMFLY-PT analog of the Khovanov complex of a tangle. Let $K^b(SBim_n)$ denote the homotopy category of bounded complexes over $SBim_n$ (not to be confused with the Grothendieck group $K(SBim_n)$.)

Theorem 5.8.1 (Rouquier) [71]). *There is a well-defined map* $C : Br_n \to K^b(SBim_n)$ *satisfying*

$$C(\sigma_i) = \underline{q\mathbb{1}} \xrightarrow{S_i} \mathbb{B}_i \qquad C(\sigma_i^{-1}) = \mathbb{B}_i \xrightarrow{S_i'} \underline{q^{-1}\mathbb{1}}$$

and $C(\sigma\sigma') \sim C(\sigma) \otimes C(\sigma')$. *(The underlined terms in each complex are in homological grading* 0.*)*

Sketch of proof. The statement of the theorem clearly determines the complex assigned to a given braid word. To check that the theorem holds, we must show that braid words which correspond to the same braid have homotopy equivalent complexes. There are three things to check: $C(\sigma_i) \otimes C(\sigma_i^{-1}) \sim \mathbb{1}$, $C(\sigma_i\sigma_j) \sim C(\sigma_j\sigma_i)$ if $|i-j| > 1$, and $C(\sigma_1\sigma_2\sigma_1) \sim C(\sigma_2\sigma_1\sigma_2)$. The first relation corresponds to the second Reidemeister move, and its proof is entirely analogous to the proof of Proposition 4.9.3.

The second relation (far commutativity) is easy: if $|i-j| > 1$, then $B_iB_j = B_jB_i$, so $\mathbb{B}_i\mathbb{B}_j \sim \mathbb{B}_j\mathbb{B}_i$. The morphism spaces $Hom(q\mathbb{B}_i, \mathbb{B}_i\mathbb{B}_j)$ and $Hom(q\mathbb{B}_j, \mathbb{B}_i\mathbb{B}_j)$ are one-dimensional, so one easily checks that $C(\sigma_j\sigma_i) \simeq C(\sigma_i\sigma_j)$.

The proof of the braid relation is more complicated. Using Proposition 5.6.5, we see that $C(\sigma_1\sigma_2\sigma_1)$ is isomorphic to a complex of the form

$$q^3\mathbb{1} \to q^2(\mathbb{B}_1 \oplus \mathbb{B}_1 \oplus \mathbb{B}_2) \to q(\mathbb{B}_{12} \oplus \mathbb{B}_{21}) \oplus (1+q^2)\mathbb{B}_1 \to \mathbb{B}_w \oplus \mathbb{B}_1.$$

cancelling, we see this is homotopy equivalent to a complex

$$q^3\mathbb{1} \to q^2(\mathbb{B}_1 \oplus \mathbb{B}_2) \to q(\mathbb{B}_{12} \oplus \mathbb{B}_{21}) \to \mathbb{B}_w.$$

Using exercise 5.7.8, we see that all components of the differential belong to 1-dimensional morphism spaces. In fact, all possible components of the differential are nonzero. (This is best proved by using invariance under the second Reidemeister move to see that $C(\sigma_1\sigma_2\sigma_1\sigma_1^{-1}\sigma_2^{-1}\sigma_1^{-1}) \sim \mathbb{1}$.)

Finally, the relation $d^2 = 0$ can be used to show that any two complexes of the form above are isomorphic. Since the form of this complex is completely symmetric under the operation of permuting 1 and 2, it must be homotopy equivalent to $C(\sigma_2\sigma_1\sigma_2)$ as well. □

The Rouquier complex categorifies the homomorphism $\Psi : \mathrm{Br}_n \to H_n$. If C is a chain complex over an additive category $\mathcal{C}$, we define $\chi(C) = \sum_i (-1)^i [C_i] \in K(\mathcal{C})$, where as usual $[C_i]$ denotes the image of the object $[C_i]$ in the Grothendieck group $K(\mathcal{C})$.

Proposition 5.8.2. *If* $\sigma \in \mathrm{Br}_n$, $\chi(C(\sigma)) = \Psi(\sigma) \in H_n$.

Proof. Comparing with equation (5.4.1), we see that $\chi(C(\sigma_i^{\pm 1})) = \Psi(\sigma_i^{\pm 1})$. Since $\chi(C \otimes C') = \chi(C)\chi(C')$, the statement holds for all σ. □

Just as with the Khovanov complex of a tangle, positivity of the category SBim ensures that $C(\sigma)$ has a unique minimal representative which is well-defined up to isomorphism.

5.9. HOMFLY-PT homology The Hochschild homology HH_* is a covariant functor from SBim_n to the category of modules over R_n, so $HH_*(C(\sigma))$ is a complex of R_n-modules. Its homology $H_*(HH_*(C(\sigma)))$ will have three gradings: the homological gradings on the complex and on Hochschild homology, and the q-grading. In keeping with our previous notation for the Hochschild homology and the Bar-Natan complex, we denote a shift in the homological grading by multiplication by t, and in the Hochschild grading by multiplication by a^2.

Definition 5.9.1. If $\sigma \in \mathrm{Br}_n$, the HOMFLY-PT homology of its closure $\overline{\sigma}$ is defined to be $HHH(\overline{\sigma}) = t^{-n_+(\sigma)}(aq)^{-w(\sigma)}H_*(HH_*(C(\sigma)))$.

We index the homological gradings so that

$$HHH_{i,j}(\overline{\sigma}) = H_{i-n_+(\sigma)}(HH_{j-w(\sigma)}(C(\sigma))).$$

Theorem 5.9.2 (Khovanov-Rozansky [40, 44]). HHH *is a well-defined link invariant which categorifies the HOMFLY-PT polynomial, in the sense that*

$$\sum_{i,j} (-1)^i(-a^2q^2)^j \,\mathrm{qdim}\, HHH_{i,j}(L) = \overline{P}(L).$$

Sketch of proof. The relation with the Euler characteristic follows from Proposition 5.8.2, Theorem 5.7.5, and Definition 5.2.3. To prove that HHH is a link invariant, we must check that it is invariant under the Markov moves. Invariance under Markov 1) follows from Proposition 5.7.4. For invariance under Markov 2), we use the computation of $\mathrm{HH}(\iota(B))$ and $\mathrm{HH}(\iota(B)B_n)$ in the proof of Theorem 5.7.5 to see that $\mathrm{HH}(\sigma\sigma_n^{\pm 1}) \sim \mathrm{HH}(\sigma) \otimes (aq)^{-1}C_{\pm}$, where $C_{\pm}$ are the bicomplexes shown below, and $R = \mathbb{C}[X_{n+1}]$.

$$
\begin{array}{ccc}
qR & \xrightarrow{X_{n+1}} & q^{-1}R \\
{\scriptstyle 0}\uparrow & & {\scriptstyle 0}\uparrow \\
q^3R & \xrightarrow{1} & q^3R
\end{array}
\qquad\qquad
\begin{array}{ccc}
q^{-1}R & \xrightarrow{1} & q^{-1}R \\
{\scriptstyle 0}\uparrow & & {\scriptstyle 0}\uparrow \\
q^3R & \xrightarrow{X_{n+1}} & qR
\end{array}
$$

It is easy to see that $H_*(C_{\pm}) \simeq \mathbb{C}$. The homology is supported in the right-hand column of both bicomplexes, which is homological grading 1 for the positive crossing, but homological grading 0 for the negative one. This is accounted for by the shift by $t^{-n_+(\sigma)}$ in the definition of HHH. □

Alternate description: Expanding out the chain complex used to compute the Hochschild homology yields the following alternate description of the HOMFLY-PT homology. This is the original definition of Khovanov and Rozansky [40].

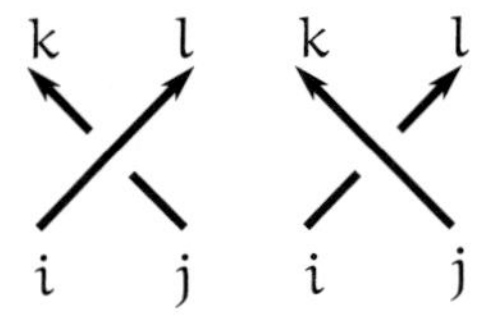

Figure 5.9.3. Positive and negative crossings

Suppose D is a planar braid diagram. Let $R = R(D) = \mathbb{C}[X_e]/(L_c)$, where e runs over the edges of D and (L_c) is an ideal generated by one linear relation L_c for each crossing c. If we let c_+ and c_- be positive and negative crossings with edges labeled as in the figure $L_{c_+} = L_{c_-} = X_i + X_j - X_k - X_l$. Since the L_c's are all linear polynomials, $R(D)$ is isomorphic to a polynomial ring.

We form a bicomplex

$$C(D) = \bigotimes_c C(c),$$

where the tensor product runs over all crossings of D. The bicomplexes $C(c)$ associated to the crossings c takes one of two forms, depending on whether c is positive or negative. Omitting the grading shifts, the diagram below shows the bicomplexes associated to c_+ (left) and c_- (right).

$$
\begin{array}{ccc}
R & \xrightarrow{(X_i-X_k)(X_i-X_l)} & R \\
{\scriptstyle 1}\uparrow & & {\scriptstyle X_i-X_l}\uparrow \\
R & \xrightarrow{X_i-X_k} & R
\end{array}
\qquad\qquad
\begin{array}{ccc}
R & \xrightarrow{X_i-X_k} & R \\
{\scriptstyle X_i-X_l}\uparrow & & {\scriptstyle X_i-X_l}\uparrow \\
R & \xrightarrow{(X_i-X_k)(X_i-X_l)} & R
\end{array}
$$

The maps in the horizontal direction are the Hochschild differentials, and the maps in the vertical direction correspond to differentials in the cube of resolutions. Up to an appropriate grading shift, we have

$$\mathrm{HHH}(\overline{\sigma}) = \mathrm{H}(\mathrm{H}(\mathrm{C}(\mathrm{D}), d_h), d_v).$$

In this formulation, we can form a reduced homology HHH^r by working over the ring $R_r(D) \subset R(D)$ generated by the differences $X_i - X_j$. If K is a knot, it can be shown that $\mathrm{HHH}^r(K)$ is finite dimensional, and that

$$\mathrm{HHH}(K) = \mathrm{HHH}^r(K) \otimes \mathrm{HHH}(\bigcirc).$$

5.10. Connections and Further Reading Khovanov's paper [44] is well worth reading. Elias and Williamson [21,22] developed a calculus for morphisms in the Soergel category which led to a proof of the Kazhdan-Lusztig conjectures in all types. More generally, the Kazhdan-Lusztig basis and the Soergel category paly a key role in the development of geometric representation theory and relate to many topics that are both beautiful and attractive to the geometrically inclined. The books by Humphreys [34] and Bjorner and Brenti [8] are good starting places.

In a different direction, we can consider the relation between HOMFLY-PT homology and Khovanov homology. The Jones polynomial is obtained from the HOMFLY-PT polynomial by substituting $a = q^2$. The categorified analog of such a substitution is a spectral sequence from HHH Kh [68]. In fact there is an entire family of such spectral sequences, corresponding to substituting $a = q^n$. The Alexander polynomial is obtained by substituting $a = q^0$, so it is natural to ask if there is a spectral sequence from HHH^r to $\widehat{HFK}$. Indeed Dowlin [18] recently proved that there is a spectral sequence from Kh_r to $\widehat{\mathrm{HFK}}$, which provides a (somewhat roundabout) way of constructing one.

The HOMFLY-PT homology of torus knots has many interesting connections to other areas of geometry and representation theory, including Hilbert schemes of plane curve singularities [58,60], Cherednik algebras [14,27,29], and the Hilbert scheme of $\mathbb{C}^2$, [28,59]. Their HHH was finally calculated through some remarkable work of Elias, Hogancamp and Mellit [20,32,53]

6. Λ^k colored polynomials

Witten [83] famously reformulated the Jones polynomial using quantum field theory. If Y is a 3-manifold equipped with a principal G-bundle E, the action of Witten's theory is given by the Chern-Simons functional on the space of connections on E. To get knot polynomials, we treat the knots as "Wilson loops" labeled by representations of G and take the expected value of a function based on the trace of the representation along the knot. To recover the Jones polynomial, we take $G = SU(2)$ and color every component with the vector (2-dimensional) representation. However Witten's formulation makes it clear that there are many other possibilities.

In mathematical terms, the theory that Witten created can be described as a relative 2+1 dimensional TQFT. (Indeed, Witten's work was a principle motivation for the definition of a TQFT.) Given a simple Lie algebra $\mathfrak{g}$, we form a category whose objects are closed surfaces Σ containing a set of points P which are *colored* (*i.e.* labeled) by representations of $\mathfrak{g}$. and whose morphisms are given by pairs $(Y, T) : (\Sigma_0, P_0) \to (\Sigma_1, P_1)$, where $Y : \Sigma_0 \to \Sigma_1$ is a cobordism and $T \subset Y$ is a colored tangle with ends on P_0 and P_1. We seek a functor from this category to the category of vector spaces and linear maps between them. So far, mathematicians are unable to give a rigorous definition of the path integral used in Witten's work, but Reshetikhin and Turaev [69] gave an algebraic construction of the invariants Witten described. They are now known as the WRT invariants.

Khovanov's program of categorification aims to upgrade these $2+1$ dimensional TQFT's to $2+1+1$ dimensional TQFT's. So far, most of its successes have been for polynomial invariants of knots and tangles in S^3 in type A (when $\mathfrak{g} = \mathfrak{sl}_n$.) We understand this process very well when all the colors are minuscule representations, *i.e.* exterior powers of the vector representation. In this situation, there are many different ways to categorify, and they are more or less all known to agree. For other colors, there are several different possible categorifications, not all of which are the same. Outside of type A, very little is known, and essentially nothing is known about categorifications for knot and tangles in 3-manifolds other than S^3. The exception comes from the world of Floer homology, where knot Floer homology, which categorifies the Alexander polynomial, fits into the 3+1 dimensional TQFT provided by Heegaard Floer homology. (For more information on this topic, see Hom's lectures in this volume.)

6.1. The yoga of WRT In order to define the WRT invariants, we must specify another parameter, known as the *level*. In general, the dependence of the invariants on the level is complicated, but in the case of tangles in B^3, it is quite simple. Instead of working over $\mathbb{C}$, we work over the ring $\mathbb{Z}[q^{\pm 1}]$. To recover the invariants at a particular level, we substitute $q = \omega$, where ω is some root of unity determined by the level.

We outline the basic properties of the WRT invariants in this setting. Fix a simple Lie algebra $\mathfrak{g}$. We consider the category $\mathfrak{g}\mathbf{Tan}$ of $\mathfrak{g}$-colored, oriented tangles in $\mathbb{R}^2 \times [0,1]$. The objects of this category are triples of the form $\mathbf{X} = (X_n, \mathbf{s}, \mathbf{V})$, where X_n is our preferred set of n points in $\mathbb{R}^2$, $\mathbf{s} \in \{\pm\}^n$ is an n-tuple of orientations on these points, and $\mathbf{V} \in (\mathrm{Rep}(\mathfrak{g}))^n$ is an n-tuple of representations labeling the points of X_n. We divide out by the equivalence relation which allows us to reverse the orientation s_i on the point x_i at the cost of replacing its label V_i with the dual representation V_i^*.

The morphisms in $\mathfrak{g}\mathbf{Tan}$ are oriented, colored tangles $T \subset \mathbb{R}^2 \times [0,1]$. If T is such a tangle, then $\partial T = \partial_0 T \amalg \partial_1 T$, where $\partial_i T \subset \mathbb{R}^2 \times i$. The points of ∂T inherit the orientations and colors of the tangle components which they lie on. In this way, we can view T as an element of $\mathrm{Mor}(\partial_0 T, \partial_1 T)$. Again, we divide out be the relation which allows us to replace the orientation on a component of T at the cost of replacing the color with the dual color. As usual, we also divide out by the action of isotopies on $\mathbb{R}^2 \times I$ which fix $\mathbb{R}^2 \times \partial I$.

To a pair of objects $\mathbf{X}$ and $\mathbf{X}'$ of $\mathfrak{g}\mathbf{Tan}$, the WRT invariant associates a free $\mathbb{Z}[q^{\pm 1}]$ module $\mathbf{W}_{\mathbf{X},\mathbf{X}'}$. To a tangle $T : \mathbf{X} \to \mathbf{X}'$ it should associate an element $\langle T \rangle_{\mathfrak{g}} \in \mathbf{W}_{\partial_0 T, \partial_1 T}$, which is well-defined up to multiplication by $\pm q^k$. (As with the Kauffman bracket, this ambiguity can be fixed by considering framed tangles.) If $T : \mathbf{X}' \to \mathbf{X}''$ is another tangle, there should be a composition map

$$\cdot : \mathbf{W}_{\mathbf{X},\mathbf{X}'} \times \mathbf{W}_{\mathbf{X}',\mathbf{X}''} \to \mathbf{W}_{\mathbf{X},\mathbf{X}''}$$

which satisfies

$$\langle TT' \rangle_{\mathfrak{g}} \sim \langle T \rangle_{\mathfrak{g}} \cdot \langle T' \rangle_{\mathfrak{g}}$$

If $\mathbf{X}_\emptyset := (X_0, \emptyset, \emptyset)$, then $\mathbf{W}_{\mathbf{X}_\emptyset,\mathbf{X}_\emptyset} = \mathbb{Z}[q^{\pm 1}]$. If $L \subset \mathbb{R}^3$ is a closed, oriented, colored link, then we can view L as a tangle $L : \mathbf{X}_\emptyset \to \mathbf{X}_\emptyset$, so $\langle L \rangle_{\mathfrak{g}} \in \mathbb{Z}[q^{\pm 1}]$. We have already seen examples of these invariants. The polynomial $\overline{P}_N(L) := \overline{P}(L)|_{a=q^N}$ is the WRT invariant corresponding to $\mathfrak{g} = \mathfrak{sl}_N$ in which all components of L are colored with the vector representation of $\mathfrak{sl}_N$.

Example 6.1.1. Suppose $\mathfrak{g} = \mathfrak{sl}_2$ and $\mathbf{X}_n$ is X_n, with all points labeled with the vector representation V. The choice of orientation is irrelevant, since $V \simeq V^*$. In this case, $\mathbf{W}_{\mathbf{X}_n,\mathbf{X}_m} = V_{n,m}$ is the linear space spanned by crossingless planar tangles (see section 4.3).

If T is a tangle whose components are labeled with the vector representation, then $\langle T \rangle_{\mathfrak{sl}_2}$ is the Kauffman bracket of T.

The dimension of $\mathbf{W}_{\mathbf{X},\mathbf{X}'}$ can be determined from the representation theory of $\mathfrak{g}$. To the object $\mathbf{X} = (X_n, \mathbf{s}, \mathbf{V})$, we assign

$$R(\mathbf{X}) = \bigotimes_{i=1}^{n} V_i^{s_i} \in \mathrm{Rep}(\mathfrak{g}),$$

where $V_i^+ = V$ and $V_i^- = V^*$. Then

$$\dim_{\mathbb{Z}[q^{\pm 1}]} \mathbf{W}_{\mathbf{X},\mathbf{X}'} = \dim_{\mathbb{C}} \mathrm{Hom}(R(\mathbf{X}), R(\mathbf{X}')).$$

If V and V' are irreps of $\mathfrak{g}$, then

$$\dim(V, V') = \begin{cases} 1 & \text{if } V \simeq V' \\ 0 & \text{if } V \not\simeq V' \end{cases}$$

Hence this dimension can be computed by decomposing $R(\mathbf{X})$ and $R(\mathbf{X}')$ into their irreducible components.

Example 6.1.2. Let $\mathfrak{g} = \mathfrak{sl}_N$, and let $\mathbf{X}_2$ consist of two positively oriented points, each labeled with V. Then $R(\mathbf{X}_2) = V \otimes V \simeq \mathrm{Sym}^2 V \oplus \Lambda^2 V$ splits as a sum of two different irreps, so $\dim \mathrm{Hom}(R(\mathbf{X}_2), R(\mathbf{X}_2)) = 2$. Hence there must be a linear relation between $\langle$$\rangle_{\mathfrak{sl}_N}$, $\langle$$\rangle_{\mathfrak{sl}_N}$, and $\langle$$\rangle_{\mathfrak{sl}_N}$. This is the source of the HOMFLY-PT skein relation.

Exercise 6.1.3. Let $\mathfrak{g} = \mathfrak{sl}_2$, and let V be the vector representation. Show that

$$\dim \operatorname{Hom}(V^{\otimes n}, V^{\otimes m}) = \begin{cases} C_{n+m/2} & \text{if } n+m \text{ is even} \\ 0 & \text{if } n+m \text{ is odd} \end{cases},$$

where

$$C_k = \frac{1}{k+1}\binom{2k}{k}$$

is the kth Catalan number. Compare with Exercise 4.2.2.

Exercise 6.1.4. Let $\mathfrak{g} = \mathfrak{sl}_N$, where $N \geqslant 4$. If you know some representation theory, show that $\Lambda^2 V \otimes \Lambda^2 V$ has three irreducible summands. More generally, show that $\Lambda^k V \otimes \Lambda^l V$ has $\min(k+1, l+1)$ irreducible summands when $N \geqslant k+l$?

6.2. Webs We now discuss the *MOY state model*, which was introduced by Murakami, Ohtsuki, and Yamada in [55]. It gives a simple way of constructing the invariants of the previous section when $\mathfrak{g} = \mathfrak{sl}_N$ and all colors are exterior powers of the vector representation. It should be viewed as a generalization of the construction of the Kauffman bracket in section 4.

We start by defining a diagrammatic category $\mathcal{W}$ which will play the same role that the category of planar tangles played in the definition of the Kauffman bracket. Let $X_n = \{x_1, \dots, x_n\}$ be our standard set of n points in $\mathbb{R}$.

Definition 6.2.1. An (n, n')–*web* is an oriented graph embedded in $\mathbb{R} \times I$ such that

(1) $\Gamma \cap (\mathbb{R} \times 0) = X_n$, $\Gamma \cap (\mathbb{R} \times 1) = X_{n'}$.
(2) Each edge e of Γ is labeled by a non-negative integer $k(e)$.
(3) $X_n \cup X_{n'}$ is the set of univalent vertices of Γ.
(4) The other vertices of Γ are locally modeled on one of the two pictures shown in the figure below.

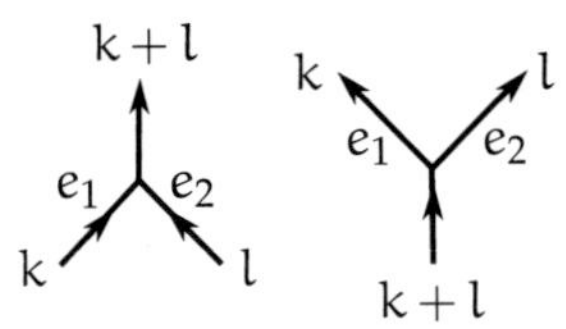

Figure 6.2.2. Internal vertices of a web

Example 6.2.3. The MOY graphs considered in the previous lecture are webs, where $k(e) = 1$ if e is a thin edge, and 2 if e is a thick edge.

If W is an (n, n')–web, its restriction to $\mathbb{R} \times 0$ determines a triple $\mathbf{X} = (X_n, \mathbf{s}, \mathbf{k})$ where s_i is the induced orientation at x_i, and $k_i = k(e_i)$, where e_i is the unique edge adjacent to x_i. The objects of $\mathcal{W}$ are such triples, and the morphisms are webs between them. Composition is given by horizontal stacking, just as in the category of planar tangles.

6.3. The MOY bracket A *closed web* W is a web with no univalent vertices; it can be viewed as a morphism $W : \mathbf{X}_\emptyset \to \mathbf{X}_\emptyset$. The key construction of [55] is a state model for evaluating the $\mathfrak{sl}_N$ bracket of a closed web.

Definition 6.3.1. Let Γ be a closed web. A $\mathfrak{sl}_N$*-state* of Γ is a function which assigns to each edge e of Γ a subset $A_e \subset \{-N+1, -N+3, \dots, N-1\}$ such that

(1) $|A_e| = k(e)$ is the label on e.
(2) At each vertex v of Γ,

$$\bigcup_{v\to e} A_e = \bigcup_{e\to v} A_e.$$

One side of the equation in condition (2) is a union of two sets, while the other side is a single set. From the balancing condition on the labels at trivalent vertices, we see that the two sets appearing in the union must be disjoint.

The definition of a state has a simple physical interpretation. We think of Γ as being a wiring diagram for a collection of electric cables, where the number of cables passing through e is exactly $k(e)$. Each cable is labeled by an element of $\{-N+1, -N+3, \dots, N-1\}$; property (2) tells us that no cables are created or destroyed as we pass through a vertex.

A state σ of Γ determines a resolution Γ_σ, which is a collection of closed oriented circles in $\mathbb{R}^2$. Each circle C is labeled by $\sigma_C \in \{-N+1, -N+3, \dots, N-1\}$. In the physical model above, the circles are the closed loops formed by individual cables.

We now define two weights associated to a state σ. The first is

$$R(\sigma) = \sum_{C\in\Gamma_\sigma} \sigma_C \operatorname{rot} C$$

where $\operatorname{rot} C$ is the rotation number: $+1$ if C is counterclockwise, and -1 if C is clockwise.

To define the second quantity, suppose that v is a vertex of Γ. After rotating the plane of the paper, v has one of the two forms shown in Figure 6.2.2. We define

$$w(\sigma, v) = \sum_{a_1\in A_{e_1}, a_2\in A_{e_2}} \frac{1}{2} s(a_1, a_2),$$

where

$$s(a,b) = \begin{cases} 1 & a > b \\ -1 & a < b \end{cases}.$$

The second weight is $W(\sigma) = \sum_{v\in\Gamma_\sigma} w(\sigma, v)$.

Definition 6.3.2. If Γ is a closed web, we define the $\mathfrak{sl}_N$ bracket

$$\langle\Gamma\rangle_N = \sum_\sigma q^{W(\sigma)+R(\sigma)}$$

where the sum runs over all $\mathfrak{sl}_N$–states of Γ.

Exercise 6.3.3. Show that if Γ is a single circle labeled by k, $\langle \Gamma \rangle_N = \begin{bmatrix} N \\ k \end{bmatrix}$, where $\begin{bmatrix} N \\ k \end{bmatrix}$ is the *quantum binomial coefficient*:

$$\begin{bmatrix} n \\ m \end{bmatrix} = \frac{[n]!}{[m]![n-m]!}$$

and $[n]! = [1][2]\cdots[n]$.

Proposition 6.3.4 ([55]). *The MOY bracket satisfies the following relations:*

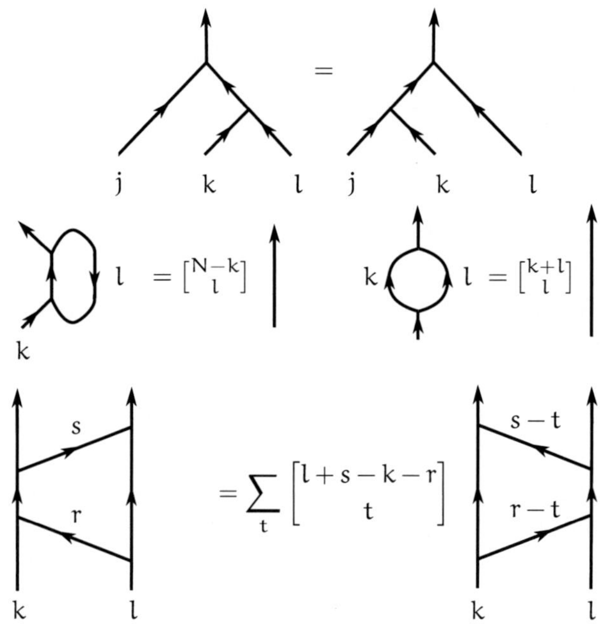

As an example, we prove the right-hand relation in the middle row, which is a generalization of the the MOY 2) move. Let $\Gamma_{k,l}$ and Γ be the graphs on the left and right-hand sides of the equation. We prove the statement by induction on l. For the base case, take $l = 1$. Let e be the edge of Γ shown in the figure and let e_1, e_2 bet the edges of $\Gamma_{k,l}$ labeled k and l. A state σ' of Γ' is specified by a state σ of Γ together with an element $a_{e_2} \in A_e$. We have $W(\sigma') = W(\sigma) + n - (k - n)$, where n is the number of elements of the set $\{a \in A_e \mid a < a_{e_2}\}$. Since $R(\sigma') = R(\sigma)$, it follows that $\langle \Gamma_{k,1} \rangle_N = [k] \langle \Gamma \rangle_N$.

For the general case, consider the two webs $\Gamma_{k,1,l}$ and $\widehat{\Gamma}_{k,1,l}$ shown below.

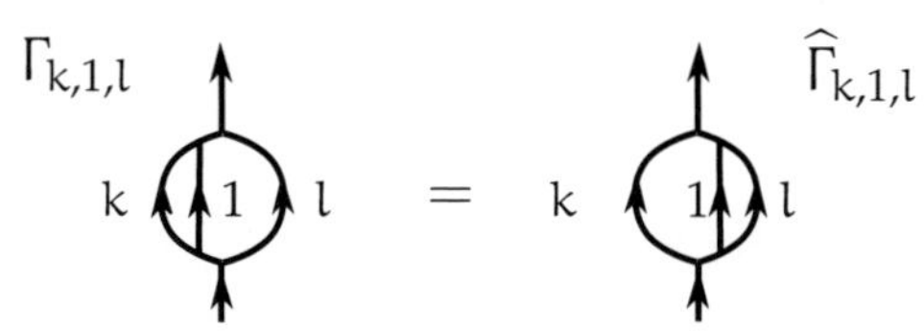

Using the MOY relation in the top row above, we see that $\langle \Gamma_{k,1,l} \rangle_N = \langle \widehat{\Gamma}_{k,1,l} \rangle_N$. Next, we use the relation for $l = 1$ to see that

$$[k+1] \langle \Gamma_{k+1,l} \rangle_N = \langle \Gamma_{k,1,l} \rangle_N = \langle \widehat{\Gamma}_{k,1,l} \rangle_N = [l+1] \langle \Gamma_{k,l+1} \rangle_N .$$

By the induction hypothesis, we see that

$$\langle \Gamma_{k,l+1}\rangle_N = \frac{[k+1]}{[l+1]}\langle \Gamma_{k+1,l}\rangle_N = \frac{[k+1]}{[l+1]}\begin{bmatrix} k+l+1 \\ l \end{bmatrix}\langle \Gamma\rangle_N = \begin{bmatrix} k+l+1 \\ l+1 \end{bmatrix}\langle \Gamma\rangle_N.$$

The relations of Proposition 6.3.4 are generalizations of the MOY relations used in section 5. Note that the coefficients that appear can all be expressed as rational functions of $a := q^N$ and q. For example

$$\begin{bmatrix} N \\ k \end{bmatrix} = \frac{\{0\}\{1\}\cdots\{k-1\}}{[1][2]\dots[k]}.$$

This can be used to prove that as N varies, the polynomials $\langle \Gamma\rangle_N$ are all specializations of a rational function of a and q.

Proposition 6.3.5 ([84]). *If Γ is a closed web, there is a rational function $\langle \Gamma\rangle \in \mathbb{Z}[a^{\pm 1}](q)$ such that $\langle \Gamma\rangle_N = \langle \Gamma\rangle|_{a=q^N}$ for all $N > 0$.*

6.4. The web category For motivation, we return to the Kauffman bracket. Associated to the objects X_n, X_m, we had a vector space $V_{n,m}$ generated by simple planar tangles. As in exercise 4.4.3, we can view $V_{n,m} = \mathrm{Hom}_{\mathbf{P}}(X_n, X_m)$, where $\mathbf{P}$ is the quotient of the category of planar tangles up to isotopy by the ideal generated by the local relation $\bigcirc = (q+q^{-1})$.

We would like to do something similar with the category $\mathcal{W}$ to obtain a category **Web** for which $\mathrm{Hom}(\mathbf{X}, \mathbf{X}')$ is a finitely generated R-module. One approach might be to try to take the quotient of $\mathrm{Add}(\mathcal{W}) \otimes R$ by the ideal generated by the MOY relations. A more elegant method is to use the following universal construction, which is due to Blanchet, Habegger, Masbaum, and Vogel [9].

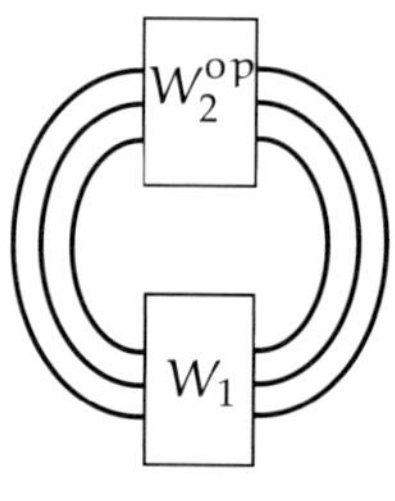

Let $\widetilde{\mathbf{W}}_{\mathbf{X},\mathbf{X}'} := \mathrm{Hom}_{\mathrm{Add}(\mathcal{W})\otimes R}(\mathbf{X}, \mathbf{X}')$ be the free R-module generated by all webs $W : \mathbf{X} \to \mathbf{X}'$. We define a pairing $(\cdot,\cdot) : \widetilde{\mathbf{W}}_{\mathbf{X},\mathbf{X}'} \times \widetilde{\mathbf{W}}_{\mathbf{X},\mathbf{X}'} \to R$ by

$$(W_1, W_2) = \langle \overline{W_1 W_2^{op}}\rangle$$

where $\overline{W_1 W_2^{op}}$ is the closed web shown in the figure above, and extending linearly to $\widetilde{\mathbf{W}}$. Let

$$I_{\mathbf{X},\mathbf{X}'} = \{W \in \widetilde{\mathbf{W}}_{\mathbf{X},\mathbf{X}'} \mid (W, W') = 0 \text{ for all } W' \in \widetilde{\mathbf{W}}_{\mathbf{X},\mathbf{X}'}\}.$$

Lemma 6.4.1. *The $I_{\mathbf{X},\mathbf{X}'}$ form an ideal $\mathcal{I}$ in* $\mathrm{Add}(\mathcal{W}) \otimes R$.

Proof. From the figure, we see that $(W_1W_2, W_3) = (W_1, W_3W_2^{r})$. If $W_1 \in I_{\mathbf{X},\mathbf{X}'}$, and $W_2 : \mathbf{X}' \to \mathbf{X}''$, then $(W_1W_2, W_3) = (W_1, W_3W_2^{r}) = 0$, so $W_1W_2 \in I_{\mathbf{X},\mathbf{X}'}$. A similar argument shows that $\mathcal{I}$ is also a right ideal. □

Definition 6.4.2. **Web** $= \mathcal{W} \otimes R/\mathcal{I}$.

Let $\mathbf{X}(k,l)$ be the object consisting of two positively oriented points with colors k and l respectively, and let $\mathbf{W}(k,l) = \mathrm{Hom}_{\mathbf{Web}}(\mathbf{X}(k,l), \mathbf{X}(k,l))$.

Proposition 6.4.3. $\dim \mathbf{W}(k,l) = \min(k+1, l+1)$.

6.5. The Λ^k colored HOMFLY-PT polynomial We defined the Kauffman bracket for tangles by assigning to a tangle diagram $D : X_n \to X_m$ its bracket $\langle D \rangle \in V_{n,m}$. Similarly, a colored oriented tangle diagram $D : \mathbf{X} \to \mathbf{X}'$ should determine an element $\langle D \rangle \in \mathbf{W}_{\mathbf{X},\mathbf{X}'}$.

As we did with the Jones polynomial, we start with the diagram of a single crossing. Let $\nwarrow\!\!\!\!\nearrow_{k,l}$ be the diagram of a positive crossing, where the overstrand is labeled by k, and the understrand is labeled by l. Similarly, let $\nearrow\!\!\!\!\nwarrow_{k,l}$ be the diagram of a negative crossing where the understrand is labeled by k and the overstrand is labeled by l.

We should have $\langle \nwarrow\!\!\!\!\nearrow_{k,l} \rangle \in \mathbf{W}(k,l)$. By Proposition 6.4.3, we know this space has dimension $\min(k+1, l+1)$. Our first task is to write down a nice basis for this space.

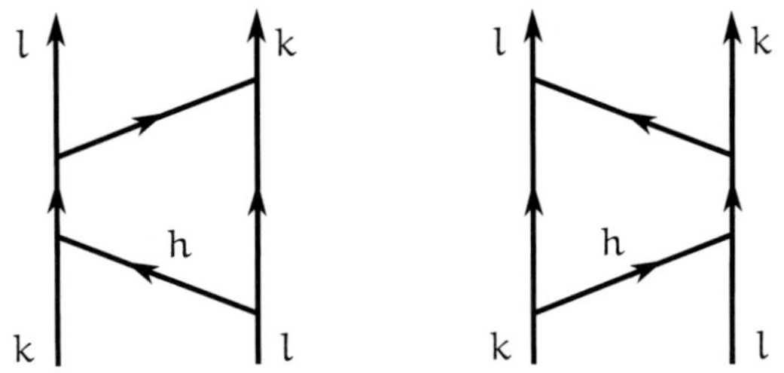

Figure 6.5.1. Basis webs for $\mathbf{W}(k,l)$

If $k \geqslant l \geqslant h \geqslant 0$, let $W_{k,l;h}$ be the open web shown in the figure to the left; while if $l \geqslant k \geqslant h \geqslant 0$, let $W'_{k,l;h}$ be the open web shown in the figure to the right.

Definition 6.5.2. If $k \geqslant l$, we define

$$\langle \nwarrow\!\!\!\!\nearrow_{k,l} \rangle = \sum_{h=0}^{l} (-q)^{h-l} \langle W_{k,l;h} \rangle \qquad \langle \nearrow\!\!\!\!\nwarrow_{k,l} \rangle = \sum_{h=0}^{l} (-q)^{l-h} \langle W_{k,l;h} \rangle$$

Similarly, if $k \leqslant l$, we define

$$\langle \nwarrow\!\!\!\!\nearrow_{k,l} \rangle = \sum_{h=0}^{k} (-q)^{h-k} \langle W'_{k,l;h} \rangle \qquad \langle \nearrow\!\!\!\!\nwarrow_{k,l} \rangle = \sum_{h=0}^{k} (-q)^{k-h} \langle W'_{k,l;h} \rangle$$

In the case where $k = l = 1$, these relations reduce to the previously known ones for the HOMFLY-PT polynomial:

$$\langle \nwarrow\!\!\!\!\nearrow \rangle = q \langle)(\rangle - \langle \Upsilon \rangle \quad \text{and} \quad \langle \nearrow\!\!\!\!\nwarrow \rangle = q^{-1} \langle)(\rangle - \langle \Upsilon \rangle.$$

Exercise 6.5.3. If $k = l$, use MOY relations to show that the two definitions are equal.

If $D : \mathbf{X} \to \mathbf{X}'$ is a colored oriented tangle diagram Definition 6.5.2 expresses $\langle D \rangle$ as a sum of webs. The MOY bracket $\langle D \rangle$ is the image of this sum in $\mathbf{W}_{\mathbf{X},\mathbf{X}'}$. Just like the Kauffman bracket, the MOY bracket is invariant under Reidmeister moves 2) and 3), but not Reidemeister 1).

If D is a colored link diagram, we define $\mathbf{w}(D) = \sum_i k_i w(L_i)$, where the sum runs over the components L_i of D.

Theorem 6.5.4 ([55]). *If* D *and* D′ *are related by a Reidemeister move, then we have* $a^{-k\mathbf{w}(D)}\langle D \rangle = a^{-k\mathbf{w}(D')}\langle D' \rangle$.

Exercise 6.5.5. Check that the theorem holds for the Reidemeister 1) move.

Definition 6.5.6. If L is an integer-colored link, $\overline{P}(L) := a^{-k\mathbf{w}(D)}\langle D \rangle \in \mathbb{Z}[a^{\pm 1}](q)$ is the unnormalized Λ-colored HOMFLY-PT polynomial of L.

When all $k_i = 1$ $\overline{P}(L)$ is the HOMFLY-PT polynomial as defined in the previous section. $\overline{P}(L)|_{a=q^N}$ is the WRT invariant of L associated to the Lie algebra $\mathfrak{sl}_N$, where each component of L is colored by the representation $\Lambda^{k_i}V$. If K is a knot with color k, we write $\overline{P}(K) = \overline{P}^{\Lambda^k}(K)$. In this case, we can define the normalized polynomial

$$P^{\Lambda^k}(K) = \frac{\overline{P}^{\Lambda^k}(K)}{\overline{P}^{\Lambda^k}(\bigcirc)},$$

and $P^{\Lambda^k}(K) \in \mathbb{Z}[q^{\pm 1}, q^{\pm 1}]$.

6.6. Categorification Having defined the Λ^k-colored $\mathfrak{sl}_N$ polynomials, we briefly discuss the problem of categorifying them. Again, we seek to imitate the constructions of section 4. The first step is to find the analog of the 2-category **Cob**.

Definition 6.6.1. A *pre-foam* is a 2-dimensional finite cell complex F embedded in $\mathbb{R} \times I \times I$ together with a labeling of every two-dimensional face f by an orientation, a non-negative integer $k(f)$, and a polynomial $p(f)$, which is a symmetric polynomial in $k(f)$ variables. F should satisfy the following conditions:

(1) $F \cap \mathbb{R} \times I \times 0 = W_0 : \mathcal{X} \to \mathbf{X}'$ and $F \cap \mathbb{R} \times I \times 1 = W_1 : \mathcal{X} \to \mathbf{X}'$ are webs.
(2) $F \cap \mathbb{R} \times 0 \times I = \mathbf{X} \times I$ and $F \cap \mathbb{R} \times 1 \times I = \mathbf{X}' \times I$
(3) In the interior of $\mathbb{R} \times I \times I$, F is locally modeled on the following pictures:
  (a) A plane labeled by $j \in \mathbb{N}$.
  (b) $Y \times I$, where Y is a trivalent vertex with edges labeled by j, k and $j+k$.
  (c) The cone on the 1-skeleton of a tetrahedron, with the six edges labeled by $j, k, l, j+k, j+l, j+k+l$.

The polynomial $p(f)$ fills the role of the "dots" in **Cob**. There is a 2-category **PFoam** of pre-foams, in which the objects are colored oriented points **X**, in which the 1-morphisms are webs $W : \mathbf{X} \to \mathbf{X}'$, and in which the 2-morphisms are prefoams $F : W_0 \to W_1$. Compositions are given by horizontal and vertical stacking.

In **PFoam**, there are foams $F_+ : W_{k,l,h} \to W_{k,l,h+1}$ and $F_- : W_{k,l,h} \to W_{k,l,h-1}$. The foam F_+ is illustrated by the movie in the figure below. To get the foam F_-, we run the movie in reverse. Reflecting across a vertical line gives foams $F'_+ : W'_{k,l,h} \to W'_{k,l,h+1}$ and $F'_- : W'_{k,l,h} \to W'_{k,l,h-1}$.

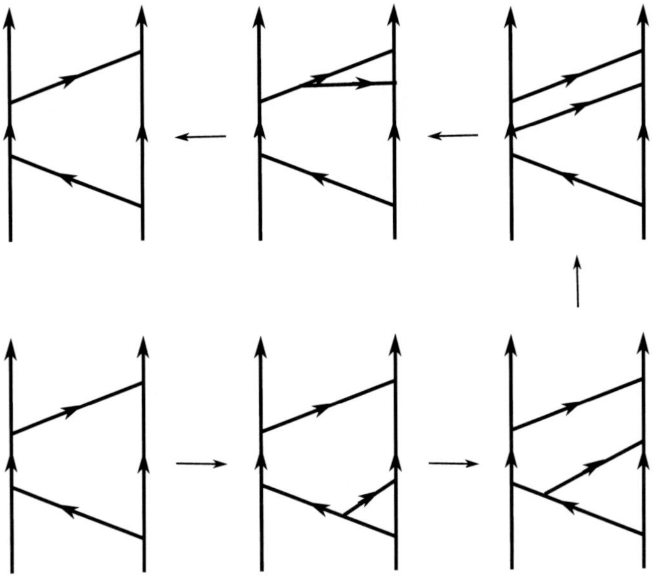

Definition 6.6.2. If $g \geqslant l$, the *Rickard complex* associated to the crossing $\nwarrow\!\!\!\nearrow_{k,l}$ ($k \geqslant l$) is

$$\langle \nwarrow\!\!\!\nearrow_{k,l} \rangle = q^l W_{k,l;0} \xrightarrow{F_+} q^{(l-1)} W_{k,l;1} \xrightarrow{F_+} \dots \xrightarrow{F_+} W_{k,l;l}.$$

There are similar formulas for the other configurations of crossings.

Note that the Rickard complex is not actually a complex in **PFoam**; there is no reason that $d^2 = 0$. To make it into one, we must pass to a new category **Foam**, which is a quotient of **PFoam** — just as we passed from **Cob** to Bar-Natan's category. The resulting category should also have enough relations to ensure that the MOY relations categorify to isomorphisms, in the same way that we were able to prove $\bigcirc \simeq (1 + q^{-1})$ in $\mathcal{C}^{BN}$. A careful list of local relations which must hold in **Foam** may be found in [65]. There are many ways to construct such a category: using matrix factorizations [45], [84], representation theory [51], [78], derived categories of sheaves [13], or categorified quantum groups [65]. Perhaps the simplest construction is due to Robert and Wagner [70]. They define a state sum formula for evaluating a closed foam, analogous to the MOY state sum formula for evaluating a closed web, and show that it satisfies the relations in [65]. The quotient category **Foam** can then be constructed using the BHMV method [9], just as we constructed **Web** from **PWeb**.

6.7. Connections and Further Reading Just as the reduced $\mathfrak{sl}_N$ homologies of a knot limit to its reduced HOMFLY-PT homology as $N \to \infty$, the reduced Λ^k colored homologies limit to a well-defined Λ^k-colored HOMFLY-PT homology $H^{\Lambda^k}(K)$. There are many conjectures in the physics literature about the structure of these invariants. We mention two of the most interesting. The first is the exponential growth conjecture of Gukov and Stosic [31], which states that $\dim H^{\Lambda^k}(K) \geqslant \dim(HHH^r(K))^k$. This was proved by Wedrich [81], who studied deformations and differentials on the Λ^k colored homology. The second is the knots-quivers correspondence [49]. This remarkable conjecture states that all the polynomials $P^{\Lambda^k}(K)$ can be recovered from a finite amount of data, in the form of a vector space equipped with a quadratic form and two linear forms. The conjecture is known to hold for all arborescent knots [76].

In a different direction, we could consider categorifying the WRT invariants for arbitrary colors in type A. A standard method for defining the WRT invariants is via Jones-Wenzl projectors. A Jones-Wenzl projector is a central idempotent in the Hecke algebra H_n; the set of such idempotents is naturally in bijection with the set of partitions of n. As is well-known, any such partition λ determines a representation V_λ of $\mathfrak{sl}_N$. If L is a link whose components L_i are labeled by partitions λ_i, we form a new link $\widehat{L}$ by taking the $|\lambda_i|$-strand cable of each L_i and "inserting the projector" e_{λ_i} into it. (More precisely, we take a braid representative of $\widehat{L}$, look at its image in the Hecke algebra, and insert a factor of e_{λ_i} into each clutch of $|\lambda_i|$ strands.) The trace of the resulting Hecke algebra element is the $\vec{\lambda}$ colored HOMFLY-PT polynomial of L.

A beautiful theorem of Rozansky [72] says that if we take an approriate limit of $CKh(T(n, m))$ as $m \to \infty$, the result is a categorification of the Jones-Wenzl projector in TL_n which corresponds to the representation $\mathrm{Sym}^n(V)$. Similar categorified projectors exist in the Soergel category and were studied by Hogancamp [33] and Abel-Hogancamp [1], who computed the stable HOMFLY-PT homology of the torus knots, as first conjectured in [19]. They can be used to define colored versions of the HOMFLY-PT homology but be careful: using this approach to define a Λ^k colored homology group results in an invariant which is different from those described above.

References

[1] M. Abel and M. Hogancamp, *Categorified Young symmetrizers and stable homology of torus links II*, Selecta Math. (N.S.) **23** (2017), no. 3, 1739–1801, DOI 10.1007/s00029-017-0336-4. MR3663594 ↑166

[2] M. Abouzaid and I. Smith, *Khovanov homology from Floer cohomology*, J. Amer. Math. Soc. **32** (2019), no. 1, 1–79, DOI 10.1090/jams/902. MR3867999 ↑122

[3] D. Bar-Natan, *On Khovanov's categorification of the Jones polynomial*, Algebr. Geom. Topol. **2** (2002), 337–370, DOI 10.2140/agt.2002.2.337. MR1917056 ↑105, 122

[4] D. Bar-Natan and J. Green, *FastKh and JavaKh*, 2005. in Mathematica package KnotTheory. Available at http://katlas.math.toronto.edu. ↑137

[5] D. Bar-Natan, *Khovanov's homology for tangles and cobordisms*, Geom. Topol. **9** (2005), 1443–1499, DOI 10.2140/gt.2005.9.1443. MR2174270 ↑122, 139

[6] J. Batson and C. Seed, *A link-splitting spectral sequence in Khovanov homology*, Duke Math. J. **164** (2015), no. 5, 801–841, DOI 10.1215/00127094-2881374. MR3332892 ↑122

[7] A. Beĭlinson and J. Bernstein, *Localisation de* $\mathfrak{g}$*-modules* (French, with English summary), C. R. Acad. Sci. Paris Sér. I Math. **292** (1981), no. 1, 15–18. MR610137 ↑147

[8] A. Björner and F. Brenti, *Combinatorics of Coxeter groups*, Graduate Texts in Mathematics, vol. 231, Springer, New York, 2005. MR2133266 ↑147, 156

[9] C. Blanchet, N. Habegger, G. Masbaum, and P. Vogel, *Topological quantum field theories derived from the Kauffman bracket*, Topology **34** (1995), no. 4, 883–927, DOI 10.1016/0040-9383(94)00051-4. MR1362791 ↑162, 165

[10] C. Blanchet, *An oriented model for Khovanov homology*, J. Knot Theory Ramifications **19** (2010), no. 2, 291–312, DOI 10.1142/S0218216510007863. MR2647055 ↑115

[11] J.-L. Brylinski and M. Kashiwara, *Kazhdan-Lusztig conjecture and holonomic systems*, Invent. Math. **64** (1981), no. 3, 387–410, DOI 10.1007/BF01389272. MR632980 ↑147

[12] J. S. Carter and M. Saito, *Reidemeister moves for surface isotopies and their interpretation as moves to movies*, J. Knot Theory Ramifications **2** (1993), no. 3, 251–284, DOI 10.1142/S0218216593000167. MR1238875 ↑114

[13] S. Cautis and J. Kamnitzer, *Knot homology via derived categories of coherent sheaves. II.* $\mathfrak{sl}_m$ *case*, Invent. Math. **174** (2008), no. 1, 165–232, DOI 10.1007/s00222-008-0138-6. MR2430980 ↑165

[14] I. Cherednik, *DAHA-Jones polynomials of torus knots*, Selecta Math. (N.S.) **22** (2016), no. 2, 1013–1053, DOI 10.1007/s00029-015-0210-1. MR3477341 ↑156

[15] D. Clark, S. Morrison, and K. Walker, *Fixing the functoriality of Khovanov homology*, Geom. Topol. **13** (2009), no. 3, 1499–1582, DOI 10.2140/gt.2009.13.1499. MR2496052 ↑115

[16] R. Crowell, *Genus of alternating link types*, Ann. of Math. (2) **69** (1959), 258–275, DOI 10.2307/1970181. MR99665 ↑100

[17] R. H. Crowell and R. H. Fox, *Introduction to knot theory*, Ginn and Co., Boston, Mass., 1963. Based upon lectures given at Haverford College under the Philips Lecture Program. MR0146828 ↑95

[18] N. Dowlin, *Knot Floer homology and Khovanov-Rozansky homology for singular links*, Algebr. Geom. Topol. **18** (2018), no. 7, 3839–3885, DOI 10.2140/agt.2018.18.3839. MR3892233 ↑156

[19] N. M. Dunfield, S. Gukov, and J. Rasmussen, *The superpolynomial for knot homologies*, Experiment. Math. **15** (2006), no. 2, 129–159. MR2253002 ↑166

[20] B. Elias and M. Hogancamp, *On the computation of torus link homology*, Compos. Math. **155** (2019), no. 1, 164–205, DOI 10.1112/s0010437x18007571. MR3880028 ↑156

[21] B. Elias and G. Williamson, *The Hodge theory of Soergel bimodules*, Ann. of Math. (2) **180** (2014), no. 3, 1089–1136, DOI 10.4007/annals.2014.180.3.6. MR3245013 ↑156

[22] B. Elias and G. Williamson, *Soergel calculus*, Represent. Theory **20** (2016), 295–374, DOI 10.1090/ert/481. MR3555156 ↑156

[23] C. D. Feustel, *On the torus theorem and its applications*, Trans. Amer. Math. Soc. **217** (1976), 1–43, DOI 10.2307/1997556. MR394666 ↑90

[24] P. Freyd, D. Yetter, J. Hoste, W. B. R. Lickorish, K. Millett, and A. Ocneanu, *A new polynomial invariant of knots and links*, Bull. Amer. Math. Soc. (N.S.) **12** (1985), no. 2, 239–246, DOI 10.1090/S0273-0979-1985-15361-3. MR776477 ↑139

[25] R. E. Gompf and A. I. Stipsicz, *4-manifolds and Kirby calculus*, Graduate Studies in Mathematics, vol. 20, American Mathematical Society, Providence, RI, 1999, DOI 10.1090/gsm/020. MR1707327 ↑121

[26] C. McA. Gordon and J. Luecke, *Knots are determined by their complements*, J. Amer. Math. Soc. **2** (1989), no. 2, 371–415, DOI 10.2307/1990979. MR965210 ↑90

[27] E. Gorsky and A. Neguţ, *Refined knot invariants and Hilbert schemes* (English, with English and French summaries), J. Math. Pures Appl. (9) **104** (2015), no. 3, 403–435, DOI 10.1016/j.matpur.2015.03.003. MR3383172 ↑156

[28] E. Gorsky, A. Neguţ, and J. Rasmussen, *Flag Hilbert schemes, colored projectors and Khovanov-Rozansky homology*, Adv. Math. **378** (2021), 107542, 115, DOI 10.1016/j.aim.2020.107542. MR4192994 ↑156

[29] E. Gorsky, A. Oblomkov, J. Rasmussen, and V. Shende, *Torus knots and the rational DAHA*, Duke Math. J. **163** (2014), no. 14, 2709–2794, DOI 10.1215/00127094-2827126. MR3273582 ↑156

[30] J. E. Grigsby, A. M. Licata, and S. M. Wehrli, *Annular Khovanov homology and knotted Schur-Weyl representations*, Compos. Math. **154** (2018), no. 3, 459–502, DOI 10.1112/S0010437X17007540. MR3731256 ↑122

[31] S. Gukov and M. Stošić, *Homological algebra of knots and BPS states*, Proceedings of the Freedman Fest, Geom. Topol. Monogr., vol. 18, Geom. Topol. Publ., Coventry, 2012, pp. 309–367, DOI 10.2140/gtm.2012.18.309. MR3084243 ↑166

[32] M. Hogancamp, *Khovanov-Rozansky homology and higher Catalan sequences*, 2017. arXiv:1704.01562. ↑156

[33] M. Hogancamp, *Categorified Young symmetrizers and stable homology of torus links*, Geom. Topol. **22** (2018), no. 5, 2943–3002, DOI 10.2140/gt.2018.22.2943. MR3811775 ↑166

[34] J. E. Humphreys, *Reflection groups and Coxeter groups*, Cambridge Studies in Advanced Mathematics, vol. 29, Cambridge University Press, Cambridge, 1990, DOI 10.1017/CBO9780511623646. MR1066460 ↑156

[35] M. Jacobsson, *An invariant of link cobordisms from Khovanov homology*, Algebr. Geom. Topol. **4** (2004), 1211–1251, DOI 10.2140/agt.2004.4.1211. MR2113903 ↑115

[36] V. F. R. Jones, *Hecke algebra representations of braid groups and link polynomials*, Ann. of Math. (2) **126** (1987), no. 2, 335–388, DOI 10.2307/1971403. MR908150 ↑87, 140

[37] L. H. Kauffman, *State models and the Jones polynomial*, Topology **26** (1987), no. 3, 395–407, DOI 10.1016/0040-9383(87)90009-7. MR899057 ↑87

[38] D. Kazhdan and G. Lusztig, *Representations of Coxeter groups and Hecke algebras*, Invent. Math. **53** (1979), no. 2, 165–184, DOI 10.1007/BF01390031. MR560412 ↑147

[39] M. Khovanov, *A categorification of the Jones polynomial*, Duke Math. J. **101** (2000), no. 3, 359–426, DOI 10.1215/S0012-7094-00-10131-7. MR1740682 ↑113, 115

[40] M. Khovanov and L. Rozansky, *Matrix factorizations and link homology II*, 2005. math.QA/0505056. ↑154, 155

[41] M. Khovanov, *A functor-valued invariant of tangles*, Algebr. Geom. Topol. **2** (2002), 665–741, DOI 10.2140/agt.2002.2.665. MR1928174 ↑122, 139

[42] M. Khovanov, *Patterns in knot cohomology. I*, Experiment. Math. **12** (2003), no. 3, 365–374. MR2034399 ↑121

[43] M. Khovanov, *Triply-graded link homology and Hochschild homology of Soergel bimodules*, Internat. J. Math. **18** (2007), no. 8, 869–885, DOI 10.1142/S0129167X07004400. MR2339573 ↑152

[44] M. Khovanov, *Triply-graded link homology and Hochschild homology of Soergel bimodules*, Internat. J. Math. **18** (2007), no. 8, 869–885, DOI 10.1142/S0129167X07004400. MR2339573 ↑154, 156

[45] M. Khovanov and L. Rozansky, *Matrix factorizations and link homology*, Fund. Math. **199** (2008), no. 1, 1–91. MR2391017 ↑122, 165

[46] J. Kock, *Frobenius algebras and 2D topological quantum field theories*, London Mathematical Society Student Texts, vol. 59, Cambridge University Press, Cambridge, 2004. MR2037238 ↑111

[47] A. Kotelskiy, L. Watson, and C. Zibrowius, *Immersed curves in Khovanov homology*, 2019. arXiv:1910.14584. ↑139

[48] P. B. Kronheimer and T. S. Mrowka, *Khovanov homology is an unknot-detector*, Publ. Math. Inst. Hautes Études Sci. **113** (2011), 97–208, DOI 10.1007/s10240-010-0030-y. MR2805599 ↑122

[49] P. Kucharski, M. Reineke, M. Stošić, and P. Sułkowski, *Knots-quivers correspondence*, Adv. Theor. Math. Phys. **23** (2019), no. 7, 1849–1902, DOI 10.4310/atmp.2019.v23.n7.a4. MR4101645 ↑166

[50] E. S. Lee, *An endomorphism of the Khovanov invariant*, Adv. Math. **197** (2005), no. 2, 554–586. MR2173845 ↑115, 117, 119, 122

[51] V. Mazorchuk and C. Stroppel, *A combinatorial approach to functorial quantum $\mathfrak{sl}_k$ knot invariants*, Amer. J. Math. **131** (2009), no. 6, 1679–1713, DOI 10.1353/ajm.0.0082. MR2567504 ↑165

[52] C. T. McMullen, *The Alexander polynomial of a 3-manifold and the Thurston norm on cohomology* (English, with English and French summaries), Ann. Sci. École Norm. Sup. (4) **35** (2002), no. 2, 153–171, DOI 10.1016/S0012-9593(02)01086-8. MR1914929 ↑104

[53] A. Mellit, *Homology of torus knots*, 2017. arXiv:1704.07630. ↑156

[54] J. Milnor, *Whitehead torsion*, Bull. Amer. Math. Soc. **72** (1966), 358–426, DOI 10.1090/S0002-9904-1966-11484-2. MR196736 ↑104

[55] T. Mrowka, P. Ozsváth, and B. Yu, *Seiberg-Witten monopoles on Seifert fibered spaces*, Comm. Anal. Geom. **5** (1997), no. 4, 685–791, DOI 10.4310/CAG.1997.v5.n4.a3. MR1611061 ↑159, 160, 161, 164

[56] K. Murasugi, *On the genus of the alternating knot. I, II*, J. Math. Soc. Japan **10** (1958), 94–105, 235–248, DOI 10.2969/jmsj/01010094. MR99664 ↑100

[57] K. Murasugi, *On alternating knots*, Osaka Math. J. **12** (1960), 277–303. MR137107 ↑97

[58] A. Oblomkov, J. Rasmussen, and V. Shende, *The Hilbert scheme of a plane curve singularity and the HOMFLY homology of its link*, Geom. Topol. **22** (2018), no. 2, 645–691, DOI 10.2140/gt.2018.22.645. With an appendix by Eugene Gorsky. MR3748677 ↑156

[59] A. Oblomkov and L. Rozansky, *Knot homology and sheaves on the Hilbert scheme of points on the plane*, Selecta Math. (N.S.) **24** (2018), no. 3, 2351–2454, DOI 10.1007/s00029-017-0385-8. MR3816507 ↑156

[60] A. Oblomkov and V. Shende, *The Hilbert scheme of a plane curve singularity and the HOMFLY polynomial of its link*, Duke Math. J. **161** (2012), no. 7, 1277–1303, DOI 10.1215/00127094-1593281. MR2922375 ↑156

[61] P. Ozsváth and Z. Szabó, *On the Heegaard Floer homology of branched double-covers*, Adv. Math. **194** (2005), no. 1, 1–33. MR2141852 ↑122

[62] P. Ozsváth and Z. Szabó, *Kauffman states, bordered algebras, and a bigraded knot invariant*, Adv. Math. **328** (2018), 1088–1198, DOI 10.1016/j.aim.2018.02.017. MR3771149 ↑139

[63] P. Ozsváth and Z. Szabó, *Algebras with matchings and knot Floer homology*, 2019. arXiv:1912.01657. ↑139

[64] J. H. Przytycki and P. Traczyk, *Invariants of links of Conway type*, Kobe J. Math. **4** (1988), no. 2, 115–139. MR945888 ↑139

[65] H. Queffelec and D. E. V. Rose, *The $\mathfrak{sl}_n$ foam 2-category: a combinatorial formulation of Khovanov-Rozansky homology via categorical skew Howe duality*, Adv. Math. **302** (2016), 1251–1339, DOI 10.1016/j.aim.2016.07.027. MR3545951 ↑165

[66] J. Rasmussen, *Khovanov homology and the slice genus*, Invent. Math. **182** (2010), no. 2, 419–447, DOI 10.1007/s00222-010-0275-6. MR2729272 ↑119, 120

[67] J. Rasmussen, *Some differentials on Khovanov-Rozansky homology*, Geom. Topol. **19** (2015), no. 6, 3031–3104, DOI 10.2140/gt.2015.19.3031. MR3447099 ↑152

[68] J. Rasmussen, *Some differentials on Khovanov-Rozansky homology*, Geom. Topol. **19** (2015), no. 6, 3031–3104, DOI 10.2140/gt.2015.19.3031. MR3447099 ↑156

[69] N. Reshetikhin and V. G. Turaev, *Invariants of 3-manifolds via link polynomials and quantum groups*, Invent. Math. **103** (1991), no. 3, 547–597. MR1091619 ↑157

[70] L.-H. Robert and E. Wagner, *A closed formula for the evaluation of foams*, Quantum Topol. **11** (2020), no. 3, 411–487, DOI 10.4171/qt/139. MR4164001 ↑165

[71] R. Rouquier, *Categorification of $\mathfrak{sl}_2$ and braid groups*, Trends in representation theory of algebras and related topics, Contemp. Math., vol. 406, Amer. Math. Soc., Providence, RI, 2006, pp. 137–167, DOI 10.1090/conm/406/07657. MR2258045 ↑153

[72] L. Rozansky, *An infinite torus braid yields a categorified Jones-Wenzl projector*, Fund. Math. **225** (2014), no. 1, 305–326, DOI 10.4064/fm225-1-14. MR3205575 ↑166

[73] P. Seidel and I. Smith, *A link invariant from the symplectic geometry of nilpotent slices*, Duke Math. J. **134** (2006), no. 3, 453–514, DOI 10.1215/S0012-7094-06-13432-4. MR2254624 ↑112

[74] P. Seidel and I. Smith, *A link invariant from the symplectic geometry of nilpotent slices*, Duke Math. J. **134** (2006), no. 3, 453–514, DOI 10.1215/S0012-7094-06-13432-4. MR2254624 ↑122

[75] W. Soergel, *Kategorie $\mathcal{O}$, perverse Garben und Moduln über den Koinvarianten zur Weylgruppe* (German, with English summary), J. Amer. Math. Soc. **3** (1990), no. 2, 421–445, DOI 10.2307/1990960. MR1029692 ↑147, 150

[76] M. Stosic and P. Wedrich, *Tangle addition and the knots-quivers correspondence*, 2020. arXiv:2004.10837. ↑166

[77] C. Stroppel, *Categorification of the Temperley-Lieb category, tangles, and cobordisms via projective functors*, Duke Math. J. **126** (2005), no. 3, 547–596, DOI 10.1215/S0012-7094-04-12634-X. MR2120117 ↑122

[78] J. Sussan, *Category O and sl(k) link invariants*, ProQuest LLC, Ann Arbor, MI, 2007. Thesis (Ph.D.)–Yale University. MR2710319 ↑165

[79] V. Turaev, *Introduction to combinatorial torsions*, Lectures in Mathematics ETH Zürich, Birkhäuser Verlag, Basel, 2001. Notes taken by Felix Schlenk, DOI 10.1007/978-3-0348-8321-4. MR1809561 ↑104

[80] P. R. Turner, *Calculating Bar-Natan's characteristic two Khovanov homology*, J. Knot Theory Ramifications **15** (2006), no. 10, 1335–1356, DOI 10.1142/S0218216506005111. MR2286127 ↑115, 117

[81] P. Wedrich, *Exponential growth of colored HOMFLY-PT homology*, Adv. Math. **353** (2019), 471–525, DOI 10.1016/j.aim.2019.06.023. MR3982970 ↑166

[82] W. Whitten, *Knot complements and groups*, Topology **26** (1987), no. 1, 41–44, DOI 10.1016/0040-9383(87)90019-X. MR880506 ↑90

[83] E. Witten, *Quantum field theory and the Jones polynomial*, Comm. Math. Phys. **121** (1989), no. 3, 351–399. MR990772 ↑88, 156
[84] H. Wu, *A colored* $\mathfrak{sl}(N)$ *homology for links in* S^3, Dissertationes Math. **499** (2014), 217, DOI 10.4064/dm499-0-1. MR3234803 ↑162, 165

DPMMS, University of Cambridge, Cambridge, UK
Email address: J.Rasmussen@dpmms.cam.ac.uk

IAS/Park City Mathematics Series
Volume 28, Pages 171–200
https://doi.org/10.1090/pcms/028/00876

Lecture notes on Heegaard Floer homology

Jennifer Hom

Abstract. These are the lecture notes for a course on Heegaard Floer homology given at the Park City Mathematics Institute Graduate Summer School in 2019. We describe Heegaard diagrams, Heegaard Floer homology, knot Floer homology, and the relationship between the knot and 3-manifold invariants.

Contents

Introduction

Heegaard Floer homology, defined by Ozsváth-Szabó [25], and knot Floer homology, defined by Ozsváth-Szabó [23] and independently Rasmussen [38], are

2020 *Mathematics Subject Classification.* Primary 57K10, 57K18, 57K30.
Key words and phrases. Park City Mathematics Institute, Heegaard Floer homology.
The author was partially supported by NSF grant DMS-1552285 and a Sloan Research Fellowship.

invariants of 3-manifolds and knots inside of them. These notes aim to provide an overview of these invariants and the relationship between them.

We will describe our manifolds and knots via Heegaard diagrams, which we define in Section 1. Any two Heegaard diagrams for the same manifold or knot are related by a sequence of moves, called Heegaard moves, much like any two projections for a knot are related by a sequence of Reidemeister moves.

From such diagrams, we will construct chain complexes for 3-manifolds (Section 2) and knots (Section 3) whose chain homotopy type (and in particular, homology) is independent of the choice of Heegaard diagram. The proof that these invariants do not depend on the choice of Heegaard diagram relies on showing invariance under Heegaard moves.

From the knot invariant associated to a knot K in S^3, one can compute the 3-manifold invariant for any Dehn surgery along K; we discuss this relationship in the case of integral surgery in Section 4.

Acknowledgements. I would like to thank the organizers of the 2019 Park City Mathematics Institute for the opportunity to give the lecture series associated with these notes, and the participants of PCMI for being an attentive audience. I would also like to thank Miriam Kuzbary and Robert Lipshitz for helpful comments on earlier versions of these notes.

1. Heegaard splittings and diagrams

1.1. Heegaard splittings Our goal is to define an invariant of closed 3-manifolds and knots inside of them. In order to do this, we will need a way to describe our manifolds and knots. We will do this using Heegaard diagrams. Throughout, we will assume that all of our 3-manifolds are closed and oriented, and that all of our homeomorphisms are orientation preserving, unless stated otherwise. Much of what follows comes from [41, Lecture 1] and [30, Sections 2 and 3].

Definition 1.1.1. A *handlebody of genus* g is a closed regular neighborhood of a wedge of g circles in $\mathbb{R}^3$.

Definition 1.1.2. Let Y be a 3-manifold. A *Heegaard splitting* of Y is a decomposition of $Y = H_1 \cup_f H_2$ where H_1, H_2 are handlebodies and f is an orientation reversing homeomorphism from ∂H_1 to ∂H_2. The *genus* of the Heegaard splitting is the genus of the surface ∂H_1 or equivalently ∂H_2.

Example 1.1.3. Note that $S^3 = B^3 \cup B^3$. This is a genus 0 Heegaard splitting of S^3.

Example 1.1.4. Consider S^3 as the 1-point compactification of $\mathbb{R}^3$. Consider the circle consisting of the z-axis and the point at infinity. A regular neighborhood of this union is a handlebody H_1 of genus 1. The complement of H_1 is also a handlebody of genus 1. Together, these two handlebodies form a genus 1 Heegaard splitting of S^3.

Theorem 1.1.5. *Any closed, orientable 3-manifold Y admits a Heegaard splitting.*

Proof. Consider a triangulation of Y. Let H_1 be a closed regular neighborhood of the 1-skeleton of the triangulation; since H_1 is a neighborhood of a graph, it is a handlebody. The complement $H_2 = Y - H_1$ is also a handlebody; namely it is a neighborhood of the dual 1-skeleton, that is, the graph whose vertices are the centers of the tetrahedra and whose edges are segments perpendicular to the faces of the tetrahedra. □

Remark 1.1.6. Alternatively, a self-indexing Morse function on Y gives rise to a Heegaard splitting of Y; see [30, Section 3], in particular, Exercise 3.5.

Definition 1.1.7. Two Heegaard splittings $Y = H_1 \cup_f H_2$ and $Y = H_1' \cup_{f'} H_2'$ of Y are *homeomorphic* if there exists a homeomorphism $\phi \colon Y \to Y$ taking H_i to H_i'.

Remark 1.1.8. One may also consider the following stricter notion of equivalence: two Heegaard splittings $Y = H_1 \cup_f H_2$ and $Y = H_1' \cup_{f'} H_2'$ of Y are *isotopic* if there exists a map $\psi \colon Y \times [0,1] \to Y$ such that

(1) $\psi|_{Y\times\{0\}} = \mathrm{id}_Y$,
(2) $\psi|_{Y\times t}$ is a homeomorphism for all t,
(3) $\psi|_{Y\times\{1\}}$ sends H_i to H_i'.

Note that if $Y = H_1 \cup_f H_2$ and $Y = H_1' \cup_{f'} H_2'$ are isotopic, then they are homeomorphic (via $\psi|_{Y\times\{1\}}$). The converse is false, since the homeomorphism $\phi \colon Y \to Y$ need not be isotopic to the identity.

Definition 1.1.9. Let $Y = H_1 \cup_f H_2$ be a genus g Heegaard splitting of Y. A *stablization* $H_1' \cup_{f'} H_2'$ of $H_1 \cup_f H_2$ is the genus $g+1$ Heegaard splitting of Y where H_1' consists of H_1 together with a neighborhood N of a properly embedded unknotted arc γ in H_2, and H_2' consists of $H_2 - N$.

Exercise 1.1.10. Prove that the homeomorphism type (in fact, isotopy type) of a stabilization of $Y = H_1 \cup_f H_2$ is independent of the choice of γ.

The following theorem of Reidemeister and Singer highlights the importance of stabilizations.

Theorem 1.1.11 ([39, 42]). *Any two Heegaard splittings of Y become isotopic after sufficiently many stablizations.*

1.2. Heegaard diagrams

We will describe a Heegaard splitting via a Heegaard diagram $\mathcal{H}$, as defined below.

Definition 1.2.1. Let H be a handlebody of genus g. A *set of attaching circles* for H is a set $\{\gamma_1, \dots, \gamma_g\}$ of simple closed curves in $\Sigma = \partial H$ such that

(1) the curves are pairwise disjoint,
(2) $\Sigma - \gamma_1 - \cdots - \gamma_g$ is connected,
(3) each γ_i bounds a disk in H.

Exercise 1.2.2. Show that $\Sigma - \gamma_1 - \cdots - \gamma_g$ is connected if and only if $[\gamma_1], \ldots, [\gamma_g]$ are linearly independent in $H_1(\Sigma; \mathbb{Z})$.

See Figure 1.2.3 for an example of a set of attaching circles.

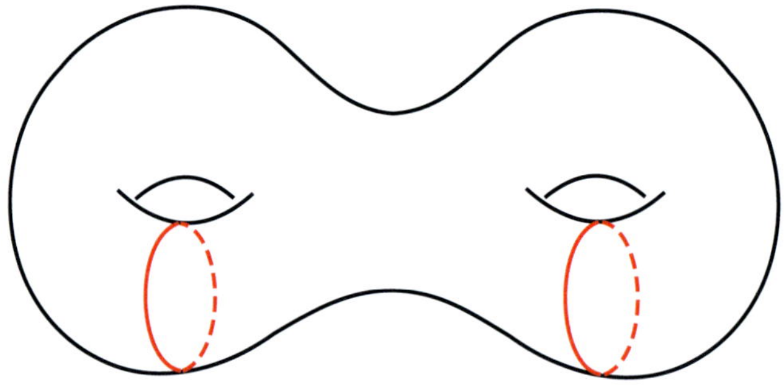

Figure 1.2.3. A set of attaching circles for the "obvious" handlebody in $\mathbb{R}^3$ bounded by Σ.

Definition 1.2.4. A *Heegaard diagram* compatible with $Y = H_1 \cup_f H_2$ is a triple $\mathcal{H} = (\Sigma, \boldsymbol{\alpha}, \boldsymbol{\beta})$ such that

(1) Σ is closed oriented surface of genus g,
(2) $\boldsymbol{\alpha} = \{\alpha_1, \ldots, \alpha_g\}$ is a set of attaching circles for H_1,
(3) $\boldsymbol{\beta} = \{\beta_1, \ldots, \beta_g\}$ is a set of attaching circles for H_2.

We call $(\Sigma, \boldsymbol{\alpha}, \boldsymbol{\beta})$ a *Heegaard diagram for* Y.

Remark 1.2.5. The convention in the field is to draw α-circles in red and β-circles in blue.

See Figure 1.2.6 for examples of Heegaard diagrams for $\mathbb{R}P^3$. (Another example of a Heegaard diagram is given in Figure 1.3.5 below.)

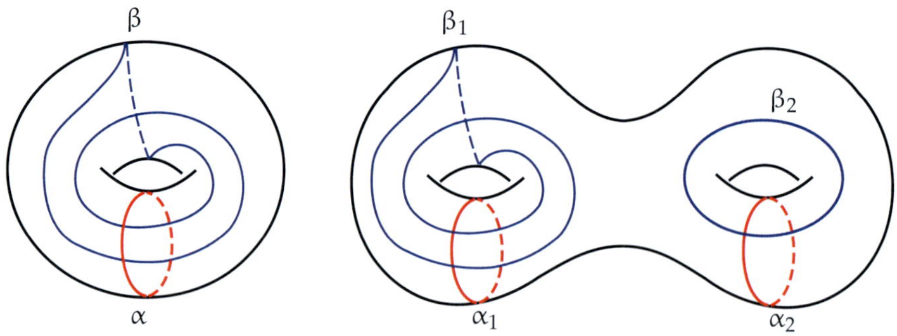

Figure 1.2.6. Left, a Heegaard diagram for $\mathbb{R}P^3$. Right, a stabilization.

Given a Heegaard diagram $(\Sigma, \boldsymbol{\alpha}, \boldsymbol{\beta})$ for Y, we can build a 3-manifold as follows. Thicken Σ to $\Sigma \times [0,1]$. Attach thickened disks along $\alpha_i \times \{0\}$, $1 \leqslant i \leqslant g$, and along $\beta_i \times \{1\}$, $1 \leqslant i \leqslant g$.

Exercise 1.2.7. Verify that since $\boldsymbol{\alpha}$ and $\boldsymbol{\beta}$ are sets of attaching circles, the boundary of the resulting 3-manifold is homeomorphic to $S^2 \sqcup S^2$.

Now fill in each of these boundary components with a copy of B^3; there is a unique way to do so, since any orientation preserving homeomorphism from S^2 to itself is isotopic to the identity. The resulting 3-manifold is homeomorphic to Y.

Exercise 1.2.8. Show that $H_1(Y, \mathbb{Z}) \cong H_1(\Sigma; \mathbb{Z})/\langle[\alpha_1], \dots [\alpha_g], [\beta_1], \dots, [\beta_g]\rangle$.

Exercise 1.2.9. Let $\mathcal{H} = (\Sigma, \boldsymbol{\alpha}, \boldsymbol{\beta})$ be a Heegaard diagram for Y. Compute $H_1(Y; \mathbb{Z})$ from $\mathcal{H}$ as follows. Choose an order and orientation on the α- and β-circles and form the matrix $M = (M_{ij})$ where M_{ij} is the algebraic intersection number between the ith α-circle and the jth β-circle. Show that M is a presentation matrix for $H_1(Y; \mathbb{Z})$.

We now consider three ways to alter a Heegaard diagram that do not change the associated 3-manifold. These moves are called *Heegaard moves* and consist of isotopies, handleslides, and stabilizations/destabilizations. Isotopies and handleslides do not change the genus of the Heegaard diagram, while stabilizations (respectively destablizations) increase the genus by one (respectively decrease the genus by one).

Let $\{\gamma_1, \dots, \gamma_g\}$ be a set of attaching circles for a handlebody H where $\partial H = \Sigma$. The set $\{\gamma_1, \dots, \gamma_g\}$ is *isotopic* to $\{\gamma_1', \dots, \gamma_g'\}$ if there is a 1-parameter family of disjoint simple closed curves starting at $\{\gamma_1, \dots, \gamma_g\}$ and ending at $\{\gamma_1', \dots, \gamma_g'\}$.

A *handleslide* of, say, γ_1 over γ_2 produces a new set $\{\gamma_1', \gamma_2, \dots, \gamma_g\}$ of attaching circles where γ_1' is any simple closed curve disjoint from $\gamma_1, \dots, \gamma_g$ such that γ_1', γ_1, and γ_2 cobound an embedded pair of pants in $\Sigma - \gamma_3 - \dots - \gamma_g$. See Figure 1.2.10.

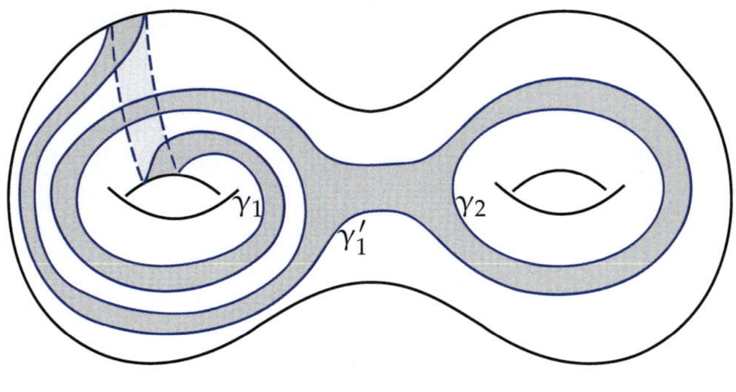

Figure 1.2.10. A handleslide.

Here is another way to think of a handleslide. Suppose that γ_1 and γ_2 can be connected by an arc δ in $\Sigma - \gamma_3 - \dots - \gamma_g$. Let γ_1' be the connected sum of γ_1 with a parallel copy of γ_2, where the connected sum is taken along a neigborhood of δ. See Figure 1.2.12.

Exercise 1.2.11. Prove that the two descriptions of handleslides agree (up to isotopy).

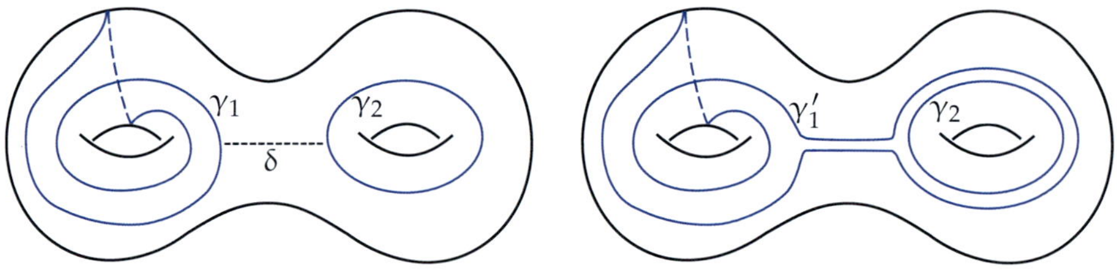

Figure 1.2.12. Another way to view a handleslide.

A *stabilization* of a Heegaard diagram (Σ, α, β) results in a Heegaard diagram $(\Sigma', \alpha', \beta')$ where

(1) $\Sigma' = \Sigma \# T^2$, where $T^2 = S^1 \times S^1$,
(2) $\alpha' = \alpha \cup \{\alpha_{g+1}\}$ and $\beta' = \beta \cup \{\beta_{g+1}\}$, where α_{g+1} and β_{g+1} are two simple closed curves supported in T^2 intersecting transversally in a single point.

We say that (Σ, α, β) is a *destablilization* of $(\Sigma', \alpha', \beta')$.

Exercise 1.2.13. Show that a stabilization of a Heegaard diagram corresponds to a stabilization of the corresponding Heegaard splitting.

We have the following standard fact about Heegaard diagrams (see, for example, [25, Proposition 2.2]).

Theorem 1.2.14. *Let (Σ, α, β) and $(\Sigma', \alpha', \beta')$ be two Heegaard diagrams for Y. Then after applying a finite sequence of isotopies, handleslides, and stabilizations to each of them, the two diagrams become homeomorphic (i.e., there is a homeomorphism $\Sigma \to \Sigma'$ taking α to α' and β to β', setwise).*

We will be interested in *pointed Heegaard diagrams*, that is, tuples $(\Sigma, \alpha, \beta, w)$ where w is a basepoint in $\Sigma - \alpha - \beta$. We now consider pointed isotopies, where the isotopies are not allowed to pass over w, and pointed handleslides, where w is not allowed to be in the pair of pants involved in the handleslide. We have the following upgraded version of Theorem 1.2.14.

Theorem 1.2.15 ([25, Proposition 7.1])**.** *Let $(\Sigma, \alpha, \beta, w)$ and $(\Sigma', \alpha', \beta', w')$ be two pointed Heegaard diagrams for Y. Then after applying a finite sequence of pointed isotopies, pointed handleslides, and stabilizations to each of them, the two diagrams become homeomorphic.*

1.3. Doubly pointed Heegaard diagrams We will also be interested in describing knots inside our 3-manifolds. For simplicity, we will focus on the case where the ambient 3-manifold is S^3.

Definition 1.3.1. A *doubly pointed Heegaard diagram* for a knot $K \subset S^3$ is a tuple $(\Sigma, \alpha, \beta, w, z)$, where w, z are basepoints in $\Sigma - \alpha - \beta$, such that

(1) $(\Sigma, \boldsymbol{\alpha}, \boldsymbol{\beta})$ is a Heegaard diagram for S^3,

(2) K is the union of arcs a and b where a is an arc in $\Sigma - \boldsymbol{\alpha}$ connecting w to z, pushed slightly into H_1 and b is an arc in $\Sigma - \boldsymbol{\beta}$ connecting z to w, pushed slightly into H_2.

See Figure 1.3.2 for an example of a doubly pointed Heegaard diagram for the left-handed trefoil.

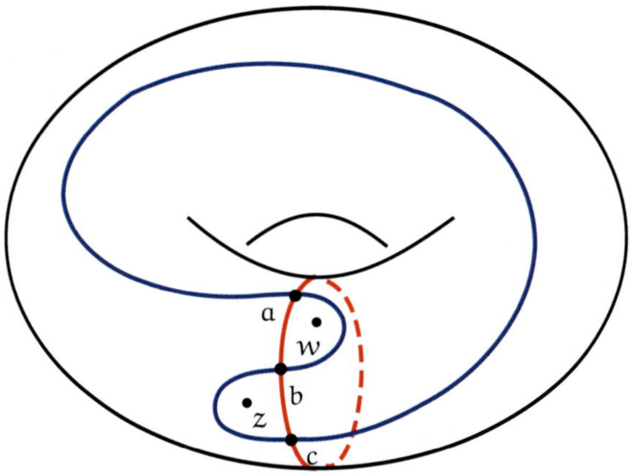

Figure 1.3.2. A doubly pointed Heegaard diagram for the left-handed trefoil, $-T_{2,3}$. (The labelled points a, b, and c will be used in Section 3.)

Given a knot diagram D for a knot K, one can obtain a doubly pointed Heegaard diagram for K as follows. Suppose that D has c crossings. Forgetting the crossing data of the diagram D yields an immersed curve C in the plane. The complement of C is $c+2$ regions in the plane, one of which is unbounded. Let Σ be the boundary of a regular neighborhood of C in $\mathbb{R}^3$; note that Σ is a surface of genus $c+1$. For each of the bounded regions in the complement of C, we put a β-circle on Σ. For each crossing of D, we put an α-circle on Σ as in Figure 1.3.3. Lastly, we add an α-circle, say α_{c+1}, corresponding to a meridan of K. Place a w-basepoint on one side of α_{c+1} and a z-basepoint on the other side. (Note that which side one choose for w determines the orientation of K.) See Figure 1.3.4 for an example.

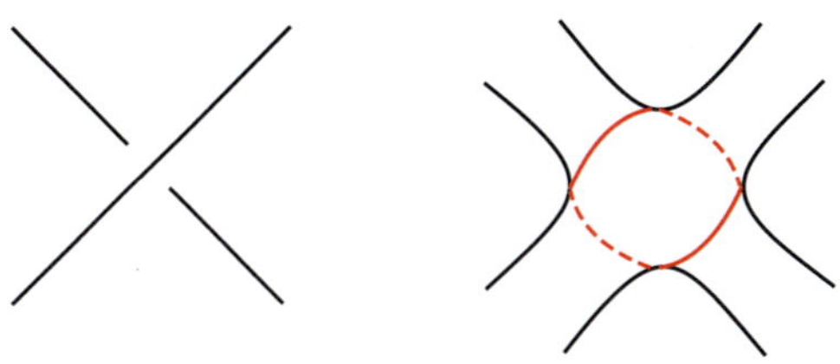

Figure 1.3.3. A knot crossing and the corresponding portion of the associated doubly pointed Heegaard diagram.

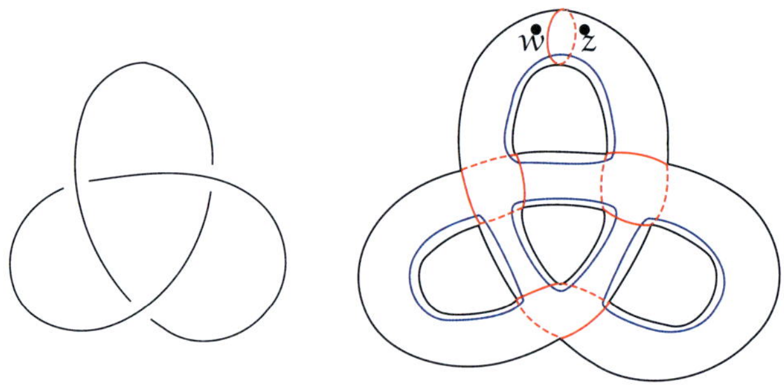

Figure 1.3.4. Another doubly pointed Heegaard diagram for the left-handed trefoil, $-T_{2,3}$.

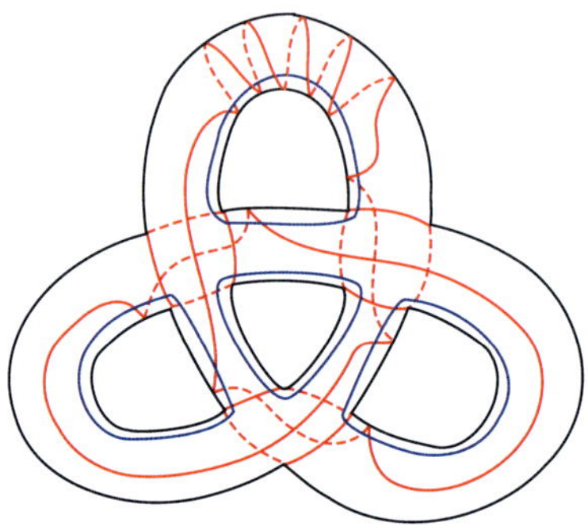

Figure 1.3.5. A Heegaard diagram for $S^3_{+5}(-T_{2,3})$.

Exercise 1.3.6. Show that the above construction yields a doubly pointed Heegaard diagram for $K \subset S^3$.

In the construction above, if we replace the circle α_{c+1} with an n-framed longitude and remove the basepoint z, we obtain a pointed Heegaard diagram for $S^3_n(K)$. See Figure 1.3.5 for an example.

We now consider doubly pointed isotopies, which are required to miss both w and z, and doubly pointed handleslides, where neither w nor z is allowed to be in the pair of pants involved in the handleslide. We have the following analog of Theorem 1.2.15.

Theorem 1.3.7 ([23, Proposition 3.5]). *Let* $(\Sigma, \boldsymbol{\alpha}, \boldsymbol{\beta}, w, z)$ *and* $(\Sigma', \boldsymbol{\alpha}', \boldsymbol{\beta}', w', z')$ *be two doubly pointed Heegaard diagrams for* $K \subset S^3$. *Then after applying a finite sequence of doubly pointed isotopies, doubly pointed handleslides, and stabilizations to each of them, the two diagrams become homeomorphic.*

Exercise 1.3.8. Find a sequence of doubly pointed Heegaard moves passing from Figure 1.3.2 to Figure 1.3.4.

In Sections 2 and 3, we will define invariants of 3-manifolds and knots in S^3. These invariants will be defined in terms of pointed and doubly pointed

Heegaard diagrams, and invariance will follow from the fact that the invariants remain unchanged under pointed and doubly pointed Heegaard moves.

We conclude this section with some additional exercises.

Exercise 1.3.9. Find a genus 1 doubly pointed Heegaard diagram for the figure eight knot.

Exercise 1.3.10. Find a genus 1 doubly pointed Heegaard diagram for the torus knot $T_{p,q}$.

Exercise 1.3.11. Let (Σ, α, β) be a Heegaard diagram for Y. What is the manifold described by $(-\Sigma, \alpha, \beta)$? By (Σ, β, α)?

Exercise 1.3.12. Let $(\Sigma, \alpha, \beta, w, z)$ be a Heegaard diagram for a knot K in S^3. What is the knot described by $(\Sigma, \alpha, \beta, z, w)$? By $(-\Sigma, \beta, \alpha, z, w)$?

2. Heegaard Floer homology

2.1. Overview From a Heegaard diagram $\mathcal{H}$ for Y, we will build chain complexes $\widehat{CF}(\mathcal{H})$ and $CF^-(\mathcal{H})$ whose chain homotopy types are invariants of Y; the former is a finitely generated chain complex over $\mathbb{F}$ and the latter is a finitely generated graded chain complex over $\mathbb{F}[U]$. Throughout, $\mathbb{F}$ denotes the field $\mathbb{Z}/2\mathbb{Z}$ and U is a formal variable of degree -2. We denote the homology of these complexes by $\widehat{HF}(Y)$ and $HF^-(Y)$ respectively; the former is a graded vector space over $\mathbb{F}$ and the latter is a graded module over $\mathbb{F}[U]$. When discussing properties that apply to either flavor of Heegaard Floer homology, will write HF° rather than $\widehat{HF}$, HF^-, HF^+, or HF^∞ (the latter two of which are defined below). The material in this section draws from [30, Sections 4-8] and [25, Sections 3 and 4].

For simplicity, we will consider the case where Y is a rational homology sphere. The case $b_1(Y) > 0$ requires some additional admissibility assumptions on $\mathcal{H}$; see [25, Section 4.2]. Before defining these invariants, we discuss certain aspects of their formal structure.

The Heegaard Floer homology of Y splits as a direct sum over spin^c structures on Y:

$$HF^\circ(Y) = \bigoplus_{\mathfrak{s}\in \text{spin}^c(Y)} HF^\circ(Y, \mathfrak{s})$$

See [30, Section 6] for a discussion of spin^c structures in terms of homotopy classes of non-vanishing vector fields. Note that spin^c structures on Y are in (non-canonical) bijection with elements in $H^2(Y;\mathbb{Z}) \cong H_1(Y;\mathbb{Z})$. The above splitting on the level of homology comes from a splitting on the chain level.

Both $\widehat{HF}(Y)$ and $HF^-(Y)$ are finitely generated and graded. Finitely generated graded vector spaces are simply a direct sum of graded copies of $\mathbb{F}$. Finitely generated graded modules over a PID are also completely characterized. (See, for example, [36, Proposition A.4.3].) Since the only homogenously graded polynomials in $\mathbb{F}[U]$ are the monomials U^n, any finitely generated graded module over

$\mathbb{F}[U]$ is isomorphic to

$$\bigoplus_i \mathbb{F}[U]_{(d_i)} \oplus \bigoplus_j \mathbb{F}[U]_{(c_j)}/(U^{n_j}), \tag{2.1.1}$$

where $\mathbb{F}[U]_{(d)}$ denotes the ring $\mathbb{F}[U]$ where the element 1 has grading d.

Moreover, by [24, Theorem 10.1], we have that for a rational homology sphere Y, for all $\mathfrak{s} \in \mathrm{spin}^c(Y)$,

$$HF^-(Y, \mathfrak{s}) \cong \mathbb{F}[U]_{(d)} \oplus \bigoplus_j \mathbb{F}[U]_{(c_j)}/U^{n_j}, \tag{2.1.2}$$

that is, there is exactly one free summand in $HF^-(Y, \mathfrak{s})$. We define the d-*invariant* of $(Y, \mathfrak{s})$ to be

$$d(Y, \mathfrak{s}) = \max\{\mathrm{gr}(x) \mid x \in HF^-(Y, \mathfrak{s}),\ U^N x \neq 0\ \forall\ N > 0\}.$$

Note that $d(Y, \mathfrak{s})$ is equal to d in Equation (2.1.2). The U-torsion part is called $HF_{\mathrm{red}}(Y, \mathfrak{s})$; that is, using the notation in Equation (2.1.2),

$$HF_{\mathrm{red}}(Y, \mathfrak{s}) = \bigoplus_j \mathbb{F}[U]_{(c_j)}/U^{n_j}.$$

A rational homology sphere Y with $HF_{\mathrm{red}}(Y, \mathfrak{s}) = 0$ for all $\mathfrak{s} \in \mathrm{spin}^c(Y)$ is called an *L-space*.

Remark 2.1.3. Different grading conventions exist in the literature. We have chosen our grading convention so that $HF^-(S^3) \cong \mathbb{F}[U]_{(0)}$, as opposed to the perhaps more common $\mathbb{F}[U]_{(-2)}$. Our grading convention choice simplifies certain formulas, such as the Künneth formula [24, Theorem 1.5]:

$$CF^-(Y_1 \# Y_2, \mathfrak{s}_1 \# \mathfrak{s}_2) \simeq CF^-(Y_1, \mathfrak{s}_1) \otimes_{\mathbb{F}[U]} CF^-(Y_2, \mathfrak{s}_2).$$

Note that our choice of grading convention in turn impacts the gradings on $HF_{\mathrm{red}}(Y, \mathfrak{s})$.

The chain complexes $\widehat{CF}(\mathcal{H}, \mathfrak{s})$ and $CF^-(\mathcal{H}, \mathfrak{s})$ fit in the following U-equivariant short exact sequence:

$$0 \to CF^-(\mathcal{H}, \mathfrak{s}) \xrightarrow{\cdot U} CF^-(\mathcal{H}, \mathfrak{s}) \to \widehat{CF}(\mathcal{H}, \mathfrak{s}) \to 0, \tag{2.1.4}$$

yielding the U-equivariant exact triangle:

$$\begin{array}{ccc} HF^-(Y, \mathfrak{s}) & \xrightarrow{\quad \cdot U \quad} & HF^-(Y, \mathfrak{s}) \\ & \nwarrow \quad \swarrow & \\ & \widehat{HF}(Y, \mathfrak{s}). & \end{array}$$

Remark 2.1.5. Equation (2.1.2) and the above exact triangle imply that for a rational homology sphere Y, we have $\dim \widehat{HF}(Y) \geqslant |H_1(Y; \mathbb{Z})|$; cf. Exercise 2.2.6. It follows that $\dim \widehat{HF}(Y) = |H_1(Y; \mathbb{Z})|$ if and only if Y is an L-space.

We may also consider $HF^\infty(Y, \mathfrak{s}) = H_*(CF^\infty(Y, \mathfrak{s}))$ where

$$CF^\infty(Y, \mathfrak{s}) = CF^-(Y, \mathfrak{s}) \otimes_{\mathbb{F}[U]} \mathbb{F}[U, U^{-1}].$$

Note that $CF^-(Y,\mathfrak{s}) \subset CF^\infty(Y,\mathfrak{s})$ and define $CF^+(Y,\mathfrak{s})$ to be the quotient

$$CF^\infty(Y,\mathfrak{s})/CF^-(Y,\mathfrak{s}).$$

We have the following short exact sequence:

$$0 \to CF^-(Y,\mathfrak{s}) \to CF^\infty(Y,\mathfrak{s}) \to CF^+(Y,\mathfrak{s}) \to 0.$$

The above short exact sequence on the chain level induces the U-equivariant exact triangle shown below:

$$\begin{array}{ccc} HF^-(Y,\mathfrak{s}) & \longrightarrow & HF^\infty(Y,\mathfrak{s}) \\ & \nwarrow \quad \swarrow & \\ & HF^+(Y,\mathfrak{s}). & \end{array} \tag{2.1.6}$$

Exercise 2.1.7. Use Equation (2.1.2) to show that when Y is a rational homology sphere, $HF^\infty(Y,\mathfrak{s}) \cong \mathbb{F}[U,U^{-1}]$. Use this fact combined with Equation (2.1.6) to show that if

$$HF^-(Y,\mathfrak{s}) \cong \mathbb{F}[U]_{(d)} \oplus \bigoplus_j \mathbb{F}[U]_{(c_j)}/U^{n_j},$$

then

$$HF^+(Y,\mathfrak{s}) \cong \mathcal{T}^+_{(d+2)} \oplus \bigoplus_j \mathbb{F}[U]_{(c_j+1)}/U^{n_j},$$

where $\mathcal{T}^+ = \mathbb{F}[U,U^{-1}]/U\mathbb{F}[U]$.

Remark 2.1.8. In light of Exercise 2.1.7, one may either define $HF_{\text{red}}(Y,\mathfrak{s})$ in terms of $HF^-(Y,\mathfrak{s})$ or $HF^+(Y,\mathfrak{s})$; this choice also affects the grading of $HF_{\text{red}}(Y,\mathfrak{s})$. The most common convention in the literature is to define HF_{red} in terms of HF^+.

A smooth cobordism W from Y_0 to Y_1 (that is, a compact smooth 4-manifold W with $\partial W = -Y_0 \sqcup Y_1$) induces a map from the Heegaard Floer homology of Y_0 to the Heegaard Floer homology of Y_1. More specifically, given a spin^c structure $\mathfrak{t}$ on W, we have a homomorphism

$$F^\circ_{W,\mathfrak{t}}\colon HF^\circ(Y_0, \mathfrak{t}|_{Y_0}) \to HF^\circ(Y_1, \mathfrak{t}|_{Y_1}),$$

which is defined via a handle decomposition of W (but does not depend on the choice of handle decomposition). The cobordism map $F^\circ_{W,\mathfrak{t}}$ has a grading shift depending only on W and $\mathfrak{t}$. See [28] for further details or [31, Section 3.2] for an expository overview.

Let Z be a compact 3-manifold with torus boundary, and γ_0, γ_1, and γ_∞ three simple closed curves in ∂Z such that

$$\#(\gamma_0 \cap \gamma_1) = \#(\gamma_1 \cap \gamma_\infty) = \#(\gamma_\infty \cap \gamma_0) = -1.$$

See Figure 2.1.10 for an example of such a triple of curves (which justifies the choice of subscripts on γ). Let Y_i be the result of Dehn filling Z along γ_i for $i = 0, 1, \infty$; that is, Y_i is the union of Z and a solid torus, where γ_i is a meridian of the solid torus. Then by [24, Theorem 9.12], we have an exact triangle

$$\begin{array}{ccc} \widehat{HF}(Y_0) & \xrightarrow{\widehat{F}_0} & \widehat{HF}(Y_1) \\ & \nwarrow_{\widehat{F}_\infty} \quad\quad {}_{\widehat{F}_1}\swarrow & \\ & \widehat{HF}(Y_\infty) & \end{array} \tag{2.1.9}$$

where $\widehat{F}_i$ is the cobordism map associated to the corresponding 2-handle cobordism. The analogous exact triangle also holds for HF^+.

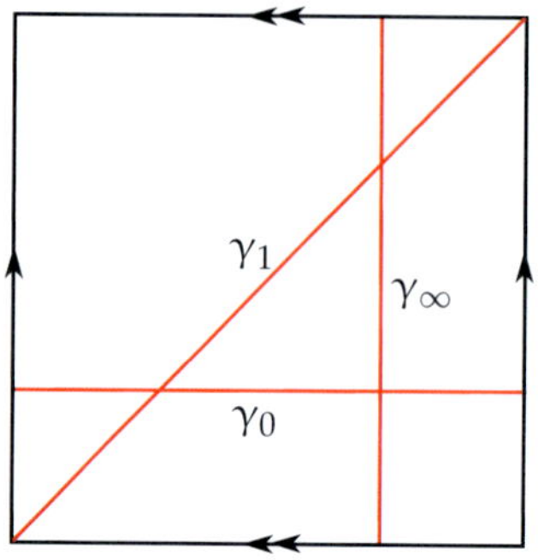

Figure 2.1.10

Remark 2.1.11. For the analogous exact triangle for the minus or infinity flavors, one must work over the formal power series ring $\mathbb{F}[[U]]$ and semi-infinite Laurent polynomials $\mathbb{F}[[U, U^{-1}]$ respectively; see [14, Section 2].

2.2. The Heegaard Floer chain complex Having completed our overview of the structural properties of Heegaard Floer homology, we turn to the construction of the Heegaard Floer chain complex associated to a Heegaard diagram.

Let $\mathcal{H} = (\Sigma, \boldsymbol{\alpha}, \boldsymbol{\beta}, w)$ be a pointed Heegaard diagram for Y where Σ has genus g, and as usual $\boldsymbol{\alpha} = \{\alpha_1, \ldots, \alpha_g\}$ and $\boldsymbol{\beta} = \{\beta_1, \ldots, \beta_g\}$. We further require that the α- and β-circles intersect transversally.

Consider the g-fold symmetric product

$$\mathrm{Sym}^g(\Sigma) = \Sigma^{\times g}/S_g,$$

where S_g denotes the symmetric group on g-elements. Points in $\mathrm{Sym}^g(\Sigma)$ consist of unordered g-tuples of points in Σ.

Exercise 2.2.1. Even though the action of S_g on $\Sigma^{\times g}$ is not free, show that the quotient $\mathrm{Sym}^g(\Sigma)$ is a smooth manifold. (Hint: Fix a complex structure on Σ. Then use the Fundamental Theorem of Algebra to define a map between ordered and unordered g-tuples of complex numbers.)

We have two half-dimensional subspaces of $\mathrm{Sym}^g(\Sigma)$:

$$\mathbb{T}_\alpha = \alpha_1 \times \cdots \times \alpha_g \qquad \text{and} \qquad \mathbb{T}_\beta = \beta_1 \times \cdots \times \beta_g.$$

The chain complex $\widehat{CF}(\mathcal{H})$ is freely generated over $\mathbb{F}$ by $\mathbb{T}_\alpha \cap \mathbb{T}_\beta$, that is, intersection points between $\mathbb{T}_\alpha$ and $\mathbb{T}_\beta$. Note that points in $\mathbb{T}_\alpha \cap \mathbb{T}_\beta$ can be viewed in

Σ as unordered g-tuples of intersection points between the α- and β-circles such that each α-circle, respectively β-circle, is used exactly once.

Exercise 2.2.2. Find all of the generators $\mathbb{T}_\alpha \cap \mathbb{T}_\beta$ in Figure 1.3.5, viewed as unordered g-tuples of points in Σ.

We will also be interested in the following subspace of $\mathrm{Sym}^g(\Sigma)$:

$$V_w = \{w\} \times \mathrm{Sym}^{g-1}(\Sigma).$$

This subspace can be viewed as unordered g-tuples of points in Σ such that at least one point is w.

The differential $\partial\colon \widehat{CF}(\mathcal{H}) \to \widehat{CF}(\mathcal{H})$ will count certain holomorphic disks in $\mathrm{Sym}^g(\Sigma)$. Let $\mathbb{D}$ denote the unit disk in $\mathbb{C}$, and let e_α, respectively e_β, denote the arc in $\partial\mathbb{D}$ with $\mathrm{Re}(z) \geqslant 0$, respectively $\mathrm{Re}(z) \leqslant 0$.

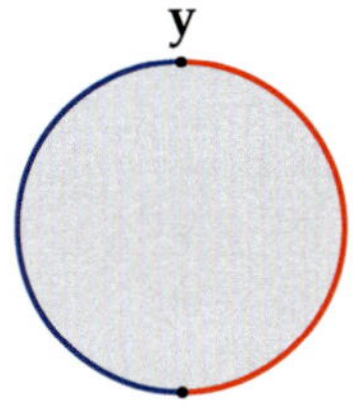

Figure 2.2.3. A schematic of a Whitney disk.

A *Whitney disk* from $\mathbf{x}$ to $\mathbf{y}$, shownn schematically in Figure 2.2.3, is a continuous map $\phi\colon \mathbb{D} \to \mathrm{Sym}^g(\Sigma)$ such that

(1) $\phi(-i) = \mathbf{x}$,
(2) $\phi(i) = \mathbf{y}$,
(3) $\phi(e_\alpha) \subset \mathbb{T}_\alpha$,
(4) $\phi(e_\beta) \subset \mathbb{T}_\beta$.

Let $\pi_2(\mathbf{x}, \mathbf{y})$ denote the set of homotopy classes of Whitney disks from $\mathbf{x}$ to $\mathbf{y}$.

Exercise 2.2.4. Following [30, Section 5.1], there is a rather straightforward obstruction to the existence of a Whitney disk from $\mathbf{x}$ to $\mathbf{y}$.

(1) Viewing $\mathbf{x}$, respectively $\mathbf{y}$, as an unordered g-tuple $\{x_1, \ldots, x_g\}$, respectively $\{y_1, \ldots, y_g\}$, of points in $\boldsymbol{\alpha} \cap \boldsymbol{\beta}$, we may choose a collection of arcs $a \subset \boldsymbol{\alpha}$ such that $\partial a = y_1 + \cdots + y_g - x_1 - \cdots - x_g$ and a collection of arcs $b \subset \boldsymbol{\beta}$ such that $\partial b = x_1 + \cdots + x_g - y_1 - \cdots - y_g$. Then $a - b$ is a 1-cycle in Σ. Using Exercise 1.2.8, verify that the image of $\epsilon(\mathbf{x}, \mathbf{y}) = [a - b] \in H_1(Y; \mathbb{Z})$ is well-defined.

(2) Show that $H_1(\mathrm{Sym}^g(\Sigma); \mathbb{Z}) \cong H_1(\Sigma; \mathbb{Z})$. (Hint: See [25, Lemma 2.6].) Combined with Exercise 1.2.8, conclude that

$$\frac{H_1(\mathrm{Sym}^g(\Sigma); \mathbb{Z})}{H_1(\mathbb{T}_\alpha; \mathbb{Z}) \oplus H_1(\mathbb{T}_\beta; \mathbb{Z})} \cong H_1(Y; \mathbb{Z}),$$

and that if $\epsilon(\mathbf{x}, \mathbf{y}) \neq 0 \in H_1(Y; \mathbb{Z})$, then there cannot exist a Whitney disk in $\mathrm{Sym}^g(\Sigma)$ from $\mathbf{x}$ to $\mathbf{y}$.

A complex structure on Σ induces one on $\mathrm{Sym}^g(\Sigma)$. Given $\phi \in \pi_2(\mathbf{x}, \mathbf{y})$, let $\mathcal{M}(\phi)$ denote the moduli space of holomorphic representatives of ϕ. ([25, Proposition 3.9] ensures that under suitably generic conditions, $\mathcal{M}(\phi)$ is smooth.) The expected dimension of $\mathcal{M}(\phi)$ is called the *Maslov index* of ϕ and denoted $\mu(\phi)$. There is an $\mathbb{R}$-action on $\mathcal{M}(\phi)$ coming from complex automorphisms of $\mathbb{D}$ that preserve i and $-i$. (One way to see this $\mathbb{R}$-action is to use the Riemann mapping theorem to change the unit disk to the infinite strip $[0,1] \times i\mathbb{R} \subset \mathbb{C}$ where e_α corresponds to $\{1\} \times i\mathbb{R}$ and e_β to $\{0\} \times i\mathbb{R}$. Then the $\mathbb{R}$-action corresponds to vertical translations.) Let $\widehat{\mathcal{M}}(\phi) = \mathcal{M}(\phi)/\mathbb{R}$. If $\mu(\phi) = 1$, then $\widehat{\mathcal{M}}(\phi)$ is a compact zero dimensional manifold [25, Theorem 3.18]. Let $n_w(\phi)$ denote the algebraic intersection between $\phi(\mathbb{D})$ and V_w.

We define a relative $\mathbb{Z}$-grading, called the *Maslov grading*, on $\widehat{CF}(\mathcal{H})$ as follows. (Recall that a relative grading defines the difference between the gradings of two elements.) Let $\phi \in \pi_2(\mathbf{x}, \mathbf{y})$. Define

$$\mathrm{gr}(\mathbf{x}) - \mathrm{gr}(\mathbf{y}) = \mu(\phi) - 2n_w(\phi)$$

By [25, Proposition 2.15], this relative grading is well-defined whenever we have $\epsilon(\mathbf{x}, \mathbf{y}) = 0$, where ϵ is the function defined in Exercise 2.2.4.

We may use the function ϵ to partition the intersection points in $\mathbb{T}_\alpha \cap \mathbb{T}_\beta$. These equivalence classes are in bijection with $H_1(Y;\mathbb{Z})$ and hence are in bijection with $\mathrm{spin}^c(Y)$; see [25, Section 2.6] for more details. In particular, we have a splitting $\widehat{CF}(\mathcal{H}) = \oplus_{\mathfrak{s}\in \mathrm{spin}^c(Y)} \widehat{CF}(\mathcal{H}, \mathfrak{s})$.

The differential $\partial\colon \widehat{CF}(\mathcal{H}, \mathfrak{s}) \to \widehat{CF}(\mathcal{H}, \mathfrak{s})$ is defined to be

$$\partial \mathbf{x} = \sum_{\mathbf{y}\in \mathbb{T}_\alpha \cap \mathbb{T}_\beta} \sum_{\substack{\phi \in \pi_2(\mathbf{x},\mathbf{y}) \\ \mu(\phi)=1 \\ n_w(\phi)=0}} \#\widehat{\mathcal{M}}(\phi)\, \mathbf{y}.$$

Note that in the first summation, it suffices to only consider $\mathbf{y}$ with $\epsilon(\mathbf{x}, \mathbf{y}) = 0$; that is, the differential respects the splitting $\widehat{CF}(\mathcal{H}) = \oplus_{\mathfrak{s}\in \mathrm{spin}^c(Y)} \widehat{CF}(\mathcal{H}, \mathfrak{s})$. It follows from the definition of the relative Maslov grading that the differential ∂ lowers the Maslov grading by one. By [25, Theorem 4.1], $\partial^2 = 0$. Let

$$\widehat{HF}(\mathcal{H}, \mathfrak{s}) = H_*(\widehat{CF}(\mathcal{H}, \mathfrak{s})).$$

Remark 2.2.5. We may define a relative $\mathbb{Z}/2\mathbb{Z}$ grading on $\widehat{CF}(\mathcal{H})$ that agrees with the mod 2 reduction of the relative Maslov grading as follows. Following [24, Section 5], choose an orientation of $\mathbb{T}_\alpha$ and $\mathbb{T}_\beta$ and define the relative $\mathbb{Z}/2\mathbb{Z}$ grading between $\mathbf{x}, \mathbf{y} \in \mathbb{T}_\alpha \cap \mathbb{T}_\beta$ to be the product of their local intersection numbers. (Here, we are identifying $\mathbb{Z}/2\mathbb{Z}$ with $\{\pm 1\}$ under multiplication.)

Exercise 2.2.6. Let $\mathcal{H}$ be a Heegaard diagam for a rational homology sphere Y. Use Remark 2.2.5 to prove that $\chi(\widehat{HF}(\mathcal{H})) = \pm |H_1(Y;\mathbb{Z})|$ and then conclude that $\dim \widehat{HF}(\mathcal{H}) \geqslant |H_1(Y;\mathbb{Z})|$. (Hint: Use Exercise 1.2.9.) Compare with Remark 2.1.5.

We now define the chain complex $CF^-(\mathcal{H})$, which is freely generated over $\mathbb{F}[U]$ by $\mathbb{T}_\alpha \cap \mathbb{T}_\beta$. Here, U is a formal variable with $\mathrm{gr}(U) = -2$. We no longer require that Whitney disks miss the basepoint (i.e., we remove the $n_w(\phi) = 0$ requirement), and instead use the variable U to count the algebraic intersection number $n_w(\phi)$ of $\phi(\mathbb{D})$ and V_w:

$$\partial \mathbf{x} = \sum_{\mathbf{y} \in \mathbb{T}_\alpha \cap \mathbb{T}_\beta} \sum_{\substack{\phi \in \pi_2(\mathbf{x},\mathbf{y}) \\ \mu(\phi)=1}} \#\widehat{\mathcal{M}}(\phi) U^{n_w(\phi)} \, \mathbf{y}.$$

The definition of ∂ is extended to all elements of $CF^-(\mathcal{H})$ by $\mathbb{F}[U]$-linearity. As in the case of the complex $\widehat{CF}$, the chain complex $CF^-(\mathcal{H})$ splits over $\text{spin}^c(Y)$, that is, $CF^-(\mathcal{H}) = \oplus_{\mathfrak{s} \in \text{spin}^c(Y)} CF^-(\mathcal{H}, \mathfrak{s})$. By [25, Theorem 4.3], $\partial^2 = 0$. Let

$$HF^-(\mathcal{H}, \mathfrak{s}) = H_*(CF^-(\mathcal{H}, \mathfrak{s})).$$

Theorem 2.2.7 ([25, Theorem 1.1]). *Let $\mathcal{H}$ be a pointed Heegaard diagram for Y. Then the isomorphism type of $\widehat{HF}(\mathcal{H}, \mathfrak{s})$ and $HF^-(\mathcal{H}, \mathfrak{s})$ is an invariant of Y and $\mathfrak{s}$.*

In order to prove the above theorem, one must show that $\widehat{HF}(\mathcal{H}, \mathfrak{s})$ is independent of the choice of Heegaard diagram, basepoint, and complex structure. Indeed, Ozsváth-Szabó show that Heegaard moves induce homotopy equivalences on the associated chain complexes, as do changes in complex structure.

Remark 2.2.8. Note that $\widehat{CF}(\mathcal{H}) = CF^-(\mathcal{H})/(U = 0)$, giving rise to the short exact sequence in (2.1.4).

In [28, Theorem 7.1], Ozsváth-Szabó prove that the relative $\mathbb{Z}$-grading may be lifted to a well-defined absolute $\mathbb{Q}$-grading; this is done by considering a cobordism from S^3 to Y and considering holomorphic triangles associated to a Heegaard triple.

We conclude this section with some computations.

Exercise 2.2.9. Compute $\widehat{HF}(L(p,q), \mathfrak{s})$ and $HF^-(L(p,q), \mathfrak{s})$ for all $\mathfrak{s} \in \text{spin}^c(L(p,q))$.

In general, computing $\widehat{HF}(Y)$ and $HF^-(Y)$ from a Heegaard diagram is not easy. In Section 4, we will see how to compute $\widehat{HF}(Y)$ and $HF^-(Y)$ when Y is surgery on a knot in S^3.

Exercise 2.2.10. Recall that $pq \pm 1$ surgery on the torus knot $T_{p,q}$ is a lens space. Use Exercise 2.2.9 and Equation (2.1.9) to compute $\widehat{HF}(S^3_n(T_{p,q}))$ for $n \geqslant pq - 1$.

Exercise 2.2.11. If $\mathcal{H} = (\Sigma, \boldsymbol{\alpha}, \boldsymbol{\beta}, w)$ a Heegaard diagram for Y and $\mathcal{H}' = (-\Sigma, \boldsymbol{\beta}, \boldsymbol{\alpha}, w)$, show that the chain complex $CF^-(\mathcal{H})$ is isomorphic to $CF^-(\mathcal{H}')$.

Exercise 2.2.12. Use Exercise 1.3.11 and the fact that $d(-Y, \mathfrak{s}) = -d(Y, \mathfrak{s})$ [24, Proposition 4.2] to determine the relationship between $HF^-(Y)$ and $HF^-(-Y)$ as absolutely graded modules.

In general, direct computation of Heegaard Floer homology is difficult. Sarkar and Wang [40] proved that $\widehat{HF}$ can be computed combinatorially using nice diagrams. Lipshitz-Ozsváth-Thurston [13] give an alternative combinatorial method for computing $\widehat{HF}$ using bordered Floer homology, a version of Heegaard Floer homology for 3-manifolds with boundary. Manolescu-Ozsváth-Thurston [17] show that all flavors of Heegaard Floer homology are algorithmically computable in terms of a link surgery description for a 3-manifold and a grid diagram for that link. In Section 4, we will see how to compute HF^- of surgery on a knot in S^3 in terms of the knot Floer complex, a knot invariant which we described in Section 3 below.

Lastly, we note that Theorem 2.2.7 merely states that the isomorphism type of $HF^-(Y)$ is an invariant of Y. Juhász-Thurston-Zemke [10] prove that Heegaard Floer homology in fact assigns a concrete group (that is, not just a group up to isomorphism) to a 3-manifold Y equipped with a basepoint.

3. Knot Floer homology

3.1. Overview Let K be a knot in S^3. The simplest version of the knot invariant is $\widehat{HFK}(K)$, a bigraded vector space over the field $\mathbb{F} = \mathbb{Z}/2\mathbb{Z}$:

$$\widehat{HFK}(K) = \bigoplus_{m,s\in\mathbb{Z}} \widehat{HFK}_m(K,s),$$

where $\widehat{HFK}(K)$ is supported in finitely many bigradings. Knot Floer homology categorifies the Alexander polynomial [23, Equation (1)] in the sense that the graded Euler characteristic of $\widehat{HFK}(K)$ is $\Delta_K(t)$,

$$\Delta_K(t) = \sum_{m,s} (-1)^m \dim \widehat{HFK}_m(K,s)\, t^s.$$

and strengthens two key properties of the Alexander polynomial. Let

$$\Delta_K(t) = a_0 + \sum_{s>0} a_s(t^s + t^{-s})$$

denote the symmetrized Alexander polynomial. While the Alexander polynomial gives a lower bound on the genus of K in the following manner:

$$g(K) \geqslant \max\{s \mid a_s \neq 0\},$$

knot Floer homology actually detects $g(K)$ [22]:

$$g(K) = \max\{s \mid \widehat{HFK}(K,s) \neq 0\}. \tag{3.1.1}$$

Similarly, while the Alexander polynomial obstructs fiberedness in that

$$K \text{ is fibered } \Rightarrow a_{g(K)} = \pm 1,$$

knot Floer homology actually detects fiberedness [3, 19]:

$$K \text{ is fibered } \Leftrightarrow \widehat{HFK}(K, g(K)) = \mathbb{F}.$$

3.2. The knot Floer complex We now modify the constructions in Section 2 to the case of doubly pointed Heegaard diagrams in order to define knot invariants. As mentioned above, for simplicity, we restrict ourselves to knots in S^3.

(With mild modifications, the constructions described here apply to any null-homologous knot in a rational homology sphere.) Some of this material comes from [30, Section 10]; see [23] for more details and proofs. Many of our conventions and notations come from [45]; see especially [45, Section 1.5]. Knot Floer homology was independently defined by Rasmussen in [38].

Recall that in the differential of the chain complex CF^-, the variable U recorded information about the basepoint w. Now that we have two basepoints, we will work with the two variable polynomial ring $\mathbb{F}[U,V]$. We endow this ring with a bigrading $\mathrm{gr} = (\mathrm{gr}_U, \mathrm{gr}_V)$. We call gr_U the U-*grading* and gr_V the V-*grading*. The variables U and V have grading

$$\mathrm{gr}(U) = (-2,0) \qquad \text{and} \qquad \mathrm{gr}(V) = (0,-2).$$

The following linear combination of gr_U and gr_V will often be convenient:

$$A = \frac{1}{2}(\mathrm{gr}_U - \mathrm{gr}_V),$$

called the *Alexander grading*. Note that $A(U) = -1$ and $A(V) = 1$.

For $\mathcal{H}$ a doubly pointed Heegaard diagram for a knot $K \subset S^3$, let $CFK_{\mathbb{F}[U,V]}(\mathcal{H})$ be the free $\mathbb{F}[U,V]$-module generated by $\mathbb{T}_\alpha \cap \mathbb{T}_\beta$. This module is relatively bigraded as follows. Let $\phi \in \pi_2(\mathbf{x},\mathbf{y})$ and define

$$\begin{aligned} \mathrm{gr}_U(\mathbf{x}) - \mathrm{gr}_U(\mathbf{y}) &= \mu(\phi) - 2n_w(\phi) \\ \mathrm{gr}_V(\mathbf{x}) - \mathrm{gr}_V(\mathbf{y}) &= \mu(\phi) - 2n_z(\phi). \end{aligned}$$

By [24, Proposition 7.5] (see also [45, Section 5.1]), this relative grading is well-defined. The relative gradings gr_U and gr_V can be lifted to absolute $\mathbb{Z}$-gradings, using the absolute grading on $HF^-(S^3)$; we describe this process below.

The differential $\partial\colon CFK_{\mathbb{F}[U,V]}(\mathcal{H}) \to CFK_{\mathbb{F}[U,V]}(\mathcal{H})$ is defined to be

$$\partial \mathbf{x} = \sum_{\mathbf{y} \in \mathbb{T}_\alpha \cap \mathbb{T}_\beta} \sum_{\substack{\phi \in \pi_2(\mathbf{x},\mathbf{y}) \\ \mu(\phi)=1}} \#\widehat{\mathcal{M}}(\phi) U^{n_w(\phi)} V^{n_z(\phi)}\, \mathbf{y},$$

and is extended to all elements of $CFK_{\mathbb{F}[U,V]}(\mathcal{H})$ by $\mathbb{F}[U,V]$-linearity. Note that the differential preserves the Alexander grading.

Setting $V = 1$ and forgetting gr_V (that is, only considering the grading gr_U), we recover $CF^-(S^3)$, whose homology is isomorphic to $\mathbb{F}[U]_{(0)}$, where the subscript (0) now denotes $\mathrm{gr}_U(1)$; this determines the absolute U-grading. Note that setting $V = 1$ corresponds to forgetting the z-basepoint. To find the absolute V-grading, we simply reverse the roles of U and V. That is, we set $U = 1$, forget gr_U, and only consider gr_V, recovering $CF^-(S^3)$, whose homology is isomorphic to $\mathbb{F}[V]$ where $\mathrm{gr}_V(1) = 0$. This corresponds to forgetting the w-basepoint.

Theorem 3.2.1 ([23, Theorem 3.1]). *Let $\mathcal{H}$ be a doubly pointed Heegaard diagram for a knot $K \subset S^3$. The chain homotopy type of $CFK_{\mathbb{F}[U,V]}(\mathcal{H})$ is an invariant of $K \subset S^3$.*

Note that [23, Theorem 3.1] is phrased in terms of filtered chain complexes; see [45, Section 1.5] for a description of the translation between filtered chain complexes and modules over $\mathbb{F}[U,V]$. We will often abuse notation and write $CFK_{\mathbb{F}[U,V]}(K)$ rather than $CFK_{\mathbb{F}[U,V]}(\mathcal{H})$.

Example 3.2.2. Figure 1.3.2 shows a doubly pointed Heegaard diagram for the left-handed trefoil. We have

$$\partial a = Ub$$
$$\partial b = 0$$
$$\partial c = Vb.$$

Setting $V = 1$, we see that the homology, which is isomorphic to $\mathbb{F}[U]$, is generated by $[a + Uc]$, implying that $gr_U(a) = gr_U(Uc) = 0$. Setting $U = 1$, we see that the homology, which is isomorphic to $\mathbb{F}[V]$, is generated by $[c + Va]$, implying that $gr_V(c) = gr_V(Va) = 0$. Hence, the generators a, b, c have the following gradings:

	gr_U	gr_V	A
a	0	2	-1
b	1	1	0
c	2	0	1

Then $H_*(CFK_{\mathbb{F}[U,V]}(-T_{2,3})) \cong \mathbb{F}[U,V]_{(0,0)} \oplus \mathbb{F}_{(1,1)}$, where the subscript denotes $gr = (gr_U, gr_V)$ of 1. The $\mathbb{F}[U,V]$ summand is generated by $Va + Uc$ and the $\mathbb{F}$ summand by b.

Exercise 3.2.3. Let $\mathcal{H} = (\Sigma, \alpha, \beta, w, z)$ be a doubly pointed Heegaard diagram for $K \subset S^3$.

(1) Let $(CFK_{\mathbb{F}[U,V]}(\mathcal{H}) \otimes_{\mathbb{F}[V]} \mathbb{F}[V, V^{-1}], s)$ denote the summand, in Alexander grading s, of $CFK_{\mathbb{F}[U,V]}(\mathcal{H}) \otimes_{\mathbb{F}[V]} \mathbb{F}[V, V^{-1}]$ thought of as an $\mathbb{F}[W]$-module, where $W = UV$. Note that multiplication by UV preserves the Alexander grading on $CFK_{\mathbb{F}[U,V]}(\mathcal{H})$. Let $CF^-(\mathcal{H}) = CF^-(\Sigma, \alpha, \beta, w)$. Show that there is an isomorphism of $\mathbb{F}[W]$-modules
$$(CFK_{\mathbb{F}[U,V]}(\mathcal{H}) \otimes_{\mathbb{F}[V]} \mathbb{F}[V, V^{-1}], s) \cong CF^-(\mathcal{H}),$$
by viewing the right-hand side as a module over $\mathbb{F}[W]$, rather than $\mathbb{F}[U]$. Hint: Consider the map $CF^-(\mathcal{H}) \to (CFK_{\mathbb{F}[U,V]}(\mathcal{H}) \otimes_{\mathbb{F}[V]} \mathbb{F}[V, V^{-1}], s)$ given by $W^n\mathbf{x} \mapsto U^n V^{n+s-A(\mathbf{x})}\mathbf{x}$ and conclude that
$$H_*(CFK_{\mathbb{F}[U,V]}(\mathcal{H}) \otimes_{\mathbb{F}[V]} \mathbb{F}[V, V^{-1}]) \cong \mathbb{F}[U, V, V^{-1}].$$

(2) Repeat part (1) reversing the roles of U and V.

The knot Floer complex behaves nicely under connected sum, reversal, and mirroring. By [23, Theorem 7.1], we have that
$$CFK_{\mathbb{F}[U,V]}(K_1 \# K_2) \simeq CFK_{\mathbb{F}[U,V]}(K_1) \otimes_{\mathbb{F}[U,V]} CFK_{\mathbb{F}[U,V]}(K_2).$$

Let K^r denote the reverse of K and mK the mirror. By [23, Section 3], we have that
$$CFK_{\mathbb{F}[U,V]}(mK) \simeq CFK_{\mathbb{F}[U,V]}(K)^*, \tag{3.2.4}$$

where $C^* = \mathrm{Hom}_{\mathbb{F}[U,V]}(C, \mathbb{F}[U,V])$ and

$$CFK_{\mathbb{F}[U,V]}(K^r) \simeq CFK_{\mathbb{F}[U,V]}(K). \tag{3.2.5}$$

Exercise 3.2.6. Let $\mathcal{H} = (\Sigma, \boldsymbol{\alpha}, \boldsymbol{\beta}, w, z)$ be a doubly pointed Heegaard diagram for $K \subset S^3$. Show that $\mathcal{H}_1 = (\Sigma, \boldsymbol{\alpha}, \boldsymbol{\beta}, z, w)$ and $\mathcal{H}_2 = (-\Sigma, \boldsymbol{\beta}, \boldsymbol{\alpha}, w, z)$ are both diagrams for K^r. Show that $CFK_{\mathbb{F}[U,V]}(\mathcal{H})$ is isomorphic to $CFK_{\mathbb{F}[U,V]}(\mathcal{H}_2)$ (cf. Exercise 2.2.11). Conclude that Equation (3.2.5) holds.

Remark 3.2.7. It follows from Equation (3.2.5) and Exercise 3.2.6 that $CFK_{\mathbb{F}[U,V]}(K)$ is chain homotopy equivalent to complex C' obtained from $CFK_{\mathbb{F}[U,V]}(K)$ by exchanging the roles of U and V. (Note that one should then also exchange the roles of gr_U and gr_V.)

Exercise 3.2.8. Compute $CFK_{\mathbb{F}[U,V]}(T_{2,3})$ two ways: from a doubly pointed Heegaard diagram and by applying Equation (3.2.4) to Example 3.2.2, and confirm that the two answers agree.

Exercise 3.2.9. Compute $CFK_{\mathbb{F}[U,V]}$ for the figure eight knot and for the torus knot $T_{3,4}$ using the doubly pointed Heegaard diagrams from Exercises 1.3.9 and 1.3.10.

3.3. Algebraic variations As usual, let $\mathcal{H}$ be a doubly pointed Heegaard diagram for a knot $K \subset S^3$. There are several algebraic modifications we may make to $CFK_{\mathbb{F}[U,V]}(\mathcal{H})$. Since the chain homotopy type of $CFK_{\mathbb{F}[U,V]}(\mathcal{H})$ is an invariant of the knot K, it follows that these algebraic modifications also yield knot invariants.

The first modification we want to consider is setting both $U = 0$ and $V = 0$, resulting in a bigraded chain complex $CFK_{\mathbb{F}}(\mathcal{H})$ over the field $\mathbb{F}$. Choosing to set $U = V = 0$ is equivalent to requiring that $n_w(\phi) = n_z(\phi) = 0$ in the definition of the differential. We denote the homology of $CFK_{\mathbb{F}}(\mathcal{H})$ by $\widehat{HFK}(K)$, which is a bigraded vector space.

It is common to use gr_U and A as the bigrading on $\widehat{HFK}(K)$. (Of course, this is the same information as gr_U and gr_V, since $A = \frac{1}{2}(\mathrm{gr}_U - \mathrm{gr}_V)$.) We write $\widehat{HFK}_m(K, s)$ to denote the summand of $\widehat{HFK}(K)$ with $\mathrm{gr}_U = m$ and $A = s$. The grading gr_U is often called the *Maslov grading*.

Example 3.3.1. Setting $U = V = 0$ in Example 3.2.2 results in

$$\partial a = \partial b = \partial c = 0.$$

Thus, $\widehat{HFK}(K)$ has dimension 3, generated by a, b, and c, with gradings given in the table in Example 3.2.2. That is,

$$\widehat{HFK}_m(K, s) = \begin{cases} \mathbb{F} & \text{if } (m, s) = (0, -1), (1, 0), \text{ or } (2, 1) \\ 0 & \text{otherwise.} \end{cases}$$

Theorem 3.3.2 ([23, Equation (1)]). *The graded Euler characteristic of $\widehat{HFK}(K)$ is equal to the Alexander polynomial of K:*

$$\Delta_K(t) = \sum_{m,s} (-1)^m \dim \widehat{HFK}_m(K, s)\, t^s.$$

Recall from [11] (see also [30, Theorem 11.3]) that the Alexander polynomial of K can be computed in terms of the Kauffman states of a diagram for K. Note that the Kauffman states of the left diagram in Figure 1.3.4 are in bijection with the Heegaard Floer generators of the right diagram. This observation, together with a computation of the bigradings, is at the heart of the proof of Theorem 3.3.2; see [30, Sections 11-13] for details.

Another algebraic modification is to set a single variable, say V, equal to zero, resulting in a chain complex $CFK_{\mathbb{F}[U]}(\mathcal{H})$ over the PID $\mathbb{F}[U]$. This corresponds to requiring that $n_z(\phi) = 0$ in the definition of the differential. The homology of $CFK_{\mathbb{F}[U]}$ is a $\mathbb{F}[U]$-module, denoted $HFK^-(K)$. As a finitely generated graded module over a PID, $HFK^-(K)$ is isomorphic to a direct sum of free summands and U-torsion summands as in Equation (2.1.1).

It is common to view $HFK^-(K)$ as bigraded by gr_U and A. The action of U lowers gr_U by 2 and A by 1.

Exercise 3.3.3. Let $K \subset S^3$. Prove that

$$\begin{aligned} HFK^-(K) \otimes_{\mathbb{F}} \mathbb{F}[U, U^{-1}] &\cong H_*(CFK_{\mathbb{F}[U]}(K) \otimes_{\mathbb{F}[U]} \mathbb{F}[U, U^{-1}]) \\ &\cong \widehat{HF}(S^3) \otimes_{\mathbb{F}} \mathbb{F}[U, U^{-1}] \\ &\cong \mathbb{F}[U, U^{-1}] \end{aligned}$$

and conclude that there is a unique free summand in $HFK^-(K)$.

Example 3.3.4. Setting $V = 0$ in Example 3.2.2 results in the free $\mathbb{F}[U]$-module generated by a, b, and c with differential

$$\begin{aligned} \partial a &= Ub \\ \partial b &= 0 \\ \partial c &= 0. \end{aligned}$$

Hence

$$HFK^-(K) \cong \mathbb{F}[U]_{(2)} \oplus \mathbb{F}_{(1)}$$

where $[c]$ is a generator for the $\mathbb{F}[U]$-summand, $[b]$ is a generator for the $\mathbb{F}$-summand, and the subscript denotes gr_U.

Other algebraic modifications one may consider are setting $U^n = 0$ or $UV = 0$.

3.4. Computations How does one compute $CFK_{\mathbb{F}[U,V]}$ in practice? For small crossing knots, $CFK_{\mathbb{F}[U,V]}$ can be computed via grid diagrams [15, 16]; see [2] for a table of $\widehat{HFK}$ for knots up to 12 crossings computed using grid diagrams. For an excellent textbook on the subject of grid diagrams, see [36]. The invariant $CFK_{\mathbb{F}[U,V]/(UV=0)}$ can be algorithmically computed following [34] (see also [35]); such computations are significantly faster than computations with grid diagrams. At the time of writing, a computer implementation of this algorithm is available at `https://web.math.princeton.edu/~szabo/HFKcalc.html`.

For certain special families of knots, we can compute $CFK_{\mathbb{F}[U,V]}$ directly from the definition (in the case of $(1,1)$-knots, described below) or from other easier to compute knot invariants such as the Alexander polynomial and signature (in the case of alternating knots and knots admitting L-space surgeries).

A knot in S^3 that admits a genus 1 doubly pointed Heegaard diagram is called a $(1,1)$-*knot*. For a $(1,1)$-knot K, the complex $CFK_{\mathbb{F}[U,V]}(K)$ can be computed by counting embedded disks in the universal cover of $\Sigma = T^2$, similar to Example 3.2.2; see [4].

If K is alternating (or more generally, quasi-alternating; see [27, Definition 3.1]), then [21, Theorem 1.3] states that $\widehat{HFK}(K)$ is completely determined by the Alexander polynomial and signature of K. Moreover, [37, Lemma 7] (which is completely algebraic) states that if K is alternating, then $\widehat{HFK}(K)$ completely determines the chain homotopy type of $CFK_{\mathbb{F}[U,V]}(K)$.

Exercise 3.4.1. Let K be an alternating knot. A key ingredient in the proof of [21, Theorem 1.3] is that if $\widehat{HFK}_m(K,s) \neq 0$, then $m = s + \frac{\sigma(K)}{2}$, where $\sigma(K)$ denotes the signature of K. (If $\widehat{HFK}(K)$ is supported on a single diagonal with respect to the Maslov and Alexander gradings, we say K is *homologically thin*.) Show that this fact combined with Theorem 3.3.2 completely determines the bigraded vector space $\widehat{HFK}(K)$ when K is an alternating knot.

If K admits a lens space surgery (or more generally, an L-space surgery), it follows from [26, Theorem 1.2] that $CFK_{\mathbb{F}[U,V]}(K)$ is completely determined by the Alexander polynomial of K, as follows. If K admits an L-space surgery, then the non-zero coefficients in $\Delta_K(t)$ are all ± 1 and they alternate in sign [26, Corollary 1.3]. Let

$$\Delta_K(t) = \sum_{i=0}^{n} (-1)^i t^{a_i},$$

for some decreasing sequence (a_i) and even n. Let $b_i = a_i - a_{i-1}$. If K admits a positive L-space surgery, then $CFK_{\mathbb{F}[U,V]}(K)$ is generated by $x_0, \dots, x_n$ where for i odd,

$$\partial x_i = U^{b_i} x_{i-1} + V^{b_{i+1}} x_{i+1},$$

and for i even, $\partial x_i = 0$. The absolute grading is determined by $\mathrm{gr}_U(x_0) = 0$ and $\mathrm{gr}_V(x_n) = 0$. (Note that there is no loss of generality in considering only positive L-space surgeries. Indeed, if K admits a negative L-space surgery, then mK admits a positive L-space surgery and one can apply Equation (3.2.4).)

Exercise 3.4.2. Compute $CFK_{\mathbb{F}[U,V]}(T_{3,4})$ using the fact that $T_{3,4}$ admits a positive L-space surgery and the above description of $CFK_{\mathbb{F}[U,V]}(K)$ in terms $\Delta_K(t)$ for knots admitting L-space surgeries. Compare with Exercise 3.2.9.

Exercise 3.4.3. Suppose K admits a positive L-space surgery. Express $\widehat{HFK}(K)$ in terms of $\Delta_K(t)$, and verify that $\widehat{HFK}(K)$ satisfies Theorem 3.3.2.

(For the relationship between $(1,1)$-knots and L-space knots, see [5].)

4. Heegaard Floer homology of knot surgery

4.1. Large surgery In this section, we discuss the relationship between the knot Floer complex $\mathit{CFK}_{\mathbb{F}[U,V]}(K)$ and $\mathit{HF}^-(S^3_n(K))$, where $S^3_n(K)$ denotes n-surgery on $K \subset S^3$.

We begin with some observations about $\mathit{CFK}_{\mathbb{F}[U,V]}(K)$, which is a chain complex over $\mathbb{F}[U,V]$. Let $W = UV$. Since multiplication by W preserves the Alexander grading, as a chain complex over $\mathbb{F}[W]$, the complex $\mathit{CFK}_{\mathbb{F}[U,V]}(K)$ splits as a direct sum over the Alexander grading. (However, note that neither multiplication by U nor by V respects this splitting.)

Following [23, Section 4], one may identify spin^c-structures on $S^3_n(K)$ with $\mathbb{Z}/n\mathbb{Z}$. Recall that $\mathit{HF}^-(S^3_n(K))$ is a module over a polynomial ring in a single variable.

Theorem 4.1.1 ([23, Theorem 4.4],[38, Section 4]). *Let $n \geqslant 2g(K)-1$ and $|s| \leqslant \lfloor \frac{n}{2} \rfloor$. Then*

$$\mathit{HF}^-(S^3_n(K),[s]) \cong H_*(\mathit{CFK}_{\mathbb{F}[U,V]}(K,s))$$

as relatively $\mathbb{Z}$-graded modules over a polynomial ring in a single variable W. That is, on the left-hand side, $W = U$ while on the right-hand side, $W = UV$. The relative grading on the right-hand side may be taken to be either gr_U or gr_V.

Remark 4.1.2. See [23, Corollary 4.2] for the absolutely graded version of Theorem 4.1.1.

Example 4.1.3. Let $Y = S^3_{+1}(-T_{2,3})$. (It follows from [18, Proposition 3.1] that $S^3_{+1}(-T_{2,3}) \cong -\Sigma(2,3,7)$.) We will use Theorem 4.1.1 to compute $\mathit{HF}^-(Y)$. Since Y is an integer homology sphere, there is a unique spin^c-structure on Y. From Example 3.2.2, we have that $\mathit{CFK}_{\mathbb{F}[U,V]}(-T_{2,3},0)$ is generated over $\mathbb{F}[W]$, where $W = UV$, by

$$Va, b, Uc.$$

The differential is given by

$$\begin{aligned} \partial(Va) &= W \cdot b \\ \partial b &= 0 \\ \partial(Uc) &= W \cdot b. \end{aligned}$$

Note that the elements Va and Uc are in the same relative grading, while the relative grading of b is one greater than the relative grading of Va. We have that

$$\mathit{HF}^-(Y) \cong H_*(\mathit{CFK}_{\mathbb{F}[U,V]}(-T_{2,3},0)) \cong \mathbb{F}[W]_{(0)} \oplus \mathbb{F}_{(1)},$$

where the $\mathbb{F}[W]$-summand is generated by $Va + Uc$ and the $\mathbb{F}$-summand is generated by b. (The absolute gradings are computed following [23, Corollary 4.2].)

Example 4.1.4. Let $Y = S^3_{+3}(-T_{2,3})$. By Theorem 4.1.1,

$$HF^-(Y,[s]) \cong H_*(CFK_{\mathbb{F}[U,V]}(-T_{2,3}, s)$$

for $s = -1, 0, 1$. By Example 4.1.3, we have

$$HF^-(Y,[0]) \cong H_*(CFK_{\mathbb{F}[U,V]}(-T_{2,3}, 0)) \cong \mathbb{F}[W] \oplus \mathbb{F}.$$

From Example 3.2.2, we have that $CFK_{\mathbb{F}[U,V]}(-T_{2,3}, -1)$ is generated over $\mathbb{F}[W]$ by

$$a, Ub, U^2c.$$

The differential is given by

$$\begin{aligned} \partial a &= Ub \\ \partial(Ub) &= 0 \\ \partial(U^2c) &= W \cdot Ub. \end{aligned}$$

Hence $H_*(CFK_{\mathbb{F}[U,V]}(-T_{2,3}, -1)) \cong \mathbb{F}[W]$, generated by $U^2c + Wa$. Similarly, $H_*(CFK_{\mathbb{F}[U,V]}(-T_{2,3}, 1)) \cong \mathbb{F}[W]$, generated by $V^2a + Wc$; we leave this calculation to the reader. See Figure 4.1.5.

The proof of Theorem 4.1.1 relies on relating the Heegaard diagrams for (S^3, K) and $S^3_n(K)$; see, for example, Figures 1.3.4 and 1.3.5. The rough idea is that for each generator of the Heegaard diagram for (S^3, K) and for each spin^c-structure on $S^3_n(K)$, there is a canonical "nearest" generator obtained by replacing the intersection point on the meridian with a nearby intersection point on the n-framed longitude. The remainder of the proof relies on the relationship between spin^c-structures and the Alexander grading, as well as a count of holomorphic triangles in a Heegaard triple. See [23, Section 4] for more details.

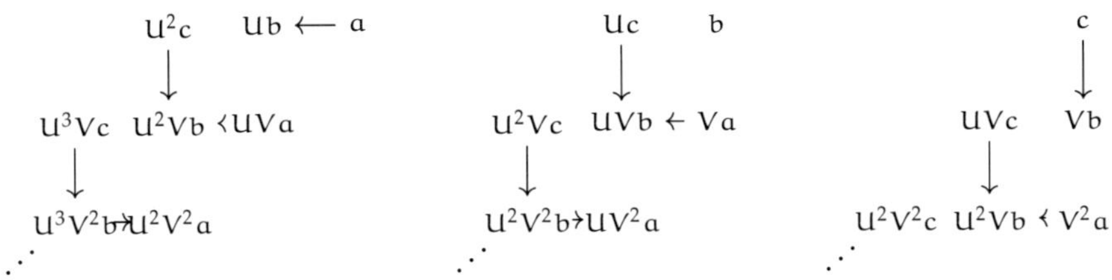

Figure 4.1.5. Three Alexander graded summands of $CFK_{\mathbb{F}[U,V]}(-T_{2,3})$. Left: $A_{-1} = CFK_{\mathbb{F}[U,V]}(-T_{2,3}, -1)$. Center: $A_0 = CFK_{\mathbb{F}[U,V]}(-T_{2,3}, 0)$. Right: $A_1 = CFK_{\mathbb{F}[U,V]}(-T_{2,3}, 1)$.

Exercise 4.1.6. Compute $HF^-(S^3_{+1}(4_1))$, where 4_1 denotes the figure eight knot, using $CFK_{\mathbb{F}[U,V]}(4_1)$ as computed in Exercise 3.2.9.

Exercise 4.1.7. Compute $HF^-(S^3_{+5}(T_{3,4}))$ using $CFK_{\mathbb{F}[U,V]}(T_{3,4})$ as computed in Exercise 3.2.9.

4.2. Integer surgery Note that Theorem 4.1.1 requires that surgery coefficient n to be greater than or equal to $2g(K)-1$. In [32, Theorem 1.1], Ozsváth-Szabó provide a recipe for computing the Heegaard Floer homology of any integer surgery along $K \subset S^3$; they improve this to a formula for rational surgery in [33].

In these notes, we will work with the minus flavor, as in [14, Theorem 1.1]. One disadvantage to working with the minus flavor is that one must work with completed coefficients (cf. Remark 2.1.11), that is, we work over the power series rings $\mathbb{F}[[U]]$ and $\mathbb{F}[[U,V]]$. To this end, for a pointed Heegaard diagram $\mathcal{H}$ for Y, let

$$CF^-_{\mathbb{F}[[U]]}(\mathcal{H}) = CF^-(\mathcal{H}) \otimes_{\mathbb{F}[U]} \mathbb{F}[[U]]$$

and let $HF^-_{\mathbb{F}[[U]]}(Y) = H_*(CF^-_{\mathbb{F}[[U]]}(\mathcal{H}))$. Note that since $\mathbb{F}[[U]]$ is flat over $\mathbb{F}[U]$, we have that $HF^-_{\mathbb{F}[[U]]}(Y) \cong HF^-(Y) \otimes_{\mathbb{F}[U]} \mathbb{F}[[U]]$. Similarly, let

$$CFK_{\mathbb{F}[[U,V]]}(K) = CFK_{\mathbb{F}[U,V]}(K) \otimes_{\mathbb{F}[U,V]} \mathbb{F}[[U,V]].$$

Let

$$A_s = CFK_{\mathbb{F}[[U,V]]}(K,s)$$

denote the part of $CFK_{\mathbb{F}[[U,V]]}(K)$ in Alexander grading s. (Note that this is not quite a grading in the usual sense, as $CFK_{\mathbb{F}[[U,V]]}(K)$ is a direct product rather than a direct sum of its homogenously graded pieces. We will abuse notation and still refer to s as the Alexander grading, and similarly for gr_U and gr_V.) We have that

$$CFK_{\mathbb{F}[[U,V]]}(K) \cong \prod_s A_s.$$

Let

$$B_s = (CFK_{\mathbb{F}[U,V]}(K) \otimes_{\mathbb{F}[U,V]} \mathbb{F}[[U,V,V^{-1}]], s)$$

denote the part of $CFK_{\mathbb{F}[U,V]}(K) \otimes_{\mathbb{F}[U,V]} \mathbb{F}[[U,V,V^{-1}]]$ in Alexander grading s. By Exercise 3.2.3, we have that $B_s \simeq CF^-_{\mathbb{F}[[W]]}(S^3)$ for every s, where $W = UV$. Note that

$$V\colon B_s \to B_{s+1} \qquad \text{and} \qquad V^{-1}\colon B_{s+1} \to B_s.$$

Since the composition of these two maps is the identity, we see that V maps B_s isomorphically to B_{s+1}. We have

$$CFK_{\mathbb{F}[U,V]}(K) \otimes_{\mathbb{F}[U,V]} \mathbb{F}[[U,V,V^{-1}]] \cong \prod_s B_s.$$

with V, V^{-1}, and W acting as follows

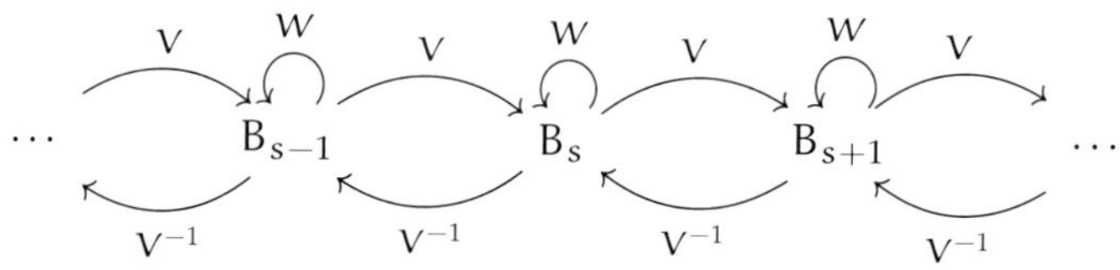

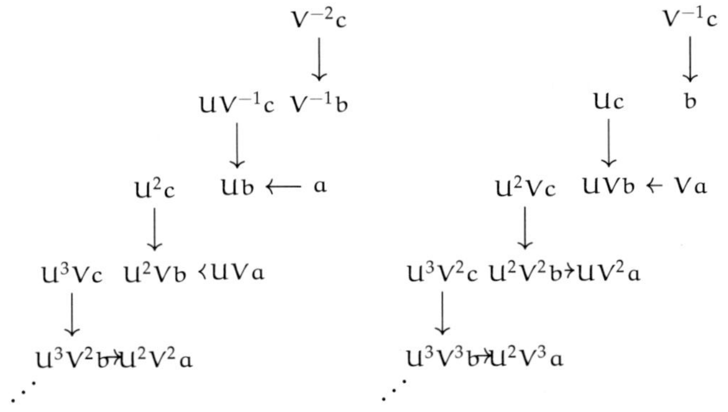

Figure 4.2.1. Different Alexander graded pieces of $CFK_{\mathbb{F}[U,V]}(-T_{2,3}) \otimes_{\mathbb{F}[U,V]} \mathbb{F}[[U,V,V^{-1}]] \cong \prod_s B_s$. Left, B_{-1}. Right, B_0. Note that the two complexes are isomorphic, with the isomorphism provided by multiplication by $V^{\pm 1}$.

By Remark 3.2.7, we have that $CFK_{\mathbb{F}[U,V]}(K) \otimes_{\mathbb{F}[U,V]} \mathbb{F}[[U,V,V^{-1}]]$ is chain homotopy equivalent to $CFK_{\mathbb{F}[U,V]}(K) \otimes_{\mathbb{F}[U,V]} \mathbb{F}[[U,U^{-1},V]]$ after exchanging the roles of U and V. (Note that this chain homotopy equivalence reverses the Alexander grading.) Moreover, in any fixed Alexander grading, both complexes are homotopy equivalent to B_s, so let

$$(CFK_{\mathbb{F}[U,V]}(K) \otimes_{\mathbb{F}[U,V]} \mathbb{F}[[U,U^{-1},V]], s) \xrightarrow{\phi_s} (CFK_{\mathbb{F}[U,V]}(K) \otimes_{\mathbb{F}[U,V]} \mathbb{F}[[U,V,V^{-1}]], s)$$

denote an Alexander grading-preserving chain homotopy equivalence between these two $\mathbb{F}[UV]$-modules. Let $\phi = \prod_s \phi_s$. (Note that ϕ is not $\mathbb{F}[U,V]$-equivariant, although it is $\mathbb{F}[UV]$-equivariant.)

Let

$$\iota_V\colon CFK_{\mathbb{F}[[U,V]]}(K) \to CFK_{\mathbb{F}[U,V]}(K) \otimes_{\mathbb{F}[U,V]} \mathbb{F}[[U,V,V^{-1}]]$$

and

$$\iota_U\colon CFK_{\mathbb{F}[[U,V]]}(K) \to CFK_{\mathbb{F}[U,V]}(K) \otimes_{\mathbb{F}[U,V]} \mathbb{F}[[U,U^{-1},V]]$$

denote inclusion. Since multiplication by V is invertible in $CFK_{\mathbb{F}[U,V]}(K) \otimes_{\mathbb{F}[U,V]} \mathbb{F}[[U,V,V^{-1}]]$, we have that

$$V^n\colon CFK_{\mathbb{F}[U,V]}(K) \otimes_{\mathbb{F}[U,V]} \mathbb{F}[[U,V,V^{-1}]] \to CFK_{\mathbb{F}[U,V]}(K) \otimes_{\mathbb{F}[U,V]} \mathbb{F}[[U,V,V^{-1}]]$$

is a (relatively graded) isomorphism. Note that $V^n|_{B_s}\colon B_s \to B_{s+n}$.

Consider the chain map

$$D_n\colon CFK_{\mathbb{F}[[U,V]]}(K) \to CFK_{\mathbb{F}[U,V]}(K) \otimes_{\mathbb{F}[U,V]} \mathbb{F}[[U,V,V^{-1}]]$$

where $D_n = \iota_V + V^n \circ \phi \circ \iota_U$.

Recall that given two chain complexes $(X, \partial_X), (Y, \partial_Y)$ and a chain map $f\colon X \to Y$ between them, the *mapping cone* of f is the chain complex $\mathrm{Cone}(f) = X \oplus Y$ with the differential $\partial(x,y) = (\partial_X x, f(x) + \partial_Y y)$. (Note that we are working over a ring of characteristic two; otherwise, one needs to insert some appropriate minus signs in the definition of the mapping cone.)

Exercise 4.2.2. Let $f\colon X \to Y$ be a chain map. Show that $(\mathrm{Cone}(f), \partial)$ is a chain complex.

The following surgery formula expresses $HF^-(S^3_n(K))$ in terms of the mapping cone of D_n.

Theorem 4.2.3 ([14, Theorem 1.1], cf. [32, Theorem 1.1]). *We have the following isomorphism of* $\mathbb{F}[W]$*-modules*

$$HF^-(S^3_n(K)) \otimes_{\mathbb{F}[W]} \mathbb{F}[[W]] \cong H_*(\mathrm{Cone}(D_n)),$$

where on the left-hand side $HF^-(S^3_n(K)$ *is viewed as a module over* $W = U$ *and on the right-hand side,* $W = UV$.

There is a similar formula for rational surgeries; see [33]. Note that

$$\iota_V|_{A_s}\colon A_s \to B_s \qquad \text{and} \qquad V^n \circ \phi \circ \iota_U|_{A_s}\colon A_s \to B_{s+n}.$$

Hence

$$D_n|_{A_s}\colon A_s \to B_s \oplus B_{s+n},$$

and so D_n may be depicted as follows:

$$\begin{array}{ccccccc} & A_{s-n} & & A_s & & A_{s+n} & \\ \cdots & & \searrow^{V^n\phi\iota_U} & \downarrow{\scriptstyle \iota_V} & \searrow^{V^n\phi\iota_U} & \downarrow{\scriptstyle \iota_V} & \cdots \\ & & & B_s & & B_{s+n} & \end{array}$$

Exercise 4.2.4. Use (3.1.1) to conclude that for $s \geqslant g(K)$, the maps

$$\iota_V\colon A_s \to B_s \qquad \text{and} \qquad V^n \circ \phi \circ \iota_U\colon A_{-s} \to B_{-s+n}$$

are homotopy equivalences. Use this to conclude that $\mathrm{Cone}(D_1)$ is homotopy equivalent to

$$\begin{array}{ccccccccc} A_{1-g} & & A_{2-g} & & & & A_{g-2} & & A_{g-1} \\ & \searrow^{V\phi\iota_U} & \downarrow{\scriptstyle \iota_V} & \searrow & \cdots & & \downarrow & \searrow^{V\phi\iota_U} & \downarrow{\scriptstyle \iota_V} \\ & & B_{2-g} & & & & & & B_{g-1} \end{array}$$

and that $\mathrm{Cone}(D_{-1})$ is homotopy equivalent to

$$\begin{array}{ccccccc} & & A_{1-g} & & & & A_{g-1} \\ & \swarrow^{V^{-1}\phi\iota_U} & \downarrow{\scriptstyle \iota_V} & \swarrow & \cdots & \downarrow \quad \swarrow^{V^{-1}\phi\iota_U} & \downarrow{\scriptstyle \iota_V} \\ B_{-g} & & B_{1-g} & & & B_{g-2} & B_{g-1}. \end{array}$$

Similar statements hold for arbitrary positive (respectively negative) surgery coefficients.

Remark 4.2.5. We consider the knot Floer complex as a module over the bigraded ring $\mathbb{F}[U, V]$, whereas Ozsváth-Szabó originally defined the knot Floer complex as a $\mathbb{Z}$-filtered chain complex over $\mathbb{F}[U]$. The two formulations are equivalent, as described in [45, Section 1.5]. Indeed, our notation A_s and B_s in this section lines up with the (minus flavor over the power series ring) of the complexes A_s and B_s in [32], while our ι_V (respectively $V^n \phi \iota_U$) is denoted v (respectively h) in [32].

Remark 4.2.6. Theorem 4.2.3 has analogous versions for the plus and hat flavors of Heegaard Floer homology. Both the plus and hat flavors have the advantage that one may pass to homology before taking the mapping cone. For the hat flavor, this is because the chain complexes are all vector spaces, while for the plus flavor, this is because the maps $\iota_{V,*}$ and $(V^n \circ \phi \circ \iota_U)_*$ are surjective for knots in S^3. See [32, Section 5] for some sample calculations.

Exercise 4.2.7. Compute $HF^-(S^3_{+1}(T_{3,4}))$ using Exercise 4.2.4 and the computation from Exercise 4.1.7.

4.3. Applications We conclude with some applications of the mapping cone formula.

Recall the *Cosmetic Surgery Conjecture*, which posits that for a nontrivial knot $K \subset S^3$ and $r, r' \in \mathbb{Q}$, if $S^3_r(K)$ and $S^3_{r'}(K)$ are homeomorphic as oriented manifolds, then $r = r'$. Using the mapping cone formula, Ni-Wu [20] prove the following:

Theorem 4.3.1 ([20, Theorem 1.2]). *Suppose K is a nontrivial knot in S^3 such that $S^3_r(K) \cong S^3_{r'}(K)$ as oriented manifolds where r, r' are distinct rational numbers. Then $r = -r'$ and r is of the form p/q where p, q are relatively prime integers with $q^2 \equiv -1 \pmod{p}$.*

Remark 4.3.2. Hanselman [6, Theorem 2] recently improved this result to $r = \pm 2$ or $\pm 1/q$, using bordered Floer homology, a version of Heegaard Floer homology for 3-manifolds with parametrized boundary. Hanselman's result relies on a reinterpretation of bordered Floer homology as immersed curves [7].

Heegaard Floer homology can also be used to obstruct manifolds from being obtained by surgery on a knot in S^3. Lickorish [12] and Wallace [43] proved that every closed oriented 3-manifold can be obtained by surgery on a link in S^3. However, since $H_1(S^3_{p/q}(K)) \cong \mathbb{Z}/p\mathbb{Z}$, it follows that if $H_1(Y; \mathbb{Z})$ is not cyclic, then Y cannot be obtained by surgery on a knot in S^3. For example, since we know that $H_1(\mathbb{RP}^3 \# \mathbb{RP}^3; \mathbb{Z}) \cong \mathbb{Z}/2\mathbb{Z} \oplus \mathbb{Z}/2\mathbb{Z}$, this shows that $\mathbb{RP}^3 \# \mathbb{RP}^3$ is not surgery on any knot in S^3.

This H_1 obstruction vanishes for integer homology spheres. Auckly [1] gave an example of a hyperbolic integer homology sphere that cannot be obtained via surgery on a knot in S^3; moreover, his techniques can be used to give infinitely many such examples. Using the mapping cone formula, Hom-Karakurt-Lidman [8] prove the following:

Theorem 4.3.3 ([8, Theorem 1.1]). *The Brieskorn homology spheres* $\Sigma(n, 2n-1, 2n+1)$, $n \geqslant 8$, n *even, cannot be obtained by surgery on a knot in* S^3.

The proof of Theorem 4.3.3 relies on a surgery obstruction which relates the d-invariant of surgery on a knot in S^3 with HF_{red}; see [8, Theorem 1.2] for the precise statement.

The mapping cone formula has also played a role in results that make no explicit reference to Dehn surgery. For example, one could ask whether there exist simply-connected, positive definite symplectic 4-manifolds. Recall that a manifold is *geometrically simply-connected* if it admits a handle decomposition with no 1-handles.

Theorem 4.3.4 ([9, Theorem 1.1]). *Let* X *be a closed, geometrically simply-connected 4-manifold with* $b_2^+(X) \geqslant 2$. *If the intersection form of* X *is positive definite, then* X *is not symplectic.*

Remark 4.3.5. Yasui [44] has recently obtained another proof of Theorem 4.3.4, using Seiberg-Witten theory.

The proof of Theorem 4.3.4 relies on the Ozsváth-Szabó closed 4-manifolds invariant [29]. Their invariant in defined as a composition of certain cobordism maps, which, in certain special cases, may be computed using the mapping cone formula.

References

[1] D. Auckly, *Surgery numbers of 3-manifolds: a hyperbolic example*, Geometric topology (Athens, GA, 1993), AMS/IP Stud. Adv. Math., vol. 2, Amer. Math. Soc., Providence, RI, 1997, pp. 21–34. MR1470719 ↑197

[2] J. A. Baldwin and W. D. Gillam, *Computations of Heegaard-Floer knot homology*, J. Knot Theory Ramifications **21** (2012), no. 8, 1250075, 65, DOI 10.1142/S0218216512500757. MR2925428 ↑190

[3] P. Ghiggini, *Knot Floer homology detects genus-one fibred knots*, Amer. J. Math. **130** (2008), no. 5, 1151–1169, DOI 10.1353/ajm.0.0016. MR2450204 ↑186

[4] H. Goda, H. Matsuda, and T. Morifuji, *Knot Floer homology of* $(1,1)$*-knots*, Geom. Dedicata **112** (2005), 197–214, DOI 10.1007/s10711-004-5403-2. MR2163899 ↑191

[5] J. E. Greene, S. Lewallen, and F. Vafaee, $(1,1)$ *L-space knots*, Compos. Math. **154** (2018), no. 5, 918–933, DOI 10.1112/S0010437X17007989. MR3798589 ↑192

[6] J. Hanselman, J. Rasmussen, S. D. Rasmussen, and L. Watson, *L-spaces, taut foliations, and graph manifolds*, Compos. Math. **156** (2020), no. 3, 604–612, DOI 10.1112/s0010437x19007814. MR4055962 ↑197

[7] J. Hanselman, J. Rasmussen, and L. Watson, *Bordered Floer homology for manifolds with torus boundary via immersed curves*, 2016. preprint, arXiv:1604.03466. ↑197

[8] J. Hom, Ç. Karakurt, and T. Lidman, *Surgery obstructions and Heegaard Floer homology*, Geom. Topol. **20** (2016), no. 4, 2219–2251, DOI 10.2140/gt.2016.20.2219. MR3548466 ↑198

[9] J. Hom and T. Lidman, *A note on positive-definite, symplectic four-manifolds*, J. Eur. Math. Soc. (JEMS) **21** (2019), no. 1, 257–270, DOI 10.4171/JEMS/835. MR3880209 ↑198

[10] A. Juhász, D. Thurston, and I. Zemke, *Naturality and mapping class groups in Heegaard Floer homology*, 2012. preprint, arXiv:1210.4996v4. ↑186

[11] L. H. Kauffman, *Formal knot theory*, Mathematical Notes, vol. 30, Princeton University Press, Princeton, NJ, 1983. MR712133 ↑190

[12] W. B. R. Lickorish, *A representation of orientable combinatorial 3-manifolds*, Ann. of Math. (2) **76** (1962), 531–540, DOI 10.2307/1970373. MR151948 ↑197

[13] R. Lipshitz, P. S. Ozsváth, and D. P. Thurston, *Computing $\widehat{\mathrm{HF}}$ by factoring mapping classes*, Geom. Topol. **18** (2014), no. 5, 2547–2681, DOI 10.2140/gt.2014.18.2547. MR3285222 ↑185

[14] P. S. Ozsváth and Z. Szabó, *Knot Floer homology and integer surgeries*, Algebr. Geom. Topol. **8** (2008), no. 1, 101–153, DOI 10.2140/agt.2008.8.101. MR2377279 ↑182, 194, 196

[15] C. Manolescu, P. Ozsváth, and S. Sarkar, *A combinatorial description of knot Floer homology*, Ann. of Math. (2) **169** (2009), no. 2, 633–660. ↑190

[16] C. Manolescu, P. Ozsváth, Z. Szabó, and D. Thurston, *On combinatorial link Floer homology*, Geom. Topol. **11** (2007), 2339–2412, DOI 10.2140/gt.2007.11.2339. MR2372850 ↑190

[17] C. Manolescu, P. Ozsváth, and D. Thurston, *Grid diagrams and Heegaard Floer invariants*, 2009. preprint, arXiv:0910.0078v3. ↑186

[18] L. Moser, *Elementary surgery along a torus knot*, Pacific J. Math. **38** (1971), 737–745. MR383406 ↑192

[19] Y. Ni, *Knot Floer homology detects fibred knots*, Invent. Math. **170** (2007), no. 3, 577–608. ↑186

[20] Y. Ni and Z. Wu, *Cosmetic surgeries on knots in* S^3, J. Reine Angew. Math. **706** (2015), 1–17. ↑197

[21] P. Ozsváth and Z. Szabó, *Heegaard Floer homology and alternating knots*, Geom. Topol. **7** (2003), 225–254, DOI 10.2140/gt.2003.7.225. MR1988285 ↑191

[22] P. Ozsváth and Z. Szabó, *Holomorphic disks and genus bounds*, Geom. Topol. **8** (2004), 311–334. ↑186

[23] P. Ozsváth and Z. Szabó, *Holomorphic disks and knot invariants*, Adv. Math. **186** (2004), no. 1, 58–116. ↑171, 178, 186, 187, 188, 189, 192, 193

[24] P. Ozsváth and Z. Szabó, *Holomorphic disks and three-manifold invariants: properties and applications*, Ann. of Math. (2) **159** (2004), no. 3, 1159–1245, DOI 10.4007/annals.2004.159.1159. MR2113020 ↑180, 182, 184, 185, 187

[25] P. Ozsváth and Z. Szabó, *Holomorphic disks and topological invariants for closed three-manifolds*, Ann. of Math. (2) **159** (2004), no. 3, 1027–1158, DOI 10.4007/annals.2004.159.1027. MR2113019 ↑171, 176, 179, 183, 184, 185

[26] P. Ozsváth and Z. Szabó, *On knot Floer homology and lens space surgeries*, Topology **44** (2005), no. 6, 1281–1300, DOI 10.1016/j.top.2005.05.001. MR2168576 ↑191

[27] P. Ozsváth and Z. Szabó, *On the Heegaard Floer homology of branched double-covers*, Adv. Math. **194** (2005), no. 1, 1–33. ↑191

[28] P. Ozsváth and Z. Szabó, *Holomorphic triangles and invariants for smooth four-manifolds*, Adv. Math. **202** (2006), no. 2, 326–400. ↑181, 185

[29] P. Ozsváth and Z. Szabó, *Holomorphic triangles and invariants for smooth four-manifolds*, Adv. Math. **202** (2006), no. 2, 326–400, DOI 10.1016/j.aim.2005.03.014. MR2222356 ↑198

[30] P. Ozsváth and Z. Szabó, *An introduction to Heegaard Floer homology*, Floer homology, gauge theory, and low-dimensional topology, Clay Math. Proc., vol. 5, Amer. Math. Soc., Providence, RI, 2006, pp. 3–27. MR2249247 ↑172, 173, 179, 183, 187, 190

[31] P. Ozsváth and Z. Szabó, *Lectures on Heegaard Floer homology*, Floer homology, gauge theory, and low-dimensional topology, Clay Math. Proc., vol. 5, Amer. Math. Soc., Providence, RI, 2006, pp. 29–70. MR2249248 ↑181

[32] P. S. Ozsváth and Z. Szabó, *Knot Floer homology and integer surgeries*, Algebr. Geom. Topol. **8** (2008), no. 1, 101–153, DOI 10.2140/agt.2008.8.101. MR2377279 ↑194, 196, 197

[33] P. Ozsváth and Z. Szabó, *Knot Floer homology and rational surgeries*, Algebr. Geom. Topol. **11** (2011), no. 1, 1–68. ↑194, 196

[34] P. Ozsváth and Z. Szabó, *Bordered knot algebras with matchings*, Quantum Topol. **10** (2019), no. 3, 481–592, DOI 10.4171/QT/127. MR4002230 ↑190

[35] P. Ozsváth and Z. Szabó, *Algebras with matchings and knot Floer homology*, 2019. preprint, arXiv:1912.01657. ↑190

[36] P. S. Ozsváth, A. I. Stipsicz, and Z. Szabó, *Grid homology for knots and links*, Mathematical Surveys and Monographs, vol. 208, American Mathematical Society, Providence, RI, 2015, DOI 10.1090/surv/208. MR3381987 ↑179, 190

[37] I. Petkova, *Cables of thin knots and bordered Heegaard Floer homology*, Quantum Topol. **4** (2013), no. 4, 377–409, DOI 10.4171/QT/43. MR3134023 ↑191

[38] J. A. Rasmussen, *Floer homology and knot complements*, ProQuest LLC, Ann Arbor, MI, 2003. Thesis (Ph.D.)–Harvard University. MR2704683 ↑171, 187, 192

[39] K. Reidemeister, *Zur dreidimensionalen Topologie* (German), Abh. Math. Sem. Univ. Hamburg **9** (1933), no. 1, 189–194, DOI 10.1007/BF02940644. MR3069596 ↑173

[40] S. Sarkar and J. Wang, *An algorithm for computing some Heegaard Floer homologies*, Ann. of Math. (2) **171** (2010), no. 2, 1213–1236, DOI 10.4007/annals.2010.171.1213. MR2630063 ↑185

[41] N. Saveliev, *Lectures on the topology of 3-manifolds*, De Gruyter Textbook, Walter de Gruyter & Co., Berlin, 1999. An introduction to the Casson invariant, DOI 10.1515/9783110806359. MR1712769 ↑172

[42] J. Singer, *Three-dimensional manifolds and their Heegaard diagrams*, Trans. Amer. Math. Soc. **35** (1933), no. 1, 88–111, DOI 10.2307/1989314. MR1501673 ↑173

[43] A. H. Wallace, *Modifications and cobounding manifolds*, Canadian J. Math. **12** (1960), 503–528, DOI 10.4153/CJM-1960-045-7. MR125588 ↑197

[44] K. Yasui, *Geometrically simply connected 4-manifolds and stable cohomotopy Seiberg-Witten invariants*, Geom. Topol. **23** (2019), no. 5, 2685–2697, DOI 10.2140/gt.2019.23.2685. MR4019901 ↑198

[45] I. Zemke, *Link cobordisms and absolute gradings on link Floer homology*, Quantum Topol. **10** (2019), no. 2, 207–323, DOI 10.4171/QT/124. MR3950650 ↑187, 197

School of Mathematics, Georgia Institute of Technology, Atlanta, GA 30332
Email address: hom@math.gatech.edu

IAS/Park City Mathematics Series
Volume 28, Pages 201–252
https://doi.org/10.1090/pcms/028/00877

Advanced topics in gauge theory: mathematics and physics of Higgs bundles

Laura P. Schaposnik

Abstract. These notes have been prepared as reading material for the mini-course given by the author at the *2019 Graduate Summer School* at the Park City Mathematics Institute. We begin by introducing Higgs bundles and their main properties (Lecture 1), and then we discuss the Hitchin fibration and its different uses (Lecture 2). The second half of the course is dedicated to studying different types of subspaces (branes) of the moduli space of complex Higgs bundles, their appearances in terms of flat connections and representations (Lecture 3), as well as correspondences between them (Lecture 4).

Contents

2020 *Mathematics Subject Classification.* Primary 14D21, 14J33; Secondary 53D30, 53 D37.
Key words and phrases. Higgs bundles, Hitchin systems, Mirror Symmetry, Langlands duality, branes.

1. Introduction

Higgs bundles provide a unifying structure for diverse branches of mathematics and physics. The *Dolbeault Moduli space* of $G_\mathbb{C}$-Higgs bundles $\mathcal{M}_{G_\mathbb{C}}$ has a hyperkähler structure, and through different complex structures it can be understood as different moduli spaces:

- Via the non-abelian Hodge correspondence the moduli space is diffeomorphic as a real manifold to the *de Rham moduli space* of flat connections [39,44,64,94,98].
- Via the Riemann-Hilbert correspondence there is an analytic correspondence between the de Rham moduli space and the *Betti moduli space* of surface group representations.

Through these correspondances, Higgs bundles manifest as both flat connections and representations, fundamental objects which are ubiquitous in contemporary mathematics, and closely related to theoretical physics. Some examples are:

- Through the Hitchin fibration, $\mathcal{M}_{G_\mathbb{C}}$ gives examples of hyperkähler manifolds which are *integrable systems* [66], leading to remarkable applications in physics, for instance in the works of Gukov, Hitchin, Donagi, Pantev, and Witten (e.g., see [40–42,58,59,67,101]).
- Hausel-Thaddeus [62] related Higgs bundles to *mirror symmetry*, and with Donagi-Pantev [40], presented $\mathcal{M}_{G_\mathbb{C}}$ as a fundamental example of mirror symmetry for Calabi-Yau manifolds, whose geometry and topology continues to be studied.
- Kapustin-Witten [72] used Higgs bundles to obtain a physical derivation of the *geometric Langlands correspondence* through mirror symmetry. Soon after, Ngô found Higgs bundles key ingredients when proving the fundamental lemma in [82].

In these Lecture Notes we will focus on Higgs bundles and their moduli spaces, by considering three main aspects of the theory:

- **The Hitchin fibration**. We shall review the Hitchin fibration as a tool to understand Higgs bundles, and describe the abelianization introduced by Hitchin in [66], and the nonabelianization appearing for certain subspaces of Higgs bundles as in [69].
- **Construction of branes**. We shall construct and study families of branes in the moduli space of Higgs bundles. In particular, branes are obtained by imposing different structures on the Riemann surface and the gauge group. We shall also consider geometric structures appearing through branes, among which are for example *the spaces of hyperpolygons*.
- **Correspondences**. Finally, we will also consider correspondences between Higgs bundles and between branes, in relation to different areas of maths and physics: e.g. mirror symmetry, correspondences arising through group automorphisms, and 3-manifolds.

For a longer introduction to spectral data for Higgs bundles the reader may refer to the lecture notes in [92] and [88]. Other references shall be mentioned across this Lecture Notes.

Bibliography. We shall highlight the main references considered, as well as the precise places where the methods used were developed. Since it proves to be very difficult to give a comprehensive and exhaustive account of research in tangential areas, we restrain ourselves to mentioning related work only when it directly involves methods mentioned in the lectures. The reader should refer to references within the bibliography for further information (e.g., see references in [6, 84, 88]).

2. The geometry of the moduli space of Higgs bundles

Proof is the end product of a long interaction between creative imagination and critical reasoning. Without proof the program remains incomplete, but without the imaginative input it never gets started.

Sir Michael Atiyah

We shall dedicate this first chapter to review the basic concepts of Higgs bundles and some of their generalizations to principal $G_{\mathbb{C}}$-Higgs bundles, real Higgs bundles, parabolic Higgs bundles and wild Higgs bundles (for more details refer to [39, 44, 64, 66, 83, 94, 95]).

2.1. Higgs bundles for complex groups. Unless specified otherwise, we shall let Σ be a compact Riemann surface of genus $g \geqslant 2$ with canonical bundle $K = T^*\Sigma$.

Definition 2.1.1. A *Higgs bundle* is a pair (E, Φ) for E a holomorphic vector bundle on Σ, and the *Higgs field* $\Phi \in H^0(\Sigma, \mathrm{End}(E) \otimes K)$.

Along these notes, we shall refer to *classical Higgs bundles* as the above Higgs bundles in Definition 2.1.1. Recall that holomorphic vector bundles E on Σ are topologically classified by their rank $\mathrm{rk}(E)$ and their degree $\deg(E)$, though which one may define their *slope* as

$$\mu(E) := \deg(E)/\mathrm{rk}(E).$$

Then, a vector bundle E is *stable* (*or semi-stable*) if for any proper, non-zero subbundle $F \subset E$ one has $\mu(F) < \mu(E)$ (or $\mu(F) \leqslant \mu(E)$). It is *polystable* if it is a direct sum of stable bundles whose slope is $\mu(E)$. The moduli space of stable bundles of fixed degree d and rank n is an algebraic variety $\mathcal{N}(n, d)$ which can be constructed via Mumford's Geometric Invariant Theory. Moreover, for coprime n and d, the space is a smooth projective algebraic variety of dimension $n^2(g-1)+1$.

Example 2.1.2. Since line bundles are stable, the space $\mathcal{N}(1, d)$ contains all line bundles of degree d, and is isomorphic to the Jacobian variety $\mathrm{Jac}^d(\Sigma)$, an abelian variety of dimension g.

Generalising the stability condition of vector bundles to Higgs bundles, we shall consider the following:

Definition 2.1.3. A vector subbundle F of E for which $\Phi(F) \subset F \otimes K$ is said to be a Φ-*invariant subbundle* of E. A Higgs bundle (E, Φ) is said to be

- *stable* (*semi-stable*) if for each proper Φ-invariant $F \subset E$ one has
$$\mu(F) < \mu(E) \text{ (equiv. } \leqslant);$$
- *polystable* if
$$(E, \Phi) = (E_1, \Phi_1) \oplus (E_2, \Phi_2) \oplus \ldots \oplus (E_r, \Phi_r),$$
where (E_i, Φ_i) is stable with $\mu(E_i) = \mu(E)$ for all i.

Remark 2.1.4. The characteristic polynomial of Φ restricted to an invariant subbundle divides the characteristic polynomial of Φ. Hence, if the characteristic polynomial of the Higgs field is irreducible, one knows that the corresponding Higgs bundle is automatically stable.

Example 2.1.5. Choose a square root $K^{1/2}$ of the canonical bundle K, and a section ω of K^2. A family of classical Higgs bundles (E, Φ_ω) (as in Definition 2.1.1) may be obtained by considering the vector bundle $E = K^{\frac{1}{2}} \oplus K^{-\frac{1}{2}}$ and the Higgs bundle Φ_ω given by
$$\Phi_\omega = \begin{pmatrix} 0 & \omega \\ 1 & 0 \end{pmatrix} \in H^0(\Sigma, \mathrm{End}(E) \otimes K)$$

Consider a strictly semi-stable Higgs bundle (E, Φ). Since E is not stable, it admits a minimal Φ-invariant subbundle $F \subset E$ for which $\mu(F) = \mu(E)$. Then, the induced pair (F, Φ) is stable and the induced quotient the quotient $(E/F, \Phi')$ is semistable. By induction, one obtains a flag of subbundles
$$F_0 = 0 \subset F_1 \subset \ldots \subset F_r = E$$
where $\mu(F_i/F_{i-1}) = \mu(E)$ for $1 \leqslant i \leqslant r$, and where the induced Higgs bundles $(F_i/F_{i-1}, \Phi_i)$ are stable. This is the *Jordan-Hölder filtration* of E, and whilst it is not unique, there is an induced graded object which is unique up to isomorphism
$$\mathrm{Gr}(E, \Phi) := \bigoplus_{i=1}^{r} (F_i/F_{i-1}, \Phi_i)$$

Two semi-stable Higgs bundles (E, Φ) and (E', Φ') are said to be *S-equivalent* if $\mathrm{Gr}(E, \Phi) \cong \mathrm{Gr}(E', \Phi')$. Following [83, Theorem 5.10] we let $\mathcal{M}(n, d)$ be the moduli space of S-equivalence classes of semi-stable Higgs bundles of fixed degree d and fixed rank n. The moduli space $\mathcal{M}(n, d)$ is a quasi-projective scheme, and has an open subscheme $\mathcal{M}'(n, d)$ which is the moduli scheme of stable pairs. Thus, every point is represented by either a stable or a polystable Higgs bundle. When d and n are coprime, the moduli space $\mathcal{M}(n, d)$ is a smooth non-compact variety which has complex dimension $2n^2(g-1)+2$ and the cotangent space of $\mathcal{N}(n, d)$ over the stable locus is contained in $\mathcal{M}(n, d)$ as a Zariski open subset.

One of the most important characterisations of stable Higgs bundles on a compact Riemann surface Σ of genus $g \geqslant 2$ is given in the work of Hitchin [66] and Simpson [94], and which carries through to more general settings:

Theorem 2.1.6. *If a Higgs bundle* (E, Φ) *is stable and deg* $E = 0$*, then there is a unique unitary connection* A *on* E*, compatible with the holomorphic structure, such that*

$$F_A + [\Phi, \Phi^*] = 0 \in \Omega^{1,1}(\Sigma, End\, E), \tag{2.1.7}$$

where F_A *is the curvature of the connection.*

Equation (2.1.7) together with the holomorphicity condition

$$d''_A \Phi = 0 \tag{2.1.8}$$

are called the *Hitchin equations*, where $d''_A\Phi$ is the anti-holomorphic part of the covariant derivative of the Higgs field Φ. Moreover, Hitchin showed that the moduli space of Higgs bundles is a hyperkähler manifold with natural symplectic form ω defined on the infinitesimal deformations $(\dot{A}, \dot{\Phi})$ of a Higgs bundle

$$\omega((\dot{A}_1, \dot{\Phi}_1), (\dot{A}_2, \dot{\Phi}_2)) = \int_\Sigma \mathrm{tr}(\dot{A}_1\dot{\Phi}_2 - \dot{A}_2\dot{\Phi}_1),$$

where $\dot{A} \in \Omega^{0,1}(\mathrm{End}_0 E)$ and $\dot{\Phi} \in \Omega^{1,0}(\mathrm{End} E)$ (see [64, 66] for details). For simplicity, we shall fix n and d and write $\mathcal{M}$ for $\mathcal{M}(n, d)$.

The notion of Higgs bundle can be generalized to encompass principal $G_{\mathbb{C}}$-bundles, for $G_{\mathbb{C}}$ a complex semi-simple Lie group. For more details, the reader should refer to [66].

Definition 2.1.9. A $G_{\mathbb{C}}$-*Higgs bundle* is a pair (P, Φ) where P is a principal $G_{\mathbb{C}}$-bundle over Σ, and the Higgs field Φ is a holomorphic section of the vector bundle $adP \otimes_{\mathbb{C}} K$, for adP the vector bundle associated to the adjoint representation.

When $G_{\mathbb{C}} \subset GL(n, \mathbb{C})$, a $G_{\mathbb{C}}$-Higgs bundle gives rise to a Higgs bundle in the classical sense, with some extra structure reflecting the definition of $G_{\mathbb{C}}$. In particular, classical Higgs bundles are given by $GL(n, \mathbb{C})$-Higgs bundles.

Example 2.1.10. The Higgs bundles in Example 2.1.5 have traceless Higgs field, and the determinant $\Lambda^2 E$ is trivial. Hence, for each quadratic differential ω one has an $SL(2, \mathbb{C})$-Higgs bundle (E, Φ_ω).

For $G_{\mathbb{C}}$ be a complex semisimple Lie group, we denote by $\mathcal{M}_{G_{\mathbb{C}}}$ the moduli space of S-equivalence classes of polystable $G_{\mathbb{C}}$-Higgs bundles. Following [85] one can define stability for principal $G_{\mathbb{C}}$-bundles, which when considering groups of type A, B, C, D, can be expressed in terms of stability for classical Higgs bundles (see [6, Section 1.1] for a comprehensive study). Moreover, the notion of polystability may be carried over to principal $G_{\mathbb{C}}$-bundles, leading to the construction of the moduli space of isomorphism classes of polystable principal $G_{\mathbb{C}}$-bundles of fixed topological type over the compact Riemann surface Σ. Since in these notes we shall be working with Higgs pairs which are automatically stable (as in Remark 2.1.4), we shall not dedicate time to recall the main study of stability for principal Higgs bundles. For details about the corresponding constructions, the reader should refer for example to [21, Section 3] and [6, Section 1].

Example 2.1.11. An $SO(2n,\mathbb{C})$-Higgs bundle is a pair (E,Φ), for E a holomorphic vector bundle of rank $2n$ with a non-degenerate symmetric bilinear form $(\ ,\ )$, and $\Phi \in H^0(\Sigma, \mathrm{End}_0(E)\otimes K)$ the Higgs field satisfying

$$(\Phi v, w) = -(v, \Phi w). \tag{2.1.12}$$

Equivalently, an $SO(2n+1,\mathbb{C})$-Higgs bundle is a pair (E,Φ) for E a holomorphic vector bundle of rank $2n+1$ with a non-degenerate symmetric bilinear form (v,w), and Φ a Higgs field in $H^0(\Sigma, \mathrm{End}_0(E)\otimes K)$ which satisfies (2.1.12).

When considering the structure of an orthogonal Higgs field Φ, one can see that for a generic matrix $A \in \mathfrak{so}(2n+1,\mathbb{C})$ (and also for $A \in \mathfrak{so}(2n,\mathbb{C})$), its distinct eigenvalues occur in $\pm\lambda_i$ pairs. Thus, the characteristic polynomial of A, which shall become important in the next chapters, must be of the form

$$\det(x-A) = x(x^{2n} + a_1 x^{2n-2} + \ldots + a_{n-1}x^2 + a_n).$$

2.2. Real Higgs bundles. Higgs bundles for real forms were first studied by N. Hitchin in [64], and the results for $SL(2,\mathbb{R})$ were generalised in [65], where Hitchin studied the case of $G = SL(n,\mathbb{R})$ appearing through involutions on the Lie algebra, a perspective we shall come back to in later chapters. Using Higgs bundles he counted the number of connected components and, in the case of split real forms, he identified a component homeomorphic to $\mathbb{R}^{\dim G(2g-2)}$ and which naturally contains a copy of a Teichmüller space.

Following [66], in order to obtain a G-Higgs bundle for a non-compact real form G of $G_\mathbb{C}$ and real structure τ, the connection A which solves Hitchin equations (2.1.7)-(2.1.8) needs to satisfy

$$\nabla = \nabla_A + \Phi - \rho(\Phi)$$

to have holonomy in G, where ρ is the compact real structure of $G_\mathbb{C}$.

Since the connection A has holonomy in the compact real form of $G_\mathbb{C}$, we have $\rho(\nabla_A) = \nabla_A$. Hence, requiring $\nabla = \tau(\nabla)$ is equivalent to $\nabla_A = \tau(\nabla_A)$ and $\Phi - \rho(\Phi) = \tau(\Phi - \rho(\Phi))$. In terms of the involution $\sigma := \rho\tau$, these two equalities are given by

$$\sigma(\nabla_A) = \nabla_A$$
$$\sigma(\rho(\Phi) - \Phi) = \Phi - \rho(\Phi).$$

From the above, ∇ has holonomy in the real form G if ∇_A is invariant under σ, and Φ anti-invariant. Hence for G a real form of a complex semisimple Lie group $G_\mathbb{C}$, we may construct G-Higgs bundles as follows. For H the maximal compact subgroup of G, we have seen that the Cartan decomposition of $\mathfrak{g}$ is given by $\mathfrak{g} = \mathfrak{h} \oplus \mathfrak{m}$, for $\mathfrak{h}$ the Lie algebra of H, and $\mathfrak{m}$ its orthogonal complement. This induces a decomposition for the complex Lie algebras as $\mathfrak{g}_\mathbb{C} = \mathfrak{h}_\mathbb{C} \oplus \mathfrak{m}_\mathbb{C}$. Through the induced isotropy representation $\mathrm{Ad}|_{H^{\mathbb{C}}} : H^{\mathbb{C}} \to GL(\mathfrak{m}^{\mathbb{C}})$, one has a concrete description of G-Higgs bundles (for more details, see for example [30]):

Definition 2.2.1. A principal G-Higgs bundle is a pair (P, Φ) where P is a holomorphic principal $H^{\mathbb{C}}$-bundle on Σ, and Φ is a holomorphic section of the bundle $P \times_{Ad} \mathfrak{m}^{\mathbb{C}} \otimes K$.

Example 2.2.2. Consider the non-compact real form $SU(p, q)$ of $SL(p+q, \mathbb{C})$ with Lie algebra

$$\mathfrak{su}(p, q) = \left\{ \begin{pmatrix} Z_1 & Z_2 \\ \overline{Z}_2^t & Z_3 \end{pmatrix} \middle| \begin{array}{c} Z_1, Z_3 \text{ skew Hermitian of order } p \text{ and } q, \\ \mathrm{Tr} Z_1 + \mathrm{Tr} Z_3 = 0\ ,\ Z_2 \text{ arbitrary} \end{array} \right\},$$

whose complexification has Cartan decomposition

$$\mathfrak{su}(p, q)^{\mathbb{C}} = \mathfrak{sl}(p+q, \mathbb{C}) = \mathfrak{h}^{\mathbb{C}} \oplus \mathfrak{m}^{\mathbb{C}},$$

where $\mathfrak{m}^{\mathbb{C}}$ corresponds to the off diagonal elements of $\mathfrak{sl}(p+q, \mathbb{C})$. The centre of $SU(p, q)$ is $U(1)$, and its maximal compact subgroup is $H = S(U(p) \times U(q))$, whose complexified Lie group is

$$H^{\mathbb{C}} = \{(X, Y) \in GL(p, \mathbb{C}) \times GL(q, \mathbb{C}) : \det Y = (\det X)^{-1}\}.$$

Hence, an $SU(p, q)$-Higgs bundle over Σ is a pair (E, Φ) where $E = V_p \oplus V_q$ for V_p, V_q vector bundles over Σ of rank p and q such that $\Lambda^p V_p \cong \Lambda^q V_q^*$, and the Higgs field Φ is given by

$$\Phi = \begin{pmatrix} 0 & \beta \\ \gamma & 0 \end{pmatrix},$$

for $\beta : V_q \to V_p \otimes K$ and $\gamma : V_p \to V_q \otimes K$.

For more details about the construction of real Higgs bundles and the study of their geometric properties, the reader might be interested to see the lecture notes and thesis [6,54,84,88,92], as well as the following – certainly non-exhaustive- list of papers which we have found useful in the past (and references therein): [7,12, 16,29–31,37,38,49,51,52,65,69,76,90,91,99,102] In particular, the reader should refer to [50] for the Hitchin-Kobayashi type correspondence for real forms.

2.3. Parabolic Higgs bundles. The notion of Higgs bundles can be generalized in several directions, and one which has caught the attention of mathematicians and physicists during the last decades is the case of *parabolic Higgs bundle*, following the generalization of vector bundles to parabolic bundles [77]. On a compact Riemann surface Σ of genus g with marked points $p_1, \ldots, p_n$ satisfying $2g - 2 + n \geqslant 0$, we denote by D the effective divisor $D = p_1 + \ldots + p_n$. Then, from [77] a *parabolic vector bundle* E on Σ is an algebraic rank r vector bundle E on Σ together with a *parabolic structure*, which is a (not necessarily full) flag for the fibre of E over the marked points

$$0 = E_{p_i,0} \subset E_{p_i,1} \subset E_{p_i,2} \subset \ldots \subset E_{p_i,r_i} = E|_{p_i},$$

together with a set $\alpha_i = (\alpha_{i,0}, \ldots, \alpha_{i,r_i})$ of *parabolic weights* associated to each marked point p_i given by real numbers satisfying

$$1 = \alpha_{i,0} > \alpha_{i,1} > \ldots > \alpha_{i,r_i} \geqslant 0.$$

A *parabolic* holomorphic map is a map $f : E \to E'$ between parabolic bundles for which $\alpha_i(p_j) > \alpha'_k(p_j)$ implies $f(E_{p_j,i}) \subset E'_{p_j,k+1}$ for all $p_j \in D$, where $\alpha'_k(p_j)$ are the weights on E'. The parabolic degree of a parabolic bundle E are then defined as

$$\operatorname{par}\deg(E) = \deg(E) + \sum_{j=1}^{n} \sum_{i=1}^{r_j} m_{j,i}\alpha_{j,i},$$

where $m_{j,i} = \dim(E_{p_j,i}/E_{p_j,i+1})$, and the parabolic slope of the parabolic bundle is defined as $\operatorname{par}\mu(E) = \operatorname{par}\deg(E)/\mathrm{rk}(E)$. A parabolic bundle is called (semi)-stable if for every parabolic subbundle F of E, the parabolic slope satisfies $\operatorname{par}\mu(F) \leqslant \operatorname{par}\mu(E)$ (resp. $\operatorname{par}\mu(F) < \operatorname{par}\mu(E)$). Through the parabolic stability for E one can construct the moduli space of semi-stable parabolic vector bundles [77], usually denoted by $\mathcal{N}_\alpha$, which is a normal projective variety, non-singular when the weights are generic.

One can generalize the notion of parabolic vector bundles and consider *parabolic Higgs bundles*, or PHG, as was first done by C. Simpson in [36].

Definition 2.3.1. A *(classical) parabolic Higgs bundle* is a pair (E, Φ) where

- E is a parabolic vector bundle as defined above,
- the Higgs field $\Phi \in H^0(\Sigma, \mathrm{End}(E) \otimes K_\Sigma(D))$ satisfies $\Phi|_{p_i}(E_{i,j}) \subset E_{i,j} \otimes K_\Sigma(D)|_{p_i}$.

The notions of stability extend to parabolic Higgs bundles, and through these one obtains the moduli space $\mathcal{P}_\alpha$ of α-semi-stable parabolic Higgs bundles of degree d and rank r, which was first constructed by Yokogawa in [103] (see also [27]). This space hasdimension $(2g-2)r^2 + 2 + nr(r-1)$, is a normal quasi-projective variety, and was constructed as a hyperkähler quotient using gauge theory in [73], and topological properties of this moduli space for rank $r = 2$ parabolic Higgs bundles were studied through the Morse theoretic approach of Hitchin in [103].

Example 2.3.2. A $U(p, q)$*-parabolic Higgs bundle* on Σ is a parabolic Higgs bundle (E, Φ) such that $E = V \oplus W$, where V and W are parabolic vector bundles of rank p and q respectively, and

$$\Phi = \begin{pmatrix} 0 & \beta \\ \gamma & 0 \end{pmatrix} : (V \oplus W) \to (V \oplus W) \otimes K(D),$$

where the non-zero components $\beta : W \to V \otimes K(D)$ and $\gamma : V \to W \otimes K(D)$ are parabolic morphisms and D is an effective divisor on Σ (the reader should compare this to Example 2.2.2).

Parabolic Hitchin systems can be considered to allow simple poles in the gauge and Higgs fields at marked points on a Riemann surface, whilst higher order poles lead us *Wild Higgs bundles*, which we shall consider next.

2.4. Wild Higgs bundles. Across these notes we shall limit ourselves to the study of Wild Higgs bundles (E, Φ) over the projective line $\mathbb{P}^1$. For a fixed effective divisor $D = p_1 + \ldots + p_n$ in $\mathbb{P}^1$, we take these pairs as composed of a vector bundle E and a $\mathcal{O}_{\mathbb{P}^1}(D)$ valued endomorphism $\Phi : \mathcal{E} \to \mathcal{E} \otimes \mathcal{O}_{\mathbb{P}^1}(D)$ called the Higgs field. The divisor D controls where Φ is allowed to have poles, so it is sometimes referred to as the *polar divisor*, and in this setting one considers stable, integrable connections with irregular singularities of the form

$$\mathcal{A} = d + T_n \frac{dz}{z^n} + \ldots + T_1 \frac{dz}{z},$$

for $n > 1$, and stable parabolic Higgs bundles (E, Φ) as in the previous section, where the Higgs field has polar parts, e.g.,

$$T_n \frac{dz}{z^n} + \ldots + T_1 \frac{dz}{z}, \tag{2.4.1}$$

and is defined such that the fibre of E at p_j is preserved by the residue $\operatorname{res}(\Phi, p_j)$, and where $T_1, T_2, \ldots, T_n$ are elements of the Lie algebra of the group[1]. When considering Higgs bundles of low rank, we shall sometimes take the notation of [25] and write a complexified connection as:

$$\mathcal{A} = dQ + T_1 \frac{dz}{z}, \tag{2.4.2}$$

for Q a square diagonal matrix of meromorphic connections[2], where varying Q in 2.4.2, is equivalent to varying $(T_n, T_{n-1}, \ldots, T_2)$ in (2.4.1).

Definition 2.4.3. The **formal monodromy** of $\mathcal{A}$ is $M_0 := e^{2\pi i T_1}$. Moreover, we say that T_1 is the **exponent** of the formal monodromy.

Following Simpson's construction in [95] for parabolic Higgs bundles, it is natural to assume that the connections and Higgs fields are holomorphically gauge equivalent to ones with diagonal polar parts. Similar to the regular case, the moduli space of these wild Higgs bundles can be obtained as a hyperkähler quotient [23, 26].

In these notes we shall follow the notation of [24] where a wild Higgs bundle of type $(m, r_1, r_2, \ldots)$ is a Higgs bundle with m poles of orders $k_i := r_i + 1$. To understand the moduli spaces of Higgs bundles, in what follows we shall restrict our attention to rank 2 Higgs bundles with poles over $\mathbb{P}^1$.

Remark 2.4.4. It is interesting to consider Higgs bundles whose number of poles with multiplicities is exactly 4 since those are closely related to solutions to Painlevé equations: (A) type $(1, 3)$ gives Painlevé II – see Problem I.(5).iii.; (B) type $(2, 1, 1)$ gives Painlevé III – see Example 2.4; (C) type $(4, 0, 0, 0, 0)$ gives Painlevé VI – see Section 2.3).

[1]One usually assumes that the matrix T_n is diagonalizable with distinct eigenvalues; this rules out nilpotence which rules out the possibility of a higher order regular singularity and also allows one to define the formal diagonalizing gauge transformation taking a connection to its irregular type given by (1.15).

[2]One typically applies a formal diagonalizing procedure before defining Q.

For rank two Higgs bundles, from [24, Remark 9.12] one obtains moduli spaces of complex dimension two when $(m, r_1, r_2, \ldots) = (4, 0, 0, 0, 0)$ as in (C) of Remark 2.4.4, $(3, 1, 0, 0), (2, 1, 1)$ as in (B) of Remark 2.4.4, $(2, 2, 0)$, and $(1, 3)$ as in (A) of Remark 2.4.4. In order to illustrate the main ingredients in the study of wild Higgs bundles, in what remains of the chapter we shall consider some particular examples. The reader should refer to Boalch's papers mentioned above for more details and complete proofs, as well as [5, Appendix A] for further examples appearing in relation to String theory. Finally, for open problems in the area, a non-exhaustive list of questions relating Wild Higgs bundles to singular geometry is given in [4], and for more on the relation with Painlevé's equations see for example [96] and references therein.

Higgs bundles with one pole. In what follows we shall give a construction of the main ingredients introduced in [25] by first studying rank 2 Higgs bundles on $\mathbb{P}^1$ with one double pole on a fixed point $p_1 = 0$ (see also [5, Appendix A]). In this setting, the connection in (2.4.2) is

$$\mathcal{A} = dz\left(\frac{T_2}{z^2} + \frac{T_1}{z}\right), \tag{2.4.5}$$

where $T_2 = \frac{dQ}{dz}z^2$ and T_1 is a constant diagonal matrix. In particular, we write

$$Q = \begin{pmatrix} q_1(z) & 0 \\ 0 & q_2(z) \end{pmatrix},$$

and define $q_{ij}(z)$ as the leading term of $q_i(z) - q_j(z)$. For $q_1(z) - q_2(z) = a/z + b$,

$$q_{12}(z) = a/z, \text{ and } q_{21}(z) = -a/z.$$

The set of *anti-Stokes directions* $\mathbb{A} \subset S^1$ is composed of the directions $d \in S^1$ for which either $q_{12}(z) \in \mathbb{R}_{<0}$, or $q_{21}(z) \in \mathbb{R}_{<0}$ for z on the ray specified by d. These are the directions along which $e^{(a/z+b)}$ (respectively $e^{-(a/z+b)}$) decays most rapidly as z approaches 0, and since if $\pm\frac{a}{z} \in \mathbb{R}_{<0}$ then $\mp\frac{a}{z}e^{-i\pi} \in \mathbb{R}_{<0}$, and thus $\#\mathbb{A} := r = 2$. Note that if $q_{12}(z) \in \mathbb{R}_{<0}$ for a direction d, then $q_{21}(z) \notin \mathbb{R}_{<0}$ for that direction (and vice versa). We shall refer to a direction as a *half-period*: for $l := r/2$, a *half-period* is an l-tuple $d = (d_1, \ldots, d_l) \subset \mathbb{A}$ of consecutive anti-Stokes directions. The *multiplicity* $\mathrm{Mult}(d)$ of d is the number of roots supporting d. When weighted by their multiplicities, the number of anti-Stokes directions in any half-period is $1 = \mathrm{Mult}(d_1) + \ldots + \mathrm{Mult}(d_l)$. In our setting of (2.4.5), letting $a = \tilde{r}e^{i\tilde{\theta}}$ there are only two half periods:

- From q_{12} a direction $d_1 := e^{i(\tilde{\theta}-\pi)}$, for which taking $z = r'e^{i(\tilde{\theta}-\pi)}$ one has $q_{12}(z) = -\frac{\tilde{r}}{r'} \in \mathbb{R}_{<0}$;
- From q_{21} a direction $d_2 := e^{-i\pi}d_1 = e^{i\tilde{\theta}}$, for which taking $z = r'e^{i\tilde{\theta}}$ gives $q_{21}(z) = -\frac{\tilde{r}}{r'} \in \mathbb{R}_{<0}$.

For d an anti-Stokes direction, the *roots* of d are

$$\mathrm{Roots}(d) := \{(ij) \mid q_{ij}(z) \in \mathbb{R}_{<0} \text{ along } d\}.$$

Note that the Stokes directions d_1 and d_2 define the orderings $q_1 <_{d_1} q_2$, and $q_2 <_{d_2} q_1$. Moreover, the directions define the *Stokes sectors* $\mathrm{Sect}_i := (d_i, d_{i+1})$ for $i \bmod \ell$. For the Higgs bundle arising from (2.4.5) the roots are $\mathrm{Roots}(d_1) = \{(12)\}$ and $\mathrm{Roots}(d_2) = \{(21)\}$.

Remark 2.4.6. For a Higgs bundle arising from (2.4.2) with one pole of order 3 at zero, writing $q_1 - q_2 = \frac{a}{z^4} + \frac{b}{z^3} + \frac{c}{z^2} + \frac{d}{z} + e$, leads to $q_{12}(z) = \frac{a}{z^3}$ and $q_{21}(z) = -\frac{a}{z^3}$. This system has 6 anti-Stokes directions, and we shall come back to it in the problem set.

The *group of Stokes factors* associated to a direction d is the group

$$\mathrm{Sto}_d(\mathcal{A}) := \{K \in G \mid (K)_{ij} = \delta_{ij} \ \text{ unless } \ (ij) \ \text{ is a root of } d\},$$

which for rank 2 Higgs bundles is a unipotent subgroup of $G = GL(2,\mathbb{C})$ of dimension 1. For all directions d_1 and d_2 that we have, the group of *Stokes matrices*[3] are 1-dimensional subgroup

(2.4.7)

$$\mathrm{Sto}_{d_1}(\mathcal{A}) = \left\{ \begin{pmatrix} 1 & \tilde{z} \\ 0 & 1 \end{pmatrix} \text{ for } \tilde{z} \in \mathbb{C} \right\} \text{ and } \mathrm{Sto}_{d_2}(\mathcal{A}) = \left\{ \begin{pmatrix} 1 & 0 \\ \tilde{z} & 1 \end{pmatrix} \text{ for } \tilde{z} \in \mathbb{C} \right\}.$$

To study the moduli space of Wild Higgs bundles one needs to consider fundamental solutions in each Stokes sector, which are then extended using the Stokes matrices, which can then be thought of as the transition matrices between the canonical fundamental solutions. However, we shan't deepen into this, and rather refer the reader to Boalch's papers mentioned before (e.g. see [19, 23–26]).

Higgs bundles with multiple poles. When considering Higgs bundles with different poles over a divisor D, more data needs to be taken into account in order to build their moduli space. In what follows we shall mention a few ingredients to give the reader a flavour of the theory by studying rank 2 Higgs bundles with four poles. We shall consider here a divisor $D = \{p_1, \ldots, p_4\}$, take V a homomorphically trivial bundle over $\mathbb{P}^1$, and let $\nabla = d - \mathcal{A}$ be a meromorphic connections with poles on D, for

$$\mathcal{A} = \sum_{i=1}^{4} \left(T^i_{k_i} \frac{dz}{(z-p_i)^{k_i}} + \cdots + T^i_1 \frac{dz}{(z-p_i)^1} \right),$$

where T^i_j are diagonal 2×2 matrices. Here the connection ∇ has a pole of order k_i at p_i, and assuming that $T^1_1 + \cdots + T^4_1 = 0$, it has no other poles. To deal with this type of Higgs bundles we choose for each i disjoint open discs D_i on $\mathbb{P}^1$ with center p_i, and coordinate z_i on D_i vanishing at p_i. Then, the local theory described before for one single pole is repeated on each such disc: Stokes matrices corresponding to distinct poles are related to each other by *connection matrices*,

[3] It should be noted that Boalch and other authors (e.g. see [2]) call *Stokes matrices* a product of a half-period's worth of *Stokes factors*, the objects in (2.4.7) – in the rank 2 case these two definitions coincide.

which are transition matrices defined in terms of local fundamental solutions. These connection matrices are associated to what is known as a *tentacle*.

To complete the description of the monodromy data of an irregular connection with multiple poles, one needs the Stokes matrices and formal monodromy at each pole along with the connection matrices relating fundamental solutions in neighborhoods of distinct poles, making choices of points and paths between them. These choices are encoded in a choice of the so-called ***tentacle***:

(1) A point a_j in some sector at p_j between two anti-Stokes rays[4], for each $j = 1, \dots, 4$.
(2) A base-point $p_0 \in \mathbb{P}^1/\{p_1, \dots, p_4\}$ and a path
$$\gamma_j : [0,1] \to \mathbb{P}^1/\{p_1, \dots, p_4\},$$
from p_0 to a_j for each j, such that the loop
$$(\gamma_4^{-1}\beta_4\gamma_4)(\gamma_3^{-1}\beta_3\gamma_3)(\gamma_2^{-1}\beta_2\gamma_2)(\gamma_1^{-1}\beta_1\gamma_1)$$
based at p_0 is contractible in $\mathbb{P}^1/\{p_1, \dots, p_4\}$, for β_j a loop in $D_j/\{p_j\}$ based at a_j around p_j once in a positive sense.

As an example, consider a rank 2 Higgs bundles with 2 poles of order 2 and its tentacle in Figure 2.4.8, where the grey labels and dotted paths have been chosen to define the theory:

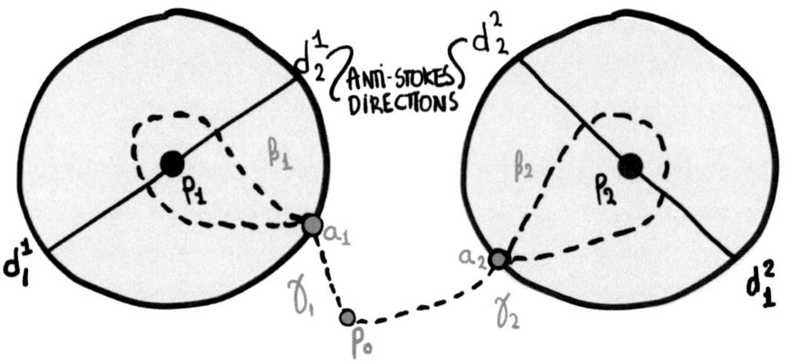

Figure 2.4.8. Tentacles for two poles of order 2.

We shall conclude by noting that considering the collision of points, one has the following conjecture about the relation of Wild Higgs bundles and parabolic Higgs bundles:

Conjecture 2.4.9 ([5]). *There exists a flat morphism between the parabolic moduli space and the wild moduli space in the case of singular parabolic Higgs bundle with n-simple poles in their connection, and that of wild Higgs bundles with a single, higher order pole of order n.*

---

[4]Note that a choice of tentacle implies a labelling P_j of the permutation associated to p_j (which for simple poles is the identity matrix). Hence, one must either fix various conventions about how those choices are to be made, or sacrifice some of the geometry by allowing for cyclic permutations of the monodromy data.

2.5. Problem set I. This problem set is meant to make the reader be acquainted with the different ways of encoding the data defining Higgs bundles in their various forms. They provide a mixture of problems from the lectures as well as some additional ones.

Problem 1. Complex Higgs bundles. As mentioned before, through Definition 2.1.9 one can define principal $G_{\mathbb{C}}$-Higgs bundles which may be described in terms of a pair (E, Φ) of a vector bundle E and a holomorphic Higgs filed $\Phi : E \to E \otimes K$ with some extra conditions reflecting the nature of the group:

i. Give a description of $SO(2p+1, \mathbb{C})$-Higgs bundles in terms of pairs (E, Φ) as in Definition 2.1.1 and list the extra conditions that need to be satisfied.
ii. Show that if λ is an eigenvalue of an $Sp(2p, \mathbb{C})$-Higgs bundle (E, Φ), then necessarily so is $-\lambda$.

Problem 2. Real Higgs bundles. In order to understand $Sp(2p, 2q)$-Higgs bundles, we shall begin by considering the structure of the corresponding Lie algebra:

i. Give a 4×4 matrix expression (with matrix entries) of the non-compact real form $\mathfrak{sp}(2p, 2q)$ of the complex symplectic Lie algebra.
ii. Show that $\mathfrak{m}^{\mathbb{C}}$ for $\mathfrak{sp}(2p, 2q) = \mathfrak{h} \oplus \mathfrak{m}$ as in Example 2.2.2 can be expressed as subset of certain off-diagonal matrices.
iii. Prove that if a Higgs bundle (E, Φ) is stable, then for $\lambda \in \mathbb{C}^*$ and α a holomorphic automorphism of E, the induced Higgs bundles $(E, \lambda\Phi)$ and $(E, \alpha^*\Phi)$ are stable.
iv. A definition of $Sp(2p, 2q)$-Higgs bundles as pairs (E, Φ) satisfying extra conditions which reflect the nature of the group.
v. A description of the characteristic polynomial of a generic $Sp(2p, 2q)$-Higgs field.

Problem 3. Real Higgs bundles. Let $\mathfrak{g}_c$ be a complex Lie algebra with complex structure i, whose Lie group is $G_{\mathbb{C}}$. Recall that a *real form* of $\mathfrak{g}_{\mathbb{C}}$ is a real Lie algebra which satisfies $\mathfrak{g}_{\mathbb{C}} = \mathfrak{g} \oplus i\mathfrak{g}$. Given a real form $\mathfrak{g}$ of $\mathfrak{g}_{\mathbb{C}}$, an element $Z \in \mathfrak{g}_{\mathbb{C}}$ may be written as $Z = X + iY$ for $X, Y \in \mathfrak{g}$. The mapping $X + iY \mapsto X - iY$ is called the *conjugation* with respect to $\mathfrak{g}$. Moreover, any real form $\mathfrak{g}$ of $\mathfrak{g}_{\mathbb{C}}$ is given by the fixed points set of an antilinear involution τ on $\mathfrak{g}_{\mathbb{C}}$. i.e., a map satisfying

$$\tau(\tau(X)) = X, \qquad \tau(zX) = \bar{z}\tau(X),$$
$$\tau(X+Y) = \tau(X) + \tau(Y), \qquad \tau([X,Y]) = [\tau(X), \tau(Y)]$$

for $X, Y \in \mathfrak{g}_{\mathbb{C}}$ and $z \in \mathbb{C}$. In particular the conjugation with respect to $\mathfrak{g}$ satisfies these properties. From the above definitions, obtain the following:

i. Obtain an involution σ on $\mathfrak{g}_{\mathbb{C}}$ such that:
   - the fixed point set $\mathfrak{g}_{\mathbb{C}}^{\sigma}$ of σ is given by the complexification of the maximal compact subalgebra $\mathfrak{h}$ of $\mathfrak{g}$;
   - the anti-invariant set under the involution σ is given by $\mathfrak{m}^{\mathbb{C}}$, where $\mathfrak{g} = \mathfrak{h} \oplus \mathfrak{m}$ is the Cartan decomposition.

ii. A description of the involution on $\mathfrak{gl}(n, \mathbb{C})$ and $\mathfrak{sl}(n, \mathbb{C})$ fixing the unitary groups with signature $\mathfrak{u}(p, q)$ and $\mathfrak{su}(p, q)$.
iii. A description of $Sp(2p, 2q)$-Higgs bundles as the fixed point set of some involution on the whole moduli space of $Sp(2p+2q, \mathbb{C})$-Higgs bundles, induced by actions on the Lie algebra, using the Hitchin equations and question **Problem 2** above.

Problem 4. Parabolic Higgs bundles. When considering $GL(2, \mathbb{C})$-Higgs bundles, recall that Hitchin established in [66] a correspondence between equivalence classes of irreducible $GL(2, \mathbb{C})$ representations of $\pi_1(\Sigma)$ and isomorphism classes of $GL(2, \mathbb{C})$-Higgs bundles of degree zero. A similar correspondence was later obtained by Simpson in [36] between parabolic Higgs bundles and and the fundamental group of $\Sigma/\{p_1, \ldots, p_n\}$. Though those identifications, particular properties of parabolic Higgs bundles and their corresponding representations can be obtained, as was done for example in [20].

i. Give an example of a parabolic stable Higgs bundle (E, Φ) of rank 2 which has parabolic degree zero.
ii. Give an example of a parabolic stable $U(3, 1)$-Higgs bundle (E, Φ), using the descriptions of Example 2.3.2.
iii. Give an expression for the flat connection corresponding to the Higgs bundle (E, Φ) in **Problem 4.i.**, and modify the pair until it can be shown that the holonomy of the flat connection is contained (after conjugation) in $SL(2, \mathbb{R})$.
iv. From [45], recall that a *minimal parabolic structure* on E is a choice at each p_i of a partial flag $0 \subset L_i \subset E_{p_i}$, where L_i is a line. A *minimal parabolic Higgs field* is an $\mathcal{O}_X$-linear map $\phi \in H^0(X, \mathrm{End}(E) \otimes K_X(D))$, such that at each point p_i, the residue of ϕ_i of ϕ is strictly triangular with respect to the flag: $\phi(E_{p_i}) \subseteq L_i, \phi_i(L) = 0$. Construct an example of a rank 3 *minimal parabolic Higgs field*.
v. Show that if minimal parabolic Higgs field ϕ as in **Problem 4.iv.** is strictly parabolic, then the residues ϕ_i are nilpotent of order 2, and they lie in the closure of the minimal nilpotent orbit of $\mathfrak{sl}_r$.

Problem 5. Wild Higgs bundles. Consider a rank 2 Higgs bundle with 1 pole of order 3 at p_2 and one of order 1 at p_1. For this, take diagonal matrices ${}^jA_i$ and

$$\nabla := d - \left({}^1A_1 \frac{dz}{(z-p_1)}\right) - \left({}^2A_3 \frac{dz}{(z-p_2)^3} + {}^2A_2 \frac{dz}{(z-p_2)^2} + {}^2A_1 \frac{dz}{(z-p_2)}\right).$$

Using the examples given in this chapter, consider the following:

i. Describe the sets of anti-Stokes directions and identify whether the Stokes matrices are upper or lower triangular;
ii. Give a description of the tentacle as in Figure 2.4.8.
iii. For Higgs bundles with one pole of order 4 as in Remark 2.4.6, show that the system has 6 anti-stokes directions and give a description of them.

3. The geometry of the Hitchin fibration

Integrability of a system of differential equations should manifest itself through some generally recognizable features:
⋆ the existence of many conserved quantities;
⋆ the presence of algebraic geometry;
⋆ the ability to give explicit solutions.
These guidelines should be interpreted in a very broad sense."

Nigel Hitchin

The moduli space $\mathcal{M}_{G_\mathbb{C}}$ of $G_\mathbb{C}$-Higgs bundles can be naturally studied through the Hitchin fibration, as introduced in [66]. We shall dedicate this chapter to the study of this fibration and the description of some of its geometric properties.

3.1. The Hitchin fibration and the Teichmüller component. This Hitchin fibration is defined through a homogeneous basis p_i, for $i = 1, \ldots, k$ for the algebra of invariant polynomials of the Lie algebra $\mathfrak{g}_c$ of $G_\mathbb{C}$, whose degrees shall be denoted by d_i. Then, the *Hitchin fibration* is given by

$$h : \mathcal{M}_{G_\mathbb{C}} \longrightarrow \mathcal{A}_{G_\mathbb{C}} := \bigoplus_{i=1}^{k} H^0(\Sigma, K^{d_i}),$$
$$(E, \Phi) \mapsto (p_1(\Phi), \ldots, p_k(\Phi)).$$

The map h is referred to as the *Hitchin map*, and is a proper map for any choice of basis (see [66, Section 4] for details). Furthermore, the Hitchin base $\mathcal{A}_{G_\mathbb{C}}$ always satisfies

$$\dim\mathcal{A}_{G_\mathbb{C}} = \dim\mathcal{M}_{G_\mathbb{C}}/2,$$

making the Higgs bundle moduli space into an integrable system.

By considering different homogenous basis of invariant polynomials, one can make different aspects of the geometry of the Hitchin fibration explicit. In particular, one may sometimes ask whether certain property of the fibration (or of a fibre) is preserved under a change of basis, and we shall see in later sections examples of this. Across these notes, for classical Higgs bundles (E, Φ) of rank n we shall often take the basis given by $\mathrm{tr}(\Phi^i)$ for $1 \leqslant i \leqslant n$, or given by the coefficients of the characteristic polynomial of the Higgs field Φ.

Remark 3.1.1. Through changes of basis, one obtains correspondences between different expressions of the Hitchin fibration for the same group. For instance, the coefficients of the characteristic polynomial can be expressed in terms of traces as follows:

$$p(x) = x^n + \mathrm{tr}(\Phi)x^{n-1} + \sum_{k=2}^{n-1} x^{n-k}\,(-1)^k\,\mathrm{Tr}(\Lambda^k\Phi) + (-1)^n \det(\Phi) \tag{3.1.2}$$

where $\mathrm{Tr}(\Lambda^k\Phi)$ is the trace of the kth exterior power of Φ, with dimension $\frac{n!}{k!(n-k)!}$.

The *Teichmüller component* [65], or so-called *Hitchin section* of the Hitchin fibration, is induced by the Kostant slice $\mathcal{K}$ of $\mathfrak{g}_{\mathbb{C}}$, which is the space given by

$$\mathcal{K} := \{f \in \mathfrak{g}_{\mathbb{C}} \mid f = f_0 + a_1e_1 + a_2e_2 + \ldots + a_re_r\},$$

for e_i the highest weight vectors of $\mathfrak{g}_{\mathbb{C}}$ and a_i complex numbers, and f_0 a nilpotent element of a three-dimensional subalgebra $\mathfrak{s}$ of $\mathfrak{g}_{\mathbb{C}}$. Recall that this is a subalgebra of $\mathfrak{g}_{\mathbb{C}}$ generated by a semisimple element h_0, and nilpotent elements e_0 and f_0 of $\mathfrak{g}_{\mathbb{C}}$ satisfying the relations

$$[h_0, e_0] = e_0 \ ; \ [h_0, f_0] = -f_0 \ ; \ [e_0, f_0] = h_0.$$

From the decomposition of $\mathfrak{g}_{\mathbb{C}}$ into eigenspaces of $\mathrm{ad}h_0$, consider the vector bundle

$$E = \mathrm{ad}P \otimes \mathbb{C} = \bigoplus_{j=-k_r}^{k_r} \mathfrak{g}_j \otimes K^j,$$

where $k_r = \max\{k_i\}_i$ is the maximal exponent of $\mathfrak{g}_{\mathbb{C}}$. This is the adjoint bundle of $\mathfrak{g}_{\mathbb{C}}$ associated to the principal $\mathrm{Ad}G_{\mathbb{C}}$-bundle $P = P_1 \times_i G$, where P_1 is the holomorphic principal $SL(2,\mathbb{C})$-bundle associated to $K^{-1/2} \oplus K^{1/2}$, and the inclusion $i : SL(2,\mathbb{C}) \hookrightarrow \mathrm{Ad}G_{\mathbb{C}}$ corresponds to the principal three dimensional subalgebra $\mathfrak{s}_0$. Although P_1 involves a choice of square root $K^{1/2}$, the bundle E is independent of it. The $\mathrm{Ad}G_{\mathbb{C}}$-Higgs bundle (E, Φ), for

$$\Phi = f_0 + a_1e_1 + a_2e_2 + \ldots + a_re_r$$

and $a_i \in H^0(\Sigma, K^{k_i+1})$, is stable. Note that $(\mathrm{ad}h_0)f_0 = -f_0$ and thus we may regard f_0 as a section of $(\mathfrak{g}_{-1} \otimes K^{-1}) \otimes K$. Furthermore, the highest weight vectors $e_j \in \mathfrak{g}_{k_j}$ and thus a_je_j is a section of $\mathfrak{g}_{k_j} \otimes K^{k_j+1}$, making Φ a well defined holomorphic section of $E \otimes K$.

Since $p_i(\Phi) = a_i$, the above construction defines a section

$$s : (a_1, \ldots, a_r) \mapsto (E, \Phi) \tag{3.1.3}$$

of $\mathcal{M}_{\mathrm{Ad}G_{\mathbb{C}}}$ whose image is the *Teichmüller component*. This component defines an origin in the smooth fibres of h, and the Hitchin fibration becomes a fibration of abelian varieties (for more details the reader should refer to [66] where Hitchin introduced the *Teichmüller component*) The following result ([65, Theorem 7.5]) relates the Teichmüller component to the space of representations of the split real form of $\mathrm{Ad}G_{\mathbb{C}}$:

Theorem 3.1.4. *The section s of $\mathcal{M}_{\mathrm{Ad}G_{\mathbb{C}}}$ defines a smooth connected component of the moduli space of reductive representations of $\pi_1(\Sigma)$ into the split real form of $\mathrm{Ad}G_{\mathbb{C}}$.*

3.2. The regular fibres of the Hitchin fibration. We shall first consider the regular fibres of the Hitchin fibration, which are those fibres over regular – generic – points in the Hitchin base. Formally, we shall be considering fibres over points $a \in \mathcal{A}_{G_{\mathbb{C}}}$ defining *smooth* spectral curves S. As before, we consider $K = T^*\Sigma$, and $X = \mathrm{Tot}(K)$ its total space with projection $\pi : X \to \Sigma$. We shall denote by η

the tautological section of the pull back π^*K on X. Abusing notation we denote with the same symbols the sections of powers K^i on Σ and their pull backs to X (for more details see the equations in [17] where the distinction is made when defining spectral data).

The characteristic polynomial of a Higgs bundle (E, Φ) in a generic fibre $h^{-1}(a)$ defines a smooth curve $\pi : S_a \to \Sigma$ in X, the *spectral curve* of Φ, whose equation is

$$\det(\eta - \Phi) = \eta^n + a_1\eta^{n-1} + a_2\eta^{n-2} + \ldots + a_{n-1}\eta + a_n = 0, \tag{3.2.1}$$

for $a_i \in H^0(\Sigma, K^i)$ (for simplicity, we will drop the subscript a of S_a). By the adjunction formula on X (see e.g. [56]), The cotangent bundle of Σ is a symplectic manifold and hence has trivial canonical bundle, so $K_S \otimes \pi^*K^{-m}$ is trivial and $K_S = \pi^*K^n$. Taking degrees in both sides gives the genus

$$g_S = 1 + n^2(g-1).$$

The *spectral data* associated to a Higgs bundle (E, Φ) provides a geometric description of the fibres of the Hitchin fibration as abelian varieties, and is then given by:

- the spectral curve $\pi : S \to \Sigma$, and
- a vector bundle on S, satisfying appropriate conditions reflecting the nature of $G_{\mathbb{C}}$.

In the case of classical Higgs bundles, the smooth fibres are Jac(S). For η the tautological section of π^*K, one recovers (E, Φ) from the curve $(S, L \in \mathrm{Jac}(S))$ by taking $E = \pi_*L$ and $\Phi = \pi_*\eta$. Indeed, recall that by the definition of direct image sheaf, given an open set $\mathcal{U} \subset \Sigma$, one has that $H^0(\pi^{-1}(\mathcal{U}), L) = H^0(\mathcal{U}, \pi_*L)$, and multiplication by η induces the map

$$H^0(\pi^{-1}(\mathcal{U}), L) \xrightarrow{\eta} H^0(\pi^{-1}(\mathcal{U}), L \otimes \pi^*K).$$

By considering the direct image of the map, one obtains

$$\Phi : \pi_*L \to \pi_*L \otimes K,$$

giving a Higgs field $\Phi \in H^0(\Sigma, \mathrm{End}E \otimes K)$ for $E := \pi_*L$. Moreover, the Higgs field satisfies its characteristic equation, which by construction is given by the equation $\eta^n + a_1\eta^{n-1} + a_2\eta^{n-2} + \ldots + a_{n-1}\eta + a_n = 0$. Furthermore, since S is irreducible, from Remark 2.1.4 there are no invariant subbundles of the Higgs field, making the induced Higgs bundle (E, Φ) stable. Conversely, let (E, Φ) be a classical Higgs bundle. The characteristic polynomial is given by (3.2.1) and its coefficients define the spectral curve S in the total space X.

From [17, Proposition 3.6], there is a bijective correspondence between Higgs bundles (E, Φ) and the line bundles L on the spectral curve S described previously. This correspondence identifies the fibre of the Hitchin map with the Picard variety of line bundles of the appropriate degree. By tensoring the line bundles L with a chosen line bundle of degree $-\deg(L)$, one obtains a point in the Jacobian Jac(S), the abelian variety of line bundles of degree zero on S, which has dimension g_S as in (3.2). In particular, the Jacobian variety is the connected component of the identity in the Picard group $H^1(S, \mathcal{O}_S^*)$. Thus, the fibre of the classical Hitchin fibration $h : \mathcal{M} \to \mathcal{A}$ is isomorphic to the Jacobian of the spectral curve S. For more details, the reader should refer for example to [67, Section 2].

Example 3.2.2. In the case of a classical rank 2 Higgs bundle (E, Φ), the characteristic polynomial of Φ defines a spectral curve $\pi : S \to \Sigma$. This is a 2-fold cover of Σ in the total space of K, and has equation $\eta^2 + a_2 = 0$, for a_2 a quadratic differential and η the tautological section of $\pi^* K$. By [17, Remark 3.5] the curve is smooth when a_2 has simple zeros, and in this case the ramification points are given by the divisor of a_2. For z a local coordinate near a ramification point, the covering is given by $z \mapsto z^2 := w$. In a neighbourhood of $z = 0$, a section of the line bundle M looks like $f(w) = f_0(w) + z f_1(w)$. Since the Higgs field is obtained via multiplication by η, one has

$$\Phi(f_0(w) + z f_1(w)) = w f_1(w) + z f_0(w),$$

and thus a local form of the Higgs field Φ is given by

$$\Phi = \begin{pmatrix} 0 & w \\ 1 & 0 \end{pmatrix}.$$

By considering additional conditions on the Higgs field Σ (e.g., requiring it to have a $U(p, q)$-structure as in Example 2.2.2) the characteristic polynomial, and thus the curve it defines, acquires further structure, and examples of these will be described below. Hence, when $G_\mathbb{C} \subset GL(n, \mathbb{C})$, the spectral data of a $G_\mathbb{C}$-Higgs bundle is given by the spectral data of the pair as a classical Higgs bundle, satisfying extra conditions. An example of this is how the spectral data for real G-Higgs bundles can be expressed when G is the split real form of the complex Lie group $G_\mathbb{C}$.

Theorem 3.2.3 ([88, Theorem 4.12])**.** *The spectral data for G-Higgs bundles when G is the split real form of the complex Lie group $G_\mathbb{C}$ is given by the $G_\mathbb{C}$ spectral data of order two. That is, the intersection of the subspace of the Higgs bundle moduli space $\mathcal{M}_{G_\mathbb{C}}$ corresponding to the split real form of $\mathfrak{g}_\mathbb{C}$ with the smooth fibres of the Hitchin fibration*

$$h : \mathcal{M}_{G_\mathbb{C}} \to \mathcal{A}_{G_\mathbb{C}},$$

is given by the elements of order two in those fibres.

Through the above theorem, one can certain real Higgs bundles as a covering space in the Hitchin fibration, and thus their geometry can be studied through the properties of covering spaces – for instance, through the monodromy action for the fibration. To define the monodromy action for the Hitchin fibration, consider a generic fibration $p : Y \to B$ which is locally trivial, i.e. for any point $b \in B$ there is an open neighbourhood $U_b \in B$ such that $p^{-1}(U_b) \cong U_b \times Y_b$ where Y_b denotes the fibre at b. The nth homologies of the fibres Y_b form a locally trivial vector bundle over B, which we denote $\mathcal{H}_n(B)$. This bundle carries a canonical flat connection, the *Gauss-Manin connection*.

To define this connection we identify the fibres of $\mathcal{H}_n(B)$ at nearby points $b_1, b_2 \in B$, i.e. $H_n(Y_{b_1})$ and $H_n(Y_{b_2})$. Consider $N \subset B$ a contractible open set which includes b_1 and b_2. The inclusion of the fibres $Y_{b_1} \hookrightarrow p^{-1}(N)$ and $Y_{b_2} \hookrightarrow p^{-1}(N)$ are homotopy equivalences sowe obtain an isomorphism between the homology of a fibre over a point in a contractible open set N and $H_n(p^{-1}(N))$:

$$H_n(Y_{b_1}) \cong H_n(p^{-1}(N)) \cong H_n(Y_{b_2}).$$

This means that the vector bundle $\mathcal{H}_n(B)$ over B has a flat connection, the Gauss-Manin connection. The *monodromy* of $\mathcal{H}_n(B)$ is the holonomy of this connection, i.e. a homomorphism $\pi_1(B) \to \mathrm{Aut}(H_n(Y))$ as an action of $\pi_1(B)$ on $H_n(Y)$. By applying these results to the fibration $\mathcal{M}_{reg} \to \mathcal{A}_{reg}$ one has that the Gauss-Manin connection on the cohomology of the fibres of $h : \mathcal{M}_{reg} \to \mathcal{A}_{reg}$ defines the monodromy action for the Hitchin fibration. This monodromy action as been studied for some groups in [10, 15, 87, 88].

The Hitchin fibration and $SL(n,\mathbb{C})$-Higgs bundles. To give an example of the conditions needed, we shall consider $G_\mathbb{C} = SL(n,\mathbb{C})$. Here, through Definition 2.1.9 an $SL(n,\mathbb{C})$-Higgs bundle is a classical Higgs bundle (E,Φ) where the rank n vector bundle E has trivial determinant and the Higgs field has zero trace. A basis for the invariant polynomials on the Lie algebra $\mathfrak{sl}(n,\mathbb{C})$ is given by the coefficients of the characteristic polynomial of a trace-free matrix $A \in \mathfrak{sl}(n,\mathbb{C})$, and thus the spectral curve $\pi : S \to \Sigma$ defined by the Higgs field has equation (3.2.1) with $a_1 = 0$. In this case the generic fibres of the Hitchin fibration are given by the subset of Jac(S) of line bundles L on S for which $\pi_* L = E$ and $\Lambda^n \pi_* L$ is trivial: the generic fibre of the $SL(n,\mathbb{C})$ Hitchin fibration is biholomorphically equivalent to the $\mathrm{Prym}(S,\Sigma)$, for S the spectral curve defined as in (3.2.1) with the coefficient $a_1 = 0$.

In order to give the flavour of some of the geometry arising through the Hitchin fibration, we shall consider next the case of $n = 2$ and let $\mathcal{A}$ be the $SL(2,\mathbb{C})$ Hitchin base and by $\mathcal{M}$ the moduli space of $SL(2,\mathbb{C})$-Higgs bundles. For any $\omega \in \mathcal{A} - \{0\}$ it is shown in [55, Theorem 8.1] that the fibre $\mathcal{M}_\omega$ is connected. For any isomorphism class of (E,Φ) in $\mathcal{M}$, one may consider the zero set of its characteristic polynomial

$$\det(\Phi - \eta I) = \eta^2 + \omega = 0,$$

where $\omega = \det(\Phi) \in \mathcal{A}$. This defines a spectral curve $\rho : S \to \Sigma$ in the total space X on K, for η the tautological section of the pull back of K on X. We shall denote by $\mathcal{M}_{reg}$ the regular fibres of the Hitchin map h, and let $\mathcal{A}_{reg}$ be the regular locus of the base, which is given by quadratic differentials with simple zeros. Note that the curve S is non-singular over the regular locus $\mathcal{A}_{reg}$, and the ramification points are given by the intersection of S with the zero section. The curve S has a natural involution $\sigma(\eta) = -\sigma$ and thus we can define the Prym variety $\mathrm{Prym}(S, \Sigma)$ as the set of line bundles $M \in \mathrm{Jac}(S)$ which satisfy

$$\sigma^* M \cong M^*.$$

This definition is consistent with the one given for $n > 2$ via the Norm map.

Since S is a 2-fold cover in the total space X of K, the direct image of the trivial bundle $\mathcal{O}$ in $\mathrm{Prym}(S, \Sigma)$ is given by $\pi_* \mathcal{O} = \mathcal{O} \oplus K^{-1}$ (e.g. see [17, Remark 3.1]). Moreover, from the natural involution σ, the sections of $\pi_* \mathcal{O}$ can be separated into invariant and anti-invariant ones. As $\mathcal{O}$ corresponds to the invariant sections, and $\Lambda^2 \pi_* \mathcal{O} \cong K^{-1}$, necessarily $\pi_* \mathcal{O} = \mathcal{O} \oplus K^{-1}$.

The *Hitchin section* as introduced in (3.1.3) of the Hitchin fibration can be obtained naturally from the line bundle $L = \pi^* K^{1/2}$ on S, for which one has

$$\pi_* \pi^* K^{1/2} = K^{1/2} \otimes \pi_* \mathcal{O} = K^{1/2} \oplus K^{-1/2}.$$

Hence, the line bundle $\mathcal{O} \in \mathrm{Prym}(S, \Sigma)$ has associated Higgs bundle given by $(K^{1/2} \oplus K^{-1/2}, \Phi_\omega)$, and Higgs field Φ_ω, as in Examples 2.1.5 and 3.2.2 given by

$$\Phi_\omega = \begin{pmatrix} 0 & \omega \\ 1 & 0 \end{pmatrix}, \quad \text{for } \omega \in H^2(\Sigma, K^2).$$

For generic $G_{\mathbb{C}}$, a description of the fibres can be obtained by means of Cameral covers [43] (see also [41]) which is equivalent to the one given in the next sections for classical Lie groups. For a comprehensive description of Cameral covers for real Higgs bundles, the reader may refer to [84].

The split real forms. In order to study Higgs bundles for the split real form $SL(2, \mathbb{R})$, one should remember that their spectral data corresponds to elements of order 2 in the abelian varieties $\mathrm{Prym}(S, \Sigma)$ giving the generic fibres of the Hitchin fibration. Moreover, in this case the monodromy is generated by the action of $\pi_1(\mathcal{A}_{reg})$ on $H^1(\mathrm{Prym}(S, \Sigma), \mathbb{Z})$, a space which in turn can be shown to give the set $P[2]$ of torsion two points of the fibre (this can be generalized to higher rank, and the reader should refer to [15] for a more detailed study).

In what follows we shall see how the space $H^1(\mathrm{Prym}(S, \Sigma), \mathbb{Z})$ is isomorphic to the torsion two points $P[2]$ in the generic fibre $\mathrm{Prym}(S, \Sigma)$. Since a generic fibre of h is given by the abelian variety $\mathrm{Prym}(S, \Sigma)$, i.e., by the quotient of a complex vector space V by some lattice $\wedge$, one has an associated exact sequence of homology groups

$$\ldots \to \pi_1(\wedge) \to \pi_1(V) \to \pi_1(\mathrm{Prym}(S, \Sigma)) \to \pi_0(\wedge) \to \ldots.$$

Hence, there is a short exact sequence $0 \to \pi_1(\mathrm{Prym}(S, \Sigma)) \to \wedge \to 0$, from where $\wedge \cong \pi_1(\mathrm{Prym}(S, \Sigma))$. Therefore, $\pi_1(\mathrm{Prym}(S, \Sigma))$ is an abelian group, i.e.,

$$\pi_1(\mathrm{Prym}(S, \Sigma)) \cong H_1(\mathrm{Prym}(S, \Sigma), \mathbb{Z}).$$

We shall denote by $P[2]$ the elements of order 2 in Prym (S, Σ), which are equivalent classes in V of points x such that $2x \in \wedge$. Then, $P[2]$ is given by $\frac{1}{2}\wedge$ modulo

$\wedge$, and as $\wedge$ is torsion free,

$$P[2] \cong \wedge/2\wedge \cong H_1(\mathrm{Prym}(S,\Sigma),\mathbb{Z}_2).$$

Moreover, $H^1(\mathrm{Prym}(S,\Sigma),\mathbb{Z}_2) \cong \mathrm{Hom}(H_1(\mathrm{Prym}(S,\Sigma),\mathbb{Z}),\mathbb{Z}_2)$ and thus

$$H^1(\mathrm{Prym}(S,\Sigma),\mathbb{Z}_2) \cong \mathrm{Hom}(\wedge,\mathbb{Z}_2) \cong \wedge/2\wedge \cong P[2].$$

Hence, the covering space $P[2]$ is determined by the action of $\pi_1(\mathcal{A}_{reg})$ on the first cohomology of the fibres with $\mathbb{Z}_2$ coefficients. This action is studied in [87] for $n = 2$, and in [10, 15] for $n > 2$.

Real forms with signature. The locus of $SU(p, p+q)$-Higgs bundles is fixed by an involution $\Theta_{SU(p,p+q)} : (E,\Phi) \mapsto (E,-\Phi)$ on $SL(2p+q,\mathbb{C})$-Higgs bundles corresponding to bundles E which have an automorphism conjugate to $I_{p,p+q}$ sending Φ to $-\Phi$, and whose ± 1 eigenspaces have dimensions p and $p+q$ (see also Example 2.2.2). The involution $-\sigma$ acts trivially on the polynomials of even degree. In this case, the characteristic polynomial is given by

$$\det(\eta - \Phi) = \eta^q(\eta^{2p} + a_2\eta^{2p-2} + \ldots + a_{2p-2}\eta^2 + a_{2p}), \tag{3.2.4}$$

defines a spectral curve $\pi : S \to \Sigma$ as

$$S = \{0 = \eta^{2p} + a_2\eta^{2p-2} + \ldots + a_{2p-2}\eta^2 + a_{2p}\} \subset \mathrm{Tot}(K),$$

where η is the tautological section of π^*K and $a_i \in H^0(\Sigma, K^i)$.

Whilst the spectral data is not known for $q \neq 0, 1$, in the case of $q = 0$ it has been described in [88, Chapter 6] and [90] by looking at $U(p,p)$-Higgs bundles $(W_1 \oplus W_2, \Phi)$, which when satisfying $\Lambda^p W_1 \cong \Lambda^q W_2^*$ correspond to $SU(p,p)$-Higgs bundles (the case of $q = 1$ is studied in [84] through Cameral data). Moreover, for $q = 0$, the curve S is a $2p$-fold cover of the Riemann surface Σ, ramified over the $4p(g-1)$ zeros of a_{2p}, and has a natural involution

$$\sigma : \eta \mapsto -\eta \tag{3.2.5}$$

which has as fixed points the ramification points of the cover. The quotient of S by the action of σ defines a p-fold cover $\overline{\rho} : \overline{S} \to \Sigma$ in the total space of K^2, whose equation is given by $\xi^p + a_1\xi^{p-1} + \ldots + a_{p-1}\xi + a_p = 0$, where $\xi = \eta^2$ is the tautological section of $\overline{\rho}^*K^2$. Since $\overline{S}$ is the quotient of a smooth curve, it is also smooth. We let $\pi : S \to \overline{S}$ be the double cover given by the above quotient:

$$\begin{array}{ccc} S & \xrightarrow[\pi]{2:1} & \overline{S}. \\ & {\scriptstyle 2p:1}\searrow^{\rho} \quad {}^{\overline{\rho}}\swarrow{\scriptstyle p:1} & \\ & \Sigma & \end{array}$$

We shall denote by g_S and $g_{\overline{S}}$ the genus of S and $\overline{S}$, respectively. Since the cotangent bundle has trivial canonical bundle, and the canonical bundle of K^2 is $\overline{\rho}^*K^{-1}$, the adjunction formula gives $K_S \cong \rho^*K^{2p}$ and $K_{\overline{S}} \cong \overline{\rho}^*K^{2p} \otimes \overline{\rho}^*K^{-1}$. Thus,

$$g_S = 4p^2(g-1) + 1,$$

$$g_{\overline{S}} = (2p^2 - p)(g-1) + 1.$$

The spectral data for $SU(p,p)$-Higgs bundles can then be deduced from the spectral data of $U(p,p)$-Higgs bundles described as follows:

Theorem 3.2.6 ([90]). *Given a $U(p,p)$-Higgs bundle with non-singular spectral curve one can construct a pair (S, M) where*

(a) *the curve $\rho : S \to \Sigma$ is an irreducible non-singular $2p$-fold cover of Σ given by the equation $\eta^{2p} + a_1\eta^{2p-2} + \ldots + a_{p-1}\eta^2 + a_p = 0$, in the total space of K, where $a_i \in H^0(\Sigma, K^{2i})$, and η is the tautological section of ρ^*K. The curve S has an involution σ acting by $\sigma(\eta) = -\eta$;*

(b) *M is a line bundle on S such that $\sigma^*M \cong M$.*

Conversely, given a pair (S, M) satisfying (a) and (b), there is an associated stable $U(p,p)$-Higgs bundle whose spectral curve is as in (a).

From the structure of the characteristic polynomials of $SU(p,p+q)$-Higgs bundles described in (3.2.4) one can see that for $q \geqslant 1$ generically the polynomial defines a reducible curve, and thus all of the $SU(p,p+q)$-Higgs bundles lie in the fibres over the discriminant locus of the Hitchin fibration. Most other non-split real forms also give Higgs bundles which lie over the non-regular locus of the Hitchin fibration, as shown in [88].

More generically, one may ask whether a given set of Higgs bundles can be ever expressed as elements over the regular locus of the Hitchin fibration. For the case of rank 2 Higgs bundles, Wentworth and Wolf proved in [100] that for every non-elementary representation of a surface group into $SL(2,\mathbb{C})$ there is a Riemann surface structure such that the Higgs bundle associated to the representation lies outside the discriminant locus of the Hitchin fibration For $n \geqslant 3$, and for any other group, the question remains open: indeed, from [100, Remark 3.iv.] there will be obstructions in any generalization of their theorem, and some of these will come from other real forms of $SL(n,\mathbb{C})$ which always lie in the discriminant locus. In particular, their work involves a result of Gallo-Kapovich-Marden which would need to be generalized for higher rank.

3.3. The singular locus of the Hitchin fibration. Across these notes we have said that a point $\omega \in \mathcal{A}_{G_\mathbb{C}}$ in the Hitchin base is regular when the spectra curve S it defines is smooth – equivalently, we said it is singular when the corresponding spectral curve defined through the characteristic polynomial of the Higgs field (3.2.1) is singular. There are, however, more elaborated and useful ways of stratifying the Hitchin base, as for instance those used in [82] (see [33] for further descriptions of those stratifications). However, we shall restrict ourselves here to the simpler definition of singular point and make some comments on some of the arising geometry and open problems.

The most singular fiber is the fibre over $\mathbf{0} \in \mathcal{A}_{G_\mathbb{C}}$, named *the nilpotent cone* by Laumon [74], to emphasise the analogy with the nilpotent cone in Lie algebra.

This fibre has been studied from different perspectives, one of which is via the moment map μ of the S^1 action $(E, \Phi) \to (E, e^{i\theta}\Phi)$ as done in [61,64]. In particular, note that given a stable bundle E, take $\Phi = 0$, then the Higgs bundle $(E, 0)$ is stable and trivially fixed by the S^1-action.

As shown in [60, Theorem 5.2], the nilpotent cone is preserved by the flow by μ and it encodes the topology of the moduli space since points of $\mathcal{M}_{G_C}$ flow towards the nilpotent cone. For a overview of work done concerning singular Higgs bundles and Higgs bundles over the singular locus of the Hitchin fibration the reader should refer to [4]. In particular, the nilpotent cone has primarily been studied for $SL(n, \mathbb{C})$ and $GL(n, \mathbb{C})$, and much of its geometry remains unknown for the moduli spaces of G_C-Higgs bundles. For $SL(n, \mathbb{C})$ and $GL(n, \mathbb{C})$-Higgs bundles, the irreducible components of the nilpotent cone are labeled by connected components of the fixed point set of the S^1 action. The nilpotent cone can be also considered for other types of Higgs bundles, and we shall make some comments in Section 5 about the nilpotent cone appearing through hyperpolygons. Among the components of the nilpotent cone for classical Higgs bundles is the moduli space $\mathcal{N}$ of semistable bundles.

Other fibers over the singular locus of the Hitchin fibration have been the subject of more recent research (e.g., see [16,28,33,55,69,91]). Since we will return to singular fibres when studying branes of Higgs bundles, we shall conclude this chapter with few short comments:

- In the case of $GL(n, \mathbb{C})$-Higgs bundles we mentioned that for generic points of the Hitchin base, the corresponding fiber can be identified with the Jacobian $Jac(S)$ of the spectral curve S. When the spectral curve S is not smooth but is integral, the corresponding fiber is seen to be the *compactified Jacobian* [17,93–95] (see also [78, Fact 10.3] for a clear explanation). The compactified Jacobian $\overline{Jac}(S)$ is the moduli space of all torsion-free rank-1 sheaves on S, where the usual "locally-free" condition is missing. Intuitively, the compactified Jacobian can be obtained by considering a path of smooth curves S_t approaching a singular curve S_0 which is the base point of the nilpotent cone; since the limit of $Jac(S_t)$ does not depend on the choice of smooth family [71], this limit is $\overline{Jac}(S)$.
- For $GL(n, \mathbb{C})$-Higgs bundles whose spectral curve S is not integral, the fine moduli space needs to be considered.
- For $SL(2, \mathbb{C})$-Higgs bundles, connectedness of the fiber of $\mathcal{M}_{SL(2,\mathbb{C})}$ when S is irreducible and has only simple nodes has been considered in [48, §5.2.2], whilst a full description of the singular fibers is given in [55].

3.4. Problem set II. This problem set is meant to make the reader used to *spectral data* for complex Higgs bundles, and understand how it can be expressed in an abelian manner. Moreover, it should provide an introduction to ways in which one can study singular fibres of the Hitchin fibration. They provide a mixture of problems from the lectures as well as some additional ones.

Problem 6. Spectral curves. Through its characteristic polynomial, a Higgs field defines a natural spectral curve in the total space of the canonical bundle K. Moreover, the extra conditions satisfied by the Higgs field are reflected, in particular, in the properties of this curve.

i. For which complex Lie groups among the types A, B, C, D, the spectral curves defined by the corresponding Higgs field carry an involution?
ii. Considering low rank groups, find the conditions on a Higgs bundle (E, Φ) for the Higgs field to define a spectral curve which has an order 4 automorphism (e.g. see [11]).
iii. By considering properties of characteristic polynomials from a Linear Algebra perspective, give a description of two particular types of Higgs bundles (E, Φ) for which the spectral curve $S = \{(\det \Phi - \lambda) = 0\}$ has equation $\{q^3(\lambda) = 0\}$ for some irreducible polynomial $q(\lambda)$.
iv. Following on **Problem 6.iii.**, for which Higgs bundles one has
$$(\det \Phi - \lambda) = q^n(\lambda)$$
for some irreducible polynomial $q(\lambda)$?

Problem 7. Generic fibres. We have mentioned that by adding conditions to a Higgs pair (E, Φ), one can obtain a description of a principal $G_{\mathbb{C}}$-Higgs bundle. Equivalently, one may add conditions to the classical spectral curve S, and data on it $\mathrm{Jac}(S)$ to recover principal Higgs bundles.

i. Use Grothendieck-Riemann-Roch to $\deg(E) = \deg(L) + (n^2 - n)(1 - g)$ when considering spectral data for classical Higgs bundles.
ii. Show that when considering $SL(n, \mathbb{C})$-Higgs bundles, the condition
$$\Lambda^n E \cong \mathcal{O}$$
is equivalent to requiring the spectral line bundle L to be in $\mathrm{Prym}(S, \Sigma)$.
iii. Show that the smooth $SO(2n, \mathbb{C})$-Hitchin fibres are $\mathrm{Prym}(\hat{S}, \hat{S}/\hat{\sigma})$, for $\hat{S}$ the desingularization of the curve S associated to the regular base point.
iv. Show that when considering generic $SO(2n+1, \mathbb{C})$-Higgs bundles, the spectral curve S defined via the characteristic polynomial is never smooth.

Problem 8. The Hitchin fibration. When considering a subspace of Higgs bundles which has finite intersection with the regular locus, one can study its geometry and topology though the monodromy action as well as through other involutions.

i. There is a natural involution σ_s on the Lie algebra $\mathfrak{g}_{\mathbb{C}}$ given by
$$\sigma_s([\mathrm{ad}\, f_0]^n e_i) = (-1)^{n+1} [\mathrm{ad}\, f_0]^n e_i,$$
uniquely defined by $\sigma_s(e_i) = -e_i$ and $\sigma_s(f_0) = -f_0$. Show that $-\sigma_s$ on $\mathfrak{g}_{\mathbb{C}}$ acts trivially on the ring of invariant polynomials of the Lie algebra $\mathfrak{g}_{\mathbb{C}}$.
ii. Show that for non-split real forms G, the characteristic polynomial of a G-Higgs field defines a reducible curve.

iii. **((*))** Generalize [18] to define generalized parabolic $G_{\mathbb{C}}$-Higgs bundles on X, as well as a Hitchin fibration.

Problem 9. Singular fibres. Consider $SU(p, p+q)$-Higgs bundles $(V \oplus W, \Phi)$, whose characteristic polynomial is

$$\det(\eta - \Phi) = \eta^q(\eta^{2p} + a_2\eta^{2p-2} + \ldots + a_{2p-2}\eta^2 + a_{2p}),$$

with natural involution $\sigma : \eta \mapsto -\eta$ as in (3.2.5).

i. For $q = 0$, consider line bundles U_1 and U_2 on

$$\bar{\pi} : \overline{S} = S/\sigma \to \Sigma,$$

such that $\bar{\pi}_* U_1 = V$ and $\bar{\pi}_* U_1 = W$, and express $\Lambda^p V$ and $\Lambda^p W$ and $\Lambda^p V \cong \Lambda^p W^*$ in terms of U_i.

ii. For $p = 2$ and $q = 0$ the quotient curve $\bar{S}$ of **Problem 9.i.** is a double cover of the Riemann surface Σ. Describe any additional geometric structure appearing from the fact that the Higgs bundle has low rank.

iii. **((*))** How can the structure of an $SO(2,4)$-Higgs bundle appear through the data described in **Problem 9.ii.**?

iv. **((*))** Consider $p = 2$ and $q = 2$ and give a description of spectral data that could be associated to $U(2,4)$-Higgs bundles – see [16] for details on spectral data for orthogonal Higgs bundles with signature.

Problem 10. Parabolic Hitchin fibration As explained in [9], the moduli space of parahoric Higgs bundles, and in particular of parabolic Higgs bundles as defined in the previous chapter, carry a Hitchin map which is a Poisson map whose generic fibres are abelian varieties. Using the paper [9], think of parabolic Higgs bundles in terms of parahoric Higgs bundles:

i. Explain the relations between parahoric Higgs bundles and parabolic Higgs bundles;

ii. By considering the parahoric global nilpotent cone, describe the fibre over 0 for parabolic $\mathcal{G}$-Higgs bundles in terms the cotangent bundle of $\mathrm{Bun}_{\mathcal{G}}$.

iii. Give a summery of how to establish that the Hitchin map on the moduli of polystable parabolic Higgs bundles is proper.

iv. Consider parabolic $SL(4, \mathbb{C})$-Higgs bundles, and give a comparison of the Hitchin maps defined in [75] and [9].

4. Branes in the moduli space of Higgs bundles

If people do not believe that mathematics is simple, it is only because they do not realize how complicated life is.

John von Neumann

The appearance of Higgs bundles (and flat connections) within string theory and the geometric Langlands program has led researchers to study the *derived category of coherent sheaves* and the *Fukaya category* of these moduli spaces. Therefore, it has become fundamental to understand Lagrangian submanifolds of the moduli space of Higgs bundles supporting holomorphic sheaves (A-branes), and their dual objects (B-branes). The space of solutions to Hitchin's equations (2.1.7)–(2.1.8) is a hyperkähler manifold, and thus there is a family of complex structures from which we shall fix I, J, K obeying quaternionic equations, following the notation of [14, 64, 72]. In particular, under this convention the smooth locus of $\mathcal{M}_{G_\mathbb{C}}$ corresponds to the space of solutions to Hitchin's equations with structure I. Throughout this notes we shall adopt the physicists' language in which a Lagrangian submanifold supporting a flat connection is called an *A-brane*, and a complex submanifold supporting a complex sheaf is a *B-brane*. By considering the support of branes, we shall refer to a submanifold of a hyperkähler manifold as being of type A or B with respect to each of the structures, and hence one may consider branes of type

$$(B, B, B),\ (B, A, A),\ (A, B, A) \text{ and } (A, A, B).$$

We shall dedicate this chapter to the construction, description, and study of such branes, by considering additional structure on the Riemann surface Σ and on the complex Lie group $G_\mathbb{C}$:

- **Branes through finite group actions on Riemann surface Σ.** Finite group actions
  $\Gamma \times \Sigma \to \Sigma$ on compact connected Riemann surfaces have long been considered, and through these actions we shall consider (B, B, B)-branes appearing through Γ-equivariant representations following [22, 63].
- **Branes through anti-holomorphic involutions.** We have seen before that anti-holomorphic involutions can be used to define real Higgs bundles, as initiated in [65]. By considering real structures on the Riemann surface and their compositions with group involutions, we shall define (B, A, A), (A, B, A) and (A, A, B)-branes following [13, 14] (see also [89]). These constructions will play a special role later on when considering hyperpolygons following [86] in the final chapter of these notes.

4.1. Branes through finite group actions. As shown in [63], a natural way to define (B, B, B)-branes in the moduli spaces of complex Higgs bundles is though the action of a finite group Γ on the base Riemann surface of genus g. When the genus is 2 or 3, a complete classification of all finite group actions appears in [34, Tables 4, 5], which allows one to perform explicit calculations in those cases. Moreover, in the case of actions induced on rank 2 bundles through automorphisms of Σ, a very concrete description of the fixed points in terms of parabolic structures is given in [3].

Within this setting, a much studied question is the appearance of flat Γ-equivariant $G_\mathbb{C}$-connections on Σ, for which one needs to fix a C^∞ trivialization $\underline{\mathbb{C}}^n = \Sigma \times \mathbb{C}^n \to \Sigma$ of the underlying vector bundle. There are a few different perspectives on Γ-equivariant flat connection on a Riemann surface Σ, and here we shall consider the definition of [63], through which a Γ-equivariant flat connection is a flat connection ∇ on $\underline{\mathbb{C}}^n \to \Sigma$ such that for every $\phi \in \Gamma$ there exist a $GL(n,\mathbb{C})$-gauge transformation $g_\phi \colon \Sigma \to GL(n,\mathbb{C})$ for which

- $\phi^*\nabla = \nabla \cdot g_\phi$,
- and $\phi \mapsto g_\phi$ a *generalized group homomorphism*, i.e., such that

$$g_{id} = id$$
$$g_{(\phi\circ\tau)}(p) = g_{\tau(p)} \circ g_\phi(\tau(p)).$$

Consider $G_\mathbb{C} = GL(n,\mathbb{C}), SL(n,\mathbb{C})$, and $\Gamma \times \Sigma \to \Sigma$ a finite group action by holomorphic automorphisms without fix points. Then, any Γ-equivariant flat $G_\mathbb{C}$-connection is given by the pull-back of a flat $G_\mathbb{C}(n,\mathbb{C})$-connection on Σ/Γ (e.g. see [63, Proposition 1]). When the action has fixed points, the image $B \subset \Sigma/\Gamma$ of the fixed points gives the branch points of the ramified cover

$$\pi_\Gamma : \Sigma \to \Sigma/\Gamma,$$

and thus any Γ-equivariant flat $G_\mathbb{C}$-connection is given by the pull-back of a flat $G_\mathbb{C}$-connection on $\Sigma/\Gamma - B$ (e.g. see [63, Proposition 2]). From the definition of Γ-equivariant flat connections it follows that a Γ-equivariant flat irreducible $G_\mathbb{C}$-connection ∇ on Σ, gives rise to a Γ-equivariant connection $\nabla \cdot g$ for $g \colon \Sigma \to G_\mathbb{C}$ a gauge transformation, which corresponds to the same point in the moduli space of flat (possibly singular) $G_\mathbb{C}$-connections on Σ/Γ (e.g. see [63, Lemma 1]).

Example 4.1.1. To illustrate the appearance of equivariant connections (and thus of equivariant Higgs bundles) consider the following example from [63]. Let Σ be a compact Riemann surface of genus 3 admitting a fixed point free involution $\phi \colon \Sigma \to \Sigma$. Through this action, one has a double cover to the quotient Riemann surface $M := \Sigma/\mathbb{Z}_2$ of genus 2.

Let ∇ be a flat unitary line bundle connection which is not self-dual, and such that $\phi^*\nabla = \nabla^*$. Then,

$$\nabla \oplus \nabla^*$$

is a $\mathbb{Z}_2$-equivariant flat connection on Σ, with a corresponding irreducible flat connection on $\Sigma/\mathbb{Z}_2$.

By considering different finite group actions of a group Γ on a Riemann surface Σ of genus $g \geqslant 2$, in [63, Theorem 8] it is shown that the connected components of the space of gauge equivalence classes of irreducible Γ-equivariant flat $G_\mathbb{C}$-connections are hyperkähler submanifolds of the moduli space of flat irreducible $G_\mathbb{C}$-connections on Σ, and hence give (B, B, B)-branes in the moduli space of $G_\mathbb{C}$-Higgs bundles.

When $G_\mathbb{C} = SL(2,\mathbb{C})$, from the above, any component of the moduli space of flat Γ-equivariant $SL(2,\mathbb{C})$-connections on Σ can be identified with the moduli space of flat $SL(2,\mathbb{C})$-connections or $PSL(2,\mathbb{C})$-connections on a n-punctured compact Riemann surface of genus γ, for $n \in \mathbb{N}$ satisfying some compatibility conditions. furthermore, more can be said about the type of fixed point sets of the induced involution on the moduli spaces of Higgs bundles by the finite group:

Theorem 4.1.2 ([63] Theorem 11). *Let Σ be a compact Riemann surface of genus $g \geqslant 2$ and Γ be a finite group acting on Σ by holomorphic automorphisms such that a component of the moduli space of Γ-equivariant flat $SL(2, \mathbb{C})$-connections on Σ has half the dimension of the moduli space $\mathcal{M}_g$. Then one of the following holds:*

(I) $\Gamma = \mathbb{Z}_2$ *acts by a fix-point free involution on* Σ, *or*

(II) Σ *is hyperelliptic of genus* 3 *and* $\Gamma = \mathbb{Z}_2 \times \mathbb{Z}_2$.

In the later case (II), one of the $\mathbb{Z}_2$-factors corresponds to the hyperelliptic involution, whilst the other $\mathbb{Z}_2$-factor corresponds to an involution with 4 fixed points.

It is interesting to note that all genus 3 Riemann surfaces with fixed point free actions must be hyperelliptic (see [63, Proposition 21]) and thus a hyperelliptic Riemann surface of genus 3 with an additional involution with 4 fixed points is the same as Riemann surface of genus 3 with a fixed point free involution.

From the above, fixed point free involutions become of particular interest. In particular, for $SL(2, \mathbb{C})$-Higgs bundles, it is shown in [63, Proposition 23] that given a fixed point free involution τ, a flat and irreducible τ-invariant $SL(2, \mathbb{C})$-connection on a Riemann surface of genus 3 is equivariant with respect to the hyperelliptic involution. It should be noted that irreducibility is not a necessary condition to be in an equivariant brane, since there exist flat reducible connections on Riemann surfaces which correspond to irreducible connections on the hyperelliptic genus 2 quotient Riemann surface surface.

The geometry of these branes can be studied, in particular, by considering their intersection with the Hitchin fibration – as was done in the last chapter when considering real Higgs bundles. For the branes within the moduli space of $SL(2, \mathbb{C})$-Higgs bundles, one can see them expressed in terms of Prym varieties:

Theorem 4.1.3 ([63] Theorem 14). *Let τ be a fix point free involution. The τ-equivariant (B, B, B)-brane intersects a generic fibre of the corresponding Hitchin fibration over a point defining the spectral curve S in the abelian variety $Prym(S/\tau, \Sigma/\tau)/\mathbb{Z}_2$.*

Following [63] and [22], one may define *anti-equivariant Higgs bundles* $(\bar{\partial}, \Phi)$ with respect to a fix point free involution τ, as pairs given by a τ-equivariant holomorphic structure $\bar{\partial}$ and a Higgs field Φ which satisfies

$$\tau^* \Phi = -g^{-1} \circ \Phi \circ g,$$

where g is the gauge transformation such that $\tau^* \bar{\partial} = \bar{\partial}.g$ and $(g \circ \tau) g = \text{id}$. These Higgs bundles form examples of Lagrangians subspaces in the moduli space of Higgs bundles, and whilst they can be defined for compact Riemann surfaces of any genus, we shall restrict our attention here to odd genus following [63] (the more general setting is treated in [22]). Indeed, it is shown in [63, Proposition 15] that anti-equivariant $SL(2, \mathbb{C})$-Higgs bundles for connected components in

$\mathcal{M}_{SL(2,\mathbb{C})}$ which are complex submanifolds with respect to the complex structure I and Lagrangian with respect to the complex structures J and K, and thus define (B, A, A)-branes. Finally, we should mention that interesting research directions arise from [63] when considering generalizations of the results appearing there for more general groups.

4.2. Branes through anti-holomorphic involutions. Branes of different types can be obtained by considering anti-holomorphic involutions both on the compact Riemann surface as well as on the complex Lie group, and this is the perspective of [13, 14] and which we shall describe in what remains of the chapter. Following [65] and recalling the previous analysis in the first chapter about Higgs bundles for real groups, one has the following description of G-Higgs bundles considered in [88]:

Proposition 4.2.1. *Let* G *be a real form of a complex semi-simple Lie group* $G_{\mathbb{C}}$*, whose real structure is* σ*. Then,* G-Higgs bundles *are given by the fixed points in* $\mathcal{M}_{G_{\mathbb{C}}}$ *of the involution* Θ_G *acting by*

$$\Theta_G : (P, \Phi) \mapsto (\theta(P), -\theta(\Phi)),$$

where $\theta = \rho\sigma$*, for* ρ *the compact real form of* $G_{\mathbb{C}}$*.*

One should note that a fixed point of Θ_G in $\mathcal{M}_{G_{\mathbb{C}}}$ in Proposition 4.2.1 gives a representation of $\pi_1(\Sigma)$ into the real form G up to the equivalence of conjugation by the normalizer of G in $G_{\mathbb{C}}$. This may be bigger than the Lie group G itself, and thus two distinct classes in $\mathcal{M}_G$ could be isomorphic in $\mathcal{M}_{G_{\mathbb{C}}}$ via a complex map. Hence, although there is a map from $\mathcal{M}_G$ to the fixed point subvarieties in $\mathcal{M}_{G_{\mathbb{C}}}$, this might not be an embedding. A description of this phenomena in the case of rank 2 Higgs bundles is given in [87, Section 10], where one can see how the $SL(2,\mathbb{R})$-Higgs bundles which have different topological invariants lie in the same connected component as $SL(2,\mathbb{C})$-Higgs bundles.

The point of view of Proposition 4.2.1, which is considered throughout [88], fits into a more global picture described in [13, 14], through which families of (B, A, A), (A, B, A) and (A, A, B) branes in $\mathcal{M}_{G_{\mathbb{C}}}$ appear as the fixed point sets involutions arising from a real form σ on a complex Lie group and a real structure f on a Riemann surface. A real structure[5] on a compact connected Riemann surface Σ is an anti-holomorphic involution $f : \Sigma \to \Sigma$. The classification of real structures on compact Riemann surfaces is a classical result of Klein, who showed that all such involutions on Σ may be characterised by two integer invariants (n, a): the fixed point set of f is known to be a disjoint union of n copies of the circle embedded in Σ; the complement of the fixed point set has one or two components, corresponding to $a = 1$ or $a = 0$ respectively. A real structure f on the Riemann surface Σ induces involutions on the moduli space of representations $\pi_1(\Sigma) \to G_{\mathbb{C}}$, of flat connections and of $G_{\mathbb{C}}$-Higgs bundles on Σ.

[5]Anti-involutions on Σ are also found in the literature as an anti-conformal maps (e.g. see [35])

- From the Cartan involution θ of G one obtains a (B, A, A)-brane fixed by

$$i_1(\bar{\partial}_A, \Phi) = (\theta(\bar{\partial}_A), -\theta(\Phi)). \tag{4.2.2}$$

- From a real structure $f : \Sigma \to \Sigma$ one obtains an (A, B, A)-brane fixed by

$$i_2(\bar{\partial}_A, \Phi) = (f^*(\partial_A), f^*(\Phi^*)) = (f^*(\rho(\bar{\partial}_A)), -f^*(\rho(\Phi))). \tag{4.2.3}$$

- Lastly, by looking at $i_3 = i_1 \circ i_2$, one obtains an (A, A, B)-brane fixed by

$$i_3(\bar{\partial}_A, \Phi) = (f^*\sigma(\bar{\partial}_A), f^*\sigma(\Phi)).$$

In what follows we shall give an overview of the above branes in terms of their geometry and topology, and describe how they can be studied though the Hitchin fibration. Generalizations of the above methods from [13, 14] have now been obtained for many other moduli spaces – e.g. see [46, 47, 70, 86] among others.

The (B, A, A)-brane of G-Higgs bundles. As mentioned in the first Lecture, the moduli spaces $\mathcal{M}_{G_\mathbb{C}}$ have a natural symplectic structure, which we denoted by ω. Moreover, following [66], an involution Θ_G from Proposition 4.2.1, which corresponds to i_1 in (4.2.2), sends $\omega \mapsto -\omega$. Thus, at a smooth point, the fixed point set must be Lagrangian and thus the expected dimension of $\mathcal{M}_G$ is half the dimension of $\mathcal{M}_{G_\mathbb{C}}$. These branes of real G-Higgs bundles can be seen sitting inside $\mathcal{M}_{G_\mathbb{C}}$ as fixed points of the involution Θ_G in the preserved fibres over the Hitchin base $\mathcal{A}_{G_\mathbb{C}}$.

By considering Cartan's classification of classical Lie algebras, we can describe the different involutions i_1 that can be considered on the moduli spaces of $G_\mathbb{C}$-Higgs bundles. To do so concretely, it is helpful to denote by $I_{p,q}$, J_n and $K_{p,q}$ the matrices

$$I_{p,q} = \begin{pmatrix} -I_p & 0 \\ 0 & I_q \end{pmatrix}, \; J_n = \begin{pmatrix} 0 & I_n \\ -I_n & 0 \end{pmatrix}, \; K_{p,q} = \begin{pmatrix} -I_p & 0 & 0 & 0 \\ 0 & I_q & 0 & 0 \\ 0 & 0 & -I_p & 0 \\ 0 & 0 & 0 & I_q \end{pmatrix}.$$

Recalling that the cartan involution Θ_G of Proposition 4.2.1 is obtained $\theta = \rho\sigma$, for ρ the compact real form of $G_\mathbb{C}$ and σ an anti-holomorphic involution of $G_\mathbb{C}$ fixing a real form G, we consider the following compact and split real forms:

$\mathfrak{g}_c$ $G_\mathbb{C}$	Split form	Compact form $\mathfrak{u}$	anti-involution ρ fixing $\mathfrak{u}$	dim $\mathfrak{u}$
$\mathfrak{a}_n$	$\mathfrak{sl}(n, \mathbb{R})$	$\mathfrak{su}(n)$	$\rho(X) = -\overline{X}^t$	$n(n+1)$
$\mathfrak{b}_n$	$\mathfrak{so}(n, n+1)$	$\mathfrak{so}(2n+1)$	$\rho(X) = \overline{X}$	$n(2n+1)$
$\mathfrak{c}_n$	$\mathfrak{sp}(2n, \mathbb{R})$	$\mathfrak{sp}(n)$	$\rho(X) = J_n\overline{X}J_n^{-1}$	$n(2n+1)$
$\mathfrak{d}_n$	$\mathfrak{so}(n, n)$	$\mathfrak{so}(2n)$	$\rho(X) = \overline{X}$	$n(2n-1)$

Table 4.2.4. Compact forms $\mathfrak{u}$ of classical Lie algebras

Then, the following table allows one to see the different possible values that θ may take in terms of the anti-holomorphic involutions on the complex Lie algebras.

$\mathfrak{g}_c$	Real form G	σ fixing G	$\theta = \rho\sigma$
$\mathfrak{a}_n$	$SL(n,\mathbb{R})$	$\sigma(X) = \overline{X}$	$\theta(X) = -X^t$
	$SU^*(2m)$	$\sigma(X) = J_m\overline{X}J_m^{-1}$	$\theta(X) = -J_mX^tJ_m^{-1}$
	$SU(p,q)$	$\sigma(X) = -I_{p,q}\overline{X}^tI_{p,q}$	$\theta(X) = I_{p,q}XI_{p,q}$
$\mathfrak{b}_n$	$SO(p,q)$	$\sigma(X) = I_{p,q}\overline{X}I_{p,q}.$	$\theta(X) = I_{p,q}XI_{p,q}$
$\mathfrak{c}_n$	$Sp(2n,\mathbb{R})$	$\sigma(X) = \overline{X}$	$\theta(X) = J_nXJ_n^{-1}$
	$Sp(2p,2q)$	$\sigma(X) = -K_{p,q}X^*K_{p,q}.$	$\theta(X) = K_{p,q}XK_{p,q}$
$\mathfrak{d}_n$	$SO(p,q)$	$\sigma(X) = I_{p,q}\overline{X}I_{p,q}.$	$\theta(X) = I_{p,q}XI_{p,q}$
	$SO^*(2m)$	$\sigma(X) = J_m\overline{X}J_m^{-1}.$	$\theta(X) = J_mXJ_m^{-1}$

Table 4.2.5. Non-compact forms G of classical Lie algebras $G_{\mathbb{C}}$

Example 4.2.6. One can see $SO(p, p+q)$-Higgs bundles as fixed points of the involution $\Theta_{SO(p,p+q)} : (E,\Phi) \mapsto (E,-\Phi)$ on the moduli space of $SO(2p+q,\mathbb{C})$ corresponding to vector bundles E which have an automorphism f conjugate to $I_{p,p+q}$ sending Φ to $-\Phi$ and whose ± 1 eigenspaces have dimensions p and $p+q$.

In what follows we shall give a summary of different appearances of the (B,A,A)-brane of G-Higgs bundles as the fixed point set of the induced action of i_1 from (4.2.2) on the Hitchin fibration. In particular, these branes may intersect the regular fibres of the Hitchin fibration in very different ways:

- **Finite intersection with the regular fibres**. As mentioned before, from [88, Theorem 4.12], the (B,A,A)-brane of G-Higgs bundles for G a split real form of $G_{\mathbb{C}}$ intersects the fibres of the Hitchin fibration in points of order two. Since we have given an overview of this case in previous chapters, we shall not consider them any further here (see [16, 28, 32, 87, 88, 91] for papers treating the spectral data for split real forms).
- **Positive dimensional intersection with the regular fibres**. The (B,A,A)-brane of $U(p,p)$-Higgs bundles in $\mathcal{M}_{GL(2p,\mathbb{C})}$, as seen in Exercise **II.(4)**, and of $SU(p,p)$ in $\mathcal{M}_{SL(2p,\mathbb{C})}$ intersects the regular fibres in finite dimensional abelian varieties given by Jacobians or Prym varieties of quotient spectral curves, as shown in [88, 90]. Having mentioned the spectral data for these objects in previous chapters, we shall dedicate the reminder of this section to the third possibility described below.
- **Empty intersection with the regular fibres**. From the structure of real Higgs bundles, as shown in [88], the characteristic polynomial for non-split Higgs bundles for real forms other than $U(p,p)$ and $SU(p,p)$ will

define reducible spectral curves and thus their (B, A, A)-brane will lie completely over the discriminant locus of the Hitchin fibration. These branes can be shown to intersect the non-regular fibres of the Hitchin fibration in different ways, and below we shall consider three different cases:

- **Abelian spectral data.** Examples of these come from quasi-split but not split real forms. For these groups, it is shown that the Cameral data for the groups $SU(p, p+1)$ and $SO(p, p+2)$ should be abelian in [84], and the abelian spectral data for $SO(p, p+2)$-Higgs bundles is described in [16].
- **Non-Abelian spectral data**. Higgs bundles for certain groups will carry natural non-abelian spectral data describing the intersection of their (B, A, A)-brane with the non-regular fibres of the Hitchin fibration. To study this situation, one can consider Higgs bundles (E, Φ) whose generic eigenspace is r-dimensional, and $\det(xI - \Phi) = p(x)^r$ for some polynomial $p(x)$: since the Hitchin map is surjective, these cases certainly occur. For $r = 2$, by extending the approach from [88], it is shown in [69] a *"nonabelianization"* procedure appearing when considering Higgs bundles corresponding to flat connections on Σ with holonomy in the real Lie groups $G = Sp(2p, 2p)$, $SL(p, \mathbb{H})$ and $SO(2p, \mathbb{H})$ (the latter two also known as $SU^*(2p)$ and $SO^*(2p)$ respectively), spectral data is given by certain subspace of rank 2 Higgs bundles on a spectral curve.
- **Abelian and Non-Abelian spectral data**. Interestingly, the generic intersection of certain (B, A, A)-branes with the Hitchin fibration needs abelian and non-abelian data to be described. This is the case, in particular, of $SO(p+q, p)$-Higgs bundles and $Sp(2p+2q, 2p)$-Higgs bundles described in [16]. In this setting, the Higgs field has generically a kernel of dimension $2q$, and its characteristic polynomial has a degree $4p$ polynomial which generically defines a smooth curve. By considering this smooth curve and abelian data on it, as well as an auxiliary spectral cover with non-abelian data on it, a geometric description of the brane through *"Cayley and Langlands type correspondences"* is obtained in [16] and will be described below.

In what follows we shall give some examples of branes which lie completely over the discriminant locus of the Hitchin fibration, mostly following the work in [69] and [16].

Nonabelianization. Following the Definition 2.2.1 consider the following:

- $SL(n, \mathbb{H})$-Higgs bundles consisting of a rank $2n$ symplectic vector bundle (V, ω) and a Higgs field satisfying $\Phi = \Phi^T$ for the symplectic transpose. Using ω to identify V and V^*, this means that $\Phi = \omega^{-1}\phi$ for $\phi \in H^0(\Sigma, \Lambda^2 V \otimes K)$.

- $SO(2p, \mathbb{H})$-Higgs bundle whose vector bundles $V = W \oplus W^*$ for a rank 2p vector bundle W. The inner product is defined by the natural pairing between W and W^*. The Higgs field is of the form $\Phi(w, \xi) = (\beta(\xi), \gamma(w))$ where $\beta : W^* \to W \otimes K$ and $\gamma : W \to W^* \otimes K$ are skew-symmetric.
- $Sp(2p, 2p)$-Higgs bundles, which have vector bundle $W_1 \oplus W_2$ for symplectic rank 2p vector bundles $(W_1, \omega_1), (W_2, \omega_2)$ and the Higgs field is of the form $\Phi(v, w) = (\beta(w), \gamma(v))$ where $\beta : W_2 \to W_1 \otimes K$ and $\gamma : W_1 \to W_2 \otimes K$ with $\beta = -\gamma^T$, using the symplectic transpose.

The spectral data for the above Higgs bundles is described in [69] by means of rank 2 vector bundles, and can be summarised as follows by considering rank an $SL(2n, \mathbb{C})$-Higgs bundles (V, Φ) on a compact Riemann surface Σ of genus $g \geqslant 2$, whose Higgs field has characteristic polynomial

$$\det(\Phi - \eta) = (\eta^n + a_2 x^{n-2} + \cdots + a_n)^2. \tag{4.2.7}$$

Higgs bundles for the groups $SL(n, \mathbb{H})$, $SO(2p, \mathbb{H})$ and $Sp(2p, 2p)$ lie within this setting, by letting $n = 2p$. By considering $a_i \in H^0(\Sigma, K^i)$ and η the tautological section as before, (4.2.7) defines a generically smooth spectral curve

$$S = \{\sqrt{\det(\Phi - \eta)} = 0\}. \tag{4.2.8}$$

Theorem 4.2.9 ([69]). *Consider a rank an $SL(2n, \mathbb{C})$-Higgs bundle (E, Φ) with characteristic polynomial as in (4.2.7). Given a rank 2 vector bundle E on the curve S defined as in (4.2.8), the direct image of $\eta : E \to E \otimes \pi^* K$ defines a semi-stable $SL(n, \mathbb{H})$-Higgs bundle on Σ if and only if*

(i) $\Lambda^2 E \cong \pi^* K^{n-1}$

(ii) *E is semi-stable.*

When $n = 2p$, the curve S in (4.2.8) has a natural involution $\sigma : \eta \mapsto -\eta$. In this case, the direct image of $\eta : E \to E \otimes \pi^* K$ defines a semi-stable $SO(2p, \mathbb{H})$-Higgs bundle on Σ if and only if **(i)** *and* **(ii)** *hold, as well as*

(iii) *$\sigma^* E \cong E$ where the induced action on $\Lambda^2 E = \pi^* K^{4p-1}$ is trivial.*

Finally, the direct image of $\eta : E \to E \otimes \pi^* K$ defines a semi-stable $Sp(2p, 2p)$-Higgs bundle if and only if **(i)** *and* **(ii)** *hold, as well as*

(iv) *$\sigma^* E \cong E$ where the induced action on $\Lambda^2 E = \pi^* K^{2m-1}$ is -1.*

Interestingly, as noted in [88] and [69], when considering $Sp(2p, 2p)$-Higgs bundle one has an action of σ by -1 on $\Lambda^2 E$, and thus at a fixed point one has distinct $+1$ and -1 eigenspaces. Following [3] this defines a rank 2 bundle on the quotient curve $\pi' : \bar{S} := S/\sigma \to S$ with a parabolic structure at the fixed points defined by the flag given by the -1 eigenspace and the parabolic weight $1/2$, thus closely related to parabolic Higgs bundles as described in Chapter 1. The direct image of E on the quotient curve gives a rank 2 vector bundle which can be decomposed into line bundles $E^+ \oplus E^-$ through the invariant and anti-invariant sections under the induced action of σ. rank 4p vector bundle can be recovered through the quotient curve by taking $\bar{\pi}(E^+) = W_1$ and $\bar{\pi}(E^+) = W_2$ for $\bar{\pi}$ the rank

2p-projection $\bar{\pi} : \bar{S} \to \Sigma$. Moreover, through the Lefschetz fixed point formula, the degrees of W_1, W_2 can be expressed in terms of those of $E^{\pm}$.

Cayley and Langlands type correspondences. Recall that $SO(p+q,p)$-Higgs bundles give a (B,A,A)-branes which for most values of $p, q \in \mathbb{N}$, require both abelian and non-abelian data to be described within the Hitchin fibration. From Definition 2.2.1, the characteristic polynomial of a generic $SO(p+q,p)$-Higgs bundle $(V \oplus W, \Phi)$ can be written as

$$\det(\Phi - \eta) = \eta^q(\eta^{2p} + a_1 x^{2p-2} + \cdots + a_p),$$

and defines a generically smooth spectral curve $\pi : S \to \Sigma$ given by

$$S = \{0 = \eta^{2p} + a_1 x^{2p-2} + \cdots + a_p\}, \tag{4.2.10}$$

taking $a_i \in H^0(\Sigma, K^{2i})$ and η the tautological section of $\pi^* K$. Since the polynomial defining S in (4.2.10) only has even coefficients, there is a natural involution $\sigma : \eta \mapsto -\eta$ through which one can define the quotient curve $\bar{\pi} : \bar{S} := S/\sigma \to \Sigma$.

From [16, Theorem 4.11], there is a one-to-one correspondence between $SO(p+q,p)$-Higgs bundles and the triple of spectral data (L, M, τ) given by:

(i) A line line bundle $L \in \mathrm{Jac}(\bar{S})[2]$, space of torsion two line bundles of degree zero on $\bar{S}$;
(ii) A rank q equivariant orthogonal bundle M of type $(q-1, 1)$ on the 2-fold auxiliary curve $C := \{0 = \eta^2 - a_p\} \subset \mathrm{Tot}(K^p)$, for a_p as in (4.2.10), with a choice of orientation, and satisfying the following semistability condition: all invariant isotropic subbundles $M' \subset M$ have $\deg(M') \leqslant 0$;
(iii) A choice of isometry τ over each zero x of a_1, between the corresponding fibres of W and those of the -1-eigenspace M^- of M, giving the extension data needed between (i) and (ii). This data may be identified with a collection $\{\tau_x\}$ of unit vectors $\tau_x \in L_x$ for each ramification point x of C.

In the above description, the spectral curve C has a sheet swapping involution $\sigma_C : C \to C$ defined by $\sigma_C(\eta) = -\eta$. Then, the equivariant orthogonal bundle $(M, Q_M, \tilde{\sigma}_C)$ is said to have type $(q-1, 1)$ if the lift of the involution to M has ± 1 eigenspaces of dimensions $(q-1, 1)$.

From the above, one can see the intersection of the (B,A,A)-brane of $SO(p+q,p)$-Higgs bundles with a fibre of the $SO(2p+q,\mathbb{C})$ Hitchin fibration over a point a is identified with a covering of the product of two moduli spaces:

$$\mathcal{M}_{\mathrm{Cay}}(a) \times \mathcal{M}_{\mathrm{Lan}}(a).$$

The covering in question corresponds to extension data τ in (iii) above. The Cayley and Langlands moduli spaces $\mathcal{M}_{\mathrm{Cay}}(a)$ and $\mathcal{M}_{\mathrm{Lan}}(a)$ are given as follows:

- $\mathcal{M}_{\mathrm{Cay}}(a)$ is a fiber of the Hitchin map for the moduli space of K^2-twisted $GL(p,\mathbb{R})$-Higgs bundles on Σ, which we identify with $\mathrm{Jac}(\bar{S})[2]$ in (i) above.
- $\mathcal{M}_{\mathrm{Lan}}(a)$ is the moduli space of equivariant $SO(q)$-bundles M on C in (ii) above

As explained in [16], the reconstruction of the $SO(p+q,p)$-Higgs bundle (E,Φ) from the spectral data involves taking an extension of the form

$$0 \to V_0 \to E \to F \otimes K^{1/2} \to 0,$$

where F is the $Sp(2p,\mathbb{R})$-Higgs bundle associated to the Cayley moduli space, and V_0 is the invariant direct image under $\pi_C : C \to \Sigma$ of the equivariant orthogonal bundle M on C. The *extension data* τ is given by the extension class defining the above extension, leading to the following conjecture.

Conjecture 4.2.11. *The extra components in the moduli space of* $SO(p+q,q)$*-Higgs bundles conjectured to contain positive representations by Guichard and Wienhard* [57, Conjecture 5.6] *are those connected components containing Higgs bundles whose spectral data* (L,M,τ) *has the form* $(\overline{\pi}^* I, \mathcal{O}^{\oplus q}, \tau)$*, where* I *is a non-trivial* $\mathbb{Z}_2$*-line bundle on the Riemann surface* Σ*.*

The (A,B,A)-brane from a real structure on the Riemann surface. As noted before, given a real structure $f : \Sigma \to \Sigma$ the natural (A,B,A)-brane of Higgs bundles described in [13, 14] appears as the fixed point set of the induced involution

$$i_2(\bar{\partial}_A, \Phi) = (f^*(\partial_A), f^*(\Phi^*)) = (f^*(\rho(\bar{\partial}_A)), -f^*(\rho(\Phi))).$$

These branes can be studied through the spectral data associated to $G_\mathbb{C}$-Higgs bundles introduced in [66], and considering the (A,B,A)-brane as sitting inside the corresponding Hitchin fibration. By doing so, new examples of real integrable systems are obtained in [13] over a real subspace of the base of the Hitchin fibration:

Theorem 4.2.12 ([13] Theorem 17)**.** *If* (A,B,A)*-brane contains smooth points then the intersection of the brane with the Hitchin fibration gives a Lagrangian fibration with singularities. The generic fibre is smooth and consists of a finite number of tori.*

In the case of $G_\mathbb{C} = GL(n,\mathbb{C})$, the number of components that each fibre in Theorem 4.2.12 has can be expressed in terms of the number of components of the fixed point set of an induced involution on the spectral curves ([13, Proposition 32]). Moreover in the rank 2 case the calculations lead to an exact formula in terms of the associated quadratic differential:

Theorem 4.2.13 ([13])**.** *Let* $q \in H^0(\Sigma, K^2)$ *be a quadratic differential which has only simple zeros and such that* $f^*(\overline{q}) = q$*, and consider the spectral curve*

$$S = \{0 = \eta^2 - q\} \subset Tot(K).$$

Let n_+ *be the number of fixed components of* f *on which* q *is non-vanishing and positive, and* u *the number of zeros of* q *which are fixed by* f*. Then the number of components of the induced involution on the generic fibre* $Jac(S)$ *of the* $GL(2,\mathbb{C})$ *Hitchin fibration, giving the components of the* (A,B,A)*-brane, is* 2^d*, where*

$$d = \begin{cases} 2n_+ + \frac{u}{2} - 1 & \text{if } \; 2n_+ + \frac{u}{2} > 0 \\ 1 & \text{otherwise} \end{cases}.$$

Moreover, for $SL(2,\mathbb{C})$ can be specified further when considering the Prym variety $Prym(S,\Sigma)$, as seen in [13, Theorem 38].

Theorem 4.2.14 ([13]). *Considering the setting of Theorem 4.2.13 and suppose that at least one branch point of* S *is fixed by* f. *Then the fixed point set of the action of* $\tilde{f}$ *on* $Prym(S,\Sigma)$ *has* $2^{n_0+ufrm-e-1}$ *connected components, where* n_0 *is the number of fixed components of* $f:\Sigma\to\Sigma$ *which do not contain branch points, and* u *is the number of branch points which are fixed by* f.

The study of these (A,B,A)-branes allows one to consider the following problem: given a 3-manifold M with boundary Σ, by restricting representations $\pi_1(M)$ to the boundary gives a natural map

$$\mathrm{Rep}(\pi_1(M),G_{\mathbb{C}})\to\mathrm{Rep}(\pi_1(\Sigma),G_{\mathbb{C}}),$$

and one can show that representations that extend from the boundary to the three manifold always define an (A,B,A)-brane inside $\mathcal{M}_{G_{\mathbb{C}}}$. Whilst a general description of these (A,B,A)-branes appearing though 3-manifolds is not available, a first step towards understanding them is to consider the (A,B,A)-branes constructed in [14] as part of a triple through (4.2.3), fixed by the involution i_2 in (4.2.3). Indeed, consider the the 3-manifolds

$$M=\Sigma\times[-1,1]/(x,t)\mapsto(f(x),-t),$$

a form which any handle body can be put in, one can see that the quotient M is a 3-manifold with boundary $\partial M=\Sigma$.

Theorem 4.2.15 ([14]). *Let* (E,Φ) *be a fixed point of* i_2 *with simple holonomy. Then the associated connection extends over* M *as a flat connection iff the class* $[E]\in\tilde{K}^0_{\mathbb{Z}_2}(\Sigma)$ *in the reduced equivariant* K*-theory is trivial.*

4.3. Problem set III. This problem set is meant to make the reader be acquainted with actions on Riemann surfaces and on groups, and their induced actions on the moduli space of Higgs bundles. They provide a mixture of problems from the lectures as well as some additional ones.

Problem 11. Topological invariants. The topological invariants associated to $(W_1\oplus W_2,\Phi)$ in the (B,A,A)-brane of $U(p,q)$-Higgs bundles are the degrees $\deg(W_1)$ and $\deg(W_2)$. In the case of $p=q$, from [90] the invariants can be seen in terms of the degree of the line bundle L on S and the number of ramification points of S over which the linear action of σ on the fibre of L is -1. These invariants satisfy several properties we shall see here:

(i) Use the Lefschetz fixed point formula in [8, Theorem 4.12] to see that the parity of the degree of L and the number of points over which σ acts as -1 needs to be the same.

(ii) As in [29], the Toledo invariant $\tau(\deg(W_1),\deg(W_2))$ associated to $U(p,p)$-Higgs bundles is defined as $\tau(\deg(W_1),\deg(W_2)):=\deg(W_1)-\deg(W_2)$.

Use Exercise 11 to express the invariant in terms of fixed points of σ and obtain natural bounds.

(iii) **((*))** In the case of $SU(p,p)$-Higgs bundles with maximal Toledo Toledo invariant the spectral data is given by a covering of $\mathrm{Prym}(S/\sigma,\Sigma)$. For $SU(p,p+1)$-Higgs bundles, the Cameral data can be used to obtain the spectral data as seen in [84]. Use the tools of [90] together with those of [102] to obtain a description of the $SU(p,1)$ spectral data.

Problem 12. (B,B,B)**-branes from finite group.** Recall that the dimension of the moduli space of flat space is $SL(2,\mathbb{C})$-connections on a Riemann surface Σ of genus g is $\dim \mathcal{M}_g = 6g-6$, and the dimension of the moduli space $\mathcal{M}_{\gamma,n}$ of flat $SL(2,\mathbb{C})$-connections on a n-punctured Riemann surface of genus γ with fixed local monodromies (with simple eigenvalues) is $\dim \mathcal{M}_{\gamma,n} = 6\gamma - 6 + 2n$, provided that either $\gamma \geqslant 2$, or that $\gamma \geqslant 1$ and $n \geqslant 1$, or finally that $\gamma = 0$ and $n \geqslant 4$. Otherwise, the dimension of $\mathcal{M}_{\gamma,n}$ is either 0 or 2.

i. Use Riemann-Hurwitz to give a relation between the genus g and γ of the Riemann surfaces Σ and $\Sigma/\mathbb{Z}_p$.
ii. Let $g \geqslant 2$, and Γ be a finite group of order h acting on Σ by holomorphic automorphisms. Using **Problem 12.i.**, show that if a component of the moduli space of equivariant flat $SL(2,\mathbb{C})$-connections on Σ has half the dimension of $\mathcal{M}_g$, then $h = 2^k$ for some natural number $k \in \mathbb{N}$.
iii. What can you say about the genus $g = 1$ case in **Problem 12.ii.**?

Problem 13. Equivariant and anti-equivariant spectral data. Given a compact Riemann surface Σ, double covers of Σ can be constructed through holomorphic sections $s \in H^0(\Sigma, L^2)$, for L a holomorphic line bundle on Σ. Restricting to sections which have simple zeros, there is a double cover $\pi\colon S \to \Sigma$ branched over the zeros of s such that a square root $t \in H^0(S,\pi^*L)$ satisfying $t^2 = s$ exists.

i. When considering the Hitchin fibration, the smooth loci in the Hitchin base is given by sections in $H^0(\Sigma, K^2)$ with simple zeros. Consider instead $L = K^2$ and prove the there is a (unique) double cover $\bar{\pi} : \bar{S} \to \Sigma$ branched over the zeros of $s \in H^0(\Sigma, L^2)$ such that a square root $t \in H^0(S,\pi^*L)$ satisfying $t^2 = s$ exists.
ii. Note that for Φ an $SL(2,\mathbb{C})$-Higgs field with spectral line bundle L, the Higgs field Φ has spectral line bundle σ^*L, where σ is the natural involution switching the sheets of the 2-fold spectral curve. Show that anti-equivariant $SL(2,\mathbb{C})$-Higgs bundles with respect to a fixed point free involution τ acting on Σ have spectral data determined by
$$\mathcal{P}^\vee := \{L \in \mathrm{Jac}(S) \mid \sigma^*L = L^*, \tau^*L = L^*\} \subset \mathrm{Prym}(S,\Sigma).$$
iii. Recalling that Higgs bundles for split real forms correspond to torsion two points in the Hitchin fibres for complex Higgs bundles, show a similar result to **Problem 13.ii.** for $SL(2,\mathbb{R})$-Higgs bundles.

Problem 14. (B,A,A)**-branes from real structures.** We have seen how real Higgs bundles can be defined through involutions on the Lie algebras. These actions can also be considered to understand properties of the spectral data description of the brane, as well as of the topological invariants labelling components of the brane.

i. The associated involution θ from Table 4.2.5 decomposes $\mathfrak{u} = \mathfrak{h} \oplus i\mathfrak{m}$. Give an explicit description of $\mathfrak{m}$ and $\mathfrak{h}$ and of the real form $\mathfrak{g} = \mathfrak{h} \oplus \mathfrak{m}$.
ii. **((*))** Nonabelianization of the fixed point set of $\Theta_{Sp(2p,2p)}$ can also be seen through Cameral covers, as shown in [84]. Realise the action of σ in terms of Cameral covers.
iii. **((*))** Considering the notion of "strong real form" from [1], describe the corresponding Higgs bundles and give a definition of Θ_G for which one does not have the problem described in the above paragraph.

Problem 15. (A,B,A)**-branes from real structures.** Let Σ be of genus 2 with a real structure f whose invariants are $(n,a) = (3,0)$.

Consider an $SL(2,\mathbb{C})$-Higgs bundle with spectral curve $S = \{0 = \eta^2 - q\}$ for q a quadratic differential with simple zeros. For $z \in \Sigma$, the involution f induces an involution $\tilde{f}(\eta, z) = (f^*(\overline{\eta}), f(z))$.

i. What is the effect of replacing q by αq for $\alpha \in \mathbb{R}_+$? What about replacing q by $-q$?
ii. Suppose Σ is hyperelliptic and describe the system of quadratic differentials on Σ through two points $a_1, a_2 \in \mathbb{P}^1$.
iii. What are the conditions on a_1 and a_2 from **Problem 15.ii.** to obtain a quadratic differential q with $f^*(\overline{q}) = q$?

5. Higgs bundles and correspondences

In recent times, a good number of the most striking resolutions of longstanding problems have come about from an unexpected combination of ideas from different areas of mathematics.

Sir Michael Atiyah

When considering the moduli space of Higgs bundles for complex groups, one may ask which correspondences it has with other moduli spaces (of Higgs bundles or of other mathematical objects), and whether any of its subspaces have interesting correspondences and interpretations. In what fallows we shall describe a few of these for the whole moduli spaces of Higgs bundles, as well as for branes of Higgs bundles as defined in Chapter 4.

5.1. Group homomorphisms. Given a fixed Riemann surface Σ and a homomorphism between two Lie groups

$$\Psi : G_{\mathbb{C}} \to G'_{\mathbb{C}}, \tag{5.1.1}$$

there is a naturally induced map between representation spaces

$$\Psi : \mathrm{Rep}(\Sigma, G_{\mathbb{C}}) \to \mathrm{Rep}(\Sigma, G'_{\mathbb{C}}),$$

for $\mathrm{Rep}(\Sigma, G)$ the space of representations modulo conjugation. As mentioned in the beginning of the notes, there is a correspondence between surface group representations and Higgs bundles, and hence there should be similar induced morphism of $G_{\mathbb{C}}$-Higgs bundle moduli spaces. Therefore, one may ask which correspondences there are between Higgs bundles from group homomorphism, and which of those maps descend to correspondences between subgroups.

Through Definition 2.1.9, one can see that the map in (5.1.1) induces a map on principal $G_{\mathbb{C}}$-Higgs bundles $(P_{G_{\mathbb{C}}}, \Phi)$ given by

$$\Psi : (P_{G_{\mathbb{C}}}, \Phi) \mapsto (P_{G_{\mathbb{C}}} \times_{\Psi} G'_{\mathbb{C}}, d\Psi(\Phi)), \tag{5.1.2}$$

where $P_{G'_{\mathbb{C}}} := P_{G_{\mathbb{C}}} \times_{\Psi} G'_{\mathbb{C}}$, and the image of the Higgs field is $\Phi' := d\Psi\Phi$ for

$$d\Psi : \mathfrak{ad}(P_{G_{\mathbb{C}}}) \to \mathfrak{ad}(P'_{G_{\mathbb{C}}})$$

defined by the derivative at the identity of the map Ψ.

Induced maps from isogenies. A first step towards tackling the above question is to consider the case where the group homomorphism is given by the isogenies of Lie groups, as we shall see next. Through correspondences between low rank Lie algebras, one can discover some interesting relations between different Higgs bundles. Some of the low rank isogenies between complex Lie algebras are

$$\begin{aligned} SO(3,\mathbb{C}) &\cong Sp(2,\mathbb{C}) \cong SL(2,\mathbb{C}); \\ SO(4,\mathbb{C}) &\cong SL(2,\mathbb{C}) \times SL(2,\mathbb{C}); \\ SO(5,\mathbb{C}) &\cong Sp(4,\mathbb{C}); \\ SO(6,\mathbb{C}) &\cong SL(4,\mathbb{C}). \end{aligned}$$

These induce maps between real Lie algebras, which can be described as follows:

$$\begin{aligned} &(a)\ \mathfrak{so}(2,1) \cong \mathfrak{sl}(2,\mathbb{R}), && (f)\ \mathfrak{so}(3,3) \cong \mathfrak{sl}(4,\mathbb{R}), \\ &(b)\ \mathfrak{so}(2,2) \cong \mathfrak{sl}(2,\mathbb{R}) \times \mathfrak{sl}(2,\mathbb{R}), && (g)\ \mathfrak{so}(2,4) \cong \mathfrak{su}(2,2), \\ &(c)\ \mathfrak{so}^*(4) \cong \mathfrak{su}(2) \times \mathfrak{sl}(2,\mathbb{R}), && (h)\ \mathfrak{so}(1,5) \cong \mathfrak{su}(4), \\ &(d)\ \mathfrak{so}(1,3) \cong \mathfrak{sl}(2,\mathbb{C}), && (i)\ \mathfrak{so}(2,3) \cong \mathfrak{sp}(4,\mathbb{R}). \\ &(e)\ \mathfrak{so}(1,4) \cong \mathfrak{sp}(2,2), && \end{aligned}$$

The rank 2 and 3 isogenies in (b) and (f) inducing maps between branes were described in [32] through the regular fibres of the Hitchin fibration, whereas [28] will take care of the correspondences in (c) and (d) between branes inside singular fibres. All of the other correspondences have never been studied through spectral data, and thus they provide a rich list of open questions one may ask. In what follows, we shall give an overview of this area and the questions one may ask, based on [32] and [28]. For further background, the reader may want to refer to [6] where some further explanations about isogenies in terms of Higgs bundles are given.

From (5.1.2), it is not hard to see that the correspondence between Higgs bundles at the level of the vector bundle and Higgs field can be deduced naturally (see [6,32]). Recall that an $\mathrm{SL}(n,\mathbb{C})$-Higgs bundle can be seen as is a pair (E,Φ) for E a rank n holomorphic bundle on Σ with trivial determinant, and Φ a traceless Higgs field. Moreover, an $\mathrm{SO}(n,\mathbb{C})$-Higgs bundle can be written as a pair (E,Φ) where E is a rank n holomorphic bundle on Σ with an orthogonal structure Q and a compatible trivialization of its determinant bundle, where a trivialization $\delta : \det(E) \simeq \mathcal{O}_\Sigma$ is compatible with Q if δ^2 agrees with the trivialization of $(\Lambda^n E)^2$ given by the discriminant $Q : \Lambda^n E \to \Lambda^n E^*$, and the Higgs field Φ satisfies $Q(u,\Phi v) = -Q(\Phi u, w)$.

Proposition 5.1.3. *The isogenies between Lie groups of rank 2 and 3 given by*

$$\mathcal{I}_2 : SL(2,\mathbb{C}) \times SL(2,\mathbb{C}) \to SO(4,\mathbb{C});$$
$$\mathcal{I}_3 : SL(4,\mathbb{C}) \to SO(6,\mathbb{C});$$

induce maps on $SL(2,\mathbb{C})$-Higgs bundles and $SL(4,\mathbb{C})$-Higgs bundles given by

$$\mathcal{I}_2 : (E_1,\Phi_1),(E_2,\Phi_2) \mapsto [E_1 \otimes E_2, \Phi_1 \otimes 1 + 1 \otimes \Phi_2] \in \mathcal{M}_{SO(4,\mathbb{C})},$$
$$\mathcal{I}_3 : (E,\Phi) \mapsto \left[\Lambda^2 E, \Phi \otimes 1 + 1 \otimes \Phi\right] \in \mathcal{M}_{SO(6,\mathbb{C})}.$$

Proof. This follows directly from (5.1.1) and (5.1.2) applied to the particular $G_{\mathbb{C}}$-Higgs bundles. □

Our interest is in the understanding of the maps described in Proposition 5.1.3 through spectral data. For this, one needs to study the correspondence between eigenvalues of the Higgs field, which leads to the correspondence between spectral curves, and between the eigenspaces, and in turn to the line bundles on the spectral curves.

In order to understand the relation between eigenvalues for Higgs bundles through group homomorphisms, it is useful to recall that the characteristic polynomial of a Higgs field Φ defines a curve in the total space of the canonical bundle K whose equation can be obtained by considering

$$p(x) = x^n + \mathrm{tr}(\Phi)x^{n-1} + \sum_{k=2}^{n-1} x^{n-k}\,(-1)^k\,\mathrm{Tr}(\Lambda^k\Phi) + (-1)^n \det(\Phi)$$

where $\mathrm{Tr}(\Lambda^k\Phi)$ is the trace of the kth exterior power of Φ, with dimension $\binom{n}{k}$, as in (3.1.2) before.

As seen before, and explained in [66], the spectral curve $\pi : S \to \Sigma$ associated to an $\mathrm{SL}(n,\mathbb{C})$-Higgs bundle (E,Φ) has equation

$$S := \{0 = \det(I\eta - \Phi)\} = \{0 = \eta^n + a_2\eta^{n-2} + \ldots + a_{n-1}\eta + a_n\}, \tag{5.1.4}$$

for $a_i \in H^0(\Sigma, K^i)$, and η the tautological section of π^*K. The spectral curve S is generically smooth and the fiber over a generic point in the Hitchin base can be identified with the Prym variety $\mathrm{Prym}(S,\Sigma)$.For a fixed choice of $K^{1/2}$, the vector bundle E is recovered from $L \in \mathrm{Prym}(S,\Sigma)$ as $E := \pi_*(L \otimes \pi^*(K^{(n-1)/2}))$.

For $SO(2n,\mathbb{C})$-Higgs bundles, the Higgs field Φ defines a cover $\pi : S \to \Sigma$ given by (5.1.4) where $a_n = p_n \in H^0(\Sigma, K^n)$ is the Pfaffian of Φ. The curve is singular at $\eta = p_n = 0$, which are the fixed points of a natural involution $\sigma : \eta \mapsto -\eta$. The normalization $\hat{\pi} : \hat{S} \to \Sigma$ of S, is what we shall refer to as the spectral curve. This curve $\hat{S}$ is generically smooth, and the involution σ extends to an involution $\hat{\sigma}$ which does not have fixed points. The generic $SO(2n,\mathbb{C})$ fibers by a connected component of $\mathrm{Prym}(\hat{S}, \hat{S}/\sigma)$. Similarly to the previous case, for a choice of $K^{1/2}$, the vector bundle E is recovered from $L \in \mathrm{Prym}(\hat{S}, \hat{S}/\sigma)$ as

$$E := \pi_*(L \otimes (K_{\hat{S}} \otimes \pi^*(K^*))^{1/2}).$$

Through the above spectral data, together with the correspondence given in Proposition 5.1.3, the isogenies induce the following map on the Hitchin fibrations from [32, Theorem 1 & Theorem 2]. Moreover, the morphisms from Theorem 5.1.5 restrict to real (e.g. see Problems 16 and 17).

Theorem 5.1.5 ([32]). *Consider two generic* $SL(2,\mathbb{C})$*-Higgs bundles* (E_i, Φ_i)*, whose spectral data is* S_i *and* $L_i \in \mathrm{Prym}(S_i, \Sigma)$*. Then* $\mathcal{I}_2$ *is a* 2^{2g} *fold-covering onto its images and the spectral data for the image* $\mathcal{I}_2((E_1, \Phi_1), (E_2, \Phi_2))$ *is given by* $(\hat{S}_4, \mathcal{L}_4)$ *where*

- $\hat{S}_4 := S_1 \times_\Sigma S_2$ *is a smooth ramified fourfold cover, and*
- $\mathcal{L}_4 := p_1^*(L_1) \otimes p_2^*(L_2)$ *where* $p_i : S_1 \times_\Sigma S_2 \to S_i$ *are the projection maps.*

Moreover, for a generic $SL(4,\mathbb{C})$*-Higgs bundle* (E', Φ') *with spectral data* (S', L')*, the* $SO(6,\mathbb{C})$*-Higgs bundle given by the image* $\mathcal{I}_3(E, \Phi)$ *has spectral data* $(\hat{S}_6, \mathcal{I}_3(L))$*, where*

- $\hat{S}_6$ *is the symmetrization of the non-diagonal component in* $S' \times_\Sigma S'$*;*
- $\mathcal{I}_3(L)$ *is a canonical twist generated by local sections of* $\mathcal{L}' := p_1^*(L) \otimes p_2^*(L)$ *that are anti-invariant with respect to the symmetry of* $S' \times_\Sigma S'$*.*

It should be noted that the Higgs bundles considered in Theorem 5.1.5 lie all within the generic fibres of the Hitchin fibration. However, the isogeny descends to all the real forms of those complex Lie groups, and thus one should be able to obtain the morphisms between real Higgs bundles which lie completely over the singular locus of the Hitchin fibration. The first cases studied, which involve the geometry of ribbons, appear in [28].

Remark 5.1.6. The approach of [32] was also of use when studying problems in F-theory related to degenerations of elliptically fibred CY 3-folds as done in [5].

5.2. Polygons and Hyperpolygons. An interesting relation between mathematical objects appears when considering *parabolic Higgs bundles* (as defined in Section 2.3) and *quiver varieties* (representation varieties of quivers, for which the base manifold is a point), as studied in [80, 81] and subsequent work of many researchers.

Definition 5.2.1. A quiver $Q = (V, A, h, t)$ is an oriented graph, consisting of a finite vertex set V, a finite arrow set A, and head and tail maps $h, t : A \to V$.

Example 5.2.2. Quivers may be represented through oriented graphs, and an example that will appear useful in coming sections is that of a star shaped quiver as bellow, where the set of vertices is $V = \{v_1, \ldots, v_3\}$, and the set of arrows is $A = \{a_i : v_i \to v_0 \mid 1 \leqslant i \leqslant 3\}$:

(5.2.3)

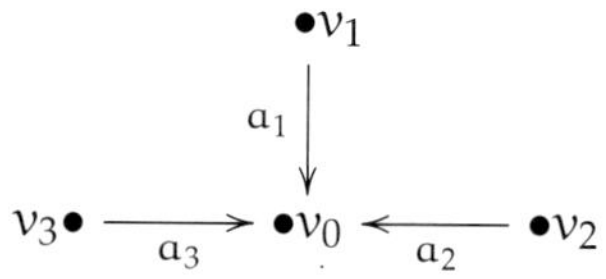

Here, the head map is given by $h(a_i) = v_0$, and the tail map by $t(a_i) = v_i$.

The *dimension vector* of Q is the $(n+1)$-tuple of positive integers $d = (r, 1, \ldots, 1)$, and this vector determines Q uniquely as a star quiver. Moreover, quivers carry naturally defined spaces of representations:

$$\mathrm{Rep}(Q) = \bigoplus_{i=1}^{n} \mathrm{Hom}(\mathbb{C}, \mathbb{C}^r) \cong \mathrm{Hom}(\mathbb{C}^n, \mathbb{C}^r) = \mathbb{C}^{n \times r}.$$

Of particular interest for us are star-shaped quivers Q_n, with one central vertex v_0 and n outer vertices $v_0, \ldots, v_n$ and arrows $a_i : v_i \to v_0$, as the one in Example 5.2.2 which has $n = 3$. In what follows we shall consider fixed dimension vector $d = (r, 1, \cdots, 1) \in \mathbb{N}^{n+1}$, and recall the value by writing the star-shaped quiver as Q_n^r.

Remark 5.2.4. A representation of a quiver Q can also be written as

$$W := ((W_v)_{v \in V}, (\phi_a)_{a \in A})$$

where W_v is finite dimensional vector space; and $\phi_a : W_{t(a)} \to W_{h(a)}$ is a linear map for all $a \in A$.

Similar to when working with parabolic Higgs bundles, one may define the slope and stability conditions of quivers. For a stability parameter $\alpha := (\alpha_v)_{v \in V}$ in $\mathbb{Z}^V$, the α-slope of W is

$$\mu_\alpha(W) := \frac{\sum_{v \in V} \alpha_v \dim W_v}{\sum_{v \in V} \dim W_v} \in \mathbb{Q}.$$

Then, the representation W is said to be: *α-semistable* if $\mu_\alpha(W') \leqslant \mu_\alpha(W)$ for all sub-representation $0 \neq W' \subset W$; *α-stable* if $\mu_\alpha(W') < \mu_\alpha(W)$ for all subrepresentations $0 \neq W' \subsetneq W$. There are natural notions of Jordan-Hölder filtrations (as those considered in Section 2.1.9), and two α-semistable representations of Q are said to be S-equivalent if their associated graded objects for their Jordan–Hölder filtrations are isomorphic. For more details, the reader may refer, for example, to [70] whose notation we will follow for most of this section.

Our interest in quiver varieties comes from their relation to vector bundles and Higgs bundles which can be obtained, in particular, considering the space of polygons and hyperpolygons (e.g., see [53] and the references therein).

The polygon space $\mathcal{M}_{poly}(n, \ell)$ consisting of n-gons in the Euclidean space $\mathbb{E}^3$ with lengths given by $\ell = (\ell_1, \dots, \ell_n)$ modulo orientation-preserving isometries is homeomorphic the moduli space of ℓ-semistable n ordered points on $\mathbb{P}^1$ modulo the automorphisms of the projective line (see [70] and references therein). Moreover, it is also isomorphic to a moduli space of representations of a star-shaped quiver Q_n^ℓ.

In order to define the moduli space of hyperpolygons, we shall consider a quiver $Q = (V, A, h, t)$ and define the doubled quiver as $\overline{Q} := (V, \overline{A}, h, t)$ where $\overline{A} = A \cup A^*$ for

$$A^* := \{a^* : h(a) \to t(a)\}_{a \in A}.$$

Through the double quiver of a star shaped quiver, the authors of [45] constructed the hyperpolygon spaces for any value of r, which are the hyperkähler analogues of polygon spaces (see references therein for previous work appearing in a less general manner).

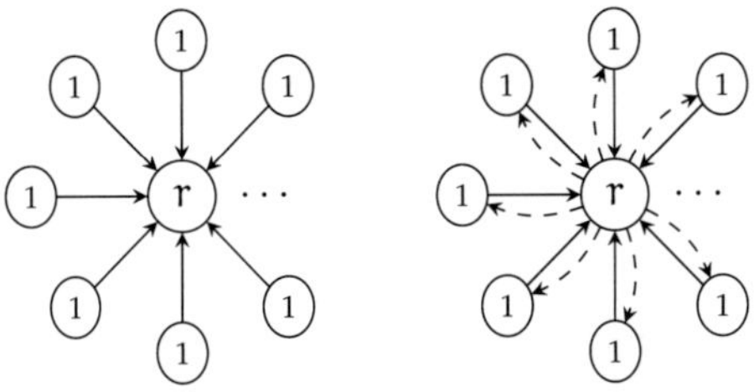

Figure 5.2.5. Left, the quiver whose moduli of representations is given by polygon space; Right, the doubled quiver used to construct hyperpolygon space.

Carrying on with the work of [45], through the natural action of the group $G := P(U(r) \times (S^1)^n)$ on the star-shaped quivers in Figure 5.2.5, there are two natural moment maps for the action of G on $T^*\mathrm{Rep}(Q)$. Denoting by$(\cdot)_0$ the trace-free part of a matrix, the scalar multiples of these maps can be written:

$$\text{Real moment map } \mu_r(x, y) = ((xx^* - y^*y)_0, |x_1|^2 - |y_1|^2, \dots, |x_n|^2 - |y_n|^2)$$

$$\text{Complex moment map } \mu_c(x, y) = ((xy)_0, y_1x_1, \dots, y_nx_n)$$

for . The *polygon space* $\mathcal{P}_n^r(\alpha)$ is the symplectic quotient

$$\mathcal{P}_n^r(\alpha) = \mathrm{Rep}(Q)//_\alpha G.$$

Similarly, the *hyperpolygon space* $\mathcal{X}_n^r(\alpha)$ can be defined be the hyperkähler quotient

$$\mathcal{X}_n^r(\alpha) = T^*\mathrm{Rep}(Q)///_{(\alpha,0)} G.$$

Nakajima showed that polygon and hyperpolygon spaces are examples of Kähler and hyperkähler quiver varieties, which can be then used to understand parabolic Higgs bundles. Although relations between quiver varieties and parabolic Higgs bundles have been long studied, many interesting problems are still open. Building on a natural analogy between hyper-kähler versions of moduli spaces of semistable n-gons in complex projective space and certain parabolic Higgs bundles, Rayan and Fisher identified in [45] the spaces $\mathcal{X}_n^r(\alpha)$ with degenerate Hitchin systems: for a hyperpolygon (x,y) in $\mathcal{X}_n^r(\alpha)$, and $D = \sum_{i=1}^n p_i$ a fixed divisor on Σ for which no two p_i are the same, a Higgs field on $\mathbb{P}^1$ with vector bundle $E = \mathbb{C}^r \times \mathbb{P}^1$ is given by

$$\phi(z) = \sum_{i=1}^{n} \frac{\phi_i dz}{z - p_i},$$

for z an affine coordinate on $\mathbb{P}^1$, and where $\phi_i = x_i y_i$. More precisely, for parabolic $SL(n,\mathbb{C})$-Higgs bundle, one has the following general result (see [45, Section 4] for a clear explanation of how parabolic Higgs bundles correspond to hyperpolygons):

Theorem 5.2.6 ([45]). *The space $\mathcal{X}_n^r(\alpha)$ may be identified with a moduli space of rank r parabolic Higgs bundles on $\mathbb{P}^1$, such that the residue at each marked point lies in the closure of the minimal nilpotent orbit of $\mathfrak{sl}_r$, and whose underlying bundle is trivial.*

Moreover, under the identification above as shown in [45] there is a natural Hitchin map whose base is an affine space of half the total dimension of hyperpolygon space.

Remark 5.2.7. The mathematical construction of the above general hyperpolygon spaces should lead to interesting applications in string theory. In particular, from [5], these quivers can be generated by an intersecting brane configuration in F-theory. The central node corresponds to the contribution from the 7-brane wrapped over the gauge theory curve, and the satellite nodes to the flavour branes of the system.

5.3. Langlands duality. A final correspondence between subspaces of the moduli space of $G_\mathbb{C}$-Higgs bundle comes from Langlands duality. One of the main reasons for considering branes is the connection to mirror symmetry and the geometric Langlands program. For ${}^L G_\mathbb{C}$ the Langlands dual group of $G_\mathbb{C}$, there is a correspondence between invariant polynomials for $G_\mathbb{C}$ and ${}^L G_\mathbb{C}$ giving an identification $\mathcal{A}_{G_\mathbb{C}} \simeq \mathcal{A}_{{}^L G_\mathbb{C}}$ of the Hitchin basis. The two moduli spaces $\mathcal{M}_{G_\mathbb{C}}$ and $\mathcal{M}_{{}^L G_\mathbb{C}}$ are then torus fibrations over a common base and their non-singular fibres are dual abelian varieties [40]. Kapustin and Witten give a physical interpretation of this in terms of S-duality, using it as the basis for their approach to the geometric Langlands program [72].

In this approach a crucial role is played by the various types of branes and their transformation under mirror symmetry. Langlands duality exchanges branes types according to

$$(B,B,B) \leftrightarrow (B,A,A),$$
$$(A,B,A) \leftrightarrow (A,B,A),$$
$$(A,A,B) \leftrightarrow (A,A,B),$$

but the exact correspondence is not yet known. Moreover, by studying branes of Higgs bundles through the Hitchin fibration, one may consider a Fourier-Mukai type transform to obtain the duality in terms of subspaces of the dual fibres, and there are a few things one can say about the branes constructed in Chapter 4 and Langlands duality:

- **3-manifold branes.** Recall that there is a natural (A, B, A)-brane obtained in [14] as fixed points of $i_2(\bar{\partial}_A, \Phi) = (f^*(\partial_A), f^*(\Phi^*))$where $f : \Sigma \to \Sigma$ is a real structure on Σ, and ρ is the compact real form of $G_{\mathbb{C}}$. By means of the compact form ${}^L\rho$ of the Langlands dual group, these spaces have a corresponding involution on $\mathcal{M}_{{}^LG_{\mathbb{C}}}$ given by
$$ {}^L i_2(\bar{\partial}_A, \Phi) = (f^*(\partial_A), f^*(\Phi^*)) = (f^*({}^L\rho(\bar{\partial}_A)), -f^*({}^L\rho(\Phi))), $$
whose fixed point set in $\mathcal{M}_{{}^LG_{\mathbb{C}}}$ is conjectured in [14] to be the dual brane (for details see also [89, Conjecture 1]).
- **Pseudo-real Higgs bundles.** As seen in [14], pseudo-real Higgs bundles in $\mathcal{M}_{G_{\mathbb{C}}}$ form an (A, A, B)-brane and can be seen as the fixed point set of the involution $i_3(\bar{\partial}_A, \Phi) = (f^*\sigma(\bar{\partial}_A), f^*\sigma(\Phi))$, for σ a real form of the complex Lie group $G_{\mathbb{C}}$. In this case, it is conjectured in [14] that the dual brane is the fixed point set of a corresponding involution defined as
$$ {}^L i_3(\bar{\partial}_A, \Phi) = (f^*\hat{\sigma}(\bar{\partial}_A), f^*\hat{\sigma}(\Phi)), $$
for a real structure $\hat{\sigma}$ on ${}^LG_{\mathbb{C}}$ which is conjectured to be the Nadler dual of σ (see below for details).
- **Hyperpolygon spaces.** The branes arising in terms of quivers may carry group structures, and Langlands duality can be considered for them. General results about duality between branes arising through quivers representations appear in [70], and geometric properties of hyperpolygon branes in the spirit of [14] appear in [86].

One of the most interesting cases of Langlands duality for branes of Higgs bundles appears when considering the (B, A, A)-brane of real Higgs bundles $\mathcal{M}_G \subset \mathcal{M}_{G_{\mathbb{C}}}$, corresponding to real surface group representations of the compact Riemann surface Σ. Recall that these branes can be seen through the Cartan involution θ of a real form G of G_c, as the fixed point set of the involution
$$ i_1(\bar{\partial}_A, \Phi) = (\theta(\bar{\partial}_A), -\theta(\Phi)). $$

The dual $\mathcal{M}^L_G \subset^L \mathcal{M}_{{}^LG_{\mathbb{C}}}$ must be of type (B, B, B), a submanifold which is complex with respect to I, J, K, supporting a hyperholomorphic sheaf. From [14], the proposed dual brane should be taken as follows (for details see also [89, Conjecture 2]):

Conjecture 5.3.1. *The support of the dual brane to the fixed point set of i_1 is the moduli space $\mathcal{M}_{\check{H}} \subset \mathcal{M}_{{}^LG_c}$ of $\check{H}$-Higgs bundles for $\check{H}$ the group associated to the Lie algebra $\check{\mathfrak{h}}$ appearing in* [79, Table 1].

Whilst the above conjecture has not been proven yet, a lot of work as appeared in different directions with progress towards it. In what follows we shall mention some of this work

- **Branes with non-abelian data.** As mentioned before, the (B, A, A)-branes of $SL(m, \mathbb{H})$, $SO(2n, \mathbb{H})$ and $Sp(2m, 2m)$-Higgs bundles were shown to have non-abelian intersection with the Hitchin fibration [69]. The duality for these branes was studied in Branco's DPhil thesis recently, where the geometry of the whole singular fibres containing these branes was described before obtaining the dual spaces [33].
- **Quasi split real forms.** The intersection of the (B, A, A)-brane of $U(m, m)$-Higgs bundles with the regular fibres of the Hitchin fibration was studied in [90], and through this description a candidate for the dual brane, agreeing with Conjecture 5.3.1, is presented in [68]. Moreover, for $m = 1$, further support of the conjecture using cohomology classes and topological invariants recently appeared in [97].
- **Split real forms.** In the case of a split real form G of $G_\mathbb{C}$, it is shown in [88, Theorem 4.12] that the (B, A, A)-brane obtained through the involution i_1 lies over the Hitchin base as 2-torsion points, and hence a 0-dimensional space in each fibre. The explicit description of the branes (as well as the monodromy action for most of the classical split real groups [10,15,87,88]) should present a good starting point to study mirror symmetry, noting the similarities with the case studied by Hitchin in [68].

5.4. Problem set IV. This problem set is meant to make the reader be acquainted with the space of polygons and hyperpolygons, and their interpretation in terms of Higgs bundles. Moreover, the reader will find the problems helpful to understand the different correspondences between Lie groups and their induced correspondences between Hitchin systems.

Problem 16. Isogenies. To understand the tensor product of an $SL(4, \mathbb{R})$-Higgs bundles with itself, start with an $SL(4, \mathbb{R})$-Higgs bundle (E, φ) where E has rank 4 and trivial determinant, and Φ is trace free and symmetric with respect to a local frame $\{e_1, \ldots, e_4\}$. Then the Higgs field on $E \otimes E$ is

$$\Phi = \varphi \otimes I + I \otimes \varphi.$$

i. Recall that for any vector bundle $\mathbf{V}$ of rank n, and for all $k < n$, there is an isomorphism $\Lambda^k(\mathbf{V}^*) \otimes \Lambda^n(\mathbf{V}) \longrightarrow \Lambda^{n-k}(\mathbf{V})$.

   Which structure arises when considering $n = 4$, $k = 2$, the determinant $\Lambda^4 \mathbf{V} \simeq \mathcal{O}$, and such that $\mathbf{V}$ has an orthogonal structure given by an isomorphism $q : \mathbf{V} \simeq \mathbf{V}^*$.

ii. With respect to the local frame for

$$E \otimes E = \Lambda^2 E \oplus Sym^2(E),$$

how does the restriction of Φ to $\Lambda^2 E$ look like? does it have the structure of a G-Higgs field for some real group G?

iii. Compute how the whole Higgs field Φ looks like with respect to the frame in Problem 16.ii., this is, with respect to

$$\{e_i \wedge e_j \; |i < j\} \cup \{e_i \otimes e_i \; |i = 1,2,3,4\} \cup \{e_i \otimes e_j + e_j \otimes e_i \; |i < j\}.$$

Problem 17. Isogenies. If the homomorphism Ψ from (5.1.2) restricts to a map between real forms which respects the Cartan decompositions of the Lie algebras, then it induces a map between real Higgs bundles. Considering the morphisms between complex groups, calculate the following:

i. Show what the isogeny between $SL(4,\mathbb{C})$ and $SO(6,\mathbb{C})$ implies when considered over real symplectic Higgs bundles, this over pairs (E,φ) where the bundle is $V \oplus V^*$ and the Higgs field has symmetric entries

$$\varphi = \begin{pmatrix} 0 & \beta \\ \gamma & 0 \end{pmatrix}.$$

ii. What are the relations between the eigenvalues of an $SL(4,\mathbb{R})$-Higgs bundle and those of an $SO(3,3)$-Higgs bundle obtained thought the isogeny between complex Lie groups?

Problem 18. Polygons. We say that n ordered points $(p_1,\ldots,p_n)$ on $\mathbb{P}^1$ are r-semistable if, for all $p_0 \in \mathbb{P}^1$ one has that

$$\sum_{i \,|\, p_i = p_0} r_i \leqslant \sum_{i \,|\, p_i \neq p_0} r_i$$

i. Show that the moduli space of r-semistable n ordered points on $\mathbb{P}^1$ modulo the automorphisms of the projective line moduli space can be constructed via GIT as a quotient of the SL_2-action on $(\mathbb{P}^1)^n$ with respect to an ample linearisation associated to the weights (e.g. see [70, Section 5.2])

ii. Describe the space of representations corresponding to polygon space for a fixed star shaped quiver and values of d, n.

iii. Show that every automorphism of the star shaped quiver Q_n must fix v_0, so $\mathrm{textrmAut}(Q_n) = \mathrm{Aut}^+(Q_n) \cong S_n$.

iv. Construct two subgroups of $\mathrm{Aut}(Q_n)$ and describe their actions on Q_n, as well as the quotient quivers under those group actions.

Problem 19. Hyperpolygons. Consider the quiver Q given by $1\bullet \overset{x}{\underset{y}{\rightleftarrows}} \bullet 2$

i. For σ the covariant involution which sending the vertex 1 to the vertex 2, and sending the arrow x to the arrow y, when is $d = (d_1, d_2)$ the dimension vector σ-compatible in the sense of [70]?

ii. Encode the information of a $G_{\mathbb{C}}$-Higgs bundle (E,Φ) on $\mathbb{P}^1$ within the above quiver, for some particular complex or real Lie group $G_{\mathbb{C}}$.

iii. In terms of hyperpolygons, describe the moduli space of parabolic Higgs bundles associated to Example 5.2.3 and its hyperkähler structure?

Problem 20. Hyperpolygons. Having seen how polygons and hyperpolygons can be understood in terms of Higgs bundles, consider the framed Jordan quiver

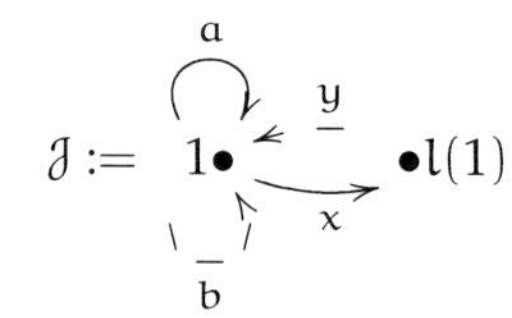

i. Give an expression of the real and complex moment maps (up to scale) associated to such quivers following [45].
ii. Let $\sigma : \mathcal{J} \to \mathcal{J}$ be the involution of the Jordan quiver which exchanges the vertices, and exchanges the arrows. Through the induced action on the moduli space, give a description of the fixed point set.
iii. As mentioned before, the space of hyperpolygons studied in the previous sections inherits a hyperkähler structure and thus one may consider three fixed complex structures and their associated symplectic forms, and how these relate to a given subspace. Following [70] and [86], decide whether the fixed point set in Problem **19.ii.** is a brane, and if so, give its type.

Acknowledgments. The author would like to thank the Institute of Advance Studies and the organisers of the *2019 Graduate Summer School*, as well as the director of PCMI Rafe Mazzeo, for providing an ideal environment for research and collaboration. We are also thankful to Steve Rayan, Charles Alley, Melissa Zhang, Mengxue Yang, Szilard Szabo, Ciprian Manolescu and Richard Derryberry, for useful comments on a draft of the notes; to Siqi He and Richard Derryberry for their help running the course, and to Ian Morrison for his help with the notes. The author also acknowledges support from the Humboldt Foundation as well as from the U.S. National Science Foundation grants DMS-1509693 and CAREER DMS 1749013.

References

[1] J. Adams, *Strong real forms and the Kac classification*, At. of Lie Groups and Rep (2005). ↑238

[2] C. Alley, *On the Monodromy of Meromorphic Cyclic Opers on the Riemann Sphere*, ProQuest LLC, Ann Arbor, MI, 2019. Thesis (Ph.D.)–University of Illinois at Chicago. MR4060655 ↑211

[3] J. E. Andersen and J. Grove, *Automorphism fixed points in the moduli space of semi-stable bundles*, Q. J. Math. **57** (2006), no. 1, 1–35, DOI 10.1093/qmath/hai008. MR2204258 ↑226, 233

[4] L. B. Anderson, M. Esole, L. Fredrickson, and L. P. Schaposnik, *Singular geometry and Higgs bundles in string theory*, SIGMA Symmetry Integrability Geom. Methods Appl. **14** (2018), Paper No. 037, 27, DOI 10.3842/SIGMA.2018.037. MR3787793 ↑210, 223

[5] L. B. Anderson, J. J. Heckman, S. Katz, and L. P. Schaposnik, *T-branes at the limits of geometry*, J. High Energy Phys. **10** (2017), 058, front matter+55, DOI 10.1007/jhep10(2017)058. MR3731155 ↑210, 212, 241, 244

[6] M. A. Arroyo, *The geometry of* $SO(p, q)$*-Higgs bundles*, Ph.D. Thesis, Salamanca, Spain (2009). ↑203, 205, 206, 207, 239, 240

[7] M. Aparicio-Arroyo, S. Bradlow, B. Collier, O. García-Prada, P. B. Gothen, and A. Oliveira, $SO(p,q)$*-Higgs bundles and higher Teichmüller components*, Invent. Math. **218** (2019), no. 1, 197–299, DOI 10.1007/s00222-019-00885-2. MR3994589 ↑207

[8] M. F. Atiyah and R. Bott, *A Lefschetz fixed point formula for elliptic complexes. II. Applications*, Ann. of Math. (2) **88** (1968), 451–491, DOI 10.2307/1970721. MR232406 ↑236

[9] D. Baraglia, M. Kamgarpour, and R. Varma, *Complete integrability of the parahoric Hitchin system*, Int. Math. Res. Not. IMRN **21** (2019), 6499–6528, DOI 10.1093/imrn/rnx313. MR4027558 ↑225

[10] D. Baraglia, *Monodromy of the* $SL(n)$ *and* $GL(n)$ *Hitchin fibrations*, Math. Ann. **370** (2018), no. 3-4, 1681–1716, DOI 10.1007/s00208-017-1611-6. MR3770177 ↑219, 221, 246

[11] D. Baraglia, I. Biswas, and L. P. Schaposnik, *Automorphisms of* $\mathbb{C}^*$ *moduli spaces associated to a Riemann surface*, SIGMA Symmetry Integrability Geom. Methods Appl. **12** (2016), Paper No. 007, 14, DOI 10.3842/SIGMA.2016.007. MR3449917 ↑224

[12] D. Baraglia, I. Biswas, and L. P. Schaposnik, *Brauer group of moduli of Higgs bundles and connections*, Geometry and physics. Vol. II, Oxford Univ. Press, Oxford, 2018, pp. 387–398. MR3931778 ↑207

[13] D. Baraglia and L. P. Schaposnik, *Higgs bundles and* (A,B,A)*-branes*, Comm. Math. Phys. **331** (2014), no. 3, 1271–1300, DOI 10.1007/s00220-014-2053-6. MR3248065 ↑226, 229, 230, 235, 236

[14] D. Baraglia and L. P. Schaposnik, *Real structures on moduli spaces of Higgs bundles*, Adv. Theor. Math. Phys. **20** (2016), no. 3, 525–551, DOI 10.4310/ATMP.2016.v20.n3.a2. MR3565861 ↑226, 229, 230, 235, 236, 245

[15] D. Baraglia and L. P. Schaposnik, *Monodromy of rank 2 twisted Hitchin systems and real character varieties*, Trans. Amer. Math. Soc. **370** (2018), no. 8, 5491–5534, DOI 10.1090/tran/7144. MR3812111 ↑219, 220, 221, 246

[16] D. Baraglia and L. P. Schaposnik, *Cayley and Langlands type correspondences for orthogonal Higgs bundles*, Trans. Amer. Math. Soc. **371** (2019), no. 10, 7451–7492, DOI 10.1090/tran/7587. MR3939583 ↑207, 223, 225, 231, 232, 234, 235

[17] A. Beauville, M. S. Narasimhan, and S. Ramanan, *Spectral curves and the generalised theta divisor*, J. Reine Angew. Math. **398** (1989), 169–179, DOI 10.1515/crll.1989.398.169. MR998478 ↑217, 218, 220, 223

[18] U. N. Bhosle, *Generalized parabolic Hitchin pairs*, J. Lond. Math. Soc. (2) **89** (2014), no. 1, 1–23, DOI 10.1112/jlms/jdt058. MR3174731 ↑225

[19] O. Biquard and P. Boalch, *Wild non-abelian Hodge theory on curves*, Compos. Math. **140** (2004), no. 1, 179–204, DOI 10.1112/S0010437X03000010. MR2004129 ↑211

[20] I. Biswas, P. Arés-Gastesi, and S. Govindarajan, *Parabolic Higgs bundles and Teichmüller spaces for punctured surfaces*, Trans. Amer. Math. Soc. **349** (1997), no. 4, 1551–1560, DOI 10.1090/S0002-9947-97-01870-9. MR1407481 ↑214

[21] I. Biswas and T. L. Gómez, *Connections and Higgs fields on a principal bundle*, Ann. Global Anal. Geom. **33** (2008), no. 1, 19–46, DOI 10.1007/s10455-007-9072-x. MR2369185 ↑206

[22] I. Biswas, S. Heller, and L. P. Schaposnik, *Branes and moduli spaces of Higgs bundles on smooth projective varieties* (2020). ↑226, 228

[23] P. Boalch, *Hyperkähler manifolds and nonabelian Hodge theory of (irregular) curves* (2012), available at 1203.6607. ↑209, 211

[24] P. P. Boalch, *Geometry and braiding of Stokes data; fission and wild character varieties*, Ann. of Math. (2) **179** (2014), no. 1, 301–365, DOI 10.4007/annals.2014.179.1.5. MR3126570 ↑209, 210, 211

[25] P. Boalch, *Symplectic manifolds and isomonodromic deformations*, Adv. Math. **163** (2001), no. 2, 137–205, DOI 10.1006/aima.2001.1998. MR1864833 ↑209, 210, 211

[26] P. Boalch, *Poisson varieties from Riemann surfaces*, Indag. Math. (N.S.) **25** (2014), no. 5, 872–900, DOI 10.1016/j.indag.2014.07.004. MR3264779 ↑209, 211

[27] H. U. Boden and K. Yokogawa, *Moduli spaces of parabolic Higgs bundles and parabolic* $K(D)$ *pairs over smooth curves. I*, Internat. J. Math. **7** (1996), no. 5, 573–598, DOI 10.1142/S0129167X96000311. MR1411302 ↑208

[28] S. Bradlow, L. Branco, and L. P. Schaposnik, *Orthogonal Higgs bundles with singular spectral curves*, Comm. Anal. Geom. **28** (2020), no. 8, 1895–1931, DOI 10.4310/CAG.2020.v28.n8.a6. MR4201792 ↑223, 231, 239, 241

[29] S. B. Bradlow, O. García-Prada, and P. B. Gothen, *Surface group representations and* $U(p,q)$*-Higgs bundles*, J. Differential Geom. **64** (2003), no. 1, 111–170. MR2015045 ↑207, 236

[30] S. B. Bradlow, O. García-Prada, and P. B. Gothen, *Maximal surface group representations in isometry groups of classical Hermitian symmetric spaces*, Geom. Dedicata **122** (2006), 185–213, DOI 10.1007/s10711-007-9127-y. MR2295550 ↑207

[31] S. B. Bradlow, O. García-Prada, and P. B. Gothen, *Higgs bundles for the non-compact dual of the special orthogonal group*, Geom. Dedicata **175** (2015), 1–48, DOI 10.1007/s10711-014-0026-8. MR3323627 ↑207

[32] S. B. Bradlow and L. P. Schaposnik, *Higgs bundles and exceptional isogenies*, Res. Math. Sci. **3** (2016), Paper No. 14, 28, DOI 10.1186/s40687-016-0062-0. MR3531373 ↑231, 239, 240, 241

[33] L. Branco, *Higgs bundles, lagrangians and mirror symmetry*, DPhil. Thesis, University of Oxford (2017). ↑222, 223, 246

[34] S. A. Broughton, *Classifying finite group actions on surfaces of low genus*, J. Pure Appl. Algebra **69** (1991), no. 3, 233–270, DOI 10.1016/0022-4049(91)90021-S. MR1090743 ↑226

[35] E. Bujalance, F.-J. Cirre, J.-M. Gamboa, and G. Gromadzki, *Symmetry types of hyperelliptic Riemann surfaces* (English, with English and French summaries), Mém. Soc. Math. Fr. (N.S.) **86** (2001), vi+122, DOI 10.24033/msmf.399. MR1891804 ↑229

[36] A. C. Casimiro, S. Ferreira, and C. Florentino, *Principal Schottky Bundles over Riemann surfaces* (2016), available at 1612.08662. ↑208, 214

[37] B. Collier, *Finite order automorphisms of Higgs Bsundles: Theory and application*, ProQuest LLC, Ann Arbor, MI, 2016. Thesis (Ph.D.)–University of Illinois at Urbana-Champaign. MR3624308 ↑207

[38] B. Collier, $SO(n, n+1)$*-surface group representations and their Higgs bundles* (2017), available at 1710.01287. ↑207

[39] K. Corlette, *Flat* G*-bundles with canonical metrics*, J. Differential Geom. **28** (1988), no. 3, 361–382. MR965220 ↑202, 203

[40] R. Donagi and T. Pantev, *Langlands duality for Hitchin systems*, Invent. Math. **189** (2012), no. 3, 653–735. MR2957305 ↑202, 244

[41] R. Donagi, *Decomposition of spectral covers*, Astérisque **218** (1993), 145–175. Journées de Géométrie Algébrique d'Orsay (Orsay, 1992). MR1265312 ↑202, 220

[42] R. Donagi, *Spectral covers*, Current topics in complex algebraic geometry (Berkeley, CA, 1992/93), Math. Sci. Res. Inst. Publ., vol. 28, Cambridge Univ. Press, Cambridge, 1995, pp. 65–86. MR1397059 ↑202

[43] R. Donagi and E. Markman, *Spectral covers, algebraically completely integrable, Hamiltonian systems, and moduli of bundles*, Integrable systems and quantum groups (Montecatini Terme, 1993), 1996, pp. 1–119. MR1397273 ↑220

[44] S. K. Donaldson, *Twisted harmonic maps and the self-duality equations*, Proc. London Math. Soc. (3) **55** (1987), no. 1, 127–131, DOI 10.1112/plms/s3-55.1.127. MR887285 ↑202, 203

[45] J. Fisher and S. Rayan, *Hyperpolygons and Hitchin systems*, Int. Math. Res. Not. IMRN **6** (2016), 1839–1870, DOI 10.1093/imrn/rnv172. MR3509942 ↑214, 243, 244, 248

[46] E. Franco, M. Jardim, and S. Marchesi, *Branes in the moduli space of framed sheaves*, Bull. Sci. Math. **141** (2017), no. 4, 353–383, DOI 10.1016/j.bulsci.2017.04.002. MR3657959 ↑230

[47] E. Franco, M. Jardim, and G. Menet, *Brane involutions on irreducible holomorphic symplectic manifolds*, Kyoto J. Math. **59** (2019), no. 1, 195–235, DOI 10.1215/21562261-2018-0009. MR3934628 ↑230

[48] E. Frenkel and E. Witten, *Geometric endoscopy and mirror symmetry*, Commun. Number Theory Phys. **2** (2008), no. 1, 113–283, DOI 10.4310/CNTP.2008.v2.n1.a3. MR2417848 ↑223

[49] O. Garcia-Prada, P. B. Gothen, and I. M. i Riera, *Representations of surface groups in real symplectic groups* (2008). ↑207

[50] O. García-Prada, P. B. Gothen, and I. M. i Riera, *The Hitchin-Kobayashi correspondence, Higgs pairs and surface group representations* (2009). ↑207

[51] O. García-Prada and A. G. Oliveira, *Higgs bundles for the non-compact dual of the unitary group*, Illinois J. Math. **55** (2011), no. 3, 1155–1181 (2013). MR3069300 ↑207

[52] O. García-Prada and A. G. Oliveira, *Connectedness of the moduli of* $Sp(2p, 2q)$*-Higgs bundles*, Q. J. Math. **65** (2014), no. 3, 931–956. MR3261975 ↑207

[53] V. Ginzburg, *Lectures on Nakajima's quiver varieties* (English, with English and French summaries), Geometric methods in representation theory. I, Sémin. Congr., vol. 24, Soc. Math. France, Paris, 2012, pp. 145–219. MR3202703 ↑242

[54] P. B. Gothen, *The topology of Higgs bundle moduli spaces*, Ph.D. thesis, Mathematics Institute, University of Warwick (1995). ↑207

[55] P. B. Gothen and A. G. Oliveira, *The singular fiber of the Hitchin map*, Int. Math. Res. Not. IMRN **5** (2013), 1079–1121, DOI 10.1093/imrn/rns020. MR3031827 ↑219, 223

[56] P. Griffiths and J. Harris, *Principles of algebraic geometry*, Wiley Classics Library, John Wiley & Sons, Inc., New York, 1994. Reprint of the 1978 original, DOI 10.1002/9781118032527. MR1288523 ↑217

[57] O. Guichard and A. Wienhard, *Positivity and higher Teichmüller theory*, European Congress of Mathematics, Eur. Math. Soc., Zürich, 2018, pp. 289–310. MR3887772 ↑235

[58] S. Gukov and E. Witten, *Gauge theory, ramification, and the geometric Langlands program*, Current developments in mathematics, 2006, Int. Press, Somerville, MA, 2008, pp. 35–180. MR2459305 ↑202

[59] S. Gukov and E. Witten, *Branes and quantization*, Adv. Theor. Math. Phys. **13** (2009), no. 5, 1445–1518. MR2672467 ↑202

[60] T. Hausel, *Compactification of moduli of Higgs bundles*, J. Reine Angew. Math. **503** (1998), 169–192, DOI 10.1515/crll.1998.096. MR1650276 ↑223

[61] T. Hausel, *Geometry of the moduli space of Higgs bundles* (2001), available at `math/0107040`. ↑223

[62] T. Hausel and M. Thaddeus, *Mirror symmetry, Langlands duality, and the Hitchin system*, Invent. Math. **153** (2003), no. 1, 197–229, DOI 10.1007/s00222-003-0286-7. MR1990670 ↑202

[63] S. Heller and L. P. Schaposnik, *Branes through finite group actions*, J. Geom. Phys. **129** (2018), 279–293, DOI 10.1016/j.geomphys.2018.03.014. MR3789253 ↑226, 227, 228, 229

[64] N. J. Hitchin, *The self-duality equations on a Riemann surface*, Proc. London Math. Soc. (3) **55** (1987), no. 1, 59–126. MR887284 ↑202, 203, 205, 206, 223, 226

[65] N. J. Hitchin, *Lie groups and Teichmüller space*, Topology **31** (1992), no. 3, 449–473, DOI 10.1016/0040-9383(92)90044-I. MR1174252 ↑206, 207, 216, 226, 229

[66] N. Hitchin, *Stable bundles and integrable systems*, Duke Math. J. **54** (1987), no. 1, 91–114, DOI 10.1215/S0012-7094-87-05408-1. MR885778 ↑202, 203, 205, 206, 214, 215, 216, 230, 235, 240

[67] N. Hitchin, *Langlands duality and* G_2 *spectral curves*, Q. J. Math. **58** (2007), no. 3, 319–344, DOI 10.1093/qmath/ham016. MR2354922 ↑202, 218

[68] N. Hitchin, *Higgs bundles and characteristic classes*, Arbeitstagung Bonn 2013, Progr. Math., vol. 319, Birkhäuser/Springer, Cham, 2016, pp. 247–264. MR3618052 ↑246

[69] N. Hitchin and L. P. Schaposnik, *Nonabelianization of Higgs bundles*, J. Differential Geom. **97** (2014), no. 1, 79–89. MR3229050 ↑202, 207, 223, 232, 233, 246

[70] V. Hoskins and F. Schaffhauser, *Group actions on quiver varieties and applications*, Internat. J. Math. **30** (2019), no. 2, 1950007, 46, DOI 10.1142/S0129167X19500071. MR3922303 ↑230, 242, 243, 245, 247, 248

[71] J.-i. Igusa, *Fibre systems of Jacobian varieties*, Amer. J. Math. **78** (1956), 171–199, DOI 10.2307/2372489. MR84848 ↑223

[72] A. Kapustin and E. Witten, *Electric-magnetic duality and the geometric Langlands program*, Commun. Number Theory Phys. **1** (2007), no. 1, 1–236, DOI 10.4310/CNTP.2007.v1.n1.a1. MR2306566 ↑202, 226, 244

[73] H. Konno, *Construction of the moduli space of stable parabolic Higgs bundles on a Riemann surface*, J. Math. Soc. Japan **45** (1993), no. 2, 253–276, DOI 10.2969/jmsj/04520253. MR1206652 ↑208

[74] G. Laumon, *Un analogue global du cône nilpotent* (French), Duke Math. J. **57** (1988), no. 2, 647–671, DOI 10.1215/S0012-7094-88-05729-8. MR962524 ↑222

[75] M. Logares and J. Martens, *Moduli of parabolic Higgs bundles and Atiyah algebroids*, J. Reine Angew. Math. **649** (2010), 89–116, DOI 10.1515/CRELLE.2010.090. MR2746468 ↑225

[76] E. Markman and E. Z. Xia, *The moduli of flat* $\mathrm{PU}(p, p)$*-structures with large Toledo invariants*, Math. Z. **240** (2002), no. 1, 95–109. MR1906709 ↑207

[77] V. B. Mehta and C. S. Seshadri, *Moduli of vector bundles on curves with parabolic structures*, Math. Ann. **248** (1980), no. 3, 205–239. MR575939 ↑207, 208

[78] M. Melo, A. Rapagnetta, and F. Viviani, *Fourier-Mukai and autoduality for compactified Jacobians, II*, Geom. Topol. **23** (2019), no. 5, 2335–2395, DOI 10.2140/gt.2019.23.2335. MR4019895 ↑223

[79] D. Nadler, *Perverse sheaves on real loop Grassmannians*, Invent. Math. **159** (2005), no. 1, 1–73. MR2142332 ↑245

[80] H. Nakajima, *Varieties associated with quivers*, Representation theory of algebras and related topics (Mexico City, 1994), CMS Conf. Proc., vol. 19, Amer. Math. Soc., Providence, RI, 1996, pp. 139–157. MR1388562 ↑241

[81] H. Nakajima, *Quiver varieties and Kac-Moody algebras*, Duke Math. J. **91** (1998), no. 3, 515–560, DOI 10.1215/S0012-7094-98-09120-7. MR1604167 ↑241

[82] B. C. Ngô, *Le lemme fondamental pour les algèbres de Lie* (French), Publ. Math. Inst. Hautes Études Sci. **111** (2010), 1–169, DOI 10.1007/s10240-010-0026-7. MR2653248 ↑202, 222

[83] N. Nitsure, *Moduli space of semistable pairs on a curve*, Proc. London Math. Soc. (3) **62** (1991), no. 2, 275–300. MR1085642 ↑203, 204

[84] A. Peón-Nieto, *Cameral data for* $SU(p, p+1)$*-Higgs bundles* (2015). ↑203, 207, 220, 221, 232, 237, 238

[85] A. Ramanathan, *Stable principal bundles on a compact Riemann surface*, Math. Ann. **213** (1975), 129–152. MR0369747 ↑205

[86] S. Rayan and L. P. Schaposnik, *Moduli spaces of generalized hyperpolygons*, The Quarterly Journal of Mathematics **71** (2021). ↑226, 230, 245, 248

[87] L. P. Schaposnik, *Monodromy of the* SL_2 *Hitchin fibration*, Internat. J. Math. **24** (2013), no. 2, 1350013, 21, DOI 10.1142/S0129167X13500134. MR3045345 ↑219, 221, 229, 231, 246

[88] L. P. Schaposnik, *Spectral data for* $U(m, m)$*-Higgs bundles*, Int. Math. Res. Not. IMRN **11** (2015), 3486–3498. MR3373057 ↑203, 207, 218, 219, 221, 222, 229, 231, 232, 233, 246

[89] L. P. Schaposnik, *Higgs bundles and applications*, Oberwolfach Reports **31** (2015), 31–33, available at 1603.06691. ↑226, 245

[90] L. P. Schaposnik, *Spectral data for* $U(m, m)$*-Higgs bundles*, Int. Math. Res. Not. IMRN **11** (2015), 3486–3498. MR3373057 ↑207, 221, 222, 231, 236, 237, 246

[91] L. P. Schaposnik, *A geometric approach to orthogonal Higgs bundles*, Eur. J. Math. **4** (2018), no. 4, 1390–1411, DOI 10.1007/s40879-017-0206-9. MR3866701 ↑207, 223, 231

[92] L. P. Schaposnik, *An introduction to spectral data for Higgs bundles*, The geometry, topology and physics of moduli spaces of Higgs bundles, Lect. Notes Ser. Inst. Math. Sci. Natl. Univ. Singap., vol. 36, World Sci. Publ., Hackensack, NJ, 2018, pp. 65–101. MR3837869 ↑203, 207

[93] D. Schaub, *Courbes spectrales et compactifications de jacobiennes* (French), Math. Z. **227** (1998), no. 2, 295–312, DOI 10.1007/PL00004377. MR1609069 ↑223

[94] C. T. Simpson, *Constructing variations of Hodge structure using Yang-Mills theory and applications to uniformization*, J. Amer. Math. Soc. **1** (1988), no. 4, 867–918, DOI 10.2307/1990994. MR944577 ↑202, 203, 205, 223

[95] C. T. Simpson, *Higgs bundles and local systems*, Inst. Hautes Études Sci. Publ. Math. **75** (1992), 5–95. MR1179076 ↑203, 209, 223

[96] S. Szabo, *Simpson's geometric* $P = W$ *conjecture in the Painlevé VI case via abelianization* (2019). ↑210

[97] T. Hausel, A. Mellit, and D. Pei, *Mirror symmetry with branes by equivariant Verlinde formulas*, Geometry and physics. Vol. I, Oxford Univ. Press, Oxford, 2018, pp. 189–218. MR3932262 ↑246

[98] K. Uhlenbeck and S.-T. Yau, *On the existence of Hermitian-Yang-Mills connections in stable vector bundles*, Comm. Pure Appl. Math. **39** (1986), no. S, suppl., S257–S293. Frontiers of the mathematical sciences: 1985 (New York, 1985). MR861491 ↑202

[99] R. A. Wentworth and G. Wilkin, *Cohomology of* $U(2,1)$ *representation varieties of surface groups*, Proc. Lond. Math. Soc. (3) **106** (2013), no. 2, 445–476, DOI 10.1112/plms/pds048. MR3021468 ↑207

[100] R. A. Wentworth and M. Wolf, *Surface group representations to* $SL(2,\mathbb{C})$ *and Higgs bundles with smooth spectral data*, Geom. Topol. **20** (2016), no. 5, 3019–3032, DOI 10.2140/gt.2016.20.3019. MR3556354 ↑222

[101] E. Witten, *Gauge theory and wild ramification*, Anal. Appl. (Singap.) **6** (2008), no. 4, 429–501, DOI 10.1142/S0219530508001195. MR2459120 ↑202

[102] E. Z. Xia, *The moduli of flat* $U(p,1)$ *structures on Riemann surfaces*, Geom. Dedicata **97** (2003), 33–43, DOI 10.1023/A:1023675007667. Special volume dedicated to the memory of Hanna Miriam Sandler (1960–1999). MR2003688 ↑207, 237

[103] K. Yokogawa, *Compactification of moduli of parabolic sheaves and moduli of parabolic Higgs sheaves*, J. Math. Kyoto Univ. **33** (1993), no. 2, 451–504, DOI 10.1215/kjm/1250519269. MR1231753 ↑208

Department of Mathematics, University of Illinois at Chicago, IL 60607, USA
Email address: schapos@uic.edu

IAS/Park City Mathematics Series
Volume 28, Pages 253–346
https://doi.org/10.1090/pcms/028/00878

Gauge theory and a few applications to knot theory

Tomasz S. Mrowka and Donghao Wang

Abstract. These lectures give a brief introduction to some applications of gauge theory to the topology of three manifolds and knot theory.

Contents

2020 *Mathematics Subject Classification.* Primary 57K18, 57K41, 58J28; Secondary 58J90, 58J30, 58J32.
Key words and phrases. Park City Mathematics Institute, gauge theory, knot theory, Chern-Simons invariants, Floer homology.
This material is based upon work supported by the National Science Foundation under Grant DMS-1808794. Any opinions, findings, and conclusions or recommendations expressed in this material are those of the author(s) and do not necessarily reflect the views of the National Science Foundation.

1. Preamble: Some background material not included in the lectures

This chapter contains background material not included in the lectures. The first section sketches in some detail the parts of Morse theory that will be important for later purposes. The second section is a crash course in principal bundles, Chern-Weil and Chern-Simons theory.

1.1. A brief introduction to Morse homology The invariants we will discuss, the Floer homology groups of three manifolds, knots and spatial graphs in three manifolds are all based on the idea of *Morse homology*. This sub-section aims to introduce a student to the principal ideas of Morse theory in the finite dimensional setting.

Morse invented Morse theory to understand the interplay between functions on a manifold and its algebraic topology. The initial impetus of Morse was to understand the existence of closed geodesics on manifolds by viewing them as critical points of the energy function on paths on a Riemannian manifold; then the homology of loop space of a manifold would imply the existence of closed geodesics. Bott exploited Morse's ideas to study the homotopy type of the loop space of manifolds and worked out the case of Lie groups (or more generally symmetric spaces) very concretely leading to the Bott periodicity theorem. In the hands of Bott's student Smale, Morse theory finally became a tool for differential topology. Morse's initial foray already envisaged Morse theory as a tool for the study of infinite dimensional manifolds for functions like the energy function on the loop space of manifolds. A key result that makes the energy functional behave sufficiently like a function on a finite dimensional manifold is the Morse index theorem stating that the dimension of a maximal negative subspace for the Hessian of the energy functional is finite. Floer's key insight is that the Morse homology exists for more subtle functions on infinite dimensional spaces where the Hessian has unbounded spectrum in both the positive and negative directions so the Morse index is no longer defined. Key examples are the Chern-Simons function on connections on three manifolds or the symplectic action function on the loop space of a symplectic manifold [18, 19]. The resulting homology groups no longer reflect the algebraic topology of the infinite dimensional manifold and rather give rise to more subtle kinds of invariants of symplectic manifolds and three dimensional manifolds.

The figure on the next page depicts a 2-torus as embedded in $\mathbb{R}^3$, with 4 critical points associated to the height function: a maximum, a minimum and two saddle points. For some choice of Riemannian metric on the torus, we draw local stable and unstable manifolds for the top saddle point.

The gradient and Hessian We recall some standard terminology. Let M be a smooth manifold of dimension n and $f : M \to \mathbb{R}$ be a smooth function. Let $\mathrm{Crit}(f) = \{p \in M : d_p f = 0\}$ be the set of critical points of f where $d_p f$ denotes the differential of f. If we choose a Riemannian metric g and local coordinates

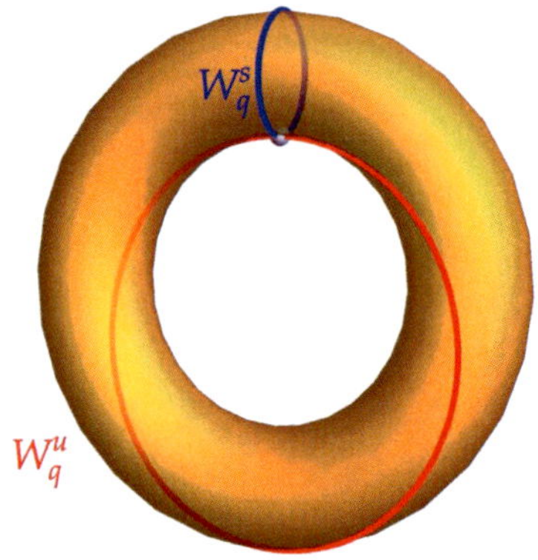

$x^1, \ldots x^n$, then we may write

$$g = g_{ij}\, dx^i \otimes dx^j$$

where $\{g_{ij}\}$ is a family of positive definite symmetric matrices. In local coordinates, we write a vector field as

$$X = \sum_{i=1}^{n} X^i \frac{\partial}{\partial x^i}.$$

The Riemannian metric induces an inner product on the fibers of TM

$$\langle X, Y \rangle = \sum_{i,j=1}^{n} g_{ij} X^i Y^j,$$

which allows us to convert df, the differential of the function f, into the gradient vector field ∇f defined locally as

$$\nabla f = \sum_{i,j=1}^{n} g^{ij} \frac{\partial f}{\partial x^i} \frac{\partial}{\partial x^j}.$$

Here g^{ij} is the inverse matrix of g_{ji}. The *Hessian* of f is initially defined as a quadratic form at each critical point in local coordinates via the matrix of second derivatives of f

$$\mathrm{Hess}_p^* f \doteq \frac{\partial^2 f}{\partial x^i \partial x^j}.$$

It is easy to check that resulting quadratic form

$$\mathrm{Hess}_p^* f(X, Y) = \sum_{i,j=1}^{n} \frac{\partial^2 f}{\partial x^i \partial x^j} X^i Y^j$$

defined by $\mathrm{Hess}_p^*(f) : T_pM \otimes T_pM \to \mathbb{R}$ is a symmetric quadratic form independent of the coordinate chart. A critical point is called *non-degenerate* if the Hessian is a non-degenerate quadratic form. The *Morse index* of a critical point p is the number of negative eigenvalues (counted with multiplicities) of the Hessian of f at p, denoted by μ_p. If p is non-degenerate, then the number of positive eigenvalues of the Hessian at p is $n - \mu_p$.

Definition 1.1.1. A function f is called a *Morse* function if all critical points are non-degenerate. ◊

The Riemannian metric allows us to globalize the Hessian. We can take the covariant derivative of the vector field ∇f. Then

$$v \mapsto \nabla_v(\nabla f)$$

gives rise to a bundle map

$$\mathrm{Hess} f : TM \to TM$$

which is fiberwise symmetric and at a critical point p satisfies

$$\langle \mathrm{Hess}_p f(X), Y\rangle = \mathrm{Hess}_p^* f(X, Y).$$

Example. A wonderful example of Morse functions on the real projective space $\mathbb{RP}^n$ is obtained by taking a symmetric $(n+1) \times (n+1)$ matrix A and by considering the function

$$f : \mathbb{RP}^n \to \mathbb{R}$$
$$[x] \mapsto \frac{\langle x, Ax\rangle}{\|x\|^2}.$$

Here $[x] \in \mathbb{R}^{n+1}$ denotes the equivalence class of $x \in \mathbb{R}^{n+1}\backslash 0$. If we diagonalize A we can write

$$f([x_0 : x_1 : \ldots : x_n]) = \frac{\sum_{i=0}^n \lambda_i |x_i|^2}{\sum_{i=0}^n |x_i|^2}$$

where $[x_0 : x_1 : \ldots : x_n]$ are homogeneous coordinates on $\mathbb{RP}^n$ corresponding to an orthonormal basis of eigenvectors and the λ_i's are non-decreasing. The function f is induced from $\tilde{f} : S^n \to \mathbb{R}$ with

$$\tilde{f}(x_0, x_1, \ldots, x_n) = \sum_{i=0}^n \lambda_i |x_i|^2.$$

It is easy to see the following.

- The critical set of f is the projectivization $[E_\lambda]$ of the eigenspaces E_λ of A. In particular, if all eigenvalues have multiplicity one, then the critical points are isolated and we denote the projectivization of E_{λ_i} by ϵ_i.
- The critical values of $[E_\lambda]$ is λ.
- The Hessian of f using the metric induced from the sphere at ϵ_i is the diagonal matrix with diagonal entries

$$\lambda_1 - \lambda_i, \lambda_2 - \lambda_i, \ldots, \widehat{\lambda_i - \lambda_i}, \ldots, \lambda_n - \lambda_i. \tag{1.1.2}$$

1.2. The negative gradient flow and (un)stable manifolds The relation between the topology of M and the Morse function f is manifested by the negative gradient flow of f. We study solutions $\gamma : \mathbb{R} \to M$ to the ODE

$$\dot{\gamma}(t) = -\nabla_{\gamma(t)} f.$$

A key role in this study is played by the stable and unstable manifolds of the critical points. Given $p \in \mathrm{Crit}(f)$, define the unstable (resp: stable) set of p to be

$$\mathcal{W}^u_p = \{x \in M \mid \exists \gamma,\ \gamma(0) = x,\ \dot\gamma(t) = -\nabla_{\gamma(t)} f,\ \lim_{t\to-\infty} \gamma(t) = p\},$$

$$\mathcal{W}^s_p = \{x \in M \mid \exists \gamma,\ \gamma(0) = x,\ \dot\gamma(t) = -\nabla_{\gamma(t)} f,\ \lim_{t\to\infty} \gamma(t) = p\}.$$

By the unique continuation theorem for solutions to ODEs, these can be equivalently described as a set of flow lines. For example, the unstable manifold can also be characterized as follows.

$$\mathcal{W}^u_p = \{\gamma : [0,\infty) \to M \mid \dot\gamma(t) = -\nabla_{\gamma(t)} f,\ \lim_{t\to-\infty} \gamma(t) = p\}. \tag{1.2.1}$$

The first basic theorem of Morse theory tells us that for a non-degenerate critical point, the stable and unstable sets are manifolds. More precisely, we have:

Theorem 1.2.2. *Let $p \in \mathrm{Crit}(f)$ be a non-degenerate critical point. Then $\mathcal{W}^u_p$ and $\mathcal{W}^s_p$ are smoothly embedded submanifolds diffeomorphic to $\mathbb{R}^{\mu_p}$ and $\mathbb{R}^{n-\mu_p}$ respectively. They intersect only at the critical point p, where the intersection is also transverse.*

The local version of this theorem is due independently to Grobman [23] and Hartman [25]. The global version is due to Smale [46,47]

For any $p, q \in \mathrm{Crit}(f)$, consider the intersection of the stable and unstable manifolds $\mathcal{M}(p,q) \doteq \mathcal{W}^u_p \cap \mathcal{W}^s_q$, which is also called *the moduli space* of trajectories from p to q. By definition, each point of $\mathcal{M}(p,q)$ lies in a trajectory γ connecting p to q.

$$\mathcal{M}(p,q) = \{x \in M \mid \exists \gamma, \gamma(0) = x,\ \lim_{t\to-\infty} \gamma(t) = p,\ \lim_{t\to\infty} \gamma(t) = q\}.$$

If $\mathcal{W}^u_p$ and $\mathcal{W}^s_q$ intersect transversely, then

$$\dim M(p,q) = \mu_p + (n - \mu_q) - n = \mu_p - \mu_q.$$

Definition 1.2.3. A pair of a Morse function $f : M \to \mathbb{R}$ and Riemannian metric g on M is called *Morse-Smale* if for all pairs of critical points the intersections $\mathcal{W}^u_p \cap \mathcal{W}^s_q$ are transverse. ◇

Example: continued The negative gradient flow of $f : \mathbb{RP}^n \to \mathbb{R}$ above is induced by a flow $\phi_t : \mathbb{R}^{n+1} \to \mathbb{R}^{n+1}$ with

$$\phi_t(x_0, x_1, \ldots, x_n) = (e^{-\lambda_0 t} x_0, e^{-\lambda_1 t} x_1, \ldots, e^{-\lambda_n t} x_n).$$

In the case that all eigenvalues have multiplicity one, the Morse index of $[E_{\lambda_i}]$ is easily seen to be i. We can also describe the stable and unstable manifolds of the critical points. The unstable manifold of $[E_{\lambda_i}]$ is

$$\mathcal{W}^u_{\epsilon_i} = \mathrm{Proj}(e_{\lambda_i} + \sum_{l=0}^{i-1} E_{\lambda_l})$$

while the stable manifold is

$$\mathcal{W}^s_{\epsilon_i} = \mathrm{Proj}(e_{\lambda_i} + \sum_{l=i+1}^{n} E_{\lambda_l})$$

Here Proj denotes projectivization and e_{λ_i} is an eigenvector of A corresponding to λ_i. Then for $i < j$, the moduli space of trajectories is the intersection of the stable and unstable manifolds

$$\mathcal{W}^s_{\epsilon_i} \cap \mathcal{W}^u_{\epsilon_j} = \mathrm{Proj}(\mu e_{\lambda_i} + \nu e_{\lambda_j} + \sum_{i<l<j} E_{\lambda_l}) \tag{1.2.4}$$

where both μ and ν are non-zero. Thus $W_s(\epsilon_i) \cap W_u(\epsilon_j)$ has two components according to whether the signs of μ and ν agree or not. Each component is diffeomorphic in this case to $\mathbb{R}^{j-i}$. For example, if $j = n = 1$ and $i = 0$, then this recovers the familiar fact that for the height function on the circle, the trajectories from the top to the bottom are two copies of the real line.

The length and energy of paths in a Riemannian manifold For any path $\gamma : [t_0, t_1] \to M$ in a Riemannian manifold, the length and energy of γ are naturally defined as

$$l(\gamma) \doteq \int_{t_0}^{t_1} |\dot{\gamma}(t)|dt, \ E(\gamma) \doteq \int_{t_0}^{t_1} |\dot{\gamma}(t)|^2 dt.$$

Recall that geodesics are characterized as critical points of the energy while fixing the end points. A key fact is the following relation between the length, energy and the length of parameterizing interval.

Lemma 1.2.5. *Let $\gamma : [t_0, t_1] \to M$ be a path. If the energy $E(\gamma)$ is finite, so is the length $l(\gamma)$, and we have the following estimate*

$$l(\gamma) \leqslant E(\gamma)^{1/2}(t_1 - t_0)^{1/2}.$$

Proof. Using Hölder's inequality we argue as follows.

$$l(\gamma) = \int_{t_0}^{t_1} |\dot{\gamma}(t)|dt \leqslant (\int_{t_0}^{t_1} |\dot{\gamma}(t)|^2 dt)^{\frac{1}{2}} (\int_{t_0}^{t_1} 1 \cdot dt)^{\frac{1}{2}} \leqslant E(\gamma)^{\frac{1}{2}}(t_1 - t_0)^{\frac{1}{2}}. \qquad \square$$

Next observe that the negative gradient flowlines of f are characterized by their energy.

Lemma 1.2.6. *For any path $\gamma : [t_0, t_1] \to M$, we have the following identity:*

$$f(\gamma(t_1)) - f(\gamma(t_2)) = \frac{1}{2}\int_{t_1}^{t_2} (|\dot{\gamma}(t)|^2 + |\nabla_\gamma f|^2 - |\dot{\gamma} + \nabla_\gamma f|^2)dt.$$

Proof. Computing directly, we find

$$\begin{aligned} f(\gamma(t_1)) - f(\gamma(t_2)) &= -\int_{t_1}^{t_2} \frac{d}{dt} f(\gamma(t))dt = -\int_{t_1}^{t_2} \langle \nabla_\gamma f, \dot{\gamma}\rangle dt \\ &= \frac{1}{2}\int_{t_1}^{t_2} (|\dot{\gamma}(t)|^2 + |\nabla_\gamma f|^2 - |\dot{\gamma} + \nabla_\gamma f|^2)dt. \qquad \square \end{aligned}$$

In particular, for negative gradient flowlines, we have the following corollary.

Corollary 1.2.7. *For any negative gradient flowline $\gamma : [t_0, t_1] \to M$, we have*

$$E(\gamma) = f(\gamma(t_1)) - f(\gamma(t_2)).$$

Combining Corollary 1.2.7 and Lemma 1.2.5, we obtain the following.

Proposition 1.2.8. *For any negative gradient flowline* $\gamma : [t_0, t_1] \to M$ *of* f, *we have*

$$l(\gamma) \leqslant (f(\gamma(t_0)) - f(\gamma(t_1))^{\frac{1}{2}} (t_1 - t_0)^{\frac{1}{2}}. \tag{1.2.9}$$

Thus given a bound on the length of the interval, the length of the flowline is controlled by the change in the function along the flow line and the length of the parameterizing interval.

Another consequence of Corollary 1.2.7 is that for any flowline γ in $\mathcal{M}(p, q)$, the Energy agrees with the total drop of the Morse function f:

$$E(\gamma) = \lim_{t \to -\infty} f(\gamma(t)) - \lim_{t \to \infty} f(\gamma(t)) = f(p) - f(q),$$

which is *independent* of the particular flowline.

With the idea of compactifying the moduli spaces of flow lines, let us make a simple observation.

Lemma 1.2.10. *For any sequence of flowlines*

$$\gamma_i : [t_0, t_1] \to M, i \in \mathbb{N}$$

such that the drop of f *along each of* γ_i *has a uniform upper bound:*

$$f(\gamma_i(t_0)) - f(\gamma_i(t_1)) \leqslant C \textit{ for all } i,$$

we can find a convergent subsequence among γ_i *(convergence is initially* C^0 *but can easily be upgraded to* C^∞*).*

Proof. We combine the uniform drop with Proposition 1.2.8 applied to any subinterval $[s_0, s_1] \subset [t_0, t_1]$ to derive that

$$l(\gamma_i|_{[s_0, s_1]}) \leqslant \frac{C}{2} (s_1 - s_0)^{\frac{1}{2}}$$

so $\{\gamma_i\}$ a bounded sequence in the Hölder space $C^{\frac{1}{2}}([t_0, t_1], M)$. The compact inclusion $C^{\frac{1}{2}}([t_0, t_1], M) \to C^0([t_0, t_1], M)$ implies that there is convergent subsequence. This can be easily upgraded to C^∞ convergence using the equation

$$\dot{\gamma} = -\nabla_\gamma f.$$

□

The behavior of the flow near a critical point In this section, we will show that the length of a flow line that remains within a sufficiently small neighborhood of a critical point is bounded by the drop of (the square root of) the function along the flowline, i.e. a version of Inequality (1.2.9) holds *without* any dependence on the length of the parameterizing interval. The key is to show that the function approaches the critical value exponentially in forward time along the flow as long as the function value is above the critical value and similarly approaches the critical value exponentially in backwards time along the flow as long as the function value is below the critical value, as indicated in the following graph. The blue curve is the upper bound, the yellow curve is the lower bound, the red curve represents the function value along the flow and the green line is the critical value.

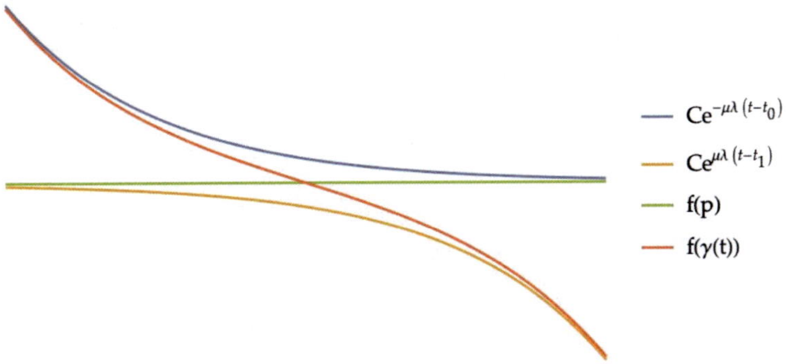

Proposition 1.2.11. *Let* p *be a non-degenerate critical point of* f*. There is a neighborhood* U *of* p *and some* $\mu > 0$ *such that for any flowline* $\gamma\colon [t_0, t_1] \to U \subset M$*, we have*

- *If* $f(\gamma(t)) > f(p)$ *for any* $t \in [t_0, s)$*, then*
$$(f(\gamma(t_0)) - f(p))e^{-\mu(t-t_0)} \geqslant f(\gamma(t)) - f(p);$$
- *If* $f(\gamma(t)) < f(p)$ *for any* $t \in (s, t_1]$*, then*
$$f(\gamma(t)) - f(p) \geqslant (f(\gamma(t_1)) - f(p))e^{\mu(t-t_1)}.$$

Proof. Consider a non-degenerate critical point p of a function f. For simplicity, assume that a coordinate chart is chosen so that we can view $f : \mathbb{R}^n \to \mathbb{R}$ with the origin being the non-degenerate critical point and $f(0) = 0$. Trivializing the tangent bundle of $\mathbb{R}^n$, we view $\nabla_x f : \mathbb{R}^n \to \mathbb{R}^n$ a family of linear maps with $\nabla_0 f = 0$.

Since $\mathrm{Hess}_0 f$ is invertible, the map $x \mapsto \nabla_x f$ is a local diffeomorphism. Thus for some neighborhood U of the origin we can find constants $c, C > 0$, so that

$$c|\nabla_x f|^2 \leqslant |x|^2 \leqslant C|\nabla_x f|^2$$
$$|f(x)| \leqslant C|x|^2 \leqslant C^2|\nabla_x f|^2. \tag{1.2.12}$$

These imply that

$$\frac{d}{dt} f(\gamma(t)) = -|\nabla_\gamma f|^2 \leqslant -\mu|f(\gamma(t))|. \tag{1.2.13}$$

If $f(\gamma(t)) > 0$ for any $t \in [t_0, s]$, then this inequality reads:

$$\frac{d}{dt} f(\gamma(t)) \leqslant -\mu f(\gamma(t))$$

or

$$\frac{d}{dt} e^{\mu t} f(\gamma(t)) \leqslant 0.$$

Integrating this inequality from t_0 to t, we obtain that

$$e^{\mu t} f(\gamma(t)) - e^{\mu t_0} f(\gamma(t_0)) \leqslant 0,$$

so

$$f(\gamma(t)) \leqslant f(\gamma(t_0))e^{-\mu(t-t_0)}$$

for any $t \in [t_0, s]$. Thus we have exponential decay of f along the flowline until the sign of f changes. If $f(\gamma(t)) < 0$ for $t \in [s, t_1]$, Inequality (1.2.13) reads

$$\frac{d}{dt} f(\gamma(t)) \leqslant \mu f(\gamma(t))$$

or

$$\frac{d}{dt} e^{-\mu t} f(\gamma(t)) \leqslant 0.$$

Integrating the inequality from t to t_1 yields

$$e^{-\mu t_1} f(\gamma(t_1)) - e^{-\mu t} f(\gamma(t)) \leqslant 0$$

so

$$f(\gamma(t)) \geqslant f(\gamma(t_1)) e^{-\mu(t_1 - t)}$$

for $s < t \leqslant t_1$. Thus we have exponential decay in backwards time until f changes sign. □

We use this to prove the length estimate.

Theorem 1.2.14. *Let* p *be a non-degenerate critical point of* f. *There is a neighborhood* U_p *of* p *and a constant* $C > 0$ *so that the length of a flowline* $\gamma : [t_0, t_1] \to U_p \subset M$ *satisfies the bound*

$$l(\gamma) \leqslant C(f(\gamma(t_1))^{\frac{1}{2}} - f(\gamma(t_0))^{\frac{1}{2}})$$

if $f(\gamma(t_0)) - f(p)$ *and* $f(\gamma(t_1)) - f(p)$ *have the same sign and*

$$l(\gamma) \leqslant C((f(\gamma(t_0)) - f(p))^{\frac{1}{2}} + (f(p) - f(\gamma(t_1))^{\frac{1}{2}}$$

if they have different signs.

Proof. We assume the same setup as in the proof of Proposition 1.2.11. Suppose $\gamma(t) \in U$ for any $t \in [t_0, t_1]$ and f is defined in a neighborhood of 0 in $\mathbb{R}^n$. The origin is the unique critical point and $f(0) = 0$. Suppose that $f(\gamma(t)) > 0$ on $[t_0, t_1]$, where t_1 is allowed to be ∞. Using Inequality (1.2.12) in the form $1 \leqslant C|f(\gamma(t))|^{-\frac{1}{2}} |\nabla_{\gamma(t)} f|$, we deduce that

$$\begin{aligned} l(\gamma) = \int_{t_0}^{t_1} |\dot{\gamma}(t)| dt = \int_{t_0}^{t_1} |\nabla_{\gamma(t)} f| dt \leqslant C \int_{t_0}^{t_1} f(\gamma(t))^{-\frac{1}{2}} |\nabla_{\gamma(t)} f|^2 dt \\ \leqslant C \int_{t_0}^{t_1} f(\gamma(t))^{-\frac{1}{2}} (-\frac{d}{dt}(f(\gamma(t))) dt \\ = -C \int_{t_0}^{t_1} \frac{d}{dt} f(\gamma(t))^{\frac{1}{2}} dt \\ = f(\gamma(t_0))^{\frac{1}{2}} - f(\gamma(t_1))^{\frac{1}{2}}. \end{aligned}$$

The remaining cases are treated similarly. □

Let us summarize three key properties we proved so far for gradient flow lines.

- The length of flowlines where $|\nabla_x f|$ is large (bounded away from 0) is bounded by the energy.

- On uniformly bounded intervals, flowlines with small energy are short (contained in some neighborhood $\mathcal{U}_p$ for some $p \in \mathrm{Crit}(f)$).
- The length of flowlines where $|\nabla_x f|$ is small is independent of the time spent in the region $(\bigcup_{p\in\mathrm{Crit}(f)} \mathcal{U}_p)$.

1.3. Palais-Smale condition C In order for f to be sensitive to the algebraic topology of M, we need certain assumptions on the Morse function f (automatically satisfied if M is compact). A standard key assumption is the Palais-Smale condition C [42] (where C is for compactness).

Definition 1.3.1. A function $f : M \to \mathbb{R}$ is said to satisfy *Condition* C if every sequence $\{x_i\}$ with values $\{f(x_i)\}$ bounded (above and below) and with $\nabla_{x_i} f$ converging to zero has a subsequence converging to a critical point p of f. ♢

For a finite dimensional manifold, there is no ambiguity in this definition. For an infinite dimensional manifold, we need to take extra care to specify the topologies in which the above convergence is happening.

1.4. Towards a compactification of the space of unparameterized trajectories A central object of study in Morse theory is the quotient of $\mathcal{M}(p, q)$ by the action of translation:

$$\check{\mathcal{M}}(p, q) = \mathcal{M}(p, q)/\mathbb{R}.$$

This is the moduli space of unparameterized flowlines. We can exploit the previous work to begin to understand sequences of flowlines in $\check{\mathcal{M}}(p, q)$. To do this, let $I_n = [n-1, n+1]$. Then for any $\epsilon > 0$, the estimate

$$\int_{I_n} |\dot{\gamma}(t)|^2 + |\nabla_\gamma f|^2 > \epsilon \tag{1.4.1}$$

holds only for finitely many intervals I_n. The number m of such intervals is bounded above by $(f(p) - f(q))/\epsilon$. We choose ϵ small so that when the inequality of equation (1.4.1) is violated for I_n, the flow $\gamma|_{I_n}$ is contained in some small neighborhood of $\mathrm{Crit}(f)$. To see this happen, note that if

$$\int_{I_n} |\dot{\gamma}(t)|^2 + |\nabla_\gamma f|^2 < \epsilon \tag{1.4.2}$$

then there must be a time $t \in I_n$ where $|\nabla_{\gamma(t)} f|^2 < \epsilon$, so by Condition C, $\gamma(t)$ lies in a neighborhood of critical points. Now Theorem 1.2.14 implies that the total length of the flow line on this interval is controlled by ϵ, and so it must remain in a slightly larger neighborhood.

We now seek a compactification of $\check{\mathcal{M}}(p, q)$. Consider a sequence of flowlines $\{[\gamma_k]\}_k \subset \check{\mathcal{M}}(p, q)$ and choose a representative γ_k for each of them. The hypothesis implies that $\dot{\gamma}_k$ is bounded uniformly. Thus this is an equicontinuous family and hence has a C^0 and indeed $C^{0,\alpha}$ convergent subsequence. Convergence in higher regularity follows from the differential equation. Thus in the case of the compact manifold, any sequence of flow lines has a convergent subsequence on any compact subsets of $\mathbb{R}$.

For each γ_k, suppose that the property (1.4.1) is satisfied for

$$J_i^k = I_{n_i}^k, 1 \leqslant i \leqslant m_k, n_i < n_{i+1}.$$

Since m_k is bounded uniformly, by taking a subsequence, we may assume m_k is the same for each k. There are two potential scenarios for each i:

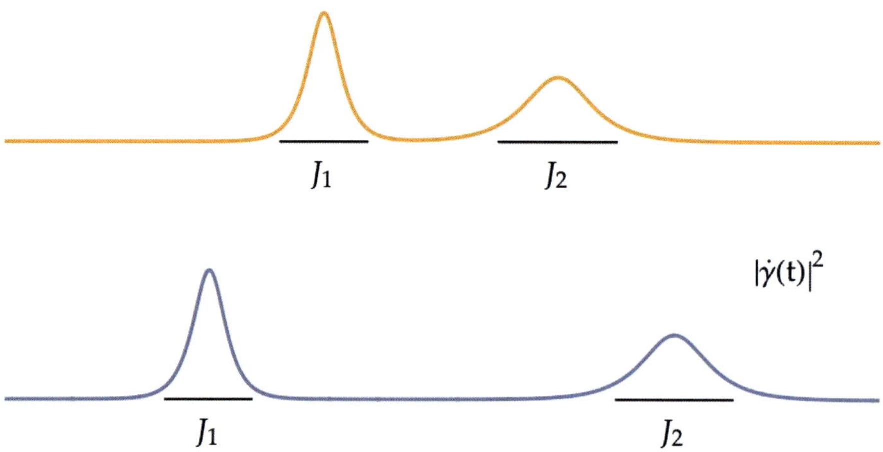

(1) The distance between J_i^k and J_{i+1}^k remains bounded as $k \to \infty$ (for a subsequence). Then we combine J_i^k, J_{i+1}^k and the intervals in between.
(2) The distance between J_i^k and J_{i+1}^k diverge to infinity as $k \to \infty$. Then we would have a broken trajectory.

This dichotomy leads to the idea of compactifying of $\check{\mathcal{M}}(p,q)$ by incorporating broken flowlines:

$$\begin{aligned}
\check{\mathcal{M}}^+(p,q) =& \check{\mathcal{M}}(p,q) && \dim = \mu_p - \mu_q - 1\\
&\cup \check{\mathcal{M}}(p,q_1) \times \check{\mathcal{M}}(q_1,q) && \dim = \mu_p - \mu_q - 2\\
&\cup \check{\mathcal{M}}(p,q_1) \times \check{\mathcal{M}}(q_1,q_2) \times \check{\mathcal{M}}(q_2,q) && \dim = \mu_p - \mu_q - 3\\
&\cdots \text{ lower dimensional strata}
\end{aligned} \tag{1.4.3}$$

Theorem 1.4.4. *The moduli space of unparameterized broken flowlines $\check{\mathcal{M}}^+(p,q)$ is compact and is a compactification of $\check{\mathcal{M}}(p,q)$.*

By construction, when (f,g) is a Morse-Smale pair the union in equation (1.4.3) above is a union of manifolds of decreasing dimensions and the frontier of a given "stratum" is contained in the union of lower strata. This union is an example of a topological space *stratified by manifolds*. It is important to understand the behavior in neighborhoods of the lower strata. What one can show is that a neighborhood of a broken trajectory

$$\check{\gamma}^+ = (\check{\gamma}_1, \check{\gamma}_1, \dots \check{\gamma}_r) \in \check{\mathcal{M}}(p_0,p_1) \times \check{\mathcal{M}}(p_1,p_2) \times \dots \times \check{\mathcal{M}}(p_r,p_{r+1})$$

is homeomorphic to $(\epsilon,0]^r \times U_1 \times \dots U_{r+1}$ where $U_i \subset \check{\mathcal{M}}(p_i,p_{i+1})$ are neighborhoods of $\check{\gamma}_i$. This homeomorphism can be chosen to be a diffeomorphism on the strata of $(\epsilon,0]^r \times U_1 \times \dots U_{r+1}$ where some set of the coordinates of the $(\epsilon,0]$

are zero and others are non-zero (see [30]). This makes this compactification very nearly a smooth manifold with corners (topologically it is). Some authors give $\check{\mathcal{M}}^+(p,q)$ a smooth structure for which it actually is, but this smooth structure is not the natural one defined as follows. Given p and q, let $c_0 < c_1 < \ldots < c_l$ denote the set of critical values of f in the interval $[f(q), f(p)]$ where $f(p) = c_l$ and $f(q) = c_0$. Choose an interlacing sequence

$$c_0 < b_1 < c_1 < \ldots b_l < c_l.$$

Then define a map $\iota\colon \check{\mathcal{M}}^+(p,q) \to \prod_{k=1}^{l} f^{-1}(b_k)$ by sending a broken flowline $\check{\gamma}^+$ to its intersection with $\prod_{k=1}^{l} f^{-1}(b_k)$. The compactification $\check{\mathcal{M}}^+(p,q)$ can be seen as the closure of the image $\iota(\check{\mathcal{M}}(p,q))$.

1.5. The Morse Homology When $\mu_p - \mu_q = 1$, $\check{\mathcal{M}}(p,q)$ is a compact 0-dim manifold, i.e., a finite collection of points. Let

$$C = C(M,f,g) = \bigoplus_{p \in \mathrm{Crit}(f)} \mathbb{Z}_2\langle p\rangle$$

$$\partial : C \to C,\ p \mapsto \sum_{q \mid \mu_p - \mu_q = 1} \#\check{\mathcal{M}}^+(p,q)\cdot q$$

When $\mu_p - \mu_q = 2$, $\dim \check{\mathcal{M}}^+(p,q) = 1$ and its boundary is described as

$$\partial\check{\mathcal{M}}^+(p,q) = \bigcup_{q_1} \check{\mathcal{M}}(p,q_1)\times\check{\mathcal{M}}(q_1,q).$$

which has 0 (mod 2) points. Therefore,

$$\partial^2 p = \sum_{\substack{q,r\\ \mu_p-\mu_q=1\\ \mu_q-\mu_r=1}} \#\check{\mathcal{M}}(p,q)\#\check{\mathcal{M}}(q,r)\cdot r = 0.$$

Finally, the Morse homology is defined as $H(M,f,g) = \mathrm{Ker}(\partial)/\mathrm{Im}(\partial)$. If M is compact, then one can show $H(M,f,g) \cong H_*(M;\mathbb{Z}_2)$. Indeed f induces a CW structure on M and the chain complex (C,∂) coincides with the CW complex that computes $H_*(M;\mathbb{Z}_2)$. See for example Smale's paper [46].

Remark. When X is non-compact, it is essential to have some analogue of the Palias-Smale condition C in order to define the Morse homology. If is it satisfied one can show that the Morse complex computes a limit of the homology of X. Note that if $c < c'$ then there is a map

$$H_*(f^{-1}(-\infty,c]) \to H_*(f^{-1}(-\infty,c']).$$

Taking the limit of these maps

$$\varinjlim_{c} H_*(f^{-1}(-\infty,c]) \tag{1.5.1}$$

1.6. Orientations If $\mathcal{W}_p^u \cap \mathcal{W}_q^s = \mathcal{M}(p,q)$, the negative gradient flow will stretch the normal bundle of $\mathcal{M}(p,q)$ within $\mathcal{W}_p^u$ which is identified with $T_q\mathcal{W}_q^u$ by taking $t\to\infty$. Thus for any $x \in \mathcal{M}(p,q)$, we have a short exact sequence

$$0 \to T_x\mathcal{M}(p,q) \to T_x\mathcal{W}_p^u \to T_q\mathcal{W}_q^u \to 0.$$

When $\mu_p - \mu_q = 1$, $\mathcal{M}(p, q)$ consists of finitely many copies of $\mathbb{R}$ and the flow orients $M(p, q)$ in a natural way. For any $\gamma \in \check{\mathcal{M}}(p, q)$, the short exact sequence induces an identification:

$$\epsilon_\gamma : \Lambda(p) \to \Lambda(q)$$

where $\Lambda(p)$ is the 2-element set of orientations of $T_p \mathcal{W}^u_p$. Now we can refine our

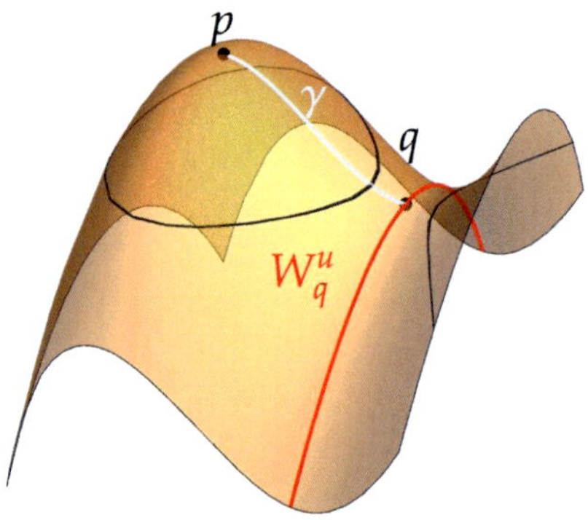

Figure 1.6.1. Critical points of index differing by 1 and a flowline between them.

differential by setting $\mathbb{Z}\Lambda(p) = \mathbb{Z} \times_{\mathbb{Z}/2\mathbb{Z}} \Lambda(p)$ and

$$C(M, f, g) = \bigoplus_{p \in \mathrm{Crit}(f)} \mathbb{Z}\Lambda(p),$$

$$\partial = \sum_{q | \mu_p - \mu_q = 1} \sum_{\gamma \in \check{\mathcal{M}}(p,q)} \epsilon_\gamma.$$

If the orientation is ignored, this chain complex reduces to the previous case.

1.7. Homology with Local Coefficients Let $\Gamma \to M$ be a flat bundle over M with fiber a module S over a commutative ring R. Using the flat structure, we can define the holonomy map by parallel transportation along a path $\gamma : [a, b] \to X$,

$$h_\gamma : \Gamma_{\gamma(a)} \to \Gamma_{\gamma(b)},$$

that depends only on the homotopy class (rel boundary) of γ. Now given a Morse function f and a Riemannian metric g, define the chain group by

$$C(f, g, \Gamma) = \bigoplus_{c \in \mathrm{Crit}(f)} \Gamma_c \cdot \Lambda(c)$$

and differentials by

$$\partial : C(M, f, g, \Gamma) \to C(M, f, g, \Gamma)$$

$$\partial = \sum_{\substack{a,b \in \mathrm{Crit}(f) \\ \mu(a) - \mu(b) = 1}} \sum_{\gamma \in M(a,b)} h_\gamma \otimes \epsilon_\gamma.$$

We can prove $\partial^2 = 0$ as usual.

Example 1.7.1. Suppose X is not orientable.

Consider the real line bundle $\bigwedge^n(T^*X)$ where $n = \dim X$. It admits an $\mathbb{R}^+$ action. The quotient bundle $\lambda \to X$ is the orientation bundle with fiber $\lambda_x \cong \{\pm 1\}$. It is not a trivial bundle when X is not orientable. ◊

How do we see Poincaré duality in terms of Morse homology? We can simply reverse the direction of the flow by taking the negative of f:

$$(X, f, g) \mapsto (X, -f, g)$$

but this duality is not canonical because of λ. Recall that

$$C(f, g) = \bigoplus_{c \in \mathrm{Crit}(f)} \mathbb{Z}\Lambda(c) \qquad \text{Orientations of } \textbf{unstable submanifolds} \text{ of } f,$$

$$C(-f, g) = \bigoplus_{c \in \mathrm{Crit}(f)} \mathbb{Z}\Lambda'(c) \qquad \text{Orientations of unstable submanifolds of } -f$$

$$= \text{Orientations of } \textbf{stable submanifolds} \text{ of } f.$$

To identify $\Lambda(c)$ with $\Lambda'(c)$, we need a section of λ, which does not exist when λ is non trivial. This could be resolved if we use local coefficients:

$$C(-f, g, \lambda) = \bigoplus_{c \in \mathrm{Crit}(f)} \mathbb{Z}\Lambda'(c) \otimes \lambda_c$$

and exploit the fact that

$$\Lambda(c) \otimes \Lambda'(c) \cong \lambda_c.$$

In this case, the Poincaré duality is formulated as

$$H^*(X; \mathbb{Z}) \cong H_*(X; \underline{\mathbb{Z}}_\lambda).$$

where $\underline{\mathbb{Z}}_\lambda$ is the local coefficient system determined by λ.

Note that a flat bundle is in bijection with isomorphism classes of representations of $\pi_1(X) \to \mathrm{Aut}(S)$. In the example above, $R \cong \mathbb{Z}/2\mathbb{Z}$ and $M \cong \mathbb{Z}$. The action of R on M is given by sending -1 to the reflection of $\mathbb{Z}$.

1.8. Novikov-Morse Homology Let θ be a closed 1-form and write $\theta = \theta^i dx_i$ in a local coordinate chart. The dual vector field is $\theta^\# = g^{ij}\theta_j\partial_i$. We investigate the flowlines driven by $\theta^\#$:

$$\dot{\gamma} = -\theta^\#_{\gamma(t)}.$$

Since θ is closed, we can locally write $\theta = df$ for some real-valued function f, so γ follows a negative gradient flow within a local coordinate chart; this property is not usually true globally.

The story goes through *mutatis mutandis* with minor two differences. The first comes with the energy identity. For a flowline $\gamma : [a, b] \to X$, we have instead

$$\int_a^b |\dot{\gamma}|^2 = \int_a^b |\theta^\#_{\gamma(t)}|^2 = -\int_a^b \langle \theta^\#_{\gamma(t)}, \dot{\gamma} \rangle = -\int_a^b \gamma^*\theta.$$

The second difference is that closed or periodic trajectories appear. They turn out to be less important in the development of the theory. If $[\theta] \neq 0 \in H^1(X, \mathbb{R})$, then $\int_a^b \gamma^*\theta$ depends on the relative homotopy class of γ so we refine the moduli space $\mathcal{M}(a,b)$ using relative homotopy classes $\pi_1(X,a,b)$ of paths from a to b:

$$\mathcal{M}(a,b) = \bigcup_{[\gamma]\in\pi_1(X,a,b)} \mathcal{M}_{[\gamma]}(a,b)$$

and subdivide the unparameterized space $\check{\mathcal{M}}(a,b)$:

$$\check{\mathcal{M}}(a,b) = \bigcup_{[\gamma]\in\pi_1(X,a,b)} \check{\mathcal{M}}_{[\gamma]}(a,b) = \bigcup_{[\gamma]\in\pi_1(X,a,b)} \mathcal{M}_{[\gamma]}(a,b)/\mathbb{R}.$$

Define

$$\check{\mathcal{M}}^+_{[\gamma]}(a,b) = \check{\mathcal{M}}_{[\gamma]}(a,b) \bigcup_{c\in\mathrm{Crit}(f)} \bigcup_{\substack{[\gamma_1]\in\pi_1(a,c)\\ [\gamma_2]\in\pi_1(c,b)\\ [\gamma_1]+[\gamma_2]=[\gamma]}} \check{\mathcal{M}}_{[\gamma_1]}(a,c)\times\check{\mathcal{M}}_{[\gamma_2]}(c,b).$$

Each $\check{\mathcal{M}}^+_{[\gamma]}(a,b)$ is still compact as usual. If θ has integral periods so that it is viewed as coming from a map $\Theta : X \to S^1$, this picture reduces to that of the $\mathbb{Z}$-cover of X determined by Θ.

Example 1.8.1. Let $X = S^1$ and $\theta = d\theta^0$. In this case, the Novikov-Morse homology is trivial, because there are no critical points at all. ♢

To define the Novikov-Morse homology with local coefficients, note that the differential

$$\partial = \sum_{\substack{a,b\in\mathrm{Crit}(f)\\ \mu(a)-\mu(b)=1}} \sum_{\gamma\in M(a,b)} h_\gamma \otimes \epsilon_\gamma.$$

may involve an infinite sum. To make sense of this, we need the fiber of Γ to carry a suitable topology so that we can speak of the convergence of this sum. The previous computation shows that $E(\gamma_k)\to\infty$ only if

$$\int_a^b \gamma_k^*\theta \to -\infty. \tag{1.8.2}$$

Now we introduce the Novikov ring $\mathcal{R}$:

$$\mathcal{R} := \{f:\mathbb{R}\to\mathbb{R} : \mathrm{supp}(f)|_{[n,+\infty)} \text{ is finite for all } n\}.$$

An element f is identified with a formal power series:

$$f \mapsto F(t) = \sum_{\lambda\in\mathrm{supp} f} f(\lambda)t^\lambda.$$

$F(t)$ is like a Laurent series that is finite in the positive direction (an element of $\mathbb{R}[[t^{-1},t])$, but the exponents are allowed to be any real numbers. Choose a base point x_0. Then the $\pi_1(X,x_0)$-representation

$$\gamma \mapsto t^{\int_\gamma \theta} \in \mathrm{Aut}(\mathcal{R},\mathcal{R})$$

determines the local coefficient system. To specify the flat bundle, consider the trivial bundle $X\times\mathcal{R}$. For each $\gamma\in\pi_1(a,b)$, we define the parallel transportation to be

$$\Gamma_\gamma = t^{\int_\gamma \theta}.$$

and form the differential of the Novikov-Morse chain complex as

$$\partial = \sum_{\substack{a,b \in \mathrm{Crit}(f) \\ \mu(a)-\mu(b)=1}} \sum_{\gamma \in M(a,b)} t^{\int_\gamma \theta} \otimes \epsilon_\gamma .$$

The infinite sum converges in $\mathrm{End}(\mathcal{R}, \mathcal{R})$ by equation (1.8.2).

2. Principal Bundles, connections, Chern-Weil theory and the Chern-Simons invariant

From Section 2.1 to 2.9, we shall go over the basics on connections and curvatures setting up the framework for instanton Floer homology – the Morse homology defined for the Chern-Simons functional. There are many excellent introductions to this material, for example, Atiyah and Bott's paper [3] or the thorough treatment in Kobayashi and Nomizu's book [29].

2.1. Principal bundles Let G be a Lie group. A principal G-bundle over a smooth manifold X is a smooth fiber bundle $\pi : P \to X$ together with a free G-action acting on the right:

$$P \times G \to P$$
$$(p, g) \mapsto pg,$$

such that each fiber $\pi^{-1}(x)$ is a G-orbit. Two principal G-bundles P, P' over X are said to be isomorphic if there is a bundle map η

$$\begin{array}{ccc} P & \xrightarrow{\eta} & P' \\ & \searrow \quad \swarrow & \\ & X & \end{array}$$

that covers the identity map on X and commutes with the right G-action.

Besides this highbrow point of view, a lowbrow point of view characterizes a principal bundle in terms of transition functions.

A principal G-bundle is determined by an open over $\{U_\alpha\}$ of X with transition functions defined on each intersection:

$$g_{\alpha\beta} : U_\alpha \cap U_\beta \to G.$$

such that the co-cycle condition is satisfied:

$$g_{\alpha\beta} g_{\beta\gamma} g_{\gamma\alpha} = e \in G \text{ on } U_\alpha \cap U_\beta \cap U_\gamma.$$

for any α, β and γ. Two bundles $\{g_{\alpha\beta}\}$ and $\{g'_{\alpha\beta}\}$ are equivalent if one can find intertwining maps:

$$\eta_\alpha : U_\alpha \to G$$

such that

$$g_{\alpha\beta}\eta_\beta = \eta_\alpha g'_{\alpha\beta} \text{ on } U_\alpha \cap U_\beta.$$

It is also necessary to define equivalence with respect to refinements of open covers $\{U_\alpha\}$, which is omitted here.

2.2. The big and little adjoint bundles Two associated bundles arise constantly in the discussion of principal bundles. The "little" adjoint bundle

$$\operatorname{ad} P = P \times_{\operatorname{ad}} \mathfrak{g}.$$

where $\operatorname{ad} : G \to \operatorname{Aut}(\mathfrak{g})$ is the adjoint action of the Lie group on it Lie algebra, i.e. for a matrix group $\operatorname{ad}_g(\xi) = g\xi g^{-1}$. Thus sections of $s \colon X \to \operatorname{ad} P$ can viewed as equivariant maps

$$\hat{s} \colon P \to \mathfrak{g}.$$

We may also think of sections of $\operatorname{ad} P$ as infinitesimal automorphisms of P. The pull back $\pi^*(\operatorname{ad} P)$ to P can be identified with the vertical tangent bundle of P:

$$\pi^*(\operatorname{ad} P) \simeq \ker(d\pi) = VTP.$$

Then we can view $\operatorname{ad} P$ as the quotient

$$\operatorname{ad} P \simeq VTP/G.$$

The "big" adjoint bundle is the associated bundle

$$\operatorname{Ad} P = P \times_{\operatorname{Ad}} G.$$

where $\operatorname{Ad} : G \to \operatorname{Aut}(G)$ is the adjoint action of G on itself. Note that $\operatorname{Ad} P$ is a bundle of groups. Sections of $\operatorname{Ad} P$ are automorphisms of P that commute with the G-action. Indeed, given such an automorphism η,

$$\begin{array}{ccc} P & \xrightarrow{\eta} & P \\ & \searrow \quad \swarrow & \\ & X & \end{array}$$

and some $p \in P$, we can find $\tilde{\eta}(p) \in G$ such that

$$\eta(p) = p \cdot \tilde{\eta}(p).$$

Since η preserves the right-action of G, $\eta(p \cdot g) = \eta(p) \cdot g$ and hence,

$$(p \cdot g) \cdot \tilde{\eta}(p \cdot g) = \eta(p \cdot g) = \eta(p) \cdot g = (p \cdot \tilde{\eta}(p)) \cdot g.$$

This shows that $\tilde{\eta}(p \cdot g) = g^{-1}\tilde{\eta}(p)g$ and the assignment

$$p \mapsto (p, \tilde{\eta}(p)) \in P \times G$$

descends to a section $X \to \operatorname{Ad} P$.

Definition 2.2.1. The space of sections is often called the *gauge group* and we will denote it by $\mathcal{G}_P$ or $\mathcal{G}$ when the bundle is obvious from context. ◇

2.3. Connections We give five definitions of connections on a principal bundle P:

(1) A connection A is a G-invariant sub-bundle of TP, denoted by $H_A \subset TP$, such that under the projection map $\pi : P \to X$,

$$d\pi|_{H_{A,p}} : H_{A,p} \xrightarrow{\cong} T_{\pi(p)}X.$$

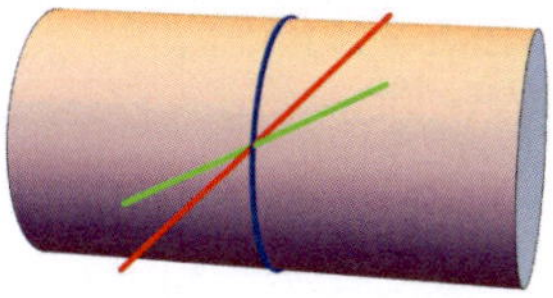

Figure 2.3.1. Two horizontal spaces (red and green). A blue fiber is indicated.

(2) The projection map π induces a short exact sequence of bundles:

$$\begin{array}{ccccccccc} 0 & \longrightarrow & \ker(d\pi) & \underset{\curvearrowleft A}{\longrightarrow} & TP & \xrightarrow{d\pi} & \pi^*TX & \longrightarrow & 0 \\ & & & \nwarrow^{0} & \uparrow & & & & \\ & & & & H_A & & & & \end{array}$$

A connection $A : TP \to \ker(d\pi)$ is a left inverse to $\ker(d\pi)$ such that this sequence splits G-equivariantly. The horizontal space H_A is $\ker(A)$. For any fixed $p \in P$, the inclusion map

$$\begin{aligned} \iota_p : G &\to P \\ g &\mapsto p \cdot g \end{aligned}$$

identifies $\mathfrak{g} = T_e G$ with $\ker(d\pi_p) \subset T_p P$ by taking the tangent map.

(3) Covariant derivative:

$$\nabla_A : \mathcal{C}^\infty(X, \operatorname{ad} P) \to \mathcal{C}^\infty(X, T^*X \otimes \operatorname{ad} P).$$

(4) Holonomy (Parallel Transportation). Given a C^1-path $\gamma : [a, b] \to X$ and any $p \in P$ with $\pi(p) = \gamma(a)$, we can find a unique path lifting γ upstairs:

$$\tilde{\gamma}_{A,p} : [a, b] \to P$$

such that $\pi \circ \tilde{\gamma} = \gamma$ and $\tilde{\gamma}_{A,p}(a) = p$. We also require this lift to be equivariant: if $p' = p \cdot g$ for some $g \in G$, then

$$\tilde{\gamma}_{A,p'}(t) = \tilde{\gamma}_{A,p}(t) \cdot g.$$

for any $t \in [a, b]$.

(5) This is the lowbrow point of view. A connection is equivalent to a collection of 1-forms:

$$a_\alpha \in \Gamma(T^*U_\alpha \otimes \mathfrak{g})$$

such that the overlapping condition is satisfied

$$a_\beta = g_{\alpha\beta}^{-1} a_\alpha g_{\alpha\beta} + g_{\alpha\beta}^{-1} dg_{\alpha\beta}$$

on $U_\alpha \cap U_\beta$.

The bundle $\bigwedge^*(T^*X) \otimes \operatorname{ad} P$ inherits a *super Lie algebra* structure. For any two sections ω, η, we combine the Lie bracket on $\operatorname{ad} P$ with the wedge product of forms to obtain a "bracket wedge" operation

$$[\omega \wedge \eta],$$

which satisfies

(1) Super anti-commutativity:

$$[\omega^p \wedge \eta^q] = (-1)^{pq+1}[\eta^q \wedge \omega^p];$$

(2) Super Jacobi identity:

$$(-1)^{pr}[\omega^p \wedge [\omega^q \wedge \omega^r]] + (-1)^{qp}[\omega^q \wedge [\omega^r \wedge \omega^p]] + (-1)^{rq}[\omega^r \wedge [\omega^p \wedge \omega^q]] = 0.$$

In terms of covariant derivatives, it is easy to see that two connections A, A' of P differ by a Lie algebra valued 1-form:

$$A' - A = a \in \Gamma(X, T^*X \otimes \operatorname{ad} P) = \Omega^1(X, \operatorname{ad} P).$$

The second definition of connections also allows us to think of A as an 1-form on P taking value in $\mathfrak{g}$ (since $\ker(d\pi) = P \times \mathfrak{g}$ is trivial). Recall that a form ω (possibly Lie algebra valued) on P is called **basic** if for any vertical vector field v:

- (Contraction) $\iota_v(\omega) = 0$;
- (Lie derivative) $\mathcal{L}_v\omega = 0$;

and a basic form ω is necessarily (at least locally) pulled back from $\pi : P \to X$. One can verify that $A - A'$ satisfies (Contraction) and a variant of (Lie derivative) namely

$$\mathcal{L}_v\omega = -\operatorname{ad}(v)\omega$$

for any vertical v, so it is pulled back from $\Omega^1(X, \operatorname{ad} P)$.

If we define $\mathcal{A}_P$ to be the space of connections in P (or $\mathcal{A}$ when the bundle is obvious from context), then $\mathcal{A}_P$ is an affine space over $\Omega^1(X, \operatorname{ad} P)$ — fixing a base connection $A_0 \in \mathcal{A}_P$, we have

(2.3.2) $$\mathcal{A}_P = A_0 + \Omega^1(X, \operatorname{ad} P).$$

2.4. The action of gauge transformations on connections Gauge transformations act on connections by pullback. If we view a connection A as a sub-bundle of TP or as vertical valued 1-form on P and a gauge transformation g as a diffeomorphism $g : P \to P$, then one can check that

$$g^*A$$

is another connection. To make gauge transformations act via a *left*-action we will use the convention that

(2.4.1) $$g \cdot A = g^{-1*}(A).$$

Since connections form an affine space, we should look for a formula for the difference $A - g^*(A)$. Each definition of connections leads to a different sort of expression for this difference. For our later purposes, a very concrete one is preferred.

Given a faithful linear representation $\rho : G \to \mathrm{End}(V)$, we view our connections as covariant derivatives in the induced bundle $E = P \times_\rho V$. When $G = SO(n), U(n)$ or $Sp(n)$, one may take ρ to be the defining representation.

Then both Ad P and ad P are sub-bundles of

$$\mathrm{End}(E) = E^\vee \otimes E.$$

The latter has a connection induced from the one on E.

Proposition 2.4.2. *The pull-back connection* g^*A *is given by*

$$g^*A = A + g d_A g^{-1} = A - (d_A g) g^{-1} \tag{2.4.3}$$

where $d_A = \nabla_A$ *is the covariant derivative induced on* $\mathrm{End}(E)$.

Proof. The induced connection ∇_A on End(V) is defined by the property:

$$(\nabla_A \phi)v + \phi(\nabla_A v) = \nabla_A(\phi(v)).$$

The pull-back connection acts on sections of E by $g(\nabla_A v) = g^*(\nabla_A)gv$. As a result, $g^*(\nabla_A v) = g \circ \nabla_A \circ g^{-1}$ and

$$\begin{aligned} g \circ \nabla_A(g^{-1}v) &= g \circ (\nabla_A g^{-1})v + g \circ g^{-1}\nabla_A v \\ &= (\nabla_A + g\nabla_A g^{-1})v. \end{aligned}$$

The last expression follows from the identity

$$\nabla_A g^{-1} = -g^{-1}(\nabla_A g)g^{-1}$$

□

We have only used the representation ρ to make this formula more concrete; specifically, the expression $d_A g$ has an interpretation in this formula because g is viewed as a section of $\mathrm{End}(E)$, so $d_A g$ is a one form with values in $\mathrm{End}(E)$. In the general situation, only the expression $(d_A g)g^{-1}$ has a meaning.

2.5. The Curvature The curvature of A is the obstruction of H_A being integrable in the sense of Frobenius (tangent spaces of foliations) and it has several equivalent definitions.

(1) For $X, Y \in T_p P$, let $\tilde{X}, \tilde{Y}$ be local extensions of vector fields and denote by $h_A : TP \to H_A$ the horizontal projection, so that $A + h_A = \mathrm{Id}$. Define

$$F_A(X, Y) = -A([h_A\tilde{X}, h_A\tilde{Y}]).$$

One verifies that F_A is basic 2-form, so $F_A \in \Gamma(X, \bigwedge^2 T^*X \otimes \mathrm{ad}\, P)$. We may think of F_A as a (basic) 2-form either upstairs or downstairs.

(2) Since A is a 1-form on P, we can take its exterior differential, so

$$F_A = dA + \frac{1}{2}[A \wedge A].$$

(3) On the adjoint bundle ad P, A determines an exterior differential

$$d_A : \Omega^*(X, \mathrm{ad}\, P) \to \Omega^{*+1}(X, \mathrm{ad}\, P).$$

In general, $d_A^2 \neq 0$, but it is tensorial and

$$d_A^2 s = [F_A \wedge s]$$

for all $s \in \Omega^*(X, \mathrm{ad}\, P)$. When $* = 0$, this relation determines F_A.

(4) The limit of holonomy around a tiny square.

(5) (The lowbrow point of view) Using local trivializations of P,

$$(F_A)_\alpha = da_\alpha + \frac{1}{2}[a_\alpha \wedge a_\alpha]$$

The connection A is called **flat** if $F_A \equiv 0$, in which case $d_A^2 = 0$ (by the third definition) and

$$\Omega^0(X, \mathrm{ad}\, P) \xrightarrow{d_A} \Omega^1(X, \mathrm{ad}\, P) \xrightarrow{d_A} \Omega^2(X, \mathrm{ad}\, P) \to \cdots$$

is a complex. Here are a few equivalent definitions of flatness:

(1) A bundle *with a flat connection* can be trivialized so that the transition functions $g_{\alpha\beta}$ are locally constant.
(2) The end point of path lifting depends on the homotopy class of the path in the base manifold X only.

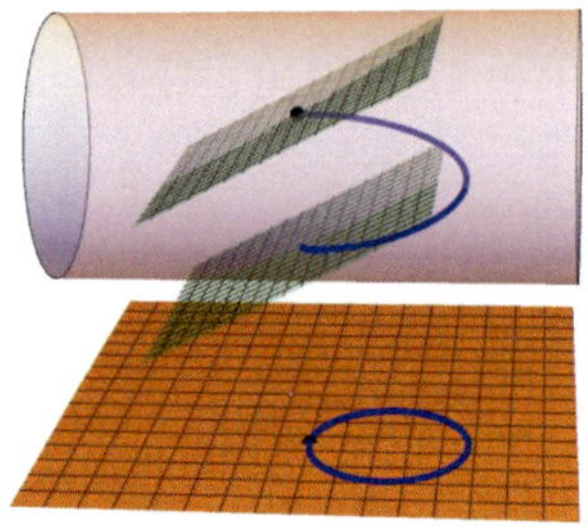

Figure 2.5.1. A loop in the base and its lift to the total space of the principal bundle.

2.6. The holonomy representation of flat connections The last property allows us to define the holonomy map. For a base point $x \in X$, choose $p \in \pi^{-1}(x)$. Define

$$\mathrm{hol}_{A,p} : \pi_1(X, x) \to G$$
$$\gamma \mapsto g_\gamma$$

if the end point of the lift $\tilde{\gamma}_{A,p}$ is $p \cdot g_\gamma$. Changing p to $p \cdot g$ will change the image of γ into

$$\mathrm{hol}_{A,pg}(\gamma) = g^{-1}\, \mathrm{hol}_{A,p}(\gamma) g.$$

So a flat connection determines a representation of $\pi_1(X)$ up to conjugacy.

On the other hand, the universal cover $\tilde{X}$ of X can be viewed as a principal $\pi_1(X)$-bundle. The right action of $\pi_1(X)$ is obtained from Deck transformations (which is a left action) by the usual trick. Thus, a representation

$$\rho : \pi_1(X, x) \to G,$$

induces a principal G-bundle

$$\begin{array}{ccc} G & \longrightarrow & \tilde{X} \times_\rho G \\ & & \downarrow \\ & & X. \end{array}$$

We declare $(\tilde{\gamma}, e) \subset \tilde{X} \times_\rho G$ to be the horizontal lift of $\gamma = \pi(\tilde{\gamma})$. This makes it a flat bundle. Different representations can lead to non-isomorphic bundles. For example, the two representations $\pi_1(S^1) \to \mathbb{Z}_2$ give the trivial and the non-trivial $\mathbb{Z}_2$-bundles over S^1. These two constructions are inverse to each other, but we need to be careful about the statement. The correct one is:

$$\{\text{Flat connections up to isomorphism in a principal bundle } P\} =$$
$$\{\pi_1\text{-representations which induce the bundle } P \text{ up to conjugacy}\}$$

Let us set

$$\mathcal{R}_P(Y) = \mathrm{Hom}_P(\pi_1(Y), G)/\text{conjugation}, \tag{2.6.1}$$

and also for later

$$\mathcal{R}_P(Y)^o = \mathrm{Hom}_P(\pi_1(Y), G). \tag{2.6.2}$$

The subscript P indicates that the principal G-bundles induced by this representation is isomorphic to P and we omit it when P is clear from the context.

This last space of representations can be identified with flat connection modulo the based gauge group:

$$\mathcal{G}_P^o = \{g \in \mathcal{G}_P |\ g|_{x_o} = \mathrm{Id}\}.$$

2.7. Deformations of connections Now we study the variation of curvatures under deformations of connections.

Lemma 2.7.1. *If* $A' = A + a$ *for* $a \in \Omega^1(X, \mathrm{ad}\, P)$, *then*

$$F_{A'} = F_A + d_A a + \frac{1}{2}[a \wedge a].$$

Proof. For a section $s \in \Omega^0(X, \mathrm{ad}\, P)$,

$$d_{A+a} s = d_A s + [a \wedge s].$$

Apply it twice:

$$\begin{aligned} d^2_{A+a} s &= (d_A + a) \circ (d_A + a) s \\ &= d^2_A s + [a \wedge d_A s] + d_A[a \wedge s] + [a \wedge [a \wedge s]] \\ &= d^2_A s + [d_A a \wedge s] + \frac{1}{2}[[a \wedge a] \wedge s] \\ &= [(F_A + d_A a + \frac{1}{2}[a \wedge a]) \wedge s]. \end{aligned}$$

In the middle, we used the super Jacobi identity to write:

$$[a \wedge [a \wedge s]] = \frac{1}{2}[[a \wedge a] \wedge s].$$ □

2.8. Classifications of $U(n)$ and $SU(n)$-bundles on 3-manifolds Since the first cell of $BSU(n)$ is in dimension 4:

$$BSU(n) = e^4 \cup \text{higher dimensional cells},$$

each map $Y^3 \to BSU(n)$ is null-homotopic by cellular approximation. This implies all $SU(n)$-bundles are trivial over Y^3.

Here is a lowbrow proof. Suppose $E \to Y^3$ is the induced n-plane bundle. Since the zero section

$$s_0 : Y \to E$$

has co-dimension $2n > 3$, we may pick a unit-length section s so that $s \cap s_0 = \emptyset$. This section reduces the structure group to $SU(n-1) \oplus \{e\}$. Since $SU(1)$ is the trivial group, all $SU(2)$-bundles are trivial over Y^3. Similarly, we can classify $U(n)$-bundles and $PU(n) = U(n)/Z(U(n))$-bundles over Y^3:

(1) $U(n)$-bundles. The lowbrow argument above allows us to write:

$$E \cong \mathbb{C}^{n-1} \oplus \Lambda^n E.$$

Since line bundles are classified by their first Chern classes c_1, P is classified by $c_1(\Lambda^n E) = c_1(E)$. The highbrow point of view is that

$$\det : U(n) \to S^1$$

induces a map.

$$\Lambda^n : [Y, BU(n)] \to [Y, BU(1)].$$

The embedding $\iota : S^1 \to U(n)$ via $e^{i\theta} \mapsto diag\{e^{i\theta}, 1, \cdots\}$ gives a map $BU(1) \to BU(n)$ that induces a right inverse to Λ^n, so Λ^n is surjective. Since each map $f : Y \to BU(n)$ is homotopic to a map with image in the 2-skeleton $BU(n)^{(2)} \subset BU(1)$, Λ^n is also injective.

(2) $PU(n)$-bundles. A principal bundle is also classified by the first Čech cohomology group

$$\check{H}^1(Y, PU(n)).$$

which fits into the long exact sequence:

$$\to \check{H}^1(Y, SU(n)) \to \check{H}^1(Y, PU(n)) \xrightarrow{\iota} \check{H}^2(Y, \mathbb{Z}_n)$$

induced by the short exact sequence

$$0 \to \mathbb{Z}_n \to SU(n) \to PU(n) \to 0.$$

Since $\check{H}^1(Y, SU(n)) = 0$, ι is injective. To show it is surjective, we need to solve the lifting problem:

$$\begin{array}{ccc} BSU(n) & \longrightarrow & BPU(n) \\ & \nearrow & \downarrow \\ Y^3 & \longrightarrow & K(\mathbb{Z}_n, 2) \end{array}$$

Since $\pi_i(BSU(n)) = 0$ for $0 \leqslant i \leqslant 3$, the first obstruction class lies in $H^5(Y^3, \pi_4(BSU(n)))$ which is zero for dimension reasons.

Note that a $PU(n)$-bundle P does not necessarily come from a $U(n)$-bundle. This is determined by whether $\iota([P]) \in H^2(Y, \mathbb{Z}_n)$ has an integral lift in $H^2(Y, \mathbb{Z})$.

2.9. Characteristic classes and the Chern-Simons invariant Let $\wp : \mathfrak{g} \to \mathbb{R}$ be an Ad-invariant polynomial function. Such a function extends to a map

$$\Omega^*(X; \operatorname{ad} P) \to \Omega^*(X).$$

Ad-invariance means that we can compute the results in any local trivializations of P. To get the value of $\wp(\omega^p)$ on some p-vectors, we simply evaluate ω^p in some trivialization to get an element of the Lie algebra and then apply the function $\wp$. For example, we can do this for the curvature form of a bundle, then

$$\wp(F_A) \in \Omega^*(X)$$

is a closed real-valued form on X. Its cohomology class

$$c_\wp(P) = [\wp(F_A)] \in H^*(X, \mathbb{R})$$

turns out to be independent of the connection A and gives rise to a characteristic class of P in de Rham cohomology. This assignment

$$\wp \mapsto [\wp(F_A)] \in H^*(X, \mathbb{R})$$

is the Chern-Weil homomorphism (see for example [12]). In particular, if A_0 and A_1 are two connections, then

$$\wp(F_{A_1}) - \wp(F_{A_0}) = dG(A_1, A_0)$$

for some form $G(A_1, A_0)$. Chern and Simons figure out a *canonical* expression for this $G(A_1, A_0)$ (see [13]), as we now explain. Suppose $\wp$ is homogeneous of degree k meaning

$$\wp(\xi) = \tilde{\wp}(\xi, \xi, \cdots, \xi)$$

for a symmetric Ad-invariant multi-linear function $\tilde{\wp} : \bigotimes_k \mathfrak{g} \to \mathbb{R}$. We can write $A_1 = A_0 + a$ and set

$$A_t = A_0 + ta.$$

Then we can compute the derivative $\frac{\partial}{\partial t}\wp(F_{A_t})$ as follows.

$$\begin{aligned}\frac{\partial}{\partial t}\tilde{\wp}(F_{A_t}, \cdots, F_{A_t}) &= k \cdot \tilde{\wp}(\frac{\partial}{\partial t}F_{A_t}, F_{A_t}, \cdots, F_{A_t}) \text{ (because } \tilde{c} \text{ is symmetric)}\\ &= k \cdot \tilde{\wp}(d_{A_t}a, F_{A_t}, \cdots, F_{A_t})\\ &= k \cdot d(\tilde{\wp}(a, F_{A_t}, \cdots, F_{A_t})).\end{aligned}$$

using the Bianchi identity $d_{A_t}F_{A_t} = 0$. Finally, we define the *Chern-Simons form*

(2.9.1) $$\mathfrak{cs}_\wp(A_1, A_0) = k \int_0^1 (\tilde{\wp}(a, F_{A_t}, \cdots, F_{A_t}))dt.$$

Reading the computation backwards, we see that

(2.9.2) $$d\mathfrak{cs}_\wp(A_1, A_0) = \int_0^1 \frac{\partial}{\partial t}\tilde{\wp}(F_{A_t}, \cdots, F_{A_t})dt = \wp(F_{A_1}) - \wp(F_{A_0})$$

An alternative view on this computation is as follows. The path of connection A_t can be viewed as defining a connection $\tilde{A}$ on $[0,1] \times M$ whose curvature

$$F_{\tilde{A}} = dt \wedge a + F_{A_t}$$

is a Lie algebra valued 2-form on $[0,1] \times M$. Then

$$\mathfrak{cs}_{\wp}(A_1, A_0) = \int_0^1 \wp(F_{\tilde{A}}) \tag{2.9.3}$$

where we integrate out the dt part. If $\wp$ is homogeneous of degree 2, this integral gives

$$\begin{aligned} \mathfrak{cs}_{\wp}(A_1, A_0) &= 2\int_0^1 (\tilde{\wp}(a, F_{A_0} + t d_{A_0} a + \frac{t^2}{2}[a \wedge a])dt \\ &= \tilde{\wp}(a, 2F_{A_0} + d_{A_0} a + \frac{t^2}{3}[a \wedge a]). \end{aligned} \tag{2.9.4}$$

3. Floer Homology overview

Morse theory was motivated by Morse's desire to prove existence of closed geodesics on Riemannian manifolds (see [38]). Thus he wasn't originally interested in functions on finite dimensional manifolds themselves. Rather he was interested in understanding critical points of the energy functional on the loop space of the manifold and how they were influenced by the topology of the underlying manifold. This energy functional shares some common features with the one in finite dimensional case:

- The negative gradient flow is well defined (it is a non-linear heat equations; the positive flow is not well-defined however).
- The number of negative eigenvalues of the Hessian is finite and critical points (if non-degenerate) have finite dimensional unstable manifolds

It was Floer [19] who first realized that Morse Homology would work for some subtler infinite dimensional settings, including the symplectic action functional and our focus, the Chern-Simons functional.

3.1. The Chern-Simons Functional Fix a principal G-bundle P over a 3-manifold Y^3, a symmetric G-invariant homogeneous polynomial $\wp$ of degree 2 and a reference connection A_0. Write $A = A_0 + a$ for $a \in \Omega^1(Y, \operatorname{ad} P)$. The Chern-Simons functional is then the integral of the Chern-Simons form in equation (2.9.1):

$$\begin{aligned} \mathcal{CS}(A) \doteq \mathcal{CS}(A, A_0) &= \int_Y \mathfrak{cs}(A, A_0) \\ &= \int_Y \wp(2F_{A_0} + d_{A_0} a + \frac{1}{3}[a \wedge a], a). \end{aligned} \tag{3.1.1}$$

In dimension 3, we can give a slightly different definition of the Chern-Simons function, since every oriented 3-manifold Y^3 bounds an oriented 4-manifold. Fix such an X^4, a principle G-bundle $\tilde{P} \to X^4$ extending P and a connection $\tilde{A}$ in $\tilde{P}$ extending A. Then we compute the Chern-Weil integral

$$\mathcal{CS}(A) \doteq \mathcal{CS}(A; \tilde{P}) = \int_X \wp(F_{\tilde{A}}).$$

In this second formulation, the choice of reference connection is replaced by the choice of $\tilde{P}$ (and hence $\tilde{X}$). In either situation, we are faced with understanding how the function depends on the choice up to isomorphism, i.e, what would happen if we choose a gauge equivalent connection A_0' or a different bundle $\tilde{P}'$ extending P. In either case, the value of $\mathcal{CS}$ is affected by the corresponding 4-dimensional characteristic class $c_\wp$ for some principal G-bundle Q over a closed oriented 4-manifold Z that contains Y:

$$\int_Z \wp(F_{\tilde{A}}) = \langle c_\wp(Q), [Z]\rangle \in \mathbb{R}.$$

This set of values is important for the development of our story: they are integral multiples of a positive real number $\mathfrak{m}_\wp$ depending on $\wp$.

$$\langle c_\wp(Q), [Z]\rangle = \mathfrak{m}_\wp k \text{ for some } k \in \mathbb{Z}. \tag{3.1.2}$$

3.2. Formal Gradients Given a Riemannian metric on Y, we can define an L^2-Riemannian metric on $\mathcal{A}_P$. Given tangent vectors $a, b \in L^2(Y, \Lambda^1 \otimes \operatorname{ad} P)$, define

$$\langle a, b\rangle = 2\int_Y \tilde{\wp}(a, \star b)$$

Here we assume that our choice of $\wp$ is positive definite; the additional factor 2 is to remove some annoying factors that would have to be carried along later. Given some $A \in \mathcal{A}_P$ and a deformation $s \mapsto A + sa$, consider a family of connections interpolating between A and $A + sa$:

$$A_{t,s} = A + tsa.$$

Viewed as connections on $[0,1]_t \times Y$, their curvatures are given by

$$F_{\tilde{A}_{t,s}} = F_A + tsd_A a + \frac{t^2s^2}{2}[a \wedge a] + s dt \wedge a,$$

so

$$\frac{\partial}{\partial s} F_{\tilde{A}_{t,s}}|_{s=0} = F_A + tsd_A a + \frac{t^2s^2}{2}[a \wedge a] + dt \wedge a\Big|_{s=0} = td_A a + dt \wedge a.$$

Now we compute the derivative of the Chern-Simons functional $\mathcal{CS}_\wp$ along the path $s \mapsto A + sa$:

$$\begin{aligned}
\frac{\partial}{\partial s}\mathcal{CS}(A + sb)\Big|_{s=0} &= \frac{\partial}{\partial s}\int_{[0,1]\times Y} \wp(F_{\tilde{A}_{t,s}})\Big|_{s=0} \\
&= 2\int_{[0,1]\times Y} \tilde{\wp}(td_A a + dt \wedge a, F_A) \\
&= 2\int_Y \tilde{\wp}(a, F_A) \\
&= 2\int_Y \tilde{\wp}(a, \star\star F_A) \\
&= \langle a, \star F_A\rangle.
\end{aligned} \tag{3.2.1}$$

Thus we have proved

Proposition 3.2.2. *The L^2-gradient of Chern-Simons is*

$$\nabla_A \mathcal{CS} = \star F_A.$$

In particular, the critical points of $\mathcal{CS}$ consist of flat connections.

Note that both the simplicity of the formula for the gradient and the conclusion about the critical points is special to dimension 3. In higher dimensions, a wedge product of F_A will appear in the first variation.

3.3. Hessians The Hessian of the Chern-Simons functional at a connection A is

$$\mathrm{Hess}_A\, \mathcal{CS}(a) = \star d_A\, a.$$

There are two notable properties of this operator.

(1) If A is flat (i.e. a critical point of $\mathcal{CS}$), then the kernel of $\mathrm{Hess}_A\, \mathcal{CS}$ contains the image of d_A by $d_A^2 = 0$.
(2) Its spectrum is unbounded in both the positive and negative directions (see Exercise 3.5.4).

The first property is a consequence of gauge invariance: the Chern-Simons functional is locally constant on orbits of the (infinite dimensional) gauge group $\mathcal{G}_P$. Nonetheless, it is a self-adjoint operator and once this gauge invariance is taken into account, $\mathrm{Hess}_A\, \mathcal{CS}$ is enhanced into an elliptic operator and hence Fredholm.

The second property hints at subtler issues making the theory far more delicate than the finite dimensional case or even the case of the energy function on the loop space of a manifold (see Milnor's Morse Theory book [37]). Property (2) indicates that we do not have a proper notion of Morse index in this case. Instead, the notion of **spectral flow** developed by Atiyah, Patodi and Singer (see [5–7]) will make sense of relative indices, which is all we need for a Floer theory.

Example 3.3.1. Consider the operator $i\frac{\partial}{\partial t} + \alpha$ acting on $L_1^2(S^1, \mathbb{C})$. Treat S^1 as $\mathbb{R}/2\pi\mathbb{Z}$. Then for any $\alpha \in [0,1]$, we have

$$\mathrm{Spec}(i\frac{\partial}{\partial t} + \alpha) = \alpha + \mathbb{Z}.$$

Indeed, eigenvectors are given by e^{-int} and

$$(i\frac{\partial}{\partial t} + \alpha)e^{-int} = (\alpha + n)e^{-int}.$$

When $\alpha = 1$, the operator is unitarily equivalent to $i\frac{\partial}{\partial t}$:

$$(i\frac{\partial}{\partial t} + 1) = e^{it}(i\frac{\partial}{\partial t})e^{-it}.$$

This is a new phenomenon for infinite dimensional spaces. ◇

More generally, the space $\mathrm{SA}_{\mathrm{Fred}}$ of bounded self-adjoint Fredholm operators on an infinite dimensional separable Hilbert space H has a rich homotopy type See Atiyah and Singer's paper [8] for more detail. The identification of the homotopy type of $\mathrm{SA}_{\mathrm{Fred}}$ hinges on Kuiper's surprising result that the unitary group of the Hilbert space H is contractible (see Kuiper's original paper [33] and also Mityagin's paper [15] for a cleaner proof). First of all, it has three connected components:

$$\mathrm{SA}_{\mathrm{Fred}} = \mathrm{SA}_+ \bigcup \mathrm{SA}_0 \bigcup \mathrm{SA}_- .$$

where $SA_{+(-)}$ denotes the space of operators with essentially positive (negative) spectrum. If an operator A is Fredholm, then it only has discrete spectrum near the origin and it makes sense to talk about the negative (positive) subspace H_- (H_+) of A by functional calculus of self-adjoint operators. Then A is called essentially positive (negative) if

$$\dim H_- \ (\dim H_+) < \infty.$$

The components SA_+ and SA_- are contractible, whereas the middle part SA_0 carries non-trivial topology. In fact,

$$\begin{aligned}\Omega\, SA_0 &\sim \mathbb{Z} \times BU \text{ if } H \text{ is over } \mathbb{C}\\ &\sim \mathbb{Z} \times BO \text{ if } H \text{ is over } \mathbb{R}.\end{aligned}$$

and the $\mathbb{Z}$-component represents the spectral flow. Here BU is the classifying space of the complex vector bundles and BO is the classifying space of the real ones. In fact, Example 3.3.1 represents a non-trivial loop in SA_0 (though they are unbounded operators on L^2):

$$S^1 \to (L^2_1(S^1, \mathbb{C}) \to L^2(S^1, \mathbb{C})).$$

Morse index does not make sense anymore. Nevertheless, it is the difference $\mu(a) - \mu(b)$ of Morse indices that matters, which can be described instead as the spectral flow.

Suppose connections $A_0, A_1 \in \mathcal{P}$ fit into a path $A_t \in \mathcal{A}$ for $t \in [0, 1]$ and assume

$$0 \notin \mathrm{Spec}(\mathrm{Hess}_{A_0} \mathcal{CS}) \cup \mathrm{Spec}(\mathrm{Hess}_{A_1} \mathcal{CS})$$

(pretend that they are actual Fredholm operators by resolving Property (1)). Then the spectral flow of this family $A_t, t \in [0, 1]$ simply counts the signed number of eigenvalues that come across 0 as t varies and we define

$$\mu(A_0) - \mu(A_1) = \mathrm{SF}(\mathrm{Hess}_{A_t} \mathcal{CS}),$$

(even though the value of $\mu(A_0)$ is not well-defined). This properly generalizes the Morse index from the finite dimensional case.

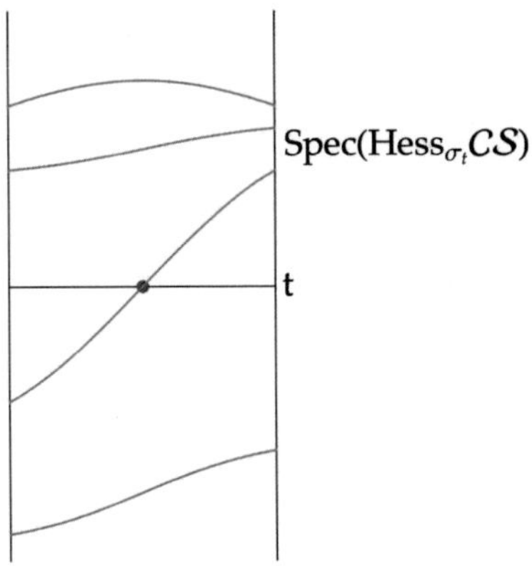

Figure 3.3.2. The spectral flow of the Hessian

3.4. Novikov Rings? In finite dimensional Morse theory, except for the case of exact 1-forms, we will require the use of a Novikov ring. This might be avoided sometimes in the infinite dimensional case and let us explain the mechanism for this here. When we try to push $\mathcal{CS}$ down to the quotient of $\mathcal{A}_P$ by gauge transformations, we get a circle valued function:

$$\begin{array}{ccc} \mathcal{A}_P & \xrightarrow{\mathcal{CS}} & \mathbb{R} \\ \downarrow & & \downarrow \\ \mathcal{A}_P/\mathcal{G}_P & \longrightarrow & \mathbb{R}/\mathfrak{m}_\wp\mathbb{Z}. \end{array}$$

This provides a homomorphism from the fundamental group

$$\Delta\mathcal{CS} : \pi_1(\mathcal{A}/\mathcal{G}) \to \mathfrak{m}_\wp\mathbb{Z} \subset \mathbb{R}. \tag{3.4.1}$$

In the meantime, the spectral flow also defines map on $\pi_1(\mathcal{A}/\mathcal{G})$:

$$\mathrm{SF}_{\mathcal{CS}} : \Omega(\mathcal{A}/\mathcal{G}) \to \pi_1(\mathcal{A}/\mathcal{G}) \to \mathbb{Z}.$$

If they are proportional, meaning, for some $\lambda \in \mathbb{R}$, $\lambda\,\mathrm{SF}_{\mathcal{CS}} = \Delta\mathcal{CS}$, then the ordinary Morse homology would just work — if we pin down the spectral flow, then we have uniform control for the drop of the functional and hence the energy of the flowline. However, if the theory is not monotone, then the Novikov-Morse homology is needed. This phenomenon is relevant with the notion of monotonicity in the context of Lagrangian Floer homology.

3.5. Exercises for the preamble

Exercise 3.5.1. Show that we can view the Grassmanian $G(k, n)$ as a subset of $\mathrm{Sym}(\mathbb{R}^n)$ as follows:

$$G(k, n) = \{A \in \mathrm{Sym}(\mathbb{R}^n) | A^2 = I_n \text{ and } \mathrm{tr}(A) = 2k - n\}.$$

Define a metric on $\mathrm{Sym}(\mathbb{R}^n)$ by

$$d(A, B) = \mathrm{tr}((A - B)^2)^{1/2}$$

Pick a random symmetric matrix $B \in \mathrm{Sym}(\mathbb{R}^n)$ and define a function on $G(k, n)$ by the formula

$$f_B(A) = d(A, B)^2.$$

- Find the critical points of f on $G(k, n)$
- Can you tell which element of $G(k, n)$ a given matrix in $\mathrm{Sym}(\mathbb{R}^n)$ is closest to?
- Try to describe the gradient flow for some choices of B.
- Generalize this to the complex Grassmanian.
- Find the homology of the complex Grassmaian.
- Can you compute the $\mathbb{Z}_2$ homology of the real Grassmanian? ♢

Exercise 3.5.2. A deep result of Łojasiewicz (see for example [10]) says that if $f : \mathbb{R}^n \to \mathbb{R}$ is a real analytic function with $f(0) = 0$ and $\nabla_0 f = 0$, then even if the origin is a degenerate critical point, there is a neighborhood U and constants C and θ with $\frac{1}{2} \leqslant \theta < 1$ so that in some neighborhood we have

$$|f(x)|^\theta \leqslant C|\nabla_x f|.$$

Check by hand that this inequality holds for

- $f(t) = t^n$ (at $t = 0$);
- $f(x_1, \dots, x_n) = x_1 x_2 \dots x_n$ at $(x_1, \dots, x_n) = (0, \dots, 0)$.

Show that the length of a gradient flowline $\gamma : [t_0, t_1] \to U \subset \mathbb{R}^n$ for such an f satisfies the bound

$$l(\gamma) \leqslant C(f(\gamma(t_0))^{(1-\theta)} - f(\gamma(t_0))^{(1-\theta)})$$

if $f(\gamma(t_0)) - f(p)$ and $f(\gamma(t_1) - f(p))$ have the same sign and

$$l(\gamma) \leqslant C((f(\gamma(t_0)) - f(p))^{(1-\theta)} + (f(p) - f(\gamma(t_1))^{(1-\theta)}$$

if they have different signs. ♢

Exercise 3.5.3. Let $S^3 \to S^2$ be the Hopf fibration (taking $S^3 \subset \mathbb{C}^2$ as the unit sphere) viewed as a principal S^1-bundle where S^1 is unit complex numbers acting diagonally by scalar multiplication. Give a formula for the connection one form for the connection whose horizontal space is the orthogonal complement to the orbit. ♢

Exercise 3.5.4. Compute the spectrum for the operator

$$\begin{bmatrix} 0 & -d^* \\ -d & *d \end{bmatrix} : \Omega^0(\mathbb{T}^3) \oplus \Omega^1(\mathbb{T}^3) \to \Omega^0(\mathbb{T}^3) \oplus \Omega^1(\mathbb{T}^3)$$ ♢

4. Some examples of representation spaces and computations of the Chern-Simons functional of three manifolds and knots

We will look in a little more detail at the critical points of the Chern-Simons functional on a three manifold as well as for certain knot complements. Fix a principal G-bunlde $P \to Y^3$. G will usually be $SU(2)$ or $PU(2) = SO(3)$. Let

$$\mathcal{A}_P = \text{the space of connections on } P \to Y^3,$$

$$\mathcal{G}_P = \text{the space of gauge transformations on } P,$$

so $\mathcal{A}_P$ is an affine space over $\Omega^1(Y; \operatorname{ad} P)$ and $\mathcal{G}_P$ is the space of sections of

$$\operatorname{Ad} P = P \times_{\operatorname{Ad}} G.$$

If P is trivial and a trivialization is fixed, then one can write

$$\mathcal{A}_P = \Omega^1(Y) \otimes \mathfrak{g} \text{ and } \mathcal{G}_P = C^\infty(Y, G).$$

For simple Lie groups, any homogeneous Ad-invariant polynomial $\wp$ of degree 2 is a multiple of the Killing form

$$B(\xi, \eta) = \operatorname{tr}(\operatorname{ad} \xi \operatorname{ad} \eta), \ \ \xi, \eta \in \mathfrak{g}.$$

For $G = SU(2)$ or $SO(3)$, we also have the fundamental representation on either $\mathbb{C}^2$ or $\mathbb{R}^3$ (with $\mathfrak{su}(2) \subset \mathrm{End}(\mathbb{C}^2)$ or $\mathfrak{so}(3) \subset \mathrm{End}(\mathbb{R}^3)$ resp.). We will take $\wp$ as

$$\wp(\xi, \eta) = -\operatorname{tr}(\xi\eta).$$

(note that this form is positive definite) where the trace is taken in the fundamental representation. The Chern-Weil theory as in Section 2.9 associates a 4-dimension characteristic class to each such choice and in this case it is proportional to second Chern class of an $SU(n)$ bundle, first Pontryagin class for an $SO(n)$ bundles and a linear combination of the first and second Chern classes for a $U(n)$ bundle. Precisely, we have (cf. for example [4]):

(1) $[\operatorname{tr}(F_A \wedge F_A)] = 8\pi^2 c_2(P) \in H^2(X, \mathbb{R})$ for $G = SU(2)$;

(2) $[\operatorname{tr}(F_A \wedge F_A)] = -2\pi^2 p_1(P) \in H^2(X, \mathbb{R})$ for $G = SO(3)$;

(3) $[\operatorname{tr}(F_A \wedge F_A)] = 4n\pi^2(c_2(P) - \frac{n-1}{2n} c_1^2(P)) \in H^2(X, \mathbb{R})$ for $G = U(n)$;

(4) $[\operatorname{tr}(F_A \wedge F_A)] = -2(n-2)\pi^2 p_1(P) \in H^2(X, \mathbb{R})$ for $G = SO(n)$.

In any case we obtain the Chern-Simons functional

$$\text{(4.0.1)}\quad \mathcal{CS}(A) = -\int_Y \int_0^1 \operatorname{tr}(a \wedge F_{A_t})dt = -\int_Y \operatorname{tr}(a \wedge (2F_{A_0} + d_{A_0}a + \frac{1}{3}[a \wedge a])).$$

where

$$A = A_0 + a \quad \text{and} \quad A_t = A_0 + ta = (1-t)A_0 + t(A_0 + a)\,.$$

Suppose $P \to Y$ is trivialized (which we can always do if $G = SU(n)$). Let Γ be the trivial connection, i.e., the ordinary differential on the adjoint bundle which also trivialized. Write $A = \Gamma + a$ and consider the path $A_t = \Gamma + ta$. In this way, we obtain a more explicit formula for the Chern-Simons functional:

$$CS(\Gamma + a) = -\int_Y \operatorname{tr}(a \wedge (da + \frac{1}{3}[a \wedge a])).$$

4.1. The first variation of Chern-Simons and its periods As we saw in Proposition 3.2.2, the first variation of $\mathcal{CS}(A) = \mathcal{CS}(A, A_0)$ is given by

$$\text{(4.1.1)}\quad \frac{\partial}{\partial s}\mathcal{CS}(A + sa)\big|_{s=0} = -2\int_Y \operatorname{tr}(a \wedge F_A),$$

or equivalently $\nabla_A \mathcal{CS} = \star F_A$. Hence, the first variation is zero if and only if A is a flat connection.

We saw in equations (3.1.2) and (3.4.1) that

$$\mathcal{CS} : \mathcal{A}_P/\mathcal{G}_P \to \mathbb{R}/m\mathbb{Z}.$$

for some $m \in \mathbb{R}^+$ depending on the polynomial function $\wp$. We can compute m from the formulas in (1) and (2). If $G = SU(2)$, then

$$[\frac{1}{8\pi^2}\operatorname{tr}(F_A \wedge F_A)] = c_2(P)_{\mathbb{R}} \in \mathrm{Im}(H^4(X; \mathbb{Z}) \hookrightarrow H^4(X; \mathbb{R}))$$

is the image of the Chern class $c_2(P)$ in the real cohomology. If $G = SO(3)$, then

$$[\frac{1}{8\pi^2}\operatorname{tr}(F_A \wedge F_A)] = -\frac{1}{4}p_1(P)_{\mathbb{R}} \in \mathrm{Im}(H^4(X; \mathbb{Z}) \hookrightarrow H^4(X; \mathbb{R}))$$

is the first Pontryagin class. With our convention on $\mathcal{CS}$, we must have $m = 8\pi^2$ in both of these cases (but note that the traces mean two different things in the two cases).

When $G = SU(2)$ and $P = SU(2) \times Y$ is the trivial $SU(2)$-bundle, one may take the reference connection A_0 to be the trivial one. In this case, a gauge transformation g is simply a map

$$g : Y^3 \to SU(2),$$

and its action on $\mathcal{CS}$ can be understood more concretely as

$$\mathcal{CS}(g \cdot A) - \mathcal{CS}(A) = \alpha \deg(g)$$

with $\alpha = -8\pi^2$. The space of flat connections and values of $\mathcal{CS}$ on them are all invariants of Y^3.

4.2. Examples of representation spaces and computations of the Chern-Simons invariant Fixing our principal bundle $P \to Y$, the gauge equivalences classes of critical points correspond to conjugacy classes of representations, $\mathcal{R}(Y)$ and we will also consider its based version (see (2.6.1) and (2.6.2)).

We will look at some relatively simple examples where computations can be done explicitly.

Lens Spaces For (p, q) co-prime, the Lens space $L(p, q)$ is the quotient space

$$S^3/(\mathbb{Z}/p\mathbb{Z})$$

where S^3 is viewed as the unit sphere of $\mathbb{C}^2$ and the $\mathbb{Z}/p\mathbb{Z}$-action is defined as follows: let $\omega = e^{\frac{2\pi i}{p}}$ be the p-th root of 1 and set

$$\omega \cdot (z_1, z_2) = (\omega z_1, \omega^q z_2).$$

This is a free action, so $\pi_1(L(p, q)) = \mathbb{Z}/p\mathbb{Z}$. This action extends to a semi-free circle action (semi-free means that all point stabilizers are finite and there are points with trivial stabilizer). The action is given by the same formula as above but ω is allowed to be an arbitrary unit length complex number. There is one "short" orbit for this circle action, which we use as the generator for the fundamental group $\gamma \in \pi_1(L(p, q))$, namely the map induced by

$$s \in [0, 1] \mapsto (0, \exp(\frac{2\pi q s}{p} i)).$$

Since $\pi_1(L(p, q)) = \mathbb{Z}/p\mathbb{Z}$ is abelian, any $SU(2)$-representation is conjugate to

$$\rho_n : \pi_1(L(p, q)) \to SU(2)$$

$$\gamma \mapsto \begin{pmatrix} e^{2\pi i n/p} & 0 \\ 0 & e^{-2\pi i n/p} \end{pmatrix}$$

for some $n \in \mathbb{Z}/p\mathbb{Z}$. Note that $\rho_n \cong \rho_{p-n}$ since

$$\begin{pmatrix} u & 0 \\ 0 & \bar{u} \end{pmatrix}, \begin{pmatrix} \bar{u} & 0 \\ 0 & u \end{pmatrix}$$

are conjugate in $SU(2)$. Thus $\mathcal{R}(L(p,q)$ is $\lfloor p/2 \rfloor + 1$ points while

$$\mathcal{R}^o(L(p,q)) = \text{pt} \cup \bigcup_{i=1}^{\lfloor p/2 \rfloor} S^2 \text{ if } p \text{ is odd,}$$

$$\mathcal{R}^o(L(p,q)) = 2\,\text{pts} \cup \bigcup_{i=1}^{\lfloor p/2 \rfloor - 1} S^2 \text{ if } p \text{ is even.}$$

The computation of the Chern-Simons invariant of Lens spaces was done by Fintushel-Stern [16] and Kirk-Klassen [28]. The result is:

Theorem 4.2.1. *The value of* $-\mathcal{CS}/8\pi^2$ *on* ρ_n *is* $\dfrac{n^2 r}{p} \mod 1$ *where* $rq = -1 \mod p$.

Proof. Let A_n be the flat connection associated to ρ_n, then the pull-back connection $\pi^* A_n$ on $\pi^* P \to S^3$ is also flat. On S^3, all flat connections are gauge-equivalent to the trivial one Γ, so

$$\pi^* A_n = g^*(\Gamma),$$

for some $g : S^3 \to SU(2)$. Let $a = A_n - A_0 \in \Omega^1(L(p,q); \operatorname{Ad} P)$, then

$$\begin{aligned}\mathcal{CS}_{L(p,q)}(A_n) &= \int_{L(p,q)} \mathfrak{cs}(A_n, A_0) = \frac{1}{p}\int_{S^3} \mathfrak{cs}(\pi^* A_n, \pi^* A_0) \\ &= \frac{1}{p}\mathcal{CS}_{S^3}(\pi^* A_n, \Gamma) \quad = -\frac{8\pi^2}{p}\deg(g).\end{aligned}$$

If g factors through $L(p,q)$, then $\deg(g)$ is divisible by p. We need to determine

$$\deg(g) \mod p.$$

We look for a $\mathbb{Z}/p\mathbb{Z}$-equivariant map $S^3 \to SU(2) \subset \mathbb{H} \cong \mathbb{C}^2$ intertwining the Deck transformations d and the action of ρ_n. Here is an instance of such a map:

$$g(\begin{bmatrix} z_1 \\ z_2 \end{bmatrix}) = \frac{1}{(|z_1|^{2n} + |z_2|^{2nr})^{\frac{1}{2}}} \begin{bmatrix} z_1^n & -\bar{z}_2^{nr} \\ z_2^{nr} & \bar{z}_1^n \end{bmatrix}$$

which has $\deg(g) = n^2 r$. □

One can check that the collection of values of the Chern-Simons functional determine the lens space up to homotopy equivalence (though not up to homeomorphism). For example, $L(5,1)$ and $L(5,2)$ are not homotopy equivalent, and the values of $\mathcal{CS}/8\pi^2$ that they take on π_1-representations are 0, $\frac{1}{5}$ and $\frac{4}{5}$ in the first case and 0, $\frac{2}{5}$, and $\frac{3}{5}$ in the second case; so the Chern-Simons functional $\mathcal{CS}$ encodes information beyond the fundamental group $\pi_1(L(p,q))$. The classification of lens spaces is a long and storied question with contributions by Reidemeister, Seifert, Franz, and de Rham. The final classification up to homeomorphism was completed by Brody [11].

Theorem 4.2.2. *The lens spaces* $L(p,q_1)$ *and* $L(p,q_2)$ *are*

- *homotopy equivalent if and only if* $q_1 q_2 \equiv \pm n^2 \mod p$
- *homeomorphic if and only it* $q_1 q_2^{\pm 1} \equiv \pm 1 \mod p$.

Representations of fundamental groups of tori If we let $Y = \mathbb{T}^n$, then since $\pi_1(Y) \cong \mathbb{Z}^n$, a representation $\rho : \pi_1(Y) \to SU(2)$ gives n-commuting matrices $A_1, \ldots, A_n$. Since these matrices can be diagonalized simultaneously in $SU(2)$, we may assume that

$$A_k = \begin{pmatrix} e^{2\pi i \alpha_k} & 0 \\ 0 & e^{-2\pi i \alpha_k} \end{pmatrix} \in S^1 \subset SU(2)$$

for $0 \leqslant k \leqslant n$. The remains of the action of $SU(2)$ by conjugation on such simultaneously diagonalized matrices is by the normalizer $N(S^1)$ of $S^1 \subset SU(2)$. More concretely, the action of the subgroup

$$N(S^1) = \mathrm{Pin}(2) = S^1 \bigcup S^1 \cdot \begin{pmatrix} 0 & -1 \\ 1 & 0 \end{pmatrix}$$

on these matrices factorizes through the quotient $\mathbb{Z}/2\mathbb{Z} = N(S^1)/S^1$, which simultaneously switches the diagonal entries. Thus

$$\mathcal{R}_{SU(2)}(\mathbb{T}^n) = (S^1)^n/(\mathbb{Z}/2\mathbb{Z}).$$

The $\mathbb{Z}/2\mathbb{Z}$-action comes from the conjugation by

$$j = \begin{pmatrix} 0 & -1 \\ 1 & 0 \end{pmatrix}.$$

Specializing to the two torus $\mathbb{T}^2$ (with $n = 2$) yields

$$\mathcal{R}_{SU(2)}(\mathbb{T}^2) = (S^1)^2/(\mathbb{Z}/2\mathbb{Z}) \cong \mathbb{CP}^1 = S^2.$$

In general, if G is a semi-simple compact Lie group with $\pi_1(G)$ trivial, an element of $\mathcal{R}_G(\mathbb{T}^2)$ is given by a pair of elements in the same maximal torus, so

$$\mathcal{R}_G(\mathbb{T}^2) = \mathbb{T}_G \times \mathbb{T}_G/\text{the action of Weyl group of } G.$$

A beautiful theorem of Looijenga [36] identifies this as a weighted projective space. For instance, if $G = SU(n)$, we have

$$\mathcal{R}_{SU(n)}(\mathbb{T}^2) = \mathbb{T}_{SU(n)} \times \mathbb{T}_{SU(n)}/S_n \cong \mathbb{CP}^{n-1}.$$

In this case, $\dim_{\mathbb{R}} \mathbb{T}_{SU(n)} = n - 1$ and the Weyl group is the permutation group S_n. This special case can be understood more concretely as follows: identify $\mathbb{T}_{SU(n)}$ as the kernel of the determinant map $\mathbb{T}_{U(n)} = (S^1)^n \xrightarrow{\det} S^1$. A pair of points in $\mathbb{T}_{U(n)}$ is equivalently a n-tuple of points $(x_1, \cdots, x_n)$ on the elliptic curve $C = S^1 \times S^1$. If $p \in C$ denotes the origin, then such a n-tuple modulo permutations (as an effective divisor on C) determines a section of $\mathcal{O}(np)$ up to scalar if $\sum_{i=1}^n x_i = np$. This map is bijective. Since $\dim_{\mathbb{C}} H^0(C, \mathcal{O}(np)) = n$, we have verified that

$$\mathcal{R}_{SU(n)}(\mathbb{T}^2) \cong \mathbb{P}H^0(C, \mathcal{O}(np)) \cong \mathbb{CP}^{n-1}.$$

The story is trickier when $G = SO(3)$, since not all representation lift to $SU(2)$. As an example, consider a representation $\rho : \pi_1(\mathbb{T}^2) \to SO(3)$. If ρ lifts to $SU(2)$, then $\operatorname{Im}\rho$ lies in $SO(2)$; but now there is the possibility of a representation with image the Klein four group K_4:

$$\rho_{K_4}\colon \pi_1(\mathbb{T}^2) \to K_4 \subset SO(3).$$

This representation has stabilizer K_4 under the action of $SO(3)$-conjugation. To construct a flat bundle exhibiting this representation, consider the three non-trivial real line bundles λ_1,λ_2,λ_3 over the two-torus $\mathbb{T}^2$. They are classified by their first Stiefel-Whitney classes in $H^1(\mathbb{T}^2, \mathbb{Z}_2)\backslash\{0\}$. Consider the sum

$$\Xi = \lambda_1 \oplus \lambda_2 \oplus \lambda_3.$$

This is an orientable real 3-plane bundle, i.e., an $SO(3)$–bundle, since

$$w_1(\Xi) = \sum w_1(\lambda_i) = 0,$$

whereas $w_2(\Xi) \neq 0$. These computations follow from the Whitney sum formula for the total Stiefel-Whitney class.

We can view this representation from an $SU(2)$—perspective. Consider the punctured torus $\mathbb{T}^2\backslash p$ and look for representations

$$\rho : \pi_1(\mathbb{T}^2\backslash p) \to SU(2).$$

To get $\pi_1(\mathbb{T}^2)$ from this group, we take standard generators x, y and add the relation that the commutator $xyx^{-1}y^{-1}$, the loop going around the puncture, is trivial. We focus on representations $\rho_{\mathbb{H}} : \pi_1(\mathbb{T}^2\backslash p) \to SU(2)$ such that

$$\rho_{\mathbb{H}}(xyx^{-1}y^{-1}) = -\operatorname{Id},$$

that is, the commutator goes to the non-trivial centeral element of $SU(2)$. For example, set

$$\rho_{\mathbb{H}}(x) = I \text{ and } \rho_{\mathbb{H}}(y) = J,$$

thinking of $SU(2) = Sp(1)$ as the unit quaternions. Since $SU(2)/\pm\operatorname{Id} \simeq SO(3)$, this defines an $SO(3)$–representation which is precisely ρ_{K_4}. If we set

$$\mathcal{R}^o(\mathbb{T}^2, p) \doteq \{\rho_{\mathbb{H}} : \pi_1(\mathbb{T}^2\backslash p) \to SU(2) | \rho_{\mathbb{H}}(xy^{-1}y^{-1}) = -\operatorname{Id}\},$$

then $SU(2)$ acts by conjugation and each $\rho_{\mathbb{H}}$ has trivial stabilizer. One may form the quotient space

$$\mathcal{R}(\mathbb{T}^2, p) \doteq \{\rho_{\mathbb{H}} : \pi_1(\mathbb{T}^2\backslash p) \to SU(2) | \rho_{\mathbb{H}}(xyx^{-1}y^{-1}) = -\operatorname{Id}\}/SU(2).$$

This construction can be generalized to any 3-manifolds. Pick a closed curve $\omega \subset Y$ and let μ be any meridional loop. Then we set

$$\mathcal{R}^o(Y, \omega) = \{\rho_{\mathbb{H}} : \pi_1(Y\backslash\omega) \to SU(2) | \rho_{\mathbb{H}}(\mu) = -\operatorname{Id}\}.$$

Exercise 4.2.3. Show that $\mathcal{R}(\mathbb{T}^2, p)$ consists only of one point. Show that

$$\mathcal{R}(\mathbb{T}^2 \times S^1, p \times S^1)$$

consists of two points. ◊

4.3. Seifert Fibered Spaces The material here is largely from Finstushel and Stern's paper [17].

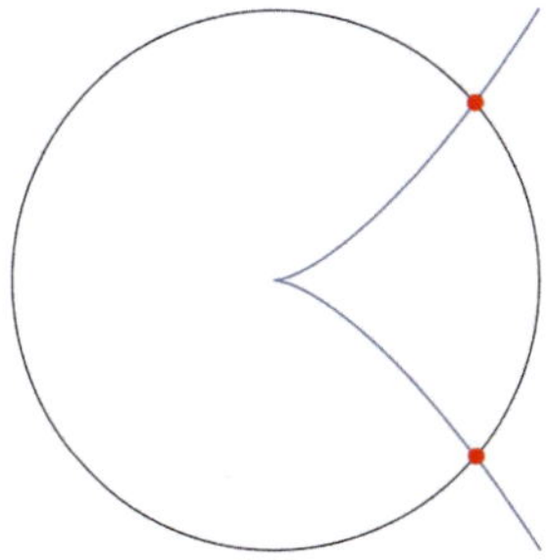

Figure 4.3.1. The sphere (black) intersecting the singular surface (in blue) at the Brieskorn sphere (in red).

Given a triple (p, q, r) of positive integers that are pairwise co-prime, consider the polynomial function:

$$f(z_1, z_2, z_3) = z_1^p + z_2^q + z_3^r$$

on $\mathbb{C}^3$. It has a unique singular point at the origin. The Brieskorn homology spheres $\Sigma(p, q, r)$ is defined by taking the intersection of the unit sphere with $f^{-1}(0)$:

$$\Sigma(p, q, r) = S^5 \cap \{z_1^p + z_2^q + z_3^r = 0\} \subset \mathbb{C}^3.$$

which admits an S^1-action:

$$u \in S^1 : (z_1, z_2, z_3) \mapsto (u^{qr} z_1, u^{pr} z_2, u^{pq} z_3).$$

The quotient space $\Sigma(p, q, r)/S^1$ is an orbifold S^2 with three orbifold points:

$$S^1 \to \Sigma(p, q, r) \to \Sigma(p, q, r)/S^1 = S^2(p, q, r).$$

To compute $\pi_1(\Sigma(p, q, r))$, one may still use the long exact sequence of homotopy groups as long as the orbifold fundamental group of $S^2(p, q, r)$ is used instead:

$$1 \to \pi_1(S^1) = \mathbb{Z} \to \pi_1(\Sigma(p, q, r)) \to \pi_1^{orb}(S^2(p, q, r)) \to 1,$$

and $\pi_1(S^1)$ lies in the center of $\pi_1(\Sigma(p, q, r))$. One presentation of the orbifold fundamental group of $S^2(p, q, r)$ is the following

$$\{t, u, v : t^p = 1, u^q = 1, v^r = 1, tuv = 1\},$$

where each of the generators t, u, and v corresponds a small loop around an orbifold point of $S^2(p, q, r)$. This is the so-called triangle group T_{pqr}.

Thus $\pi_1(\Sigma(p, q, r))$ admits a presentation as

$$\{h, \tilde{t}, \tilde{u}, \tilde{v} : h \text{ central}, \tilde{t}^p = h^{-b_1}, \tilde{u}^q = h^{-b_2}, \tilde{v}^r = h^{-b_3}, \tilde{t}\tilde{u}\tilde{v} = h^{-b}\}$$

for some $b_i, b \in \mathbb{Z}$. These numbers could be arranged (with more work required) such that

$$-b + \frac{b_1}{p} + \frac{b_2}{q} + \frac{b_3}{r} = \frac{1}{pqr},$$

from which one can easily deduce that $\Sigma(p, q, r)$ is an integral homology sphere.

Because these are homology spheres, any representation $\bar{\rho}$ of the fundamental group into SO(3) lift uniquely up to conjugacy to a representation ρ into SU(2). A representation of $\pi_1(\Sigma(p,q,r))$ into SU(2) sends its center into $Z(SU(2)) \cong \{\pm 1\}$, so $\rho(h)$ is either 1 or -1.

$$\pi_1(\Sigma(p,q,r)) \xrightarrow{\rho} SU(2) \xrightarrow{\pi} SO(3), \qquad \bar{\rho} = \pi \circ \rho.$$

This implies that $\bar{\rho}(h) = 1$ and $\bar{\rho}(\pi_1(\Sigma(p,q,r)))$ is generated by

$$T = \bar{\rho}(\tilde{t}), U = \bar{\rho}(\tilde{u}), V = \bar{\rho}(\tilde{v}).$$

This is equivalent to the representation of the triangle group T_{pqr} into SO(3). Suppose

$$\begin{aligned} T &= \text{the rotation by } 2\alpha = \frac{2\pi k}{p} \text{ about an axis } l_t, \\ U &= \text{the rotations by } 2\beta = \frac{2\pi l}{q} \text{ about an axis } l_u, \\ V &= \text{the rotations by } 2\gamma = \frac{2\pi m}{r} \text{ about an axis } l_v. \end{aligned}$$

The relation $TUV = 1 \in SO(3)$ gives a constraint for these angles and axes. Assuming $\bar{\rho}$ is irreducible (does not factorize through an S^1), these axes are not coplanar. Each two of them determine a great circle on $S^2 \subset \mathbb{R}^3$:

$$S_{uv}, S_{tv}, S_{tu}.$$

which divide S^2 into eight spherical triangles whose interior angles are given respectively by (they come in pairs):

$$\{(\alpha,\beta,\gamma),(\alpha,\pi-\beta,\pi-\gamma),(\pi-\alpha,\pi-\beta,\gamma),(\pi-\alpha,\beta,\pi-\gamma)\}.$$

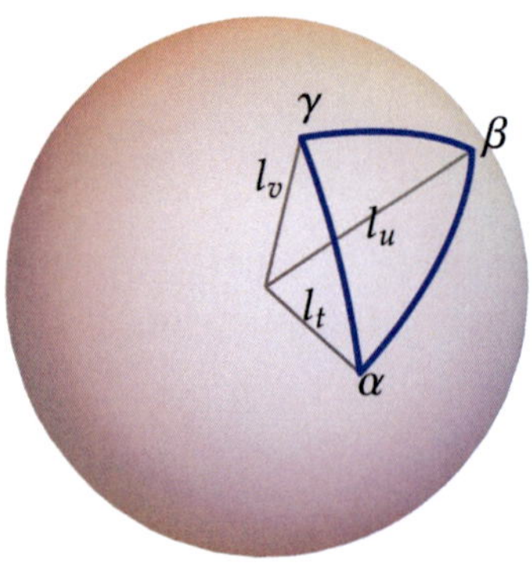

Figure 4.3.2. A spherical triangle.

Exercise 4.3.3. Prove this constraint holds. (Hint: write each rotation as product of two reflections.) ◇

When q and r are odd, Fintushel and Stern computed $\Sigma(p,q,r)$, the signed count of these representations modulo conjugacy in two ways. First they showed that this count is the number of triples (k, l, m) such that $0 < k < p$, $0 < l < q$ and $0 < m < r$ and such that,

- l, m have the same parity;
- $k/p + l/r + m/q > 1$;
- $k/p + l/r - m/q < 1$; and,
- $k/p - l/r + m/q < 1$.

Then they showed that this number, which is the Casson invariant of the Seifert-Fibered space, is equal to

$$-\frac{1}{4}\sigma(M(p,q,r))$$

where

$$M(p,q,r) = B^6 \cap \{z_1^p + z_2^q + z_3^r = \epsilon\} \subset \mathbb{C}^3.$$

is the Milnor fiber of the singularity and σ denotes the signature. Thus the $SO(3)$–representation variety of $\Sigma(p,q,r)$ is discrete and consist of finitely many points parametrized by suitable triples of integers (k, l, m).

For the quotient map $\pi : \Sigma(p,q,r) \to S^2(p,q,r)$, one may form the mapping cylinder $C\pi$. Note that if π is a genuine S^1-fiber bundle, then $C\pi$ is the associated disk bundle. This shows that near a smooth point of $S^2(p,q,r)$, $C\pi$ is a smooth manifold. Globally, $C\pi$ is an orbifold with boundary

$$\partial C\pi = \Sigma(p,q,r),$$

along with three orbifold points lying over $S^2(p,q,r)$ with links given by

$$L(p,b_1), L(q,b_2), L(r,b_3).$$

Thus, each of these orbifold points has a neighborhood homeomorphic to a cone over a Lens space. By removing these neighborhoods, we obtain a cobordism:

$$W = C\pi - (N_p \cup N_q \cup N_r) : L(p,b_1) \cup L(q,b_2) \cup L(r,b_3) \to \Sigma(p,q,r).$$

Take the trivial principal bundle P over W and a flat connection A of P. By the defining property of $\mathcal{CS}$, we have

$$\begin{aligned} 0 &= \int_W \operatorname{tr}(F_A \wedge F_A) - \int_W \operatorname{tr}(F_{A_0} \wedge F_{A_0}) \\ &= \mathcal{CS}(A|_{\Sigma(p,q,r)}) - \mathcal{CS}(A|_{L(p,b_1)}) - \mathcal{CS}(A|_{L(q,b_2)}) - \mathcal{CS}(A|_{L(r,b_3)}) \end{aligned}$$

Since all $SO(3)$ representations of $\Sigma(p,q,r)$ come from the representations of the orbifold base, they extend over W and the computation of the Chern-Simons functional on $\Sigma(p,q,r)$ is reduced to Lens spaces. This property was used by Fintushel and Stern (again in [17]) and independently Furuta (in [22]) to study the homology cobordism group Θ_3^H of oriented homology 3-spheres.

Finally, for Seifert fibered spaces over S^2 with $n \geqslant 4$ singular fibers, we will get a polygon constraint. Triangles are rigid, but spherical squares may have non-trivial moduli. When $n = 4$, it will be a union of S^2's and for $n = 5$, different rational manifolds could occur as the moduli spaces.

Exercise 4.3.4. Show that the link around each singular point of $C\pi$ is a Lens space. ◇

4.4. Representation spaces of knot groups Consider a knot $K \subset S^3$. We will look at a special class of representations of $\pi_1(S^3 \backslash K)$

$$R_I(K) = \{\rho : \pi_1(S^3 \backslash K)) \to SU(2) | \operatorname{tr}(\rho(\mu) = 0 \text{ for all meridional curves } \mu\}.$$

Thus $\rho(\mu)^2 = -1$. The set of such elements is a 2-sphere $S^2_I \subset SU(2)$. For an unlink of k-components, we get

$$R_I(K) = \times_k S^2_I.$$

For knots in S^3, it is somewhat easier to get a description of this representation space. Consider the Wirtinger presentation for the fundamental group of the knot complement.

$$\pi_1(S^3 \backslash K) = \langle x_1, x_2, \dots, x_n | r_1, \dots r_{n-1} \rangle$$

where each r_s is of the form

$$x_i x_j x_i^{-1} = x_k.$$

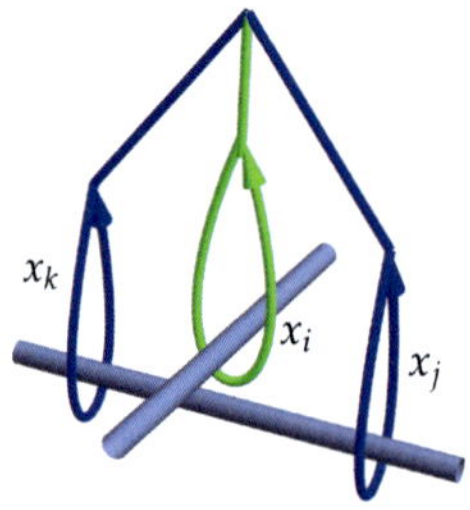

Figure 4.4.1. The Wirtinger relation.

Any $p \in S^2_I \subset SU(2)$ acts on S^2_I by conjugation as the symmetric space involution $\tau_p : S^2 \to S^2$ of reflection along geodesics through p. Since the x_i's are meridians, we see that the three points

$$\rho(x_j), \rho(x_i), \rho(x_k) \in S^2_I$$

lie on the same geodesic with $\rho(x_i)$ being a midpoint between the other two.

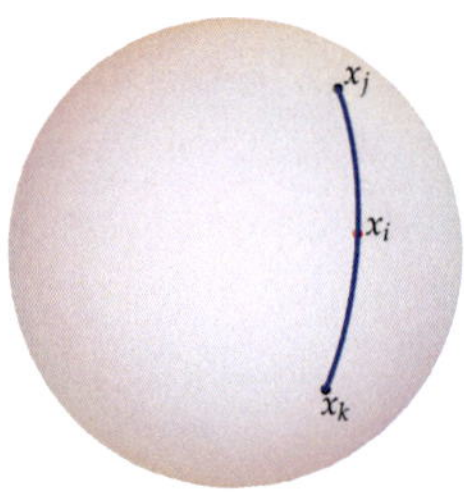

Figure 4.4.2. Three points on the 2-sphere two are interchanged by reflection about the third.

For a trefoil, the presentation is simply $\langle x, y, z | xyx^{-1} = z \text{ and cyclic}\rangle$. Thus we are looking for triples of points on S^2 lying on a great circle so that each point is a midpoint of the other two. There are two possibilities: all all three points are the same or they are equally spaced labeled points on the great circle. As a result,

$$R_I(S^3, T_{2,3}) = S^2_I \coprod \mathbb{RP}^3.$$

For a $(2, p)$-torus knot, the story is similar. The presentation is

$$\langle x_1, x_2, \ldots, x_p | x_{i+1} x_i x_{i+1}^{-1} = x_{i+2}\rangle.$$

Thus the representation space is p ordered points lying on a great circle so that $\rho(x_{i+1})$ is the midpoint of $\rho(x_i)$ and $\rho(x_{i+2})$. There are three possibilities:

- $\rho(x_1)$ and $\rho(x_2)$ (hence all the points) are the same.
- $\rho(x_1)$ and $\rho(x_2)$ are antipodal (if p is even); all odd points are mapped to $\rho(x_1)$ and all even points to $\rho(x_2)$.
- The angle between $\rho(x_1)$ and $\rho(x_2)$ is $2\pi k/p$ for some $0 < k < p/2$.

As a result, we have

$$R_I(S^3, T_{2,2m+1}) = S^2_I \coprod_m \mathbb{RP}^3,$$
$$R_I(S^3, T_{2,2m+2}) = S^2_I \coprod S^2_I \coprod_m \mathbb{RP}^3.$$

Taking the homology of R_I in any of these cases gives the Khovanov homology (without a grading) of the corresponding knot. In order to get an invariant with good properties, simply taking the homology of representation varieties turns out to be too naive. If we want an invariant with knot cobordisms inducing maps and satisfying a composition law, we need to do something more subtle. We will see that instanton Floer homology provides the appropriate scaffolding for such a rich invariant.

5. The ASD Equation: Examples and Basic Properties

Having achieved a reasonable understanding of what the critical set of the Chern-Simons function looks like in several examples we shift our attention to the study of the flowlines between critical points.

5.1. The gradient of Chern-Simons and the ASD equation In this lecture, we focus on the case when $G = SU(2)$ or $SO(3)$, but we also allow $G = U(1)$ at some point as an enlightening toy model. The configuration space $\mathcal{A}_P$ is an affine space over $\Omega^1(\operatorname{ad} P)$; so its tangent bundle

$$T\mathcal{A}_P = \mathcal{A}_P \times \Omega^1(\operatorname{ad} P)$$

is necessarily trivial. Fixing a metric on Y, we define a metric on $T\mathcal{A}_P$ as follows: given $a, b \in T_A\mathcal{A}_P = \Omega^1(\operatorname{ad} P)$, define

$$\langle a, b\rangle = -2\int_Y \operatorname{tr}(a \wedge \star_3 b),$$

where $\star_3$ denotes the Hodge $\star$-operator on forms. Using the first variation of the Chern-Simons functional (4.1.1), we compute the gradient vector field of $\mathcal{CS}$ as

$$\nabla_A \mathcal{CS} = \star_3 F_A.$$

If $A : \mathbb{R} \to \mathcal{A}_P$ is a negative gradient flowline of $\mathcal{CS}$, then it obeys the equation

$$\frac{\partial}{\partial t} A(t) = - \star_3 F_{A(t)}. \tag{5.1.1}$$

A miracle happens here. The initial value problem of this evolution equation (5.1.1) is not well-posed, that is, given an initial condition we cannot guarantee even the short time existence of a flowline much like the backward heat flow. This is because the Hessian $- \star_3 d_A$ (the linearization of the right hand side of (5.1.1)) has infinitely many positive and negative eigenvalues (acting on the space $T_{[A]}(\mathcal{A}_P/\mathcal{G}_P)$). On the other hand, (5.1.1) can be also understood as the anti-self duality equation in dimension 4 and then upgraded into an elliptic system, as we now explain.

Let X^4 be an oriented Riemannian manifold. The Hodge $\star$-operator acting on Λ^2 satisfies $\star_4^2 = 1$, so Λ^2 splits as

$$\Lambda^2 = \Lambda^+ \oplus \Lambda^-,$$

where $\Lambda^{\pm}$ is the ± 1 eigenspace of $\star_4$ respectively. Given a local orthonormal basis $\{e_1, e_2, e_3, e_4\}$ of T^*X, a framing for $\Lambda^{\pm}$ can be written as

$$e_1 \wedge e_2 \pm e_3 \wedge e_4, e_1 \wedge e_3 \pm e_4 \wedge e_2, e_1 \wedge e_4 \pm e_2 \wedge e_3.$$

Fixing a principal $SU(2)$-bundle $P \to X$ and a connection A, we impose the (anti)-self dual Yang-Mills equation:

$$F_A = \pm \star_4 F_A. \tag{5.1.2}$$

In the special case that $X = \mathbb{R} \times Y$ and $P = \mathbb{R} \times Q$ is pulled back from a principal G-bundle $Q \to Y$, a connection A of P can be written as

$$A = \tilde{B} + dt \otimes c(t)$$

where $c(t) \in \Omega^0(\operatorname{ad} Q)$ and $B(t)$ is a family of connections on Q such that

$$\tilde{B} = \frac{\partial}{\partial t} + B(t).$$

Therefore,

$$\begin{aligned} F_A &= F_{\tilde{B}} + d_{\tilde{B}}(dt \otimes c(t)) + \underbrace{\frac{1}{2}[dt \otimes c(t) \wedge dt \otimes c(t)]}_{=0} \\ &= F^{(3)}_{B(t)} + dt \wedge \frac{\partial}{\partial t} B(t) - dt \wedge d^{(3)}_{B(t)} c(t). \end{aligned} \tag{5.1.3}$$

The orientation convention is

$$dt \wedge dvol_Y = dvol_X.$$

If ω is a 2-form on Y and ν a 1-form, then

$$\star_4 \omega = dt \wedge \star_3 \omega, \; \star_4(dt \wedge \nu) = \star_3 \nu.$$

Applying $\star_4$ to (5.1.3), one obtains that:

$$\star_4 F_A = dt \wedge \star_3 F^{(3)}_{B(t)} + \star_3\left(\frac{\partial}{\partial t}B(t) - d^{(3)}_{B(t)}c(t)\right).$$

The equation (5.1.2) then becomes

$$\frac{\partial B(t)}{\partial t} = \underbrace{\pm \star_3 F^{(3)}_{B(t)}}_{\pm\nabla_{B(t)}\mathcal{CS}} + d^{(3)}_{B(t)}c(t) \tag{5.1.4}$$

We recognize the first part as the gradient of the Chern-Simons functional. Comparing with the negative gradient flow equation (5.1.1), the second part is new and it is tangential to the gauge group orbit at $B(t)$. The equation (5.1.1) is only invariant under the 3-dimensional gauge group while the (anti)-self-duality equation (5.1.2) admits a larger symmetric group: it is invariant under the 4-dimensional gauge group.

The purpose of this lecture is to understand the basic analytic properties of the ASD equation, which will allow us to construct the Morse homology of the Chern-Simons functional at a later point. To this end, we note some further analogies between the ASD equation (5.1.2) and the negative gradient flow equation (5.1.1). The first analogy is the energy identity: for a connection on $P \to X$, a natural gauge invariant notion of energy is

$$E(A) = -\int_X \mathrm{tr}(F_A \wedge \star F_A).$$

Since the Hodge $\star$-operator in dimension 4 is conformally invariant, we have:

Proposition 5.1.5. *The Yang-Mills energy is conformally invariant in dimension 4.*

At this point, another coincidence particular to dimension 4 occurs, which connects back to the energy identity from Morse theory. Let M be any Riemannian manifold and $f : M \to \mathbb{R}$ any smooth function. Then Lemma 1.2.6 provides an energy identity for any path $\gamma : [t_0, t_1] \to M$ that characterizes negative gradient flowlines:

$$f(\gamma(t_1)) - f(\gamma(t_2)) = \frac{1}{2}\int_{t_1}^{t_2} (|\dot\gamma(t)|^2 + |\nabla_\gamma f|^2 - |\dot\gamma + \nabla_\gamma f|^2)dt.$$

The generalization is the following.

Proposition 5.1.6. *Let A be any smooth connection on a compact 4-manifold X possibly with boundary. Then we have the following energy identity:*

$$\begin{aligned}\int_X \mathrm{tr}(F_A \wedge F_A) &= \int_X \mathrm{tr}(F_A^+ \wedge F_A^+) + \int_X \mathrm{tr}(F_A^- \wedge F_A^-)\\ &= 2\int_X \mathrm{tr}(F_A^+ \wedge \star F_A^+) - \int_X \mathrm{tr}(F_A^+ \wedge \star F_A^+) - \int_X \mathrm{tr}(F_A^- \wedge \star F_A^-)\\ &= 2\int_X \mathrm{tr}(F_A^+ \wedge F_A^+) + 2E(A).\end{aligned}$$

Proposition 5.1.6 has two immediate corollaries; we start with the closed case:

Corollary 5.1.7. *Let A be an ASD connection on a closed 4-manifold X. Then*

$$E(A) = 8\pi^2\langle c_2(P), [X]\rangle.$$

In particular, ASD connections exist only if $\langle c_2(P), [X]\rangle \geqslant 0$.

The next case is for manifolds with boundary:

Corollary 5.1.8. *If X is a compact oriented 4-manifold with boundary, then a smooth connection A in an* $SU(2)$*-bundle P is ASD if and only if*

$$E(A) = \mathcal{CS}(A|_{\partial X}). \tag{5.1.9}$$

As a result, the energy of ASD-connections depends only on their boundary values. Since $\partial X \neq \emptyset$*, the principal bundle* $P \to X$ *is necessarily trivial. We note that the Chern-Simons functional* $\mathcal{CS}$ *appeared in* (5.1.9) *is defined using a trivialization of* $P|_{\partial X}$ *that extends over X.*

5.2. The linear case Now we consider the case when $G = U(1)$ as a toy problem, and decide whether a complex line bundle L over a compact oriented Riemannian 4-manifold admits an ASD connection.

To this end, we recall the Hodge theory on Riemannian manifolds. On any oriented Riemannian n-manifold X, every cohomology class admits a unique harmonic representative; indeed we have the Hodge-de Rham decomposition:

$$\Omega^r(X) = d\Omega^{r-1}(X) + \mathcal{H}^r(X) + d^*\Omega^{r-1}(X),\ 0 \leqslant r \leqslant n,$$

where $\mathcal{H}^r(X)$ is the space of harmonic r-forms:

$$\alpha \in \Omega^r(X) \text{ such that } d\alpha = 0 \text{ and } d^*\alpha = 0.$$

This decomposition works for Sobolev spaces as well, with a little care of the regularity involved:

$$L^p_k(X;\Lambda^r) = dL^p_{k+1}(X;\Lambda^{r-1}) + \mathcal{H}^r(X) + d^*L^p_{k+1}(X;\Lambda^{r+1}).$$

When $n = 4$, we can also decompose $\Omega^2(X) = \Omega^+(X) \oplus \Omega^-(X)$ as well as the space of harmonic forms

$$\mathcal{H}^2(X) = \mathcal{H}^+(X) \oplus \mathcal{H}^-(X)$$

according to the eigenspaces of the Hodge $\star$-operator. Note that the fibers of $\Lambda^+(X)$ and $\Lambda^-(X)$ are point-wise orthogonal under the wedge product (since they are $\star$-invariant) and also under the Riemannian inner product on 2-forms. The signature $\sigma(X)$ of X is the difference in dimensions of these spaces and can be computed in terms of the first Pontryagin form $p_1(X, g)$:

$$\sigma(X) = \dim(\mathcal{H}^+(X)) - \dim(\mathcal{H}^-(X)) = \frac{1}{3}\int_X p_1(X, g).$$

We will run into the operator

$$d^+ \doteq \frac{1}{2}(d + \star d) : \Omega^1(X) \to \Omega^+(X).$$

Exercise 5.2.1. Show that the cokernel of d^+ consists of closed self-dual 2-forms, i.e., 2-forms ω such that $\omega = \star\omega$ and $d\omega = 0$. ◇

If A_0 is a reference connection in L, any other connection A in L can be written as $A = A_0 + a$ for some 1-form $a \in i\Omega^1(X)$; their curvatures are related by

$$F_A = F_{A_0} + da.$$

Thus L admits an ASD connection A if and only if we can solve the equation

$$0 = F^+_{A_0} + d^+ a.$$

If such a solution a exists, $F^+_{A_0}$ must be orthogonal to the cokernel of d^+. By Exercise 5.2.1, we must have

$$\int_X F^+_{A_0} \wedge \star\omega = 0$$

for any self-dual harmonic 2-form ω. Fixing any such ω, we write $F_{A_0} = F^+_{A_0} + F^-_{A_0}$. Using the self-duality of ω and the orthogonality between Λ^+ and Λ^-, we obtain that

$$\int_X F^+_{A_0} \wedge \star\omega = \int_X F_{A_0} \wedge \omega.$$

This last formula has a purely cohomological interpretation, since both F_{A_0} and ω are closed:

$$\int_X F_{A_0} \wedge \omega = \langle [F_{A_0}] \cup [\omega], [X] \rangle$$

where the brackets denote the de Rham cohomology class. The form

$$\frac{i}{2\pi} F_{A_0}$$

represents the first Chern class of L. To sum up what we have proved:

Proposition 5.2.2. *A line bundle* L *over a closed oriented 4-manifold* X *admits an ASD-connection if and only if* $\langle c_1(L) \cup [\omega], [X] \rangle = 0$ *for any self-dual harmonic 2-form* $\omega \in \mathcal{H}^+(X)$.

This is a codimension b^+-constraint on the cohomology class $c_1(L)$. Thus if $b^+ > 0$, we do not expect a generic metric would support any ASD connections on line bundles (which was proved by Donaldson). For this reason, we do not typically look for ASD-connections on $U(n)$-bundles, since taking the trace always yield a $U(1)$ ASD connection.

Exercise 5.2.3. Consider a product of two round 2-spheres of radii r and s

$$S^2(r) \times S^2(s).$$

Fixing orientations for these spheres, we can write any cohomology class as

$$a PD(S^2 \times \{pt\}) + b PD(\{pt\} \times S^2)$$

for some $a, b \in \mathbb{R}$. Here PD denotes the Poincaré duality map.

- In terms of (a, b, r, s), what is the cohomology class of a harmonic self-dual 2-form?
- Suppose $c_1(L) = k PD(S^2 \times \{pt\}) + l PD(\{pt\} \times S^2)$. When does L admit an ASD connection? ◊

5.3. The basic instanton The first interesting solution to the ASD-equation was discovered by Belavin, Polyakov, Schwartz and Tyupkin [9]. The following is a description from Atiyah's Pisa notes [2].

We will make use of quaternions $\mathbb{H} = \mathbb{R} + \mathbb{R}I + \mathbb{R}J + \mathbb{R}K$ with

$$I^2 = J^2 = K^2 = -1, IJ = -JI = K, \text{etc},$$

and think of $SU(2) = Sp(1) = \{x \in \mathbb{H} : x\bar{x} = 1\}$ as the group of unit length vectors of $\mathbb{H}$. Let

$$S^7 \subset \mathbb{H} \oplus \mathbb{H}$$

be the unit sphere of $\mathbb{H} \oplus \mathbb{H}$ endowed with the round metric. There are two distinct right $SU(2)$-actions on S^7:

$$(x, y) *_+ q = (xq, yq)$$
$$(x, y) *_- q = (\bar{q}x, \bar{q}y).$$

which make S^7 into a principal $SU(2)$-bundle over S^4 in two different ways (the quaternionic Hopf bundles). Let us call them P_+ (the right action) and P_- (the left-inverse action) respectively.

We define a connection A_+ on P_+ by declaring the horizontal space H_+ to be the orthogonal complement of the orbits of $SU(2)$. We wish to show that this connection satisfies the anti-self-dual Yang-Mills equation:

$$F^+_{A_+} = 0.$$

for the bundle

$$\begin{array}{ccc} S^3 & \longrightarrow & S^7 \\ & & \downarrow \\ & & S^4. \end{array}$$

There are many ways to verify this. One is to use the trick that the compact symplectic group $Sp(2)$ acts on $\mathbb{H}^2$:

$$Sp(2) = \{g \in SO(8) : g(v) *_+ q = g(v *_+ q) \text{ for any } q \in SU(2)\}.$$

This is the subgroup of linear transformations that preserve the metric and commute with the right action $*_+$ of $SU(2)$. More explicitly,

$$Sp(2) = \{\begin{pmatrix} q_1 \cos\theta & -q_3 \sin\theta \\ q_2 \sin\theta & q_4 \cos\theta \end{pmatrix} : q_1, q_2, q_3, q_4 \in Sp(1),\ q_4 = q_2\bar{q}_1 q_3\}.$$

The action of $Sp(2)$ on S^7 is transitive, inducing an action on the quotient space $S^4 \cong S^7/*_+$.

Thus F_{A_+} viewed as a section of $\Lambda^2 S^4 \otimes \operatorname{ad} P_+ \to S^4$ is invariant under the induced action of $Sp(2)$. To better understand this induced action, consider the subgroup of $Sp(2)$ that fixes

$$[x] = [(1, 0)] \in S^4,$$

which is a copy of $Sp(1)_+ \times Sp(1)_-$:

$$\mathtt{Stab}_{[x]} = \{\begin{pmatrix} q_1 & 0 \\ 0 & q_2 \end{pmatrix} : q_i \in Sp(1)\}.$$

The vertical tangent space at $x = (1,0) \in P_+$ (identified with the Lie algebra of $Sp(1)$ by the definition of the connection) is

$$VT_x = \mathrm{Im}(\mathbb{H}) \oplus \{0\}.$$

The horizontal space

$$H_x = \{0\} \oplus \mathbb{H}.$$

is identified with $T_{[x]}S^4$. Any $q_1 \in Sp(1)_+ \subset Sp(2)$ acts by left multiplication as a map $VT_{(1,0)} \to VT_{(q_1,0)}$. Identifying both with the Lie algebra yields the action by conjugation $q_1 \cdot \xi = q_1\xi\bar{q}_1$ on VT_x, while $Sp(1)_-$ acts trivially on VT_x. Finally, $\mathtt{Stab}_{[x]}$ acts by $(q_1, q_2) \cdot v = q_2 v \bar{q}_1$ on $T_{[x]}S^4$. This induces an action on

$$\Lambda^2 T^*_{[x]} S^4 = \Lambda^+_{[x]} \oplus \Lambda^-_{[x]}.$$

To show that $F^+_{A_+}$ takes values only in Λ^-, We have to verify that the action of $Sp(1)_-$ is non-trivial on $\Lambda^+_{[x]}$ and is trivial on $\Lambda^-_{[x]}$. The operators I, J, K give rise to different complex structures on $H_x \cong \mathbb{H}$, and we obtain a triple of symplectic forms $\omega_I, \omega_J, \omega_K$:

$$\omega_I(v, w) = \langle Iv, w\rangle, \text{etc.},$$

which are in fact self-dual:

$$\begin{aligned}
\omega_I &= dx^0 \wedge dx^1 + dx^2 \wedge dx^3, \\
\omega_J &= dx^0 \wedge dx^2 + dx^3 \wedge dx^1, \\
\omega_K &= dx^0 \wedge dx^3 + dx^1 \wedge dx^2.
\end{aligned}$$

If instead we consider $\omega'_I(v, w) = \langle vI, w\rangle$, then they are anti-self-dual. It is clear that $\omega'_I, \omega'_J, \omega'_K$ are invariant under the left action of $SU(2)$, while $\omega_I, \omega_J, \omega_K$ give rise to the adjoint-representation. This proves $F^+_{A_+} = 0$.

Similarly, we can define the connection A_- on P_-. This time we investigate the group that preserves the action $*_-$:

$$Sp(2)' = \{g \in SO(8) : g(v) *_- q = g(v *_- q)\ \forall q \in SU(2)\}.$$

Then the stabilizer C'_x at $x = (1,0)$ is

$$q \in SU(2) : \begin{pmatrix} v_1 \\ v_2 \end{pmatrix} \mapsto \begin{pmatrix} v_1 \\ v_2 \cdot \bar{q} \end{pmatrix}.$$

The induced action of C'_x on $T_{[x]}S^4$ now preserves the forms $\omega_I, \omega_J, \omega_K$. Thus, $F^-_{A_-} = 0$.

We wish to understand this canonical connection $A = A_+$ more concretely in terms of local coordinates. Let x, y be quaternion-valued coordinate functions on $\mathbb{H}^2$, so $dx, dy, d\bar{x}, d\bar{y}$ are well-defined (quaternion-valued) 1-forms on $\mathbb{H}^2$. We claim that as an $\mathfrak{su}(2)$-valued 1-form

$$A = -\operatorname{Im}(d\bar{x} \cdot x + d\bar{y} \cdot y) = \operatorname{Im}(\bar{x} \cdot dx + \bar{y} \cdot dy) \text{ on } P_+,$$

which is pulled back from the embedding $P_+ = S^7 \hookrightarrow \mathbb{H}^2$. We verify that this 1-form indeed satisfies the desired properties:

- A is the identity on vertical vectors. Given any $(x, y) \in S^7$ and any vertical vector $v_q = (x \cdot q, y \cdot q)$ with $q \in \operatorname{Im}\mathbb{H} = \mathfrak{su}(2)$, we have
$$dx(v_q) = x \cdot q,\ dy(v_q) = y \cdot q.$$
Hence, $A(v_q) = (|x|^2 + |y|^2)q = q$.
- A is invariant under $Sp(2)$. Take $g \in Sp(2)$, then
$$g = \begin{pmatrix} a & b \\ c & d \end{pmatrix}, g^{\intercal} = \begin{pmatrix} \bar{a} & \bar{c} \\ \bar{b} & \bar{d} \end{pmatrix}$$
for some $a, b, c, d \in \mathbb{H}$ and $g^{\intercal} g = \mathrm{Id}$. If $(x, y)^{\intercal}$ maps to $g(x, y)^{\intercal}$, then
$$(\bar{x}, \bar{y}) \mapsto (\bar{x}, \bar{y})g^{\intercal}, \begin{pmatrix} dx \\ dy \end{pmatrix} \mapsto g \begin{pmatrix} dx \\ dy \end{pmatrix}.$$
Since A is the imaginary part of
$$(\bar{x}, \bar{y}) \cdot \begin{pmatrix} dx \\ dy \end{pmatrix},$$
one easily sees that it is invariant under the action of g.
- A kills the horizontal space H. It suffices to check this at a point. Take $w = (1, 0) \in S^7$, where H_w is $\{(0, y) \in T_w S^7 : y \in \mathbb{H}\}$. Since $\bar{y}(w) = 0$ and $dx = 0$ on H_w, $\ker A_w = H_w$.
- $q^*A = \operatorname{Ad}(q^{-1})(A)$ for any $q \in Sp(1)$ acting on the right. Indeed,
$$\bar{x} \cdot dx \mapsto \overline{xq} \cdot d(xq) = \bar{q}(\bar{x} \cdot dx)q.$$

One can compare the similar construction for the complex Hopf bundle

$$\begin{array}{ccc} S^1 & \longrightarrow & S^3 \\ & & \downarrow \\ & & S^2. \end{array}$$

The same expression then provides the standard connection for $\mathcal{O}(-1) \to S^2$.

If we trivialize P_+ locally using the section

$$\mathbb{H} \ni x \mapsto s(x) = \frac{(x, 1)}{(1 + |x|^2)^{1/2}} \in S^7,$$

then the connection 1-form under this trivialization is

$$a = s^*A = \frac{\operatorname{Im}(\bar{x} \cdot dx)}{1 + |x|^2}$$

Exercise 5.3.1. Consider the quaternionic 2-from $d\bar{x} \wedge dx$ with

$$d\bar{x} = dx^0 - I dx^1 - J dx^2 - K dx^3,\ dx = dx^0 + I dx^1 + J dx^2 + K dx^3.$$

Show that

$$d\bar{x} \wedge dx = 2I\omega_I' + 2J\omega_J' + 2K\omega_K'$$

is anti-self-dual, and that if we take the wedge product in the reverse order, it will be a self-dual form:

$$dx \wedge d\bar{x} = -2I\omega_I - 2J\omega_J - 2K\omega_K. \qquad \diamondsuit$$

Exercise 5.3.2. Show that

$$s^*F_A = \mathrm{Im}(\frac{d\bar{x} \wedge dx}{(1+|x|^2)^2}). \qquad \diamondsuit$$

5.4. More instantons on S^4 by conformal transformations To obtain more solutions on the round sphere S^4, we exploit the conformally invariance of the (anti)-self-dual equation

$$F_A = \pm \star F_A.$$

See Proposition 5.1.5. Indeed, the Hodge $\star$-operator:

$$\star : \Lambda^2 \to \Lambda^2$$

depends only on the conformal class of the metric when $\dim X = 4$. In general, if $\dim X = n$ and $\tilde{g} = e^{2\sigma} g$ for some $\sigma : X \to \mathbb{R}$, then

$$\star_{\tilde{g}} = e^{(n-2p)\sigma} \star_g : \Lambda^p \to \Lambda^{n-p},$$

so it remains unchanged when $p = n/2$ is the middle dimension.

Consider the conformal group $SO(5,1)$ of S^4:

$$SO(5,1) = \{u : S^4 \to S^4 \text{ orientation preserving} |\ g = \text{the round metric}, \\ u^*g = e^{2\sigma} g \text{ for some } \sigma : S^4 \to \mathbb{R}\}.$$

This group is generated by rotations ($SO(5) \subset SO(5,1)$) and dilations. It is isomorphic to the special orthogonal group of type $(5,1)$ as the name indicates. Since A_+ is invariant under the $Sp(2)$-action on S^7, it is also invariant under the action of $SO(5)$ on S^4. Therefore, the moduli space of instantons on S^4 obtained by applying conformal transformations to A_+ is

$$SO(5,1)/SO(5) \cong \mathbb{R}^5.$$

Let us compute this family explicitly in terms of connection 1-forms. The dilation $\tau_\lambda(x) = \lambda x$ on $\mathbb{R}^4$ extends to a conformal transformation of S^4, and so

$$\tau_\lambda^*(a) = \frac{\lambda^2 \mathrm{Im}(\bar{x} \cdot dx)}{1 + \lambda^2|x|^2} = \frac{\mathrm{Im}(\bar{x} \cdot dx)}{1/\lambda^2 + |x|^2},$$

$$\tau_\lambda^*(F_A) = \mathrm{Im}(\frac{\lambda^2 d\bar{x} \wedge dx}{(1+\lambda^2|x|^2)^2}).$$

Initially, $\|F_{A_+}\|^2$ distributes evenly on S^4. As $\lambda \to \infty$, $\|\tau_\lambda^* F_{A_+}\|^2$ concentrates at the origin and

$$\tau_\lambda^*(a) \to a_\infty = \frac{\mathrm{Im}(\bar{x} \cdot dx)}{|x|^2}.$$

Consider the singular gauge transformation:

$$g : \mathbb{R}^4 \backslash \{0\} \to SU(2), \; x \mapsto \frac{x}{|x|}.$$

Then $g^* a_\infty = g \cdot (a_\infty - g^{-1} dg) \cdot g^{-1}$ and

$$g^{-1} dg = \frac{\bar{x}}{|x|} \cdot \left(\frac{dx}{|x|} - \frac{x d|x|^2}{2|x|^3}\right) = \frac{\bar{x} \cdot dx - d\bar{x} \cdot x}{2|x|^2} = a_\infty.$$

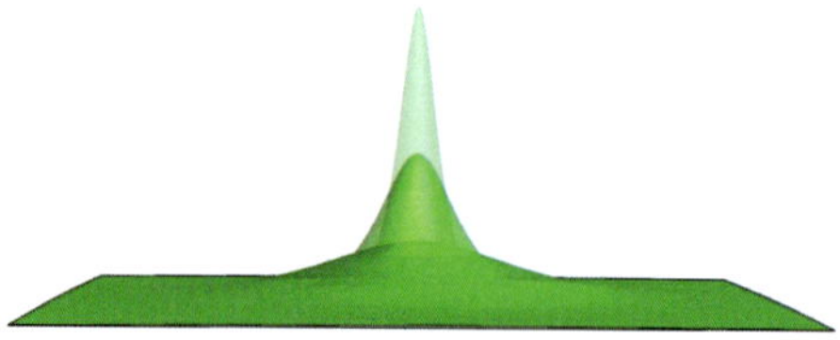

Figure 5.4.1. A cartoon suggesting how the curvature can concentrate near a point.

This shows that $g^* a_\infty$ is the trivial connection. The path of connections $\tau_\lambda^*(a)$ converges away from the origin to the trivial connection (up to gauge transformations) and the limit extends to a connection in a different principal $SU(2)$-bundle. Therefore, to compactify the moduli space of ASD connections, we have to allow the topology of the bundle to change.

This "bubbling" phenomenon also occurs in symplectic topology. If we study pseudo-holomorphic curves and fix a conformal structure on the Riemann surface, then bubbling of spheres may occur. It is important to also allow the conformal structure to change, tying this story with the Riemann moduli space of complex structures on 2-manifolds. Here is a short history of this analysis:

(1) Sacks & Uhlenbeck (see [43]) first discovered this phenomenon when studying the energy equation for a Riemann surface mapping into a Riemannian manifold.
(2) Uhlenbeck (see [51, 52]) and Taubes (see [48, 50]) worked out the case for the Yang-Mills equation.
(3) Gromov (see [24]) worked out the case for Pseudo-holomorphic curves.

5.5. Gauge Transformations We wish to take seriously the idea of doing Morse homology as outlined in the preamble for the Chern-Simons functional, and in particular, developing an analogue of the broken trajectory compactification in Theorem 1.4.4. To this end, we need to review some analytic properties of the ASD equation originally proved by Uhlenbeck, Taubes and Donaldson.

The gauge group $\mathcal{G}_P$ acts on $\mathcal{A}_P$ by the formula

$$\begin{aligned} \mathcal{A}_P \times \mathcal{G}_P &\to \mathcal{A}_P \\ (A, g) &\mapsto g \cdot A = A + g^{-1} d_A g \end{aligned}$$

from equations (2.4.1) and (2.4.3).

How bad is this gauge action? Orbits with non-trivial stabilizers give rise to "non-smooth" points in $\mathcal{B}$. Given a central element $u \in Z(G)$, the bundle map

$$\tilde{u} : P \to P,\ p \mapsto p \cdot u$$

fixes all connections on P, because a connection is always invariant under the right multiplication. This shows that

$$u^* A = A,$$

and $u \in \mathrm{Stab}_A$ for any $A \in \mathcal{A}_P$. Working with the quotient group $\mathcal{G}_P/Z(G)$ makes this action faithful, so the center $Z(G)$ does not cause problems.

If $g \cdot A = A$, then $g^{-1}\nabla_A g = 0$. This means that g is a parallel section of $\mathrm{End}(V)$; so g is determined by its value $g(p)$ at a base point $p \in P$. Hence,

$$\mathrm{Stab}_A \subset G.$$

To understand this more concretely, take a loop $\gamma : [0,1] \to X$ around $x = \pi(p) \in X$. Let $\tilde{\gamma}$ be the horizontal lift of γ with $\tilde{\gamma}(0) = p$. Then $\tilde{\gamma}(1) = p \cdot u^{-1}$ for some $u \in G$. This element

$$u = \rho_{A,p}(\gamma) \in G$$

is the holonomy of A around γ and $\rho_{A,p}$ defines a representation:

$$\rho_{A,p} : \Omega(X, x) \to G.$$

The inverse in the definition of u is to make $\rho_{A,p}$ a group homomorphism. If $g \in \mathrm{Stab}_A$, then $\tilde{g} \circ \tilde{\gamma}$ is a horizontal lift of γ with $\tilde{g} \circ \tilde{\gamma}(0) = p \cdot g(p)$, so it must agree with the path $\tilde{\gamma}(t) \cdot g(p)$ for any $t \in [0,1]$. In particular, comparing the end points gives:

$$\begin{aligned} p \cdot \underline{g(p) \cdot u^{-1}} &= \tilde{g}(p) \cdot u^{-1} = \tilde{g}(p \cdot u^{-1}) = \tilde{g}(\tilde{\gamma}(1)) \\ &= \tilde{\gamma}(1) \cdot g(p) = p \cdot \underline{u^{-1} \cdot g(p)}. \end{aligned}$$

Thus $g(p)$ lies in the commutant of $\mathrm{Im}\,\rho_{A,p}$. This is also a sufficient condition. Hence,

$$\mathrm{Stab}_A = C(\mathrm{Im}\,\rho_{A,p}) \subset G.$$

For a subgroup $H \subset G$, denote

$$C(H) \doteq \text{the commutant of } H \subset G.$$

We have to decide which subgroup of G can be realized as the commutant of other subgroups. We list some properties and examples:

(1) If $H = C(K)$ for some $K \subset G$, then

$$C\big(C(H)\big) = H.$$

This property is not true in general if the assumption $H = C(K)$ is removed.

(2) All commutants form a directed graph $\mathbb{G}$ where arrows are given by inclusions of subgroups. The center $Z(G)$ is the initial object, while G is the

final object. By (1), taking commutants identifies $\mathbb{G}$ with its dual graph (where arrows are reversed).

(3) Let $G = U(n)$. Let H consist of elements of the block matrices

$$\begin{pmatrix} x_1 I_{n_1} & & & & & \\ & \cdots & & & & \\ & & I_{n_s} & & & \\ & & & U(k_1) & & \\ & & & & \cdots & \\ & & & & & U(k_r) \end{pmatrix},$$

where $x_i \in S^1$. Then $C(H)$ is given by

$$\begin{pmatrix} U(n_1) & & & & & \\ & \cdots & & & & \\ & & U(n_s) & & & \\ & & & x_1 I_{k_1} & & \\ & & & & \cdots & \\ & & & & & x_r I_{k_r} \end{pmatrix}.$$

Up to conjugacy, all commutants in $U(n)$ are given in this form.

(4) For $SO(3)$ the graph of commutants is

$$\begin{array}{ccccccc} & & & K_4 = \mathbb{Z}_2 \times \mathbb{Z}_2 & & & \\ & & \nearrow & & \searrow & & \\ 1 \longrightarrow & \mathbb{Z}_2 & & & & O_2 & \longrightarrow SO(3), \\ & & \searrow & & \nearrow & & \\ & & & SO(2) & & & \end{array}$$

where

- $C(SO(3)) = 1$;
- $\mathbb{Z}_2 = \{e, \operatorname{diag}(-1,-1,1)\}$. $C(\mathbb{Z}_2) = O(2)$;
- $K_4 = \{A = \operatorname{diag}(\pm1, \pm1, \pm1) : \det A = 1\}$. $C(K_4) = K_4$;
- $SO(2) = \{\begin{pmatrix} A & 0 \\ 0 & 1 \end{pmatrix} : A \in SO(2)\}$. $C(SO(2)) = SO(2)$;
- $O(2) = \{\begin{pmatrix} A & 0 \\ 0 & \det A \end{pmatrix} : A \in O(2)\}$. $C(O(2)) = \mathbb{Z}_2$.

Arrows denote inclusions of subgroups. It is easily seen that this graph is dual to itself.

Definition 5.5.1. A connection $A \in \mathcal{A}_P$ is called reducible if $\mathrm{Stab}_A \neq Z(G)$. ◇

A reducible connection will make the quotient space singular.

5.6. Sobolev Completions To analyze the quotient configuration space $\mathcal{A}_P/\mathcal{G}_P$. We would like to consider this quotient space as some kind of infinite dimensional manifold, modeled on a Hilbert or Banach space. The model for this work can be found in [3, 14, 21, 35, 52] and many other places. To begin, we consider the completion with respect to some Sobolev topology. For an oriented Riemannian manifold X and a vector bundle $E \to X$, fix a smooth reference connection A. For any section $s : X \to E$, define its Sobolev norm to be

$$\|s\|^p_{L^p_{k,A}} = \int_X \sum_{l=0}^{k} |\nabla_A^{(l)} s|^p \, dvol_X.$$

By completing $\mathcal{C}^\infty(X, E)$ with respect to $\|\cdot\|_{L^p_{k,A}}$, we obtain the Sobolev space

$$L^p_k(X, E).$$

This space is independent of the reference connection A: the norms $\|\cdot\|_{L^p_{k,A}}$ induced by different sufficiently smooth connections are equivalent.

Let E be either $T^*X \otimes \operatorname{ad} P$ or $\operatorname{End}(V)$, then we obtain the Sobolev completion of $\mathcal{A}_P$ and $\mathcal{G}_P$ with respect to $\|\cdot\|_{L^p_k}$:

$$\mathcal{A}^p_k, \mathcal{G}^p_k.$$

Here $\mathcal{G}_P$ is viewed as a subset of $\mathcal{C}^\infty(X, \operatorname{End}(V))$ and V is a vector bundle induced from a faithful representation of G. If p and k are not large enough, then $L^p_k \not\hookrightarrow \mathcal{C}^0$, and it is a delicate matter to interpret the meaning of the pointwise condition:

$$s_x \in (\operatorname{Ad} P)_x \subset \operatorname{End}(V_x).$$

Theorem 5.6.1 (Sobolev Embedding Theorem). *In dimension n, we have embeddings:*

(1) $L^p_k \hookrightarrow L^q_l$ *if* $k - \frac{n}{p} \geqslant l - \frac{n}{q}$ *and* $k > l$.
(2) $L^p_k \hookrightarrow \mathcal{C}^{l,\alpha}$ *if* $k - \frac{n}{p} \geqslant l + \alpha$, $k > l$ *and* $0 < \alpha < 1$.

In particular, $k - \frac{n}{p} > 0$ implies $L^p_k(X, E) \hookrightarrow \mathcal{C}^0(X, E)$. In addition to the embedding theorem, we also have the multiplication theorem which says in this case $L^p_k(X, E)$ is a Banach algebra. We will show

Proposition 5.6.2. *If $k - \frac{n}{p} > 0$, then $\mathcal{G}^p_k$ is a Banach Lie group.*

We focus on the case when $G = U(n)$. Let V be the bundle induced by the fundamental representation of $U(n)$. Then $(\operatorname{Ad} P)_x \subset \operatorname{End}(V_x)$ is the space of unitary transformations.

Proof of Proposition 5.6.2 when $G = U(n)$. Consider the map

$$\phi : L^p_k(X, \operatorname{End}(V)) \to L^p_k(X, \operatorname{Herm}(V))$$
$$f \mapsto f^* f.$$

where $\operatorname{Herm}(V)$ denotes the bundle of Hermitian endomorphisms of V. Then $\mathcal{G}^p_k = \phi^{-1}(\mathrm{Id}_V)$. For any $g \in \mathcal{G}^p_k$, the tangent map

$$d_g\phi : L^p_k(X, \operatorname{End}(V)) \to L^p_k(X, \operatorname{Herm}(V)).$$

is surjective. By the implicit function theorem, $\ker d_g\phi$ gives a local coordinate chart of $\mathcal{G}_k^p$. It remains to check that the multiplication and the inverse map:

$$\mathcal{G}_k^p \times \mathcal{G}_k^p \xrightarrow{\times} \mathcal{G}_k^p, \; \mathcal{G}_k^p \xrightarrow{inv} \mathcal{G}_k^p$$

are smooth maps between Banach manifolds. This follows from the fact that L_k^p is an algebra. □

The action of $\mathcal{G}_{k+1}^p$ on $\mathcal{A}_k^p$ is given by

$$\begin{aligned} \mathcal{G}_{k+1}^p \times \mathcal{A}_k^p &\to \mathcal{A}_k^p \\ (g, A) &\mapsto A + g d_A g^{-1}. \end{aligned}$$

This action is smooth if

$$k + 1 - \frac{n}{p} > 0 \text{ with } n = \dim X.$$

When $n = 4$ and $p = 2$, we need $k > 1$. In the borderline case that $k = 1$, $L_2^2 \not\hookrightarrow \mathcal{C}^0$. The completion $\mathcal{G}_2^2$ is defined to be

$$\{g \in L_2^2(X^4, \operatorname{End}(V)) : g(x) \in \operatorname{Ad}P \text{ a.e. } x \in X\}.$$

For $A \in \mathcal{A}_1^2$, $A = A_0 + a$ for some $a \in L_1^2(X, \operatorname{ad} P)$ and A_0 a smooth reference connection; and we have

$$F_A = F_{A_0} + d_{A_0} a + \frac{1}{2}[a \wedge a] \in L^2.$$

The L^2-norm of F_A is the one controlled by the energy identity, so this borderline case is also important.

5.7. Local Charts The next step is to construct "orbifold" charts for the quotient space $\mathcal{B}_k^p \doteq \mathcal{A}_k^p/\mathcal{G}_{k+1}^p$. We will not give a precise definition of an orbifold but the basic idea is to locally model the quotient space as the quotient of a open subset of a Banach space by a compact Lie group acting linearly. To get a feeling for what we want to do, suppose that we have a finite dimensional manifold M with an action by a compact Lie group G. Perhaps the original idea for this construction goes back to Mostow's 1956 Annals paper [39]. Pick $x \in M$. A *transversal* to the orbit through x is a sub-manifold S_x of M that intersects transversely with the orbit $\mathcal{O}_x = G \cdot x$. To construct a transversal, pick a Riemannian metric on M. Average it under the group action to get an invariant Riemannian metric. Let B_x be a ball in the orthogonal complement to $T_x\mathcal{O}_x \subset T_xM$. Then if the ball is small enough the desired transversal is $S_x = \exp_x(B_x)$. To construct an orbifold chart in the quotient space M/G near $[x]$, consider the following two cases separately:

(1) If the action is free, consider the map

$$\phi\colon G \times S_x \to M \text{ quad by } \quad (g, s) \mapsto g \cdot s. \tag{5.7.1}$$

To show that (5.7.1) is a diffeomorphism onto its image when S_x is small enough, we apply the implicit function theorem (using the fact that $T_xS_x \oplus T_xG \cong T_xM$) and exploit the G-equivariance of the map (5.7.1).

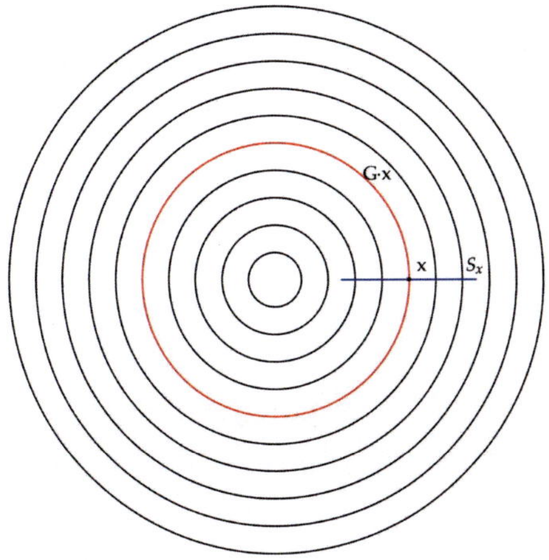

(2) If the action is not free and S_x is Stab_x-invariant, then Stab_x acts on $G \times S_x$ by the formula:

$$h(g,s) \mapsto (gh^{-1}, hs).$$

This action is free. By the first case, the quotient space

$$G \times S_x/\mathrm{Stab}_x$$

is a manifold. Consider the map

$$\begin{aligned}\phi : G \times S_x/\mathrm{Stab}_x &\to M\\ [(g,s)] &\mapsto g \cdot s.\end{aligned}$$

When S_x is small enough, ϕ is a diffeomorphism onto a neighborhood of the orbit $G \cdot x$. Then the orbifold chart comes from

$$S_x/\mathrm{Stab}_x \cong (G \times S_x/\mathrm{Stab}_x)/G \subset M/G.$$

The reader can use this construction as a model for the definition of an orbifold. To get a little more of taste of what the definition keeps track of, note that if y lies in the image of ϕ, then it is identified with

$$[(g,z)] \in G \times S_x/\mathrm{Stab}_x$$

for some $g \in G$ and $z \in S_x$ (not necessarily unique). Therefore, $\mathrm{Stab}_z \subset \mathrm{Stab}_x$. Now Stab_z and Stab_y are isomorphic as they lie in the same orbit and the element g conjugates Stab_z to Stab_y. Furthermore given z, there is a map $\hat{g} : S_y \to G$ so that the induced map

$$y' \to \hat{g}(y') \cdot y' \text{ mapping } S_y \to S_x \subset M$$

is a diffeomorphism onto its image, equivariant under the action of Stab_z. The definition keeps track of all of this data, but in general one can construct data satisfying all the axioms but which do not arise as a global quotient. One can also realize the construction of the transition map ϕ_{xy} from an orbifold chart (S_y, ϕ_y) to (S_x, ϕ_x).

We wish to define the orbifold charts for the $\mathcal{G}$-action on $\mathcal{A}$:

$$(g, A) \mapsto A + g^{-1} d_A g.$$

The differential of this action is the operator

$$T_e\mathcal{G} = \Omega^0(X, \operatorname{ad} P) \to \Omega^1(X, \operatorname{ad} P)$$
$$\xi \mapsto d_A \xi$$

and its image is the tangent space to the gauge group orbit of A. Since the L^2-inner product is $\mathcal{G}$-invariant, consider the L^2-complement of $\operatorname{Im} T_e\mathcal{G}$ and define

$$S_A = \{a \in \Omega^1(X, \operatorname{ad} P) : \langle d_A \xi, a\rangle_{L^2(X)} = 0 \ \forall \xi \in \Omega^0(X, \operatorname{ad} P)\}.$$

When X is closed,

$$(\langle d_A \xi, a\rangle_{L^2(X)} = 0 \ \forall \xi) \Leftrightarrow (d_A^* a = 0).$$

When X is not closed,

$$(\langle d_A \xi, a\rangle_{L^2(X)} = 0 \ \forall \xi) \Leftrightarrow (d_A^* a = 0, *a|_{\partial X} = 0).$$

An interesting question is how large a local orbifold chart could be. In our case, it is necessary to take a Sobolev completion. The stronger the Sobolev topology we use, the smaller the chart we get. But this is fine for our primary applications — elliptic regularity provides bijections between moduli spaces $\{F_A^+ = 0\}/\mathcal{G}$ in different Sobolev completions.

5.8. The Construction of Slices We now construct orbifold charts for $\mathcal{A}_k^p/\mathcal{G}_{k+1}^p$. For $A \in \mathcal{A}_k^p$, let

$$S_{A,\epsilon} = \{A + a : d_A^* a = 0, *a|_{\partial X} = 0, \|a\|_{L_k^p} < \epsilon\}.$$

By Hodge theory (solving a Neumann boundary value problem for $d_A^* d_A$),

$$L_k^p(X, T^*X \otimes \operatorname{ad} P) = \operatorname{Im} d_A \oplus \{a \in L_k^p(X, T^*X \otimes \operatorname{ad} P) : d_A^* a, *a|_{\partial X} = 0\} \tag{5.8.1}$$

with $d_A : L_{k+1}^p(X, \operatorname{ad} P) \to L_k^p(X, T^*X \otimes \operatorname{ad} P)$. Now consider the map

$$m : \mathcal{G}_{k+1}^p \times S_{A,\epsilon} \to \mathcal{A}_k^p.$$
$$(g, A + a) \mapsto g^*(A + a) = A + a + g d_{A+a} g^{-1}.$$

and its differential at (e, A):

$$\mathcal{D}_{(e,A)} m : L_{k+1}^p(X, \operatorname{ad} P) \times \{a \in L_k^p : d_A^* a = 0, *a|_{\partial X} = 0\} \to L_k^p$$

which maps (ξ, a) to $-d_A \xi + a$. By (5.8.1), we have

Theorem 5.8.2. *The differential $\mathcal{D}m$ is surjective at (e, A) for all $g \in \mathcal{G}_{k+1}^p$.*

We first identify $\ker \mathcal{D}_{(e,A)} m$ with the Lie algebra of Stab_A:

$$(\xi, \alpha) \in \ker \mathcal{D}_{(e,A)} m \Rightarrow -d_A \xi + \alpha = 0 \Rightarrow d_A \xi = 0, \alpha = 0.$$

Our goal is then to show that

$$\bar{m} : \mathcal{G}_{k+1}^p \times S_{A,\epsilon}/\operatorname{Stab}_A \to \mathcal{A}_k^p.$$

is a diffeomorphism onto the image when $\epsilon \ll 1$.

We start with the local version:

Theorem 5.8.3. *When* $\epsilon \ll 1$, $\bar{\mathfrak{m}}$ *is a local diffeomorphism onto its image.*

Proof. We have shown that $\mathcal{D}_{(e,A)}\bar{\mathfrak{m}}$ is a bijection. By implicit function theorem, there is a Stab_A-invariant neighborhood $\mathcal{U}$ of $\mathrm{Stab}_A \subset \mathcal{G}^p_{k+1}$ and $\epsilon \ll 1$ such that

$$\mathcal{U} \times S_{A,\epsilon}/\mathrm{Stab}_A$$

is a diffeomorphism onto its image, so the theorem is proved at the point

$$[(e, A)].$$

Now we use the fact that $\bar{\mathfrak{m}}$ is $\mathcal{G}^p_{k+1}$-equivariant:

$$\begin{array}{ccc} \mathcal{U} \times S_{A,\epsilon}/\mathrm{Stab}_A & \xrightarrow{\bar{\mathfrak{m}}} & \mathcal{G}^p_k \\ \downarrow g & & \downarrow g \\ (g \cdot \mathcal{U}) \times S_{A,\epsilon}/\mathrm{Stab}_A & \xrightarrow{\bar{\mathfrak{m}}} & \mathcal{G}^p_k. \end{array}$$

Since vertical maps are diffeomorphisms, the map $\bar{\mathfrak{m}}$ is a diffeomorphism on $(g \cdot \mathcal{U}) \times S_{A,\epsilon}/\mathrm{Stab}_A$. But this is a neighborhood of $[(g, A)]$, and so proves the theorem. □

To prove $\bar{\mathfrak{m}}$ is a diffeomorphism globally, it remains to show that $\bar{\mathfrak{m}}$ is injective when $\epsilon \ll 1$. We will allow $n = \dim X$ to be any integer $\geqslant 2$. Suppose we have

$$g^*(A + a) = A + b$$

for some $a, b \in S_{A,\epsilon}$ and $g \in \mathcal{G}^p_{k+1}$. We wish to show $g \in \mathrm{Stab}_A$. Since

$$d_A g = ga - bg$$

as sections of $\mathrm{End}(E)$, we apply d_A^* to both sides to obtain

$$\begin{aligned} d_A^* d_A g &= d_A^*(ga - bg) \\ &= -\star d_A\big(g(\star a) - (\star b)g\big) \\ &= -\star (d_A g \wedge \star a - (-1)^{n-1} \star b \wedge d_A g), \end{aligned}$$

where we used the fact that $d_A^* a = d_A^* b = 0$. This computation shows

$$d_A^* d_A g = \Phi(d_A g, a) - \Psi(b, d_A g).$$

for some bilinear bundle maps Φ, Ψ.

Lemma 5.8.4. *For some* $C > 0$,

$$\|d_A^* d_A g\|_{(L^2_1)^*} \geqslant C\|d_A g\|_{L^2},$$

for all $g \in L^2_1(X, \mathrm{End}(E))$.

Assuming this lemma, we have

$$\begin{aligned} C\|d_A g\|_{L^2} &\leqslant \|d_A^* d_A g\|_{(L^2_1)^*} \leqslant \|\Phi(d_A g, a)\|_{(L^2_1)^*} + \|\Psi(b, d_A g)\|_{(L^2_1)^*} \\ &\leqslant C_1(\|d_A g\|_{L^2}\|a\|_{L^n} + \|d_A g\|_{L^2}\|b\|_{L^n}). \end{aligned} \tag{5.8.5}$$

In the last step, we used the fact that in dimension $n > 2$, $L_1^2 \hookrightarrow L^{2n/(n-2)}$ and

$$L^{2n/(n-2)} \times L^2 \times L^n \to L^1 \xrightarrow{\int_X} \mathbb{R}$$

is bounded linear. Indeed,

$$\frac{n-2}{2n} + \frac{1}{2} + \frac{1}{n} = 1.$$

This implies that $L^2 \times L^n \to (L_1^2)^*$ is bounded linear. It is a bit subtle fact that the estimate holds also in dimension 2 because of the precise nature of the terms involved in the product (see [40]). Hence, for $n \geqslant 2$, if

$$\|a\|_{L^n} + \|b\|_{L^n} \leqslant \frac{C}{2C_1},$$

then Inequality (5.8.5) forces $\|d_A g\|_2$ to be zero and we are done. This requirement is independent of the choice of Sobolev completion of $\mathcal{G}$. Thus the injectivity of $\bar{m}$ holds for a ball of a weaker Sobolev norm L^n. We have thus proved:

Proposition 5.8.6. *Consider the slice*

$$S^{big}_{A,\epsilon} = \{a \in L^p_k : d_A^* a = 0, \star a|_{\partial X} = 0, \|a\|_{L^n} < \epsilon\}.$$

Then for ϵ sufficiently small, the induced map

$$\bar{m} : \mathcal{G}^p_{k+1} \times S^{big}_{A,\epsilon}/\operatorname{Stab}_A \to \mathcal{A}^p_k$$

is injective.

In fact, one can enhance this result in an important and subtle way.

Proposition 5.8.7. *In Proposition 5.8.6, the induced map*

$$\bar{m} : \mathcal{G}^p_{k+1} \times S^{big}_{A,\epsilon}/\operatorname{Stab}_A \to \mathcal{A}^p_k$$

is injective and is open *in the L^n-topology on $\mathcal{A}^p_k$, for ϵ sufficiently small.*

We do not prove this here. The main ideas are contained in Taubes' paper [49]. Finally, we prove Lemma 5.8.4.

Proof of Lemma 5.8.4. We assume the regularity $g \in L_1^2$ and $a, b \in L_1^{n/2}$. We think of g as a section of $\operatorname{End}(V) \to X$, instead of a gauge transformation. The equation

$$d_A g = ag - gb$$

still holds in L^2. Because of the multiplication structure

$$L_1^{n/2} \times L_1^2 \to L_1^{\frac{2n}{n+2}}$$

$d_A g \in L_1^{\frac{2n}{n+2}}$ and so its boundary value makes sense. Since $\star a, \star b = 0$ on ∂X,

$$\star d_A g|_{\partial X} = 0.$$

Therefore, we can integrate by parts without worrying about the boundary term and write

$$\|d_A^* d_A g\|_{L_1^2(X)^*} = \sup_{\|v\|_{L_1^2}=1} \langle d_A^* d_A g, v\rangle_{L^2} = \sup_{\|v\|_{L_1^2}=1} \langle d_A g, d_A v\rangle_{L^2}.$$

Write $g = g_1 + g_2$ with

$$g_1 \in \ker d_A \text{ and } \langle g_2, g_1 \rangle_{L^2} = 0.$$

Then $\|d_A g\|_2 = \|d_A g_2\|_2 \geqslant C\|g_2\|_{L^2_1}$ for some $C > 0$. Take

$$v = \frac{g_2}{\|g_2\|_{L^2_1}}$$

to be the test function. Hence,

$$\|d_A^* d_A g\|_{L^2_1(X)^*} \geqslant \frac{1}{\|g_2\|_{L^2_1}} \langle d_A g, d_A g \rangle_{L^2} \geqslant \frac{C}{\|d_A g\|_2} \|d_A g\|_2^2 \geqslant C\|d_A g\|_2.$$

The heuristic for this estimate is that d_A^* is injective on $\operatorname{Im} d_A$:

$$\Omega^0 \underset{d_A}{\overset{d_A^*}{\rightleftarrows}} \Omega^1,$$

so we expect that $\|d_A^* d_A g\|_{L^2_{k+1}} \geqslant \|d_A g\|_{L^2_k}$ for any k. When X has a non-empty boundary, it is not true that

$$L^2_{-k} = (L^2_k)^*.$$

Instead,

$$L^2_{-k} = (L^2_{k,0})^*.$$

Here $L^2_{k,0}$ denotes the space of L^2_k functions whose restriction at the boundary is zero (if $k > \frac{1}{2}$). This space allows us to do integration by parts without worrying about the boundary term. □

5.9. The local structure of the moduli space on a closed 4-manifold Let $P \to X^4$ be a principal G-bundle over a closed 4-manifold. Define the moduli space of ASD connections as

$$\mathcal{M}(X) \doteq \{A \in \mathcal{A}_P : F_A^+ = 0\}/\mathcal{G}_P.$$

If $[A] \in \mathcal{M}(X)$, then $\mathcal{M}(X)$ can be locally described as the zero locus of the map

$$G_A : \{A + a \in S_A\} \to \Omega^2_+(\operatorname{ad} P)$$
$$A + a \mapsto F^+_{A+a},$$

whose linearization at $a = 0$ is d_A^+. With appropriate Sobolev completions, G_A becomes a Fredholm map. Indeed, there is a complex of differential forms associated to any ASD connection:

$$0 \to \Omega^0(\operatorname{ad} P) \xrightarrow{d_A} \Omega^1(\operatorname{ad} P) \xrightarrow{d_A^+} \Omega^2_+(\operatorname{ad} P) \to 0 \tag{5.9.1}$$

because $d_A^+ \circ d_A = F_A^+ = 0$. In fact, the complex of equation (5.9.1) is elliptic, meaning the symbol sequence is exact. This implies the following theorem due to Atiyah-Hitchin-Singer.

Theorem 5.9.2 ([4]). *For any* $p \geqslant 1$ *and* $k \geqslant 1$, *we have a Fredholm complex*

$$L^p_{k+1}(X, \operatorname{ad} P) \xrightarrow{d_A} L^p_k(X, T^*X \otimes \operatorname{ad} P) \xrightarrow{d_A^+} L^p_{k-1}(X, \Lambda^2_+ \otimes \operatorname{ad} P)$$

which means that

(1) *it has finite dimensional cohomology groups;*
(2) *all maps have closed range.*

Furthermore the (negative) index of this complex is given by

$$8k - \frac{3}{2}(\chi(X) + \sigma(X)) = 8k - 3(1 - b^1(X) + b^+(X)).$$

Theorem 5.9.2 suggests that $\mathcal{M}(X)$ is an orbifold of corresponding dimension, if it is cut out "transversely" by the map G_A.

5.10. Uhlenbeck's Fundamental Lemma Our next step is to understand the compactness property of the ASD moduli space $\mathcal{M}(X)$. The motivating question is whether an ASD connection with curvature close to F_A can have the connection 1-form close to A after a suitable gauge transformation. As the base case, Uhlenbeck's fundamental lemma addresses this problem when $X = B^4$ is the standard flat 4-ball.

Theorem 5.10.1 (Uhlehbeck [52]). *For $n \leqslant 4$, there exist constants $C, \epsilon > 0$ such that for any L^2_1-connection A on B^n with*

$$\int_{B^n} |F_A|^2 \leqslant \epsilon,$$

we can find a gauge transformation $g \in \mathcal{G}^2_2$ such that

$$g^*A = \Gamma + a$$

where Γ is the trivial connection and

(1) $d^*_\Gamma a = 0$,
(2) $\star a|_{\partial B^4} = 0$,
(3) $\int_{B^4}(|\nabla_\Gamma a|^2 + |a|^2)dvol_{B^n} \leqslant C \int_{B^4} |F_A|^2 dvol_{B^n}$.

When $n = 4$, the hypothesis of theorem is scale invariant; so the conclusion holds over a smaller ball $B^4(r), 0 < r < 1$ (with a slight modification because $|a|^2$ is not scale invariant). When $n < 4$, if we scale $B^4(r)$ up to the unit size, then the L^2-norm of the curvature shrinks. It follows that one can alway find a cover of a compact 2 or 3 dimensional manifold on which the hypothesis holds.

We will not give the full proof of this but highlight a few key computations. To get a feeling for why it holds and where the dimension constraint comes from, we content ourselves with proving the following lemma.

Lemma 5.10.2. *There exists constants $C_1, \epsilon_1 > 0$ such that for all $A = \Gamma + b$ with*

- $\|b\|_4 \leqslant \epsilon_1$,
- $d^*b = 0$,
- $*b|_{\partial B^4} = 0$,

we have

$$\int_{B^4} |\nabla_\Gamma b|^2 + |b|^2 \leqslant C_1 \int_{B^4} |F_A|^2.$$

Proof. By the Weitzenböck formula, we have

$$\int_{B^4} |(d + d^*)b|^2 = \int_{B^4} |\nabla_\Gamma b|^2 + \int_{\partial B^4} \star b \wedge d^* b + \int_{\partial B} |b|^2. \tag{5.10.3}$$

for any $b \in \Omega^1(B^4)$. Since the Euclidean metric on B^4 is flat, no Ricci curvature term appears here. The last term is positive definite, because $S^3 = \partial B^4 \subset B^4$ has positive mean curvature.

Exercise 5.10.4. Prove the Weitzenböck formula (5.10.3). ♢

By assumption, we have $d^* b = 0$ and $*b|_{\partial B^4} = 0$. Equation (5.10.3) then implies that

$$\int_{B^4} |db|^2 \geqslant \int_{B^4} |\nabla_\Gamma b|^2. \tag{5.10.5}$$

Since $L^2_1 \hookrightarrow L^4$, for some $C_{\text{Sob}} > 0$,

$$\|b\|_4 \leqslant C_{\text{Sob}} \|b\|_{L^2_1} \doteq \left(\int_{B^4} |\nabla_\Gamma b|^2 + |b|^2\right)^{\frac{1}{2}}.$$

Putting all these together, we have

$$\begin{aligned}
\int_{B^4} |F_A|^2 = \int_{B^4} |db + b \wedge b|^2 &\gtrsim \int_{B^4} |db|^2 - |b|^4 \\
\text{(by equation (5.10.5))} &\geqslant \left(\int_{B^4} |\nabla_\Gamma b|^2\right) - \|b\|_4^4 \\
(\text{since } \|b\|_4 < \epsilon) &\geqslant \left(\int_{B^4} |\nabla_\Gamma b|^2\right) - \epsilon_1^2 C_{\text{Sob}}^2 \|b\|_{L^2_1}^2 \\
&\geqslant \frac{\lambda_1^2}{\lambda_1^2 + 1} \|b\|_{L^2_1}^2 - \epsilon_1^2 C_{\text{Sob}}^2 \|b\|_{L^2_1}^2 \\
&\geqslant \left(\frac{\lambda_1^2}{\lambda_1^2 + 1} - \epsilon_1^2 C_{\text{Sob}}^2\right) \|b\|_{L^2_1}^2 \geqslant C_1^{-1} \|b\|_{L^2_1}^2,
\end{aligned}$$

if $\epsilon_1 \ll 1$. Here we used the fact that

$$\int_{B^4} |\nabla_\Gamma b|^2 \geqslant \lambda_1^2 \int_{B^4} |b|^2.$$

for some $\lambda_1 > 0$. To follows by contradiction, since a parallel 1-form b on B^4 with the boundary condition $*b|_{\partial X} = 0$ also vanishes on the whole space. □

5.11. The Curvature Map Is Proper To get an idea where compactness for ASD connections comes from, we will show that in regions of small energy the curvature map is proper. Consider the small curvature slice:

$$SC_\epsilon = \{A = \Gamma + b : d^* b = 0, *b|_{\partial B^4} = 0, \int_{B^4} |F_A|^2 < \frac{\epsilon}{2}\}$$

with the same ϵ as in Theorem 5.10.1.

Theorem 5.11.1. *The curvature map*

$$A \in SC_\epsilon \mapsto F_A \in L^2$$

is proper on the slice SC_ϵ.

Proof. Let $A_i = \Gamma + a_i \in SC_\epsilon$. Suppose that their curvatures $\{F_{A_i}\}$ are Cauchy in L^2. We can use Uhlenbeck's fundamental lemma, Theorem 5.10.1 to aid in the following estimates:

$$\begin{aligned}\int_{B^4} |F_{A_i} - F_{A_j}|^2 &\gtrsim \int_{B^4} |d(a_i - a_j)|^2 - \int_{B^4} |(a_i - a_j) \wedge a_i + a_j \wedge (a_i - a_j)|^2 \\ &\geqslant \int_{B^4} |(d + d^*)(a_i - a_j)|^2 - C_2 \|a_i - a_j\|^2_{L^2_1} (\|a_i\|^2_{L^2_1} + \|a_j\|^2_{L^2_1}) \\ &\geqslant C_3 \|a_i - a_j\|^2_{L^2_1} - \epsilon C_2 C \|a_i - a_j\|^2_{L^2_1} \\ &= (C_3 - \epsilon C_2 C)\|a_i - a_j\|^2_{L^2_1}.\end{aligned}$$

This implies that $\{a_i\}$ is Cauchy if $\epsilon \ll 1$. □

5.12. The Uhlenbeck Compactness Theorem.

A local compactness result Let $B_r = B^4(r)$ be the ball of radius r in $\mathbb{R}^4$, and

$$\mathcal{M}(B_r) = \{[A] : F_A = - * F_A\} \subset \mathcal{A}^2_1/\mathcal{G}^2_2,$$

be the moduli space of ASD connections on B_r. When $r = 1$, the space

$$\mathcal{M}^\epsilon(B_1) = \{[A] \in \mathcal{M}(B_1) : \int_{B_1} |F_A|^2 dvol_{B_1} < \epsilon\}$$

can be understood using Uhlenbeck's fundamental lemma, Theorem 5.10.1. By restricting connections to a smaller ball, we obtain a map:

$$r : \mathcal{M}^\epsilon(B_1) \to \mathcal{M}(B_r).$$

Theorem 5.12.1 (Uhlenbeck)**.** *When $\epsilon \ll 1$, The restriction map r is compact.*

This theorem is analogous to a classical theorem in complex analysis. Consider the space of holomorphic functions on the unit disk $B(0,1) \subset \mathbb{C}$ with uniformly bounded L^2-norms, then the restriction map to a smaller disk $B(0,r)$ is compact in $\mathcal{C}^k$-topology for any $k > 0$:

$$\{f : B(0,1) \to \mathbb{C} : \bar{\partial} f = 0, \|f\|_2 < C\} \xrightarrow{r} \mathcal{C}^k(B(0,r)).$$

The global version The argument below should be compared with the sketch in Subsection 1.4 for the moduli space of unparameterized Morse trajectories. The key role is played the decomposition into small energy pieces. Taking up the global picture, we fix a principal G-bundle

$$P \to X^4$$

over a closed 4-manifold and pick a sequence $\{A_i\}$ of ASD connections. Corollary 5.1.7 then says

$$E(A) = 8\pi^2 \langle c_2(P), [X^4] \rangle$$

is a constant independent of A. We can cover

$$X = \bigcup_{i \in \Lambda} B(x_i, \delta)$$

by balls of a given radius δ such that for a fixed constant $N > 0$, any $x \in X$ lies in at most N such δ-balls. Then we can find a subsequence of $\{A_j\}$ such that except for

$$\frac{E(A)N}{\epsilon}$$

bad balls, the curvature constraint

$$\int_{B(x_i,\delta)} |F_{A_j}|^2 < \epsilon \tag{5.12.2}$$

holds for all $i \in \Lambda$ and all connections A_j in this subsequence.

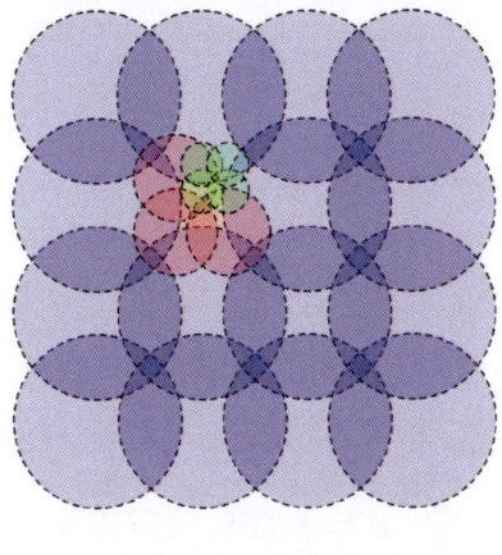

Figure 5.12.3. A cartoon indicating how the curvature may concentrate in a sequence.

Remark 5.12.4. One way to produce such a cover is as follows. Find a maximal collection of disjoint balls of radius $\frac{\delta}{3}$ inside X. Let them be

$$B(x_i, \frac{\delta}{3}), 1 \leqslant i \leqslant m.$$

Then the collection $\{B(x_i, \delta)\}$ necessarily covers X. For any $x \in X$, if $x \in B(x_i, \delta)$, then

$$B(x_i, \frac{\delta}{3}) \subset B(x, 2\delta).$$

Therefore,

$$N \leqslant N_0 \doteq \frac{\max_{x \in X} \mathrm{Vol}(B(x, 2\delta))}{\min_{1 \leqslant i \leqslant m} \mathrm{Vol}(B(x_i, \delta/3))}.$$

At this step, we can apply Gromov-Bishop comparison theorem to estimate N_0 in terms of a curvature bound of X. Alternatively, we shrink the size of δ. Locally, any metric is close to the standard Euclidean metric. Then N is bounded by the corresponding number N_0 in $\mathbb{R}^n$. ◇

We can iterate this process (by letting $\delta \to 0$) and obtain a cover of

$$X^c \doteq X - \{x_1, x_2, \cdots, x_k\}$$

such that for any ball in this cover, the constraint of equation (5.12.2) holds and we can apply Uhlenbeck's fundamental lemma. After proper gauge fixing, we have $\mathcal{C}^\infty_{loc}$-convergence

$$[A_i]|_{X^c} \xrightarrow{\mathcal{C}^\infty_{loc}} [A] \text{ on } X^c$$

to a smooth connection A on $P|_{X^c}$. However, if

$$\frac{1}{2}\int_{X^c}|F_A|^2 < \frac{1}{2}\int_X |F_{A_i}|^2 = 4\pi^2\langle c_2(P),[X^4]\rangle,$$

then we can only have weak convergence $|F_{A_i}| \xrightarrow{w-L^2} |F_A|$. To extend A to a global connection on X^4, we need to apply Uhlenbeck's Removable Singularities Theorem:

Theorem 5.12.5 (Ulhenbeck [51]). *Suppose* $A = \Gamma + a$ *is a smooth connection on* $G \times (B^4 - \{0\})$ *with* $F_A^+ = 0$. *If* $E(A) < \infty$, *then we can find a smooth gauge transformation* $g : B^4 - \{0\} \to G$ *such that*

$$g \cdot A = \Gamma + a'$$

and a' *extends to a smooth connection 1-form on* $B^4 \times G$.

The map from the punctured ball to G determines an element in $\pi_3(G)$. This theorem allows us to extend the limit A to a connection A' but in a possibly non-isomorphic bundle $P' \to X$. If $\pi_1(G) \neq \{0\}$, then P might have characteristic classes in dimension 2, but they are unaffected under the transformation g in Theorem 5.12.5. Only a 4-dimensional class might be possibly changed.

6. Lecture 3: The Construction of Floer Homology

6.1. The ASD Equation and Gauge Fixing on A Cylinder Consider an oriented closed 3-manifold Y and the trivial principal G-bundle $Q \to Y$. Let Γ be the trivial connection with respect to a fixed trivialization. Recall that for $B = \Gamma + b$, the Chern-Simons functional is defined as

$$\mathcal{CS}(b) = -\int \operatorname{tr}(b \wedge db + \frac{1}{3} b \wedge [b \wedge b]),$$

and its gradient at B is

$$\nabla \mathcal{CS}(B) = \star F_B.$$

The negative gradient flow equation

$$\frac{\partial}{\partial t} B(t) = -\star F_{B(t)}$$

is equivalent to the ASD equation on the 4-manifold $\mathbb{R} \times Y$. We would like to transfer ideas from Morse Theory to the Chern-Simons functional, and mimic the construction of Morse Homology. There are some potential problems:

- The functional $\mathcal{CS}$ is not single valued on the quotient space $\mathcal{B}_Q = \mathcal{A}_Q/\mathcal{G}_Q$. We may need the Novikov-Morse homology (See Section 6.4).
- The space of connections on Y^3 can have complicated topology. The quotient group $\mathcal{G}_Q/Z(G)$ may not act freely on $\mathcal{A}_Q$.
- The moduli space of flowlines might be non-compact due to the bubbling phenomenon.
- We have to perturb the Chern-Simons functional $\mathcal{CS}$ to make it Morse.

A general four dimensional connection on $\mathbb{R} \times Y$ can be written in terms of the pull back of a reference connection B_o on Y as

$$A = \pi^* B_o + b + c\,dt$$

where $b = b(t)$ is a t-dependent Lie algebra valued one-form c is t-dependent Lie algebra valued function. Rewriting equation (5.1.4) is these terms we have

$$0 = \frac{\partial}{\partial t} b + \star_3 (F_{B_o} + D_{B_o} b + \frac{1}{2}[b \wedge b]) - d_{B_o} c - [b \wedge c]. \tag{6.1.1}$$

The negative gradient flow equation (6.1.1) itself does not form an elliptic system. We will also need the Coulomb gauge fixing condition:

$$\begin{aligned} 0 &= -d_B^*(b + c\,dt) = \star_4 d_B \star_4 (b + c\,dt) \\ &= -\star_4 d_B (dt \wedge \star_3 b + \star_3 c) \\ &= \frac{\partial}{\partial t} c(t) - d_B^* b. \end{aligned}$$

Together with the linearization of (6.1.1) for B, we have:

$$0 = \frac{\partial}{\partial t}\begin{pmatrix} b \\ c \end{pmatrix} + \begin{pmatrix} \star_3 d_B & -d_B \\ -d_B^* & 0 \end{pmatrix}\begin{pmatrix} b \\ c \end{pmatrix}$$

This operator applied to (b, c) can be cast into the form

$$\frac{\partial}{\partial t} + D_B$$

with D_B first order self-adjoint and elliptic on $C^\infty(Y, (\Lambda_Y^1 \oplus \Lambda_Y^0) \otimes \mathrm{ad}\, Q)$. Thus,

$$\ker D_B = \operatorname{Coker} D_B.$$

We would like to understand what this kernel means and when it is trivial.

6.2. The Extended Hessian A critical point of $\mathcal{CS}$ is a flat connection B of $Q \to Y$, i.e. $F_B \equiv 0$. A flat connection also induces a π_1-representation:

$$\rho : \pi_1(Y) \to G.$$

Consider the equation $D_B(b, c) = 0$:

$$\begin{cases} \star_3 d_B b - d_B c &= 0 \\ \qquad\quad d_B^* b &= 0 \end{cases} \tag{6.2.1}$$

Note that $\operatorname{Im} d_B^*$ is orthogonal to $\operatorname{Im} d_B$ in $\Omega^1(Y, \mathrm{ad}\, Q)$. To see this, consider the twisted de Rham complex:

$$\Omega^0(Y, \mathrm{ad}\, Q) \xrightarrow{d_B} \Omega^1(Y, \mathrm{ad}\, Q) \underset{\overset{\curvearrowleft}{d_B^* = -\star d_B \star}}{\overset{\overset{\star}{\curvearrowright}}{\xrightarrow{d_B}}} \Omega^2(Y, \mathrm{ad}\, Q) \xrightarrow{d_B} \Omega^3(Y, \mathrm{ad}\, Q).$$

This is a chain complex of differential forms, because $d_B^2 = F_B = 0$. Hence, (b, c) is a solution to equation 6.2.1 if and only if

$$d_B b = d_B^* b = 0, d_B c = 0,$$

i.e. b is a harmonic form in $\mathcal{H}^1(Y, \mathrm{ad}_B)$ and $c \in \mathcal{H}^0(Y, \mathrm{ad}_B)$. Here $\mathcal{H}^*(Y, \mathrm{ad}_B)$ stands for harmonic forms with respect to d_B.

Definition 6.2.2. A critical points $[B]$ of $\mathcal{CS}$ is called non-degenerate if
(1) $\mathcal{H}^0(Y, \mathrm{ad}_B) = \{0\}$,
(2) $\mathcal{H}^1(Y, \mathrm{ad}_B) = \{0\}$.
The first condition implies that B is irreducible as a flat connection. ♢

To achieve conditions (1) and (2), we have to apply holonomy perturbations to the Chern-Simons functional $\mathcal{CS}$. This technical step will be not discussed in our lectures. We will introduce the notion of "admissible bundles" (Definition 6.6.3) in Section 6.6 to achieve (1). If B is a non-degenerate critical point, then the operator

$$D_B : L_k^2(Y, (\Lambda^0 \oplus \Lambda^1) \otimes \mathrm{ad}\, Q) \to L_{k-1}^2(Y, (\Lambda^0 \oplus \Lambda^1) \otimes \mathrm{ad}\, Q)$$

is invertible. D_B is also called the extended Hessian at B.

Theorem 6.2.3. *The operator* $\frac{\partial}{\partial t} + D_B$ *is invertible as a bounded linear operator between*

$$L_k^2(\mathbb{R} \times Y, (\Lambda_Y^0 \oplus \Lambda_Y^1) \otimes \mathrm{ad}\, Q) \to L_{k-1}^2(\mathbb{R} \times Y, (\Lambda_Y^0 \oplus \Lambda_Y^1) \otimes \mathrm{ad}\, Q),$$

if and only if D_B *is invertible.*

In this theorem, the connection B is independent of the time t.

Proof. We verify this theorem directly when $k = 1$. Let

$$u \in \mathcal{C}_c^\infty(\mathbb{R} \times Y, (\Lambda_Y^0 \oplus \Lambda_Y^1) \otimes \mathrm{ad}\, Q)$$

be any compactly supported section; then

$$\|(\frac{\partial}{\partial t} + D_B)u\|_{L^2(\mathbb{R}\times Y)}^2 = \int_{\mathbb{R}\times Y} |\frac{\partial}{\partial t}u|^2 + |D_B u|^2 + 2\langle \frac{\partial}{\partial t}u, D_B u\rangle.$$

As D_B is self-adjoint, $\frac{\partial}{\partial t}\langle u, D_B u\rangle = \langle \frac{\partial}{\partial t}u, D_B u\rangle + \langle u, \frac{\partial}{\partial t}(D_B u)\rangle = 2\langle \frac{\partial}{\partial t}u, D_B u\rangle$. Note that u is compactly support, so

$$\int_{\mathbb{R}\times Y} 2\langle \frac{\partial}{\partial t}u, D_B u\rangle = \frac{\partial}{\partial t}\langle u, D_B u\rangle \Big|_{-\infty}^{+\infty} = 0.$$

On the other hand, the invertibility of D_B implies that

$$\|D_B u\|_{L^2(Y)} \geqslant c\|u\|_{L_1^2(Y)}^2$$

for some $c > 0$. Hence,

$$\|(\frac{\partial}{\partial t} + D_B)u\|_{L^2(\mathbb{R}\times Y)}^2 \geqslant \int_{\mathbb{R}\times Y} |\frac{\partial}{\partial t}u|^2 + \|u\|_{L_1^2(Y)}^2.$$

This estimate shows that $\frac{\partial}{\partial t} + D_B : L_1^2 \to L^2$ is injective with closed range. It remains to verify the surjectivity. At this point, there are at least two ways to proceed:

(1) The formal adjoint of $\frac{\partial}{\partial t} + D_B$ is $-\frac{\partial}{\partial t} + D_B$. The same estimate shows that it is injective.

(2) Verify that the range of $\frac{\partial}{\partial t} + D_B$ is dense in L^2 by hand. D_B has a spectral decomposition, i.e., it has an L^2-complete set of eigenvectors. Denote these, for $\lambda \in \mathrm{Spec}(D_B)$, by $\{\phi_\lambda\}$.

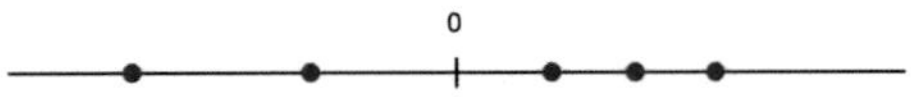

The spectrum of D_B is discrete and accumulates only at infinity so

$$\|u\|^2_{L^2_1(Y)} = \sum_{\lambda \in \mathrm{Spec}\, D_B} \lambda^2 c_\lambda^2$$

if $u = \sum c_\lambda \phi_\lambda$ and $0 \notin \mathrm{Spec}\, D_B$. With respect to the spectral decomposition, $\frac{\partial}{\partial t} + D_B$ is a diagonal matrix with diagonal entries given by

$$\frac{\partial}{\partial t} + \lambda,\ \lambda \in \mathrm{Spec}\, D_B.$$

We can show each $\frac{\partial}{\partial t} + \lambda$ has a dense range in

$$L^2(\mathbb{R}, \mathbb{R}).$$

For the easy extension higher regularities (with $k > 1$), see Dondl. □

6.3. The Spectral Flow and Fredholm Index As above, if we write

$$A = \pi^* B_o + a = \pi_o^* + (b + c\,dt)$$

with $(b, c) \in L^p_k(\mathbb{R} \times Y, (\Lambda^1_Y \oplus \Lambda^0_Y) \otimes \mathrm{ad}\, Q))$, then the linearized operator

$$D_A = \frac{\partial}{\partial t} + D_B + a\#$$

is Fredholm of index 0. (Here $a\#\cdot$ denotes some algebraic operator depending on a which is self=adjoint in this case.) Indeed if $a \equiv 0$, D_A is invertible. The error term $a\#\cdot$ is a compact operator, so $F\mathfrak{D}_A$ is Fredholm of index 0.

Here is another way to think about the Fredholmness. Since $a \to 0$ as $t \to \pm\infty$, the operator

$$D_B \cdot + a(t)\#\cdot$$

is close to D_B in the operator norm as $t \to \pm\infty$. Hence, $D_B + a(t)\#\cdot$ is invertible when $|t| \gg 1$.

We would like to understand the general relation between the spectral flow and the index of D_A when $a \notin L^p_k(Z)$. We assume that $a \in L^p_{k,loc}(Z)$ and there exist translation invariant forms a_+, a_- in $\mathcal{C}^\infty(\mathbb{R} \times Y)$ such that

$$a - a_+|_{[0,\infty) \times Y} \in L^p_k([0,\infty) \times Y),$$
$$a - a_-|_{(-\infty,0] \times Y} \in L^p_k((-\infty,0] \times Y).$$

Lemma 6.3.1. *If $\frac{\partial}{\partial t} + D_B + \alpha_+$ and $\frac{\partial}{\partial t} + D_B + \alpha_-$ are both invertible, then the operator*

$$D_A = \frac{\partial}{\partial t} + D_B + a\# : L^p_k(\mathbb{R} \times Y) \to L^p_{k+1}(\mathbb{R} \times Y)$$

is Fredholm.

Proof. This is proved by the standard parametrix patching argument. □

The next step is to compute the index of D_A.

Example 6.3.2. Consider the operator

$$D_+ = \frac{\partial}{\partial t} + \tanh(t)$$

on the real line $\mathbb{R}$. As $t \to \pm\infty$, this operator approximates

$$\frac{\partial}{\partial t} + 1 \text{ and } \frac{\partial}{\partial t} - 1$$

respectively, which are invertible. We wish to compute its index as a Fredholm operator between spaces:

$$L_1^2(\mathbb{R}) \to L^2(\mathbb{R}).$$

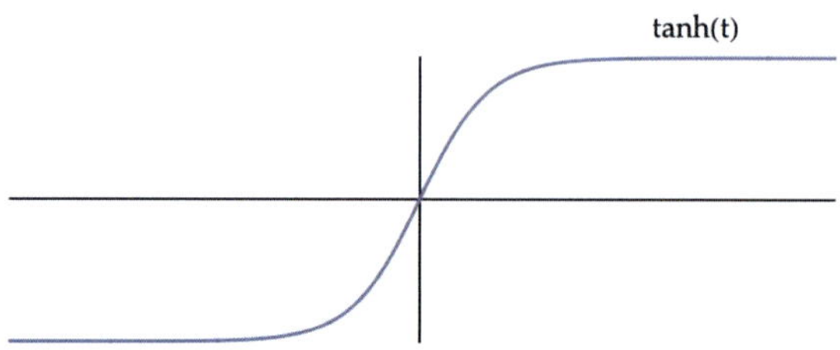

Figure 6.3.3. The graph of $\tan(t)$.

Note that $\ker D_+$ is spanned by $\operatorname{sech}(t) \in L^2(\mathbb{R})$ while $\operatorname{Coker} D_+$ is spanned by $\cosh(t) \notin L^2(\mathbb{R})$. Hence, $\operatorname{Ind} D_+ = 1$. Similarly, if we define

$$D_- = \frac{\partial}{\partial t} - \tanh(t)$$

then $\operatorname{Ind} D_- = -1$. ◇

In general, the Fredholm index is the net number of eigenvalues going from negative to positive. To prove "SF=Ind" in general, we use the fact that the index is a homotopy invariant. When fixing $a_\pm$, we are allowed to homotope the path $a(t)$. The limits $a_\pm$ are allowed to change as long as they stay invertible during a homotopy. We apply perturbations of $a_\pm$ such that

$$D_B + a_\pm$$

have the same eigenvectors. Then we reduce the problem to the model examples above. Recall from Example 3.3.1 that the spectral flow along a closed loop of operators might not be zero.

6.4. The Spectral Flow in the Instanton Floer Homology Over the trivial principal bundle $Q = SU(2) \times Y \to Y$, the Chern-Simons functional defines a circle valued functional:

$$\mathcal{CS} : \mathcal{A}/\mathcal{G} \to \mathbb{R}/\alpha\mathbb{Z} \ (\alpha > 0).$$

In this case, $\mathcal{G} = \operatorname{Map}(Y, SU(2))$ and $\pi_0(\mathcal{G})$ is classified by the degree of the map:

$$u : Y \to SU(2).$$

Therefore, $\pi_1(\mathcal{A}/\mathcal{G}) = \mathbb{Z}$. It is important to understand what is the spectral flow along a generator of π_1.

Let $\gamma : [0,1] \to \mathcal{A}/\mathcal{G}$ to be a closed loop. Let $A_0 = \gamma(0)$ and $A_1 = \gamma(1)$. Then $u \cdot A_0 = A_1$ for a gauge transformation $u : Y \to SU(2)$. Consider the mapping torus:

$$\begin{array}{c} P = [0,1] \times Q/(u \cdot p, 0) \sim (p,1) \\ \big\downarrow \pi \\ X = S^1 \times Y. \end{array}$$

The index of the operator

$$\frac{\partial}{\partial t} + D_{A(t)}$$

on the cylinder $\mathbb{R} \times Y$ equals the expected dimension of the ASD moduli space on $P \to X$. Since $\sigma(X) = \chi(X) = 0$ by Theorem 5.9.2, we have

$$SF = \text{Ind} = 8c_2(P) = 8\deg(u).$$

Another important fact is that $\mathcal{CS}$ detects π_1, i.e.,

$$\mathcal{CS}_* : \pi_1(\mathcal{A}/\mathcal{G}) \to \pi_1(S^1)$$

is an isomorphism. In addition, the total change $\Delta_\gamma \mathcal{CS}$ of the Chern-Simons functional along a closed loop is proportional to the spectral flow. When counting differentials in the Floer theory, we focus on moduli spaces of formal dimension 1. Hence,

$$\begin{aligned} &\text{pin down the dimension of moduli spaces,} \\ \Rightarrow\ &\text{pin down the homotopy classes of paths,} \\ \Rightarrow\ &\text{pin down the change of } \mathcal{CS}. \end{aligned}$$

As a result, even though $\mathcal{CS}$ is a circle valued function, we do not have to use Novikov-Morse homology. This is comparable with the monotonicity condition in the symplectic Floer theory, where

$$\mathcal{CS} \xrightarrow{\text{is replaced by}} \text{the symplectic action},$$

$$SF \xrightarrow{\text{is replaced by}} c_1(M, J).$$

The monotonic condition then states that $[\omega] = \lambda c_1(M, J) \in H^2(M, \mathbb{Z})$ for some $\lambda \in \mathbb{R}$.

6.5. Compactification of the Moduli Space Let $Q = G \times Y \to Y$ be the trivial principal G-bundle over Y and $Z = \mathbb{R} \times Y$ be the cylinder. The proof sketch of Section 1.4 largely goes through, even though we have the additional complication of bubbling.

Suppose the Chern-Simons functional $\mathcal{CS}$ descends to a non-degenerate Morse function on the quotient space:

$$\mathcal{B}_Q = \mathcal{A}_Q/\mathcal{G}_Q.$$

Let $\alpha, \beta \in \mathrm{Crit}(\mathcal{CS}) \subset \mathcal{B}_Q$ and a, b be their lifts in $\mathcal{A}_Q$ so that $[a] = \alpha$ and $[b] = \beta$. Fix a reference connection A_0 on Z such that

$$A_0\Big|_{(-\infty,-T]\times Y} = \pi^* a, \quad A_0\Big|_{[T,\infty)\times Y} = \pi^* b.$$

Consider the configuration space

$$\mathcal{A}_{P,A_0}(a, b) = \{A_0 + a' : a' \in \mathcal{C}_0^\infty(Z, iT^*Z \otimes \mathrm{ad}\, Q)\}$$

and its quotient space

$$\mathcal{B}_\gamma(\alpha, \beta) \doteq \mathcal{A}_{P,A_0}(a, b)/\mathcal{G}_Z.$$

Here $\mathcal{G}_Z$ is the subgroup of gauge transformations that approximate $e \in G$ as $t \to \pm\infty$:

$$\mathcal{G}_Z = \{u : Z \to G : u - e \in C_0^\infty(Z, \mathrm{End}(V))\}.$$

Any element in $\mathcal{A}_{P,A_0}(a, b)$ determines the same relative homotopy class of paths connecting α and β in $\mathcal{B}_Q$. Use γ to denote this homotopy class of path determined by A_0:

$$\gamma \in \pi_1(\alpha, \mathcal{B}_Q, \beta).$$

Remark 6.5.1. Note that $\pi_1(\alpha, \mathcal{B}_Q, \beta)$ is a $\mathbb{Z}$-torsor: it admits an action by integers. This is defined by "adding instantons" at some point x in Z. For the principal bundle $P_k \to S^4$ with $\langle c_2(P_k), [X]\rangle = k$, pick any connection A_k. By taking the connected sum of connections $A_0\#_x A_k$, we obtain a new homotopy class of paths: $\gamma + k$. Here k is allowed to be any integer and A_k is not necessarily an ASD connection. When $k > 0$, A_k is ASD and Z is closed, this grafting constructing produces new instantons out of a given ASD connection. $\diamondsuit$

Now we look at finite energy solutions to the ASD equation inside $\mathcal{B}_\gamma(\alpha, \beta)$:

$$\mathcal{M}_\gamma(\alpha, \beta) \subset \mathcal{B}_\gamma(\alpha, \beta).$$

Define the moduli space of ideal instantons of homotopy class γ as follows:

$$\mathcal{IM}_\gamma(\alpha, \beta) = \bigcup_{i \geqslant 0} \mathcal{M}_{\gamma - i}(\alpha, \beta) \times \mathrm{Sym}^i Z$$

and its unparameterized version $\breve{\mathcal{IM}}_\gamma(\alpha, \beta) = \mathcal{IM}_\gamma(\alpha, \beta)/\mathbb{R}$. To compactify this space, we need to add broken trajectories. Define $\mathcal{IM}_\gamma^+(\alpha, \beta)$ as the union

$$\bigcup_{r \geqslant 0} \; \bigcup_{\substack{\alpha_1, \cdots, \alpha_r \\ \in \mathrm{Crit}(\mathcal{CS})}} \; \bigcup_{\substack{\gamma_0, \cdots \gamma_r \\ \gamma_i \in \pi_1(\alpha_i, \mathcal{B}_Q, \alpha_{i+1}) \\ \sum \gamma_i = \gamma}} \breve{\mathcal{IM}}_\gamma(\alpha, \alpha_1) \times \breve{\mathcal{IM}}_\gamma(\alpha_1, \alpha_2) \times \cdots \times \breve{\mathcal{IM}}_\gamma(\alpha_r, \beta). \tag{6.5.2}$$

When r is fixed, we assumed $\alpha_0 = \alpha$ and $\alpha_{r+1} = \beta$. This union is over all homotopy classes of chains:

$$\alpha = \alpha_0 \xrightarrow{\gamma_0} \alpha_1 \xrightarrow{\gamma_1} \alpha_2 \cdots \xrightarrow{\gamma_r} \alpha_{r+1} = \beta.$$

The first union in (6.5.2) is over the length of chains. The second is over possible critical points α_i that appear in this chain. The last is over all possible homotopy classes of paths between two adjacent critical points. Finally, we require $\sum \gamma_i = \gamma$.

When $r = 0$, this union contains a single piece: $\breve{\mathcal{IM}}_\gamma(\alpha,\beta) = \mathcal{IM}_\gamma(\alpha,\beta)/\mathbb{R}$, which corresponds to unbroken trajectories. It is also the top stratum of the stratified space $\overline{\mathcal{IM}}_\gamma(\alpha,\beta)$. The length r records the number of breaks in a broken trajectory. The expected dimension will drop by 1 for each break. The top dimension of spaces in the middle union in (6.5.2) is $\dim \breve{\mathcal{IM}}_\gamma(\alpha,\beta) - r$. It also contains lower dimensional strata from the bubbling phenomenon:

$$(\mathcal{M}_{\gamma_i - k}(\alpha_i, \alpha_{i+1}) \times \mathrm{Sym}^k(Z))/\mathbb{R}.$$

Each bubble will drop the dimension by 4. The final extension of the broken-trajectories compactification keeping track of bubbles, (or Uhlenbeck compactification) is due to Floer [18].

Theorem 6.5.3. *Assuming that the Chern-Simons function (or a suitable perturbation of it) is Morse-Smale, then the space $\mathcal{IM}^+_\gamma(\alpha,\beta)$ is compact and is the closure of $\mathcal{IM}^+_\gamma$.*

The proof of this theorem requires quite a bit of work involving some (perhaps now standard) gluing analysis going back to Taubes.

6.6. A Criterion to Avoid Reducible Flat Connections We have assumed until now that the Chern-Simons functional $\mathcal{CS}$ is unperturbed and all critical points are non-degenerate. In general, a perturbation is alway needed to make $\mathcal{CS}$ Morse, and the perturbation space needs to be carefully designed so that the space of critical points will remain compact and the compactness arguments can carry over without essential changes. In the unperturbed case, critical points arise from representations of the fundamental group, so the compactness is automatic.

If our critical points are non-degenerate, then finite energy flow lines always have finite length and $\breve{\mathcal{M}}_\gamma(\alpha,\beta)$ is dense in $\overline{\mathcal{IM}}_\gamma(\alpha,\beta)$ by an argument similar to Theorem 1.2.14. When $G = SU(2)$, $\mathrm{Crit}(\mathcal{CS})$ always contains reducible connections (especially the trivial connection). The non-degeneracy condition is never satisfied for $\mathcal{CS}$.

The case when $G = PU(2) = SO(3)$ is slightly better. Here is a word of caution: we will not work with the full $SO(3)$-gauge group but only the subgroup of gauge transformations that lift to $SU(2)$.

Consider a genus-g ($g \geqslant 1$) surface Σ_g and an $SO(3)$ bundle $Q \to \Sigma_g$ with $w_2(Q) \neq 0$. A flat connection is always induced from a group homomorphism $\rho : \pi_1(\Sigma_g) \to SO(3)$. We have to determine when the induced bundle

$$\begin{array}{ccc} \mathfrak{so}(3) & \longrightarrow & \operatorname{ad}\rho \\ & & \downarrow \\ & & \Sigma_g \end{array}$$

has $w_2(\operatorname{ad}\rho) \neq 0$. Consider the standard presentation of the surface group:

$$\pi_1(\Sigma_g) = \langle a_1, b_1, \cdots, a_g, b_g \,|\, \prod_{i=1}^g [a_i, b_i] = 1 \rangle.$$

Let $A_i = \rho(a_i) \in SO(3)$ and $B_i = \rho(b_i) \in SO(3)$. Then $\prod_{i=1}^{g}[A_i, B_i] = 1 \in SO(3)$. Consider their lifts in $SU(2)$:

$$\tilde{A}_i \in \pi^{-1}(A_i), \tilde{B}_i \in \pi^{-1}(B_i),$$

and define

$$w(\rho) \doteq \prod_{i=1}^{g}[\tilde{A}_i, \tilde{B}_i] = \pm 1 \in Z(SU(2)).$$

Apparently $w(\rho)$ does not depend on the choice of $\tilde{A}_i$ and $\tilde{B}_i$. In fact,

$$w(\rho) = w_2(\operatorname{ad}\rho) \in \mathbb{Z}/2\mathbb{Z}.$$

Now consider the odd representation variety

$$\mathcal{R}^{\mathrm{odd}}(\Sigma_g) \doteq \{\{\tilde{A}_i, \tilde{B}_i\}_{1\leqslant i\leqslant g} | \tilde{A}_i, \tilde{B}_i \in SU(2), \prod_{i=1}^{g}[\tilde{A}_i, \tilde{B}_i] = -1\}/SU(2).$$

Note that $\mathcal{R}^{\mathrm{odd}}(\Sigma_g)$ is a cover of

$$\{\rho : \pi_1(\Sigma) \to SO(3) : w_2(\operatorname{ad}\rho) \neq 0\}.$$

Theorem 6.6.1. *$\mathcal{R}^{odd}(\Sigma_g)$ contains no reducible representations.*

Proof. Take $h \in SU(2)\backslash Z(SU(2))$; then the commutant of h is the great circle S_h passing through h and ± 1. If h commutes with $\tilde{A}_i, \tilde{B}_i$ for each $1 \leqslant i \leqslant g$, then $\tilde{A}_i, \tilde{B}_i$ lie in the great circle S_h.

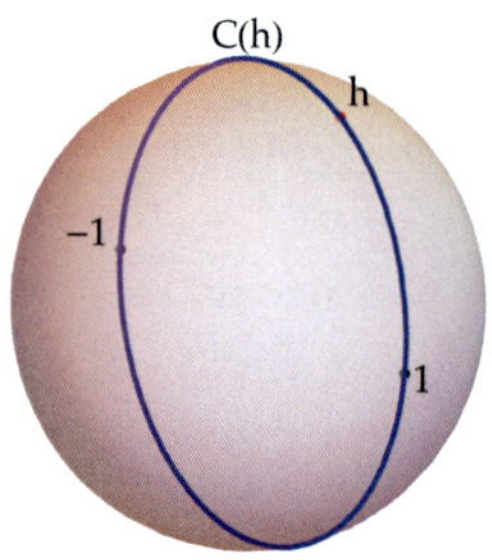

If $\{\tilde{A}_i, \tilde{B}_i\}_{1\leqslant i\leqslant g}$ has a non-trivial stabilizer (larger than $Z(SU(2))$), then $\tilde{A}_i, \tilde{B}_i$ are simultaneously diagonalizable. In particular,

$$[\tilde{A}_i, \tilde{B}_i] = 1$$

for each $1 \leqslant i \leqslant g$. This contradicts the assumption that $\prod_{i=1}^{g}[\tilde{A}_i, \tilde{B}_i] = -1$. □

Example 6.6.2. Let $g = 1$ and $\Sigma_g = \mathbb{T}^2$. Take $G = SU(2) = Sq(1)$. If $\tilde{A}_1 = I$ and $\tilde{B}_1 = J$, then

$$[\tilde{A}_1, \tilde{B}_1] = IJI^{-1}J^{-1} = -1.$$ ◇

Theorem 6.6.1 provides a convenient criterion to exclude reducible connections:

Definition 6.6.3. An $SO(3)$ bundle $Q \to Y$ with $w_2(Q) \neq 0$ is called *admissible* if we can find an oriented embedded surface $\Sigma_g \subset Y$ with $g \geqslant 1$ such that $w_2(Q)[\Sigma] \neq 0$. Then Q does not admit reducible flat connections. ◇

Remark 6.6.4. The construction of $w(\rho)$ generalizes to any $PU(n)$ bundles, where the invariant lives in

$$\beta(\rho) \in H^2(\Sigma, \mathbb{Z}/n\mathbb{Z}).$$

Using the short exact sequence

$$1 \to Z(SU(n)) = \mathbb{Z}/n\mathbb{Z} \to SU(n) \to PU(n) \to 0,$$

we choose the lift of A_i in $SU(n)$ to define $\beta(\rho)$. ◊

6.7. Instanton Floer Homology With all the machinery developed so far, we are ready to define the instanton Floer homology of (Y, Q) where $Q \to Y$ is an admissible $SO(3)$-bundle over a closed 3-manifold Y.

The chain group is freely generated by critical points of the (perturbed) Chern-Simons functional $\mathcal{CS}$:

$$C_*(Y, Q) = \bigoplus_{\alpha \in \mathrm{Crit}(\mathcal{CS})} \mathbb{Z}_2 \alpha$$

while the differential is defined by counting negative gradient flowlines of $\mathcal{CS}$:

$$\begin{aligned} \partial : C_*(Y, Q) &\to C_*(Y, Q) \\ \alpha &\mapsto \sum_{\beta \in \mathrm{Crit}(\mathcal{CS})} \sum_{\substack{\gamma \in \pi_1(\alpha, \mathcal{B}_Q, \beta) \\ sf_\gamma(\alpha, \beta) = 1}} \#\breve{\mathcal{M}}_\gamma(\alpha, \beta) \cdot \beta. \end{aligned}$$

To verify that $\partial^2 = 0$, we examine the compactification of the moduli space when $sf_\gamma(\alpha, \alpha') = 2$. In this case,

$$\mathcal{I}\breve{\mathcal{M}}_\gamma(\alpha, \alpha') = \breve{\mathcal{M}}_\gamma(\alpha, \alpha') \bigcup_{\substack{\gamma_1 \in \pi_1(\alpha, \mathcal{B}_Q, \beta) \\ \gamma_2 \in \pi_1(\beta, \mathcal{B}_Q, \alpha') \\ sf_{\gamma_1} = sf_{\gamma_2} = 1}} \breve{\mathcal{M}}_{\gamma_1}(\alpha, \beta) \times \breve{\mathcal{M}}_{\gamma_2}(\beta, \alpha'),$$

as shown schematically below.

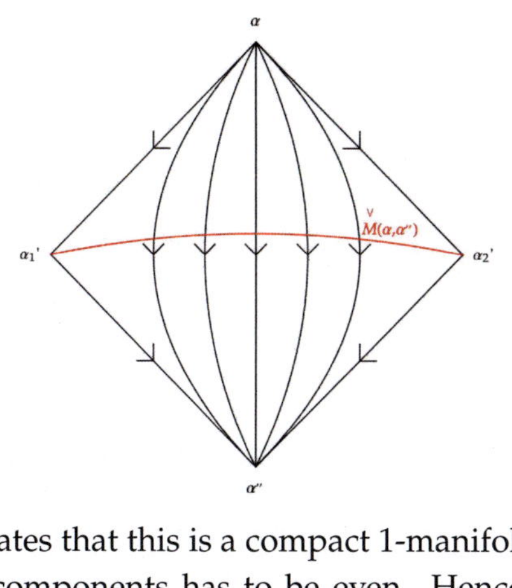

A gluing theorem states that this is a compact 1-manifold with boundaries, the number of boundary components has to be even. Hence, $\partial^2 = 0 \mod 2$. The resulting homology group is the instanton Floer homology of (Y, Q):

$$I_*(Y, Q) \doteq H_*(C_*(Y, Q), \partial).$$

7. Floer Homology for links in three manifolds, exact triangles and Khovanov Homology

7.1. Representation spaces of knot complements. We will sketch a spectral sequence abutting to the singular knot Floer homology and whose E_2-page is Khovanov homology. This yields an unexpected application of gauge theory to the classical knot theory: this allows us to show that Khovanov homology detects the unknot in S^3.

Let Y^3 be a closed 3-sphere and $L \subset Y$ a smoothly embedded link. Instead of an admissible SO(3)-bundle, the singular link Floer homology of (Y, L) is defined using an SO(3)-bundle Q singular along L: one should really treat (Y, L) as an orbifold. The local model near L is $\mathbb{C}/\mathbb{Z}_2 \times L$, and the monodromy of Q along a small meridional loop of L is of order 2 in $SO(3)$.

It is enlightening to first look at the kind of representations of $\pi_1(Y \setminus L)$ and understand what flat connections are going to be used for singular bundles. A singular π_1-representation of $\pi_1(Y \setminus L)$ sends the meridian of each link component to an order-2 element in $SO(3)$. Working instead with the 2-fold cover $SU(2)$, they are of order 4. Such elements in $SU(2)$ form a 2-sphere (the equator) $S^2_I \subset SU(2)$.

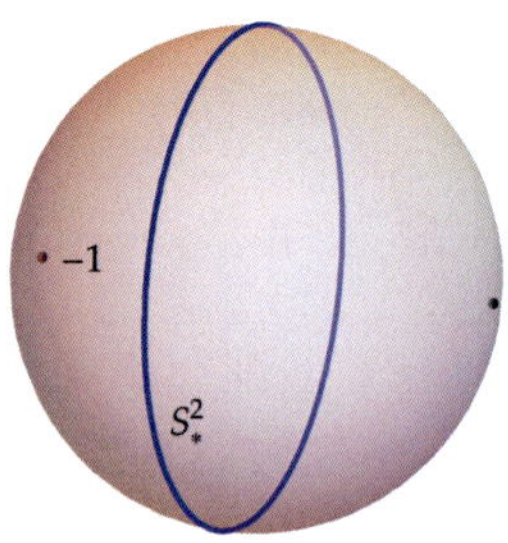

Consider the case that $Y = S^3$ and $L = H$ is the Hopf link. The complement $S^3 \setminus H$ deformation retracts to $\mathbb{T}^2$, so

$$\pi_1(S^3 \setminus H) \cong \mathbb{Z} \oplus \mathbb{Z}.$$

When $w_2(Q) = 0$, the singular $SO(3)$-representation variety of (S^3, H) is given by

$$\begin{aligned} \mathcal{R}^I(S^3, H) &= \{(\alpha, \beta) \in SU(2) \times SU(2) : [\alpha, \beta] = \mathrm{Id}, \alpha \sim \beta \sim \left(\begin{smallmatrix} i & 0 \\ 0 & -i \end{smallmatrix}\right)\}/\sim \\ &= \{(I, I), (I, -I)\} = S^2 \coprod S^2/\text{conjugation} \\ &= 2 \text{ points.} \end{aligned}$$

Each singular representation is reducible. The conjugation action of $SU(2)$ on $S^2 \coprod S^2$ is not free — the stabilizer at each point is S^1.

Let λ be an arc connecting different components of the Hopf link H and let $w_2(Q) = PD[\lambda] \neq 0 \in H^2(S^3, H)$. Then, the singular representation variety is

$$\begin{aligned} \mathcal{R}^I_\lambda(S^3, H) &= \{(\alpha, \beta) \in SU(2) \times SU(2) : [\alpha, \beta] = -1, \alpha \sim \beta \sim \left(\begin{smallmatrix} i & 0 \\ 0 & -i \end{smallmatrix}\right)\}/\sim \\ &= \{(I, J)\} = 1 \text{ point.} \end{aligned}$$

Indeed, α takes values in the equator S^2 of $SU(2)$, while β takes value in the intersection:

$$\alpha^{\perp} \cap S^2.$$

So the singular representation variety is the unit sphere bundle $S(TS^2)$ of the tangent space of S^2, on which $SU(2)/Z(SU(2))$ acts freely and transitively.

In general, define $(Y,K)^{\#} = (Y,K)\#(S^3,H)$, where we take the connected sum of 3-manifolds and the disjoint union of links, and set

$$\mathcal{R}^{\#}(Y,K) = \mathcal{R}^{I}_{\lambda}((Y,K)^{\#}).$$

We can identify this with the based representation variety $\mathcal{R}^0(Y,K)$.

The singular knot Floer homology of (Y,K) is then defined as

$$I^{\#}_{*}(Y,K) \doteq I_{*}((Y,K)^{\#}),$$

while $I_*((Y,K)^{\#})$ is defined using a singular $SO(3)$-bundle over $(Y,K)\#(S^3,H)$.

Recall from Section 4.4 that we have have the follow descriptions of the representation spaces. For the knots $U \subset S^3$ we have

$$\mathcal{R}^{\#}(S^3,U) = S^2.$$

More generally for the unlink of k components $U_k = \coprod_k U$ we have:

$$\mathcal{R}^{\#}(S^3,U_k) = \underbrace{S^2 \times \cdots \times S^2}_{\text{k copies}}.$$

For the trefoil $T(2,3) \subset S^3$ we have:

$$\mathcal{R}^{\#}(S^3,T_{(2,3)}) = SO(3) \coprod S^2.$$

More generally for the $(2,2k+1)$-torus knot $T(2,2k+1)$ we have:

$$\mathcal{R}^{\#}(S,T(2,2k+1)) = \coprod_{k} SO(3) \coprod S^2.$$

Compute the singular homology of these representation varieties:

$$\begin{aligned} H_*(\mathcal{R}^{\#}(S,U)) &= \mathbb{Z} \oplus \mathbb{Z} \\ H_*(\mathcal{R}^{\#}(S,U_k) &= (\mathbb{Z} \oplus \mathbb{Z})^{\otimes k} \\ H_*(\mathcal{R}^{\#}(S,T(2,2k+1))) &= \underbrace{(\mathbb{Z} \oplus \mathbb{Z})}_{S^2} \oplus \underbrace{(\mathbb{Z} \oplus \mathbb{Z}_2 \oplus \mathbb{Z})^{\oplus k}}_{\text{k copies of } \mathbb{RP}^3}. \end{aligned}$$

All these groups agree with Khovanov homology of the corresponding links, suggesting that there is a relation between Khovanov homology and instanton Floer homology.

Floer's original definition of instanton knot Floer homology followed a slightly different approach, and we end this section with a brief sketch. The complement of a knot $K \subset S^3$ is a homology $S^1 \times D^2$. Removing a tubular neighborhood $N(K)$ of K, a Seifert surface S_K of K defines a canonical curve in $\partial N(K)$, i.e., the longitude:

$$l = \partial S_K \subset N(K).$$

On the other hand, $(\mathbb{T}^2\backslash D^2) \times S^1$ is a 3-manifold with a torus boundary. By identifying

$$l \mapsto [\partial D^2] \times \{pt\}, m \mapsto \{pt\} \times S^1,$$

we glue $S^3\backslash N(K)$ and $(\mathbb{T}^2\backslash D^2) \times S^1$ along their common boundary to obtain a homology $\mathbb{T}^3$:

$$\mathbb{T}^3_K \doteq S^3\backslash N(K) \bigcup (\mathbb{T}^2\backslash D^2) \times S^1/\sim .$$

Consider the principal SO(3)-bundle $Q_K \to \mathbb{T}^3_K$ with

$$PD(w_2(Q_K)) = \{pt\} \times S^1 \subset (\mathbb{T}^2\backslash D^2) \times S^1$$

and the closed surface $\Sigma_K = S_K \cup (\mathbb{T}^2\backslash D^2) \times \{pt\}$. Then $\langle w_2(Q_K), [\Sigma_K]\rangle \neq 0$. The SO(3)-bundle is admissible. The original instanton knot Floer homology is defined as

$$I^{\text{Floer}}(S^3, K) \doteq I_*(\mathbb{T}^3_K, Q_K).$$

This group is indeed closely related to $I^\sharp(S^3, K)$.

In general, the boundary of Seifert surfaces of a link $L \subset S^3$ may give different homology classes in $\partial N(L)$; so Floer's construction does not provide a canonical closure for the case of a link complement.

7.2. Khovanov Homology In this section, we review the basic construction of Khovanov homology roughly following Khovanov's original paper (see [27]) and the notations therein. We follow however the conventions in [31] with regards to gradings and take homology as the primary theory (as opposed to cohomology). For any knot $K \subset S^3$, we pick a knot diagram $D(K)$ with n crossings. There are two possible ways to resolve each crossing, as indicated in the figure below; the passage from K_1 to K_0 can be realized by a concrete 2-cobordism that adds a single 1-handle to K_1.

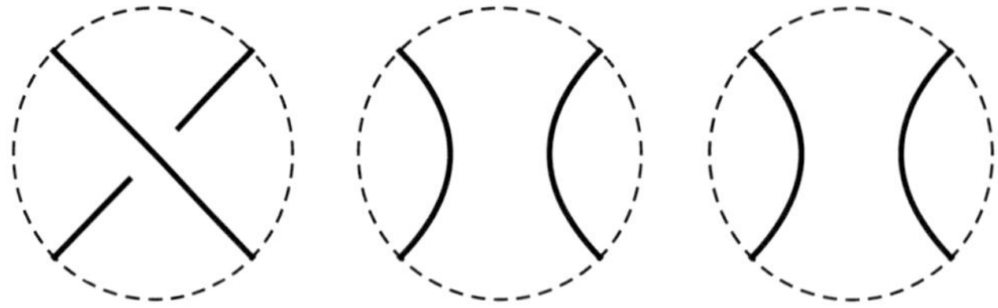

K_2 K_1 K_0

To resolve all crossings, there are 2^n different ways, each of which corresponds to a vertex of the cube:

$$\{0,1\}^n \subset [0,1]^n.$$

Each resolved digram is a disjoint union of circles embedded in $\mathbb{R}^2$. For each edge of $[0,1]^n$, we associate an arrow from 1 to 0 making the 1-skeleton of $[0,1]^n$ into a category. This cube of resolutions then provides a functor:

$$\text{1-skeleton of } [0,1]^n \to \text{Cob}_2 .$$

Example 7.2.1. Starting with the standard knot diagram of a left-handed trefoil, we arrive at a cube of resolutions in the figure below. ♢

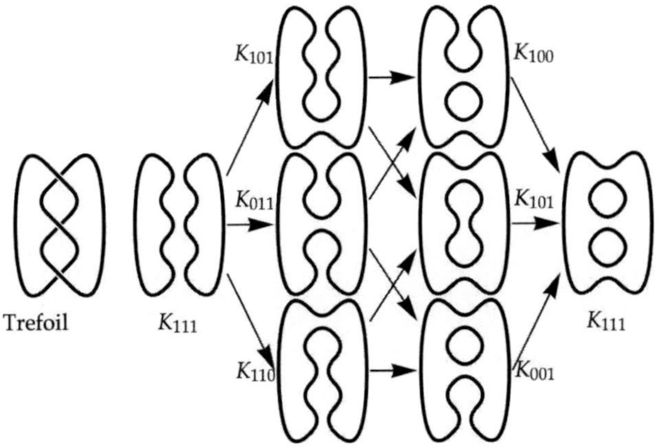

Khovanov's original ideas is to apply an $(1+1)$-dimensional TQFT to the cube of resolutions. Consider the Frobenius algebra

$$A = \mathbb{Z}[x]/(x^2) = H^*(S^2, \mathbb{Z}).$$

The ring structure of $H^*(S^2, \mathbb{Z})$ defines the multiplication map $\Delta : A \otimes A \to A$. The co-multiplication $\nabla : A \to A \otimes A$ is defined by the formulae

$$1 \mapsto 1 \otimes x + x \otimes 1,$$
$$x \mapsto x \otimes x.$$

Alternatively, one may consider the diagonal embedding $\iota : S^2 \hookrightarrow S^2 \times S^2$ and realize ∇ as the composition:

$$\begin{aligned} A = H^*(S^2, \mathbb{Z}) &\xrightarrow{PD} H_{2-*}(S^2, \mathbb{Z}) \xrightarrow{\iota *} H_{2-*}(S^2 \times S^2, \mathbb{Z}) \\ &\xrightarrow{PD} H^{2+*}(S^2 \times S^2, \mathbb{Z}) = A \otimes A. \end{aligned}$$

Define the unit and co-unit as:

$$\epsilon : \mathbb{Z} \to A,\ 1 \mapsto 1,$$
$$\eta : A \to \mathbb{Z},\ 1 \mapsto 0,\ x \mapsto 1.$$

Then the operations $(\Delta, \nabla, \epsilon, \eta)$ make A into a Frobenius algebra. To get other Frobenius algebras, one can take any compact oriented manifold X and set $A = H^*(X, \mathbb{Z})$. Then the ring structure, Poincaré duality and the diagonal embedding defines a Frobenius algebra structure on A, and one can construct others as deformations of these. However, the Frobenius algebras that give rise to knot invariants are quite restricted (see Khovanov's paper [26]).

Given the Frobenius algebra $(A, \Delta, \nabla, \epsilon, \eta)$ with $A = H^*(S^2, \mathbb{Z})$, we can construct maps from elementary cobordisms. First define

$$Z(\coprod_k S^1) = A^{\otimes k}.$$

Then for each elementary cobordism, define:

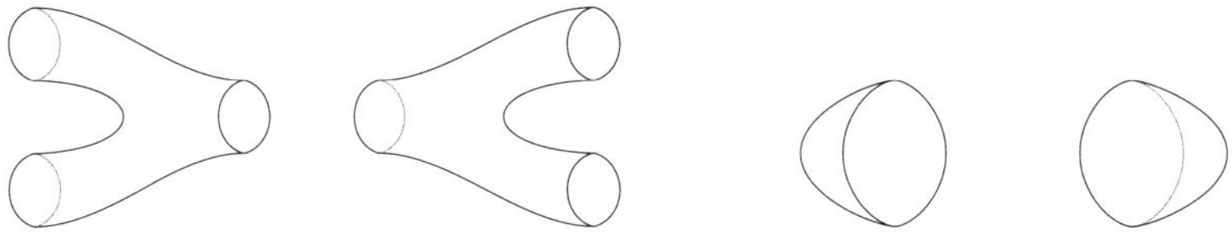

(a) $\Delta\colon A \otimes A \to A$ (b) $\nabla\colon A \to A \otimes A$ (c) $\epsilon\colon \mathbb{Z} \to A$ (d) $\eta\colon A \to \mathbb{Z}$

For a general 2-dimensional cobordism Σ, the induced map is defined by decomposing Σ into a sequence of elementary cobordisms and is independent of the particular decomposition. Apply this construction to the cube of resolutions and its edges. For each vertex $\nu \in \{0,1\}^n$, let $D(K,\nu)$ denote the corresponding planar resolved diagram. This is a union of circles, and $Z(D(K,\nu))$ is a tensor product of copies of A. Each edge of $[0,1]^n$ gives rise to an elementary cobordism which either splits a circle or merges two circles:

$$\text{each vertex} \mapsto A^{\otimes n}, \ n = \text{the number of circles},$$
$$\text{each edge} \mapsto \ \text{either } \Delta \text{ or } \nabla.$$

We form a graded chain complex by taking the direct sum

$$C_{Kh}(D(K)) = \bigoplus_{\nu \in \{0,1\}^n} Z(\nu(D(K))).$$

with grading defined by $|\nu| \doteq \sum_{1 \in \nu} 1$ (in Khovanov's original paper, he worked with cohomology; so his gradings are sorted out differently than those in this exposition).

The differential ∂_{Kh} is the direct sum of all edge morphisms, which decreases the grading by 1. Once some minus signs are assigned properly to the edge morphisms, one can assure that $\partial_{Kh}^2 = 0$. As Khovanov shows the chain homotopy type of $(C_{Kh}(D(K)), \partial_{Kh})$ is independent of the knot diagram $D(K)$. In particular, its homology $\mathrm{Kh}(K)$ is an invariant of $K \subset S^3$. For this exposition, we will ignore the gradings on these group entirely. Fix a crossing of $D(K)$. The 0- and 1-resolutions at this crossing result in two links K_0 and K_1 respectively, together with distinguished diagrams, $D(K_0)$ and $D(K_1)$. Their Khovanov chain complexes fit into a short exact sequence:

$$0 \to C_{Kh}(D(K_0)) \to C_{Kh}(D(K)) \to C_{Kh}(D(K_1)) \to 0,$$

inducing an exact triangle for Khovanov homology:

$$\begin{array}{ccc} & \mathrm{Kh}(K_2) & \\ \nearrow & & \searrow \\ \mathrm{Kh}(K_0) & \xleftarrow[\delta]{\qquad\qquad} & \mathrm{Kh}(K_1). \end{array}$$

7.3. Exact Triangles The same exact triangle holds also for singular knot Floer homology, which comes with natural cobordisms of links. Imagine that the resolution of a crossing happens within a tetrahedron T.

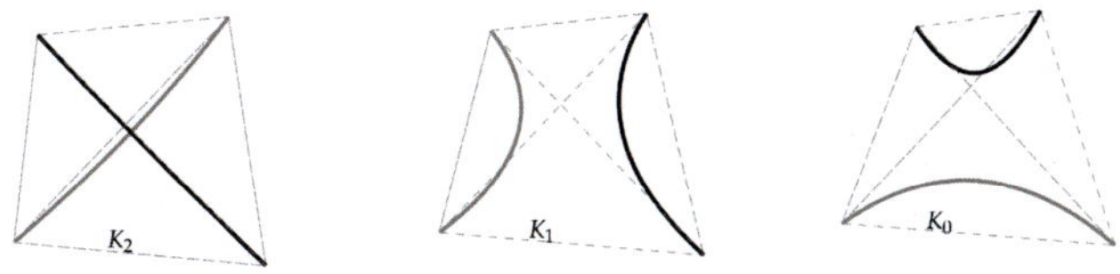

The cobordism Σ from K to K_1 happens within $T \times [0,1]$ and is obtained by attaching a handle. Consider the surface $S \subset T$:

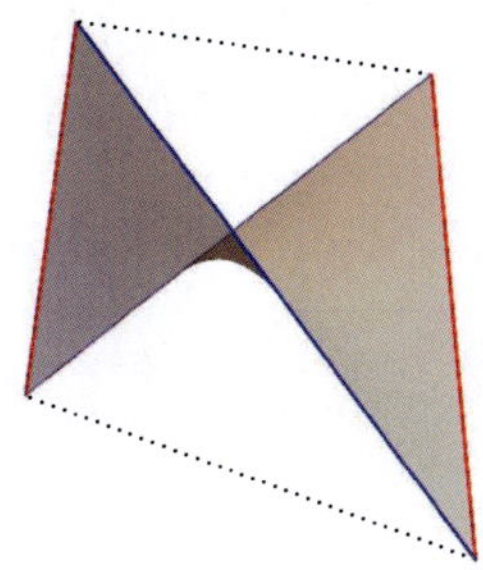

Figure 7.3.1. A projection to $\mathbb{R}^3$ of the surface S.

To obtain the surface $\Sigma \subset [0,1] \times S^3$, one extends S by a product cobordism outside $[0,1] \times T$. Recall that for each cobordism $(X, \Sigma) : (Y_-, K_-) \to (Y_+, K_+)$, there is a natural morphism

$$I^{\#}(X, \Sigma) : I^{\#}(Y_-, K_-) \to I^{\#}(Y_+, K_-)$$

defined by counting singular ASD connections on (X, Σ).

This morphism map is also defined when Σ is **non-orientable**. In this case, the normal bundle of Σ is not orientable either, so the linking holonomy is only well-defined up to $\pm\,\mathrm{Id}$. Since g is conjugate to $-g$ for any $g \in S^2_I \subset SU(2)$ (g is an element of order 4), a singular connection still makes sense.

In what follows, we will work with the following digram:

(7.3.2)

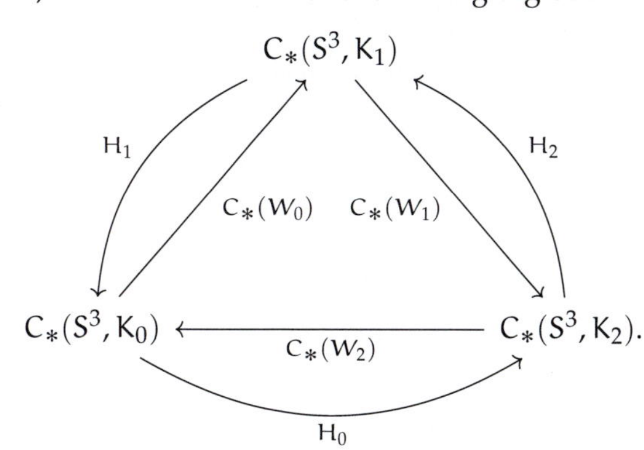

where arrows are indexed by their domains, $K_2 \doteq K$ and $W_i = (S^3 \times I, \Sigma_i)$ is the corresponding cobordism from (S^3, K_i) to (S^3, K_{i+1}). Each H_i is a concrete null-homotopy of the composition $C_*(W_{i+1}) \circ C_*(W_i)$ (to be constructed later).

To verify that (7.3.2) induces an exact triangle for singular knot Floer homology, we invoke the distinguished triangulation detection lemma (Theorem 7.6.2). The subtlety lies in the fact that each cobordism comes with an orientable surface Σ_i, but their compositions

$$\Sigma_{i+1} \circ \Sigma_i$$

are non-orientable. The co-core of the 1-handle in Σ_i and the core in Σ_{i+1} form a Möbius band in $\Sigma_{i+1} \circ \Sigma_i$:

7.4. Some Non-Orientable Surfaces in 4-Manifolds To facilitate the construction of the null-homotopy H_i in the diagram (7.3.2), we introduce some non-orientable surfaces in the standard 4-sphere S^4. Note that the self-intersection number $\Sigma \cdot \Sigma$ always makes sense, even if the surface $\Sigma \subset S^4$ is non-orientable.

Example 7.4.1. $\mathbb{RP}^2_\pm \subset S^4$. Let S^4 be the unit sphere of traceless symmetric 3×3 matrices ($\dim_{\mathbb{R}} = 5$):

$$\begin{pmatrix} x & y & z \\ y & u & v \\ z & v & -x-u \end{pmatrix}.$$

The group $SO(3)$ acts on S^4 by matrix conjugation. Define

$$\mathbb{RP}^2_+ = \{\text{matrices with a positive eigenspace of dimension } 2 \subset S^4\},$$
$$\mathbb{RP}^2_- = \{\text{matrices with a negative eigenspace of dimension } 2 \subset S^4\}.$$

As a result, any matrix $A \in \mathbb{RP}^2_+$ is conjugate to

$$\lambda \begin{pmatrix} 1 & & \\ & 1 & \\ & & -2 \end{pmatrix}$$

for some positive constant λ. The stabilizer of A is $O(2) \subset SO(3)$, so the orbit is

$$SO(3)/O(2) \cong \mathbb{RP}^2.$$

It is easier to compute the self intersection of $\mathbb{RP}^2_\pm$ using a different description. A observation of Kuiper [34], Arnol'd [1] and others states that

$$\mathbb{CP}^2/\text{complex conjugation} \cong S^4.$$

A real conic with no real roots is an $S^2 \subset \mathbb{CP}^2$ invariant under complex conjugation. For instance, we take

$$\mathbb{CP}^1 = \{x^2 + y^2 + z^2 = 0\} \subset \mathbb{CP}^2.$$

For a conic, the self intersection number is $S^2 \cdot S^2 = +4$. Passing to the quotient, the self-intersection of $\mathbb{RP}^2_+ = S^2/\mathbb{Z}_2$ is divided by 2, so

$$\mathbb{RP}^2_+ \cdot \mathbb{RP}^2_+ = 2.$$

On the other hand, the fixed point set of this action is a real projective space $\mathbb{RP}^2_- \subset \mathbb{CP}^2$. The subspace $\mathbb{RP}^2_- \subset \mathbb{CP}^2$ is a Lagrangian submanifold. Hence, $T^*\mathbb{RP}^2_- \cong N(\mathbb{RP}^2_-)$ and

$$\begin{aligned} 1 = \chi(\mathbb{RP}^2_-) &= \langle e(T\mathbb{RP}^2_-), [\mathbb{RP}^2_-, o(\mathbb{RP}^2_-)]\rangle \\ &= -\langle e(T^*\mathbb{RP}^2_-), [\mathbb{RP}^2_-, o(\mathbb{RP}^2_-)]\rangle \\ &= -\langle e(N(\mathbb{RP}^2_-)), [\mathbb{RP}^2_-, o(\mathbb{RP}^2_-)]\rangle. \end{aligned}$$

Since $\mathbb{RP}^2_-$ is fixed under conjugation, after passing to the quotient, the self-intersection number is doubled: $\mathbb{RP}^2_- \cdot \mathbb{RP}^2_- = -2$. Note that the signs of self-intersection numbers depend only on the orientation of S^4. Reversing the construction, one may take the double branched cover of S^4 along $\mathbb{RP}^2_\pm$ to obtain

$$(S^4, \mathbb{RP}^2_+) \mapsto \overline{\mathbb{CP}}^2 \text{ and } (S^4, \mathbb{RP}^2_-) \mapsto \mathbb{CP}^2. \qquad \diamondsuit$$

The geometric input required for the null-homotopy V_i in (7.3.2) is

$$W_1 \circ W_0 = (S^3 \times I, \Sigma_1) \circ (S^3 \times I, \Sigma_0) = \overline{(S^3 \times I, \Sigma_2)} \# (S^4, \mathbb{RP}^2_+).$$

where the connected sums of 4-manifolds and embedded 2-surfaces are taken at the same time. In [31], Kronheimer and the first author proved the following exact triangle.

Theorem 7.4.2. *With K_2, K_1 and K_0 as above, we have*

$$\begin{array}{ccc} & I^\#(S^3, K_2) & \\ \nearrow & & \searrow \\ I^\#(S^3, K_0) & \xleftarrow{\quad\delta\quad} & I^\#(S^3, K_1). \end{array}$$

is an exact triangle. Equivalently, one has

$$C^\#(S^3, K) \xrightarrow[\cong]{\text{chain homotopy equivalent.}} \mathrm{Cone}(C^\#(S^3, K_1) \xrightarrow{f_{1\to 0}} C^\#(S^3, K_0)). \tag{7.4.3}$$

7.5. Floer's Exact Triangle Instead of the proof of Theorem 7.4.2, we will sketch a proof of Floer's original surgery triangle. The proof here is rather different from Floer's original approach and is based on an argument developed first in the Seiberg-Witten context by Kronheimer-Mrowka-Ozsváth-Szabó in [32] and then in the instanton knot context in [31]. Another, more detailed, exposition of Floer's original triangle by this proof scheme appears in [44]. Suppose we have a 3-manifold with torus boundary $\hat{Y}$ and choose three oriented curves γ_2, γ_1 and γ_0 on $\partial\hat{Y}$ such that

$$\gamma_2 + \gamma_1 + \gamma_0 = 0 \in H_1(\mathbb{T}^2; \mathbb{Z}),$$

and

$$\gamma_2 \cdot \gamma_1 = -1;$$

then it follows that

$$\gamma_{i+1} \cdot \gamma_i = -1$$

for all i (mod 3). Consider the closed 3-manifold

$$Y_i = \hat{Y} \cup_{\phi_i} S^1 \times D^2$$

where $\phi \colon S^1 \times \partial D^2 \to \partial \hat{Y}$ is an orientation reversing diffeomorphism satisfying $\phi_i(pt \times \partial D^2) = \gamma_i$. The consecutive 3-manifolds Y_i and Y_{i+1} are naturally cobordant by a 4-manifold W_i obtained by adding a single 2-handle to Y_i. The $SO(3)$-bundle on Y_i is specified by requiring the core w_i of the solid torus to represent the second Stiefel-Whitney class w_2. For the 4-manifold W_i, this class is instead represented by ζ_i, the union of core C_i and co-core S_i of the 2-handle. Having explained the relevant context, Floer's original exact triangle is the following.

Theorem 7.5.1 (Floer [20]). *The diagram*

$$\begin{array}{ccccc} & & I_*(Y_1, w_1) & & \\ & \nearrow^{I_*(W_0;\zeta_0)} & & \searrow^{I_*(W_1;\zeta_1)} & \\ I_*(Y_0, w_0) & & \xleftarrow[I_*(W_2;\zeta_2)]{} & & I_*(Y_2, w_2) \end{array}$$

is an exact triangle.

Let us sketch a proof of Theorem 7.5.1 and point out the relevant input from geometry. The first step is to show consecutive compositions are zeros. If we compose cobordisms:

$$W_i : Y_i \to Y_{i+1} \text{ and } W_{i+1} : Y_{i+1} \to Y_{i+2},$$

then we obtain a cobordism

$$Z_i = W_{i+1} \circ W_i : Y_i \to Y_{i+2}.$$

The co-core of W_i and the core of W_{i+1} form a sphere E_i of self-intersection (-1):

$$E_i = S_i \cup C_{i+1}, \; E_i \cdot E_i = -1.$$

This means that the boundary of a tubular neighborhood $N(E_i)$ of E_i is homeomorphic to S^3:

$$\partial N(E_i) \cong S^3.$$

Moreover, $\langle w_2, E_i \rangle \neq 0 \in \mathbb{Z}_2$. If we take critical points $\alpha \in \mathrm{Crit}(\mathcal{CS}, Y_i)$ and $\beta \in \mathrm{Crit}(\mathcal{CS}, Y_{i+2})$, then the composition

$$I_*(W_{i+1}; \zeta_{i+1}) \circ I_*(W_i; \zeta_i) = I_*(Z_i; \zeta_{i+1} \cup \zeta_i).$$

is defined by counting the zero dimensional moduli space

$$\mathcal{M}_k(Z_i, \alpha, \beta).$$

This instanton number $k = \langle c_2(P_i' \to Z_i), [Z_i, \partial Z_i] \rangle \in \mathbb{Z}$ is fixed by requiring

$$\dim \mathcal{M}_k(Z_i, \alpha, \beta) = 0.$$

Now the idea is to choose a special metric on Z_i such that the moduli space $\mathcal{M}_k(Z_i, \alpha, \beta)$. is empty, and so the map $I_*(Z_i)$ is already trivial on the chain level. The special metric is constructed by stretching the neck along $\partial N(E_i) = S^3$. Consider the Riemannian manifold

$$Z_i^T = Z_i \backslash N(E_i) \bigcup [-T, T] \times S^3 \bigcup N(E_i)$$

with fixed metrics on $Z_i \backslash N(E_i)$ and $N(E_i)$. When $T \gg 0$, an ASD connection on Z_i is close to the flat connection over the cylinder $[-T, T] \times S^3$.

Lemma 7.5.2. *Since $\partial N(E_i) = S^3$, one can fill in a 4-ball to obtain a new cobordism between Y_i and Y_{i+2}. This cobordism is simply*

$$\overline{W}_{i+2} : Y_{i+2} \to Y_i.$$

and $Z_i = \overline{W}_{i+2} \# \overline{\mathbb{CP}}^2$.

Figure 7.5.3. On the left, $W_i \coprod W_{i+1}$ with cocore of W_i and core of W_{i+1} in red. On the right, $W_{i+2} \circ W_i$ creating a (red) (-1)-sphere.

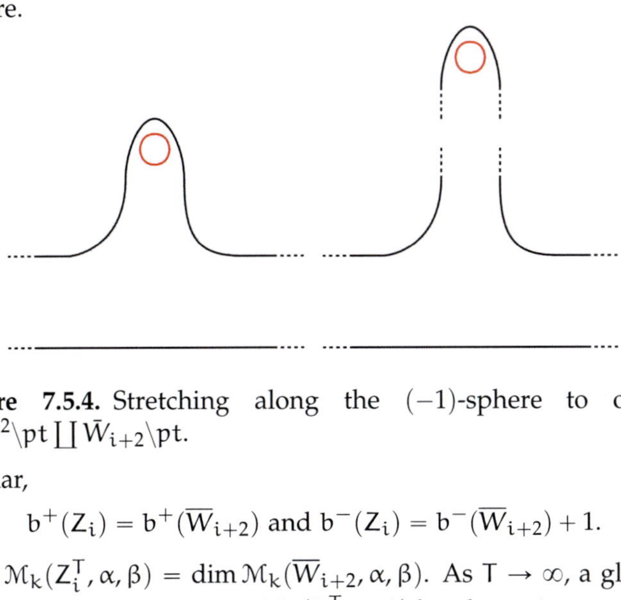

Figure 7.5.4. Stretching along the (-1)-sphere to obtain $-\mathbb{CP}^2 \backslash pt \coprod \bar{W}_{i+2} \backslash pt$.

In particular,

$$b^+(Z_i) = b^+(\overline{W}_{i+2}) \text{ and } b^-(Z_i) = b^-(\overline{W}_{i+2}) + 1.$$

and $0 = \dim \mathcal{M}_k(Z_i^T, \alpha, \beta) = \dim \mathcal{M}_k(\overline{W}_{i+2}, \alpha, \beta)$. As $T \to \infty$, a gluing theorem states that an ASD connection A in $\mathcal{M}_k(Z_i^T, \alpha, \beta)$ breaks up into

$$A \rightsquigarrow (A', A_0)$$

with $A' \in \mathcal{M}_k(\overline{W}_{i+2}, \alpha, \beta)$ and an ASD connection A_0 on $\overline{\mathbb{CP}}^2$. Moreover,

$$\mathcal{E}(A) = \mathcal{E}(A') + \mathcal{E}(A_0).$$

Since $\mathcal{E}(A)$ and $\mathcal{E}(A')$ are determined by the instanton number k, we must have $\mathcal{E}(A_0) = 0$. To draw a contradiction, it suffices to show an ASD connection on $\overline{\mathbb{CP}}^2$ with $\langle w_2, E_i \rangle \neq 0$ must have non-zero energy.

On $\overline{\mathbb{CP}}^2$, lift the $SO(3)$-bundle P to a $U(2)$-bundle $P' = P'_{U(2)}$. The condition $w_2 \neq 0$ implies that $c_1(P')$ is odd. The energy formula (see item 3 on page 284) says

$$\mathcal{E}(A) = 8\pi^2\langle c_2 - 1/4c_1^2, [\overline{\mathbb{CP}}^2]\rangle \neq 0.$$

7.6. The Second Step The second step is to show the triangle is exact. The first step provides a concrete chain homotopy between $C_*(W_{i+1}) \circ C_*(W_i)$ and 0:

$$\begin{aligned} H_i : C_*(Y_i, w_i) &\to C_*(Y_{i+2}, w_{i+2}) \\ \alpha &\mapsto \sum_\gamma \# \coprod_{T\in[0,\infty)} \mathcal{M}_k(Z_i^T, \alpha, \gamma)\gamma. \end{aligned}$$

where the instanton number $k(\alpha, \gamma)$ is chosen (for each pair (α, γ)) such that

$$\dim \mathcal{M}_k(Z_i^T, \alpha, \gamma) = -1.$$

Since the moduli space depends on an extra parameter T, the union has formal dimension 0 and there is a reasonable signed count.

To show H_i is a chain homotopy, we need to prove:

$$\partial_{i+2} \circ H_i + H_i \circ \partial_i = C_*(W_{i+1}) \circ C_*(W_i) - 0. \tag{7.6.1}$$

The idea is to compactify the moduli space

$$\coprod_{T\in[0,\infty)} \mathcal{M}_k(Z_i^T, \alpha, \beta)$$

with $\dim \mathcal{M}_k(Z_i^T, \alpha, \beta) = 0$. The formal dimension of this parametrized moduli space is 1 and its compactification is a disjoint union of compact 1-manifolds. Each term in equation (7.6.1) appears as a boundary term in the compactification:

(1) The boundary of $T \in [0, \infty]$ give rise to $C_*(W_{i+1}) \circ C_*(W_i)$ $(T = 0)$ and 0 $(T = \infty)$.
(2) The energy can slide off along ends $(-\infty, 0] \times Y_i$ and $Y_{i+2} \times [0, \infty)$, which contribute to $H_i \circ \partial_i$ and $\partial_{i+2} \circ H_i$.

Hence, we arrive at a diagram with arrows indexed by their domains:

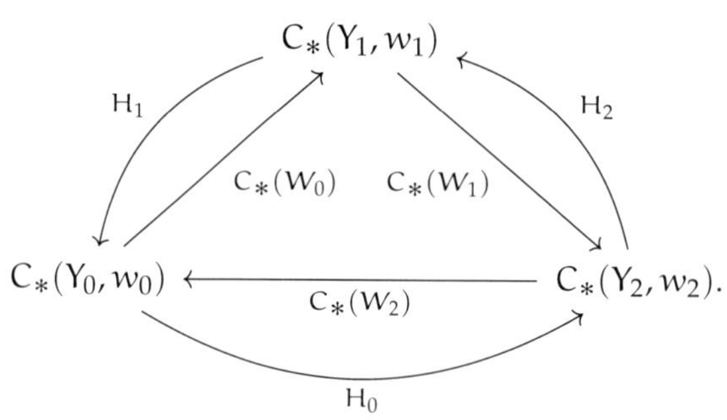

Note that $H_1W_0 + W_2H_0 : C_*(Y_0) \to C_*(Y_0)$ is a chain map (at least) over the base ring $\mathbb{Z}_2$:

$$\partial_0(H_1W_0 + W_2H_0) = H_1\partial_0W_0 + W_2W_1W_0 + W_2\partial_2H_0$$

$$= (H_1 W_0 + W_2 H_0)\partial_0 + 2W_2 W_1 W_0$$
$$= (H_1 W_0 + W_2 H_0)\partial_0 \mod 2.$$

Theorem 7.6.2 (Distinguished Triangulation Detection Lemma; see [41, Lemma 4.4] and [45, Lemma 3.7]). *If each of $H_{i+1}W_i + W_{i+2}H_i$ $(0 \leqslant i \leqslant 2)$ is a chain-homotopy equivalence, then we have a chain-homotopy equivalence.*

$$(C_*(Y_0), \partial_0) \cong \mathrm{Cone}(C_*(Y_1) \xrightarrow{W_1} C_*(Y_2)).$$

Recall that the mapping cone of a chain map defined by

$$(C_{12}, \partial_{12}) = (C_*(Y_1) \oplus C_*(Y_2), \begin{pmatrix} \partial_1 & 0 \\ W_1 & \partial_2 \end{pmatrix}).$$

The morphism in the detection theorem, Theorem 7.6.2, is constructed as

$$f : (C(Y_0), \partial_0) \to (C_{12}, \partial_{12}), \alpha \mapsto (W_0(\alpha), H_0(\alpha)).$$

whose inverse is given by

$$g : (C_{12}, \partial_{12}) \to (C(Y_0), \partial_0), (\beta, \gamma) \mapsto H_1(\beta) + W_2(\gamma).$$

so $g \circ f = H_1 W_0 + W_2 H_0$. The condition on $g \circ f$ is certainly necessary for Theorem 7.6.2 to hold.

Let us now explain why for example $H_1 W_0 + W_2 H_0$ is chain-homotopy equivalent to the identity map from the geometry. Consider the composition

$$U = W_2 \circ W_1 \circ W_0 : Y_0 \to Y_0,$$

in which we have five separating submanifolds:

(1) Y_1.
(2) Y_2.
(3) $S^3_0 = \partial N(E_0)$, where $E_0 \subset Z_0$ is the first exceptional curve.
(4) $S^3_1 = \partial N(E_1)$, where $E_1 \subset Z_1$ is the second exceptional curve.
(5) $S_* = S^1 \times S^2 = \partial N(E_1 \cup E_2)$.

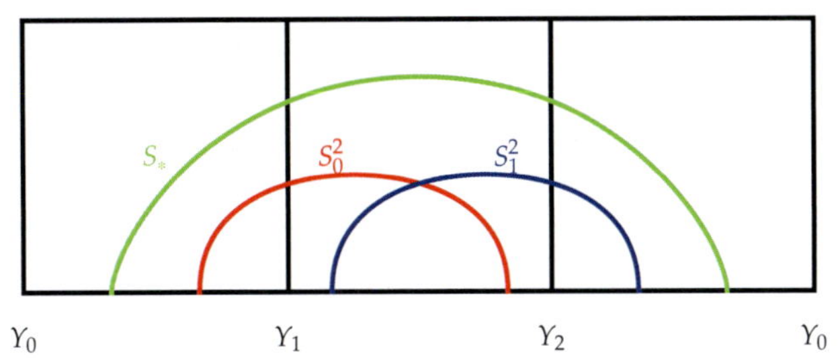

Figure 7.6.3. The triple composite cobordism with five separating submanifolds.

Starting with a fixed metric g_0 on U, we can stretch the metric g_0 in five different ways. If two submanifolds are disjoint, then they can be stretched at the same time. There are five such pairs:

$$(H_1, H_2) = (Y_1, S^3_1), (S^3_1, S_*), (S_*, S^3_0), (S^3_0, Y_2), (Y_2, Y_1).$$

For each pair (H_1, H_2), we form a family of metrics parametrized by a square:

$$(T_1, T_2) \in [0, \infty] \times [0, \infty].$$

where T_i denotes the length of the cylinder along H_i:

$$[0, T_i] \times H_i.$$

When $(T_1, T_2) = (0, 0)$, the metric is not stretched and it is just g_0. When $(T_1, T_2) = (0, \infty)$, the metric is completely stretched along H_2, but not stretched along H_1 at all. When $(T_1, T_2) = (\infty, \infty)$, it means the metric is completely stretched along both H_1 and H_2.

Figure 7.6.4. Square parameterizing metrics. The dashed lines indicate that at least one of the separating three manifolds is completely stretched.

Since we have five pairs, the five squares are combined and form a closed pentagon, denoted by P:

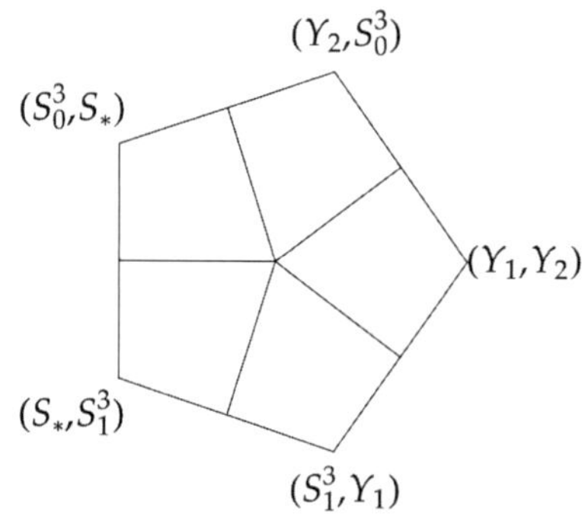

Figure 7.6.5. The pentagon P of metrics.

To construct a chain homotopy between $H_1 W_0 + W_2 H_0$ and the identity map, define

$$\begin{aligned} J : C(Y_0) &\to C(Y_0) \\ \alpha &\mapsto \sum_{\gamma} \#\Big(\coprod_{p \in P} M_k(U, g_p, \alpha, \beta) \Big) \beta \end{aligned}$$

The instanton number $k(\alpha,\gamma)$ is chosen so that the formal dimension of each individual moduli space is -2:

$$\dim M_k(U_p,\alpha,\beta) = -2.$$

Hence, the parametrized moduli space

$$\coprod_{p\in P} M_k(U,g_p,\alpha,\beta)$$

has formal dimension 0. To prove

$$\partial_0 J + J\partial_0 = H_1W_0 + W_2H_0 + \mathrm{Id},$$

consider the compactification of the 1-dimensional moduli space:

$$\coprod_{p\in P} M_{k'}(U,g_p,\alpha,\gamma)$$

where $k'(\alpha,\gamma)$ is chosen so that each individual moduli space has dimension -1. The boundary terms have two origins:

(1) The boundary ∂P of the parameter space P.
(2) The energy can slide off along the ends $(-\infty,0]\times Y_0$ and $Y_0\times[0,\infty)$, which contribute to $J\circ\partial_0$ and $\partial_0\circ J$.

In conclusion, we have

$$\partial_0 J + J\partial_0 = \text{ contributions from } \partial P.$$

There are five components of ∂P. We already know some boundary contributions:

(1) The face of Y_1: $(Y_2,Y_1)-(Y_1,S^3_1)$. The metric is completely stretched along Y_1, so we get an 1-parameter family of metrics on

$$Z_1 : Y_1 \to Y_0.$$

In one direction, the metric is completely stretched along $Y_2\subset Z_1$; in the other, it is stretched along $S^3_1\subset Z_1$. The counting (by definition) is

$$H_1W_0.$$

(2) The face of Y_2: $(S^3_0,Y_2)-(Y_2,Y_1)$. The metric is completely stretched along Y_2, so we get an 1-parameter family of metrics on

$$Z_0 : Y_0 \to Y_2.$$

In one direction, the metric is completely stretched along $Y_1\subset Z_0$; in the other, it is stretched along $S^3_0\subset Z_0$. The counting (by definition) is

$$W_2H_0.$$

(3) The face of S^3_0: $(S_*,S^3_0)-(S^3_0,Y_2)$. The metric is completely stretched along S^3_0. By the argument from the first step, we do not have any contribution from this part because ASD connections on $\overline{\mathbb{CP}}^2$ with $w_2\neq 0$ must have non-zero energy.

(4) The face of S^3_1: $(Y_1, S^3_1) - (S^3_1, S_*)$. It is similar to the face of S^3_1. Again there is no contribution.

(5) The face of S_*: $(S^3_1, S_*) - (S_*, S^3_0)$. The metric is stretched completely along S_*.

The remaining task is to show that the face of $S_* = S^1 \times S^2$ will contribute to the identity morphism on $C_*(Y_0)$.

Lemma 7.6.6. *Recall that* S_* *is the boundary of* $U_1 \doteq N(E_0 \cup E_1)$. *Write the triple composition* $U = W_2 \circ W_1 \circ W_0$ *as* $U_1 \bigcup_{S_*} (U - U_1)$. *Then*

$$U_1 = (S^2 \times D^2) \# \overline{\mathbb{CP}}^2.$$

$$[0,1] \times Y = (U - U_1) \bigcup_{S_*} S^1 \times D^3.$$

We focus on U_1 first. E_1 and E_2 are two (-1)-spheres intersecting positively at a single point. One can blow down either of them within U_1 to obtain $(S^2 \times D^2, S^2 \times S^1)$, which can be easily seen using surgery diagrams. In particular, the intersection form of U_1 is

$$\begin{pmatrix} -1 & 1 \\ 1 & -1 \end{pmatrix} \text{(represented by } E_0 \text{ and } E_1\text{).}$$

$$\sim \begin{pmatrix} -1 & 0 \\ 0 & 0 \end{pmatrix} \text{(represented by } E_0 \text{ and } E_0 + E_1\text{).}$$

The Poincaré dual of $w_2(P)$ on U_1 is the union:

$$E_0 \cup E_1 \cup D_0 \cup D_2$$

where $D_0 = U_1 \cap C_0$ (the core disk) and $D_2 = U_1 \cap S_2$ (the co-core disk). It has even evaluations on both E_1 and E_2, so $w_2(P) = 0 \in H^2(U_1, \mathbb{Z}_2)$. In particular, $w_2(P) = 0 \in H^2(S_*, \mathbb{Z}_2)$.

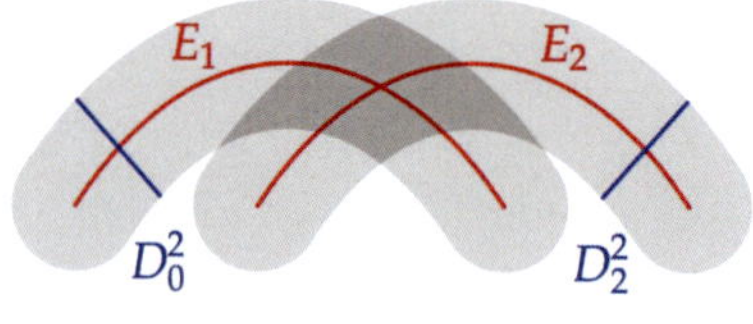

Another way to think about this is to take the intersection:

$$(E_0 \cup E_1 \cup D_0 \cup D_2) \cap S_* = (D_0 \cap S_*) \cup (D_2 \cap S_*). \tag{7.6.7}$$

One may think of $(D_0 \cap S_*)$ as in $(D^2 \times S^2, S^1 \times S^2)$ (blow down the other (-1)-sphere). Then D_0 is a fiber disk and $D_0 \cap S_*$ generates $\pi_1(S_*) \cong \mathbb{Z}$. Since we have two copies of generators in (7.6.7), $w_2(P) = 0 \in H^2(S^1 \times S^2, \mathbb{Z}_2)$.

Hence, we may think of P as the trivial $SU(2)$-bundle (instead of $SO(3)$) over $S^1 \times S^2$. The moduli space of flat connections is

$$\{\mathbb{Z} \to SU(2)\}/\text{conj} \cong [0, \pi].$$

with the isompophism pairing

$$\begin{pmatrix} e^{i\theta} & 0 \\ 0 & e^{-i\theta} \end{pmatrix} \leftrightarrow \theta.$$

The representation $\rho : \mathbb{Z} \to SU(2)$ is determined by $\rho(1) = \cos\theta + \nu\sin\theta$ with $\nu \in \operatorname{Im}\mathbb{H}$ and $|\nu| = 1$. Conjugation by $SU(2)$ will rotate ν in $\operatorname{Im}\mathbb{H}$, so the quotient space is parametrized by $\theta \in [0, \pi]$.

7.7. The Final Step Consider the 4-manifold with cylindrical ends:

$$U_1 \bigcup_{S_*} S_* \times [0, \infty).$$

Fix a flat connection $\theta_0 \in (0, \infty) \subset \mathcal{M}(S^1 \times S^2)$. We choose the instanton number k of P such that the moduli space $\mathcal{M}_{k,\theta_0}(U_1, S_*)$ of ASD connections on U with θ_0 as the limit at ∞ has dimension (-1).

Lemma 7.7.1. *Fix a family of metrics on U_1 parametrized by*

$$\mathbb{R} = (-\infty, 0] \cup [0, \infty).$$

When $T \in [0, \infty)$, the metric is stretched along S^3_1. When $T \in (-\infty, 0]$, the metric is stretched along S^3_0. Then the signed count of

$$\coprod_{T \in \mathbb{R}} \mathcal{M}_{k,\theta_0}(U_1, S_*, g_T)$$

equals 1.

Here is the heuristic for Lemma 7.7.1. As $T \to \infty$, $\mathcal{M}_k(U_1, S_*, g_T)$ contains a unique point with $\theta(T) \to \pi$. As $T \to -\infty$, $\mathcal{M}_k(U_1, S_*, g_T)$ contains a unique point with $\theta(T) \to 0$. Hence, the map

$$\theta : \coprod_{T \in \mathbb{R}} \mathcal{M}_k(U_1, S_*, g_T) \to [0, \pi]$$

has degree 1. Finally, we need to compare the gluing picture

$$U_1 \coprod_{S*} S_* \times [0, \infty) \ + \ (-\infty, 0] \times S_* \coprod_{S*} (U - U_1)$$

with

$$S^1 \times D^3 \coprod_{S*} S_* \times [0, \infty) \ + \ (-\infty, 0] \times S_* \coprod_{S*} (U - U_1).$$

The latter gives rise to the trivial cobordism

$$[0, 1] \times Y_0 : Y_0 \to Y_0,$$

which induces the identity map on Floer homology.

7.8. A Spectral Sequence A similar argument using cobordisms of knots establishes the exact triangle in Theorem 7.4.2 for singular knot Floer homology. In this section, we follow the notations from earlier sections and sketch the spectral sequence promised in the beginning of this lecture. By resolving another crossing of the knot digram $D(K)$, we can iterate the chain homotopy equivalence (7.4.3):

$$C^{\#}(S^3,K) \xrightarrow{\cong} \mathrm{Cone}(C^{\#}(S^3,K_1) \xrightarrow{f_1} C^{\#}(S^3,K_0))$$

$$\xrightarrow{\cong} \mathrm{Cone}\left(\begin{array}{ccc} C^{\#}(K_{11}) & \xrightarrow{f_{11\to 01}} & C^{\#}(K_{01}) \\ \Big\downarrow f_{11\to 10} & \searrow f_{11\to 00} & \Big\downarrow f_{01\to 00} \\ C^{\#}(K_{10}) & \xrightarrow{f_{10\to 00}} & C^{\#}(K_{00}) \end{array}\right).$$

where the chain complex $C^{\#}(S^3,K_i), i=0,1$ is replaced by

$$\mathrm{Cone}\left(\begin{array}{c} C^{\#}(K_{i1}) \\ \Big\downarrow f_{i1\to i0} \\ C^{\#}(K_{i0}) \end{array}\right) = C^{\#}(K_{i1}) \oplus C^{\#}(K_{i0}) \circlearrowright \begin{pmatrix} \partial_{i1} & 0 \\ f_{i1\to i0} & \partial_{i0} \end{pmatrix}.$$

The horizontal map f_1 in (7.4.3) is replaced by the sum of three maps:

$$f_{11\to 01},\ f_{11\to 00},\ f_{10\to 00}.$$

The diagonal map $f_{11\to 00}$ is not a chain map. It is a "canonical" chain homotopy between two compositions of chain maps:

$$f_{01\to 00} \circ f_{11\to 01} \text{ and } f_{10\to 00} \circ f_{11\to 10}.$$

To see this concretely, one may compare the right-bottom entries of the products:

$$\begin{pmatrix} \partial_{11} & 0 \\ f_{11\to 10} & \partial_{10} \end{pmatrix}\begin{pmatrix} f_{11\to 01} & 0 \\ \mathbf{f_{11\to 00}} & f_{10\to 00} \end{pmatrix} = \begin{pmatrix} f_{11\to 01} & 0 \\ \mathbf{f_{11\to 00}} & f_{10\to 00} \end{pmatrix}\begin{pmatrix} \partial_{01} & 0 \\ f_{01\to 00} & \partial_{00}. \end{pmatrix}$$

Continuing to resolve crossings and iterating, we end up with a cube of maps:

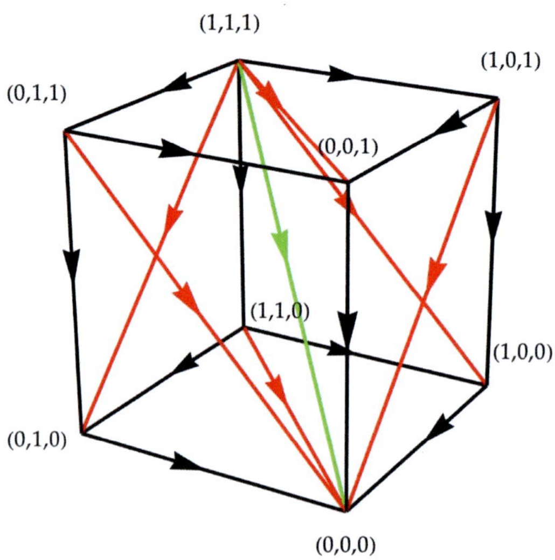

Figure 7.8.1. The 3-cube with diagonals and non-trivial maps assigned to them. There are also self-maps at each vertex.

At each vertex v of $[0,1]^n$, we have a chain complex computing the singular instanton knot homology of a unlink and each edge is associated with a chain map. A filtration on this large chain complex is given by the norm of v:

$$|v| \doteq \sum_{1 \in v} 1.$$

Hence, we obtain a spectral sequence abutting to $I^{\#}(S^3, K)$. Regarding its E_1 and E_2 pages, we have

(1) E_1-page: at each vertex $v \in \{0,1\}^n$, the resolved digram is a unlink. (E_1, d_1) computes the singular knot Floer homology at each vertex. It is isomorphic to $A^{\otimes n}$ where n is the number of link components.

(2) E_2-page: (E_2, d_2) consists of chain maps along edges of $[0,1]^n$. They agree with Khovanov differentials. Hence, the E_2-page computes the Khovanov homology of (S^3, K).

In particular, we see that $\dim \mathrm{Kh}(K) \geqslant \dim I^{\#}(S^3, K)$. To finish the proof of unknot detection, we also need ro:

(1) Compare $I^{\#}(S^3, K)$ with $I^{\natural}(S^3, K)$, the original version of knot Floer homology defined by Floer.

(2) Using sutured manifolds, show that $I^{\natural}(S^3, K)$ detects the unknot.

Now note that

$$\dim \mathrm{Kh}(U) = \min_K \dim \mathrm{Kh}(K).$$

If $\dim \mathrm{Kh}(K)$ is minimal, both $\dim I^{\#}(S^3, K)$ and $\dim I^{\natural}(S^3, K)$ are minimal as well, so K is the unknot.

References

[1] V. I. Arnol'd, *The branched covering* $\mathbf{CP}^2 \to S^4$*, hyperbolicity and projective topology* (Russian), Sibirsk. Mat. Zh. **29** (1988), no. 5, 36–47, 237, DOI 10.1007/BF00970265; English transl., Siberian Math. J. **29** (1988), no. 5, 717–726 (1989). MR971226 ↑332

[2] M. F. Atiyah, *Geometry on Yang-Mills fields*, Scuola Normale Superiore Pisa, Pisa, 1979. MR554924 ↑298

[3] M. F. Atiyah and R. Bott, *The Yang-Mills equations over Riemann surfaces*, Philos. Trans. Roy. Soc. London Ser. A **308** (1983), no. 1505, 523–615, DOI 10.1098/rsta.1983.0017. MR702806 ↑269, 305

[4] M. F. Atiyah, N. J. Hitchin, and I. M. Singer, *Self-duality in four-dimensional Riemannian geometry*, Proc. Roy. Soc. London Ser. A **362** (1978), no. 1711, 425–461, DOI 10.1098/rspa.1978.0143. MR506229 ↑284, 311

[5] M. F. Atiyah, V. K. Patodi, and I. M. Singer, *Spectral asymmetry and Riemannian geometry. I*, Math. Proc. Cambridge Philos. Soc. **77** (1975), 43–69, DOI 10.1017/S0305004100049410. MR397797 ↑280

[6] M. F. Atiyah, V. K. Patodi, and I. M. Singer, *Spectral asymmetry and Riemannian geometry. II*, Math. Proc. Cambridge Philos. Soc. **78** (1975), no. 3, 405–432, DOI 10.1017/S0305004100051872. MR397798 ↑280

[7] M. F. Atiyah, V. K. Patodi, and I. M. Singer, *Spectral asymmetry and Riemannian geometry. III*, Math. Proc. Cambridge Philos. Soc. **79** (1976), no. 1, 71–99, DOI 10.1017/S0305004100052105. MR397799 ↑280

[8] M. F. Atiyah and I. M. Singer, *Index theory for skew-adjoint Fredholm operators*, Inst. Hautes Études Sci. Publ. Math. **37** (1969), 5–26. MR285033 ↑280

[9] A. A. Belavin, A. M. Polyakov, A. S. Schwartz, and Yu. S. Tyupkin, *Pseudoparticle solutions of the Yang-Mills equations*, Phys. Lett. B **59** (1975), no. 1, 85–87, DOI 10.1016/0370-2693(75)90163-X. MR434183 ↑298

[10] E. Bierstone and P. D. Milman, *Semianalytic and subanalytic sets*, Inst. Hautes Études Sci. Publ. Math. **67** (1988), 5–42. MR972342 ↑282

[11] E. J. Brody, *The topological classification of the lens spaces*, Ann. of Math. (2) **71** (1960), 163–184, DOI 10.2307/1969884. MR116336 ↑286

[12] H. Cartan, *Notions d'algèbre différentielle; application aux groupes de Lie et aux variétés où opère un groupe de Lie* (French), Colloque de topologie (espaces fibrés), Bruxelles, 1950, Georges Thone, Liège; Masson et Cie., Paris, 1951, pp. 15–27. MR0042426 ↑277

[13] S. S. Chern and J. Simons, *Characteristic forms and geometric invariants*, Ann. of Math. (2) **99** (1974), 48–69, DOI 10.2307/1971013. MR353327 ↑277

[14] S. K. Donaldson and P. B. Kronheimer, *The geometry of four-manifolds*, Oxford Mathematical Monographs, The Clarendon Press, Oxford University Press, New York, 1990. Oxford Science Publications. MR1079726 ↑305

[15] B. S. Mitjagin and I. S. Èdel′šteĭn, *The homotopy type of linear groups of two classes of Banach spaces* (Russian), Funkcional. Anal. i Priložen. **4** (1970), no. 3, 61–72. MR0341526 ↑280

[16] R. Fintushel and R. J. Stern, *Pseudofree orbifolds*, Ann. of Math. (2) **122** (1985), no. 2, 335–364, DOI 10.2307/1971306. MR808222 ↑286

[17] R. Fintushel and R. J. Stern, *Instanton homology of Seifert fibred homology three spheres*, Proc. London Math. Soc. (3) **61** (1990), no. 1, 109–137, DOI 10.1112/plms/s3-61.1.109. MR1051101 ↑288, 291

[18] A. Floer, *An instanton-invariant for 3-manifolds*, Comm. Math. Phys. **118** (1988), no. 2, 215–240. MR956166 ↑255, 323

[19] A. Floer, *The unregularized gradient flow of the symplectic action*, Comm. Pure Appl. Math. **41** (1988), no. 6, 775–813, DOI 10.1002/cpa.3160410603. MR948771 ↑255, 278

[20] A. Floer, *Instanton homology and Dehn surgery*, The Floer memorial volume, Progr. Math., vol. 133, Birkhäuser, Basel, 1995, pp. 77–97. MR1362823 ↑334

[21] D. S. Freed and K. K. Uhlenbeck, *Instantons and four-manifolds*, 2nd ed., Mathematical Sciences Research Institute Publications, vol. 1, Springer-Verlag, New York, 1991, DOI 10.1007/978-1-4613-9703-8. MR1081321 ↑305

[22] M. Furuta, *Homology cobordism group of homology 3-spheres*, Invent. Math. **100** (1990), no. 2, 339–355, DOI 10.1007/BF01231190. MR1047138 ↑291

[23] D. M. Grobman, *Homeomorphism of systems of differential equations* (Russian), Dokl. Akad. Nauk SSSR **128** (1959), 880–881. MR0121545 ↑258

[24] M. Gromov, *Pseudo holomorphic curves in symplectic manifolds*, Invent. Math. **82** (1985), no. 2, 307–347, DOI 10.1007/BF01388806. MR809718 ↑302

[25] P. Hartman, *A lemma in the theory of structural stability of differential equations*, Proc. Amer. Math. Soc. **11** (1960), 610–620, DOI 10.2307/2034720. MR121542 ↑258

[26] M. Khovanov, *Link homology and Frobenius extensions*, Fund. Math. **190** (2006), 179–190, DOI 10.4064/fm190-0-6. MR2232858 ↑329

[27] M. Khovanov, *A categorification of the Jones polynomial*, Duke Math. J. **101** (2000), no. 3, 359–426, DOI 10.1215/S0012-7094-00-10131-7. MR1740682 ↑328

[28] P. A. Kirk and E. P. Klassen, *Chern-Simons invariants of 3-manifolds and representation spaces of knot groups*, Math. Ann. **287** (1990), no. 2, 343–367, DOI 10.1007/BF01446898. MR1054574 ↑286

[29] S. Kobayashi and K. Nomizu, *Foundations of differential geometry. Vol I*, Interscience Publishers, a division of John Wiley & Sons, New York-London, 1963. MR0152974 ↑269

[30] P. Kronheimer and T. Mrowka, *Monopoles and three-manifolds*, New Mathematical Monographs, vol. 10, Cambridge University Press, Cambridge, 2007, DOI 10.1017/CBO9780511543111. MR2388043 ↑265

[31] P. B. Kronheimer and T. S. Mrowka, *Khovanov homology is an unknot-detector*, Publ. Math. Inst. Hautes Études Sci. **113** (2011), 97–208, DOI 10.1007/s10240-010-0030-y. MR2805599 ↑328, 333

[32] P. Kronheimer, T. Mrowka, P. Ozsváth, and Z. Szabó, *Monopoles and lens space surgeries*, Ann. of Math. (2) **165** (2007), no. 2, 457–546, DOI 10.4007/annals.2007.165.457. MR2299739 ↑333

[33] N. H. Kuiper, *The homotopy type of the unitary group of Hilbert space*, Topology **3** (1965), 19–30, DOI 10.1016/0040-9383(65)90067-4. MR179792 ↑280

[34] N. H. Kuiper, *The quotient space of* $\mathbf{C}P(2)$ *by complex conjugation is the* 4*-sphere*, Math. Ann. **208** (1974), 175–177, DOI 10.1007/BF01432386. MR346817 ↑332

[35] H. B. Lawson Jr., *The theory of gauge fields in four dimensions*, CBMS Regional Conference Series in Mathematics, vol. 58, Published for the Conference Board of the Mathematical Sciences, Washington, DC; by the American Mathematical Society, Providence, RI, 1985, DOI 10.1090/cbms/058. MR799712 ↑305

[36] E. Looijenga, *Root systems and elliptic curves*, Invent. Math. **38** (1976/77), no. 1, 17–32, DOI 10.1007/BF01390167. MR466134 ↑287

[37] J. Milnor, *Morse theory*, Annals of Mathematics Studies, No. 51, Princeton University Press, Princeton, N.J., 1963. Based on lecture notes by M. Spivak and R. Wells. MR0163331 ↑280

[38] M. Morse, *The calculus of variations in the large*, American Mathematical Society Colloquium Publications, vol. 18, American Mathematical Society, Providence, RI, 1996. Reprint of the 1932 original, DOI 10.1090/coll/018. MR1451874 ↑278

[39] G. D. Mostow, *Equivariant embeddings in Euclidean space*, Ann. of Math. (2) **65** (1957), 432–446, DOI 10.2307/1970055. MR87037 ↑306

[40] T. S. Mrowka and K. Wehrheim, *L^2-topology and Lagrangians in the space of connections over a Riemann surface*, Geom. Funct. Anal. **20** (2010), no. 5, 1278–1305, DOI 10.1007/s00039-010-0086-3. MR2746955 ↑310

[41] P. Ozsváth and Z. Szabó, *On the Heegaard Floer homology of branched double-covers*, Adv. Math. **194** (2005), no. 1, 1–33, DOI 10.1016/j.aim.2004.05.008. MR2141852 ↑337

[42] R. S. Palais and S. Smale, *A generalized Morse theory*, Bull. Amer. Math. Soc. **70** (1964), 165–172, DOI 10.1090/S0002-9904-1964-11062-4. MR158411 ↑263

[43] J. Sacks and K. Uhlenbeck, *The existence of minimal immersions of 2-spheres*, Ann. of Math. (2) **113** (1981), no. 1, 1–24, DOI 10.2307/1971131. MR604040 ↑302

[44] C. W. Scaduto, *Instantons and odd Khovanov homology*, J. Topol. **8** (2015), no. 3, 744–810, DOI 10.1112/jtopol/jtv012. MR3394316 ↑333

[45] P. Seidel, *Fukaya categories and Picard-Lefschetz theory*, Zurich Lectures in Advanced Mathematics, European Mathematical Society (EMS), Zürich, 2008, DOI 10.4171/063. MR2441780 ↑337

[46] S. Smale, *Morse inequalities for a dynamical system*, Bull. Amer. Math. Soc. **66** (1960), 43–49, DOI 10.1090/S0002-9904-1960-10386-2. MR117745 ↑258, 265

[47] S. Smale, *Stable manifolds for differential equations and diffeomorphisms*, Ann. Scuola Norm. Sup. Pisa Cl. Sci. (3) **17** (1963), 97–116. MR165537 ↑258

[48] C. H. Taubes, *Self-dual Yang-Mills connections on non-self-dual 4-manifolds*, J. Differential Geometry **17** (1982), no. 1, 139–170. MR658473 ↑302

[49] C. H. Taubes, *Path-connected Yang-Mills moduli spaces*, J. Differential Geom. **19** (1984), no. 2, 337–392. MR755230 ↑310

[50] C. H. Taubes, *Self-dual connections on 4-manifolds with indefinite intersection matrix*, J. Differential Geom. **19** (1984), no. 2, 517–560. MR755237 ↑302

[51] K. K. Uhlenbeck, *Removable singularities in Yang-Mills fields*, Comm. Math. Phys. **83** (1982), no. 1, 11–29. MR648355 ↑302, 316

[52] K. K. Uhlenbeck, *Connections with L^p bounds on curvature*, Comm. Math. Phys. **83** (1982), no. 1, 31–42. MR648356 ↑302, 305, 312

Department of Mathematics, Massachusetts Institute of Technology
Email address: mrowka@mit.edu

Department of Mathematics, Massachusetts Institute of Technology
Current address: Department of Mathematics, SUNY at Stony Brook
Email address: donghaow@mit.edu

IAS/Park City Mathematics Series
Volume 28, Pages 347–402
https://doi.org/10.1090/pcms/028/00879

Lecture on Invertible Field Theories

Søren Galatius

with Exercises by Arun Debray, Søren Galatius, and Martin Palmer

Abstract. Four lectures on invertible field theories at the Park City Mathematics Institute 2019.

Cobordism categories are introduced both as plain categories and topologically enriched. We then discuss localization of categories and its relationship to classifying spaces, and state the main theorem of classification of invertible field theories in these terms. We also discuss symmetric monoidal structures and their relationship to actions of the little disk operads. In the final lecture we discuss an application of cobordism categories to characteristic classes of surface bundles.

Emphasis will be on self-contained definitions and statements, referring to original literature for proofs.

Contents

2020 *Mathematics Subject Classification.* Primary 57R90, 57R15, 57R56, 55P47.
Key words and phrases. Park City Mathematics Institute, mapping class group, surface bundles.

Introduction

The notion of *cobordism* between smooth manifolds and its relationship with *homotopy theory* goes back at least to the work of Pontryagin and Thom. In these lectures we will outline some more recent developments, including connections to *diffeomorphism groups*.

Acknowledgments I would like to thank Arun Debray and Martin Palmer for their help with the exercise sessions during the summer school including their coauthorship on the exercises at the end of these notes. I also thank David Ayala and Ulrike Tillmann for helpful conversations and feedback on draft versions of these notes. Finally, I thank the PCMI and all participants for creating a unique and inspiring atmosphere.

The author was supported by the European Research Council (ERC) under the European Union's Horizon 2020 research and innovation programme (grant agreement No. 682922), by the Danish National Research Foundation through the Centre for Symmetry and Deformation (DNRF92), and by the EliteForsk Prize.

Lecture 1: Cobordisms

Recall that a smooth manifold M is *closed* if it is compact and $\partial M = \emptyset$. We do not require M to be connected. For any $d \in \mathbb{Z}$, the empty set is a manifold of dimension d.

1.1. Classical definitions

Definition 1.1.1. Let M_0 and M_1 be smooth closed manifolds of dimension $d-1$. An (abstract) *cobordism* from M to N is a triple $(W, c_{\text{in}}, c_{\text{out}})$, where W is a compact manifold, and

$$c_{\text{in}} : [0, \infty) \times M_0 \hookrightarrow W$$
$$c_{\text{out}} : (-\infty, 0] \times M_1 \hookrightarrow W$$

are smooth embeddings with disjoint images, such that

$$\partial W = c_{\text{in}}(\{0\} \times M_0) \cup c_{\text{out}}(\{0\} \times M_1).$$

Sometimes we shall omit the two charts c_{in} and c_{out} from the notation and simply say "W is a cobordism from M_0 to M_1" and write $W : M_0 \rightsquigarrow M_1$.

Two manifolds M_0 and M_1 are *cobordant* if there exists a cobordism from M_0 to M_1.

Lemma 1.1.2. *Being cobordant is an equivalence relation.*

Proof sketch. We leave reflexivity and symmetry to the reader and focus on transitivity. If $(W, c_{\text{in}}, c_{\text{out}})$ is a cobordism from M_0 to M_1 and $(W', c'_{\text{in}}, c'_{\text{out}})$ is a cobordism from M_1 to M_2, then we may define a cobordism $(W'', c''_{\text{in}}, c''_{\text{out}})$, where the underlying space W'' is

$$W'' = \frac{W \amalg W'}{c_{\text{out}}(0, x) \sim c'_{\text{in}}(0, x),\ \forall x \in M_1}$$

in the quotient topology, i.e. the coarsest topology making the canonical maps

$$i : W \to W''$$
$$i' : W' \to W''$$

continuous. Charts near a glued point $c_{\text{out}}(0, x) = c'_{\text{in}}(0, x)$ for $x \in M_1$ are obtained as

$$(t, u) \mapsto \begin{cases} c_{\text{out}}(t, \phi(u)) & \text{for } t \leqslant 0 \\ c'_{\text{in}}(t, \phi(u)) & \text{for } t \geqslant 0 \end{cases}$$

for a chart $\phi : U \to M_1$ defined on an open $U \subset \mathbb{R}^{d-1}$. Around other points in W'' charts are obtained by composing a chart in W with i or a chart in W' with i'. It is easy to check that this defines a smooth structure on W''. We obtain a cobordism $M_0 \rightsquigarrow M_2$ by setting $c''_{\text{in}} = i \circ c_{\text{in}}$ and $c''_{\text{out}} = i' \circ c_{\text{out}}$. □

Remark 1.1.3. The set of cobordism classes of k-dimensional manifolds is often denoted $\mathfrak{N}^k$. The direct sum

$$\mathfrak{N}^* = \bigoplus_{k \geqslant 0} \mathfrak{N}^k$$

has the structure of a commutative graded ring (addition induced by disjoint union and multiplication by product of smooth manifolds) in which $2 = 0$. In his 1954 paper *Quelques propriétés globales de variétés differentiables*, René Thom completely determined this ring: there is an isomorphism of commutative graded rings

$$\mathbb{F}_2[x_k \mid k \neq 2^i - 1] \xrightarrow{\cong} \mathfrak{N}^*,$$

from the polynomial ring over $\mathbb{F}_2$ with one generator in each dimension not of the form $k = 2^i - 1$.

The citation for René Thom's Fields medal at the 1958 ICM is: "In 1954 [he] invented and developed the theory of cobordism in algebraic topology. This classification of manifolds used homotopy theory in a fundamental way and became a prime example of a general cohomology theory."

The work presented in these lectures owes an intellectual debt to Thom's profound 1954 paper, although we shall not make direct use of his calculations.

In these lectures we shall be concerned with the *reasons* that two closed manifolds M_0 and M_1 are cobordant, rather than merely *whether* they are cobordant as in Thom's 1954 paper.

1.2. Cobordism categories, first attempt

Definition 1.2.1. Let (W, c_{in}, c_{out}) and (W', c'_{in}, c'_{out}) be two cobordisms between M_0 and M_1. We say that these are *diffeomorphic as cobordisms* if there exists a diffeomorphism $\phi : W \to W'$ of underlying smooth manifolds, such that there exists an $\varepsilon > 0$ with $\phi(c_{in}(t,x)) = c'_{in}(t,x)$ for all $(t,x) \in [0,\varepsilon] \times M_0$ and $\phi(c_{out}(t,x)) = c'_{out}(t,x)$ for all $(t,x) \in [-\varepsilon, 0] \times M_1$.

Definition 1.2.2. Let W be a cobordism from M_0 to M_1 and W' a cobordism from M_1 to M_2. (Here we suppress the boundary collars from the notation, but emphasize that they are an important part of the structure.) The *composition* of W and W' is the cobordism $W'' : M_0 \rightsquigarrow M_2$ constructed in the proof of transitivity in Lemma 1.1.2. We shall also write

$$W \cup_{M_1} W'$$

for this cobordism. (This is sloppy notation, since it depends on the collars.)

Lemma 1.2.3. *Let M_0 and M_1 be closed $(d-1)$-manifolds and let $W_1, W_1' : M_0 \rightsquigarrow M_1$ be two cobordisms between the same two manifolds, and similarly $W_2, W_2' : M_1 \rightsquigarrow M_2$. If W_1 and W_1' are diffeomorphic as cobordisms and also W_2 and W_2' are diffeomorphic as cobordisms, then the compositions $W_1 \cup_{M_1} W_2$ and $W_1' \cup_{M_1} W_2'$ are diffeomorphic as cobordisms.*

Lemma 1.2.4. *Let M_0, M_1, M_2, and M_3 be closed smooth $(d-1)$-manifolds, and let $W_1 : M_0 \rightsquigarrow M_1$, $W_2 : M_1 \rightsquigarrow M_2$ and $W_3 : M_2 \rightsquigarrow M_3$ be cobordisms. Then the two cobordisms*

$$(W_1 \cup_{M_1} W_2) \cup_{M_2} W_3 : M_0 \rightsquigarrow M_3$$
$$W_1 \cup_{M_1} (W_2 \cup_{M_2} W_3) : M_0 \rightsquigarrow M_3$$

are diffeomorphic as cobordisms.

The first lemma is proved by gluing together two diffeomorphisms of cobordisms and verifying that the result is a diffeomorphism of cobordisms, the second by verifying that the canonical map of underlying sets is a diffeomorphism of cobordisms. Let us remark that the two cobordisms $(W_1 \cup_{M_1} W_2) \cup_{M_2} W_3$ and $W_1 \cup_{M_1} (W_2 \cup_{M_2} W_3)$ are likely not *equal* (their underlying sets aren't equal).

Definition 1.2.5. For closed $(d-1)$-manifolds M_0 and M_1, let $\mathrm{Cob}_d(M_0, M_1)$ denote the set of equivalence classes of cobordisms from M_0 to M_1, up to diffeomorphism of cobordisms. The induced operation

$$\begin{aligned} \mathrm{Cob}_d(M_0, M_1) \times \mathrm{Cob}_d(M_1, M_2) &\to \mathrm{Cob}_d(M_0, M_2) \\ ([W], [W']) &\mapsto [W \cup_{M_1} W'] \end{aligned}$$

is then well defined and associative, by the two lemmas above.

An easy argument shows that it is also *unital*: in fact the "cylinder" $[0,1] \times M$ gives rise to an identity element in $\mathrm{Cob}_d(M, M)$.

Definition 1.2.6. Let Cob_d be the category whose objects are closed smooth $(d-1)$-manifolds, whose morphism set $M_0 \to M_1$ is $\mathrm{Cob}_d(M_0, M_1)$, and whose composition is as in Definition 1.2.5.

We will later enhance this definition to one which for many purposes turns out to be more natural. The enhanced definition has more structure (it is not just a plain category), and therefore at first sight will seem more complicated. In the rest of this lecture we shall stick to the elementary definition above.

Remark 1.2.7. A *small category* is one whose collections of objects and morphisms form sets. For example, the category of sets is not small because there is no "set of all sets".

As it stands, the category Cob_d defined above is not small for $d \geqslant 0$. The problem is that there are too many possibilities for the underlying set of the manifolds. This issue may be handled in multiple ways, for example as follows. For any set Ω, let Cob_d^{Ω} be the subcategory of Cob_d in which the manifolds M (objects) and W (whose diffeomorphism class gives a morphism) are required to have underlying set a subset of Ω. Then Cob_d^{Ω} is a small category, and, if the cardinality of Ω is at least that of $\mathbb{R}$ then the inclusion $\mathrm{Cob}_d^{\Omega} \hookrightarrow \mathrm{Cob}_d$ is an equivalence of categories. If Ω' is another such set, then the small categories Cob_d^{Ω} and $\mathrm{Cob}_d^{\Omega'}$ are *equivalent*, in fact both inclusions of subcategories

$$\mathrm{Cob}_d^{\Omega} \hookrightarrow \mathrm{Cob}_d^{\Omega \cup \Omega'} \hookleftarrow \mathrm{Cob}_d^{\Omega'}$$

are equivalences of categories.

We shall not dwell further on this issue. To be definite, let's say we choose once and for all a set Ω of sufficiently large cardinality and henceforth by abuse of notation write Cob_d for the small category Cob_d^{Ω}.

1.3. Categories, groupoids, and spaces For $p \geqslant 0$, the set $[p] = \{0, \dots, p\}$ is totally ordered by the usual ordering of natural numbers, and may be considered the objects of a category, whose morphism set $i \to j$ is empty if $i > j$ and a singleton if $i \leqslant j$. If C is any small category, the set of all functors $f : [p] \to C$ may be identified with the set

$$N_p C = \{(f_1, \ldots, f_p) \mid \text{target}(f_i) = \text{source}(f_{i+1}) \text{ for all } i\}$$

of composable p-tuples of morphisms: f_1 is given by the value of $f : [p] \to C$ on the unique morphism $0 \to 1$, f_2 the value on the unique morphism $1 \to 2$, etc.

Definition 1.3.1. The *classifying space* of a small category C is the space

$$BC = |NC|,$$

the geometric realization of its nerve.

More explicitly, BC is the CW complex with one 0-cell for each object of C; one 1-cell for each non-identity morphism $f : x \to y$ in C, with one end attached to the 0-cell x and the other end to y; one 2-cell for each pair (f, g) of composable non-identity morphisms, etc.

In particular, the set $N_0 C = \text{Ob}(C)$, regarded as a topological space in the discrete topology, canonically sits inside BC via

$$N_0 C \hookrightarrow BC,$$

the inclusion of 0-simplices.

Let us briefly summarize some convenient properties of this construction; see e.g. [38].

- Functoriality Cat $\to$ Top: a functor $F : C \to D$ is sent to a continuous map $BF : BC \to BC$.
- A natural transformation $T : F \to G$ between functors $C \to D$ induces a homotopy between the maps $BF, BG : BC \to BD$.
- As a consequence, a functor $F : C \to D$ admitting a left or a right adjoint is sent to a homotopy equivalence. In particular, equivalences of categories are sent to homotopy equivalences. Likewise, BC is contractible if C admits a terminal or an initial object.
- Products: the canonical map $B(C \times D) \to BC \times BD$ is a homeomorphism[1].

An overarching theme in these lectures is the question of how much information about C is retained by the space BC.

Example 1.3.2. For $c_0 \in \text{Ob}(C)$, let $C_{/c_0}$ be the *over category*: its objects are morphisms $f : c \to c_0$ in C and its morphisms $(f : c \to c_0) \to (f' : c' \to c_0)$ are morphisms $g : c \to c'$ in C such that $f = f' \circ g$. There is a forgetful functor $C_{/c_0} \to C$ which induces a map of spaces

$$u : BC_{/c_0} \to BC.$$

It is not hard to show that $BC_{/c_0}$ is contractible ($\text{id} : c_0 \to c_0$ is a terminal object, and any category with a terminal object has contractible classifying space).

---

[1]Fine print: in this statement the product of BC and BD should be formed in compactly generated spaces. Otherwise we can only say it is a continuous bijection, although still a weak equivalence.

If in addition C is a *groupoid*, i.e. all morphisms are invertible, then the map u can be shown to be a covering map, and it is not hard to use this to construct an isomorphism of groups

$$\mathrm{Aut}_C(c_0) \to \pi_1(BC, c_0).$$

Since all path components of BC admit a contractible covering space, all higher homotopy groups vanish. Thus we have completely understood the homotopy type of BC when C is a groupoid: it is a disjoint union of $K(\pi, 1)$'s, one for each isomorphism class in C.

Taking its classifying space turns a category into a space. One way to go in the other direction is the *fundamental groupoid.*

Definition 1.3.3. For a space X, let $\pi_1(X)$ denote[2] the category whose objects are the points of X, whose morphisms $x \to y$ are homotopy classes of paths $[0,1] \to X$ starting at x and ending at y, up to homotopy relative to their endpoints, and whose composition is induced by concatenation of paths.

For any subset $A \subset X$ we write $\pi_1(X, A)$ for the full subcategory of $\pi_1(X)$ whose objects are the points in A.

As the name indicates, the category $\pi_1(X)$ is in fact a groupoid. It simultaneously encodes both the fundamental groups $\pi_1(X, x)$ for all $x \in X$ and the change-of-basepoint isomorphisms $\pi_1(X, x) \to \pi_1(X, x')$ induced by paths between x and x'.

Lemma 1.3.4. *Let C be a small category and let*

$$\gamma : C \to \pi_1(BC)$$

send an object $x \in C$ to the corresponding 0-simplex in BC, regarded as an object of $\pi_1(BC)$, and a morphism $f : x \to y$ to the homotopy class represented by the 1-simplex corresponding to $f : x \to y$ (and the constant path if f is an identity morphism). Then γ is a (well defined) functor.

Proof sketch. We must prove that γ preserves composition. While the 1-simplex corresponding to a composition $h = f \circ g$ may not be *equal* to the concatenation of the paths corresponding to f and g, the 2-simplex corresponding to the pair $(f, g) \in N_2C$ gives rise to a homotopy relative to end-points. □

If C is a groupoid it is not hard to show, as in Example 1.3.2 above, that γ is an equivalence of categories. In fact, if we keep track of the subset $N_0C \subset BC$ we may reconstruct C up to isomorphism of groupoids, as follows.

Proposition 1.3.5. *If C is a groupoid, then $\gamma : C \to \pi_1(BC)$ is an equivalence of categories. In fact it is an isomorphism of categories onto $\pi_1(BC, N_0C)$, the full subcategory of $\pi_1(BC)$ on the objects $N_0C \subset BC$.* □

[2]Other commonly used notation for the fundamental groupoid is $\Pi_1(X)$ or $\pi_{\leqslant 1}(X)$.

In fact γ is an equivalence of categories if and only if C is a groupoid, since $\pi_1(BC)$ is always a groupoid. In general we have the following universal property, expressing that γ is the "universal functor to a groupoid".

Proposition 1.3.6. *The functor*

$$\gamma : C \to \pi_1(BC, N_0C)$$

is the universal groupoid under C*: for any groupoid* D *and functor* $F : C \to D$*, there is a unique functor* $G : \pi_1(BC, N_0C) \to D$ *such that* $F = G \circ \gamma$.

Proof sketch. We first explain uniqueness. Since γ is a bijection of object sets it is fixed what the functor G must do on objects. A morphism in $\pi_1(BC, N_0C)$ is represented by a path $\lambda : [0, 1] \to BC$ starting and ending in $N_0C \subset BC$, and any such path is homotopic relative to its endpoints to a path running entirely in the 1-skeleton of BC. Since the 1-simplices of BC are (non-identity) morphisms in C, any morphism in $\pi_1(BC, N_0C)$ may be written as a finite composition of the form

$$(\gamma(\phi_1))^{\varepsilon_1} \dots (\gamma(\phi_n))^{\varepsilon_n},$$

where $\varepsilon_i \in \{\pm 1\}$ and ϕ_i are morphisms in C. The requirement $F = G \circ \gamma$ forces G to take this element to $(F(\phi_1))^{\varepsilon_1} \dots (F(\phi_n))^{\varepsilon_n}$.

The easiest way to check existence is to apply Proposition 1.3.5 as follows: define $G : \pi_1(BC, N_0C) \to D$ as the inverse of the isomorphism $D \to \pi_1(BD, N_0D)$, composed with $\pi_1(BF) : \pi_1(BC, N_0C) \to \pi_1(BD, N_0D)$. □

As with any universal property, $\pi_1(BC, N_0C)$ is determined up to unique isomorphism of categories by this proposition. It therefore also determines $\pi_1(BC)$ up to canonical equivalence of categories (since the inclusion $\pi_1(BC, N_0C) \hookrightarrow \pi_1(BC)$ is an equivalence).

Remark 1.3.7. The natural transformation $\gamma : C \to \pi_1(BC, N_0C)$ is a special case of *localization of categories*. More generally, given a small category C and a subcategory $S \subset C$, there exists a functor $\gamma : C \to C[S^{-1}]$ which is a bijection on object sets, sends any morphism in S to an isomorphism in $C[S^{-1}]$, and is initial with this property (i.e. if $f : C \to D$ is a functor sending morphisms in S to isomorphisms, then it factors as $f = g \circ \gamma$ for a unique $g : C[S^{-1}] \to D$). The universal functor which sends all morphisms to isomorphisms is then $C \to C[C^{-1}]$, so the result above may be summarized as an isomorphism of categories

$$C[C^{-1}] \to \pi_1(BC, N_0C)$$

and hence an equivalence of categories

$$C[C^{-1}] \xrightarrow{\simeq} \pi_1(BC).$$

Remark 1.3.8. We have explained that, up to equivalence of categories, no information is lost in the passage $C \mapsto BC$, as long as C is a groupoid.

If C is not a groupoid, the passage from C to BC should be expected to lose lots of information. For example, if C is the two-object category with object set

$\{0,1\}$, one morphism $0 \to 1$, and no other non-identity morphisms, then BC is homeomorphic to the interval $[0,1]$. If $C' = \{0\} \hookrightarrow C$ is the inclusion of the 1-object category, then $BC' \to BC$ is a homotopy equivalence (namely the inclusion $\{0\} \subset [0,1]$), even though $C' \to C$ is not an equivalence of categories.

More examples of this flavor are easily constructed. For example, to any simplicial complex X there is an associated "poset of simplices" of X, regarded as a category (i.e. $\mathrm{Hom}(\sigma,\tau)$ is a singleton if σ is a face of τ and empty otherwise). Then it is a standard result that BC is homeomorphic to $|X|$.

1.4. Invertible field theories (poor man's version) In light of what we have discussed so far, understanding functors from a small category C into groupoids is essentially the same thing as understanding the universal such $C \to C[C^{-1}]$. As we have seen, this is the same as understanding the fundamental groupoid of BC. If we want an up-to-isomorphism-of-categories answer we should also keep track of the subset $N_0C \subset BC$, but for an up-to-equivalence-of-categories answer we only need the space BC itself, and then we only need it up to homotopy equivalence. In fact we need less than the homotopy type of BC, we only need to understand the space BC up to maps that induce equivalence of fundamental groupoids (i.e. the "1-type" of BC). For example, if we can find a space X and a 2-connected map $X \to BC$ (meaning all homotopy fibers are simply connected), then the induced functor $\pi_1(X) \to \pi_1(BC)$ is an equivalence. Let us spell out this observation.

Corollary 1.4.1. *Let C be a small category and let $f : X \to BC$ be a 2-connected map. Then the functors*

$$C \xrightarrow{\gamma} \pi_1(BC) \xleftarrow{\pi_1(f)} \pi_1(X)$$

induce, after choosing an inverse to the equivalence $\pi_1(f) : \pi_1(X) \to \pi_1(BC)$, an equivalence of categories

$$C[C^{-1}] \to \pi_1(X)$$

and for any small groupoid D it induces an equivalence

$$\mathrm{Fun}(\pi_1(X), D) \to \mathrm{Fun}(C, D).$$

In turn, these groupoids are also canonically equivalent to the fundamental groupoids of the mapping spaces $\mathrm{Map}(X, BD) \simeq \mathrm{Map}(BC, BD)$, at least if X is a CW complex.

We shall be interested in the case $C = \mathrm{Cob}_d$. In this case functors from Cob_d to groupoids may be regarded as an approximation to the notion of *invertible field theories* with values in the groupoid D. The actual definition involves[3] symmetric monoidal functors into rigid symmetric monoidal groupoids, as we shall explain later in the week. The approach we shall take, namely to find a 2-connected map from an "understandable" space, also makes sense in the simpler case of plain categories and groupoids, and it seems instructive to discuss this setting first.

[3]Depending on who you ask, it may likely also involve higher categories.

Lecture 2: Topologically enriched (cobordism) categories

Recall that a (small) *topologically enriched* category consists of a category C and a specified topology on each morphism set $C(x,y)$ such that the composition map $C(x,y) \times C(y,z) \to C(x,z)$ is continuous[4] for all x, y, and z. The underlying (unenriched) category has the same object set, morphism set $x \to y$ given by $C(x,y)^\delta$, the underlying set of $C(x,y)$, and composition as in C.

A *topologically enriched* (also known as *continuous*) functor $F : C \to D$ between topologically enriched categories is a functor of underlying categories with the property that the induced map of morphism spaces is continuous.

2.1. Topologically enriched cobordism categories In this section we shall define a version of the cobordism category Cob_d which is a topologically enriched category. It is very similar in spirit to Cob_d, except that the objects and morphisms are now *submanifolds* of a fixed vector space, instead of being abstract manifolds.

If V is a finite-dimensional real vector space and $M \subset V$ is a subset, then being a d-dimensional submanifold (without boundary) is a *property* of the subset: it is the condition that it locally be the inverse image of a regular value of a smooth function to $\mathbb{R}^n$ for $n = \dim(V) - d$.

For $t > 0$ we shall say that a compact subset $W \subset [0,t] \times V$ is a *cobordism* between two compact submanifolds $M_0 \subset V$ and $M_1 \subset V$, provided there exists an $\varepsilon > 0$ such that

$$W \cap ([0,\varepsilon] \times V) = [0,\varepsilon] \times M_0$$
$$W \cap ([t-\varepsilon, t] \times V) = [t-\varepsilon, t] \times M_1,$$

and such that the subset

$$\widehat{W} = \big((-\infty, 0] \times M_0\big) \cup W \cup \big([t,\infty) \times M_1\big)$$

is a smooth submanifold of $\mathbb{R} \times V$.

Definition 2.1.1. For a finite-dimensional real vector space V, the objects of $\mathcal{C}_d^V$ are the compact subsets $M \subset V$ which are smooth $(d-1)$-dimensional submanifolds. Morphisms $M_0 \rightsquigarrow M_1$ are pairs (t, W) where $t > 0$ and $W \subset [0,t] \times V$ is a cobordism in the sense above. Composition is defined as

$$(t_1, W_1) \circ (t_2, W_2) = (t_1 + t_2, W_1 \cup (t_1 + W_2)),$$

where $t_1 + W_2 = \{(t_1, 0) + w \mid w \in W_2\}$ is the parallel translate of W_2.

Lemma 2.1.2. *The composition defined above is well defined, and makes $\mathcal{C}_d^V$ into a (small, non-unital) category.*

[4]For technical reasons it is probably better to require all morphism spaces $C(x,y)$ to be *compactly generated* topological spaces. If this is not already the case for C, it may be achieved by applying k-ification to each morphism space. Similarly, the domain of the composition map should be the product in compactly generated topological spaces. In these notes we will gloss over this and many other details.

Proof. This is in fact much easier than for the "abstract" cobordisms considered earlier. The power set of V is a set, and $\mathrm{Ob}(\mathcal{C}_d^V)$ is a subset thereof, hence it is again a set; morphisms form a set for a similar reason. Composition is given by *union of subsets* which is strictly associative (unlike the gluing operation of abstract manifolds, which is only "associative up to (preferred) diffeomorphism"). □

As defined above there are no units for composition, due to the requirement $t > 0$. We could fix this by allowing $t = 0$ with some care, but instead we shall just consider $\mathcal{C}_d(V)$ a *non-unital category*.

Next we define a topology on the morphism spaces of $\mathcal{C}_d^V$. If M_0 and M_1 are objects of $\mathcal{C}_d(V)$ and W is an *abstract* cobordism between them, we have a map

$$\begin{aligned} \mathbb{R}_{>0} \times \mathrm{Emb}_\partial(W, [0,1] \times V) &\to \mathcal{C}_d^V(M_0, M_1) \\ (t, \phi) &\mapsto (t, \lambda_t(\phi(W))), \end{aligned} \tag{2.1.3}$$

where λ_t scales the first factor by a factor of t, i.e. it is the map $\mathbb{R} \times V \to \mathbb{R} \times V$ given by $(s, v) \mapsto (st, v)$. The subscript "∂" denotes that the embeddings are required to agree with the canonical $M_0 \hookrightarrow \{0\} \times V$ and $M_1 \hookrightarrow \{1\} \times V$ when restricted to ∂W, and also that they must have a product behavior on a collar neighborhood of the boundary.

The point is now that the domain of (2.1.3) has a natural topology: the Euclidean topology on $\mathbb{R}_{>0}$ and the C^∞ topology on the embedding space. The latter is the topology in which convergence means uniform convergence of all partial derivatives of any order (we have uniform convergence of partial derivatives on compact subsets of any chart on W). We give the codomain of (2.1.3) the quotient topology (as W ranges over all abstract cobordisms).

Definition 2.1.4. Let $\mathcal{C}_d^V$ be the topologically enriched category whose objects are the closed $(d-1)$-dimensional submanifolds $M \subset V$ and whose morphism spaces are the sets of embedded cobordisms (t, W) in the quotient topology from the C^∞ topology, as defined above.

At first sight the definition of $\mathcal{C}_d^V$ may seem a bit ad hoc: for example, the particular way that "collar conditions" near incoming and outgoing boundaries is handled; one could imagine other definitions (e.g. objects could come with a germ of a d-dimensional thickening) which would lead to a topological category which is likely not isomorphic to the one defined above. However, we only want to declare this particular model to be "the" canonical model up to a suitable notion of *equivalence*, namely the so-called *DK-equivalence*. We shall return briefly to this homotopy theory question later in the lectures.

Before proceeding, let us discuss the geometric meaning of the homotopy type of the morphism spaces $\mathcal{C}_d^V(M_0, M_1)$.

2.2. Manifold bundles and bundles of cobordisms We will only give a cursory discussion of this material, referring the reader to the original sources for further details.

To simplify matters we postpone the discussion of boundaries. Hence, let W be a closed manifold, let V be a finite dimensional real vector space, and consider the orbit space

$$\mathrm{Emb}(W, V)/\mathrm{Diff}(W), \tag{2.2.1}$$

where the diffeomorphism group of W acts on the embedding space by precomposition. As a set, this quotient may be identified with a subset of the power set of V, namely those subsets of V which have the property of being submanifolds which additionally are diffeomorphic to W. The space is topologized as a quotient of the embedding space, which is topologized in the C^∞ topology.

How does one produce continuous maps *into* the space (2.2.1)? A good way is to start with an *embedded bundle*. Suppose X is a smooth manifold without boundary, but not necessarily compact, and that $E \subset X \times V$ is a subset with the following properties

- E is a smooth submanifold of $X \times V$, without boundary,
- the composition

  $$E \xrightarrow{\text{inclusion}} X \times V \xrightarrow{\text{projection}} X$$

  is proper (inverse image of compact is compact), smooth (C^∞), and a submersion (derivative at any point is surjective),
- the inverse image $E_x \subset V$ of $x \in X$ is diffeomorphic[5] to W, for all $x \in X$.

Let us say that a subset $E \subset X \times V$ with these properties is a *smooth* W*-bundle over* X *embedded in* $X \times V$. For a subset $E \subset X \times V$ with these properties, the map

$$\begin{aligned} X &\to \mathrm{Emb}(W, V)/\mathrm{Diff}(W) \\ x &\mapsto E_x \end{aligned}$$

can be shown to be continuous. Moreover, any continuous map is homotopic to one arising this way, and in fact this correspondence gives rise to a bijection

$$[X, \mathrm{Emb}(W, V)/\mathrm{Diff}(W)] \cong \frac{\{E \subset X \times V \text{ as above}\}}{\simeq} \tag{2.2.2}$$

where the equivalence relation $\simeq$ on the right denotes *concordance*: a concordance from E to E' is an embedded smooth W-bundle over $\mathbb{R} \times X$ whose restriction to $\{0\} \times X$ is $\{0\} \times E$ and to $\{1\} \times X$ is $\{1\} \times E'$.

In this sense, $\mathrm{Emb}(W, V)/\mathrm{Diff}(W)$ is a classifying space for embedded smooth W-bundles, in much the same way as Grassmannians are classifying spaces for vector bundles embedded in trivial bundles.

One way to prove (2.2.2) proceeds by giving $\mathrm{Emb}(W, V)/\mathrm{Diff}(W)$ the structure of an *infinite dimensional manifold* and proving that any continuous map from X may be homotoped to a smooth one. The usual reference seems to be [2].

The space $\mathcal{C}_d(M_0, M_1)$ may be understood similarly, as a classifying space for "embedded bundles of cobordisms".

[5] E_x is given the smooth structure arising as the inverse image of a regular point, which agrees with the smooth structure it inherits as a submanifold of V.

Definition 2.2.3. Let M_0 and M_1 be objects of $\mathcal{C}_d$, and let X be a smooth manifold without boundary (possibly non-compact). A subset $E \subset X \times ([0,1] \times V)$ is a *bundle of cobordisms* from $M_0 \subset V$ to $M_1 \subset V$ provided

(i) E is a smooth submanifold,
(ii) the projection map $\pi : E \to X$ is a proper submersion,
(iii) the boundary of E equals $X \times (\{0\} \times M_0 \cup \{1\} \times M_1)$, and for some $\varepsilon > 0$,

$$E \cap (X \times ([0,\varepsilon] \times V)) = X \times ([0,\varepsilon] \times M_0)$$
$$E \cap (X \times ([1-\varepsilon,1] \times V)) = X \times ([1-\varepsilon,1] \times M_1).$$

As for closed manifolds, such a bundle of embedded cobordisms gives rise to a continuous map

$$X \to \mathcal{C}_d^V(M_0, M_1),$$

any continuous map is homotopic to one arising this way, and this construction induces a natural bijection

$$[X, \mathcal{C}_d^V(M_0, M_1)] \cong \frac{\{E \subset X \times ([0,1] \times V) \text{ bundle of cobordisms from } M_0 \text{ to } M_1\}}{\simeq}$$

where the equivalence relation $\simeq$ on the right denotes *concordance*: a concordance from E to E' is a bundle of cobordisms over $\mathbb{R} \times X$ whose restriction to $\{0\} \times X$ is $\{0\} \times E$ and to $\{1\} \times X$ is $\{1\} \times E'$.

2.3. Infinite dimensional ambient space To get a "more universal" object we may take the colimit over V.

Definition 2.3.1. Let

$$\mathcal{C}_d = \operatorname*{colim}_{V \subset \mathbb{R}^\infty} \mathcal{C}_d^V$$

be the topologically enriched category obtained by taking the colimit over finite-dimensional linear subspaces $V \subset \mathbb{R}^\infty = \oplus_{\mathbb{N}} \mathbb{R}$.

For any other infinite-dimensional real vector space $\mathcal{U}$ we could similarly define a cobordism category $\mathcal{C}_d^{\mathcal{U}}$ by taking the colimit over finite-dimensional $V \subset \mathcal{U}$ but this would not be especially interesting: the comparison functors

$$\mathcal{C}_d^{\mathcal{U}} \to \mathcal{C}_d^{\mathcal{U} \oplus \mathbb{R}^\infty} \leftarrow \mathcal{C}_d^{\mathbb{R}^\infty} = \mathcal{C}_d.$$

are both "DK-equivalences" in the sense explained below.

For reasons of time we shall mostly stick to finite-dimensional V in these lectures, but let us record one interesting connection to diffeomorphism groups of manifolds.

Proposition 2.3.2. *There is a weak equivalence*

$$\mathcal{C}_d(M_0, M_1) \simeq \coprod_W \mathrm{BDiff}_\partial(W), \tag{2.3.3}$$

where W ranges over abstract cobordisms from M_0 to M_1, one in each diffeomorphism class, and $\mathrm{Diff}_\partial(W)$ *denotes the group of diffeomorphisms of W which act as the identity near ∂W, topologized in the C^∞ topology.*

We will again not prove this in any detail, but remark that it boils down to the embedding space $\mathrm{Emb}_\partial(W, \mathbb{R}^\infty)$ being weakly contractible and the quotient map to $\mathrm{Emb}_\partial(W, \mathbb{R}^\infty)/\mathrm{Diff}_\partial(W)$ being a fiber bundle.

2.4. Categories versus enriched categories In order to discuss the relationship between Cob_d and $\mathcal{C}_d$, we first discuss the relationship between enriched and unenriched categories in the abstract.

Definition 2.4.1. Let D be a small (ordinary) category. Then we may regard D as a topologically enriched category by giving each set $D(x, y)$ the discrete topology. This topologically enriched category shall be denoted ιD.

Let TCat denote the category of small topologically enriched categories, and topologically enriched functors between them. Thus we have defined a functor $\iota : \mathrm{Cat} \to \mathrm{TCat}$.

Definition 2.4.2. For a small topologically enriched category C, let hC denote the (unenriched) category with the same object set as C, but with morphism sets given by

$$hC(x, y) = \pi_0(C(x, y)),$$

the set of path components of the morphism space in C. Composition is induced by composition in C (using that π_0 preserves products).

Definition 2.4.3. Let C and D be topologically enriched categories. Then a functor $F : C \to D$ is a *DK-equivalence*[6] if the induced functor $(hF) : (hC) \to (hD)$ is an equivalence in the usual sense and if the maps

$$C(x, y) \to D(Fx, Fy)$$

are weak equivalences for all objects $x, y \in C$.

Example 2.4.4. Given a topologically enriched category C, we can define C' to have the same object set, but with $C'(x, y) = |\mathrm{Sing}(C(x, y))|$, where Sing denotes the total singular complex. Composition is induced by composition in C, using that both Sing and geometric realization of simplicial sets preserve finite products. The evaluation maps $|\mathrm{Sing}(C(x, y))| \to C(x, y)$ are compatible with composition and hence define a DK-equivalence $C' \to C$. In particular any topologically enriched category is equivalent to one whose morphism spaces are CW complexes. See Exercise 5.2.5 for more details.

Remark 2.4.5. The topologically enriched categories appearing in these lectures should be understood as homotopy theoretic objects. In the same way as a weak equivalence $X \to Y$ of spaces lets us consider them as "models" for the same homotopy type, we shall consider two topologically enriched categories as "models"

[6]This terminology is more commonly used in the setting of *simplicially enriched* categories, but makes sense for topologically enriched as well. If desired, a topologically enriched category can of course be turned into a simplicially enriched one by taking Sing of all morphism spaces.

for the same platonic object whenever we have given a DK-equivalence between them, or at least a zig-zag of such.

The point-set definition of $\mathcal{C}_d^V$ and $\mathcal{C}_d$ that we wrote down above should be understood in this sense, as reference models for the (embedded) cobordism category.

Topologically enriched categories form one model for $(\infty, 1)$-categories. Presumably everything in these notes could have been presented in any of the other popular settings (simplicially enriched categories, quasi-categories, complete Segal spaces, etc.).

The relationship between $\mathcal{C}_d$ and Cob_d may be expressed succinctly as an equivalence of categories $h\mathcal{C}_d \simeq \mathrm{Cob}_d$, as follows from the description (2.3.3). More precisely the following holds.

Proposition 2.4.6. *Let $\mathcal{C}_d^V \to \iota\mathrm{Cob}_d$ be the functor which sends an object $M_0 \subset V$ to the underlying abstract smooth manifold M_0 and an embedded cobordism $W \subset [0,t] \times V$ to its diffeomorphism class as a cobordism. Then the induced functor*

$$h\mathcal{C}_d^V \to \mathrm{Cob}_d$$

is an equivalence of categories, provided $\dim(V) \gg d$.

Proof sketch. It is essentially surjective if $\dim(V) \geqslant 2d - 1$, since any $(d-1)$-manifold M admits an embedding into $\mathbb{R}^{2d-1}$. It is full when $\dim(V) \geqslant 2d$, since any cobordism from M_0 to M_1 may then be embedded into $[0,t] \times \mathbb{R}^{2d}$ relative to any specified embeddings of M_0 and M_1, and it is faithful when $\dim(V) \geqslant 2d+1$ since any two embeddings of a cobordism are then isotopic, so represent the same element in $\pi_0\mathcal{C}_d^V(M_0, M_1)$. □

Since Cob_d is obtained from the topologically enriched category $\mathcal{C}_d^V$ by the functorial construction $C \mapsto hC$, we take the point of view that the latter "contains more information". Once one becomes sufficiently advanced, one might decide that the enriched category $\mathcal{C}_d$, or perhaps some even more sophisticated object, is a more natural object to consider than Cob_d. Nevertheless, we shall in the rest of this section discuss how information about $\mathcal{C}_d^V$, such as the main theorem below, has implications for Cob_d.

Under a mild assumption, the construction $C \mapsto hC$ has the following universal property.

Lemma 2.4.7. *Let C be a topologically enriched category such that all morphism spaces $C(x,y)$ are locally path connected[7]. Then, for any small (unenriched) category D there is a bijection*

$$\mathrm{Fun}(hC, D) \cong \mathrm{Fun}^{\mathrm{enr}}(C, \iota D),$$

[7]On a technical level it is often more convenient to work with categories enriched in simplicial sets, instead of in topological spaces. For example, the local path connectedness assumption here would go away: the analogue of taking the discrete topology is taking the constant simplicial set which is right adjoint to "π_0".

natural in $D \in \mathbf{Cat}$*, given by composing with the canonical functor* $C \to \iota(hC)$*. That is,* $C \to \iota(hC)$ *is the initial enriched functor to an ordinary category (more precisely, to a topologically enriched category in which all morphism spaces have the discrete topology).*

In particular, functors out of Cob_d translate to enriched functors out of $\mathcal{C}_d$.

The lemma is easily deduced from the corresponding statement about topological spaces: if X is locally path connected, then $X \to \pi_0(X)$ is the initial continuous map to a space which has the discrete topology. (Without local path connectedness, such a space might not exist! See Exercise 5.2.6.)

It also makes sense to take the classifying space of a topologically enriched category C: in this case the set $N_p(C)$ of composable p-tuples of morphisms inherits a topology from the morphism spaces of C, and hence NC is a *simplicial topological space*. The geometric realization

$$BC = |NC| = \Big(\coprod_{p \geqslant 0} \Delta^p \times N_p C \Big) / \sim$$

may be defined as for simplicial sets, except that $\Delta^p \times N_p C$ is given the product topology and the whole space BC is given the quotient topology. Actually, we shall work with the *thick* geometric realization, i.e. take only the quotient involving the face maps of NC, not the degeneracy maps. With this definition, the space BC still makes sense when C is a non-unital topologically enriched category: the nerve NC is then a *semi-simplicial space* (it has no degeneracy maps defined).

Lemma 2.4.8. *Let* C *be a topologically enriched category, possibly non-unital, with locally path connected morphism spaces. Then the map obtained by applying "B" to the canonical functor* $C \to \iota(hC)$

$$BC \to B(hC),$$

is 2-connected. Hence we get equivalences of groupoids

$$hC[(hC)^{-1}] \xrightarrow{\simeq} \pi_1(B(hC)) \xleftarrow{\simeq} \pi_1(BC).$$

Proof. We claim that the induced map of simplicial nerves has

$$N_p C \to N_p(\iota hC)$$

a $(2-p)$-connected map for all $p \geqslant 0$. For $p = 0$ it is in fact a homeomorphism since the functor $C \to (\iota hC)$ is a bijection (the identity) on object sets. For $p = 1$ it follows because the maps

$$C(x, y) \to (\iota hC)(x, y) = \pi_0(C(x, y))$$

of morphism spaces are 1-connected for all x and y. This is just the statement that for any locally path connected space the continuous map $X \to \pi_0(X)$ is 1-connected. For $p = 2$ we assert that $N_2 C \to N_2(\iota hC)$ induces a surjection on π_0, but in fact it is even surjective on the point-set level. For $p > 2$ the claim is vacuous.

It is now a general fact that the geometric realization of a map of semi-simplicial spaces is n-connected if the map of p-simplices is $(n-p)$-connected. □

The point of the lemma (and of considering topologically enriched categories in this context) is that it may be easier to describe the whole homotopy type of BC than to try to calculate $hC[(hC)^{-1}]$ "directly". With more experience, it may also turn out that one is actually interested in the extra information contained in C which is forgotten in hC: that extra information is the homotopy type of each path component of the morphism spaces $C(x, y)$.

Corollary 2.4.9. *The functors* $\mathcal{C}_d \to h\mathcal{C}_d \simeq \mathrm{Cob}_d$ *induce equivalences of groupoids*

$$\mathrm{Cob}_d[\mathrm{Cob}_d^{-1}] \xrightarrow{\simeq} \pi_1(B\mathrm{Cob}_d) \xleftarrow{\simeq} \pi_1(B\mathcal{C}_d)$$

and hence a functor $\gamma : \mathrm{Cob}_d \to \pi_1(B\mathcal{C}_d)$*, universal among functors to groupoids, up to equivalence of categories.*

2.5. The main theorem It turns out there is a nice formula for the homotopy type of $B\mathcal{C}_d^V$, the classifying space of the cobordism category. As we have seen above, such a formula implies a classification of invertible field theories (poor man's version; we return to symmetric monoidal structures later).

Definition 2.5.1. Let V be a finite dimensional real inner product space and $\mathrm{Gr}_d(V)$ denote the Grassmannian of d-planes in $\mathbb{R} \oplus V$. Let $U_{d,V}$ be the universal d-dimensional vector bundle over $\mathrm{Gr}_d(V)$, and let

$$U_{d,V}^{\perp} = \{(X, v) \in \mathrm{Gr}_d(V) \mid v \in X^{\perp}\}$$

be its orthogonal complement in V. Finally, let $T_{d,V}$ be the *Thom space* of $U_{d,V}^{\perp}$, i.e. the one-point compactification of its total space. Let S^V denote the one-point compactification of V, and let

$$\Omega^V T_{d,\mathbb{R}\oplus V} = \mathrm{Map}_*(S^V, T_{d,\mathbb{R}\oplus V})$$

denote the space of basepoint preserving maps, in the compact-open topology. (I.e., if $V = \mathbb{R}^n$ this is the n-fold based loop space.)

We can now state the main theorem of these lectures.

Theorem 2.5.2. *There is a weak equivalence*

$$B\mathcal{C}_d^V \simeq \Omega^V T_{d,\mathbb{R}\oplus V}. \tag{2.5.3}$$

Remark 2.5.4. We will not have time to discuss the proof of this theorem in these lectures, but let us make some historical remarks. There is presumably not a direct map in either direction, and in any case the known proofs construct a *zigzag* of weak equivalences. In suitable models it is compatible with taking the colimit over finite-dimensional $V \subset \mathbb{R}^\infty$ and gives a weak equivalence

$$B\mathcal{C}_d \simeq \operatorname*{colim}_{V \subset \mathbb{R}^\infty} \Omega^V T_{d,\mathbb{R}\oplus V}.$$

The object on the right hand side is an *infinite loop space*, and the corresponding spectrum is often called a Madsen–Tillmann spectrum and denoted $MTO(d)$. In this form the weak equivalence was first proved in [16, Main Theorem].

The proof for finite-dimensional V was first outlined in [9, Section 6], and in more detail in [11]. An outline is also given in my 2011 PCMI lecture notes [10].

Corollary 2.5.5. *There is*[8] *an equivalence of groupoids*

$$h\mathcal{C}_d^V[(h\mathcal{C}_d^V)^{-1}] \simeq \pi_1(\Omega^V T_{d,\mathbb{R}\oplus V}).$$

Hence for $\dim(V) \gg d$ *there is an equivalence of groupoids*

$$\mathrm{Cob}_d[\mathrm{Cob}_d^{-1}] \simeq \pi_1(\Omega^V T_{d,\mathbb{R}\oplus V}).$$

The point is that it is now purely a homotopy theoretic question to determine the fundamental groupoid of $\Omega^V T_{d,\mathbb{R}\oplus V}$, which is closely related to determining $\pi_n(T_{d,\mathbb{R}\oplus V})$ and $\pi_{n+1}(T_{d,\mathbb{R}\oplus V})$ for $n = \dim(V)$. Let us illustrate this with a low-dimensional example.

Example 2.5.6. Let us take $d = 1$ and $V = \mathbb{R}$ so that objects and morphisms may be visualized in the plane. Then $\mathrm{Gr}_1(\mathbb{R}^2) = \mathbb{R}P^1$ is homeomorphic to a circle, while the bundle $U_{1,\mathbb{R}}^{\perp}$ may be identified with the Möbius strip. Therefore $T_{1,\mathbb{R}}$ may be identified with $\mathbb{R}P^2$, so that $\Omega T_{1,\mathbb{R}}$ has two components, each of which is homotopy equivalent to ΩS^2 which has fundamental group $\pi_2(S^2) = \mathbb{Z}$. The fundamental groupoid therefore has two isomorphism classes of objects, and all objects have infinite cyclic automorphism groups.

Remark 2.5.7. In fact the references given above prove slightly more, about a more general notion of cobordism category, where the objects and morphisms are equipped with a *tangential structure*, given by specifying a space Θ with a continuous action of the Lie group $GL_d(\mathbb{R})$. One then defines a topologically enriched category $\mathcal{C}_\Theta^V$ in which morphisms are cobordisms $W \subset [0,t] \times V$ as before, equipped with a $GL_d(\mathbb{R})$-equivariant continuous map $\ell : \mathrm{Fr}(W) \to \Theta$, where $\mathrm{Fr}(W)$ denotes the frame bundle of TW. Objects are $M \subset V$ equipped with $GL_d(\mathbb{R})$-equivariant maps from the frame bundle of $\mathbb{R} \oplus TM$. A popular choice is $\Theta = \{\pm 1\}$ on which $GL_d(\mathbb{R})$ acts by multiplication by the sign of the determinant; then an equivariant map from the frame bundle is the same as an *orientation*. A version of the main theorem holds in this generality, the only difference is that $T_{d,\mathbb{R}\oplus V}$ should be replaced with a space $T_{\Theta,\mathbb{R}\oplus V}$ in which the Grassmannian $\mathrm{Gr}_d(\mathbb{R} \oplus V)$ has been replaced by a space whose points are pairs of a d-dimensional linear subspace $X \subset \mathbb{R} \oplus V$ together with an equivariant map $\ell : \mathrm{Fr}(X) \to \Theta$.

Another interesting variant concerns manifolds equipped with an action of a finite group G. In this case the embedded version of the cobordism category is defined for a finite dimensional inner-product space V, equipped with an isometric action of G; objects are then submanifolds of V invariant under the action and morphisms are submanifolds of $[0,t] \times V$, invariant when $[0,t]$ is given the trivial action. Under the assumption that $\dim(V^G) > d$, there is again a homotopy theoretic formula for the homotopy type of the classifying space of this "equivariant bordism category". Passing to the colimit over subrepresentations V of the direct

[8]The proof of course gives a specified equivalence, or at least a zig-zag of such, not just existence of an equivalence.

sum of countably many copies of the regular representation $\mathbb{R}[G]$, one gets a theorem for a cobordism category of "abstract" G-equivariant manifolds. See [15] for more details on all this, including precise statements.

To finish this section, let us state the classification result in yet another form.

Corollary 2.5.8. *Let* D *be a small groupoid, then there are equivalences of groupoids*

$$\begin{aligned}\mathrm{Fun}(h\mathcal{C}_d^V, D) &\simeq \pi_1\mathrm{Map}(B(h\mathcal{C}_d^V), BD)\\ &\simeq \pi_1\mathrm{Map}(B\mathcal{C}_d^V, BD)\\ &\simeq \pi_1\mathrm{Map}(\Omega^V T_{d,\mathbb{R}\oplus V}, BD).\end{aligned}$$

Imprecisely the poor man's field theories, namely functors from Cob_d to a groupoid D, are up to natural equivalence of functors the same as continuous maps from $\Omega^V T_{d,\mathbb{R}\oplus V}$ to BD, up to homotopy.

Lecture 3: More structure

3.1. Symmetric monoidal structures The "abstract" cobordism category Cob_d admits the structure of a *symmetric monoidal category* with respect to disjoint union. This monoidal structure is given by the *data* of

- an object $\emptyset \in \mathrm{Cob}_d$,
- a functor $\amalg : \mathrm{Cob}_d \times \mathrm{Cob}_d \to \mathrm{Cob}_d$,
- isomorphisms $M_0 \amalg (M_1 \amalg M_2) \cong (M_0 \amalg M_1) \amalg M_2$ forming a natural transformation of functors $\mathrm{Cob}_d \times \mathrm{Cob}_d \times \mathrm{Cob}_d \to \mathrm{Cob}_d$,
- isomorphisms $M_0 \amalg M_1 \cong M_1 \amalg M_0$ forming a natural transformation of functors $\mathrm{Cob}_d \times \mathrm{Cob}_d \to \mathrm{Cob}_d$,
- isomorphisms $\emptyset \amalg M_0 \cong M_0$ forming a natural transformation of functors $\mathrm{Cob}_d \to \mathrm{Cob}_d$,

and this data satisfies the axioms of a symmetric monoidal structure. See e.g. MacLane's book for the full list of axioms (or wikipedia, or many other places on the internet).

Another example of a symmetric monoidal category is that of complex vector spaces, or more generally modules over some commutative ring R, with symmetric monoidal structure given by $\otimes_R$, the trivial 1-dimensional module $1 = R$, all the usual isomorphisms $N_0 \otimes_R (N_1 \otimes_R N_2) \cong (N_0 \otimes_R N_1) \otimes_R N_2$, $R \otimes_R N \cong N$, etc. Let us write Mod_R for the category of R-modules, equipped with this symmetric monoidal structure. One version of the classical definition of a topological field theory then goes as follows.

Definition 3.1.1. An (R-linear) *topological field theory*[9] is a symmetric monoidal functor $Z : \mathrm{Cob}_d \to \mathrm{Mod}_R$.

[9]To distinguish from more elaborate notions, this is nowadays sometimes called a $(d-1, d)$-theory, because Cob_d involves manifolds of dimension $d-1$ and d.

More generally, for any symmetric monoidal category D, a D-valued topological field theory is a symmetric monoidal functor $Z : \mathrm{Cob}_d \to D$.

Let us again skip the detailed definition, but recall that a (strong) symmetric monoidal functor is a functor together with specified isomorphisms $1 \to Z(\emptyset)$ and $Z(M_0) \otimes_R Z(M_1) \to Z(M_0 \amalg M_1)$, natural in M_0 and M_1, satisfying three axioms (having to do with the associators, unitors, and symmetries in the symmetric monoidal categories).

There are also versions of the definition in which objects and cobordisms are equipped with tangential structures, as in Remark 2.5.7.

Definition 3.1.2. Let D be a symmetric monoidal category. An object x is *invertible* if there exists another object $\overline{x}$ such that $x \otimes \overline{x}$ is isomorphic to 1.

A field theory $Z : \mathrm{Cob}_d \to D$ is *invertible* if it sends all objects of Cob_d to invertible objects of D and all morphisms to isomorphisms.

In fact the condition on objects is automatic: the cylinder $[0,1] \times M$ as a cobordism from $M \amalg M$ to $\emptyset$ is sent to an (iso)-morphism $Z(M) \otimes Z(M) \to Z(\emptyset) \cong 1$, so $Z(M)$ is invertible with inverse $Z(M)$. (In tangentially structured versions, the inverse of $Z(M)$ can be taken as $Z(-M)$, the value of Z on M given the opposite tangential structure.)

There is a subcategory $D^{\sim} \subset D$ whose objects are the invertible objects and whose morphisms are the isomorphisms. Then a D-valued invertible field theory is the same thing as a symmetric monoidal functor $\mathrm{Cob}_d \to D^{\sim}$. Hence we may forget about the ambient D and just consider $D^{\sim}$-valued topological field theories. Thus an invertible topological field theory is essentially one whose target category is a rigid symmetric monoidal groupoid, defined as follows[10].

Definition 3.1.3. A symmetric monoidal groupoid D is *rigid* if all objects are invertible.

3.2. Symmetric monoidal structure on the universal groupoid under C We have already studied how to classify functors from Cob_d into the underlying groupoid of D, ignoring the symmetric monoidal structure. Indeed, they are "the same thing" as functors from $\mathrm{Cob}_d[\mathrm{Cob}_d^{-1}] \simeq \pi_1(B\mathrm{Cob}_d)$, which we identified with the fundamental groupoid of another space $\Omega^V T_{d,\nu}$ for $\dim(V) \gg d$. This answer may be enhanced to a classification of invertible topological field theories provided we have good answers to the following two (vaguely worded) questions.

- Let X be a space and $C = \pi_1(X)$ the fundamental groupoid. What extra structure on X is necessary to functorially induce a symmetric monoidal structure on C?

[10]For a general monoidal *category* the adjective "rigid" is the requirement that all objects admit *duals* but that is the same as admitting inverses up to isomorphism in the case where all morphisms are isomorphisms. Another name for "rigid symmetric monoidal groupoid" is "Picard groupoid".

- Let C be a small category and $X = BC$. What extra structure on X does a symmetric monoidal structure on C functorially induce?

There is a quite straightforward answer to these questions, based on the fact that $C \mapsto BC$ preserves products and takes natural transformations to homotopies, and conversely $X \mapsto \pi_1(X)$ takes maps to functors and homotopies to natural isomorphisms[11]. Indeed, the following list of structure on X will induce a symmetric monoidal structure on $\pi_1(X)$.

- a base point $1 \in X$,
- a "product" map $\mu : X \times X \to X$,
- a (homotopy class of a) homotopy $[0,1] \times X \times X \times X \to X$, between the maps $(x_0, x_1, x_2) \mapsto \mu(x_0, \mu(x_1, x_2))$ and $(x_0, x_1, x_2) \mapsto \mu(\mu(x_0, x_1), x_2)$,
- a (homotopy class of a) homotopy $[0,1] \times X \times X \to X$ between the maps $(x_0, x_1) \mapsto \mu(x_0, x_1)$ and $(x_0, x_1) \mapsto \mu(x_1, x_0)$,
- a (homotopy class of a) homotopy of maps $X \to X$ from the identity to the map $x \mapsto \mu(1, x)$,

subject to certain conditions, obtained by translating the axioms of a symmetric monoidal category (according to "category" $\mapsto$ "space", "functor" $\mapsto$ "continuous map", "natural transformation" $\mapsto$ "homotopy class of a homotopy", "object" $\mapsto$ "point", "vertical composition of natural transformations" $\mapsto$ "concatenation of homotopies"). Conversely if C is given a symmetric monoidal structure then $X = BC$ acquires this kind of structure. Finally, these processes are inverse in the sense that $\gamma : C \to \pi_1(BC)$ is a symmetric monoidal functor, and a symmetric monoidal equivalence if all morphisms in C are isomorphisms.

What seems to occur in nature is usually not the exact list above of data on a space X, but rather an "E_n-structure". Let us briefly recall what that is, and explain why it is sufficient for inducing a symmetric monoidal structure on $\pi_1(X)$, for $n \geqslant 3$.

3.3. Little disks Let $\mathcal{D}_n(k)$ be the space of embeddings that on each open disk is given by a translation and a scaling by a positive number. That is, $\mathcal{D}_n(k)$ is the set of $j : \{1, \dots, k\} \times \mathrm{int}(D^n) \to \mathrm{int}(D^n)$ of the form

$$(i, x) \mapsto v_i + t_i x,$$

for some $t_i \in (0,1)$ and $v_i \in \mathrm{int}(D^n)$. Use v_i and t_i to topologize this as a subspace

$$\mathcal{D}_n(k) \subset \big((0,1) \times \mathrm{int}(D^n)\big)^k.$$

These spaces come with the following structure

- an identity element $1 \in \mathcal{D}_n(1)$

[11]A possible pitfall here is that $C \mapsto BC$ does not send vertical composition of natural transformations to concatenation of homotopies, it only does so up to homotopy relative to endpoints. Therefore it is better to only consider *homotopy classes* of homotopies: identify maps $[0,1] \times X \to Y$ that are homotopic relative to $\{0,1\} \times X$. Conversely, such a homotopy class of homotopies is sufficient to induce a natural transformation of functors $\pi_1(X) \to \pi_1(Y)$.

- an action of S_k on $\mathcal{D}_n(k)$, by permuting input disks
- composition maps for $k, k' \geqslant 0$ and $i \in \{1, \dots, k\}$

$$\circ_i : \mathcal{D}_n(k) \times \mathcal{D}_n(k') \to \mathcal{D}_n(k + k' - 1),$$

where $j \circ_i j'$ is defined by inserting the codomain $\mathrm{int}(D^n)$ of j' into the ith copy of $\mathrm{int}(D^n)$ in the domain of j.

We will write $\mathcal{D}_n$ for the data of the spaces $\mathcal{D}_n(k)$ for $k \geqslant 0$ with their symmetric group actions, the identity $1 \in \mathcal{D}_n(1)$, and the compositions $\circ_i$. It satisfies the associativity axioms

$$(j \circ_i j') \circ_{i+i'-1} j'' = j \circ_i (j' \circ_{i'} j'')$$
$$(j \circ_{i'} j') \circ_{i+k'-1} j'' = (j \circ_i j') \circ_{i'} j''$$

for $j \in \mathcal{D}_n(k)$, $j' \in \mathcal{D}_n(k')$, and $j'' \in \mathcal{D}_n(k'')$, as well unit axioms

$$1 \circ_1 j = j$$
$$j \circ_i 1 = j,$$

and an axiom expressing how composition interacts with the action of symmetric groups. Together, these axioms precisely express that $\mathcal{D}_d$ is an *operad*[12] in spaces. This operad is the *little n-disk operad*.

Another example of an operad is the *endomorphism operad* End_X of a space X. Its kth space is

$$\mathrm{End}_X(k) = \mathrm{Map}(X^k, X),$$

the symmetric group acts by permuting factors in X^k, the identity $1 \in \mathrm{End}_X(1)$ is the identity map, and the composition $f \circ_i f'$ is defined by inserting the codomain X of f' into the ith factor in the domain of f, i.e.

$$(f \circ_i f')(x_1, \dots, x_{k+k'-1}) = (x_1, \dots, x_{i-1}, f(x_i, \dots, x_{i+k'-1}), x_i, \dots, x_{k+k'-1}).$$

Definition 3.3.1. An *algebra* for $\mathcal{D}_n$ is a space X together with an operad map $\mathcal{D}_n \to \mathrm{End}_X$, i.e. continuous maps

$$\mu_k : \mathcal{D}_n(k) \to \mathrm{End}_X(k) = \mathrm{Map}(X^k, X),$$

for each $k \in \mathbb{Z}_{\geqslant 0}$, preserving all the structure (symmetric group action, identity, compositions $\circ_i$).

A *morphism* of $\mathcal{D}_n$-algebras is a continuous map $f : X \to Y$ with the property that the diagrams

$$\begin{array}{ccc} \mathcal{D}_n(k) \times X^k & \xrightarrow{\mu_k} & X \\ \downarrow{\scriptstyle \mathrm{id} \times f^k} & & \downarrow{\scriptstyle f} \\ \mathcal{D}_n(k) \times Y^k & \xrightarrow{\mu_k} & Y \end{array}$$

commute for all $k \in \mathbb{Z}_{\geqslant 0}$.

[12] See e.g. [8, chapter 2] or [27] for the axioms given here, and a discussion of how they are equivalent to May's axioms.

Proposition 3.3.2. *For $n \geqslant 3$ there is a functorial way to associate a symmetric monoidal structure on $\pi_1(X)$ to a space X with E_n structure.*

Lemma 3.3.3. *The space $\mathcal{D}_n(k)$ is homotopy equivalent to the* ordered configuration space

$$\mathrm{Conf}_k(\mathbb{R}^n) = \{(x_1, \dots, x_k) \in (\mathbb{R}^n)^k \mid x_i \neq x_j \textit{ for } i \neq j\}.$$

In particular it is simply connected for all k when $n \geqslant 3$.

Proof sketch. Compose the inclusion $\mathcal{D}_n(k) \subset ((0,1) \times \mathrm{int}(D^n))^k$ with the projection to $(\mathrm{int}(D^n))^k$. Its image is the set of k-tuples of distinct elements of $\mathrm{int}(D^n) \subset \mathbb{R}^n$, and the fiber over a k-tuple of distinct elements is a non-empty convex open subset of $(0,1)^k$. After choosing a diffeomorphism $\mathrm{int}(D^n) \to \mathbb{R}^n$, we get a submersion $\mathcal{D}_n(k) \to \mathrm{Conf}_k(\mathbb{R}^n)$ with contractible fibers, which implies being a homotopy equivalence.

Forgetting the last point gives a map $\mathrm{Conf}_{k+1}(\mathbb{R}^n) \to \mathrm{Conf}_k(\mathbb{R}^n)$ which is a fiber bundle whose fiber over $(x_1, \dots, x_k)$ is

$$\mathbb{R}^n \setminus \{x_1, \dots, x_k\} \simeq \bigvee^k S^{n-1}.$$

This fiber is simply connected for $n \geqslant 3$, so in this case the long exact sequence of homotopy groups gives simple connectivity of $\mathrm{Conf}_k(\mathbb{R}^n)$ by induction on k. □

Proof of Proposition 3.3.2. Applying π_1 to the structure maps gives

$$\pi_1(\mathcal{D}_n(k)) \to \mathrm{Fun}(\pi_1(X) \times \cdots \times \pi_1(X), \pi_1(X)), \tag{3.3.4}$$

associating functors $(\pi_1(X))^k \to \pi_1(X)$ to objects of $\pi_1(\mathcal{D}_n(k))$ and natural isomorphisms to isomorphisms in $\pi_1(\mathcal{D}_n(k))$.

Choose[13] an object of $\pi_1(\mathcal{D}_n(2))$, that is, a point in the space $\mathcal{D}_n(2)$, and let $\otimes : \pi_1(X) \times \pi_1(X) \to \pi_1(X)$ be the image of m under (3.3.4).

The two points $m \circ_1 m$ and $m \circ_2 m$ of $\mathcal{D}_n(3)$ are likely not equal, but there's a unique isomorphism between them in $\pi_1(\mathcal{D}_n(3))$ by Lemma 3.3.3. Since the structure maps preserve $\circ_1$ and $\circ_2$, the image under (3.3.4) of this isomorphism $m \circ_1 m \to m \circ_2 m$ in $\pi_1(\mathcal{D}_n(3))$ is a natural transformation between the two functors

$$\begin{aligned} \pi_1(X) \times \pi_1(X) \times \pi_1(X) &\to \pi_1(X) \\ (x, y, z) \quad &\mapsto \begin{cases} x \otimes (y \otimes z) \\ (x \otimes y) \otimes z, \end{cases} \end{aligned}$$

and we take this to be the associator.

The action of the transposition $(12) \in S_2$ on m gives a new point $(12).m$ in $\mathcal{D}_n(2)$ but since $\mathcal{D}_n(2)$ is simply connected (in fact it is homotopy equivalent to S^{n-1}) there is a unique isomorphism $m \to (12).m$ in $\pi_1(\mathcal{D}_n(2))$. Take the image

[13] In case the reader feels unease about this choice, let us remind them that defining e.g. the disjoint union functor $\amalg$: Sets × Sets → Sets also involves a choice, viz. of how to replace two sets by "disjoint copies". Here and there, any two choices give naturally isomorphic functors.

under (3.3.4) of this isomorphism as the symmetry. Finally, as monoidal unit we take the image of the unique point in $\mathcal{D}_n(0)$ under (3.3.4).

It remains to verify that the axioms are satisfied, but that is in fact easy. For example, the pentagon axiom asserts that two natural transformations of functors $(\pi_1(X))^4 \to \pi_1(X)$ are equal. The two natural transformations come from the two ways of using the associator $x \otimes (y \otimes (z \otimes w)) \to ((x \otimes y) \otimes z) \otimes w$ to give a natural isomorphism. Tracing definitions, we see that they are the image under (3.3.4) of two particular isomorphisms $m \circ_2 (m \circ_2 m) \to (m \circ_1 m) \circ_1 m$ in $\pi_1(\mathcal{D}_n(4))$. But since $\mathcal{D}_n(4)$ is simply connected, *any* two isomorphisms between two objects in its fundamental groupoid must be equal. The other axioms of a symmetric monoidal category are proved by the same method. □

Remark 3.3.5. A similar argument gives a *monoidal structure* on the fundamental groupoid of an E_1 space, and a *braided monoidal structure* on the fundamental groupoid of an E_2 space.

Example 3.3.6. Finally, let us give the *standard example* of an E_n algebra, which was in fact the historical motivation for the development of operads.

Any embedding $j : \{1, \dots, k\} \times \text{int}(D^n) \to \text{int}(D^n)$ may be one-point compactified to a based map in the other direction

$$S^n \to S^n \vee \cdots \vee S^n,$$

where we regard S^n as the one-point compactification of $\text{int}(D^n)$. Hence, if Y is any based space we obtain a map

$$j_* : \Omega^n Y \times \dots \Omega^n Y \to \Omega^n Y,$$

depending continuously on $j \in \mathcal{D}_n(k)$. It is not hard to verify that this construction makes $X = \Omega^n Y$ into a $\mathcal{D}_n$-algebra. Moreover, it is *group-like* (the monoidal category $\pi_1(X)$ is rigid).

More surprisingly, it turns out that up to homotopy the space Y may be reconstructed from X and its E_d structure, at least if Y is $(n-1)$-connected. There is a functorial procedure, May's *two-sided bar construction*, which to an E_n-space X associates a pointed space $B(S^n, \mathcal{D}_n, X)$. For $X = \Omega^n Y$ there is a canonical weak equivalence between $B(S^n, \mathcal{D}_n, X)$ and the $(n-1)$-connected cover of Y. Conversely, if X is any group-like E_n space and we define $Y = B(S^n, \mathcal{D}_n, X)$, then there is a zig-zag of weak equivalences which are E_n-maps, between X and $\Omega^n Y$.

3.4. Structure on embedded cobordism categories We can "almost" give the space $B\mathcal{C}_d^V$ an E_n structure with $n = \dim(V)$, as follows. Any embedding $j : \{1, \dots, k\} \times V \to V$ gives a continuous functor

$$j_* : (\mathcal{C}_d^V)^k \to \mathcal{C}_d^V, \tag{3.4.1}$$

by taking images of objects $M_1 \subset V$ and morphisms $W \subset [0, t] \times V$ under j, and hence a continuous map

$$B(j_*) : (B\mathcal{C}_d^V)^k \to B\mathcal{C}_d^V.$$

Choose a diffeomorphism $\phi : V \approx \mathrm{int}(D^n)$ and use it to define maps

$$\mathcal{D}_n(k) \hookrightarrow \mathrm{Emb}(\{1,\dots,k\} \times \mathrm{int}(D^n), \mathrm{int}(D^n)) \xrightarrow{\phi} \mathrm{Emb}(\{1,\dots,k\} \times V, V)$$
$$\to \mathrm{Map}((B\mathcal{C}_d^V)^k, (B\mathcal{C}_d^V),$$

where the last map is $j \mapsto B(j_*)$. It is easy to verify that these maps preserve the operad structure: symmetric group actions, identity, and compositions $\circ_i$. Unfortunately they are not continuous, because the functor (3.4.1) does not "depend continuously" on j. The problem is that $\mathcal{C}_d^V$ has a set (as opposed to a space) of objects, and hence $N_0\mathcal{C}_d^V$ has the discrete topology in which the action is not continuous.

The obvious fix, which is the approach we shall take, is to topologize $\mathrm{Ob}(\mathcal{C}_d^V)$. In fact this set has a very natural topology, which does make the actions explained above continuous. The result, however, is no longer a topologically enriched category.

3.5. Topological categories A *topological category*, or a *category internal to topological spaces* is a category $\overline{C}$ together with a topology on $\mathrm{Ob}(\overline{C})$ and on the total set $\mathrm{Mor}(\overline{C})$ of all morphisms, such that the four structure maps

$$\begin{aligned} \text{identity} : \mathrm{Ob}(\overline{C}) &\to \mathrm{Mor}(\overline{C}) \\ \text{source, target} : \mathrm{Mor}(\overline{C}) &\to \mathrm{Ob}(\overline{C}) \\ \text{composition} : \mathrm{Mor}(\overline{C}) \times_{\mathrm{Ob}(\overline{C})} \mathrm{Mor}(\overline{C}) &\to \mathrm{Mor}(\overline{C}) \end{aligned}$$

are all continuous. The nerve $N\overline{C}$ is then a simplicial topological space, with $N_0\overline{C} = \mathrm{Ob}(\overline{C})$, $N_1\overline{C} = \mathrm{Mor}(\overline{C})$, etc. There is also a notion of non-unital topological categories, which have no identities; the nerve is then a semi-simplicial topological space (no degeneracy maps).

There is an easy way to extract a topologically enriched category C out of a topological category $\overline{C}$, namely take the underlying set of $\mathrm{Ob}(\overline{C})$, and then let $C(x,y) = \{f \in \mathrm{Mor}(\overline{C}) \mid \mathrm{source}(f) = x, \mathrm{target}(f) = y\}$ with the subspace topology. Composition and identity are as in $\overline{C}$. There is an induced map of simplicial topological spaces $NC \to N\overline{C}$. If we assume that $(d_0, d_1) : N_1\overline{C} \to (N_0\overline{C})^2$ is a Serre fibration, then it follows that the diagrams

$$\begin{array}{ccc} N_pC & \longrightarrow & N_p\overline{C} \\ \downarrow & & \downarrow \\ (N_0C)^{p+1} & \longrightarrow & (N_0\overline{C})^{p+1} \end{array} \tag{3.5.1}$$

are homotopy pullbacks for all p. In this case, from a homotopy theoretic point of view[14], not much is lost by this operation. For example it follows from the Bousfield–Friedlander theorem that the induced map of classifying spaces

[14] In the unital case a concise statement is the map $\mathrm{Sing}(NC) \to \mathrm{Sing}(N\overline{C})$ is a weak equivalence in the *complete Segal space* model structure [36] on simplicial spaces. Hence they may be viewed as two models for the same homotopy type of $(\infty, 1)$-categories.

$BC \to B\overline{C}$ is a weak equivalence. It is also a bijection, so $\pi_1(BC) \to \pi_1(B\overline{C})$ is an isomorphism of groupoids. If hC is unital (but C possibly not), then the canonical functors

$$hC \xrightarrow{\gamma} \pi_1 BC \xrightarrow{\cong} \pi_1(B\overline{C}) \tag{3.5.2}$$

induce an isomorphism $(hC)[(hC)^{-1}] \cong \pi_1(B\overline{C})$.

We may define a version of the embedded cobordism category which is a topological category, as follows. The set of objects of $\mathcal{C}_d^V$ is in bijection with

$$\coprod_M \mathrm{Emb}(M, V)/\mathrm{Diff}(M),$$

where the disjoint union ranges over closed $(d-1)$-manifolds, one in each diffeomorphism class. Hence we may topologize it in the quotient topology from the C^∞ topology on the embedding space. Keeping the same underlying category as $\mathcal{C}_d^V$, there is a compatible way to topologize the total set of morphisms, where each connected component is topologized as a subspace of $\mathbb{R}_{>0} \times \mathrm{Emb}_\partial(W, [0,1] \times V)/\mathrm{Diff}_\partial(W)$. See [16] for more details on such a definition. Let us temporarily write $\overline{\mathcal{C}}_d^V$ for the version as a topological category.

3.6. Conclusion Since $\mathcal{C}_d^V$ and $\overline{\mathcal{C}}_d^V$ have the same underlying (non-topologized) category, the discussion in §3.4 applies equally well to the topological category $\overline{\mathcal{C}}_d^V$, but now the functor

$$j_* : (\overline{\mathcal{C}}_d^V)^k \to \overline{\mathcal{C}}_d^V$$

depends continuously on $j \in \mathrm{Emb}(\{1, \dots, k\} \times V, V)$ and we get continuous maps

$$\begin{aligned}\mathcal{D}_n(k) \hookrightarrow \mathrm{Emb}(\{1, \dots, k\} \times \mathrm{int}(D^n), \mathrm{int}(D^n)) &\xrightarrow{\phi} \mathrm{Emb}(\{1, \dots, k\} \times V, V)\\ &\to \mathrm{Map}((B\overline{\mathcal{C}}_d^V)^k, (B\overline{\mathcal{C}}_d^V)),\end{aligned}$$

which satisfy the axioms of a map of operads. Hence they make the space $B\overline{\mathcal{C}}_d^V$ into an E_n-algebra (depending on the diffeomorphism $V \approx \mathrm{int}(D^n)$). Thus for $n = \dim(V) \geqslant 3$ the fundamental groupoid $\pi_1(B\overline{\mathcal{C}}_d^V)$ inherits a symmetric monoidal structure, from a choice of $m \in \mathcal{D}_n(2)$. This choice also promotes $h\mathcal{C}_d^V$ to a symmetric monoidal category and the universal functor (as in (3.5.2))

$$h\mathcal{C}_d^V \xrightarrow{\gamma} \pi_1(B\overline{\mathcal{C}}_d^V)$$

to a symmetric monoidal functor, factoring over a symmetric monoidal equivalence $h\mathcal{C}_d^V[(h\mathcal{C}_d^V)^{-1}] \simeq \pi_1(B\overline{\mathcal{C}}_d^V)$.

By Example 3.3.6, the space $\Omega^V T_{d,\mathbb{R}\oplus V} \approx \Omega^n T_{d,\mathbb{R}\oplus V}$ also canonically has the structure of an E_n algebra. Hence it makes sense to ask whether the weak equivalence from the main theorem, between $\Omega^V T_{d,\mathbb{R}\oplus V}$ and $B\mathcal{C}_d^V \simeq B\overline{\mathcal{C}}_d^V$ preserves this structure. The proof of this equivalence goes through a rather long zig-zag of weak equivalences. The original proofs of the main theorem did not discuss compatibility with E_n-structures, but more recently C. Schommer-Pries [37, Chapter 4] has shown how to find a zig-zag of maps, in our notation between $B\overline{\mathcal{C}}_d^V$ and

$\Omega^V T_{d,\mathbb{R}\oplus V}$, which are both maps of E_n spaces and weak equivalences of underlying spaces. (Nguyen [29] had shown this earlier in the limiting case $n \to \infty$, using Segal's Γ-spaces as a model for E_∞ spaces.) More concisely, he constructed a weak equivalence of E_n spaces $B\overline{\mathcal{C}}_d^V \simeq \Omega^V T_{d,\mathbb{R}\oplus V}$.

Combining everything, we get symmetric monoidal functors (for $\dim(V) \gg d$)

$$\mathrm{Cob}_d \xleftarrow{\simeq} h\mathcal{C}_d^V \to h\mathcal{C}_d^V[(h\mathcal{C}_d^V)^{-1}] \xrightarrow{\simeq} \pi_1(B\overline{\mathcal{C}}_d^V) \xleftrightarrow{\simeq} \pi_1(\Omega^n T_{d,\mathbb{R}\oplus V}),$$

where the symmetric monoidal structure on $\pi_1(\Omega^n T_{d,\mathbb{R}\oplus V})$ arises from the E_n-structure on $\Omega^n T_{d,\mathbb{R}\oplus V}$, as explained in §3.3.6.

Remark 3.6.1. Let X be a (based) space and let $n \geqslant 3$. Let $X \to \tau_{\leqslant n} X$ denote the Postnikov truncation[15], and $\tau_{>n} X = \tau_{\geqslant n+1} X \to X$ its homotopy fiber over the base point. Then the two maps $\tau_{\leqslant n+1}\tau_{\geqslant n} X \to \tau_{\leqslant n+1} X \leftarrow X$ induce symmetric monoidal equivalences

$$\pi_1\Omega^n(\tau_{\leqslant n+1}\tau_{\geqslant n} X) \xrightarrow{\simeq} \pi_1\Omega^n(\tau_{\leqslant n+1} X) \xleftarrow{\simeq} \pi_1\Omega^n X,$$

so the symmetric monoidal groupoid $\pi_1(\Omega^n X)$ depends only on the homotopy type of the based space $\tau_{\leqslant n+1}\tau_{\geqslant n} X$. This space is a "two-stage Postnikov tower", meaning that it has only two non-zero homotopy groups, namely $A = \pi_n(X, *)$ and $B = \pi_{n+1}(X, *)$. Moreover, it is canonically homotopy equivalent to the homotopy fiber of a map of Eilenberg–MacLane spaces

$$K(A, n) \xrightarrow{k} K(B, n+2),$$

whose homotopy class in $[K(A, n), K(B, n+2)] \cong H^{n+2}(K(A, n); B)$ is one of the k-*invariants* of X. It is a measure of how far $\tau_{\leqslant n+1}\tau_{\geqslant n} X$ is from being homotopy equivalent to the product $K(A, n) \times K(B, n+1)$.

To summarize, the universal symmetric monoidal groupoid equipped with a symmetric monoidal functor from Cob_d is entirely encoded by the two groups $A = \pi_n(T_{d,\mathbb{R}^{n+1}}, *)$ and $B = \pi_{n+1}(T_{d,\mathbb{R}^{n+1}}, *)$ as well as the k-invariant[16] connecting them. A similar discussion applies to all the other flavors of cobordism categories (e.g. with extra tangential structure).

With only a little more work, the classification of symmetric monoidal functors $\mathrm{Cob}_d \to D$ with values in a rigid symmetric monoidal groupoid D can be turned entirely into homotopy theory of spaces. The classical "infinite loop space machines" allow us to write the classifying space BD as an iterated loop space

$$BD \simeq \Omega^n(B^{n+1}D),$$

for any n, and for $n \geqslant 3$ such a delooping in turn determines the symmetric monoidal structure on $D \simeq \pi_1(BD)$, as explained in §3.3. Using the main theorem with $V = \mathbb{R}^n$ for $n \gg d$ there is then an equivalence of groupoids

$$\mathrm{Fun}^{\otimes}(\mathrm{Cob}_d, D) \simeq \pi_1 \mathrm{Map}_*(\tau_{\geqslant n} T_{d,\mathbb{R}^{n+1}}, B^{n+1}D),$$

[15] See Exercise 5.2.5, especially (d).

[16] See also Exercise 5.3.8.

where the domain denotes the category of symmetric monoidal functors and monoidal natural transformations between them and the codomain is the fundamental groupoid of the space of pointed maps.

Lecture 4: Cobordism classes and characteristic classes

As explained in Lecture 2, each path component of a morphism space in $\mathcal{C}_d$ may be interpreted as a classifying space for bundles of manifolds. In this section we explain how the main result of Lecture 2 may be used to study the cohomology of these spaces, i.e. characteristic classes of bundles of manifolds. This will make use of more of the homotopy type of $B\mathcal{C}_d$, beyond its fundamental groupoid.

4.1. Stable homology of moduli spaces of surfaces In the case $d = 2$, classification of surfaces asserts that a compact connected non-orientable smooth 2-manifold W with $\partial W \approx \amalg_1^k S^1$ must be diffeomorphic to

$$W_{h,k} = (h\mathbb{R}P^2)\#(kD^2),$$

the connected sum of $h \in \mathbb{Z}_{\geqslant 1}$ copies of the real projective plane and $k \in \mathbb{Z}_{\geqslant 0}$ copies of the 2-disk. For $h' \geqslant h$, $k \geqslant 1$, and $k' \geqslant 0$, there exist embeddings

$$W_{h,k} \overset{j}{\hookrightarrow} W_{h',k'}$$

and any such embedding gives rise to a continuous group homomorphism

$$j_* : \mathrm{Diff}_\partial(W_{h,k}) \to \mathrm{Diff}_\partial(W_{h',k'}),$$

defined by "extension by the identity": for $\varphi \in \mathrm{Diff}_\partial(W_{h,k})$ the diffeomorphism $j_*\varphi$ is given by $j \circ \varphi \circ j^{-1}$ on the image of j, and by the identity on the complement of the image of j. The resulting map $j_*\varphi$ is again a diffeomorphism, because φ acts as the identity near $\partial W_{h,k}$. Taking induced map of classifying spaces gives a map of spaces $B\mathrm{Diff}(W_{h,k}) \to B\mathrm{Diff}(W_{h',k'})$ and hence a map on homology

$$H_i(B\mathrm{Diff}(W_{h,k})) \to H_i(B\mathrm{Diff}(W_{h',k'})). \tag{4.1.1}$$

The following *homological stability* theorem was proved by Nathalie Wahl[17].

Theorem 4.1.2 ([40]). *For any choice of embedding* $j : W_{h,k} \hookrightarrow W_{h',k'}$ *the homomorphism* (4.1.1) *is an isomorphism for* $h \gg i$.

In fact Wahl proved that $h \geqslant 4i + 5$ is sufficient, which was later improved by Randal-Williams to $h \geqslant 3i + 3$ being sufficient.

For any connected, compact, and non-orientable surface W, the homology of $B\mathrm{Diff}_\partial(W)$ is therefore "independent of W" at least up to homological degree $\frac{h-3}{3}$, where h is the largest natural number for which there exists an embedding $j : W_{h,1} \hookrightarrow W$. (This number may be characterized homologically as the dimension of the cokernel of the natural map $H_1(\partial W; \mathbb{F}_2) \to H_1(W; \mathbb{F}_2)$.) By

[17]Wahl's paper works with the discrete groups $\pi_0(\mathrm{Diff}_\partial(W_{h,k}))$, but that point of view is equivalent by the earlier result [17, Théorème 1].

the homological stability theorem the group $H_i(\mathrm{BDiff}_\partial(W); A)$ may be identified with the colimit of the direct system

$$\cdots \to H_i(\mathrm{BDiff}_\partial(W_{h,1}); A) \to H_i(\mathrm{BDiff}_\partial(W_{h+1,1}); A) \to \dots,$$

which is manifestly independent[18] of W.

The range appearing in Wahl's theorem (improved by Randal-Williams) is known as the *stable range*. In this range the homology with coefficients in an abelian group A may be identified with the *stable homology* groups

$$\operatorname*{colim}_{h\to\infty} H_*(\mathrm{BDiff}_\partial(W_{h,1}); A). \tag{4.1.3}$$

In this lecture we shall discuss how cobordism categories may be used to determine the stable homology. We shall not directly employ the "field theory" language, although the interested reader may interpret some of the constructions in this lecture in terms of invertible field theories.

Remark 4.1.4. A similar homological stability result for *oriented* compact connected 2-manifolds was established earlier by Harer in [18] (see the survey [41] for a geodesic proof). Surfaces with other tangential structures were considered later by Randal-Williams in [35]. Similar homological stability results are known for manifolds of dimension higher than 2,when the dimension is even, see [13, 14].

Several homological stability results for moduli spaces of odd dimensional manifolds have been established by Perlmutter [32–34].

4.2. Cohomology of morphism spaces The connection between cobordism categories and the question of determining $H^*(\mathrm{BDiff}_\partial(W); A)$ for a (non-orientable, connected, compact) surface W is that the space $\mathrm{BDiff}_\partial(W)$ appears, up to homotopy, as a path component of a morphism space in $\mathcal{C}_2$. Indeed, if we pick an embedding $\partial W \hookrightarrow V \subset \mathbb{R}^\infty$ and denote its image by M, then M is an object of $\mathcal{C}_2$, and one of the path components of the space $\mathcal{C}_2(M, \emptyset)$ is weakly equivalent to $\mathrm{BDiff}_\partial(W)$, as explained in Section 2.2. It also appears as a path component of $\mathcal{C}_2(M_0, M_1)$ if $M_0 \amalg M_1 \approx \partial W$. We wish to deduce information about the cohomology of $\mathrm{BDiff}_\partial(W)$ from this appearance as a path component of morphism spaces in $\mathcal{C}_2$, and the knowledge provided by the main theorem cited above, which gives the weak equivalence

$$\Omega B\mathcal{C}_2 \simeq \Omega^\infty MTO(2) = \operatorname*{colim}_V \Omega^V T_{2,V}.$$

This homotopy equivalence should be regarded as a calculation of the left hand side and therefore, at least implicitly, gives information about $\mathcal{C}_2$. The right hand side is a priori more well understood—for example, standard arguments imply

[18] The diligent reader and writer ought to raise the question of whether the isomorphism is independent of choice of $j : W_{h,1} \hookrightarrow W$. This is indeed the case: assuming the image of j is contained in $W \setminus \partial W$, as may be arranged after isotoping j, it may be deduced from the classification of surfaces that the homomorphisms of diffeomorphism groups arising from two choices of j are conjugate, and hence the maps of classifying spaces are homotopic. By a similar argument the maps in the direct system are independent of choice of $W_{h,1} \hookrightarrow W_{h+1,1}$.

that each path component of this space has rational cohomology ring

$$\mathbb{Q}[x_1, x_2, \ldots], \tag{4.2.1}$$

a polynomial ring on generators x_i of cohomological degree $|x_i| = 4i$.

Let us first discuss what we are trying to do, in the abstract setting of a topologically enriched category C. There is a continuous map

$$C(x,y) \to \Omega_{x,y} BC, \tag{4.2.2}$$

where the codomain denotes the space of paths $[0,1] \to BC$ starting at the point $x \in N_0 C \subset BC$ and ending at y, in the compact-open topology. Standard arguments, generalizing the discussion in §1.3, show that this map is a weak equivalence for all x and y, if and only if hC is a groupoid. Without further assumptions the map (4.2.2) may be very far from an equivalence; for example if C is a partially ordered set the domain is either a singleton or empty, while the space BC may have any homotopy type whatsoever.

The map (4.2.2) is not quite a natural transformation in the strict sense: a morphism $f : y \to y'$ induces a map $C(x,y) \to C(x,y')$ by postcomposing with f, and a map $\Omega_{x,y} BC \to \Omega_{x,y'} BC$ by concatenating with the path given by the 1-simplex $f \in N_1(C)$, but the square

$$\begin{array}{ccc} C(x,y) & \longrightarrow & \Omega_{x,y} BC \\ \downarrow {\scriptstyle f\circ -} & & \downarrow \\ C(x,y') & \longrightarrow & \Omega_{x,y'} BC \end{array}$$

does not commute on the nose. It does commute up to homotopy though, which is sufficient to make

$$H_*(C(x,y); A) \to H_*(\Omega_{x,y} BC; A)$$

a natural transformation of functors of $(x,y) \in (hC)^{op} \times (hC)$, for any abelian group A. Concatenation with a path is a homotopy equivalence, so the codomain maps morphisms $y \to y'$ to isomorphisms $H_*(\Omega_{x,y} BC; A) \to H_*(\Omega_{x,y'} BC; A)$, and similarly for $x' \to x$.

If we have a filtered category J, an object $j_0 \in J$, and a functor $y : J \to C$, we therefore get an isomorphism

$$H_*(\Omega_{x,y(j_0)} BC; A) \xrightarrow{\cong} \operatorname*{colim}_{j \in J} H_*(\Omega_{x,y(j)} BC; A)$$

and an induced map

$$\operatorname*{colim}_{j \in J} H_*(C(x,y(j)); A) \to \operatorname*{colim}_{j \in J} H_*(\Omega_{x,y(j)} BC; A) \cong H_*(\Omega_{x,y(j_0)} BC; A) \tag{4.2.3}$$

The next result is a version[19] of the "group completion" theorem for categories.

[19] An arguably more conceptual point of view on the map (4.2.2) involves the *derived* version of the universal functor $\gamma : C \to C[C^{-1}]$ mentioned earlier for plain categories. Explicit models in simplicially enriched categories were given by Dwyer and Kan, [6,7], who also established that morphism spaces in the derived localization are path spaces in the classifying space. Another model is the localization

Proposition 4.2.4. *Let* C *and* $y : J \to C$ *be as above, let* A *be an abelian group, and assume the functor*

$$(4.2.5)\qquad \begin{aligned} hC^{op} &\to \mathrm{Ab} \\ x &\mapsto \operatorname*{colim}_{j\in J} H_*(C(x, y(j)); A) \end{aligned}$$

sends all morphisms to isomorphisms. Then the map (4.2.3) *is an isomorphism, in fact a natural isomorphism of functors of* $x \in hC^{op}$.

This abstract result (with $A = \mathbb{Z}$) is quite close to what we need: the stable homology groups (4.1.3) are (filtered) colimits of homology groups of path components of morphism spaces in the cobordism category $\mathcal{C}_2$, and we understand the space $\Omega B\mathcal{C}_2 \simeq \Omega^\infty MTO(2)$ well, including the (rational co)homology of its path components (4.2.1). Unfortunately the proposition does not apply on the nose to $C = \mathcal{C}_2$, because the many path components of $\mathcal{C}_2(x, y)$ corresponding to disconnected cobordisms prevents the assumption of the proposition from being satisfied.

Remark 4.2.6. Even though we shall not need it in the rest of this lecture, let us make a brief remark about how Proposition 4.2.4 fits into the "field theory" paradigm, under the extra assumption that C is symmetric monoidal, $y(j_0) = 1$ is the monoidal unit of C, and A is a commutative ring. Then the functor

$$\begin{aligned} hC^{op} &\to \text{graded } A\text{-modules} \\ x &\mapsto \operatorname*{colim}_{j\in J} H_*(C(x, y(j); A) \cong H_*(\Omega_{x,1}BC; A) \end{aligned}$$

may be promoted to a lax symmetric monoidal functor with codomain symmetric monoidal with respect to $\otimes_A$ and the usual symmetry $x \otimes y \mapsto (-1)^{|x||y|} y \otimes x$. The monoidality of the functor arises from applying the lax monoidal functor $H_*(-; A)$ to

$$\Omega_{x,1}(BC) \times \Omega_{x',1}(BC) \xrightarrow{\cong} \Omega_{(x,x'),(1,1)}(BC \times BC) \to \Omega_{x\otimes x', 1\otimes 1}BC \xrightarrow{\simeq} \Omega_{x\otimes x',1}BC,$$

where the middle arrow is $\Omega_{(x,x'),(1,1)}(B\otimes)$ and the last arrow is concatenating with the path corresponding to the isomorphism $1 \otimes 1 \to 1$.

Since the resulting natural transformation

$$H_*(\Omega_{x,1}BC; A) \otimes_A H_*(\Omega_{x',1}BC; A) \to H_*(\Omega_{x\otimes x',1}BC; A)$$

and also the unit $A \to H_*(\Omega_1 BC; A)$ are likely not isomorphisms, we have only constructed a *lax* monoidal functor. But if BC is path connected it lifts to a (strong) monoidal functor into the category of modules over $H_*(\Omega_1 BC; A)$, endowed with tensor product over that graded-commutative ring. The module structure is induced by the map of spaces $\Omega_{x,1}BC \times \Omega_1 BC \to \Omega_{x,1}BC$ defined by concatenation

of quasi-categories [24, §5.2.7]. Either of these settings promotes (4.2.2) to a natural transformation of (representable) functors of y. From this derived point of view, it may also be more natural to replace the "colimits of homology" by "homotopy colimits of chains" in Proposition 4.2.4; in such a formulation the assumption that J be filtered may be weakened to BJ being contractible.

of paths, and the monoidality is defined as before but factoring over the tensor product over $H_*(\Omega_1 BC; A)$. It is not hard to check that this is well defined and gives a strong monoidal functor if BC is path connected[20].

4.3. Cobordisms with connectivity restrictions It is a popular endeavour to study and/or define field theories whose domain is *larger* than Cob_d. The general flavor is to allow objects to have boundary as well: instead of only allowing closed $(d-1)$-manifolds M we could allow M to be a cobordism between $(d-2)$-manifolds. Cobordisms $M_0 \rightsquigarrow M_1$ would then be d-manifolds with corners. If we stop there, the corresponding notion of field theory is a "$(d-2, d-1, d)$-theory". Going for maximal generality, there is a notion of *fully extended* field theory. See [37] for a classification result for invertible field theories in this setting or [25] for a non-invertible classification result.

In this section we shall do the *opposite*: study functors defined on something *smaller* than $\mathcal{C}_d$. Such subcategories turn out to be very useful for understanding homology of components of the morphism spaces $\mathcal{C}_d(M_0, M_1)$ which, as we have seen, have the homotopy type of $B\mathrm{Diff}_\partial(W)$. For $d = 2$ the goal is, roughly speaking, to find a subcategory $C \subset \mathcal{C}_2$ which is large enough to contain all the $B\mathrm{Diff}_\partial(W_{h,k})$ and small enough that Proposition 4.2.4 applies to it.

Definition 4.3.1. Let $d \geqslant 0$ and $k \geqslant -1$. For objects $M_0, M_1 \in \mathrm{Cob}_d$, let

$$\mathrm{Cob}_d^k(M_0, M_1) \subset \mathrm{Cob}_d(M_0, M_1)$$

be the subset consisting of (diffeomorphism classes of abstract) cobordisms W, such that the pair (W, M_1) is k-connected. For $k = -1$ this is no condition, for $k = 0$ it means that $\pi_0(M_1) \to \pi_0(W)$ is surjective, and for $k > 0$ it additionally means that the relative homotopy groups $\pi_i(W, M_1, x)$ vanish for all $i \leqslant n$ and all basepoints $x \in M_1$.

As you will show in Exercise 5.3.5, the connectivity property is preserved under composition of cobordisms, so that the definition above gives a subcategory $\mathrm{Cob}_d^k \subset \mathrm{Cob}_d$.

Since we have functors

$$h\mathcal{C}_d^V \to h\mathcal{C}_d \xrightarrow{\simeq} \mathrm{Cob}_d,$$

we get corresponding subcategories of $\mathcal{C}_d$ and $\mathcal{C}_d^V$ by taking the inverse image. Similarly in the presence of a tangential structure Θ. Following [12] we shall write $\mathcal{C}_d^k \subset \mathcal{C}_d$ for the inverse image of $\mathrm{Cob}_d^k \subset \mathrm{Cob}_d$.

Lemma 4.3.2. $\mathrm{Cob}_d^k \subset \mathrm{Cob}_d$ *is a symmetric monoidal subcategory. Relatedly, for* $n = \dim(V)$, *the* E_n *structure on* $\overline{\mathcal{C}}_d^V$ *restricts to an action on the topological subcategory defined as the inverse image of* Cob_d^k.

[20]This construction extends to disconnected BC, at least when the commutative monoid structure on $\pi_0(BC)$ induced by $\otimes$ is a group: the functor should take values in a symmetric monoidal category of $\pi_0(BC)$-graded $H_*(\Omega_1 BC; A)$-modules, the symmetry isomorphism involves a "generalized sign"—i.e. an anti-symmetric function $c : \pi_0(BC) \times \pi_0(BC) \to \pi_1(BC, 1) \subset H_*(\Omega_1 BC; A)^\times$, and the associator involves a normalized 3-cocycle $h : (\pi_0(BC))^3 \to \pi_1(BC, 1)$.

Proof. Let $W \in \mathrm{Cob}_d(M_0, M_1)$ and $W' \in \mathrm{Cob}_d(M_0', M_1')$. If (W, M_1) and (W', M_1') are both k-connected, then so is the inclusion

$$M_1 \amalg M_1' \hookrightarrow W \amalg W'. \qquad \square$$

The case $k = 0$ is especially easy to visualize: the condition on a cobordism $W \in \mathcal{C}_d(M_0, M_1)$ is then that each path component of W must have non-empty outgoing boundary. (Symmetric monoidal) functors from $\mathcal{C}_d^k$, or from other subcategories of cobordism categories, are sometimes called *restricted field theories*.

Example 4.3.3. For $d = 1$ the subcategory $\mathrm{Cob}_1^0 \subset \mathrm{Cob}_1$ has a very simple structure. Its objects are finite sets, while morphisms from a finite set M_0 to another finite set M_1 may be viewed as injections $M_0 \hookrightarrow M_1$ together with matchings (equivalence relations where each equivalence class has cardinality 2) on the complement of the image. Symmetric monoidal functors $\mathrm{Cob}_1^0 \to \mathrm{Mod}_R$ are therefore uniquely given, up to natural isomorphism, by an object $X \in \mathrm{Mod}_R$ together with an element $\omega \in X \otimes_R X$ that is fixed under swapping the factors. For most choices of X and ω it is *not* possible to extend to a symmetric monoidal functor on all of Cob_1. In fact, it can be shown[21] that such an extension exists if and only if X is finitely generated and projective, and the adjoint $\omega : X \to X^\vee$ is an isomorphism.

Similar examples may be given of symmetric monoidal functors $\mathrm{Cob}_2^0 \to \mathrm{Mod}_R$ which do not extend to Cob_2.

The inclusion $\mathcal{C}_d^k \hookrightarrow \mathcal{C}_d$ induces a map of spaces

$$B\mathcal{C}_d^k \hookrightarrow B\mathcal{C}_d \tag{4.3.4}$$

which, surprisingly, sometimes turns out to be a weak equivalence. However, (4.3.4) can't always be a weak equivalence. For example if $k > d$, then we only allow cobordisms $W \in \mathrm{Cob}_d(M_0, M_1)$ for which the inclusion $M_1 \to W$ is a homotopy equivalence. This prevents the map (4.3.4) from inducing a bijection on π_0, much less being a weak equivalence. It is perhaps more surprising that it is *ever* a weak equivalence for $k \geqslant 0$. The following result was proved in [12, §3] (see also [16, §6] for the case $d = 2$).

Theorem 4.3.5. *Let $k \leqslant \frac{d}{2} - 1$. Then the map (4.3.4) is a weak equivalence.*

More generally let Θ be a space with $GL_d(\mathbb{R})$ action, let $\mathcal{C}_\Theta$ be the corresponding cobordism category of manifolds with Θ-structure, as in Remark 2.5.7, and let $\mathcal{C}_\Theta^k$ be the inverse image of Cob_d^k. Then the inclusion induces a weak equivalence

$$B\mathcal{C}_\Theta^k \hookrightarrow B\mathcal{C}_\Theta \tag{4.3.6}$$

for $k \leqslant \frac{d}{2} - 1$.

4.4. Cohomology of morphism spaces Let us now return to the topic of characteristic classes of smooth manifold bundles. For concreteness we will again first explain the case of non-orientable surfaces.

[21] This is the 1-dimensional *cobordism hypothesis*, [1, 25].

This is in fact just a matter of collecting together all the ingredients outlined above. Indeed, let us we apply Proposition 4.2.4 to the topologically enriched category $C = \mathcal{C}_2$ with $J = (\mathbb{N}, \leqslant)$ and $y : J \to \mathcal{C}_2$ given on objects by letting each $y(n)$ be the image of some embedding $S^1 \hookrightarrow V$ and on the generating morphisms $n < n+1$ as an embedded cobordism diffeomorphic to $([n, n+1] \times S^1)\#\mathbb{R}P^2$. Then Wahl's theorem implies that the assumption of Proposition 4.2.4 is satisfied, and the proposition implies an isomorphism between the cohomology of $B\mathrm{Diff}_\partial(W_{h,k})$ in the stable range, and the cohomology of a path component of $\Omega^\infty MTO(2)$. We deduce a map of graded $\mathbb{Q}$-algebras

$$\mathbb{Q}[x_1, x_2, \ldots] \to H^*(B\mathrm{Diff}_\partial(W_{h,k}); \mathbb{Q}),$$

which is an isomorphism in the stable range (cohomological degrees $\leqslant (h-3)/3$, by Randal-Williams' improved range).

Remark 4.4.1. The corresponding statement for oriented surfaces is the celebrated Madsen–Weiss theorem [26], establishing the Mumford conjecture. It may be proved using cobordism categories of oriented surfaces by an argument similar to the one outlined above, replacing Wahl's theorem with Harer's theorem. This method of using cobordism categories to prove Madsen and Weiss' theorem was first used in [16]. The insight that restricted cobordism categories are useful for studying stable homology goes back to Ulrike Tillmann's paper [39], in the surface case.

Remark 4.4.2. The homological stability results established in [13,14] for manifolds of even dimension $d = 2n$ may be used in a similar way to Harer's and Wahl's results, and the restricted cobordism categories $\mathcal{C}_{2n}^{n-1}$ may be used in a similar way, to determine the stable homology. See op. cit. for precise statements and detailed proofs.

The situation for odd dimensional manifolds is currently more mysterious. While Perlmutter's homological stability results [32–34] give explicit stable ranges, we do not currently know the corresponding stable homology. The statement in Theorem 4.3.5 is not quite strong enough in the odd dimensional case $d = 2n+1$: Proposition 4.2.4 does not seem to apply to $\mathcal{C}_{2n+1}^{n-1}$.

Exercises (by A. Debray, S. Galatius, M. Palmer)

The four lectures at the summer school were accompanied by three exercise sessions, TA'd by Arun Debray[22] and Martin Palmer[23]. The exercises below are edited versions of the problem sets: a few have been omitted and some have been rewritten for clarity. We thank the participants in the exercise sessions for their feedback on the original versions.

[22] University of Texas, Austin. a.debray@math.utexas.edu

[23] University of Bonn, Germany. Current affiliation: Mathematical Institute of the Romanian Academy. mpanghel@imar.ro

We have mostly preserved the ordering used at the summer school (but added a few suggestions for doing certain exercises out of order). Exercises labeled with an asterisk are likely more challenging.

5.1. Exercise set 1

Exercise 5.1.1. Recall that $\mathfrak{N}_d$ denotes the abelian group of smooth, closed d-manifolds up to cobordism, with respect to the operation of disjoint union.

(a) Show that the Klein bottle is nullbordant.

(b) Prove that the disjoint union of two manifolds of positive dimension is cobordant to their connected sum. (The connected sum depends on a choice of chart in each manifold, but the claim is true for any choice.)

(c) Deduce that every cobordism class except $0 \in \mathfrak{N}_0$ contains a connected manifold.

(d) Prove, without invoking the theorem of Thom (but you may use the classification of surfaces), that the abelian groups $\mathfrak{N}_0$, $\mathfrak{N}_1$ and $\mathfrak{N}_2$ are isomorphic to $\mathbb{Z}/2\mathbb{Z}$, 0 and $\mathbb{Z}/2\mathbb{Z}$, respectively.
(Hint: to show that $\mathfrak{N}_2 \neq 0$, check that Euler characteristic modulo 2 defines a well defined surjection $\mathfrak{N}_2 \to \mathbb{Z}/2\mathbb{Z}$.)

Write Ω_d for the analogous oriented cobordism group: the abelian group of smooth, closed, *oriented* d-manifolds up to *oriented* cobordism, with respect to the operation of disjoint union.

(e) Prove that the abelian groups Ω_0, Ω_1 and Ω_2 are isomorphic to $\mathbb{Z}$, 0 and 0 respectively.

Exercise 5.1.2. In this exercise we construct two elementary "invertible field theories" (in the naive sense, ignoring symmetric monoidal structures). Recall that a (naive) invertible field theory is a functor from the abstract cobordism category Cob_d to a groupoid; we will construct two examples where the target is a group.

(a) Construct a functor

$$F_d\colon \mathrm{Cob}_d \longrightarrow \mathfrak{N}_d$$

(where the group $\mathfrak{N}_d$ is considered as a category with one object) such that the restriction to $\mathrm{End}_{\mathrm{Cob}_d}(\varnothing)$ sends a diffeomorphism class of d-manifolds to its cobordism class.
(*Hint*: note that Cob_d is a disjoint union (coproduct) of subcategories indexed by the elements of $\mathfrak{N}_{d-1}$, so it will suffice to define F on the full subcategory of Cob_d on nullbordant $(d-1)$-manifolds.)

(b) Construct a functor

$$E_d\colon \mathrm{Cob}_d \longrightarrow \mathbb{Z}$$

(where the group $\mathbb{Z}$ is considered as a groupoid with one object) such that the restriction to $\mathrm{End}_{\mathrm{Cob}_d}(\varnothing)$ sends a diffeomorphism class of d-manifolds to its Euler characteristic.

(c) $(*)$ When $d = 2$, use the classification of surfaces to show that the functor

$$\mathrm{Cob}_2[\mathrm{Cob}_2^{-1}] \longrightarrow \mathbb{Z},$$

induced by E_2, is an equivalence.[24]

Exercise 5.1.3. In this exercise as in the text, the fundamental groupoid of an (unbased) space X is denoted $\pi_1(X)$.

(a) Let C be the poset of all non-empty, proper subsets of $\{0, 1, 2, 3\}$, considered as a category. Prove that BC is homeomorphic to S^2 and hence that $\pi_1(BC)$ is equivalent to a trivial category.
(b) Let C be the category with exactly two objects a, b and exactly two non-identity morphisms f, g, which both have source a and target b. Prove that BC is homeomorphic to S^1 and describe explicitly an equivalence of groupoids $\pi_1(BC) \to \mathbb{Z}$.
(c) Combining the above ideas, find a finite category C such that the abelian group $\pi_2(BC, x)$ is infinitely-generated for any object x.

Exercise 5.1.4. $(*)$ Another variant of the cobordism group is the *group of homotopy* d-*spheres* Θ_d. To define this, let us first denote by $\mathcal{M}_d$ the abelian monoid of isomorphism classes of smooth, closed, connected and oriented d-manifolds, up to oriented diffeomorphism, under the operation of connected sum (which is well defined up to oriented diffeomorphism). Let $\mathcal{S}_d$ be the subset of diffeomorphism classes of d-manifolds that are homotopy equivalent to the sphere S^d, which we call *homotopy spheres*.

(a) Prove that, if M and N are homotopy spheres, so is their connected sum.

Hence $\mathcal{S}_d$ is a submonoid. Two manifolds $M, N \in \mathcal{M}_d$ are called h-*cobordant* if there is a cobordism W between them such that the inclusions c_{in} and c_{out} are both homotopy equivalences.

(b) Prove that h-cobordism $\sim_h$ induces an equivalence relation on $\mathcal{S}_d$.
(c) Prove that this equivalence relation is compatible with the connected sum operation.

Hence we have a well-defined quotient monoid $\Theta_d := \mathcal{S}_d/\sim_h$ with respect to the operation of connected sum.

(d) Prove that Θ_d is a group.[25]

5.2. Exercise set 2

Exercise 5.2.1. This exercise studies the difference between being diffeomorphic and being diffeomorphic as cobordisms.

[24] This is not so easy to do by hand, in fact it is a research paper [20, Theorem 3.7]. This is a more reasonable exercise if you do Exercise 5.3.2 first.

[25] This is Theorem 1.1 of the influential paper [21], which also studies the group Θ_d in depth. It is always a finite abelian group, and the smallest d for which it is non-trivial is $\Theta_7 \cong \mathbb{Z}/28\mathbb{Z}$. In fact, except possibly for $d = 4$, the quotient $\mathcal{S}_d \to \Theta_d$ is an isomorphism, in which case $\mathcal{S}_d$ is also a (finite) group.

(a) Give examples of the following data: two closed $(d-1)$-manifolds M_0 and M_1 and two cobordisms (W, c_{in}, c_{out}) and (W', c'_{in}, c'_{out}), both between M_0 and M_1, which are *not* diffeomorphic as cobordisms, yet W is diffeomorphic to W'. (Hint: start with $d = 1$.)

(b) Fix M_0 and M_1 as above and let $M = M_0 \amalg M_1$ be the disjoint union. By sending a triple (W, c_{in}, c_{out}) to a cobordism of the form $(W, ?, \emptyset)$, construct a bijection $\mathrm{Cob}_d(M_0, M_1) \cong \mathrm{Cob}_d(M, \emptyset)$. Thus to study the difference between being diffeomorphic and being diffeomorphic as cobordisms, we need only consider the case $M_1 = \emptyset$.

(c) Construct an action of the group $\mathrm{Diff}(M)$ diffeomorphisms of M on the set $\mathrm{Cob}_d(M, \emptyset)$ by "reparametrizing the boundary". Show that two cobordisms $M \rightsquigarrow \emptyset$ are diffeomorphic (as abstract manifolds) if and only if they represent elements of $\mathrm{Cob}_d(M, \emptyset)$ which are in the same $\mathrm{Diff}(M)$-orbit.

Exercise 5.2.2. Let D be the groupoid defined abstractly by $\mathrm{Ob}(D) = \mathbb{Z}/2\mathbb{Z}$, with morphism sets

$$D(a, b) = \begin{cases} \mathbb{Z} & a = b \\ \varnothing & a \neq b, \end{cases}$$

and composition given by addition in $\mathbb{Z}$. Example 2.5.6 shows that this groupoid is equivalent to the fundamental groupoid of $\Omega T_{1,\mathbb{R}^2} \cong \Omega \mathbb{RP}^2$, so it should also be the target of a universal functor from $h\mathcal{C}_1^{\mathbb{R}}$ to a discrete groupoid. The goal of this exercise is to verify this directly, by geometric constructions.

(a) Construct a functor $f\colon h\mathcal{C}_1^{\mathbb{R}} \longrightarrow D$, by sending an object (finite subset of $\mathbb{R}$) to its cardinality modulo 2, and a morphism $W \subset [0, t] \times \mathbb{R}$ from $M_0 \subset \mathbb{R}$ to $M_1 \subset \mathbb{R}$ to the integer $\chi(X) - \chi(X \cap (\{0\} \times \mathbb{R}))$, where $X \subset [0, t] \times \mathbb{R}$ is the union of components of the complement $([0, t] \times \mathbb{R}) \setminus W$ obtained by coloring them white or gray in an alternating way, starting with white for the unbounded component in the $[0, t] \times \{-\infty\}$ direction, and setting X to be the union of the gray components.

More rigorously: a component C of $([0, t] \times \mathbb{R}) \setminus W$ is included in the union X if and only if a ray starting from the interior of C, intersecting W transversely and asymptotically equal to $s \mapsto (t/2, -s)$, has an odd number of intersections with W. For example, in the following picture, the region shaded gray is X and its boundary is W:

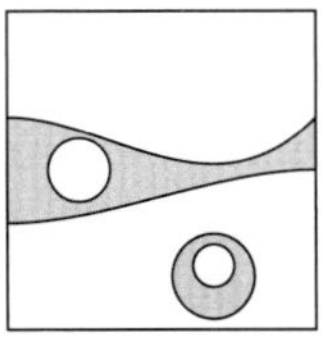

Check that this indeed defines a functor as claimed.

(b) Verify by pictures that f factors over an equivalence $h\mathcal{C}_1^{\mathbb{R}}[(h\mathcal{C}_1^{\mathbb{R}})^{-1}] \simeq D$.

(Hint: it is recommended to first look at Exercise 5.3.2 below. For a solution to this exercise see the end of these notes.)

Exercise 5.2.3 (Cobordism categories with tangential structures.).

As mentioned in the lectures, there are versions of the cobordism category for manifolds equipped with tangential structures, where a *tangential structure* is a space Θ equipped with a continuous action of $GL_d(\mathbb{R})$. A *Θ-structure* on a real vector bundle $E \to X$ is a $GL_d(\mathbb{R})$-equivariant map $Fr(E) \to \Theta$, where $Fr(E)$ is the total space of the frame bundle of the vector bundle. As explained in the lectures, the topological cobordism category $\mathcal{C}_\Theta^V$ is defined similarly to $\mathcal{C}_d^V$, except that each object M is equipped with a Θ-structure on $\mathbb{R} \oplus TM$ and each morphism W is equipped with a Θ-structure on TW.

(a) Unwind the definition of Θ-structure in the cases:
  (i) $\Theta = \{*\}$,
  (ii) $\Theta = \{\pm 1\}$, where an element A of $GL_d(\mathbb{R})$ acts by multiplication by $\det(A)/|\det(A)|$,
  (iii) $\Theta = GL_d(\mathbb{R})$, where $GL_d(\mathbb{R})$ acts on itself by left-multiplication,
  (iv) $\Theta = GL_n(\mathbb{R})/O(n)$,
  (v) $\Theta = GL_{2n}(\mathbb{R})/GL_n(\mathbb{C})$, when $d = 2n$ is even,
  (vi) $\Theta = Z$, where Z is any topological space, and $GL_d(\mathbb{R})$ acts trivially on Z.

(b) Fill in the details of the definition of $\mathcal{C}_\Theta^V$ above. As before, $\mathcal{C}_\Theta$ is then defined as the colimit of $\mathcal{C}_\Theta^V$ over all finite-dimensional linear subspaces $V \subset \mathbb{R}^\infty$.

(c) Define the abstract cobordism category Cob_Θ and show that it is equivalent to $h\mathcal{C}_\Theta$.

(d) Describe explicitly the categories $Cob_{\{\pm 1\}}$ and Cob_Z for $d = 1$ ($GL_1(\mathbb{R})$ acts trivially on Z).

Exercise 5.2.4. Let $\widetilde{\mathcal{C}}_d^V$ be the (ordinary) category obtained by giving the morphism spaces of $\mathcal{C}_d^V$ the discrete topology. What can you say about the connectivity of the map $B\widetilde{\mathcal{C}}_d^V \longrightarrow B\mathcal{C}_d^V$?

Exercise 5.2.5. Let $F : Top \to Top$ be a functor from spaces[26] to spaces. Applying F to the projections $\pi_0 : X_0 \times X_1 \to X_0$ and $\pi_1 : X_0 \times X_1 \to X_1$ gives a canonical map $F(X_0 \times X_1) \to F(X_0) \times F(X_1)$. We say that the functor *preserves finite products* if this canonical map is a homeomorphism for all spaces X_0 and X_1, and F of a one-point space is again a one-point space.

If F admits a left adjoint then it preserves all small limits and in particular finite products, but there are interesting examples not of that form.

(a) If C is a topologically enriched category and $F : Top \to Top$ preserves finite products, explain how to define a new topologically enriched category FC, with the same object set as C, but with $(FC)(x, y) = F(C(x, y))$.

[26]In this exercise it matters that "spaces" means "compactly generated topological spaces".

(b) If $F \to F'$ is a natural transformation between functors preserving finite products, explain how to define an induced topologically enriched functor $FC \to F'C$.

(c) Let $\tau_{\leqslant\infty}(X) = |\mathrm{Sing}(X)|$ denote the geometric realization of the total singular complex. Recall why $\tau_{\leqslant\infty} : \mathrm{Top} \to \mathrm{Top}$ preserves finite products, and why the evaluation map $|\mathrm{Sing}(X)| \to X$ defines a natural transformation from $\tau_{\leqslant\infty}$ to the identity functor, which is always a weak equivalence. This justifies some details of Example 2.4.4, giving a DK equivalence

$$\tau_{\leqslant\infty}C \xrightarrow{\simeq} C$$

to any topologically enriched category C from one, all of whose morphism spaces are CW complexes.

(d) Let $\mathrm{Cosk}_n\mathrm{Sing}(X)$ be the simplicial set whose p-simplices are continuous maps $\mathrm{sk}_n(\Delta^p) \to X$ from the n-skeleton of Δ^n, and let $\tau_{\leqslant n}X$ equal $|\mathrm{Cosk}_{n+1}\mathrm{Sing}(X)|$. Recall why the canonical map $\tau_{\leqslant\infty}X \to \tau_{\leqslant n}X$ is a model for *Postnikov truncation*: an $(n+1)$-connected map whose codomain has vanishing homotopy groups above degree n, with any basepoint.

Explain that we get topologically enriched functors

$$C \xleftarrow{\simeq} \tau_{\leqslant\infty}C \to \tau_{\leqslant n}C.$$

This gives a way of "Postnikov truncating each morphism space of C", retaining a topologically enriched category.

(e) For a space X, regard $\pi_0(X)$ as a topological space using the discrete topology (not the quotient topology from X). Prove that the resulting functor $\pi_0 : \mathrm{Top} \to \mathrm{Top}$ preserves finite products. Construct a natural weak equivalence $\tau_{\leqslant 0}(X) \to \pi_0(X)$ and deduce a DK-equivalence

$$\tau_{\leqslant 0}C \xrightarrow{\simeq} hC$$

for any topologically enriched category C. Thus, the construction $C \mapsto hC$ may be regarded as a special case of the Postnikov truncations

$$C \mapsto \tau_{\leqslant n}C.$$

Exercise 5.2.6.

(a) Prove that the "constant simplicial set" functor Sets $\to$ sSets is right adjoint to the functor π_0: sSets $\to$ Sets.

(b) Let us temporarily write $\mathrm{Top}^{\mathrm{lpc}}$ for the category of locally path connected topological spaces and continuous maps between such. Prove that the functor $\pi_0 : \mathrm{Top}^{\mathrm{lpc}} \to \mathrm{Sets}$ is left adjoint to the functor $\mathrm{Sets} \to \mathrm{Top}^{\mathrm{lpc}}$ assigning a set its discrete topology.

(c) Prove that the "discrete topology" functor Sets $\to$ Top does not admit a left adjoint.
(*Hint*: any functor which admits a left adjoint must preserve all small limits and in particular infinite products.)

5.3. Exercise set 3

Exercise 5.3.1. Let D be a rigid symmetric monoidal groupoid, let 1 denote the monoidal unit, and let $x \in D$ be any object. Prove that

$$\begin{aligned} \mathrm{End}_D(1) &\to \mathrm{End}_D(1 \otimes x) \\ f &\mapsto f \otimes \mathrm{id}_x \end{aligned}$$

is an isomorphism of groups. Using the unitor and its inverse, the codomain may be identified with $\mathrm{End}_D(x)$.

Does this say anything useful when $D = \mathrm{Cob}_d[\mathrm{Cob}_d^{-1}]$? (Hint: first show that D is rigid.)

Exercise 5.3.2. Let C be a small category and let $\gamma : C \to C[C^{-1}]$ be the universal functor to a groupoid. In particular we have, for each object $x \in C$,

$$\gamma : \mathrm{End}_C(x) \to \mathrm{End}_{C[C^{-1}]}(x), \tag{5.3.3}$$

a monoid homomorphism into a group. The purpose of this exercise is to work out some useful rules for determining the group $\mathrm{End}_{C[C^{-1}]}(x)$, under the additional assumption on C that for any other object y, both $C(x, y)$ and $C(y, x)$ are non-empty. Unless stated otherwise, we shall in the rest of this exercise make this assumption on C and x.

(a) Prove that the image of (5.3.3) generates.
(b) Let $y \in C$ be any object, let $w_1, w_2 \in C(y, x)$ and $w_3, w_4 \in C(x, y)$ be any morphisms, and define $a, b, c, d \in \mathrm{End}_C(x)$ by

$$a = w_1 \circ w_3, \quad b = w_2 \circ w_3 \quad c = w_1 \circ w_4, \quad d = w_2 \circ w_4.$$

Then $\gamma(a), \ldots, \gamma(d)$ are elements of the group $\mathrm{End}_{C[C^{-1}]}(x)$. Prove that[27]

$$\gamma(a) \circ (\gamma(b))^{-1} = \gamma(c) \circ (\gamma(d))^{-1}. \tag{5.3.4}$$

Let us now specialize to $C \subset h\mathcal{C}_d^V$ the full subcategory on those objects admitting a morphism from $\emptyset$, and set $x = \emptyset$. To avoid special cases, take $\dim(V) > 0$ (see Remark 6.2.29 for a discussion of that special case).

(c) Prove that this C and x satisfy the assumption above.
(d) Convince yourself that the domain of (5.3.3) is a commutative monoid, and deduce that the codomain is an abelian group.
(e) Convince yourself that, moreover, the domain of (5.3.3) is a *free* commutative monoid[28].
(f) Use the tools developed above to show that, in $\mathrm{Cob}_2[(\mathrm{Cob}_2)^{-1}]$, the endomorphisms of $\emptyset$ given by the torus, by the Klein bottle and by the empty surface are equal.
(g) Return to Exercise 5.1.2(c) and Exercise 5.2.2(b).

---

[27] See [3, Section 6] for further discussion.

[28] An easier exercise is to prove this under the assumption $\dim(V) > 2d + 1$ where the domain of (5.3.3) is free on the set of classes of path connected manifolds. For a solution in the general case see the end of these notes.

Exercise 5.3.5. In this exercise we shall study the subcategories $\mathrm{Cob}_d^k \subset \mathrm{Cob}_d$ and $\mathcal{C}_d^k$ from Section 4.3.

(a) Verify that $\mathrm{Cob}_d^k \subset \mathrm{Cob}_d$ is indeed a subcategory. Deduce that $\mathcal{C}_d^k$ is a subcategory of $\mathcal{C}_d$.

In the lectures it was stated that the inclusion $\mathcal{C}_d^k \hookrightarrow \mathcal{C}_d$ induces a weak homotopy equivalence of classifying spaces

$$B\mathcal{C}_d^k \longrightarrow B\mathcal{C}_d \tag{5.3.6}$$

as long as $k \leqslant \frac{d-2}{2}$. The purpose of this exercise is to investigate, by more elementary means, when the map (5.3.6) induces a bijection on π_0.

(b) Prove "by hand" that, when $d \geqslant 2$ and $k = 0$, the map (5.3.6) induces a bijection on π_0.

(c) Rephrase bijectivity of (5.3.6) on π_0 as follows: given any cobordism $W\colon M_0 \rightsquigarrow M_1$ in $\mathcal{C}_d$, there is a zig-zag of cobordisms between M_0 and M_1, all satisfying the k-connectivity condition on the outgoing boundary.

(d) Prove bijectivity of (5.3.6) on π_0 more generally whenever $k < d/2$.
(*Hint*: if you haven't already, learn about *elementary cobordisms* e.g. from Milnor's book on the h-cobordism theorem.)

Exercise 5.3.7. Let D be the groupoid with one object $*$ and $\mathrm{End}_D(*) = \mathbb{Z}$. Define $E : \mathrm{Cob}_d \to D$ by sending any object to $*$ and a morphism $W : M_0 \rightsquigarrow M_1$ to $\chi(W) - \chi(M_0) \in \mathbb{Z}$.

(a) Briefly explain why this is a functor.
(You may have already done this in Exercise 5.1.2(b).)

(b) Prove that E is isomorphic to the trivial functor (sending any object to $*$ and any morphism to 0) when d is odd.

(c) Prove that E is not isomorphic to the trivial functor when d is even.

(d) Can you promote E to a symmetric monoidal functor? (Hint: use addition in $\mathbb{Z}$ as $\otimes$. In this case the associator, symmetry, and unitor can all be taken to be the identity.)

This is sometimes called the "Euler TQFT".

(e) Let $V = \mathbb{R}^m$, let $e\colon T_{d,\mathbb{R}^{m+1}} \to K(\mathbb{Z}, m+1)$ be a based map, and let $\Omega^m T_{d,\mathbb{R}^{m+1}} \to \Omega^m K(\mathbb{Z}, m+1)$ be the m-fold loop of e. Prove that the fundamental groupoid of $\Omega^m K(\mathbb{Z}, m+1)$ is equivalent to D, and explain how any such map e gives rise to a symmetric monoidal invertible field theory $\mathrm{Cob}_d \to D$ if $m \geqslant 3$.

(f) $(*)$ For d even, let $e \in \widetilde{H}^{m+1}(T_{d,\mathbb{R}^{m+1}})$ be the cohomology class corresponding to the Euler class of the universal bundle under Thom isomorphism

$$H^d(\mathrm{Gr}_d(\mathbb{R} \oplus V); \widetilde{\mathbb{Z}}) \xrightarrow{\cong} H^{m+1}(T_{d,\mathbb{R}^{m+1}}; \mathbb{Z}),$$

where $\widetilde{\mathbb{Z}}$ denotes the local system corresponding to the orientation character of the canonical bundle (this same local coefficient system shows up in both the Thom isomorphism and in the Euler class). Use the same notation $e : T_{d,\mathbb{R}^{m+1}} \to K(\mathbb{Z}, m+1)$ for a pointed map classifying this cohomology class, and prove that the TQFT defined in (e) agrees with the Euler TQFT. (Hint: in this part of the exercise it may be necessary to know a geometric description of the map $\mathcal{C}_d(\emptyset, \emptyset) \to \Omega^{n+1} T_{d,\mathbb{R}^{n+1}}$. This is the *Pontryagin–Thom construction*, see e.g. [16, p. 197].)

Exercise 5.3.8. If X is a rigid symmetric monoidal groupoid, it is determined up to equivalence by three pieces of data: $\pi_0 X$ (the abelian group of isomorphism classes of objects), $\pi_1 X = \mathrm{Aut}(1_X)$, and the k-*invariant*, which we now define: see Remark 3.6.1 for the related notion of k-invariants of spaces. Given any $x \in X$, there is a canonical isomorphism $- \otimes \mathrm{id}_x \colon \mathrm{Aut}(1_X) \to \mathrm{Aut}(x)$. The k-invariant of X is the map $\pi_0 X \otimes \mathbb{Z}/2 \to \pi_1 X$ which to $x \in \pi_0 X$ assigns the image of the symmetry $\sigma \colon x \otimes x \to x \otimes x$ in $\mathrm{Aut}(x \otimes x) \cong \mathrm{Aut}(1_X) = \pi_1 X$.

Compute π_0, π_1, and k for the following rigid symmetric monoidal groupoids.

(a) The category $\mathrm{Vect}_k^{\sim}$ of invertible vector spaces over a field k.

(b) Assuming $\mathrm{char}(k) \neq 2$, the category $\mathrm{sVect}_k^{\sim}$ of invertible *super vector spaces*, i.e. the category of invertible $\mathbb{Z}/2$-graded vector spaces with the symmetry $a \otimes b \mapsto (-1)^{\deg a \deg b} b \otimes a$.

(c) $\mathrm{Cob}_1[\mathrm{Cob}_1^{-1}]$.

(d) As in (c), but with the oriented 1-dimensional cobordism category.

Exercise 5.3.9. Let us denote by $h\mathring{\mathcal{C}}_d^V$ the oriented version of the category $h\mathcal{C}_d^V$, where both $(d-1)$-manifolds and cobordisms are equipped with compatible orientations.[29] There is a functor

$$F_{d,V} \colon h\mathring{\mathcal{C}}_d^V \longrightarrow h\mathcal{C}_d^V$$

that forgets all orientations.

(a) When $d = \dim(V)$ or $d = 0$, construct a section of $F_{d,V}$. ("Section" means a functor G in the other direction, such that $F_{d,V} \circ G$ is naturally isomorphic to the identity.)

(b) When $0 < d < \dim(V)$, prove that $F_{d,V}$ does not admit a section.

(*Suggestion*: first consider the cases $(d, V) = (1, \mathbb{R})$ and $(d, V) = (1, \mathbb{R}^2)$, and look at the picture in Exercise 5.2.2.)

Solutions to selected exercises (by A. Debray, S. Galatius, M. Palmer)

6.1. Solution to Exercise 5.2.2(b) In part (a) of the exercise, we defined a functor $f \colon \mathcal{C}_1^{\mathbb{R}} \to D$ by sending a 0-manifold $M \subset \mathbb{R}$ to $\#M \bmod 2$ and a bordism W to $\chi(X) - \chi(X \cap \mathbb{R} \times \{0\}) \in \mathbb{Z}$. This is well-defined because if M and N are bordant, their cardinalities mod 2 agree, so $\mathrm{Hom}_D(f(M), f(N)) = \mathbb{Z}$; then, the additivity formula for Euler characteristic implies that f is a functor.

[29]This was called $h\mathcal{C}_{\{\pm 1\}}^V$ in Exercise 5.2.3.

Since D is a discrete category, f factors as a functor $f\colon h\mathcal{C}_1^{\mathbb{R}} \to D$; since D is a groupoid, f factors further into a functor $f\colon h\mathcal{C}_1^{\mathbb{R}}[(h\mathcal{C}_1^{\mathbb{R}})^{-1}] \to D$, where $f(W^{-1}) := f(W)^{-1}$ for a bordism W. For part (b), we show $f\colon h\mathcal{C}_1^{\mathbb{R}}[(h\mathcal{C}_1^{\mathbb{R}})^{-1}] \to D$ is an equivalence of groupoids, meaning that it is fully faithful and essentially surjective. Essential surjectivity, i.e. that f induces a surjection on the sets of isomorphism classes of objects, is clear: the empty set in $h\mathcal{C}_1^{\mathbb{R}}[(h\mathcal{C}_1^{\mathbb{R}})^{-1}]$ maps to $0 \in \mathrm{Ob}(D)$, and a single point in $h\mathcal{C}_1^{\mathbb{R}}[(h\mathcal{C}_1^{\mathbb{R}})^{-1}]$ maps to $1 \in \mathrm{Ob}(D)$.

First we reduce to considering the empty manifold, using an argument similar to Exercise 5.3.1.

Lemma 6.1.1. *Let $f\colon X \to Y$ be a map of small groupoids. If $\pi_0 f\colon \pi_0 X \to \pi_0 Y$ (i.e. the induced map on sets of isomorphism classes) is a bijection, then f is fully faithful if and only it induces isomorphisms $\mathrm{Aut}_X(x) \to \mathrm{Aut}_Y(f(x))$ for all $x \in X$.*

Proof. If $\psi\colon x \to x'$ is any map in X, composing with ψ defines a bijection $\mathrm{Aut}_X(x) \xrightarrow{\cong} \mathrm{Hom}_X(x, x')$. In the diagram

$$\begin{array}{ccc} \mathrm{Aut}_X(x) & \xrightarrow[\cong]{\psi} & \mathrm{Hom}_X(x,x') \\ \big\downarrow f & & \big\downarrow f \\ \mathrm{Aut}_Y(f(x)) & \xrightarrow[\cong]{f(\psi)} & \mathrm{Hom}_Y(f(x), f(x')), \end{array} \tag{6.1.2}$$

the horizontal arrows, given by composing with ψ and $F(\psi)$ respectively, are bijections. Hence if the left vertical arrow is an isomorphism, so is the right one. This suffices: there cannot be any x and x' such that $\mathrm{Hom}_X(x, x') = \varnothing$ and $\mathrm{Hom}_Y(f(x), f(x')) \neq \varnothing$ because $\pi_0 f$ is an isomorphism. □

Lemma 6.1.3. *Suppose the map $\mathrm{Aut}_{h\mathcal{C}_1^{\mathbb{R}}[(h\mathcal{C}_1^{\mathbb{R}})^{-1}]}(\varnothing) \to \mathrm{Aut}_D(0)$ induced by f is an isomorphism. Then f is fully faithful.*

Proof. By Lemma 6.1.1, we have only to consider automorphism groups of objects in $h\mathcal{C}_1^{\mathbb{R}}[(h\mathcal{C}_1^{\mathbb{R}})^{-1}]$; there are only two isomorphism classes of objects, given by the empty manifold and a single point, and the assumption on $\pi_0 f$ is satisfied. We will define an endofunctor F' of $h\mathcal{C}_1^{\mathbb{R}}[(h\mathcal{C}_1^{\mathbb{R}})^{-1}]$ and show that it induces an isomorphism $\mathrm{Aut}(\varnothing) \to \mathrm{Aut}(\mathrm{pt})$.

First, we define an endofunctor F of $\mathcal{C}_1^{\mathbb{R}}$. The idea of F is to disjoint union all objects with a point, and all bordisms with an interval regarded as a bordism between the new points. Fix a diffeomorphism $\phi\colon \mathbb{R} \to (-\infty, 0)$ isotopic to the identity on $\mathbb{R}$, and let $\Phi\colon [0,t] \times \mathbb{R} \to [0,t] \times (-\infty, 0)$ denote $(s, x) \mapsto (s, \phi(x))$.

- Given an object $M \subset \mathbb{R}$, let $F(M) := \phi(M) \cup \{1\}$.
- Given a bordism $W \subset [0,t] \times \mathbb{R}$, let $F(W) := \Phi(W) \cup [0,t] \times \{1\}$.

This is continuous on spaces of objects and morphisms, hence it passes to an endofunctor of $h\mathcal{C}_1^{\mathbb{R}}$, and this descends to an endofunctor F' of $h\mathcal{C}_1^{\mathbb{R}}[(h\mathcal{C}_1^{\mathbb{R}})^{-1}]$ by $F'(x^{-1}) := F(x)^{-1}$.

Choose an isotopy $H\colon [0,1] \times \mathbb{R} \to \mathbb{R}$ from $\phi \circ \phi$ to the identity. We define a natural transformation $T\colon F^2 \Rightarrow \mathrm{id}$, where F is regarded in $h\mathcal{C}_1^{\mathbb{R}}$, by "capping off" the two new points. For every object x of $\mathcal{C}_1^{\mathbb{R}}$, we produce a bordism $T_x\colon F^2x \to x$, defined to be the union of $\{H(s,p) : s \in [0,1], p \in x\}$ and a semicircle in $[0,1] \times \mathbb{R}$ containing the two new points of F^2x as its boundary, as in the following picture.

(6.1.4)

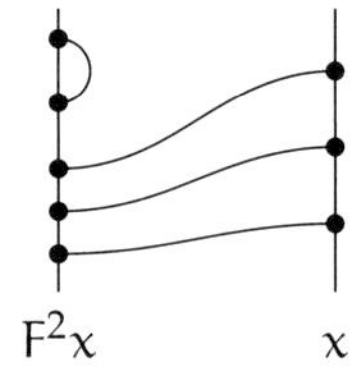

To check that this is in fact a natural transformation after passing to $h\mathcal{C}_1^{\mathbb{R}}$, we need that for every bordism $W\colon x \to y$, the diagram

$$\begin{array}{ccc} F^2x & \xrightarrow{T_x} & x \\ {\scriptstyle F^2W}\downarrow & & \downarrow{\scriptstyle W} \\ F^2y & \xrightarrow{T_y} & y \end{array} \tag{6.1.5}$$

commutes. The picture proof is that the following two bordisms are equal in $h\mathcal{C}_1^{\mathbb{R}}$.

(6.1.6)

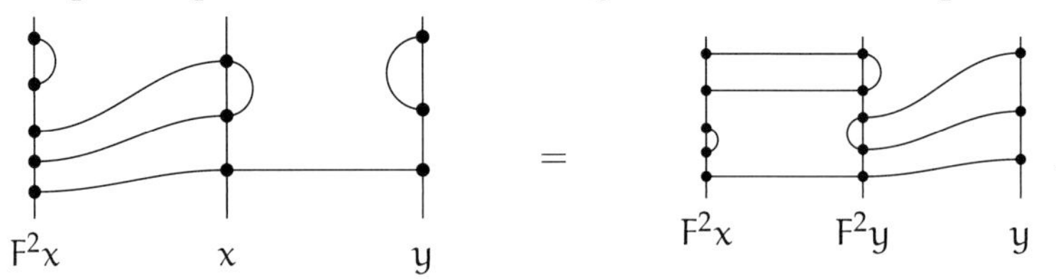

(Though we drew a specific bordism $x \to y$, the argument works in general.)

The map T also descends to a natural transformation $T'\colon (F')^2 \Rightarrow \mathrm{id}$, and every natural transformation between functors into a groupoid is a natural isomorphism. Thus applying F' defines an isomorphism

$$F'_*\colon \mathrm{Aut}_{h\mathcal{C}_1^{\mathbb{R}}[(h\mathcal{C}_1^{\mathbb{R}})^{-1}]}(\varnothing) \to \mathrm{Aut}_{h\mathcal{C}_1^{\mathbb{R}}[(h\mathcal{C}_1^{\mathbb{R}})^{-1}]}(\mathrm{pt}). \tag{6.1.7}$$

This is compatible with the identifications $\mathrm{Aut}_D(0) = \mathbb{Z} = \mathrm{Aut}_D(1)$, in that if $W \in \mathrm{Aut}_{h\mathcal{C}_1^{\mathbb{R}}[(h\mathcal{C}_1^{\mathbb{R}})^{-1}]}(\varnothing)$, then $f(W) = f(F'_*(W))$, because applying F increases the Euler characteristics of X and $X \cap (\{0\} \times \mathbb{R})$ by the same amount. Therefore if

$$f_*\colon \mathrm{Aut}_{h\mathcal{C}_1^{\mathbb{R}}[(h\mathcal{C}_1^{\mathbb{R}})^{-1}]}(\varnothing) \to \mathrm{Aut}_D(0) \tag{6.1.8}$$

is an isomorphism, so is

$$f_*\colon \mathrm{Aut}_{h\mathcal{C}_1^{\mathbb{R}}[(h\mathcal{C}_1^{\mathbb{R}})^{-1}]}(\mathrm{pt}) \to \mathrm{Aut}_D(1). \tag{6.1.9}$$

□

To show $\mathrm{Aut}_{h\mathcal{C}_1^{\mathbb{R}}[(h\mathcal{C}_1^{\mathbb{R}})^{-1}]}(\varnothing) \to \mathrm{Aut}_D(0) = \mathbb{Z}$ is surjective, it suffices to produce a preimage of the generator $1 \in \mathrm{Aut}_D(0)$. Asingle circle, interpreted as a bordism $W\colon \varnothing \to \varnothing$, works: X is a disk and does not intersect the incoming boundary, so $f(W) = 1$.

Then, we will show that for every morphism $W: \varnothing \to \varnothing$ in $h\mathcal{C}_1^{\mathbb{R}}$, the image of W in $h\mathcal{C}_1^{\mathbb{R}}[(h\mathcal{C}_1^{\mathbb{R}})^{-1}]$ is equal to a composition of n circles, where $n \in \mathbb{Z}$ (here $n < 0$ is interpreted as a composition of $-n$ formal inverses of circles). By Exercise 5.3.2, parts (a) and (c), $\mathrm{Aut}_{h\mathcal{C}_1^{\mathbb{R}}[(h\mathcal{C}_1^{\mathbb{R}})^{-1}]}(\varnothing)$ is generated by such W, so it will follow that $\mathrm{Aut}_{h\mathcal{C}_1^{\mathbb{R}}[(h\mathcal{C}_1^{\mathbb{R}})^{-1}]}(\varnothing)$ is cyclic. Since every surjective homomorphism from a cyclic group to $\mathbb{Z}$ is an isomorphism, this suffices for Lemma 6.1.3 to apply and conclude the proof.

For any bordism $W: \varnothing \to \varnothing$, the complement of W in $(0,t) \times \mathbb{R}$ has exactly one unbounded path component A: there is at least one because $[0,t] \times \mathbb{R}$ is unbounded, and at most one because W does not intersect $\partial([0,t] \times \mathbb{R})$ and is compact, hence is contained in some disk in $(0,t) \times \mathbb{R}$. Let $\overline{A}$ denote the closure of A in $(0,t) \times \mathbb{R}$. Then $\overline{A}$ is a 2-manifold with boundary and $\partial\overline{A}$ is bounded in $[0,t] \times \mathbb{R}$, so $\partial\overline{A}$ is a compact 1-manifold with empty boundary, and therefore a disjoint union of k circles. Call W simple if $k = 1$.

Lemma 6.1.10. *Every bordism $W: \varnothing \to \varnothing$ in $h\mathcal{C}_1^{\mathbb{R}}$ is a composition of simple bordisms.*

Proof. Let $C_1, \ldots, C_k$ be the components of $\partial\overline{A}$. Each C_i is diffeomorphic to a circle, so by the smooth Schoenflies theorem, the complement of C_i in $(0,t) \times \mathbb{R}$ has a unique bounded component D_i, and $\overline{D_i}$ is diffeomorphic to a closed disc. Then $Y := \overline{D_1} \cup \cdots \cup \overline{D_k}$ is the complement of A in $(0,t) \times \mathbb{R}$, so W is contained in Y. We can now apply a diffeomorphism of $[0,t] \times \mathbb{R}$ fixing the boundary and such that $\overline{D_i}$ lands in $((i-1)t/k, it/k) \times \mathbb{R}$, factoring W as the composition of the simple bordisms $W \cap \overline{D_i}$. □

Therefore we are done if we can show that for every simple bordism $W: \varnothing \to \varnothing$ in $h\mathcal{C}_1^{\mathbb{R}}$, its image in $h\mathcal{C}_1^{\mathbb{R}}[(h\mathcal{C}_1^{\mathbb{R}})^{-1}]$ is equal to a composition of n circles for some $n \in \mathbb{Z}$. Assume W has m path components; $m = 0$ would not be simple, and if $m = 1$ we are already done. We induct on m. A general simple bordism W with $m > 1$ is of the form

(6.1.11)

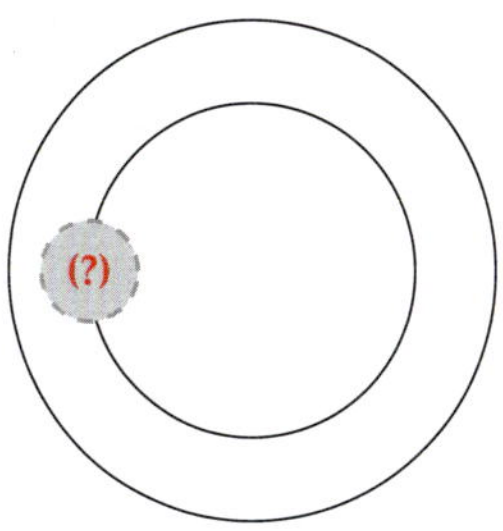

where the (?) indicates more components of W that may be present. We use Exercise 5.3.2 to simplify this picture. Recall that Exercise 5.3.2 tells us for any object y of $h\mathcal{C}_1^{\mathbb{R}}$ admitting a map from $x := \varnothing$ and any morphisms $w_1, w_2: y \to x$ and $w_3, w_4: x \to y$ in $h\mathcal{C}_1^{\mathbb{R}}$, that in $\mathrm{Aut}_{h\mathcal{C}_1^{\mathbb{R}}[(h\mathcal{C}_1^{\mathbb{R}})^{-1}]}(y)$,

$$w_1 \circ w_3 \circ (w_2 \circ w_3)^{-1} = w_1 \circ w_4 \circ (w_2 \circ w_4)^{-1}. \tag{6.1.12}$$

Apply this to $y := \mathrm{pt}^{\amalg 4}$ and the four morphisms

(6.1.13)

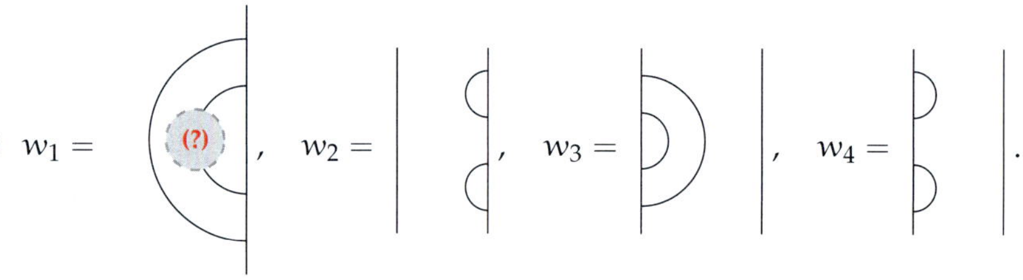

Then (6.1.12) tells us

(6.1.14)

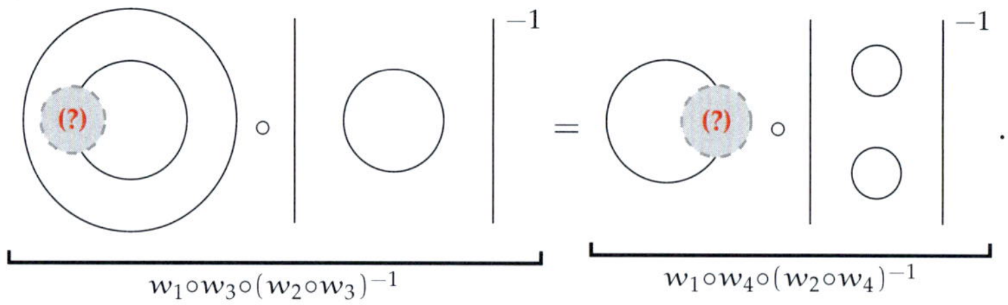

Precomposing with a single circle,

(6.1.15)

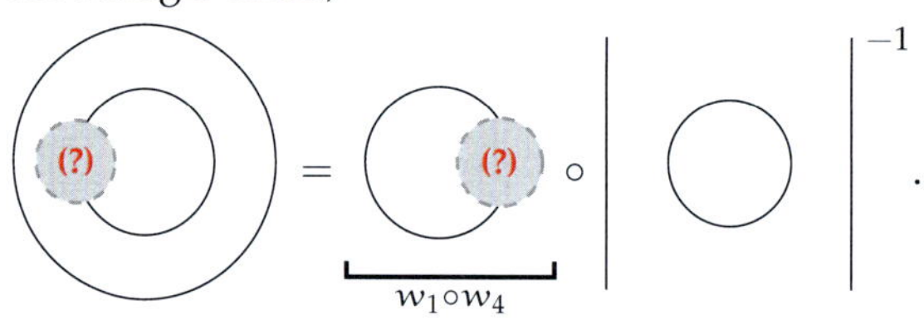

The bordism $w_1 \circ w_4$ has fewer path components than W. It is not necessarily true that $w_1 \circ w_4$ is simple, but by Lemma 6.1.10 it is a composition of simple bordisms, and each has fewer path components than W, so we induct and conclude that every simple bordism can be written as a union of circles and their inverses.

6.2. Solution to Exercise 5.3.1(e) The goal of this exercise is to show that the monoid $h\mathcal{C}_d^V(\emptyset, \emptyset)$ is *free commutative* when $\dim(V) > 0$. The fact that it is commutative (which was Exercise 5.3.1(d)) follows from Corollary 6.2.11 below, and the fact that it is *free* commutative is Corollary 6.2.26. Remark 6.2.28 describes the corresponding statements when manifolds are equipped with tangential structures and Remark 6.2.30 comments on whether the space $\mathcal{C}_d^V(\emptyset, \emptyset)$ is free as an E_m-algebra.

Description of the monoid. Let $U = \mathbb{R} \times V$ and set $m = \dim(U) = \dim(V) + 1$. Note that, unwinding the definitions, we may see that $h\mathcal{C}_d^V(\emptyset, \emptyset)$ is isomorphic to the monoid

$$\mathcal{M}_d^V = \left\{ (W, \varphi) \;\middle|\; \begin{array}{l} W \text{ is a smooth, closed } d\text{-manifold} \\ \varphi \text{ is a smooth embedding } W \hookrightarrow U \end{array} \right\} /\sim,$$

where $(W, \varphi) \sim (W', \varphi')$ if and only if

$$[\varphi' \circ \theta] = [\varphi] \in \pi_0\left(\mathrm{Emb}(W, U)/\mathrm{Diff}(W)\right)$$

for a diffeomorphism $\theta\colon W \cong W'$, and the operation in $\mathcal{M}_d^V$ is given by

$$[(W_1,\varphi_1)]\cdot[(W_2,\varphi_2)] = [(W_1 \sqcup W_2, i\circ(\varphi_1 \sqcup \varphi_2))], \tag{6.2.1}$$

where $i = j \times \mathrm{id}_V$ for any orientation-preserving embedding $j\colon \mathbb{R}\sqcup\mathbb{R} \hookrightarrow \mathbb{R}$. In other words, $\mathcal{M}_d^V$ is the monoid of isotopy classes of (unparametrized) closed, d-dimensional submanifolds of U, under the operation of "placing two submanifolds side-by-side" in U.

Embedding and diffeomorphism results. For the solution, we will need the following facts. For smooth manifolds M, N, recall that $\mathrm{Emb}(M, N)$ denotes the space of smooth embeddings of M into N and $\mathrm{Diff}(N)$ the topological group of self-diffeomorphisms of N, both equipped with the smooth Whitney topology. We also write $\mathrm{Diff}_c(N)$ for the subgroup of $\mathrm{Diff}(N)$ of those diffeomorphisms φ such that $\{x \in N \mid \varphi(x) \neq x\} \subset K$ for some compact $K \subset N$ (topologized as a colimit over larger and larger compact $K \subset N$) and $\mathrm{Diff}_c(N)_1$ for the path-component of the identity in $\mathrm{Diff}_c(N)$.

Proposition 6.2.2 (*Isotopy extension theorem*). *Let M, N be smooth manifolds without boundary, where M is compact and $\dim(M) < \dim(N)$, and let $i\colon M \hookrightarrow N$ be a smooth embedding. Then the map*

$$\mathrm{Diff}_c(N)_1 \longrightarrow \mathrm{Emb}(M,N)/\mathrm{Diff}(M), \tag{6.2.3}$$

defined by $\Phi \mapsto [\Phi\circ i]$, is a Serre fibration, and thus in particular admits path-lifting.

Proof. We may factor (6.2.3) into two maps

$$r_1\colon \mathrm{Diff}_c(N)_1 \longrightarrow \mathrm{Emb}(M,N) \quad \text{and} \quad r_2\colon \mathrm{Emb}(M,N) \longrightarrow \mathrm{Emb}(M,N)/\mathrm{Diff}(M).$$

The first map r_1 is a fibre bundle, and hence a Serre fibration, by combining Theorems A and B of [31] (see also Corollaire 2 of §II.2.2.2 of [5] and [23]). The second map r_2 is a principal $\mathrm{Diff}(M)$-bundle, and hence in particular a Serre fibration, by Theorem 13.11 of [28] (see also [2] and Theorem 44.1 of [22]). □

Proposition 6.2.4 (*Multi-disk theorem*). *Let N be a smooth, connected, oriented manifold of dimension $m \geqslant 2$, let $i_1, \ldots, i_r\colon \mathbb{D}^m \hookrightarrow N$ be a finite collection of orientation-preserving smooth embeddings of the closed m-disk into N with pairwise disjoint images, and let $j_1, \ldots, j_r$ be a second such collection. Then there exists $\Phi \in \mathrm{Diff}_c(N)_1$ such that $\Phi \circ i_s = j_s$ for all $s \in \{1, \ldots, r\}$.*

Proof. When $r = 1$ this is exactly the *disk theorem* of Palais [30, Theorem B and Corollary 1]. We show how this implies the result for general r. First, note that we may assume, without loss of generality, that the images of $i_1, \ldots, i_r$ are also disjoint from the images of the $j_1, \ldots, j_r$.

For each $s \in \{1, \ldots, r\}$, choose a smooth path γ_s in N that is disjoint from $i_t(\mathbb{D}^m)$ and $j_t(\mathbb{D}^m)$ for all $t \neq s$ and such that $\gamma_s(0) = i_s(0)$ and $\gamma_s(1) = j_s(0)$. We may also assume that the paths $\gamma_1, \ldots, \gamma_r$ have *pairwise disjoint* images.[30] We

[30]For $m \geqslant 3$ it suffices to ensure that the $\gamma_1, \ldots, \gamma_r$ are pairwise *transverse*. For $m = 2$, some extra care is needed: after ensuring that these smooth paths are pairwise transverse and therefore intersect

may then choose pairwise disjoint, connected, open subsets $A_1, \dots, A_r$ of N such that

$$i_s(\mathbb{D}^m) \cup \gamma_s([0,1]) \cup j_s(\mathbb{D}^m) \subset A_s$$

for $s \in \{1, \dots, r\}$. By the disk theorem of Palais, we may find $\Phi_s \in \mathrm{Diff}_c(A_s)_1$ such that $\Phi_s \circ i_s = j_s$ for all $s \in \{1, \dots, r\}$. We may then define Φ to be Φ_s on each A_s and the identity elsewhere. □

Another description of the monoid. The next two lemmas give a more convenient description of $\mathcal{M}_d^V$.

Lemma 6.2.5. *An alternative description of the equivalence relation in the definition of* $\mathcal{M}_d^V$ *is* $(W, \varphi) \sim (W', \varphi')$ *if and only if there is a diffeomorphism* $\Phi \in \mathrm{Diff}_c(U)_1$ *such that* $\Phi(\varphi(W)) = \varphi'(W')$.

Proof. Suppose first that there is a diffeomorphism $\Phi \in \mathrm{Diff}_c(U)_1$ such that $\Phi(\varphi(W)) = \varphi'(W')$. We then obtain a diffeomorphism $\theta\colon W \cong W'$ by setting $\theta = (\varphi')^{-1} \circ \Phi \circ \varphi$. Choose a path $\mathrm{id} \rightsquigarrow \Phi$ in $\mathrm{Diff}_c(U)_1$ and consider its composition with (6.2.3), where we set $N = U$, $M = W$ and $i = \varphi$. This is a path in $\mathrm{Emb}(W, U)/\mathrm{Diff}(W)$ from $[\varphi]$ to $[\Phi \circ \varphi] = [\varphi' \circ \theta]$. Hence $(W, \varphi) \sim (W', \varphi')$.

Conversely, suppose that $(W, \varphi) \sim (W', \varphi')$, so we have a diffeomorphism $\theta\colon W \cong W'$ and a path $[\varphi] \rightsquigarrow [\varphi' \circ \theta]$ in $\mathrm{Emb}(W, U)/\mathrm{Diff}(W)$. By Proposition 6.2.2, we may lift this to a path $\mathrm{id} \rightsquigarrow \Phi$ such that $[\Phi \circ \varphi] = [\varphi' \circ \theta]$, and hence $\Phi(\varphi(W)) = \varphi'(\theta(W)) = \varphi'(W')$. □

Lemma 6.2.6. *Choose any smooth, orientation-preserving embedding* $e\colon U \sqcup U \hookrightarrow U$. *Then the operation in* $\mathcal{M}_d^V$ *may be defined using this embedding:*

$$[(W_1, \varphi_1)] \cdot [(W_2, \varphi_2)] = [(W_1 \sqcup W_2, e \circ (\varphi_1 \sqcup \varphi_2))]. \tag{6.2.7}$$

Moreover, if $[(W_1, \varphi_1)], \dots, [(W_r, \varphi_r)]$ *is any finite collection of elements of* $\mathcal{M}_d^V$ *and* e *is a smooth, orientation-preserving embedding of* r *disjoint copies of* U *into* U, *then we have*

$$[(W_1, \varphi_1)] \cdots [(W_r, \varphi_r)] = [(W_1 \sqcup \cdots \sqcup W_r, e \circ (\varphi_1 \sqcup \cdots \sqcup \varphi_r))]. \tag{6.2.8}$$

Proof. The operation in $\mathcal{M}_d^V$ was defined exactly as in equation (6.2.7) (see equation (6.2.1)), *assuming* that the embedding e is of a particular form (namely of the form $(\mathbb{R} \sqcup \mathbb{R} \hookrightarrow \mathbb{R}) \times \mathrm{id}_V$). It follows that equation (6.2.8) also holds, again *assuming* that the embedding e is of a particular form.[31] It will therefore suffice to prove that right-hand side of (6.2.8) is independent of the choice of e, in other words, that

$$(W_1 \sqcup \cdots \sqcup W_r, e \circ (\varphi_1 \sqcup \cdots \sqcup \varphi_r)) \sim (W_1 \sqcup \cdots \sqcup W_r, e' \circ (\varphi_1 \sqcup \cdots \sqcup \varphi_r))$$

---

in finitely many points, we apply finitely many isotopies to remove these intersection points one at a time by "pushing them off the end of the arc".

[31] For example, one choice of conventions would lead to e being of the following form: let us identify U with $(0,1) \times V$ and, on the ith copy of U in the disjoint union, we define e to be the embedding $U \hookrightarrow U$ that acts by $t \mapsto (t + 2^i - 2)/2^i$ on the first coordinate and by the identity on V.

for any two smooth, orientation-preserving embeddings e, e' of r disjoint copies of U into U. By Lemma 6.2.5 this means that we need to find $\Phi \in \mathrm{Diff}_c(U)_1$ such that:

$$\Phi(e(\varphi_1(W_1) \sqcup \cdots \sqcup \varphi_r(W_r))) = e'(\varphi_1(W_1) \sqcup \cdots \sqcup \varphi_r(W_r)). \tag{6.2.9}$$

Since $\varphi_s(W_s)$ is a compact subspace of $U = \mathbb{R}^m$, we may assume that it is contained in the closed unit disk $\mathbb{D}^m$.[32] Let us write i_s for the restriction of e to the closed unit disk in the sth copy of U for $s \in \{1, \ldots, r\}$. Similarly, write j_s for the restriction of e' to the closed unit disk in the sth copy of U. We may therefore rewrite (6.2.9) as:

$$\Phi(i_1(\varphi_1(W_1))) \cup \cdots \cup \Phi(i_r(\varphi_r(W_r))) = j_1(\varphi_1(W_1)) \cup \cdots \cup j_r(\varphi_r(W_r)). \tag{6.2.10}$$

Now Proposition 6.2.4 tells us that there exists $\Phi \in \mathrm{Diff}_c(U)_1$ such that $\Phi \circ i_s = j_s$ for all $s \in \{1, \ldots, r\}$, which implies (6.2.10). □

Corollary 6.2.11. *The monoid* $h\mathcal{C}_d^V(\emptyset, \emptyset) = \mathcal{M}_d^V$ *is isomorphic to the monoid whose underlying set is the set of smooth, closed,* d*-dimensional submanifolds of* U*, up to the natural left-action of* $\mathrm{Diff}_c(U)_1$*, and whose operation is given by*

$$([W_1], [W_2]) \mapsto [e_1(W_1) \cup e_2(W_2)],$$

where (e_1, e_2) *is any choice of a pair of orientation-preserving self-embeddings* $U \hookrightarrow U$ *with disjoint images. In particular, this operation is commutative. Moreover, if* $e_1, \ldots, e_r$ *is any collection of orientation-preserving self-embeddings* $U \hookrightarrow U$ *with pairwise disjoint images, we have*

$$[W_1] \cdots [W_r] = [e_1(W_1) \cup \cdots \cup e_r(W_r)] \tag{6.2.12}$$

in this monoid.

Proof. By Lemma 6.2.5, there is a bijection from $\mathcal{M}_d^V$ to the set of smooth, closed, d-dimensional submanifolds of U, up to the natural left-action of $\mathrm{Diff}_c(U)_1$, given by sending $[(W, \varphi)]$ to $[\varphi(W)]$. Using the description of the operation of $\mathcal{M}_d^V$ from Lemma 6.2.6, we see that it corresponds under this bijection to the operation $([W_1], [W_2]) \mapsto [e_1(W_1) \cup e_2(W_2)]$ described above, which is therefore in particular well-defined. Equation (6.2.12) also follows directly from Lemma 6.2.6. □

Remark 6.2.13. Note that this in particular solves Exercise 5.3.1(d), since Corollary 6.2.11 implies that $\mathcal{C}_d^V(\emptyset, \emptyset)$ is commutative. For the rest of this solution, we identify $\mathcal{M}_d^V$ with the monoid described in Corollary 6.2.11.

A generating set. We now define a subset of $\mathcal{M}_d^V$ that will turn out to freely generate it.

Definition 6.2.14. Let us call a smooth, closed, d-dimensional submanifold W of U *inseparable* if $W \neq \emptyset$ and, if for any finite collection of smooth embeddings

[32] If it is not, we go back and choose a different representative for $[(W_s, \varphi_s)]$ by "shrinking" φ_s by an isotopy.

$i_1, \dots, i_r \colon U \hookrightarrow U$ with

$$W \subset \bigcup_{s=1}^{r} i_s(U) \qquad \text{and} \qquad i_1(U), \dots, i_r(U) \text{ are pairwise disjoint,}$$

we have that $W \subset i_s(U)$ for one $s \in \{1, \dots, r\}$.

Remark 6.2.15. Some immediate consequences of Definition 6.2.14 are as follows:

- If W is path-connected, then it is inseparable.
- Examples of inseparable but non-path-connected W are (for $d = 1, m = 2$) two nested circles in $\mathbb{R}^2$ and (for $d = 1, m = 3$) two non-trivially linked circles in $\mathbb{R}^3$.
- If $m = \dim(U) > 2d + 1$ then inseparable d-submanifolds of $U = \mathbb{R}^m$ are path-connected. To see this, let us denote the path-components of W by $W_1, \dots, W_r$ and choose any collection $i_1, \dots, i_r$ of orientation-preserving smooth embeddings $U \hookrightarrow U$ with pairwise disjoint images. Let $i \colon W \hookrightarrow U$ be the embedding that acts by i_s on W_s. Since the space of embeddings $\mathrm{Emb}(W, U)$ is path-connected for $\dim(U) > 2\dim(W) + 1$, we may choose a path from the inclusion to i and use Proposition 6.2.2 to lift this and find $\Phi \in \mathrm{Diff}_c(U)_1$ such that $\Phi|_W = i$. Hence $[i(W)] = [W]$, so by Lemma 6.2.16 below, W is inseparable if and only if $i(W)$ is inseparable. But $i(W)$ is clearly inseparable if and only if $r = 1$.

Lemma 6.2.16. *Inseparability is a well-defined property of the element $[W] \in \mathcal{M}_d^V$. In other words, if $W \sim W'$ and W is inseparable, then so is W'.*

Proof. Suppose that $i_1, \dots, i_r \colon U \hookrightarrow U$ are smooth embeddings with pairwise disjoint images and with $W' \subset i_1(U) \cup \dots \cup i_r(U)$. Since $W \sim W'$, we have (by Corollary 6.2.11) $\Phi \in \mathrm{Diff}_c(U)_1$ such that $\Phi(W) = W'$. If we define $\hat{i}_s = \Phi^{-1} \circ i_s$ for each s, then $\hat{i}_1, \dots, \hat{i}_r \colon U \hookrightarrow U$ are smooth embeddings with pairwise disjoint images and with $W \subset \hat{i}_1(U) \cup \dots \cup \hat{i}_r(U)$. Since W is inseparable we have $W \subset \hat{i}_s(U)$ for some $s \in \{1, \dots, r\}$, and hence $W' \subset i_s(U)$. □

Definition 6.2.17. Denote the set of inseparable elements of $\mathcal{M}_d^V$ by $\mathcal{I}_d^V$.[33]

The following is a consequence of Lemma 6.2.16 and the case $r = 1$ of Corollary 6.2.11.

Corollary 6.2.18. *Let $e \colon U \hookrightarrow U$ be an orientation-preserving self-embedding and let W be a smooth, closed, d-dimensional submanifold of U. Then $[e(W)] = [W]$ in $\mathcal{M}_d^V$. In particular, by Lemma 6.2.16, W is inseparable if and only if $e(W)$ is inseparable.* □

Write $\mathbb{N}[\mathcal{I}_d^V]$ for the free commutative monoid on the set $\mathcal{I}_d^V$. The inclusion $\mathcal{I}_d^V \hookrightarrow \mathcal{M}_d^V$ therefore induces a monoid homomorphism

$$\alpha \colon \mathbb{N}[\mathcal{I}_d^V] \longrightarrow \mathcal{M}_d^V. \tag{6.2.19}$$

[33]See Lemma 6.2.27 for an equivalent characterization of $\mathcal{I}_d^V$, where the r of Definition 6.2.14 is replaced by 2.

Our aim is to show that α is an isomorphism. (Surjectivity of α is the statement that $\mathcal{I}_d^V$ *generates* $\mathcal{M}_d^V$ as a commutative monoid; injectivity is the statement that it *freely* generates $\mathcal{M}_d^V$.)

Surjectivity and injectivity.

Proposition 6.2.20. *The monoid homomorphism* (6.2.19) *is surjective.*

Proof. Let $[W] \in \mathcal{M}_d^V$. We show that $[W]$ is in the image of α by induction on the number $n(W)$ of path-components of W. If $W = \emptyset$ then $[W] = \alpha(\ )$. If W is path-connected, then it must be inseparable, so $[W] = \alpha([W])$. So we may assume that $n(W) \geqslant 2$. If $[W]$ is inseparable, we are done, so let us assume that there are smooth embeddings $i_1, \ldots, i_r \colon U \hookrightarrow U$ with pairwise disjoint images ($r \geqslant 2$) such that

$$W \subset \bigcup_{s=1}^{r} i_s(U) \qquad \text{and} \qquad W \cap i_s(U) \neq \emptyset \text{ for all } s \in \{1, \ldots, r\}.$$

Without loss of generality, we may also assume that $i_1, \ldots, i_r$ are orientation-preserving. Since $\{W \cap i_s(U) \mid s \in \{1, \ldots, r\}\}$ is a partition of W into clopen subsets, each $W \cap i_s(U)$ is a union of path-components of W, and in particular compact. Hence

$$W_s' = i_s^{-1}(W)$$

are smooth, closed, d-dimensional submanifolds of U, so they represent elements $[W_s']$ of $\mathcal{M}_d^V$. Since W_s' are all non-empty and $r \geqslant 2$ we deduce that $n(W_s') < n(W)$ for all s. Thus, by the inductive hypothesis, each $[W_s']$ is in the image of α. By Corollary 6.2.11,

$$[W_1'] \cdots [W_r'] = [i_1(W_1') \cup \cdots \cup i_r(W_r')] = [W], \tag{6.2.21}$$

so $[W]$ is also in the image of α. □

Lemma 6.2.22. *Let $W_1, \ldots, W_r$ be a finite collection of inseparable manifolds embedded in U and let $e_1, \ldots, e_r \colon U \hookrightarrow U$ be a collection of smooth, orientation-preserving embeddings $U \hookrightarrow U$ with pairwise disjoint images, and let $i_1, \ldots, i_s \colon U \hookrightarrow U$ be another such collection. Assume that*

$$W = e_1(W_1) \cup \cdots \cup e_r(W_r) \subset i_1(U) \cup \cdots \cup i_s(U)$$

and that $i_t^{-1}(W)$ is compact and inseparable for each $t \in \{1, \ldots, s\}$. Then $r = s$ and $i_t(U) \cap W = e_{\sigma(t)}(W_{\sigma(t)})$ for each $t \in \{1, \ldots, r\}$ and some permutation σ of $\{1, \ldots, r\}$.

Proof. We will first prove that, for each $t \in \{1, \ldots, s\}$,

$$i_t(U) \cap W = \bigcup_{a \in A_t} e_a(W_a), \text{ for some } A_t \subset \{1, \ldots, r\}. \tag{6.2.23}$$

Given this, one may easily see that $\{A_1, \ldots, A_s\}$ is a partition of $\{1, \ldots, r\}$ into non-empty subsets. We will then prove that, for each $t \in \{1, \ldots, s\}$,

$$|A_t| = 1, \tag{6.2.24}$$

which will complete the proof.

Suppose that (6.2.23) is false. This implies that, for some $u \in \{1, \dots, r\}$, the submanifold $e_u(W_u)$ is contained in the union $i_1(U) \cup \dots \cup i_s(U)$, but it is not contained in any one of the $i_1(U), \dots, i_s(U)$. Hence $e_u(W_u)$ is separable, and so, by Corollary 6.2.18, W_u is separable, which contradicts the hypotheses.

Suppose that (6.2.24) is false. This implies that, for some $t \in \{1, \dots, s\}$, the submanifold $i_t(U) \cap W$ is contained in the union $e_1(U) \cup \dots \cup e_r(U)$, but it is not contained in any one of the $e_1(U), \dots, e_r(U)$. Hence $i_t(U) \cap W = i_t(i_t^{-1}(W))$ is separable, and so, by Corollary 6.2.18, $i_t^{-1}(W)$ is separable, which contradicts the hypotheses. □

Proposition 6.2.25. *The monoid homomorphism* (6.2.19) *is injective.*

Proof. Let $W_1, \dots, W_r$ and $X_1, \dots, X_s$ be inseparable manifolds embedded in U. Suppose that $[W_1] \cdots [W_r] = [X_1] \cdots [X_s]$ in $\mathcal{M}_d^V$. We then need to show that $r = s$ and $[X_t] = [W_{\sigma(t)}]$ for all $t \in \{1, \dots, r\}$ and some permutation σ of $\{1, \dots, r\}$.

Let $e_1, \dots, e_{\max(r,s)} : U \hookrightarrow U$ be any collection of orientation-preserving embeddings with pairwise disjoint images. Then, by Corollary 6.2.11, we have some $\Phi \in \mathrm{Diff}_c(U)_1$ such that

$$\Phi(e_1(X_1) \cup \dots \cup e_s(X_s)) = e_1(W_1) \cup \dots \cup e_r(W_r) =: W.$$

For each $t \in \{1, \dots, s\}$, we have $(\Phi \circ e_t)^{-1}(W) = X_t$, which is a compact and inseparable submanifold of U, so Lemma 6.2.22 implies that $r = s$ and

$$\Phi(e_t(U)) \cap W = e_{\sigma(t)}(W_{\sigma(t)})$$

for each $t \in \{1, \dots, r\}$ and some permutation σ of $\{1, \dots, r\}$. But then we must have $\Phi(e_t(U)) \cap W = \Phi(e_t(X_t))$, so by Corollaries 6.2.11 and 6.2.18 we conclude that

$$[X_t] = [e_t(X_t)] = [e_{\sigma(t)}(W_{\sigma(t)})] = [W_{\sigma(t)}].$$

□

Propositions 6.2.20 and 6.2.25 immediately imply:

Corollary 6.2.26. *The commutative monoid $\mathcal{M}_d^V$ is freely generated by the subset $\mathcal{I}_d^V$.*

We note that the subset $\mathcal{I}_d^V \subset \mathcal{M}_d^V$ of equivalence classes (with respect to the action of $\mathrm{Diff}_c(U)_1$) of *inseparable* submanifolds of U may be characterised in a slightly simpler way than in Definition 6.2.14, namely: a submanifold W may be separated by $r \geqslant 2$ pairwise disjoint embeddings $U \hookrightarrow U$ if and only if it may be separated by two disjoint embeddings $U \hookrightarrow U$.

Lemma 6.2.27. *An element $[W] \in \mathcal{M}_d^V$ is inseparable if and only if $W \neq \emptyset$ and, given any pair of smooth embeddings $i, j : U \hookrightarrow U$ with*

$$W \subset i(U) \cup j(U) \qquad \textit{and} \qquad i(U) \cap j(U) = \emptyset,$$

then $W \subset i(U)$ or $W \subset j(U)$.

Proof. The "only if" direction is immediate from Definition 6.2.14, setting $r = 2$. For the "if" direction, let us assume that W is separable and non-empty: we need to find a pair of smooth embeddings $i, j\colon U \hookrightarrow U$ with disjoint images, such that $W \subset i(U) \cup j(U)$ and $i(U) \cap W \neq \emptyset \neq j(U) \cap W$. Since W is separable, we have a finite collection $i_1, \ldots, i_r\colon U \hookrightarrow U$ of smooth embeddings ($r \geqslant 2$) with pairwise disjoint images, such that

$$W \subset \bigcup_{s=1}^{r} i_s(U) \qquad \text{and} \qquad i_s(U) \cap W \neq \emptyset \text{ for all } s \in \{1, \ldots, r\}.$$

Without loss of generality we may assume that each i_s is orientation-preserving. Moreover, each $i_s(U) \cap W = i_s(i_s^{-1}(W))$ is a closed subset of W, and hence compact, so $i_s^{-1}(W)$ is also compact. Thus we may also assume, without loss of generality, that $i_s^{-1}(W)$ is contained in the closed unit disk $\mathbb{D}^m \subset \mathbb{R}^m = U$. Write $j_s = i_s|_{\mathbb{D}^m}\colon \mathbb{D}^m \hookrightarrow U$ and choose another collection $k_1, \ldots, k_r\colon \mathbb{D}^m \hookrightarrow U$ of orientation-preserving embeddings with pairwise disjoint images, such that

$$k_1(\mathbb{D}^m) \subset \mathring{\mathbb{D}}^m(0, \ldots, 0) \qquad \text{and} \qquad k_2(\mathbb{D}^m), \ldots, k_r(\mathbb{D}^m) \subset \mathring{\mathbb{D}}^m(2, 0, \ldots, 0),$$

where $\mathring{\mathbb{D}}^m(x)$ is the open unit disk in $\mathbb{R}^m$ centred at x. Let $\hat{\imath}, \hat{\jmath}\colon U \hookrightarrow U$ be two embeddings with $\hat{\imath}(U) = \mathring{\mathbb{D}}^m(0, \ldots, 0)$ and $\hat{\jmath}(U) = \mathring{\mathbb{D}}^m(2, 0, \ldots, 0)$. By Proposition 6.2.4, we have $\Phi \in \mathrm{Diff}_c(U)_1$ such that $\Phi \circ j_s = k_s$ for all $s \in \{1, \ldots, r\}$. It follows that $\Phi(W)$ is contained in $\hat{\imath}(U) \cup \hat{\jmath}(U)$, but $\Phi(W) \not\subset \hat{\imath}(U)$ and $\Phi(W) \not\subset \hat{\jmath}(U)$. Now defining $i = \Phi^{-1} \circ \hat{\imath}$ and $j = \Phi^{-1} \circ \hat{\jmath}$, we obtain the required smooth embeddings $i, j\colon U \hookrightarrow U$. □

Tangential structures and E_n algebras.

Remark 6.2.28 (Tangential structures). If we fix a tangential structure Θ and consider the topological cobordism category $\mathcal{C}_\Theta^V$, we may ask whether the monoid $\mathcal{M}_\Theta^V = h\mathcal{C}_\Theta^V(\emptyset, \emptyset)$ is also free commutative when $\dim(V) > 0$. It turns out that it is, with a very similar proof to above: the result is that $\mathcal{M}_\Theta^V$ is *free commutative* on the subset $\mathcal{I}_\Theta^V$ of those Θ-manifolds whose underlying manifold (ignoring the Θ-structure) is inseparable.

Remark 6.2.29 (Dimension zero). When $\dim(V) = 0$ (and $d = 0$), the monoid $\mathcal{M}_\Theta^V$ is not commutative in general. Note that, in dimension zero, a tangential structure is just a topological space Θ, and a Θ-structure on a zero-dimensional manifold (i.e., discrete space) W is just a function $W \to \Theta$. One may then easily see that, when $d = 0$ and $\dim(V) = 0$ (so $U \cong \mathbb{R}$), the monoid $\mathcal{M}_\Theta^V$ is isomorphic to the *free non-commutative monoid* on $\pi_0(\Theta)$.[34] Since the only inseparable zero-manifolds are single points (equipped with any Θ-structure), we may identify the set $\mathcal{I}_\Theta^V$ of equivalence classes of inseparable Θ-manifolds with $\pi_0(\Theta)$, and say that $\mathcal{M}_\Theta^V$ is the free non-commutative monoid on $\mathcal{I}_\Theta^V$.

Remark 6.2.30 (E_m-algebras). The topological monoid $\mathcal{C}_\Theta^V(\emptyset, \emptyset)$ is homotopy equivalent to an E_m-algebra, where $m = \dim(V) + 1$. Defining $\mathcal{J}_\Theta^V \subset \mathcal{C}_\Theta^V(\emptyset, \emptyset)$ to be the

[34]If Θ is path-connected, this is also free commutative, but only "by accident".

subspace of *inseparable* submanifolds, the inclusion induces a map of E_m-algebras

$$F_{E_m}(\mathcal{J}_\Theta^V) \longrightarrow \mathcal{C}_\Theta^V(\emptyset, \emptyset), \tag{6.2.31}$$

where $F_{E_m}(X)$ denotes the free E_m-algebra on a space X. The result of this exercise (plus the previous two remarks) is that π_0(6.2.31) = (6.2.19) is an isomorphism (of commutative monoids if $m \geqslant 2$ and of non-commutative monoids if $m = 1$). One may naturally ask whether (6.2.31) induces isomorphisms also on higher homotopy groups. The answer is no in general.

As one counterexample, take $d = 1$, $m = 3$ and $\Theta = \{*\}$. Then the left-hand side of (6.2.31) is homotopy equivalent to the configuration space of finitely many unordered points in $\mathbb{R}^3$ each labeled by a point in the space of inseparable links in $\mathbb{R}^3$. One path-component X of this space consists of configurations of two points in $\mathbb{R}^3$ each labeled by a point in the space of unknots in $\mathbb{R}^3$. Using [19] to see that the space of unknots in $\mathbb{R}^3$ is homotopy equivalent to $\mathbb{RP}^2$, we see that $H_1(X) \cong (\mathbb{Z}/2\mathbb{Z})^2$. The corresponding[35] path-component Y of the right-hand side of (6.2.31) is the space of two-component unlinks in $\mathbb{R}^3$. Using [4, Theorem 1 and Proposition 3.7], we see that $H_1(Y) \cong (\mathbb{Z}/2\mathbb{Z})^3$. Hence (6.2.31) is not surjective on H_1, and therefore also not on π_1.

Another counterexample is given by $d = 1$, $m = 3$ and $\Theta = \{\pm 1\}$ the tangential structure of orientations. Then one path-component X of the left-hand side of (6.2.31) is homotopy equivalent to the space of configurations of two points in $\mathbb{R}^3$ each labeled by a point in the space of *oriented* unknots in $\mathbb{R}^3$. Since the space of oriented unknots in $\mathbb{R}^3$ is homotopy equivalent to S^2, we have $H_1(X) \cong \mathbb{Z}/2\mathbb{Z}$. The corresponding path-component Y of the right-hand side of (6.2.31) is the space of two-component *oriented* unlinks in $\mathbb{R}^3$. Using [4, Theorem 1 and Proposition 3.3], we see that $H_1(Y) \cong \mathbb{Z} \oplus \mathbb{Z}/2\mathbb{Z}$. So, again, (6.2.31) is not surjective on H_1, and therefore also not on π_1.

References

[1] J. C. Baez and J. Dolan, *Higher-dimensional algebra and topological quantum field theory*, J. Math. Phys. **36** (1995), no. 11, 6073–6105. ↑379

[2] E. Binz and H. R. Fischer, *The manifold of embeddings of a closed manifold*, Differential geometric methods in mathematical physics (Proc. Internat. Conf., Tech. Univ. Clausthal, Clausthal-Zellerfeld, 1978), Lecture Notes in Phys., vol. 139, Springer, Berlin-New York, 1981, pp. 310–329. With an appendix by P. Michor. MR613004 ↑358, 393

[3] M. Bökstedt and A. M. Svane, *A geometric interpretation of the homotopy groups of the cobordism category*, Algebr. Geom. Topol. **14** (2014), no. 3, 1649–1676, DOI 10.2140/agt.2014.14.1649. MR3212580 ↑386

[4] T. E. Brendle and A. Hatcher, *Configuration spaces of rings and wickets*, Comment. Math. Helv. **88** (2013), no. 1, 131–162, DOI 10.4171/CMH/280. MR3008915 ↑400

[5] J. Cerf, *Topologie de certains espaces de plongements* (French), Bull. Soc. Math. France **89** (1961), 227–380. MR140120 ↑393

[6] W. G. Dwyer and D. M. Kan, *Calculating simplicial localizations*, J. Pure Appl. Algebra **18** (1980), no. 1, 17–35. ↑376

[35] We have shown that (6.2.31) is a bijection on π_0, so "corresponding" makes sense.

[7] W. G. Dwyer and D. M. Kan, *Simplicial localizations of categories*, J. Pure Appl. Algebra **17** (1980), no. 3, 267–284. ↑376

[8] B. Fresse, *Homotopy of operads and Grothendieck-Teichmüller groups. Part 1*, Mathematical Surveys and Monographs, vol. 217, American Mathematical Society, Providence, RI, 2017. The algebraic theory and its topological background, DOI 10.1090/surv/217.1. MR3643404 ↑368

[9] S. Galatius, *Stable homology of automorphism groups of free groups*, Ann. of Math. (2) **173** (2011), no. 2, 705–768, DOI 10.4007/annals.2011.173.2.3. MR2784914 ↑363

[10] S. Galatius, *Lectures on the Madsen-Weiss theorem*, Moduli spaces of Riemann surfaces, IAS/Park City Math. Ser., vol. 20, Amer. Math. Soc., Providence, RI, 2013, pp. 139–167, DOI 10.1090/pcms/020/05. MR3114684 ↑363

[11] S. Galatius and O. Randal-Williams, *Monoids of moduli spaces of manifolds*, Geom. Topol. **14** (2010), no. 3, 1243–1302. ↑363

[12] S. Galatius and O. Randal-Williams, *Stable moduli spaces of high-dimensional manifolds*, Acta Math. **212** (2014), no. 2, 257–377, DOI 10.1007/s11511-014-0112-7. MR3207759 ↑378, 379

[13] S. Galatius and O. Randal-Williams, *Homological stability for moduli spaces of high dimensional manifolds. II*, Ann. of Math. (2) **186** (2017), no. 1, 127–204, DOI 10.4007/annals.2017.186.1.4. MR3665002 ↑375, 380

[14] S. Galatius and O. Randal-Williams, *Homological stability for moduli spaces of high dimensional manifolds. I*, J. Amer. Math. Soc. **31** (2018), no. 1, 215–264. ↑375, 380

[15] S. Galatius and G. Szűcs, *The equivariant cobordism category*, J. Topol. **14** (2021), no. 1, 215–257, DOI 10.1112/topo.12181. MR4201126 ↑365

[16] S. Galatius, U. Tillmann, I. Madsen, and M. Weiss, *The homotopy type of the cobordism category*, Acta Math. **202** (2009), no. 2, 195–239. ↑363, 372, 379, 380, 388

[17] A. Gramain, *Le type d'homotopie du groupe des difféomorphismes d'une surface compacte* (French), Ann. Sci. École Norm. Sup. (4) **6** (1973), 53–66. MR326773 ↑374

[18] J. L. Harer, *Stability of the homology of the mapping class groups of orientable surfaces*, Ann. of Math. (2) **121** (1985), no. 2, 215–249, DOI 10.2307/1971172. MR786348 ↑375

[19] A. E. Hatcher, *A proof of the Smale conjecture*, $\mathrm{Diff}(S^3) \simeq O(4)$, Ann. of Math. (2) **117** (1983), no. 3, 553–607, DOI 10.2307/2007035. MR701256 ↑400

[20] R. Juer and U. Tillmann, *Localisations of cobordism categories and invertible TFTs in dimension two*, Homology Homotopy Appl. **15** (2013), no. 2, 195–225, DOI 10.4310/HHA.2013.v15.n2.a13. MR3138377 ↑382

[21] M. A. Kervaire and J. W. Milnor, *Groups of homotopy spheres. I*, Ann. of Math. (2) **77** (1963), 504–537, DOI 10.2307/1970128. MR148075 ↑382

[22] A. Kriegl and P. W. Michor, *The convenient setting of global analysis*, Mathematical Surveys and Monographs, vol. 53, American Mathematical Society, Providence, RI, 1997, DOI 10.1090/surv/053. MR1471480 ↑393

[23] E. L. Lima, *On the local triviality of the restriction map for embeddings*, Comment. Math. Helv. **38** (1964), 163–164, DOI 10.1007/BF02566913. MR161343 ↑393

[24] J. Lurie, *Higher topos theory*, Annals of Mathematics Studies, vol. 170, Princeton University Press, Princeton, NJ, 2009, DOI 10.1515/9781400830558. MR2522659 ↑377

[25] J. Lurie, *On the classification of topological field theories*, Current developments in mathematics, 2008, Int. Press, Somerville, MA, 2009, pp. 129–280. MR2555928 ↑378, 379

[26] I. Madsen and M. Weiss, *The stable moduli space of Riemann surfaces: Mumford's conjecture*, Ann. of Math. (2) **165** (2007), no. 3, 843–941, DOI 10.4007/annals.2007.165.843. MR2335797 ↑380

[27] M. Markl, *Models for operads*, Comm. Algebra **24** (1996), no. 4, 1471–1500, DOI 10.1080/00927879608825647. MR1380606 ↑368

[28] P. W. Michor, *Manifolds of differentiable mappings*, Shiva Mathematics Series, vol. 3, Shiva Publishing Ltd., Nantwich, 1980. MR583436 ↑393

[29] H. K. Nguyen, *On the infinite loop space structure of the cobordism category*, Algebr. Geom. Topol. **17** (2017), no. 2, 1021–1040, DOI 10.2140/agt.2017.17.1021. MR3623680 ↑373

[30] R. S. Palais, *Extending diffeomorphisms*, Proc. Amer. Math. Soc. **11** (1960), 274–277, DOI 10.2307/2032968. MR117741 ↑393

[31] R. S. Palais, *Local triviality of the restriction map for embeddings*, Comment. Math. Helv. **34** (1960), 305–312, DOI 10.1007/BF02565942. MR123338 ↑393

[32] N. Perlmutter, *Homological stability for the moduli spaces of products of spheres*, Trans. Amer. Math. Soc. **368** (2016), no. 7, 5197–5228, DOI 10.1090/tran/6564. MR3456177 ↑375, 380

[33] N. Perlmutter, *Linking forms and stabilization of diffeomorphism groups of manifolds of dimension* $4n + 1$, J. Topol. **9** (2016), no. 2, 552–606, DOI 10.1112/jtopol/jtw004. MR3509973 ↑375, 380

[34] N. Perlmutter, *Homological stability for diffeomorphism groups of high-dimensional handlebodies*, Algebr. Geom. Topol. **18** (2018), no. 5, 2769–2820, DOI 10.2140/agt.2018.18.2769. MR3848399 ↑375, 380

[35] O. Randal-Williams, *Resolutions of moduli spaces and homological stability*, J. Eur. Math. Soc. (JEMS) **18** (2016), no. 1, 1–81, DOI 10.4171/JEMS/583. MR3438379 ↑375

[36] C. Rezk, *A model for the homotopy theory of homotopy theory*, Trans. Amer. Math. Soc. **353** (2001), no. 3, 973–1007, DOI 10.1090/S0002-9947-00-02653-2. MR1804411 ↑371

[37] C. Schommer-Pries, *Invertible Topological Field Theories* (2017), arXiv:1712.08029. ↑372, 378

[38] G. Segal, *Classifying spaces and spectral sequences*, Inst. Hautes Études Sci. Publ. Math. **34** (1968), 105–112. MR232393 ↑352

[39] U. Tillmann, *On the homotopy of the stable mapping class group*, Invent. Math. **130** (1997), no. 2, 257–275, DOI 10.1007/s002220050184. MR1474157 ↑380

[40] N. Wahl, *Homological stability for the mapping class groups of non-orientable surfaces*, Invent. Math. **171** (2008), no. 2, 389–424. ↑374

[41] N. Wahl, *Homological stability for mapping class groups of surfaces*, Handbook of moduli. Vol. III, Adv. Lect. Math. (ALM), vol. 26, Int. Press, Somerville, MA, 2013, pp. 547–583. MR3135444 ↑375

Department of Mathematics, University of Copenhagen, Denmark
Email address: galatius@math.ku.dk

IAS/Park City Mathematics Series
Volume 28, Pages 403–438
https://doi.org/10.1090/pcms/028/00880

Topological Quantum Field Theories, Knots and BPS states

Pavel Putrov

Abstract. The first two of the four lectures in this set of notes provide a brief introduction to the Witten-Reshetikhin-Turaev (WRT) invariant of 3-manifolds and the general notion of Topological Quantum Field Theory (TQFT). The third lecture is devoted to the asymptotic expansion conjecture, analytically continued Chern-Simons theory and the application of resurgence theory to the WRT invariant. The final, fourth lecture introduces so-called BPS invariants of 3-manifolds aimed to categorify the WRT invariant. The lectures are accompanied by sets of exercises and solutions.

Contents

2020 *Mathematics Subject Classification.* Primary 57M27; Secondary 57R56, 81T20.
Key words and phrases. Topological Quantum Field Theory; Witten-Reshetikhin-Turaev invariant.

1. Lecture I

1.1. Motivation Knots and links have an intimate relation with 3-manifolds realised via surgery procedures (to be described later in the lecture in more detail)

$$\text{Links} \quad \rightsquigarrow \quad \text{3-manifolds}$$

This leads to a close relation between invariants of 3-manifolds and links. The following table provides some examples of invariants of links and their 3-manifold counterparts. The question marks indicate the part that has yet to be mathematically defined for general 3-manifolds. This question will be addressed in more detail in Lectures III and IV.

Links	3-manifolds
Colored Jones polynomial	Witten-Reshetikhin-Turaev for $U_q(\mathfrak{sl}_2)$
Khovanov-Rozansky cohomology	??? (BPS states of 5-branes on $M^3 \times S^1 \times D^2$)
Alexander polynomial	Turaev torsion (=WRT for $U_q(\mathfrak{sl}_{1\|1})$)
Knot Floer homology	Heegaard-Floer homology

1.2. Dehn surgery and Kirby calculus Dehn surgery is a method to produce nontrivial 3-manifolds starting from a framed link in S^3 (equivalently $\mathbb{R}^3$):

$$\begin{array}{ccc} \dfrac{\{\text{Framed links in } S^3\}}{\sim \text{ambient isotopy}} & \longrightarrow & \dfrac{\{\text{Oriented closed 3-manifolds}\}}{\sim \text{homeomorphisms}}, \\ \mathcal{L} & \longmapsto & S^3(\mathcal{L}). \end{array}$$

Let $\mathcal{L} = \sqcup_{i=1}^{L} \mathcal{L}_i$ be an L-component oriented framed link in S^3 (the result will actually be independent of the choice of orientation). Its *Dehn surgery on a framed link* $\mathcal{L} \subset S^3$ is a new oriented closed 3-manifold $S^3(\mathcal{L})$ obtained by the following procedure.

- Cut out a tubular neighbourhood $\mathcal{N}(\mathcal{L})$ of $\mathcal{L}$ in S^3. The resulting 3-manifold has a boundary homeomorphic to a disjoint union of L copies of T^2 with distinguished pairs of homology classes of closed curves, *meridians* and *longitudes*, determined by the orientation and framing. The meridians m_i ($i = 1 \ldots L$) are small cycles going around the link components in the clockwise direction respectively to the link orientation. The longitudes ℓ_i are "push-offs" of the link components to the boundary in the normal direction specified by framing.

- Glue back in L copies of the solid torus $D^2 \times S^1$, with standard longitudes and meridians $m_i' = \partial D^2$ and $\ell_i' = S^1$, by identifying points on the boundary using any homeomorphism $\phi : \sqcup_{i=1}^N T^2 \to \sqcup_{i=1}^N T^2$ such that $\phi_*([m_i]) = -[\ell_i']$ and $\phi_*([\ell_i]) = [m_i']$ (i.e. meridians and longitudes are swapped). This produces a new closed manifold $S^3(\mathcal{L}) := S^3 \setminus \mathcal{N}(\mathcal{L}) \cup_\phi \sqcup_{i=1}^L D^2 \times S^1$ unique up to a homeomorphism.

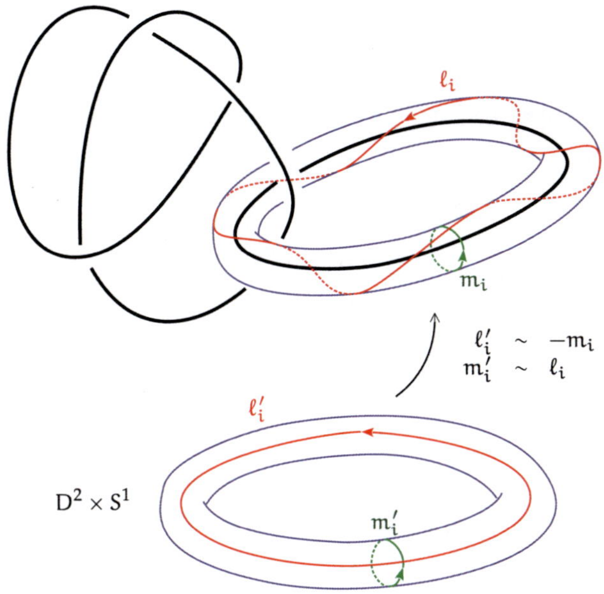

Remark 1.2.1. Framings in the pictures below are specified by integers, the linking numbers $\ell k(\mathcal{L}_i, \mathcal{L}_i')$ between the link components $\mathcal{L}_i$ and their push-offs $\mathcal{L}_i'$ toward the framing vectors.

Example 1.2.2.

S^3 (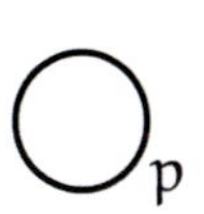) $\cong L(p,1) = \{|z_1|^2 + |z_2|^2 = 1\}/(z_1, z_2) \sim (z_1 e^{\frac{2\pi i}{p}}, z_2 e^{-\frac{2\pi i}{p}})$.

(cf. Exercises.)

Example 1.2.3.

S^3 (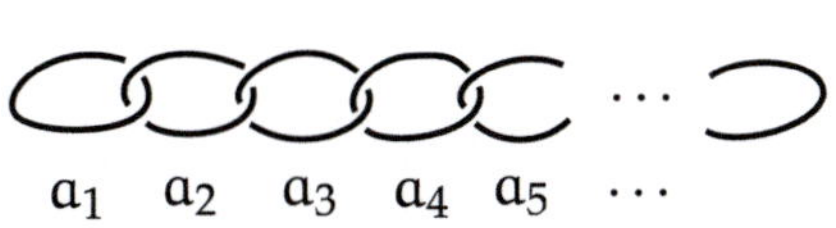) $\cong$

$$L(p,q) = \{|z_1|^2 + |z_2|^2 = 1\}/(z_1, z_2) \sim (z_1 e^{\frac{2\pi i}{p}}, z_2 e^{-\frac{2\pi i q}{p}}),$$

where p, q are coprime and $p/q = a_1 - 1/(a_2 - 1/(a_3 - \cdots) \cdots)$.

Example 1.2.4.

$$S^3\Bigg(\;$$ 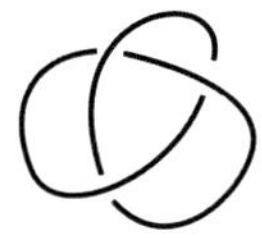 $$\;\Bigg)_1 \cong S^3/\Gamma,$$

where Γ is the binary icosohedral group (a subgroup of $SU(2) \subset SO(4)$ corresponding to E_8 via ADE classification). This is also known as a Poincaré homology sphere.

The following two theorems establish precise relation between links in a 3-sphere and closed oriented 3-manifolds.

Theorem 1.2.5 (Lickorish [18], Wallace [26]). *Any closed oriented 3-manifold can be obtained by Dehn surgery on a framed link in* S^3.

Theorem 1.2.6 (Kirby [13], Fenn-Rourke [8]). $S^3(\mathcal{L}) \cong S^3(\mathcal{L}')$ *iff* $\mathcal{L}$ *and* $\mathcal{L}'$ *can be related by a sequence of the Kirby-Fenn-Rourke moves shown below.*

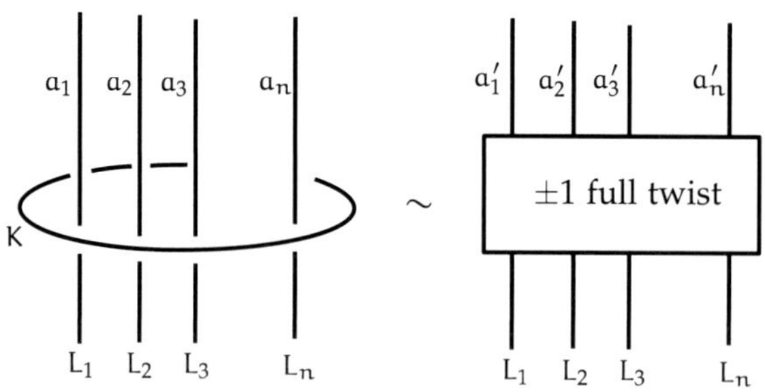

$$a_i' = a_i \mp \ell k(L_i, K)^2$$

Therefore invariants of framed links, which are also invariant under Kirby-Fenn-Rourke moves, provide invariants of closed oriented 3-manifolds.

1.3. Witten-Reshetikhin-Turaev invariant and Chern-Simons TQFT Suppose that $k \in \mathbb{Z}_+$, $\mathcal{L}$ is a framed oriented link with L components, and

$$J_{n_1,n_2,\ldots,n_L}(\mathcal{L}) \in \mathbb{Z}[q^{1/2}, q^{-1/2}]$$

is its colored Jones polynomial, normalized so that

$$J_n\Big(\;$$  $$_0\;\Big) = [n] := (q^{n/2} - q^{-n/2})/(q^{1/2} - q^{-1/2}).$$

Let M be the linking matrix ($L \times L$), i.e.

$$M_{ij} := \ell k(\mathcal{L}_i, \mathcal{L}_j), \quad i \neq j$$
$$M_{ii} := \ell k(\mathcal{L}_i, \mathcal{L}_i')$$

where $\mathcal{L}_i'$ is a push-off of $\mathcal{L}_i$ towards the framing. If b_+, b_- and b_1 are the number of positive, negative and zero eigenvalues of M then $b_+ + b_- + b_1 = L$.

Definition 1.3.1. The Witten-Reshetikhin-Turaev invariant of a Dehn surgery on a framed link $\mathcal{L}$ in S^3 is a function $\mathbb{Z}_+ \to \mathbb{C}$ defined by the following formula:

$$\begin{aligned}\tau_k(S^3(\mathcal{L})) = \Bigg(&\left(\sum_{n_i=1}^{k-1}\prod_{i=1}^{L}[n_i]J_{n_1\cdots n_L}(\mathcal{L})\right)\Bigg/ \\ &\left(\sum_{n_i=1}^{k-1}\prod_{i=1}^{b_++b_-}[n_i]J_{n_1\cdots n_{b_++b_-}}\Big(\underbrace{\bigcirc_{+1}\bigcirc_{+1}\cdots\bigcirc_{+1}}_{b_+}\underbrace{\bigcirc_{-1}\cdots\bigcirc_{-1}}_{b_-}\Big)\cdot\right. \\ &\left.\left(\sum_{n=1}^{k-1}[n]^2)^{b_1/2}\right)\right)\Bigg)\end{aligned} \tag{1.3.2}$$

where the right hand side is evaluated at $q = e^{\frac{2\pi i}{k}}$.

Theorem 1.3.3 (Reshetikhin-Turaev [22], Kirby-Melvin [14]). *$\tau_k(S^3(\mathcal{L}))$ is invariant under Kirby-Fenn-Rourke moves on $\mathcal{L}$ and thus depends only on homeomorphism class of the 3-manifold $S^3(\mathcal{L})$.*

Remark 1.3.4. The normalization in the formula above is such that

$$\tau_k(S^3) = 1$$

and the invariant behaves nicely under connected sum operation:

$$\tau_k(M\#N) = \tau_k(M)\tau_k(N).$$

The other common choice of normalization is

$$Z_{SU(2)_k}(S^3) := \tau_k(S^2)\sqrt{\frac{2}{k}}\sin\frac{\pi}{k}$$

so that

$$Z_{SU(2)_k}(S^2 \times S^1) = 1$$

(cf. Exercises). The latter is more natural from the TQFT point of view that will be discussed later.

The definition above was inspired by a more physical approach by Witten [27] where the WRT invariant of a 3-manifold M^3 is given by the following functional integral in a QFT with CS action (which is metric independent and thus the QFT is topological):

$$\begin{aligned} Z_{SU(2)_k}(M^3) &= \int_{\mathcal{A}_{SU(2)}(M^3)} DA\exp\{2\pi i(k-2)CS[A]\}, \\ CS[A] &= \frac{1}{8\pi^2}\int_{M^3}\mathrm{Tr}\left(AdA + \frac{2}{3}A^3\right) \qquad \in \mathbb{C} \mod \mathbb{Z}, \end{aligned} \tag{1.3.5}$$

where $A \in \Omega^1(M^3, \mathfrak{su}_2)$ is a connection 1-form of a $SU(2)$ principal bundle on M^3 (any $SU(2)$ bundle over a 3-manifold is trivial). The functional integral is performed over $\mathcal{A}_{SU(2)}(M^3)$, the space of all connections modulo gauge transformations $A \to gAg^{-1} + gdg^{-1}$, $g : M^3 \to SU(2)$.

In the same setting, the colored Jones polynomial of a knot $\mathcal{K}$ in S^3 is given by an expectation of a Wilson line supported on the knot in the representation corresponding to the color:

$$J_n(\mathcal{K}) = \int_{\mathcal{A}_{SU(2)}(S^3)} DA \exp\{2\pi i(k-2)CS[A]\} \operatorname{Tr}_{\text{irrep of dim } n} \operatorname{Hol}_{\mathcal{K}}(A).$$

Such formulas, however, are mathematically ill defined, because it is not clear how to rigorously define the measure DA on the infinite dimensional space $\mathcal{A}_{SU(2)}(M^3)$. One, however, can try to define this integral indirectly via Resurgence Theory (see later). The basic idea is that in practice one rarely calculates integrals directly using the definition of the measure, instead one usually deforms the contour in the complex plane and expresses them via some basic integrals for which the value is known.

1.3.1. Functorial definition of a TQFT

Definition 1.3.6 (Atiyah [2]). An n-dimensional TQFT Z is a symmetric monoidal functor

$$Z : \mathrm{Bord}_{n,n-1} \longrightarrow \mathrm{Vect}_{\mathbb{C}}.$$

The right-hand side is the category of complex (finite dimensional) vector spaces, considered as a symmetric monoidal category with the usual tensor product over $\mathbb{C}$ and the unit object being $\mathbb{C}$, and with the canonical isomorphisms $V_1 \otimes V_2 \cong V_2 \otimes V_1$. The lefthand side is the category of n-dimensional (smooth) bordisms. The objects are closed oriented $(n-1)$-dimensional manifolds. The morphisms between M_1^{n-1} and M_2^{n-1} are n-dimensional manifolds M^n with a fixed homeomorphism[1] $\partial M^n \cong \bar{M}_1^{n-1} \sqcup M_2^{n-1}$ (where bar denotes orientation reversal), modulo homeomorphisms identical on the boundaries. The composition of the morphisms is given naturally by the gluing along common boundary with the use of boundary homeomorphism. The tensor product is the disjoint union, the unit object is an empty manifold and the symmetric structure is given by the canonical isomorphisms $M_1^{n-1} \sqcup M_2^{n-1} \cong M_1^{n-1} \sqcup M_2^{n-1}$.

Remark 1.3.7. Instead of $\mathrm{Vect}_{\mathbb{C}}$ one can consider $\mathrm{Vect}_{\mathbf{k}}$ for any field $\mathbf{k}$, or, in fact, any other symmetric monoidal category.

Remark 1.3.8. Any closed n-manifold M^n can be considered as a bordism from $\emptyset$ to $\emptyset$ and thus $Z : \mathbb{C} \to \mathbb{C}$ provides a topological invariant of M^n valued in complex numbers.

2. Lecture II

2.1. 2d TQFTs 2d TQFTs are in one-to-one correspondence with Frobenius algebras over $\mathbb{C}$ (for a detailed exposition see [15]). The correspondence is realized

[1]Here we are considering the category of *topological* bordisms. Up to dimension 3 it is equivalent to the category of *smooth* bordisms.

as follows: $Z(S^1) = V$, the algebra, as a vector space and the Frobenius algebra structure is given in term of the TQFT functor as follows:

- Multiplication: $\mu : V \otimes V \to V$

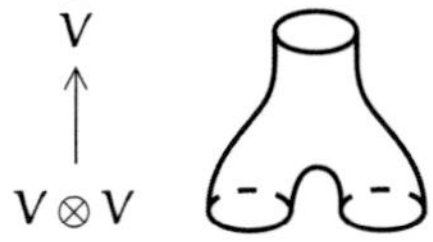

- Unit: $\mathbb{C} \to V, 1 \mapsto \mathbf{1}$

- Counit (Trace): $\mathrm{Tr} : V \to \mathbb{C}$

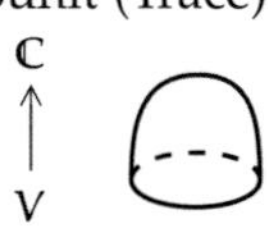

which are subject to the following conditions:

- Associativity

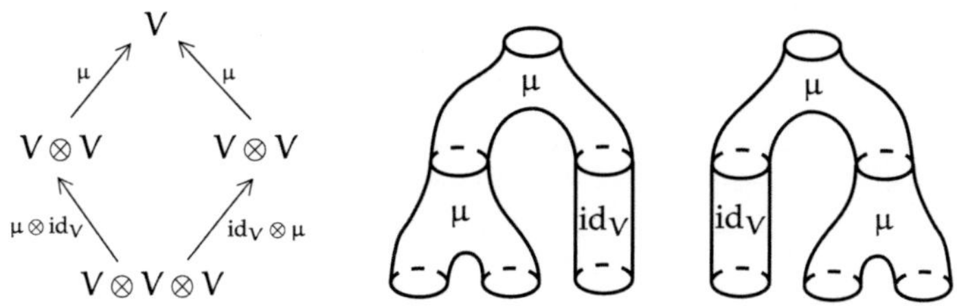

- Non-degeneracy of the Fronenis bilinear form $\mathrm{Tr} \circ \mu : V \otimes V \to \mathbb{C}$

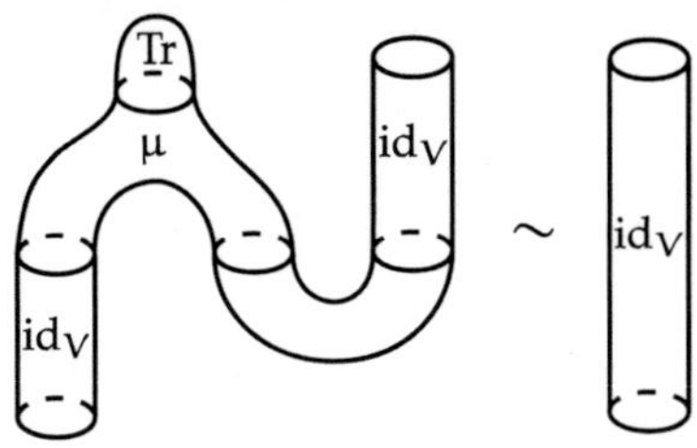

2.2. 3d TQFTs Given any Modular Tensor Category (MTC) $\mathcal{C}$ one can construct a 3d TQFT $Z_{\mathcal{C}}$ [22, 25]. The Witten-Reshetikhin-Turaev invariant considered earlier (in its "physical" normalization, i.e. $Z_{SU(2)_k}(M^3)$) of a closed 3-manifold is a value of a 3d TQFT $Z_{SU(2)_k} \equiv Z_{\mathcal{C}}$, where $\mathcal{C}$ is the category of finite-dimensional representations of $U_q(\mathfrak{sl}_2)$ at $q = e^{2\pi i/k}$ (possessing a weight decomposition), which is also equivalent to the category of the integrable representations of the affine $(\widehat{\mathfrak{sl}_2})_k$ at positive integral level k. In both such categories the set of simple objects can be identified with representations of the ordinary Lie algebra $\mathfrak{sl}_2$ of dimensions $1, 2, \ldots (k-1)$. The tensor product structure is however modified.

Without going too much into details, an MTC is a semisimple ribbon category with finite number of simple objects (up to isomorphisms) and a non-degenerate S-matrix. The ribbon structure on an abelian (linear over $\mathbb{C}$) monoidal category is, roughly, the most general structure that provides quantum invariants of framed oriented links via their braid representations [23], with link components colored by objects of $\mathcal{C}$ (for an MTC, by semi-simplicity, any object can be decomposed into a direct sum of finite number of simple objects $V_i, i \in I$). The S-matrix is the corresponding invariant of the Hopf link colored by the simple objects.

More concretely, the ribbon structure can be encoded in the following diagrammatic rules (the blackboard framing is assumed below)

- $A \in \mathrm{Hom}(V_1 \otimes \ldots V_n, W_1 \otimes \ldots W_m)$

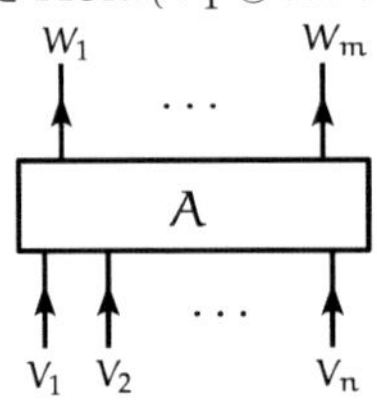

- $\mathrm{id}_V : V \to V$

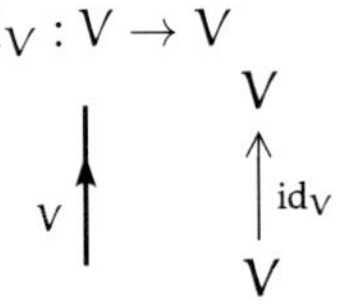

- V vs V^*:

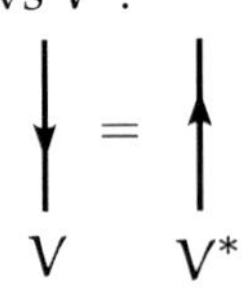

- $V \otimes V^* \to \mathbf{1}$ (the unit object w.r.t. monoidal structure)

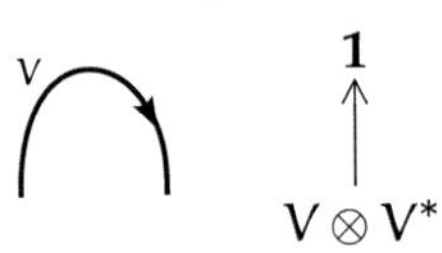

- $\mathbf{1} \to V \otimes V^*$

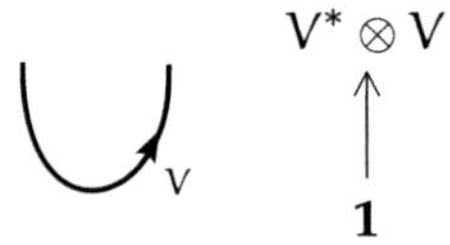

- The R-matrix $R_{V,W} : V \otimes W \to W \otimes V$

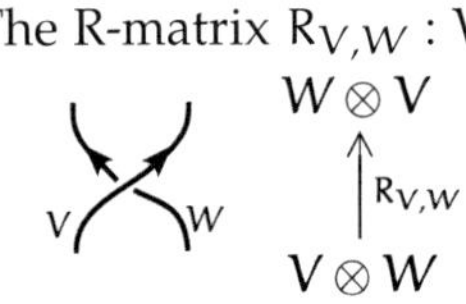

which satisfies the Yang-Baxter equation

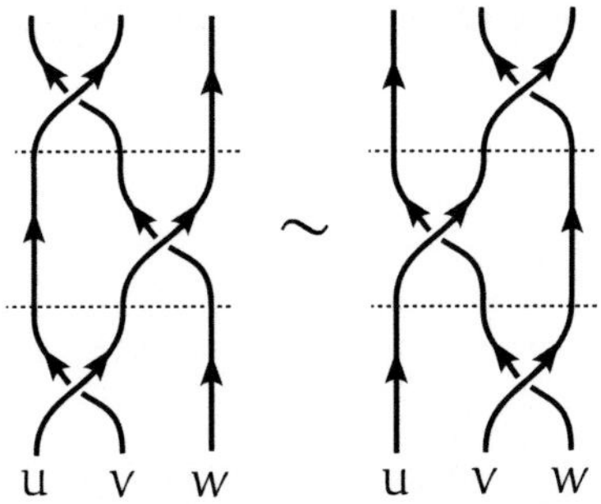

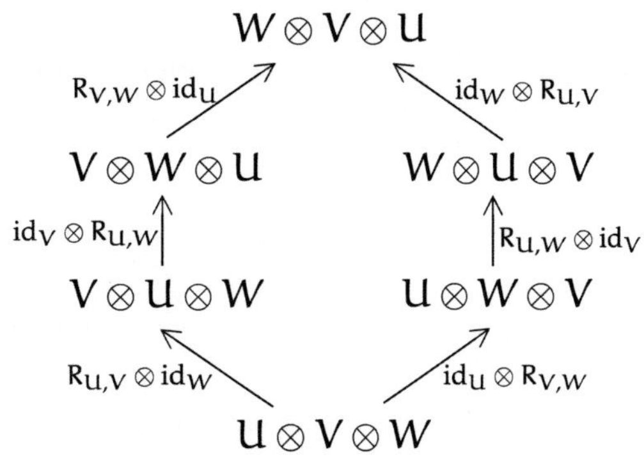

- Dependence on framing

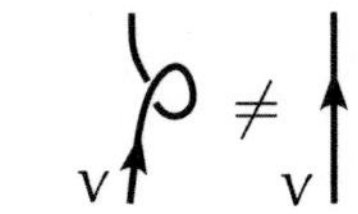

For an MTC we also have:

- Unnormalized S-matrix $\tilde{S}_{ij} \in \mathbb{C}$, required to be non-degenerate.

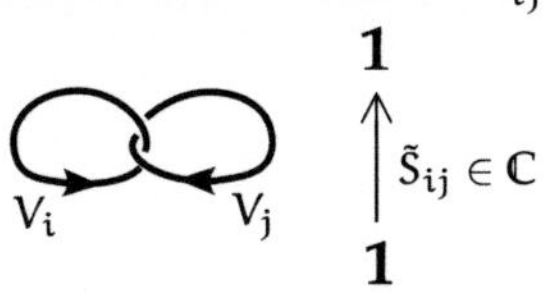

(Note that from Schur's lemma it follows that $\mathrm{Hom}(V_i, V_i) \cong \mathbb{C}$ while $\mathrm{Hom}(V_i, V_j) = 0$, for $i \neq j$.)

- Quantum dimension $d_i := \tilde{S}_{0i}$ (where $V_0 \equiv \mathbf{1}$).

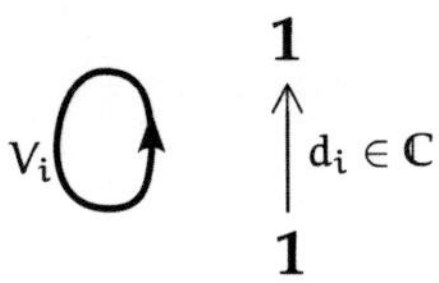

- T-matrix $T_{ij} := \delta_{ij} T_i$.

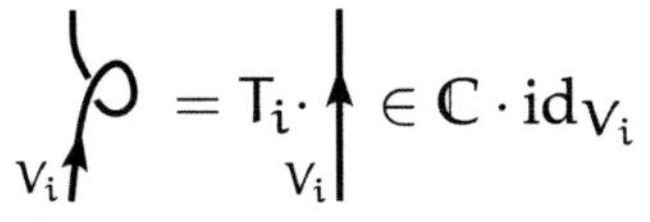

It is also useful to introduce the quantities

$$D := \sqrt{\sum_{i \in I} d_i^2}, \qquad \text{and} \qquad p_{\pm} := \sum_{i \in I} d_i^2 T_i^{\pm 1}$$

as well as the normalized S-matrix

$$S_{ij} := \tilde{S}_{ij}/D.$$

The corresponding 3d TQFT $Z_{\mathcal{C}}$ is then constructed as follows. Let Σ_g be a closed oriented Riemann surface of genus g. Then

$$Z_{\mathcal{C}}(\Sigma_g) = \mathrm{Hom}(\mathbf{1}, (\oplus_{i \in I} V_i \otimes V_i^*)^{\otimes g})$$

The Riemann surface should be understood as the boundary of a handlebody obtained by thickening of a "coupon" (a rectangle with framed links ending on one of the sides with framing vectors parallel to the side):

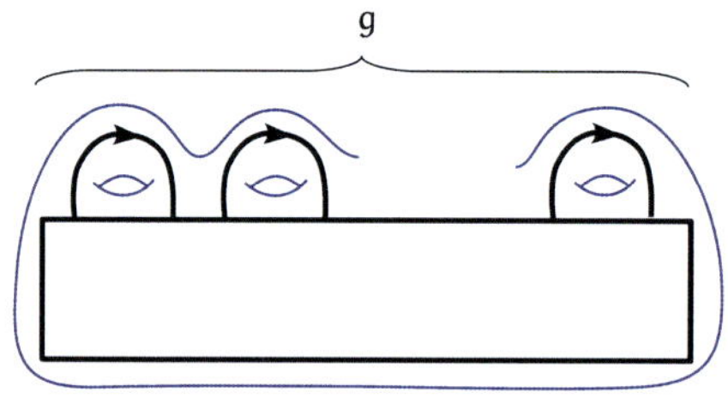

In particular $Z_{\mathcal{C}}(S^2) = \mathrm{Hom}(\mathbf{1},\mathbf{1}) \cong \mathbb{C}$ and

$$Z_{\mathcal{C}}(T^2) = \mathrm{Hom}(\mathbf{1}, \oplus_{i\in I} V_i \otimes V_i^*) \cong \oplus_{i\in I}\mathrm{Hom}(V_i, V_i) \cong \oplus_{i\in I}\mathbb{C}$$

where we used again Schur's lemma.

Let M^3 be a bordism between Σ_{g_1} and Σ_{g_2} a pair of closed oriented Riemann surfaces of genera g_1 and g_2 respectively. Then the homomorphism

$$Z_{\mathcal{C}}(M^3): \ \mathrm{Hom}(\mathbf{1}, (\oplus_{i\in I} V_i \otimes V_i^*)^{\otimes g_1}) \longrightarrow \mathrm{Hom}(\mathbf{1}, (\oplus_{i\in I} V_i \otimes V_i^*)^{\otimes g_2})$$

is determined by the values

$$\langle B, Z_{\mathcal{C}}(M^3)(A)\rangle := Z_{\mathcal{C}}(\tilde{M}^3_{A,B}) \qquad \in \mathbb{C} \tag{2.2.1}$$

where

$$A \in \mathrm{Hom}(\mathbf{1}, (\oplus_{i\in I} V_i \otimes V_i^*)^{\otimes g_2}),$$

$$B \in \mathrm{Hom}(\mathbf{1}, (\oplus_{i\in I} V_i \otimes V_i^*)^{\otimes g_1})^* \cong \mathrm{Hom}((\oplus_{i\in I} V_i \otimes V_i^*)^{\otimes g_1}, \mathbf{1}),$$

and $\tilde{M}^3_{A,B}$ is a certain closed 3-manifold with two coupons labeled by A and B embedded. It is obtained by gluing M^3 with genus g_1 and g_2 handlebodies obtained by thickening of the coupons A and B, as depicted below:

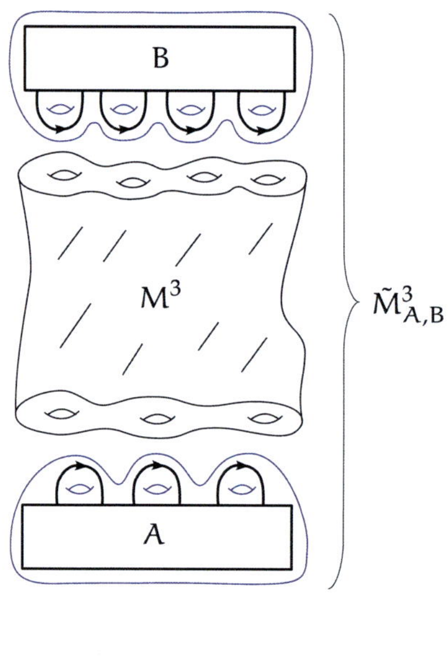

As in the case without coupons, such manifolds can be obtained by a surgery on a framed oriented link. In terms of such surgery the value of (2.2.1) is given by the generalization of the formula (1.3.2) for the WRT invariant:

$$Z_{\mathcal{C}}(\tilde{M}^3_{A,B}) := D^{-1} \sum_{i_\alpha \in I} \prod_{\substack{\alpha \text{ in the set of} \\ \text{link components} \\ \text{and arcs} \\ \text{ending on coupons}}} \frac{d_{i_\alpha}}{D}$$

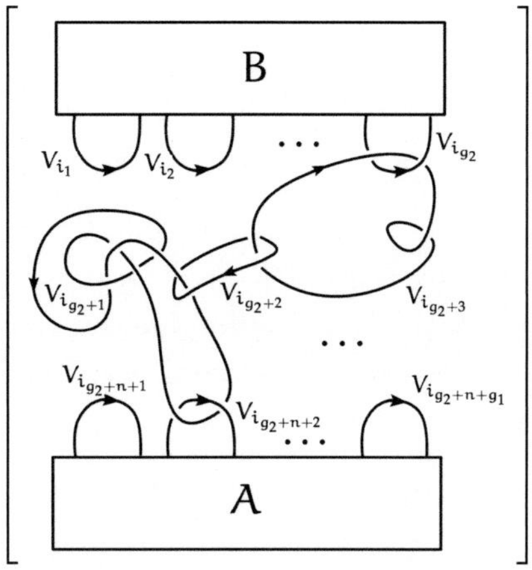

where the element in the brackets $[\cdots]$ is, according to the diagrammatic rules, understood as an element of $\mathrm{Hom}(\mathbf{1},\mathbf{1}) \cong \mathbb{C}$. In the case of WRT invariant of a closed 3-manifold, it is the value of the Jones polynomial of the link colored by the $\mathfrak{sl}_2$ representations, which are in one to one correspondence with simple objects of $\mathcal{C}$).

Remark 2.2.2. It is easy to see that the formula above is consistent with gluing of bordisms up to an overall phase which is a power of the ratio (p_+/p_-). This phase corresponds to a "framing anomaly" which can be fixed by fixing framings on 3-manifolds. I.e. one should actually consider $Z_{\mathcal{C}}$ as functor from the category of framed bordisms.

Remark 2.2.3. An MTC $\mathcal{C}$ actually provides a 3-2-1 extended TQFT $Z_{\mathcal{C}}$, that is a symmetric monoidal functor between 2-categories

$$\mathrm{Bord}_{3,2,1} \longrightarrow 2\mathrm{Vect}_{\mathbb{C}}$$

here $2\mathrm{Vect}_{\mathbb{C}}$ is a 2-cateogry of linear categories over $\mathbb{C}$ whose objects are linear categories, 1-morphisms are functors, and 2-morphisms are natural transformations and $\mathrm{Bord}_{3,2,1}$ is the 2-category of 3-dimensional bordisms whose objects are 1-manifolds, 1-morphisms are surfaces with boundaries and 2-morphisms are bordism between them (bordism "with corners"). In particular $Z_{\mathcal{C}}(S^1) = \mathcal{C}$ and $Z_{\mathcal{C}}(\emptyset) = \mathrm{Vect}_{\mathbb{C}}$. This is essentially realized by a construction similar to the one above, but with extra strands with fixed colors (which are not summed over).

3. Lecture III

3.1. Asymptotic expansion conjecture It is natural to ask if there is some prediction that one can make for the behavior of $Z_{SU(2)_k}(M^3)$ as a function of k from the formal path-integral expression (1.3.5)? If it were a finite dimensional integral of the form $\int dz e^{ik f(z)}$ one would expect its asymptotic behaviour as $k \to \infty$ to be determined by the critical points of $f(z)$.

The critical points of the Chern-Simons functional

$$CS: \mathcal{A}_{SU(2)}(M^3) \to \mathbb{C}/\mathbb{Z} \cong \mathbb{C}^*$$

are determined by the equation

$$\frac{\delta CS[A]}{\delta A} = 0 \Leftrightarrow F_A \equiv dA + \frac{1}{2}[A, A] = 0$$

where F_A is the curvature 2-form. That is critical points are in one-to-one correspondence with $SU(2)$ flat connections on M^3 modulo gauge transformations. Even though $\mathcal{A}_{SU(2)}$ is infinite dimensional, the subspace of critical points is finite dimensional (in general not connected) and can be explicitly described as follows

Definition 3.1.1. The subspace of critical points in $\mathcal{A}_{SU(2)}(M^3)$ is given by

$$\mathcal{M}^{\text{flat}}_{SU(2)}(M_3) := \{F_A = 0\}/\sim \text{gauge} \cong \text{Hom}(\pi_1(M^3), SU(2))/SU(2)$$

where the quotient in the right hand side is with respect to conjugation action by $SU(2)$. A homomorphism $\pi_1(M_1, x_0) \to SU(2)$ corresponding to a flat connection A is given by $[\gamma] \mapsto \text{Hol}_\gamma(A)$. A gauge transformation $A \to gAg^{-1} + gdg^{-1}$ changes $\text{Hol}_\gamma(A) \to g(x_0)\text{Hol}_\gamma(A)g(x_0)^{-1}$ which is in the same class of the quotient.

It is useful to distinguish flat connections by their stabilizer of $SU(2)$ action:

- central: $\{\pm 1\} \in SU(2)$
- abelian: $U(1) \subset SU(2)$
- irreducible: $SU(2)$

However we often abuse terminology and will refer to both central and abelian flat connections as just abelian flat connections.

Conjecture 3.1.2 (Asymptotic Expansion Conjecture, Witten [27], Andersen [1]). *The asymptotics of the WRT invariant of a closed manifold M^3 has the following form:*

$$Z_{SU(2)_k}(M^3) \overset{k\to\infty}{\sim} \sum_{\alpha \in \pi_0(\mathcal{M}^{flat}_{SU(2)}(M_3))} e^{2\pi i k CS[A_\alpha]} \frac{1}{k^{d_\alpha + 1}} \left(a_0^{(\alpha)} + \frac{a_1^{(\alpha)}}{k} + \frac{a_2^{(\alpha)}}{k^2} + \dots \right)$$

where A_α is any connection 1-form representing connected component α.

The coefficients $a_i^{(\alpha)}$ are known as *perturbative invariants* of 3-manifolds and can be expressed in terms of integrals over configurations of points in M^3 labeled by trivalent graphs (i.e. "Feynmann graphs"). This approach has been mathemtically defined under certain assumptions on the flat connection A_α (e.g. trivial in a rational homology sphere [5, 6, 17] or acyclic [3, 4]). In practice, one can compute $a_i^{(\text{triv})}$ in terms of Dehn surgery description of M_3 [19, 20, 24] and $a_i^{(\beta)}$, for β irreducible, in terms of triangulation [7]. The perturbative invariants can be understood as 3-manifold counterparts of Vasiliev finite-type invariants of links.

In what follows we will see that, motivated by the analysis of finite dimensional integral of the same form, one can propose a stronger version of this conjecture.

3.2. Analytically continued Chern-Simons For Khovanov-Rozansky homology it is important that Jones polynomial is a Laurent polynomial in $e^{2\pi i/k}$, and in particular it is defined for any complex q (not just a root of unity), or, equivalently, any complex non-zero k such that $q = e^{2\pi i/k}$. It is natural to ask if it also possible to extend the definition of the WRT invariant (originally defined for $k \in \mathbb{Z}_+$) to $k \in \mathbb{C}^*$.

It is actually possible to give a proposal of such analytic continuation from the path-integral approach [16, 28], which we summarize below. Let us continue to treat path integrals formally (as if they were finite dimensional integrals) and then we will arrive at a mathematically meaningful formula that can actually be used define them. This is possible because what will be actually important for us is the space of integration and the integrand (which are both well defined), not the measure of integration. In order to make sense of non-integral k one needs to also modify the space of integration in the original path integral:

$$Z^{\mathrm{CS}}_{\Gamma,k}(M^3) := \int_{\Gamma\subset\tilde{\mathcal{A}}_{SL(2,\mathbb{C})}(M^3)} DA\, e^{2\pi i k\, \mathrm{CS}[A]}$$

where

$$\tilde{\mathcal{A}}_{SL(2,\mathbb{C})}(M^3) := \frac{\{SL(2,\mathbb{C}) \text{ connection 1-forms } A\}}{\text{Based gauge transformations } (g(x_0)=1) \text{ connected to identity}}$$

Henceforth, we use the symbol $\sim_{\mathrm{bgc1}}$ to denote the quotient above. The fact that we divide only by based gauge transformations makes $\tilde{\mathcal{A}}_{SL(2,\mathbb{C})}(M^3)$ smooth (since they act freely). The fact that we divide only by gauge transformations connected to identity makes $\tilde{\mathcal{A}}_{SL(2,\mathbb{C})}(M^3)$ simply connected (otherwise it would have a fundamental group $\mathbb{Z}$ whose elements correspond to large gauge transformations given by homotopically non-trivial maps $g : M^3 \to SU(2) \cong S^3$). In particular this makes the functional $\mathrm{CS}[\cdot]$ to be valued in $\mathbb{C}$ instead of $\mathbb{C} \mod \mathbb{Z} \cong \mathbb{C}^*$.

The contour Γ is a mid-dimensional (oriented) subspace of $\tilde{\mathcal{A}}_{SL(2,\mathbb{C})}(M^3)$ with boundary lying in the subset of the boundary of $\tilde{\mathcal{A}}_{SL(2,\mathbb{C})}(M^3)$ where $\mathrm{Re}\, ik\mathrm{CS}[A]$ is infinitely small. Of course, the value of the integral should only depend on the class of Γ in the corresponding relative homology, i.e.

$$[\Gamma] \in H_{\frac{1}{2}\dim}(\tilde{\mathcal{A}}_{SL(2,\mathbb{C})}(M^3), \partial\tilde{\mathcal{A}}_{SL(2,\mathbb{C})}(M^3)|_{\mathrm{Re}\, ik\mathrm{CS}\ll 0}; \mathbb{Z}). \tag{3.2.1}$$

In order for $Z^{\mathrm{CS}}_{\Gamma,k}$ to be an analytic continuation of the WRT invariant the contour Γ should be chosen such that

$$Z^{\mathrm{CS}}_{\Gamma,k}(M^3)|_{k\in\mathbb{Z}_+} = Z_{SU(2)_k}(M^3) \tag{3.2.2}$$

This does not uniquely determine contour Γ (see more comments on this later).

The space of critical points of CS[A] on $\tilde{\mathcal{A}}_{SL(2,\mathbb{C})}(M^3)$ is given by

$$\tilde{\mathcal{M}}^{\mathrm{flat}}_{SL(2,\mathbb{C})}(M_3) := \left(\{F_A = 0\}/\sim_{\mathrm{bgc1}}\right) \cong \mathrm{Hom}(\pi_1(M^3), SL(2,\mathbb{C})) \times \mathbb{Z}.$$

There are two differences from the usual space of $SL(2,\mathbb{C})$ flat connections: the absence of the quotient w.r.t. the conjugate action, which is because we quotient over based gauge transformations in the ambient space, and an extra factor $\mathbb{Z}$, because we quotient only over gauge transformations connected to 1. The critical points which project to the same component in $\mathrm{Hom}(\pi_1(M^3), SL(2,\mathbb{C}))$ have values of the CS functional that differ by an integer.

Let $\{\mathcal{M}^{\mathbb{C}}_\alpha\}_\alpha$ be the connected components of $\tilde{\mathcal{M}}^{\text{flat}}_{SU(2)}(M_3)$, labelled by

$$\alpha \in \pi_0(\mathrm{Hom}(\pi_1(M^3), SL(2,\mathbb{C}))) \times \mathbb{Z}.$$

There is a particular choice of basis in (3.2.1) represented by *Lefschetz thimbles* Γ_α defined as the union of steepest descent paths w.r.t. $\mathrm{Re}\,(ik\mathrm{CS}:)\,\tilde{\mathcal{A}}_{SU(2)}(M^3) \to \mathbb{R}$ map from a half-dimensional compact subset $\mathcal{M}^{\mathbb{R}}_\alpha \subset \mathcal{M}^{\mathbb{C}}_\alpha$ (the homology class of Γ_α only depends on the homology class of $[\mathcal{M}^{\mathbb{R}}_\alpha]$ in $H_{\frac{1}{d}\dim}(\mathcal{M}^{\mathbb{C}}_\alpha, \mathbb{Z})$, which in our case is unique because the components of the critical subspace are of the form $\mathcal{M}^{\mathbb{C}}_\alpha \cong T^*\mathcal{M}^{\mathbb{R}}_\alpha$). For a generic value of k the flow trajectories do not encounter other critical points and go to the boundary subspace $\tilde{A}_{SL(2,\mathbb{C})}(M^3)|_{\mathrm{Re}\,ik\mathrm{CS}\ll 0}$ and thus Γ_α is a representative of a class in (3.2.1). The basic result in the Picard-Lefschetz theory (generalized to non-isolated critical points) tells us that the $[\Gamma_\alpha]$ form a basis. As one changes k one can pass through a special value of the value where the $\arg k = \theta_{\alpha\beta}$ is such that a flow from one of critical points in $\mathcal{M}^{\mathbb{C}}_\alpha$ reaches a critical point from another connected component $\mathcal{M}^{\mathbb{C}}_\beta$. The corresponding rays in the k-plane are usually referred to Stokes rays. The bases $[\Gamma_\alpha]$ on both sides are generically related by a non-trivial linear transform of triangular form:

$$[\Gamma_\alpha]|_{\arg k = \theta_{\alpha\beta} - \epsilon} = ([\Gamma_\alpha] + m^\alpha_\beta [\Gamma_\beta])|_{\arg k = \theta_{\alpha\beta} + \epsilon}, \qquad m^\alpha_\beta \in \mathbb{Z}.$$

This linear transform can also be understood as the monodromy of generators of middle-dimensional homology in the fibers of $\mathrm{CS} : \tilde{\mathcal{A}}_{SL(2,\mathbb{C})}(M^3) \to \mathbb{C}$ around the base point $\mathrm{CS}[A_\beta]$ where one of the fibers vanishes.

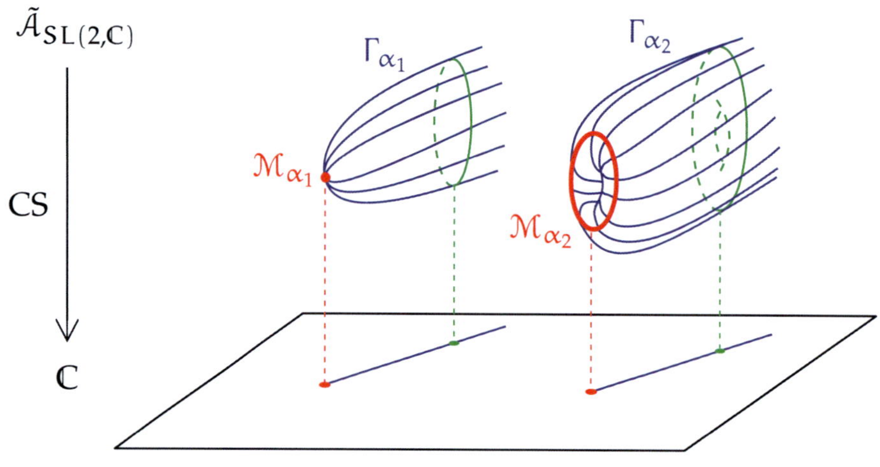

In particular, one can decompose the class of the contour Γ

$$[\Gamma] = \sum_{\alpha} n_\alpha [\Gamma_\alpha], \qquad n_\alpha \in \mathbb{Z}$$

so that

$$Z_{SU(2)_k}(M_3) = \sum_{\alpha} n_\alpha I_\alpha \qquad \text{and} \qquad I_\alpha = \int_{\Gamma_\alpha} DA e^{2\pi i k CS[A]}.$$

The condition (3.2.2) does not fix individual coefficients n_α but only fixes the partial sums

$$\sum_{\alpha \in \mathbb{Z}\text{-tower}} n_\alpha$$

performed over the connected components of the critical subspace projecting to the same component in $\mathrm{Hom}(\pi_1(M^3), SL(2,\mathbb{C})))$. The integral over Lefschetz thimbles have asymptotic expansion of the form

$$I_\alpha \overset{k\to\infty}{\sim} \sum_{n \geqslant 0} \frac{a_n^{(\alpha)}}{k^{n+d_\alpha+1}} e^{-k\xi_\alpha},$$

where

$$\xi_\alpha := -2\pi i CS[A_\alpha] \qquad \text{and} \qquad d_\alpha = (1 - \dim_{\mathbb{C}} \mathcal{M}_\alpha^{\mathbb{C}})/2,$$

where A_α is any representative of the connected component $\mathcal{M}_\alpha^{\mathbb{C}}$. Note that the coefficients $a_n^{(\alpha)}$ and numbers d_α actually depend only on the projection of the connected component α in $\tilde{\mathcal{M}}^{\text{flat}}_{SL(2,\mathbb{C})}$ to $\mathcal{M}^{\text{flat}}_{SL(2,\mathbb{C})}$. The Borel transform of the asymptotic series is a new series with coefficients related as follows

$$B^\alpha(\xi) := \sum_{n \geqslant 0} \frac{a_n^{(\alpha)}}{\Gamma(n + d_\alpha + 1)} (\xi - \xi_\alpha)^{d_\alpha + n}. \tag{3.2.3}$$

In the finite dimensional setting one can show that, unlike the asymptotic series for I_α, $B^\alpha(\xi)$ has finite radius of convergence. Moreover, it can be analytically continued to a cover of $\mathbb{C} \setminus \{\xi_\alpha\}_\alpha$ (usually called Borel plane) and its directional Laplace transform (for a generic k not on a Stokes ray) gives back I_α:

$$\int_{\gamma_\alpha = \xi_\alpha + 1/k\mathbb{R}_+} d\xi B^\alpha(\xi) e^{-k\xi} = I_\alpha.$$

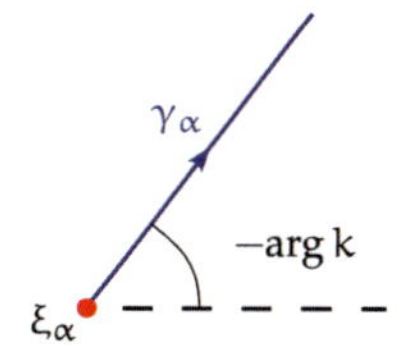

Note that the integration contour can be understood as the projection of the Lefschetz thimble Γ_α in the fibration $CS : \tilde{\mathcal{A}}_{SL(2,\mathbb{C})}(M^3) \to \mathbb{C}$, i.e. $\gamma_\alpha = -2\pi i CS[\Gamma_\alpha]$. When k is on a Stokes ray, the ray γ_α passes through a (possibly) singular point ξ_β of $B^\alpha(\xi)$. The contours in the Borel plane are related by the same triangular transform as before:

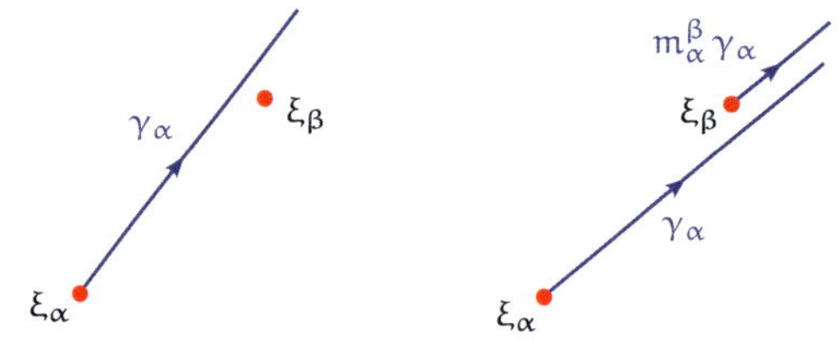

In particular,

$$I_\alpha|_{\arg k=\theta_{\alpha\beta}-\epsilon} = (I_\alpha + m^\alpha_\beta I_\beta)|_{\arg, k=\theta_{\alpha\beta}+\epsilon} \quad \text{and} \quad \theta_{\alpha\beta} = \arg(\xi_\alpha - \xi_\beta).$$

Now we are ready to formulate mathematically precise conjectures without any infinite-dimensional space shenanigans.

Conjecture 3.2.4 (A). *Suppose we are under assumptions (as discussed above) such that the perturbative invariants $a_n^{(\alpha)}$ of a 3-manifold M^3 are mathematically defined. Then the Borel series defined by the formula*

$$B^\alpha(\xi) := \sum_{n\geqslant 0} \frac{a_n^\alpha}{\Gamma(n+d_\alpha+1)}(\xi-\xi_\alpha)^{d_\alpha+n}$$

$$\xi_\alpha := -2\pi i CS[A_\alpha], \qquad d_\alpha = (1-\dim_{\mathbb{C}} \mathcal{M}^{\mathbb{C}}_\alpha)/2$$

have finite radius of convergence and can be analytically continued to a cover of $\mathbb{C}\setminus\{\xi_\alpha\}_\alpha$ *where* $\xi_\alpha \in \mathbb{C}$ *are the values of the Chern-Simons functional of* $SL(2,\mathbb{C})$ *flat connections on* M^3 *up to an integer. Moreover, the integral*

$$\int_{\gamma_\alpha=\xi_\alpha+1/k\mathbb{R}_+} d\xi B^\alpha(\xi)e^{-k\xi}$$

is convergent whenever the ray γ_α *does not pass through any* ξ_β *for* $\beta\neq\alpha$*—that is,* $\arg k \neq \arg(\xi_\alpha - \xi_\beta)$.

If this conjecture holds, then one can *define*

$$I_\alpha := \int_{\gamma_\alpha=\xi_\alpha+1/k\mathbb{R}_+} d\xi B^\alpha(\xi)e^{-k\xi}$$

without actually referring to the ill-defined path-integration over an infinite dimensional Lefschetz thimble Γ_α. This is what will be assumed in the conjectures below.

Conjecture 3.2.5 (B). *Suppose Conjecture (A) holds. Then there exist* $m^\beta_\alpha \in \mathbb{Z}$ *such that*

$$I_\alpha|_{\arg k=\theta_{\alpha\beta}-\epsilon} = (I_\alpha + m^\alpha_\beta I_\beta)|_{\arg, k=\theta_{\alpha\beta}+\epsilon} \quad \textit{and} \quad \theta_{\alpha\beta} = \arg(\xi_\alpha - \xi_\beta). \tag{3.2.6}$$

Moreover, $m^\alpha_\beta = 0$ *whenever* α *is a connected component of a lift (w.r.t. the projection* $\mathrm{Hom}(\pi_1(M^3), SL(2,\mathbb{C}))\times\mathbb{Z} \to \mathrm{Hom}(\pi_1(M^3), SL(2,\mathbb{C}))/SL(2,\mathbb{C})$*) of an irreducible flat connection and* β *is a connected component of a lift of an abelian flat connection*[2].

[2]The abelian and irreducible flat connections for $SL(2,\mathbb{C})$ gauge group are defined analogously to the $SU(2)$ case. In particular, $U(1)$ is replaced by $\mathbb{C}^*$.

On the other hand, for any abelian and irreducible flat connections there exist connected components β and α of their respective lifts such that m^{α}_{β}.

Remark 3.2.7. The motivation for the second part of the conjecture comes from the finite-dimensional analysis of integrals that model Chern-Simons path integral [27]. If it holds, it follows that one can always recover I_β for irreducible β from I_α (defined via the correpsonding perturbative invariants $a^{(\alpha)}$) for abelian α via the formula (3.2.6), but not the other way around.

Finally, the next conjecture can be understood as a "refinement" of the Asymptotic Expansion Conjecture

Conjecture 3.2.8 (C). *Suppose Conjecture (A) holds. There exist integers $N_c \in \mathbb{Z}$ such that for any $n_\alpha \in \mathbb{Q}$ satisfying the conditions*

$$\sum_{\alpha} n_\alpha$$

equals

- 1 *when α runs over lifts of a connected component a of $SU(2)$ flat connections for any a,*
- 0 *when α runs over lifts of a connected component b of non-$SU(2)$ flat connections with* $\mathrm{Im}\,(CS) \leqslant 0$ *for any b,*
- N_c *when α runs over lifts of a connected component c of $SU(2)$ flat connections with* $\mathrm{Im}\,(CS) > 0$,

then we have

$$\sum_{\alpha} n_\alpha I_\alpha|_{k\in\mathbb{Z}} = Z_{SU(2)_k}(M^3).$$

All these conjectures were (at least partially) checked for various non-trivial 3-manifolds [10, 21].

Example 3.2.9. Take $M_3 = \Sigma(2,3,7)$ which can be realized by Dehn surgery on a figure-eight know or on a 4-component link shown below:

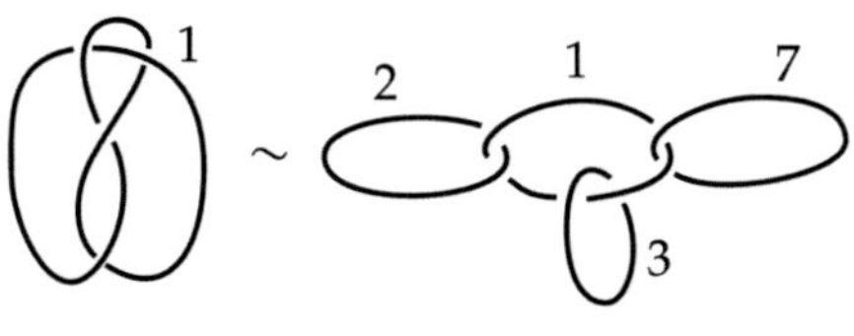

There are 4 $SL(2,\mathbb{C})$ flat connections: one a_0 is trivial and the others a_1, a_2 and a_3 are irreducible. These last three can be conjugated to $SU(2)$ flat connections (a_3 can be conjugated to an $SL(2,\mathbb{R})$ flat connection). The values of the CS functional mod 1 on them are the following:

$$CS(a_0) = 0, \qquad CS(a_1) = -\frac{25}{168}, \qquad CS(a_2) = -\frac{121}{168}, \qquad CS(a_3) = -\frac{1}{168}.$$

The Borel transforms $B^{a_i}(\xi)$ for $i = 1,2,3$ do not have any any singularities outside of the single point ξ_{a_i}. The Borel transform $B^{a_0}(\xi)$ has singularities at

$n + 2\pi i CS[a_i]$ for certain values of n. One can explicitly check that the conjectures (A), (B) and (C) hold. In particular, for $C_0 = \mathbb{R}_+$, $C_1 = \mathbb{R}_+ - \frac{25}{168}$, $C_2 = \mathbb{R}_+ - \frac{121}{168}$ and $C_3 = \mathbb{R}_+ - \frac{1}{168}$, we may write $Z_{SU(2)_k}$ as

$$\int_{C_0} B^{a_0}(\xi)e^{-k\xi} + \int_{C_1} B^{a_1}(\xi)e^{-k\xi} + \int_{C_2} B^{a_2}(\xi)e^{-k\xi} + \int_{C_3} B^{a_3}(\xi)e^{-k\xi}$$

at $k \in \mathbb{Z}_+$. A different analytic continuation (which yields the same result at integral k) can be realized as:

$$Z_{SU(2)_k} = \lim_{k \in \mathbb{Z}_+} \int_{\gamma_0^+ + \gamma_0^-} B^{a_0}(\xi)e^{-k\xi} d\xi$$

with the contours $\gamma_0^{\pm}$ shown below:

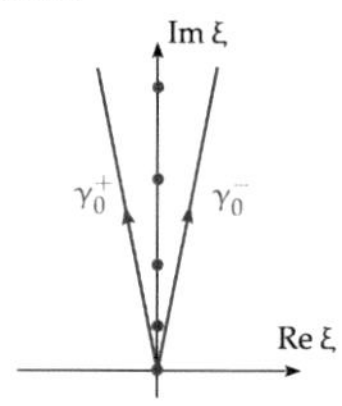

These two different realizations are consistent with the Stokes relations between abelian and irreducible flat connections.

4. Lecture IV

4.1. Problems in categorifing WRT invariant Unlike the Jones polynomial, the WRT invariant is only defined for integral k, not any generic complex k, or, equivalently, a generic value of $q = e^{2\pi i/k}$. Suppose we overcame this problem and found a proper analytic continuation in k. One can pose the question whether it has an expansion in q with integral coefficients that can be interpreted as Euler characteristics of some cohomologies, as happens for the Jones polynomial?

Consider for example the Lens space $L(5,1)$ (see the Exercises):

$$\begin{aligned} &\sqrt{2ki}q^{-1/2}Z_{SU(2)_k}(L(5,1)) \\ &= \frac{1}{\sqrt{5}}(q^{1/5}-1) + e^{2\pi i/k}\frac{(-1-\sqrt{5})q^{1/5}-4}{2\sqrt{5}} + e^{-2\pi i/k}\frac{(-1+\sqrt{5})q^{1/5}-4}{2\sqrt{5}} \end{aligned}$$

Even in this simple example we see that the answer is no! We have to do something about it.

4.2. Abelian flat connections and linking pairing In resurgent analysis we saw that, in a certain sense, abelian flat connections are more important than non-abelian ones, since the latter can be recovered from the former via the Stokes phenomenon. The abelian flat connections form a subspace of all flat connections where all holonomies can be conjugated to $U(1)$, the maximal torus in $SU(2)$. Therefore this subspace can be explicitly described as follows,

$$\begin{aligned} \mathcal{M}^{\text{flat,Ab}}_{SU(2)}(M^3) &\cong \text{Hom}(\pi_1(M^3), U(1))/\mathbb{Z}_2 \\ &\subset \text{Hom}(\pi_1(M^3), SU(2))/SU(2) \cong \mathcal{M}^{\text{flat}}_{SU(2)}(M^3) \end{aligned}$$

where SU(2) action by conjugation is reduced to a $\mathbb{Z}_2$ Weyl group action on the maximal torus. Using the fact that U(1) is abelian we can replace $\pi_1(M^3)$ by its abelianization, $H_1(M^3, \mathbb{Z})$:

$$\mathcal{M}^{\text{flat,Ab}}_{SU(2)}(M^3) \cong \text{Hom}(\pi_1(M^3), U(1))/\mathbb{Z}_2 \cong \text{Hom}(H_1(M^3, \mathbb{Z}), U(1))/\mathbb{Z}_2$$

Moreover, as in the resurgence analysis, we will be interested in the connected components of $\mathcal{M}^{\text{flat,Ab}}_{SU(2)}(M^3)$:

$$\pi_0\mathcal{M}^{\text{flat,Ab}}_{SU(2)}(M^3) \cong \text{Hom}(\text{Tor}\, H_1(M^3, \mathbb{Z}), U(1))/\mathbb{Z}_2.$$

where we used that $H_1(M^3, \mathbb{Z}) \cong \text{Tor}\, H_1 \times \mathbb{Z}^{b_1}$ (non-canonically) and

$$\begin{aligned} &\text{Hom}(H_1(M^3, \mathbb{Z}), U(1)) \\ \cong\ &\text{Hom}(\text{Tor}\, H_1(M^3, \mathbb{Z}), U(1)) \times \text{Hom}(\mathbb{Z}^{b_1}, U(1)) \\ \cong\ &\text{Hom}(\text{Tor}\, H_1(M^3, \mathbb{Z}), U(1)) \times U(1)^{b_1}. \end{aligned}$$

Definition 4.2.1. The *linking pairing* is a bilinear form on $\text{Tor}\, H_1(M^3, \mathbb{Z})$ defined as follows:

$$\begin{aligned} \ell k:\ \text{Tor}\, H_1(M_3) \otimes \text{Tor}\, H_1(M_3) &\longrightarrow \mathbb{Q}/\mathbb{Z}, \\ [a] \otimes [b] &\longmapsto \#(a \cap b')/n, \end{aligned}$$

where b' is a 2-chain such that $\partial b' = nb$ for an integer n. Such b' and n exist because $[b]$ is torsion. As usual, $\#(a \cap b')$ denotes the number of intersection points counted with signs determined by the orientation.

Remark 4.2.2. The linking pairing is a homotopy invariant of 3-manifolds.

Example 4.2.3. Consider lens space

$$L(p, q) := \{|z_1|^2 + |z_2|^2 = 1\}/(z_1, z_2) \sim (e^{2\pi i/p} z_2, e^{-2\pi i q/p})$$

for a pair of coprime integers (p, q). We have $H_1(L(p, q), \mathbb{Z}) \cong \mathbb{Z}_p$ and

$$\ell k\,(a, b) = \frac{qab}{p} \mod 1$$

Unlike the fundamental group, the linking pairing can distinguish lens spaces with the same p and different values of q.

The linking pairing provides an isomorphism between $\text{Tor}\, H_1(M_3)$ and its Pontryagin dual $(\text{Tor}\, H_1(M_3))^* \equiv \text{Hom}(\text{Tor}\, H_1(M_3), U(1))$ via the pairing $e^{2\pi i \ell k\,(\cdot,\cdot)}$. Using this isomorphism we have

$$\pi_0\mathcal{M}^{\text{flat,Ab}}_{SU(2)}(M^3) \cong \text{Tor}\, H_1(M^3, \mathbb{Z})/\mathbb{Z}_2$$

with, from the $\mathbb{Z}_2$ action, the equivalence relation $a \sim -a$ on the elements of $\text{Tor}\, H_1(M^3)$.

4.3. WRT invariant and q-series Now we are ready to formulate the following conjecture [11, 12] (also see [9] for a mathematically oriented review as well as generalization for the manifolds with a torus boundary):

Conjecture 4.3.1. *The WRT invariant can be decomposed into the following form:*

$$Z_{SU(2)_k}[M_3] = (i\sqrt{2k})^{b_1(M_3)-1} \sum_{\substack{a,b\in \\ \mathrm{Tor}\, H_1(M_3,\mathbb{Z})/\mathbb{Z}_2}} e^{2\pi i k \ell k(a,a)} S_{ab}\, \hat{Z}_b(M^3)|_{q\to e^{\frac{2\pi i}{k}}} \tag{4.3.2}$$

with $\hat{Z}_b(M^3) \in 2^{-c} q^{\Delta_b}\mathbb{Z}[[q]], \Delta_b \in \mathbb{Q}$, *and* $c \in \mathbb{Z}_+$, *convergent in* $|q| < 1$ *and*

$$S_{ab} = \frac{e^{4\pi i \ell k(a,b)} + e^{-4\pi i \ell k(a,b)}}{|\mathrm{Stab}_{\mathbb{Z}_2}(a)|\sqrt{|\mathrm{Tor}\, H_1(M_3,\mathbb{Z})|}}.$$

Remark 4.3.3. $CS(A_a) = \ell k(a,a) \mod 1$ (cf. Exercises)

Remark 4.3.4. The conjecture above is consistent with the asymptotic expansion conjecture and its stronger resurgence version. Namely, in the notations of the previous lecture

$$(i\sqrt{2k})^{b_1(M_3)-1} e^{2\pi i k \ell k(a,a)} \sum_{\substack{b\in \\ \mathrm{Tor}\, H_1(M_3,\mathbb{Z})/\mathbb{Z}_2}} S_{ab}\, \hat{Z}_b(M^3)\Big|_{q\to e^{\frac{2\pi i}{k}}} = I_{a'} + \sum_{\beta\in \mathrm{irr}} N_a^\beta I_\beta$$

where a' is any lift of a and the numbers N_a^β satisfy $\sum_{a\in \mathrm{Tor}\, H_1(M^3,\mathbb{Z})} N_a^\beta = n_\beta$ with n_β as in Conjecture (C) from the last lecture.

Example 4.3.5. The conjecture works for $L(5,1)$ with (cf. Exercises)

$$\hat{Z}(L(5,1)) = q^{-1/2}\begin{pmatrix} 1 \\ -q^{1/5} \\ 0 \end{pmatrix}.$$

Remark 4.3.6. In principle the formula (4.3.2), if it holds, does not uniquely fix $\hat{Z}_a(M^3)$. However from physics we expect that there is a canonical choice of $\hat{Z}_a(M^3)$, i.e. they should be an invariant of 3-manifolds that refines the WRT invariant. Physics predicts a categorification similar to Khovanov homology

$$\hat{Z}_a(M^3) = \sum_{i,j} (-1)^i q^j \dim \mathcal{H}_a^{ij}(M^3)$$

where for each $a \in \mathrm{Tor}\, H_1(M^3,\mathbb{Z})$ there are doubly graded vector spaces $\mathcal{H}_a^{ij}(M^3)$. Physically they are BPS states of a stack of two M5-branes (i.e. 6d (2,0) type A_1 superconformal field theory) on $D^2 \times M^3$ with topological twist along M^3 and a certain boundary condition labelled by a. The two gradings correspond to charges of $U(1)$ symmetry rotating the disc D^2 and $U(1)$ R-symmetry (part of the superalgebra acting on the Hilbert space).

Example 4.3.7. For $M^3 = L(5,1)$ physics predicts the following Poincare polynomials $P_a(q,t) := \sum_{ij} t^i q^j \dim \mathcal{H}_a^{ij}(M^3)$:

$$P(q,t) = q^{-1/2}\begin{pmatrix} 1 + q(1+t) + q^2(2+3t+t^2) + \ldots \\ q^{1/5} t(1 + q(2+2t) + q^2(3+5t+2t^2) + \ldots) \\ q^{4/5}(0 + q(1+t) + q^2(1+3t+2t^2) + \ldots) \end{pmatrix}.$$

4.4. q-series invariant for plumbed 3-manifolds Consider a 3-manifold $M_3(\Gamma)$ associated to a plumbing graph Γ. Namely, Γ is a graph with vertices labelled by integers and M_3 is obtained by a Dehn surgery on a link $\mathcal{L}(\Gamma)$ of framed unknots, associated to the plumbing graph Γ as illustrated in the example below:

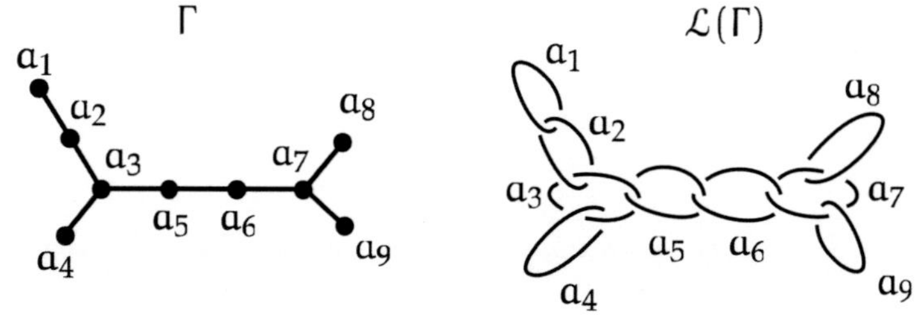

Denote by L the number of vertices of Γ. It is equal to the number of components of the link $\mathcal{L}(\Gamma)$. We will also need the $L \times L$ linking matrix of $\mathcal{L}(\Gamma)$ which will be denoted by M:

$$M_{v_1,v_2} = \begin{cases} 1, & v_1, v_2 \text{ connected}, \\ a_v, & v_1 = v_2 = v, \\ 0, & \text{otherwise}. \end{cases} \qquad \text{for } v_1 \text{ and } v_2 \text{ vertices of } \Gamma. \tag{4.4.1}$$

The first homology group of $M_3(\Gamma)$ is given by the cokernel (over $\mathbb{Z}$) of the linking matrix:

$$H_1(M_3, \mathbb{Z}) \cong \text{Coker} M = \mathbb{Z}^L / M\mathbb{Z}^L.$$

Assume for simplicity that $M_3(\Gamma)$ is a rational homology sphere (i.e. $b_1(M_3) = 0$), which implies that Γ has no loops (i.e. it is a tree). The linking form on $H_1(M_3, \mathbb{Z})$ is then given by

$$\ell k\,(a, b) = (a, M^{-1}b) \mod \mathbb{Z}, \qquad a, b \in \mathbb{Z}^L / M\mathbb{Z}^L.$$

Also assume that M is negative definite (although this condition can be weakened). Then [11]:

$$\begin{aligned} &\hat{Z}_b(q) \\ &= q^{-\frac{3L+\sum_v a_v}{4}} \cdot \text{v.p.} \int\limits_{|z_v|=1} \prod_{v \,\in\, \text{Vertices}} \frac{dz_v}{2\pi i z_v} \,(z_v - 1/z_v)^{2-\deg(v)} \cdot \Theta_b^{-M}(z) \\ &\equiv q^{-\frac{3L+\sum_v a_v}{4}} \cdot \text{v.p.} \int\limits_{|z_v|=1} \prod_{v \,\in\, \text{Vertices}} \frac{dz_v}{2\pi i z_v} \,(z_v - 1/z_v)^{2} \times \\ &\qquad \times \prod_{(v_1,v_2)\in \text{Edges}} \frac{1}{(z_{v_1} - 1/z_{v_1})(z_{v_2} - 1/z_{v_2})} \cdot \Theta_b^{-M}(z) \end{aligned} \tag{4.4.2}$$

where $\deg(v)$ is the degree of a vertex v and "v.p." means taking the principle value integral (i.e. take half-sum of contours "dodging" poles from both sides) and $\Theta_b^{-M}(x)$ is the theta function of the lattice corresponding to minus the linking form M given by

$$\Theta_b^{-M}(x) = \sum_{\ell \in 2M\mathbb{Z}^L + b} q^{-\frac{(\ell, M^{-1}\ell)}{4}} \prod_{i=1}^{L} x_i^{\ell_i}.$$

We have

$$\hat{Z}_b(q) \in 2^{-c} q^{\Delta_b} \mathbb{Z}[[q]]$$

where

$$c \in \mathbb{Z}_+, \qquad c \leqslant L,$$

$$b \in (2\mathrm{Coker}\, M + \delta)/\mathbb{Z}_2 \overset{\mathrm{Set}}{\cong} H_1(M_3, \mathbb{Z})/\mathbb{Z}_2,$$

$$\delta \in \mathbb{Z}^L/2\mathbb{Z}^L, \qquad \delta_v = \deg(v) \mod 2.$$

$$\Delta_b = -\frac{3L + \sum_v a_v}{4} - \max_{\ell \in 2M\mathbb{Z}^L + b} \frac{(\ell, M^{-1}\ell)}{4} \in \mathbb{Q}$$

and where $\mathbb{Z}_2$ acts as

$$b \mapsto -b$$

and is a symmetry of (4.4.2).

Remark 4.4.3. The formula (4.4.2) defines homological blocks $\hat{Z}_a(q)$ in terms of a plumbing graph Γ. Different plumbing graphs can give homeomorphic 3-manifolds — $M_3(\Gamma) \cong M_3(\Gamma')$ *iff* Γ and Γ' can be connected by a sequence of 3d Kirby moves:

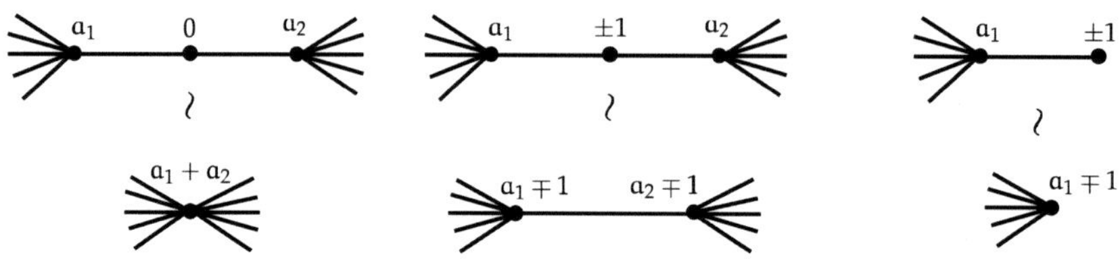

One can show that the right hand side of (4.4.2) is invariant under Kirby moves acting on Γ. Therefore, $\hat{Z}_a(q)$ actually depends only on the homeomorphism class of $M_3(\Gamma)$, and is indeed a topological invariant.

Example 4.4.4.

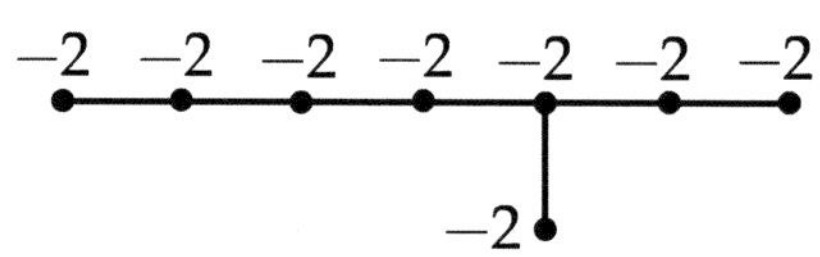

$$H_1(M_3) = 0$$

$$\hat{Z}_0 = q^{-3/2}(1 - q - q^3 - q^7 + q^8 + q^{14} + q^{20} + q^{29} - q^{31} - q^{42} + \cdots)$$

Example 4.4.5.

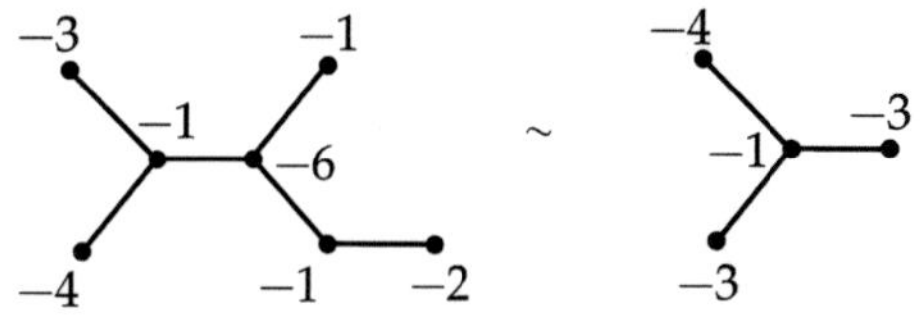

$$H_1(M_3) = \mathbb{Z}_3$$

$$\hat{Z} = \begin{pmatrix} 1 - q + q^6 - q^{11} + q^{13} - q^{20} + q^{35} + O\left(q^{41}\right) \\ q^{5/3}\left(-1 + q^3 - q^{21} + q^{30} + O\left(q^{41}\right)\right) \end{pmatrix}$$

5. Exercises

5.1. Lecture I

I.1 Show that

$$S^3\left(\bigcirc_p \right) \cong L(p,1) = \{|z_1|^2 + |z_2|^2 = 1\}/(z_1, z_2) \sim (z_1 e^{\frac{2\pi i}{p}}, z_2 e^{-\frac{2\pi i}{p}})$$

I.2 Check that the following framed links can be related by a sequence of Kirby-Fenn-Rourke moves

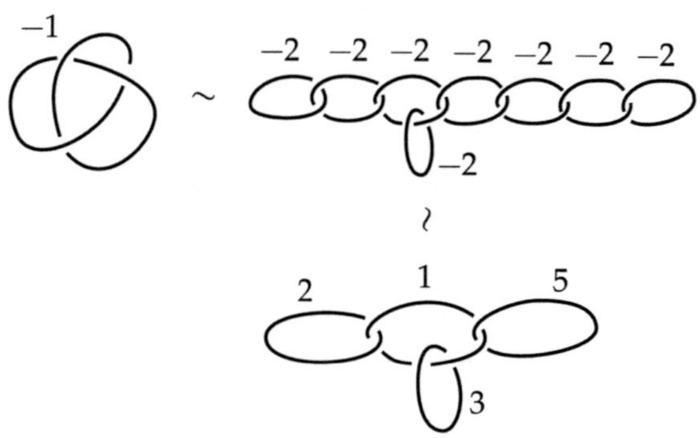

I.3 Show that indeed $\tau_k(S^3) = 1$ and $\tau_k(S^2 \times S^1) = \left(\sqrt{\frac{2}{k}} \sin \frac{\pi}{k}\right)^{-1}$. *Hint*: realize S^3 and $S^2 \times S^1$ as surgeries on an unknot with framings 1 and 0 respectively.

I.4 Let Z be an n-dimensional TQFT and M^{n-1} is a closed oriented n-manifold. Show that $Z(S^1 \times M^n) = \dim Z(M^{n-1})$. *Hint*: Use the fact that $M^{n-1} \times [0,1]$ with the diffeomorphism of its boundary to $\bar{M}^{n-1} \sqcup \bar{M}^{n-1}$ given restricting the identity map.

5.2. Lecture II

II.1 Let Γ be a finite group. Consider 2d TQFT Z_Γ corresponding to the Frobenius algebra $\mathbb{C}[\Gamma]$, i.e. the group algebra of Γ (as a vector space it is freely generated by elements of Γ over $\mathbb{C}$ and the multiplication is determined by the group law in

Γ) with

$$\text{for } g \in \Gamma, \ \operatorname{Tr}(g) = \begin{cases} 1, & g = \mathbf{1}, \\ 0, & \text{otherwise} \end{cases}$$

Let Σ_g be a closed oriented Riemann surface of genus g. Compute $Z_\Gamma(\Sigma_g)$.

II.2 Consider a mapping class cylinder M_g of a torus corresponding to an element $g \in SL(2,\mathbb{Z})$, i.e. a bordism $T^2 \times [0,1]$ with diffeomorphisms $\mathrm{id} : T^2 \times \{0\} \to T^2$ and $\phi : T^2 \times \{1\} \to T^2$, so that its pullback $\phi_* : \pi_1(T^2) \to \pi_1(T^2) \cong \mathbb{Z}^2$ is given by $g \in SL(2,\mathbb{Z})$. It is generated by $T = \left(\begin{smallmatrix} 1 & 1 \\ 1 & 0 \end{smallmatrix}\right)$ and $S = \left(\begin{smallmatrix} 0 & 1 \\ -1 & 0 \end{smallmatrix}\right)$. Let $Z_{\mathcal{C}}$ be a 3d TQFT functor corresponding to an MTC $\mathcal{C}$. As was mentioned in the lectures, $Z_{\mathcal{C}}(T^2) = \operatorname{Hom}(\mathbf{1}, \oplus_{i\in I} V_i \otimes V_i^*)$ with natural basis labelled by the simple objects of $\mathbb{C}$:

$$A_i = 0 \oplus \ldots \oplus 0 \oplus \frac{D}{d_i}$$ 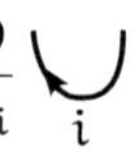 $$\oplus 0 \oplus \ldots \oplus 0$$

Similarly, for the dual space $Z_{\mathcal{C}}(T^2)^* = \operatorname{Hom}(\oplus_{i\in I} V_i \otimes V_i^*, \mathbf{1})$ we choose the basis

$$B_i = \frac{D}{d_i}$$  $$\circ \operatorname{Proj}_{V_i^* \otimes V_j}$$

(The normalization coefficient D/d_i is chosen so that it cancels out in the surgery formula)

Questions:

(1) Express $\langle B_j, Z_{\mathcal{C}}(M_g)(A_i)\rangle$ for $g = \mathbf{1}$, S and T using the basic MTC data $\tilde{S}_{ij}$ and T_i and, when needed, the quantities $d_i := \tilde{S}_{0i}$, $D := \sqrt{\sum_{i\in I} d_i^2}$, $p_\pm := \sum_{i\in I} T_i^{\pm} d_i^2$ and $S_{ij} = \tilde{S}_{ij}/D$.

(2) Using the fact that $\sum_{\ell\in I} S_{i\ell} S_{\ell j} = \delta_{ij^*}$ (where $V_{j^*} \equiv V_j^*$), show that the basis B_j is dual to A_i.

(3) Assuming that $p_\pm = D$ (this is true when there is no framing anomaly) simplify further the formulas.

5.3. Lecture III

III.1

a) Show that there are exactly 2 $SU(2)$ flat connections (modulo gauge transformations) on $L(3,1)$.

b) Obtain an explicit formula for $Z_{SU(2)_k}(L(3,1))$ which is analytic in k.

Hint: Use the Gauss reciprocity formula

$$\sum_{m\in\mathbb{Z}_{2k}} q^{-\frac{pm^2}{4}} \cdot q^{\frac{n\ell}{2}} = \frac{e^{-\frac{\pi i \,\mathrm{sign}(p)}{4}} (2k)^{1/2}}{|p|^{1/2}} \sum_{a\in\mathbb{Z}_p} e^{2\pi i k \frac{a^2}{p}} \cdot e^{2\pi i \frac{a\ell}{p}} \cdot q^{\frac{\ell^2}{4p}},$$

where, as usual, $q = e^{2\pi i/k}$, $k \in \mathbb{Z}_+$.

c) Check that asymptotic expansion conjecture holds for $L(3,1)$ (use the fact that the values of the Chern-Simons functional on flat connections are 0 and 1/3) and that in this special case the asymptotic series in $1/k$ for each flat connection are actually convergent.

III.2 Consider $M_3 = S^3_{+1}(\mathbf{4}_1)$, i.e. +1-surgery on figure-8 knot. The perturbative invariants for the trivial flat connection are explicitly known to each order. Up to an overall rational power of q (depending on the framing):

$$\sum_{n\geqslant 0} \frac{a_n^{(\mathrm{triv})}}{k^{n+3/2}} = \frac{1}{i\sqrt{2k}} \sum_{n\geqslant 0} \frac{c_n}{n!}\left(\frac{\pi i}{84k}\right)^n$$

where the coefficients c_n have the following generating function:

$$\sum_{n\geqslant 0} \frac{c_n}{(2n)!} z^{2n} = \frac{2\sinh(6z)\sinh(14z)\sinh(21z)}{\sinh(42z)}$$

a) Give a closed form expression for its Borel transform

$$B^{(\mathrm{triv})}(\xi) := \sum_{n\geqslant 0} \frac{a_n^{(\mathrm{triv})}}{\Gamma(n+3/2)} \xi^{n+1/2}$$

b) Show that it has singularities only at values of CS functional on $SL(2,\mathbb{C})$ flat connections (namely $0, -\frac{1}{168}, -\frac{25}{168}, -\frac{121}{168}$) modulo $\mathbb{Z}$

c) Show that

$$\int_{e^{i(\pi/2+\epsilon)}\mathbb{R}_+} d\xi B^{(\mathrm{triv})}(\xi) + \int_{e^{i(\pi/2-\epsilon)}\mathbb{R}_+} d\xi B^{(\mathrm{triv})}(\xi)$$

(i.e. sum of integrals along the rays infinitesimally shifted from the positive half of the imaginary axis), up to an overall constant and a power of k, can be expressed as q-series with integral coefficients.

5.4. Lecture IV

IV.1 Let $L(p,q) := \{|z_1|^2 + |z_2|^2 = 1\}/(z_1,z_2) \sim (e^{2\pi i/p}z_2, e^{-2\pi i q/p})$ be lens spaces for a pair of coprime integers (p,q). Show that $H_1 \cong \mathbb{Z}_p$ and

$$\ell k\,(a,b) = \frac{qab}{p} \mod 1$$

Hint: As in the solution of the exercise I.1 represent $L(p,q)$ as gluing of two copies of $D^2 \times S^1$ so that the standard meridians and longitudes are identified according to the matrix $\left(\begin{smallmatrix} q & p \\ r & s \end{smallmatrix}\right) \in SL(2,\mathbb{Z})$.

Remark: Unlike the fundamental group, the linking pairing can distinguish lens spaces with the same p and different values of q.

IV.2 Assume, for simplicity, that M_3 is a rational homology sphere, so that $\mathrm{Tor}\, H_1(M^3,\mathbb{Z}) \cong H_1(M^3,\mathbb{Z})$. Let A_a be representative of the a $U(1)$ flat connection labelled by an element $a \in \mathrm{Tor}\, H_1(M^3,\mathbb{Z}) \cong \mathrm{Hom}(\mathrm{Tor}\, H_1(M^3,\mathbb{Z}), U(1))$ (i.e. so that $\mathrm{Hol}_b(A_a) = e^{2\pi i \ell k\,(a,b)}$). Show that $CS[A_a] = \ell k\,(a,a) \mod 1$.

Remark: Since there can be non-trivial $U(1)$ bundles, and the connection 1-form is only locally defined, one cannot define the CS functional as just

$$CS[A] = \frac{1}{4\pi^2}\int_{M^3} AdA \mod 1.$$

Instead one needs to pick any 4-manifold M^4 such that $\partial M^4 = M^3$, extend the $U(1)$ gauge field to M^4 and define

$$\mathrm{CS}[A] := \frac{1}{4\pi^2}\int_{M^4} F_A \wedge F_A \mod 1$$

where F_A is a curvature 2-form of the connection, locally expressed as $F_A := dA$, so that $F_A \wedge F_A = d(AdA)$ (if the connection 1-form were globally defined one could use the Stokes theorem to obtain $\frac{1}{4\pi^2}\int_{M^4} F_A \wedge F_A = \frac{1}{4\pi^2}\int_{M^3} AdA$).
Hint: Take M^4 to be simply connected[3], so that from the long exact sequence of relative homology and Poincaré-Lefschetz duality we have the isomorphism $H^2(M^4,\mathbb{Z})/H_2(M^4,\mathbb{Z}) \cong H_1(M^3,\mathbb{Z})$.

IV.3

a) Using the general formula for plumbed 3-manifolds, calculate $\hat{Z}_a$ for $L(5,1)$, e.g. taking the plumbing graph

$$\bullet\, 5$$

b) Calculate the first few terms of $\hat{Z}_a$ for the plumbing

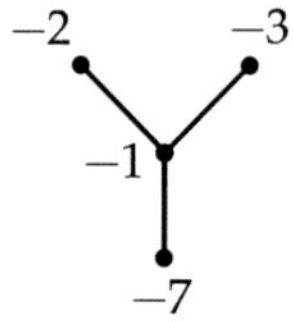

6. Solutions

6.1. Lecture I

I.1 This is a more analytic solution; one can present a more geometric argument, e.g. by starting with the Hopf fibration of S^3 and dividing it by $\mathbb{Z}_p$. Let us split $L(p,1) = \{|z_1|^2 + |z_2|^2 = 1\}/(z_1,z_2) \sim (z_1 e^{\frac{2\pi i}{p}}, z_2 e^{-\frac{2\pi i}{p}})$ into a union of submanifolds meeting along a common boundary: $L(p,1) \cong U_1 \cup U_2$, where $U_1 \cong \{|z_1|^2 \geqslant 1/2\}$ and $U_2 \cong \{|z_2|^2 \geqslant 1/2\}$. It is easy to see that $U_{1,2} \cong S^1 \times D^2$, in coordinates, $\{(u,w) \mid |u| = 1, |w| \leqslant 1| \mid z,w \in \mathbb{C}\}$, e.g. by presenting explicit diffeomorphisms:

$$U_1 := \frac{\{|z_1|^2 + |z_2|^2 = 1\} \cap \{|z_1|^2 \geqslant 1/2\}}{(z_1,z_1) \sim (e^{\frac{2\pi i}{p}} z_1, e^{-\frac{2\pi i}{p}} z_2)} \underset{f_1^{-1}}{\overset{f_1}{\rightleftarrows}} S^1 \times D^2$$

(6.1.1)

$$[(z_1,z_2)] \xmapsto{\quad f_1 \quad} (z_1^p/|z_1|^p, \sqrt{2} z_1 z_2/|z_1|)$$

$$[(\sqrt{1-|w|^2/2}\, u^{1/p}, u^{-1/p} w/\sqrt{2})] \xleftarrow{\quad f_1^{-1} \quad} (u,w)$$

[3]Suppose M^3 is obtained by Dehn surgery on a framed link in S^3. Then M^4 can be obtained by attaching 2-handles $D^2 \times D^2$ to the B^4 by the same gluing as in Dehn surgery

and the same for U_2 with indices $1 \leftrightarrow 2$. Let m and ℓ be the standard meridian and longitude of $S^1 \times D^2 = \{|u| = 1, |w| \leqslant 1 \mid z, w \in \mathbb{C}\}$, i.e. if understood as loops $[0,1] \to S^1 \times D^2$ they are given by $m(t) = (1, e^{2\pi i t})$, $\ell(t) = (e^{2\pi i t}, 1)$.

From the explicit formulas one can see that pushforwards (to π_1) of the diffeomorphisms restricted to the boundaries act as follows

$$\begin{aligned}(f_2 f_1^{-1}|_{S^1\times S^1})_*[m] &= \;\; [m] + p[\ell] \\ (f_2 f_1^{-1}|_{S^1\times S^1})_*[\ell] &= \;\; -[\ell]\end{aligned}$$

It is left to notice that the complement of tubular neighborhood of unframed unknot in S^3 is another solid torus with meridian and longitude swapped, and that p-framing shifts the class in the fundamental group of the standard longitude of the tubular neighborhood of unknot by p copies of the class of the meridian

I.2

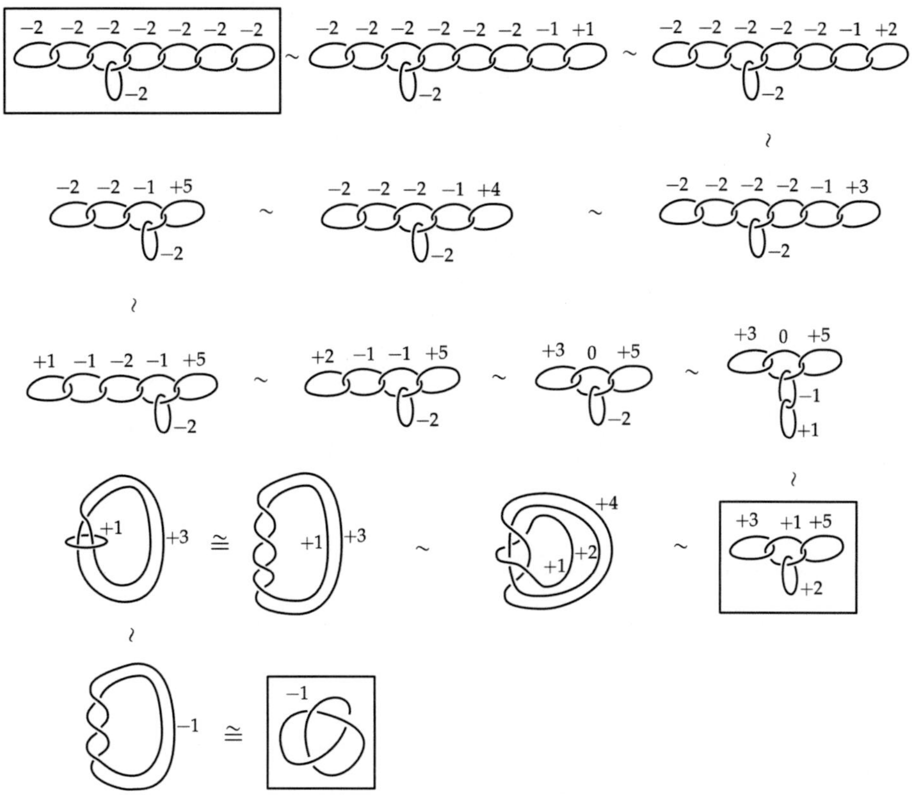

I.3 The first statement follows immediately from the definition

$$\tau_k(S^3) = \tau_k(S^3(\bigcirc_{+1})) = \frac{\sum_{n=1}^{k-1}[n]J_n(\bigcirc_{+1})}{\sum_{n=1}^{k-1}[n]J_n(\bigcirc_{+1})} = 1$$

The second statement can be shown as follows:

$$
\begin{aligned}
\tau_k(S^2 \times S^1) &= \tau_k\left(S^3\left(\bigcirc_0\right)\right) \\
&= \frac{\sum_{n=1}^{k-1}[n]J_n(\bigcirc_0)}{(\sum_{n=1}^{k-1}[n]^2)^{1/2}} \\
&= \left(\sum_{n=1}^{k-1}[n]^2\right)^{1/2} \\
&= \left(\sum_{n=1}^{k}[n]^2\right)^{1/2} \\
&= \left(\sum_{n=1}^{k}\frac{q^n+q^{-n}-2}{(q^{1/2}-q^{-1/2})^2}\right)^{1/2} \\
&= \left(\frac{-2k}{(q^{1/2}-q^{-1/2})^2}\right)^{1/2} \\
&= \sqrt{\frac{2}{k}}\frac{1}{\sin\frac{\pi}{k}}
\end{aligned}
$$

I.4 Note that the cylinder $M^{n-1} \times [0,1]$ can be considered as a bordism either from M^{n-1} to M^{n-1}, or from $M^{n-1} \sqcup \bar{M}^{n-1}$ to $\emptyset$, or from $\emptyset$ to $\bar{M}^{n-1} \sqcup M^{n-1}$. Consider a bilinear pairing

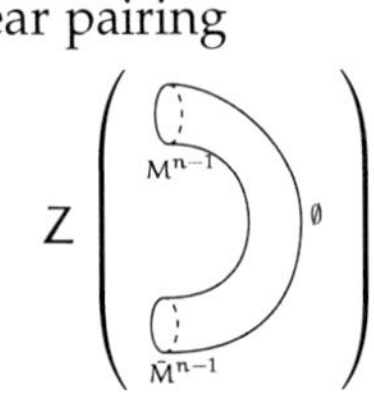

$: Z(M^{n-1}) \otimes Z(\bar{M}^{n-1}) \to \mathbb{C}$

It is non-degenerate, because it has an inverse

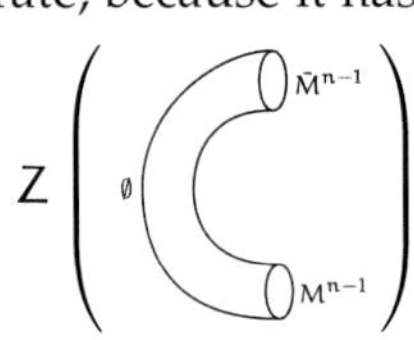

$: \mathbb{C} \to Z(\bar{M}^{n-1}) \otimes Z(M^{n-1})$

since

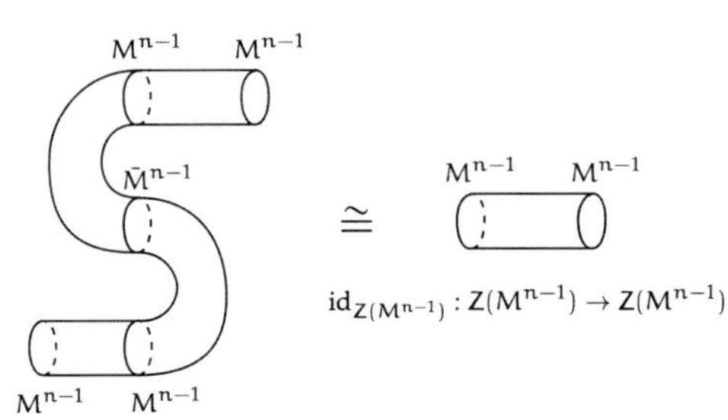

Therefore $Z(\bar{M}^{n-1})$ is a vector space dual to $Z(M^{n-1})$. Choose a basis e_i, $i = 1 \dots N$ in $Z(M^{n-1})$ and its dual e^j, $j = 1 \dots N$ in $Z(\bar{M}^{n-1})$. That is

Z (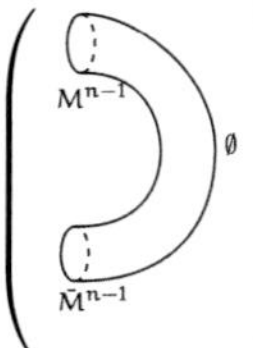) $: e_i \otimes e^j \longmapsto \delta^i_j$

Z (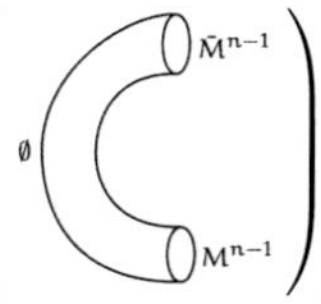) $: 1 \longmapsto \sum_{i=1}^{N} e^i \otimes e_i$

so that

Z (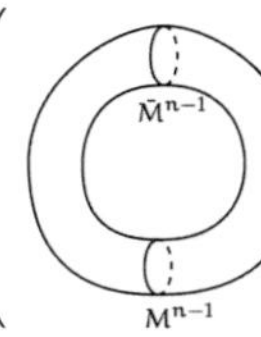) $: 1 \longmapsto \sum_{i=1}^{N} \delta^i_i = N$

6.2. Lecture II

II.1

Z_Γ (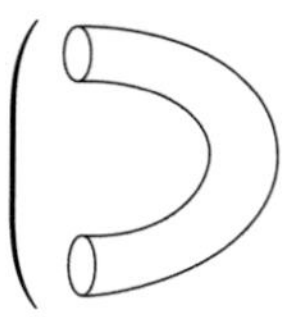) $= Z_\Gamma$ (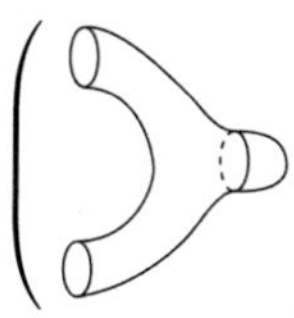) $: \ g \otimes h \longmapsto \operatorname{Tr}(gh) = \begin{cases} 1, & g = h^{-1}, \\ 0, & \text{otherwise} \end{cases}$.

Its inverse is then (using the same argument as in Problem I.4):

Z_Γ (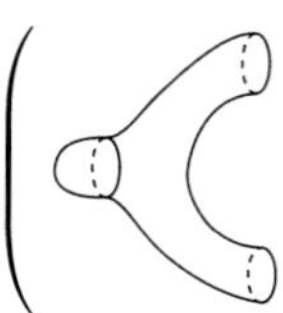) $= Z_\Gamma$ (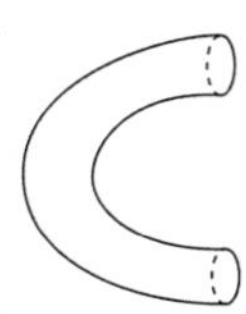) $: \ 1 \longmapsto \sum_{g \in \Gamma} g \otimes g^{-1}.$

Using the fact that

Z_Γ (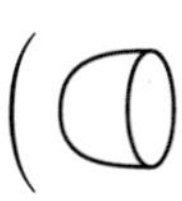) $: \quad 1 \longmapsto \mathbf{1}$

we then determine that

Z_Γ (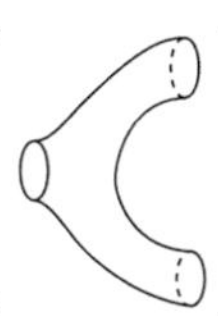) $: \mathbf{1} \longmapsto \sum_{g \in \Gamma} g \otimes g^{-1}$

and therefore

Z_Γ () $: \ \mathbf{1} \longmapsto |\Gamma| \cdot \mathbf{1}$

Finally

$$Z_\Gamma\Bigg($$ 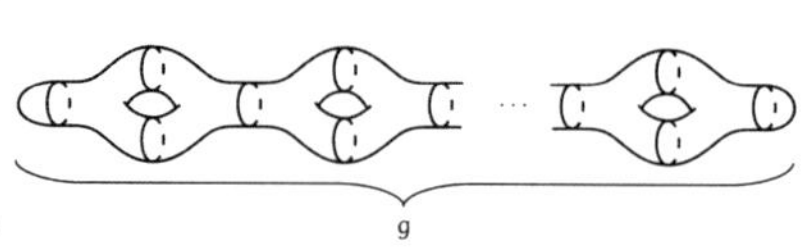 $$\Bigg) : 1 \longmapsto |\Gamma|^g \cdot 1$$

II.2

(1) Using the general approach described in the lectures, $\langle B, Z_{\mathcal{C}}(M_g)(A)\rangle$ is given in terms gluing of 2 copies of a solid tori, which are thickenings of coupons (colored by morphisms A and B) with arcs attached, according to the $SL(2,\mathbb{Z})$ element g. When $A = A_i$ and $B = B_j$, and after summing over the colors of the arcs connected to coupons, we end up having just unknots colored with colors i and j placed at $S^1 \times \{0\}$ in $S^1 \times D^2$.

For $g = \mathbf{1}$ the gluing produces $S^2 \times S^1$ (which can be realized as 0-surgery on unknot) with unknots placed at two different points of S^2 (we are assuming blackboard framing below):

$$\langle B_j, Z_{\mathcal{C}}(M_{id})(A_i)\rangle = \frac{1}{D}\sum_{\ell\in I}\frac{d_\ell}{D}$$  $$= \sum_{\ell\in I} S_{i\ell}S_{\ell j^*}$$

where we used that

 $$= \frac{\tilde{S}_{ij}}{d_i}$$  $$\in \mathrm{Hom}(V_i, V_i) \cong \mathbb{C}\cdot \mathrm{id}_{V_i}$$

which follows from the fact that $\mathrm{Hom}(V_i, V_i) \cong \mathbb{C}$ generated by id_{V_i} (Schur's lemma) and that closing the open strand above produces the Hopf link.

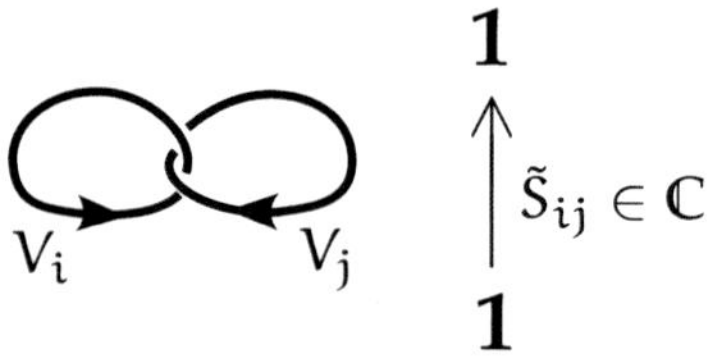

The case $g = T$ differs just by changing the framing of one of the unknots by 1.

$$\langle B_j, Z_{\mathcal{C}}(M_T)(A_i)\rangle = \frac{1}{D}\sum_{\ell\in I}\frac{d_\ell}{D}$$  $$= \sum_{\ell\in I} T_j S_{i\ell}S_{\ell j^*}$$

For $g = S$ the gluing produces S^3 (which can be realized as 1-surgery on unknot) with unknots forming the Hopf link.

$$\langle B_j, Z_{\mathcal{C}}(M_S)(A_i)\rangle = \frac{1}{D}\sum_{\ell\in I}\frac{d_\ell}{D}$$

$$= \frac{p_+}{D} S_{ij}$$

(2) This follows immediately from the fact that $Z_{\mathcal{C}}(M_{\mathrm{id}}) = \mathrm{id}_{Z_{\mathcal{C}}(T^2)}$ and the formula obtained above.

(3)

$$\langle B_j, Z_{\mathcal{C}}(M_{\mathrm{id}})(A_i)\rangle = \delta_{ij},$$

$$\langle B_j, Z_{\mathcal{C}}(M_T)(A_i)\rangle = T_j\delta_{ij},$$

$$\langle B_j, Z_{\mathcal{C}}(M_S)(A_i)\rangle = S_{ij}.$$

6.3. Lecture III

III.1

a) The fundamental group is $\pi_1(L(3,1)) = \pi_1(S^3/\mathbb{Z}_3) = \mathbb{Z}_3$. It is an abelian group with a single generator.
$\mathcal{M}^{\mathrm{flat}}_{SU(2)}(L(3,1)) = \mathrm{Hom}(\mathbb{Z}_3, SU(2))/SU(2) = \{A^3 = 1 | A \in SU(2)\}/SU(2)$.
Diagonalizing the matrix $A = \begin{pmatrix} \alpha & 0 \\ 0 & \alpha^{-1}\end{pmatrix}$ we get a set of 2 elements:

$$\mathcal{M}^{\mathrm{flat}}_{SU(2)}(L(3,1)) = \{\alpha^3 = 1\}/(\alpha \sim 1/\alpha)$$

b) Representing $L(3,1)$ as Dehn surgery on an unknot with framing 3 and using the explicit formula we have

$$Z_{SU(2)_k}(L(3,1)) = \sqrt{\frac{2}{k}}\sin\frac{\pi}{k}\,\frac{\sum_{n=1}^{k-1}\left(\frac{q^{n/2}-q^{-n/2}}{q^{1/2}-q^{-1/2}}\right)^2 q^{3\,\frac{n^2-1}{4}}}{\sum_{n=1}^{k-1}\left(\frac{q^{n/2}-q^{-n/2}}{q^{1/2}-q^{-1/2}}\right)^2 q^{\frac{n^2-1}{4}}}$$

$$= \sqrt{\frac{2}{k}}\sin\frac{\pi}{k}\,\frac{\sum_{n\in\mathbb{Z}_{2k}}(q^n+q^{-n}-2)\,q^{3\,\frac{n^2-1}{4}}}{\sum_{n\in\mathbb{Z}_{2k}}(q^n+q^{-n}-2)\,q^{\frac{n^2-1}{4}}}$$

$$= \sqrt{\frac{2}{k}}\sin\frac{\pi}{k}\,\frac{1}{\sqrt{3}}\,\frac{(2q^{1/3}-2)+2e^{\frac{2\pi i k}{3}}(2q^{1/3}\cos\frac{2\pi}{3}-2)}{q-1}$$

$$= \sqrt{\frac{2}{3k}}\,q^{-1/2}\left((q^{1/3}-1)+2e^{\frac{2\pi i k}{3}}(q^{1/3}\cos\frac{2\pi}{3}-1)\right)$$

c) This immediately follows from the expression above (the asymptotic expansion is obtained by expanding $q = e^{\frac{2\pi i}{k}}$)

III.2

a) Using the fact that

$$a_n^{(\mathrm{triv})} = \frac{1}{\sqrt{2i}}\,\frac{c_n}{n!}\left(\frac{\pi i}{84}\right)^n$$

and

$$\Gamma(n+1/2) = \frac{\sqrt{\pi}}{2^{2n}}\frac{(2n)!}{n!}$$

we have

$$\begin{aligned} B^{(\mathrm{triv})}(\xi) &:= \sum_{n\geqslant 0} \frac{a_n^{(\mathrm{triv})}}{\Gamma(n+3/2)}\xi^{n+1/2} \\ &= \frac{1}{\sqrt{2i}} \sum_{n\geqslant 0} \frac{c_n}{\Gamma(n+1/2)} \left(\frac{\pi i}{84}\right)^n \xi^{n-1/2} \\ &= \frac{2}{\sqrt{2\pi\xi i}} \left.\frac{\sinh(6z)\sinh(14z)\sinh(21z)}{\sinh(42z)}\right|_{z=\sqrt{\frac{\pi i \xi}{21}}} \end{aligned}$$

b) This immediately follows from the direct analysis of the singularities of the explicit expression above

c) Let us make the change of variable $z = \sqrt{\frac{\pi i \xi}{21}}$ in the sum of integrals (due to the nature of the question, we can work up to an overall k-independent factor):

$$\int_{e^{i(\pi/2+\epsilon)}\mathbb{R}_+} d\xi B^{(\mathrm{triv})}(\xi) + \int_{e^{i(\pi/2-\epsilon)}\mathbb{R}_+} d\xi B^{(\mathrm{triv})}(\xi) \propto \int_{i\mathbb{R}+\epsilon} F(z) e^{-\frac{21kz^2}{\pi i}}$$

where

$$F(z) = \frac{4\sinh(6z)\sinh(14z)\sinh(21z)}{\sinh(42z)}.$$

(For convergence of the integral we assume that $\operatorname{Im} k > 0$) When $z \in i\mathbb{R}+\epsilon$ it has a (uniformly) convergent expansion

$$F(z) = \sum_{n\geqslant 0} L_n e^{-nz}$$

where $L_n \in \mathbb{Z}$. Plugging this into the integral and performing gaussian integration we obtain

$$\int_{i\mathbb{R}+\epsilon} F(z) e^{-\frac{21kz^2}{\pi i}} \propto \frac{1}{\sqrt{k}} \sum_{n\geqslant 0} L_n q^{\frac{n^2}{168}} \in \frac{1}{\sqrt{k}} q^{\frac{1}{168}} \mathbb{Z}[[q]]$$

where the last statement follows from the fact that

$$L_n \neq 0 \implies n^2 = 1 \mod 168.$$

6.4. Lecture IV

IV.1 The first statement follows from the fact that $L(p,q)$ is a quotient of a simply-connected space by $\mathbb{Z}_p$:

$$\pi_1(L(p,q)) = \pi_1(S^3/\mathbb{Z}_p) = \mathbb{Z}_p$$

and its abelianization is the same: $H_1(L(p,q)) = \mathbb{Z}_p$.

Following the *hint* we have $L(p,q) \cong D^2 \times S^1 \cup_\phi D^2 \times S^1$

$$\begin{aligned} \phi_*[m'] &= q[m] + p[\ell] \\ \phi_*[\ell'] &= r[m] + s[\ell] \end{aligned}.$$

One can see that $H_1(L(p,q))$ can be generated by $[\ell]$. Indeed, $p[\ell]$ is trivial in $H_1(L(p,q))$ because $[m]$ and $[m']$ become trivial in $H_1(L(p,q))$. To determine $\ell k\,([\ell],[\ell])$ consider cycles $\ell_{1,2}$ placed at $x_{1,2}\times S^1$, $x_{1,2}\in D^2$, inside the second copy of $D^2\times S^1$. Obviously $[\ell_1]=[\ell_2]=[\ell]$. Deform the path $p\ell_2$ to the composition of two paths: one is the cycle on the boundary $\partial(D^2\times S^1)$ with a slope p/q and another which winds q times along the meridian. The first path bounds a disk in the first copy of $D^2\times S^1$ and does not intersect with ℓ_1. The second path bounds q copies of $D^2\times \mathrm{pt}$ in the second copy of $D^2\times S^1$ and intersects q times with ℓ_1. By definition we then have

$$\ell k\,([\ell],[\ell])=\frac{q}{p}\mod 1$$

And by linearity

$$\ell k\,(a[\ell],b[\ell])=\frac{qab}{p}\mod 1$$

IV.2 Take M^4 to be a simply-connected 4-manifold that bounds a rational homology sphere M^3. By Poincaré-Lefschetz duality and the universal coefficient theorem we have $\mathrm{Tor}\,H_2(M^4,\mathbb{Z})=\mathrm{Tor}\,H_2(M^4,M^3,\mathbb{Z})=0$ and $H_2(M^3,\mathbb{Z})=0$. We have the following short exact sequence:

(6.4.1)
$$\begin{array}{ccccccccc} & & & & H^2(M_4,\mathbb{Z}) & & & & \\ & & & & \uparrow\sim & & & & \\ 0 & \longrightarrow & H_2(M^4,\mathbb{Z}) & \longrightarrow & H_2(M^4,M^3,\mathbb{Z}) & \xrightarrow{\partial} & H_1(M^3,\mathbb{Z}) & \longrightarrow & 0. \end{array}$$

Moreover, using the de Rham theorem and the fact that there is no torsion, let us identify $H^2(M_4,\mathbb{Z})=\{C\in H^2_{\mathrm{dR}}(M)\mid \int_\Sigma C\in\mathbb{Z},\ \forall\Sigma\in H_2(M^4,\mathbb{Z})\}$. Then Poincaré-Lefschetz duality $H_2(M^4,M^3,\mathbb{Z})\xrightarrow{\sim}H^2(M^4,\mathbb{Z})$ is explicitly realized by $\Sigma\mapsto C_\Sigma$, such that $\int_\Xi C_\Sigma=\#(\Sigma\cap\Xi)$, for all $\Xi\in H_2(M^4,\mathbb{Z})$.

First, let us show that

(6.4.2)
$$\ell k\,(a,b)\overset{?}{=}\int_{M^4}C_{\Sigma_a}\wedge C_{\Sigma_b}\mod 1$$

for any $\Sigma_{a,b}\in H_2(M^4,M^3,\mathbb{Z})$ such that $\partial\Sigma_a=a$, $\partial\Sigma_b=b$. Let $n\in\mathbb{Z}$ such that $nb=0$. In particular this means that $n\Sigma_b$ can be lifted to a certain $\Xi\in H_2(M^4,\mathbb{Z})$.

$$n\int_{[M^4]}C_{\Sigma_a}\wedge C_{\Sigma_b}=\int_{[M^4]}C_{\Sigma_a}\wedge C_{n\Sigma_b}=\#(\Sigma_a\cap\Xi)$$

Let $c_{a,b}$ be 2-chains in M^4 representing $\Sigma_{a,b}$, and b' be a 2-chain in $M^3=\partial M^4$ such that $\partial b'=n\tilde{b}$, $[\tilde{b}]=b$. Then Ξ can be represented by a 2-chain $-b'+nc_b$. We then have

$$\begin{aligned}\#(\Sigma_a\cap\Xi)\mod n &=\#(c_a\cap(-b'+nc_b))\mod n\\ &=-\#(c_a\cap b'))\mod n\\ &=-\#(a\cap b')\mod n=-n\ell k\,(a,b)\mod n\end{aligned}$$

So indeed (6.4.2) is true (up to a sign).

Suppose the flat connection A_a on M^3 is extended to a connection on M^4.

$$e^{2\pi i \ell k(a,b)} = \mathrm{Hol}_b(A_a) = e^{i\int_{\Sigma_b}[F_{A_a}]} = e^{i\int C_{\Sigma_b}\wedge[F_{A_a}]}$$

It then follows that $[F_{A_a}/2\pi i] = C_{\Sigma_a}$ for some Σ_a (unique up to an element of $H_2(M^4,\mathbb{Z})$) such that $\partial\Sigma_a = a$. We then have

$$CS(A_a) := \frac{1}{4\pi^2}\int_{M^4} F_{A_a}\wedge F_{A_a} = \int_{[M^4]} C_{\Sigma_a}\wedge C_{\Sigma_a} = \ell k(a,a) \mod 1.$$

IV.3 a) The linking matrix is $M = (5)$ so that $\mathrm{Coker}(M)/\mathbb{Z}_2 = \mathbb{Z}_5/\mathbb{Z}_2 = \{0,1,2\}$. Using directly the formula we have

$$\hat{Z} = 2q^{-1/2}\begin{pmatrix} -1 \\ q^{-1/5} \\ 0 \end{pmatrix}$$

b) The linking matrix is

$$M = \begin{pmatrix} -1 & 1 & 1 & 1 \\ 1 & -2 & 0 & 0 \\ 1 & 0 & -3 & 0 \\ 1 & 0 & 0 & -7 \end{pmatrix}$$

so that $\mathrm{Coker}(M) = 0$. Using directly the formula we have

$$\hat{Z}_0 = q^{1/2}(1 - q - q^5 + q^{10} - q^{11} + q^{18} + q^{30} - \dots)$$

References

[1] Jørgen Ellegaard Andersen, *The asymptotic expansion conjecture* (English), Invariants of knots and 3–manifolds (kyoto 2001), 20046, pp. 474–480. ↑414

[2] M. Atiyah, *Topological quantum field theories*, Inst. Hautes Études Sci. Publ. Math. **68** (1988), 175–186 (1989). MR1001453 ↑408

[3] S. Axelrod and I. M. Singer, *Chern-Simons perturbation theory*, Proceedings of the XXth International Conference on Differential Geometric Methods in Theoretical Physics, Vol. 1, 2 (New York, 1991), World Sci. Publ., River Edge, NJ, 1992, pp. 3–45. MR1225107 ↑414

[4] S. Axelrod and I. M. Singer, *Chern-Simons perturbation theory. II*, J. Differential Geom. **39** (1994), no. 1, 173–213. MR1258919 ↑414

[5] R. Bott and A. S. Cattaneo, *Integral invariants of 3-manifolds*, J. Differential Geom. **48** (1998), no. 1, 91–133. MR1622602 ↑414

[6] R. Bott and A. S. Cattaneo, *Integral invariants of 3-manifolds. II*, J. Differential Geom. **53** (1999), no. 1, 1–13. MR1776090 ↑414

[7] T. Dimofte, S. Gukov, J. Lenells, and D. Zagier, *Exact results for perturbative Chern-Simons theory with complex gauge group*, Commun. Number Theory Phys. **3** (2009), no. 2, 363–443, DOI 10.4310/CNTP.2009.v3.n2.a4. MR2551896 ↑414

[8] R. Fenn and C. Rourke, *On Kirby's calculus of links*, Topology **18** (1979), no. 1, 1–15, DOI 10.1016/0040-9383(79)90010-7. MR528232 ↑406

[9] S. Gukov and C. Manolescu, *A two-variable series for knot complements*, Quantum Topol. **12** (2021), no. 1, 1–109, DOI 10.4171/qt/145. MR4233201 ↑421

[10] S. Gukov, M. Marino, and P. Putrov, *Resurgence in complex Chern-Simons theory* (2016), available at 1605.07615. ↑419

[11] S. Gukov, D. Pei, P. Putrov, and C. Vafa, *BPS spectra and 3-manifold invariants* (2017), available at 1701.06567. ↑421, 423

[12] S. Gukov, P. Putrov, and C. Vafa, *Fivebranes and 3-manifold homology*, J. High Energy Phys. **7** (2017), 071, front matter+80, DOI 10.1007/JHEP07(2017)071. MR3686727 ↑421

[13] R. Kirby, *A calculus for framed links in* S^3, Invent. Math. **45** (1978), no. 1, 35–56, DOI 10.1007/BF01406222. MR467753 ↑406

[14] R. Kirby and P. Melvin, *The 3-manifold invariants of Witten and Reshetikhin-Turaev for* $\mathrm{sl}(2, \mathbf{C})$, Invent. Math. **105** (1991), no. 3, 473–545, DOI 10.1007/BF01232277. MR1117149 ↑407

[15] J. Kock, *Frobenius algebras and 2D topological quantum field theories*, London Mathematical Society Student Texts, vol. 59, Cambridge University Press, Cambridge, 2004. MR2037238 ↑408

[16] M. Kontsevich, *Resurgence from the path integral perspective (Perimeter Institute, 2012); Exponential integrals (SCGP and at IHES, 2014 and 2015); Resurgence and wall-crossing via complexified path integral (TFC Sendai, 2016)*, Lectures. ↑415

[17] M. Kontsevich, *Feynman diagrams and low-dimensional topology*, First European Congress of Mathematics, Vol. II (Paris, 1992), Progr. Math., vol. 120, Birkhäuser, Basel, 1994, pp. 97–121. MR1341841 ↑414

[18] W. B. R. Lickorish, *A representation of orientable combinatorial 3-manifolds*, Ann. of Math. (2) **76** (1962), 531–540, DOI 10.2307/1970373. MR151948 ↑406

[19] T. Ohtsuki, *A polynomial invariant of integral homology 3-spheres*, Math. Proc. Cambridge Philos. Soc. **117** (1995), no. 1, 83–112, DOI 10.1017/S0305004100072935. MR1297897 ↑414

[20] T. Ohtsuki, *A polynomial invariant of rational homology 3-spheres*, Invent. Math. **123** (1996), no. 2, 241–257, DOI 10.1007/s002220050025. MR1374199 ↑414

[21] W. E. Petersen and J. E. Andersen, *Resurgence analysis of quantum invariants: Seifert manifolds and surgeries of the figure eight knot* (2018), available at `1904.06057`. ↑419

[22] N. Reshetikhin and V. G. Turaev, *Invariants of 3-manifolds via link polynomials and quantum groups*, Invent. Math. **103** (1991), no. 3, 547–597, DOI 10.1007/BF01239527. MR1091619 ↑407, 409

[23] N. Yu. Reshetikhin and V. G. Turaev, *Ribbon graphs and their invariants derived from quantum groups*, Comm. Math. Phys. **127** (1990), no. 1, 1–26. MR1036112 ↑410

[24] L. Rozansky, *A contribution of the trivial connection to the Jones polynomial and Witten's invariant of 3d manifolds. I, II*, Comm. Math. Phys. **175** (1996), no. 2, 275–296, 297–318. MR1370097 ↑414

[25] V. G. Turaev, *Quantum invariants of knots and 3-manifolds*, De Gruyter Studies in Mathematics, vol. 18, De Gruyter, Berlin, 2016. Third edition [of MR1292673], DOI 10.1515/9783110435221. MR3617439 ↑409

[26] A. H. Wallace, *Modifications and cobounding manifolds*, Canadian J. Math. **12** (1960), 503–528, DOI 10.4153/CJM-1960-045-7. MR125588 ↑406

[27] E. Witten, *Quantum field theory and the Jones polynomial*, Comm. Math. Phys. **121** (1989), no. 3, 351–399. MR990772 ↑407, 414, 419

[28] E. Witten, *Analytic continuation of Chern-Simons theory*, Chern-Simons gauge theory: 20 years after, AMS/IP Stud. Adv. Math., vol. 50, Amer. Math. Soc., Providence, RI, 2011, pp. 347–446. MR2809462 ↑415

ICTP, Strada Costiera 11, 34151 Trieste, Italy
Email address: `putrov@ictp.it`

IAS/Park City Mathematics Series
Volume 28, Pages 439–476
https://doi.org/10.1090/pcms/028/00881

Lectures on BPS states and spectral networks

Andrew Neitzke

Abstract. These are notes for a lecture series on BPS states and spectral networks, delivered at Park City Mathematics Institute, July 2019. The first part is a general review of the notions of BPS state and BPS index. The second part discusses the specific case of BPS states in $\mathcal{N} = (2,2)$ supersymmetric field theories in two dimensions, and introduces the notion of spectral network as a way of computing the BPS indices in that context. The last part discusses the more general case of 2d and 4d BPS indices associated to surface defects in four-dimensional field theories of class S.

Contents

2020 *Mathematics Subject Classification.* Primary 81T60.
Key words and phrases. BPS states, spectral networks.

1. Lecture 1: What is a BPS state?

BPS states appear very frequently in geometric applications of quantum field theory. The aim of this lecture is to explain rather generally what a BPS state is and some of their basic properties.

In one sentence: we'll study a representation $\mathcal{H} = \mathcal{H}^0 \oplus \mathcal{H}^1$ of a certain super Lie algebra $\mathcal{A} = \mathcal{A}^0 \oplus \mathcal{A}^1$, and the BPS states are the ones in irreps annihilated by nontrivial subspaces of $\mathcal{A}^1$.

1.1. Quantum mechanics A time-independent quantum system involves the following data:

- A Hilbert space $\mathcal{H}$,
- A formally self-adjoint operator $H : \mathcal{H} \to \mathcal{H}$.

We think of iH as generating the abelian Lie algebra

$$\text{Lie}(\text{ISO}(0,1)) = \text{Lie}(\text{Isom}(\mathbb{R}^{0,1})) \simeq \mathbb{R}. \tag{1.1.1}$$

Eigenvectors of H are called *bound states*; each bound states thus spans a 1-dimensional irreducible representation of $\text{ISO}(0,1)$. The fact that H is formally self-adjoint implies that this representation is unitary. The eigenvalues of H are called *bound state energies*.

Example 1.1.2 (Particle on the line)**.** In your first course on quantum mechanics you study the "particle on the line." For this example you need to fix a function $V : \mathbb{R} \to \mathbb{R}$. Then there is a time-independent quantum system $T_1[\mathbb{R}, V]$ with:

- $\mathcal{H} = L^2(\mathbb{R})$,
- $H = -\frac{1}{2}\frac{d^2}{dx^2} + V$.

Fact 1.1.3. If $V(x)$ is smooth and $V(x) \to \infty$ as $|x| \to \infty$, then H has discrete spectrum with no accumulation points, and $\mathcal{H}$ has a (Schauder) basis consisting of eigenvectors of H. There is a unique eigenvector with smallest eigenvalue, which we call the *ground state*; we call the other eigenvectors *excited states*.

Example 1.1.4 (Harmonic oscillator)**.** If we take $V(x) = \frac{1}{2}x^2$ in the above then the ground state turns out to be $\psi(x) = e^{-\frac{1}{2}x^2}$, which has $H\psi(x) = \frac{1}{2}\psi(x)$, so that the ground state energy is $\frac{1}{2}$. The bound state energies are $n + \frac{1}{2}$, $n \in \mathbb{N}$.

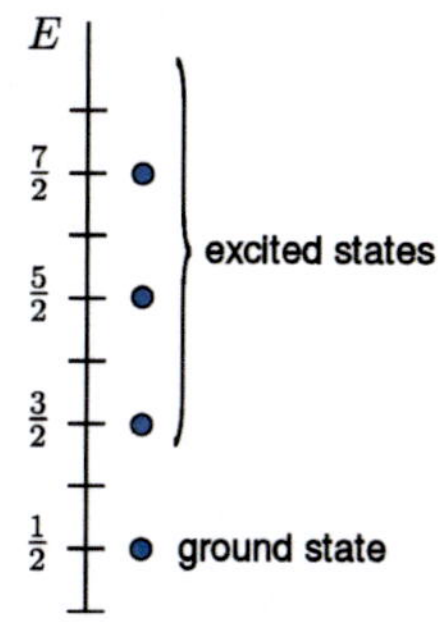

Remark 1.1.5. If $V(x)$ is more general, then $\mathcal{H}$ need not have a basis consisting of eigenvectors of H. For an extreme example, if $V(x) = 0$ there are no L^2 eigenvectors at all. For some purposes the functions $\psi_p(x) = e^{ipx}$, parameterized by $p \in \mathbb{R}$, can stand in. These functions clearly have formally $H\psi_p = \frac{p^2}{2}\psi_p$. Since $\frac{p^2}{2}$ takes all nonnegative real values, one might then expect that H has continuous spectrum consisting of all nonnegative real numbers, and this is indeed true. For more general choices of $V(x)$, H could have some discrete and some continuous spectrum, as indicated in the figure below.

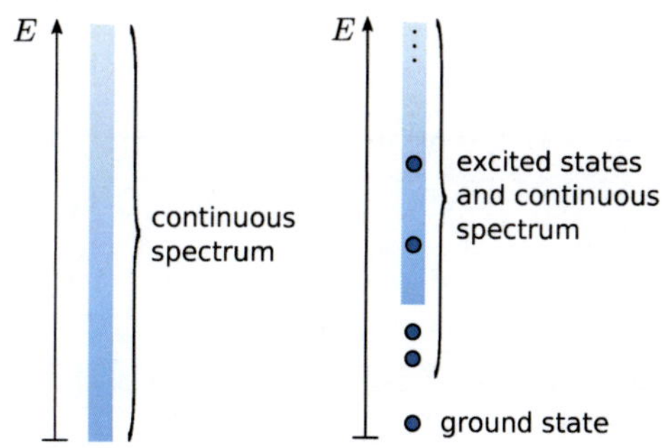

Next we consider replacing the domain $\mathbb{R}$ by a general Riemannian manifold:

Definition 1.1.6 (Laplace operator)**.** For M a Riemannian manifold, the Laplace operator $\Delta : L^2(M) \to L^2(M)$ is given by

$$\Delta = d^* d \tag{1.1.7}$$

where d^* is the formal adjoint of $d : L^2(M) \to \Omega^1_{L^2}(M)$.

Example 1.1.8 (Particle on a Riemannian manifold)**.** Fix a Riemannian manifold M and a function $V : M \to \mathbb{R}$. Then there is a quantum mechanical system $T_1[M, V]$, the "particle on M with potential V," with:

- $\mathcal{H} = L^2(M)$,
- $H = \frac{1}{2}\Delta + V$.

In particular, suppose M is compact and we take $V = 0$. In this case, $H = \frac{1}{2}\Delta$ has only discrete spectrum, with eigenvalues bounded below by 0, and $H\psi = 0$ iff ψ is a constant function.

1.2. The superparticle Now we come to our first supersymmetric example, following [51].

Definition 1.2.1 (Laplace operator on differential forms). The Laplace operator $\Delta : \Omega^*_{L^2}(M) \to \Omega^*_{L^2}(M)$ is given by

$$\Delta = [d, d^*] = dd^* + d^*d \tag{1.2.2}$$

where d^* is the formal adjoint of $d : \Omega^*_{L^2}(M) \to \Omega^*_{L^2}(M)$.

Example 1.2.3 (Superparticle on a Riemannian manifold). Again fix a Riemannian manifold M. Then there is a quantum mechanical system $T_1[M]$, the "superparticle on M", with:

- $\mathcal{H} = \Omega^*_{L^2}(M)$,
- $H = \frac{1}{2}\Delta$.

The key new feature is that in this case there are more operators around than just H: $\mathcal{H}$ is a unitary $\mathbb{Z}/2\mathbb{Z}$-graded representation of a Lie superalgebra, defined as follows.

Definition 1.2.4 ($\mathcal{N} = 2$ supersymmetry in dimension $d = 1$). [1] We define a Lie superalgebra $\mathcal{A}$ by

$$\mathcal{A} = \mathcal{A}^0 \oplus \mathcal{A}^1, \quad \mathcal{A}^0 = \mathbb{C} \cdot H, \quad \mathcal{A}^1 = \mathbb{C} \cdot Q \oplus \mathbb{C} \cdot \overline{Q}, \tag{1.2.5}$$

i.e. $\mathcal{A}$ has 2 odd generators $Q, \overline{Q}$ and one even generator H, with the brackets[2]

$$[Q, \overline{Q}] = 2H, \quad [Q, Q] = 0, \quad [\overline{Q}, \overline{Q}] = 0, \quad [Q, H] = 0, \quad [\overline{Q}, H] = 0. \tag{1.2.6}$$

A $\mathbb{Z}/2\mathbb{Z}$-*graded* representation of $\mathcal{A}$ is a representation of $\mathcal{A}$ on a $\mathbb{Z}/2\mathbb{Z}$-graded vector space $\mathcal{H} = \mathcal{H}^0 \oplus \mathcal{H}^1$, where $\mathcal{A}^i$ maps $\mathcal{H}^j \to \mathcal{H}^{j+i}$. A *unitary* representation of $\mathcal{A}$ is a representation in which $\mathcal{H}$ is a Hilbert space, H acts by a formally self-adjoint operator, and $Q, \overline{Q}$ act by operators which are formally adjoint to one another.

To realize $\mathcal{H}$ as a representation of $\mathcal{A}$ we just take

$$Q = d, \qquad \overline{Q} = d^*, \qquad H = \frac{1}{2}\Delta. \tag{1.2.7}$$

Then $\mathcal{H}$ is a unitary representation of $\mathcal{A}$. It is also $\mathbb{Z}/2\mathbb{Z}$-graded:

$$\mathcal{H} = \mathcal{H}^0 \oplus \mathcal{H}^1, \quad \mathcal{H}^0 = \bigoplus_k \Omega^{2k}_{L^2}(M), \quad \mathcal{H}^1 = \bigoplus_k \Omega^{2k+1}_{L^2}(M). \tag{1.2.8}$$

Now let us explore a bit of the unitary representation theory of $\mathcal{A}$, following [50]. The unitarity implies that all eigenvalues of H are nonnegative, because

$$2\langle\psi, H\psi\rangle = \langle\psi, Q\overline{Q}\psi\rangle + \langle\psi, \overline{Q}Q\psi\rangle = \|Q\psi\|^2 + \|\overline{Q}\psi\|^2 \geqslant 0. \tag{1.2.9}$$

---

[1]The $\mathcal{N} = 2$ refers to the fact that we have 2 odd generators. The $d = 1$ refers to the fact that H generates $ISO(0,1)$, the isometries of $\mathbb{R}^{0,1}$ — it does *not* have to do with the dimension of M, which can be anything.

[2]Our convention is that $[,]$ means the *graded* bracket, i.e. for objects x, y which are in grade n_x, n_y respectively, $[x, y] = xy - (-1)^{n_x n_y} yx$. So this bracket is a commutator unless both x and y are odd, in which case it is an anticommutator.

Moreover the norm $\|\cdot\|$ is nondegenerate, so we conclude that

$$H\psi = 0 \quad \Leftrightarrow \quad Q\psi = 0, \overline{Q}\psi = 0. \tag{1.2.10}$$

In particular, each state with $H\psi = 0$ generates a 1-dimensional (trivial) representation of $\mathcal{A}$. We call this representation V_0^0 or V_0^1 depending whether the single state is in $\mathcal{H}^0$ or $\mathcal{H}^1$. These representations are called "short." The only other possibility for a unitary irreducible $\mathbb{Z}/2\mathbb{Z}$-graded representation is a 2-dimensional representation, with one state in $\mathcal{H}^0$ and one in $\mathcal{H}^1$, both with $H\psi = E\psi$ for some $E > 0$; these representations are called "long."

Fact 1.2.11. If M is compact, then $\mathcal{H}$ of Example 1.2.3 is a countable orthogonal direct sum of unitary irreducible representations of $\mathcal{A}$. (We say $\mathcal{H}$ "contains only discrete spectrum.") See the figure below, where each dot represents one state; note that the states with $E > 0$ come paired up into long representations, while those with $E = 0$ are in short representations by themselves.

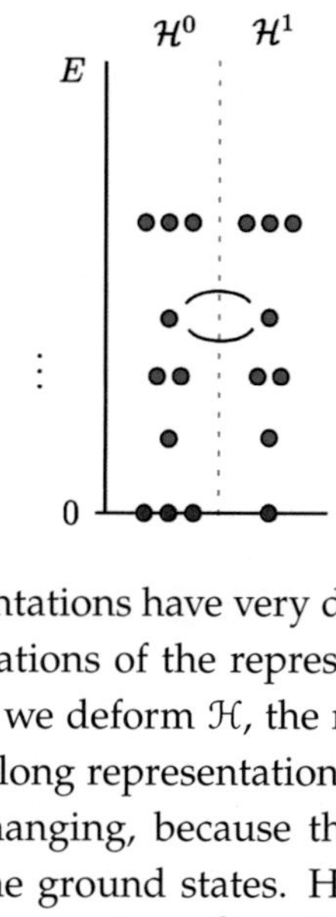

The short and long representations have very different character, which we see clearly if we consider deformations of the representation $\mathcal{H}$, e.g. by varying the Riemannian metric on M. As we deform $\mathcal{H}$, the nonzero eigenvalues $E > 0$ of H can change continuously: the long representations are not rigid. The eigenvalues $E = 0$ have a harder time changing, because the representations V_0^i are rigid. This fact helps to "protect" the ground states. However, it doesn't protect them absolutely: the reducible representation $V_0^0 \oplus V_0^1$ is not rigid, since it can deform to a long representation with $E = \epsilon > 0$. See the figure below.

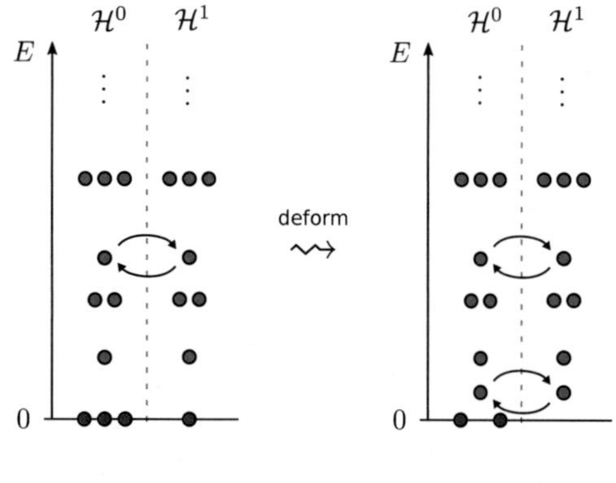

With this deformation process in mind we consider the following quantity.

Definition 1.2.12 (Index for representations of $\mathcal{N} = 2$ supersymmetry in $d = 1$). The *index*, or (signed) *ground state degeneracy*, of a representation $\mathcal{H}$ of $\mathcal{A}$ is

$$\chi(\mathcal{H}) = (\text{\# copies of } V_0^0 \text{ in } \mathcal{H}) - (\text{\# copies of } V_0^1 \text{ in } \mathcal{H}). \tag{1.2.13}$$

The key property of the index is that it is invariant under deformations of $\mathcal{H}$, as long as $\mathcal{H}$ contains only discrete spectrum: if inside $\mathcal{H}$ a copy of $V_0^0 \oplus V_0^1$ deforms into a long representation, then χ changes by $1 - 1 = 0$.

The main lesson here is: while the full Hilbert space $\mathcal{H}$ depends strongly on every little detail of the system, by using a little bit of the representation theory of the supersymmetry algebra $\mathcal{A}$ — looking at representations which are particularly rigid — we are able to extract a more robust and invariant quantity. A second lesson is that the rigid representations are the ones which are smaller than usual, by virtue of being annihilated by part of $\mathcal{A}$ (in this case actually all of $\mathcal{A}$).

1.3. Q-cohomology Here is another viewpoint on the ground states. In any representation of $\mathcal{A}$, say $W = W^0 \oplus W^1$, we can define the Q-cohomology,

$$H_Q^i(W) = \ker Q|_{W^i} / \operatorname{Im} Q|_{W^{i-1}}. \tag{1.3.1}$$

In the case of Example 1.2.3 the Q-cohomology is something well known: it is the de Rham cohomology of M (with the $\mathbb{Z}$-grading collapsed to a $\mathbb{Z}/2\mathbb{Z}$-grading),

$$H_Q^i(\mathcal{H}) = H_{dR}^i(M) = \ker d|_{W^i} / \operatorname{Im} d|_{W^{i-1}}. \tag{1.3.2}$$

Now $H_Q^i(\bigoplus_\alpha W_\alpha) \simeq \bigoplus_\alpha H_Q^i(W_\alpha)$, so to compute $H_Q^i(W)$ it's enough to understand the cohomology of each irreducible constituent of W. When W is irreducible,

$$H_Q^i(W) = \begin{cases} W & \text{if } W \simeq V_0^i, \\ 0 & \text{otherwise.} \end{cases} \tag{1.3.3}$$

Thus only the short representations in the decomposition of W contribute to $H_Q^i(W)$, and each contributes a 1-dimensional space. (This is a $\mathbb{Z}/2\mathbb{Z}$-graded version of the Hodge theorem for compact Riemannian manifolds.)

In particular, the deformation invariant index χ which we discussed has a simple interpretation:

$$\chi(\mathcal{H}) = \dim H_Q^0(\mathcal{H}) - \dim H_Q^1(\mathcal{H}) = \dim H_{dR}^0(M) - \dim H_{dR}^1(M) \tag{1.3.4}$$

i.e. it is the Euler characteristic of M!

This supersymmetric-quantum-mechanics viewpoint on the cohomology of M was introduced by Witten in [51]. There he also considered various deformations of the story, which we won't have time to treat here. One modification involves adding a Morse function on M (this becomes especially interesting in the limit where the Morse function is very large; one gets a description of the cohomology of M in terms of critical points of the Morse function and gradient flows between them.) Another involves passing to *equivariant* cohomology for a G-action on M.

1.4. Richer examples So far we've discussed how cohomology of a compact Riemannian manifold M arises as a space of supersymmetric ground states of a time-independent quantum system $T_1[M]$. This was essentially a reinterpretation of the usual story of Hodge theory of compact Riemannian manifolds.

We briefly mention a few richer examples of the same structure, to make contact with other lectures at this school. These are not rigorous statements, but rather accounts of what is claimed in the physics literature.

Physics Example 1.4.1 (Lagrangian Floer homology). Suppose M is a Kähler manifold. There is a 2-dimensional quantum field theory $T_2[M]$ known as the supersymmetric sigma model into M, with $\mathcal{N} = (2,2)$ supersymmetry. A Lagrangian submanifold $L \subset M$ gives a supersymmetric boundary condition in $T_2[M]$. Now consider $T_2[M]$ in the 2-dimensional spacetime $X = [0,1] \times \mathbb{R}$, with boundary conditions at the two sides induced by L and L'.

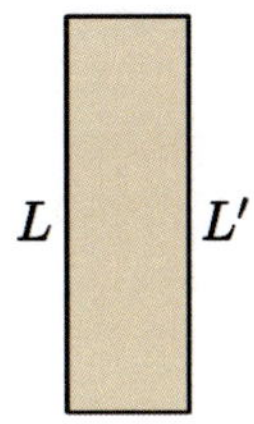

This system is effectively 1-dimensional: in particular its geometric symmetry is just $\mathrm{ISO}(0,1)$. It gives a time-independent quantum system $T_1[M, L, L']$ which is $\mathcal{N} = 2$ supersymmetric, so that its Hilbert space $\mathcal{H}$ is a representation of $\mathcal{A}$. The space of supersymmetric ground states is expected to be isomorphic to (some appropriate version of) Lagrangian Floer homology $HF(L, L')$.

Physics Example 1.4.2 (Khovanov-Rozansky homology). Let $\mathfrak{g}$ be a Lie algebra of ADE type (eg $\mathfrak{g} = \mathfrak{sl}(N)$.) Let L be a link in $\mathbb{R}^3$, and R an irreducible finite-dimensional representation of $\mathfrak{g}$. There is a 6-dimensional quantum field theory $T_6[\mathfrak{g}]$, the $(2,0)$ superconformal field theory of type $\mathfrak{g}$, which admits 2-dimensional defects of various types, one for each R. The theory $T_6[\mathfrak{g}]$ can be formulated on the spacetime $\mathbb{R}^{5,1}$ with a defect inserted along a 2-dimensional submanifold $S \subset \mathbb{R}^{5,1}$:

$$\mathbb{R}^{5,1} = \mathbb{R}^3 \times \mathbb{R}^2 \times \mathbb{R}^{0,1} \tag{1.4.3}$$

$$S = L \times \{pt\} \times \mathbb{R}^{0,1} \tag{1.4.4}$$

The geometric symmetry remaining is $\mathrm{ISO}(0,1) \times \mathrm{SO}(2)$. This gives a time-independent quantum mechanical system $T_1[\mathfrak{g}, R, L]$ which is $\mathcal{N} = 2$ supersymmetric,[3] so that its Hilbert space $\mathcal{H}$ is a representation of $\mathcal{A}$.

[3]For physicists: a cheap way of getting this counting is to remember from [57] that in $\mathcal{N} = 4$ super Yang-Mills a Wilson loop constrained to a 3-dimensional subspace is $\frac{1}{8}$-BPS, i.e. it preserves 2 of the 16 supercharges.

The full $\mathcal{H}$ probably depends on every little detail of L, but optimistically, the space of supersymmetric ground states should be well defined, finite-dimensional and depend only on the isotopy class of L; in fact it should be isomorphic to the Khovanov-Rozansky homology of L. (This is more or less conjectured in [56] Section 5.1.6, in turn a reinterpretation of a proposal in [26].)

In this case there are two $U(1)$ symmetries in the theory, so the Hilbert space $\mathcal{H}$ is $\mathbb{Z} \times \mathbb{Z}$-graded, as the Khovanov-Rozansky homology is. One of these symmetries comes from the rotational SO(2) mentioned above, the other comes from an internal "R-symmetry" of the $(2,0)$ theory.[4]

1.5. Field theory In the remainder of these lectures we will be interested in a more complicated class of examples, which arise from quantum field theory (QFT) in Minkowski space $\mathbb{R}^{d-1,1}$.

Such a QFT is an example of a time-independent quantum system, and so it determines a Hilbert space $\mathcal{H}$, as before. However, if we study this Hilbert space the way we did before, we will find that everything is unpleasantly infinite: in particular, H has no discrete spectrum except maybe the vacuum state. To make further progress we need to organize $\mathcal{H}$ better. We use the fact that instead of $ISO(0,1)$ which appeared before, now $\mathcal{H}$ is a unitary representation of a larger group, the Poincare group

$$ISO(d-1,1) = SO(d-1,1) \ltimes \mathbb{R}^{d-1,1} \subset \mathrm{Isom}(\mathbb{R}^{d-1,1}) \tag{1.5.1}$$

or more precisely its universal cover $ISpin(d-1,1)$. Thus, we need to know a little bit about the unitary representations of $ISpin(d-1,1)$.

The Lie algebra of $ISpin(d-1,1)$ is spanned by $\mathfrak{so}(d-1,1)$ plus translation generators P^i, $i \in \{0,1,\ldots,d-1\}$, with $H = P^0$. In the universal enveloping algebra there is a quadratic Casimir operator

$$\rho = (P^0)^2 - (P^1)^2 - (P^2)^2 - \cdots - (P^{d-1})^2 \tag{1.5.2}$$

which thus acts by a scalar in every irreducible representation. We will only be interested in representations where this scalar is nonnegative. It will be convenient to write M for a nonnegative square root of this scalar, i.e. $M = \sqrt{\rho}$.

A key device for understanding unitary representations of $ISpin(d-1,1)$ is to first diagonalize the translation generators, then focus on a single eigenspace, as follows.

Definition 1.5.3 (Rest subspace). Given a representation V of $ISpin(d-1,1)$, in which ρ acts by a scalar, the *rest subspace* of V is

$$V^{\mathrm{rest}} = \{\psi \in V : P^0\psi = M\psi,\ P^i\psi = 0 \text{ for } i \in \{1,\ldots,d-1\}\}. \tag{1.5.4}$$

Given V, the subspace V^{rest} is a representation of the subgroup

$$G^{\mathrm{rest}} = \mathrm{Spin}(d-1) \ltimes \mathbb{R}^{d-1,1}. \tag{1.5.5}$$

[4]The $(2,0)$ theory itself has SO(5) R-symmetry, but the surface defect on $L \times \mathbb{R}^{0,1}$ breaks that down to $U(1)$.

The action of the translation group $\mathbb{R}^{d-1,1}$ on V^{rest} is by the fixed character $(M, 0, \ldots, 0)$. In contrast, the structure of V^{rest} as a representation of $\text{Spin}(d-1)$ can be arbitrary (though unitary). Conversely, suppose given an irreducible representation S of $\text{Spin}(d-1)$ and a scalar $M > 0$; then we may extend S to a representation V^{rest} of G^{rest} by letting the translations act by the character $(M, 0, \ldots, 0)$, and there is a canonical procedure ("induction") for extending V^{rest} to a unitary irreducible representation V of the full $\text{ISpin}(d-1,1)$. Thus we have:

Fact 1.5.6 (1-particle representations of Poincare group). Given a unitary irreducible representation S of the compact group $\text{Spin}(d-1)$ and a scalar $M > 0$, there is a corresponding unitary irreducible representation $V_{S,M}$ of $\text{ISpin}(d-1,1)$, such that $V^{\text{rest}}_{S,M}$ is identical to S as a representation of $\text{Spin}(d-1)$.

Physically $V_{S,M}$ should be thought of as representing the space of states of a massive particle propagating in $\mathbb{R}^{d-1,1}$, with rest mass M, and spin governed by the representation S.

Our main focus in what follows will be theories with a mass gap. In such theories a typical structure for the full Hilbert space $\mathcal{H}$ is as follows:

- $\mathcal{H}$ contains one copy of the trivial representation of $\text{ISpin}(d-1,1)$; this represents the quantum vacuum of the theory.
- $\mathcal{H}$ contains a collection of direct summands, of the form $V_{S,M}$, with masses M_n where $0 < M_1 \leqslant M_2 \leqslant M_3 \leqslant \cdots$. These summands represent 1-particle states. M_1 is called the mass gap.
- The full $\mathcal{H}$ is not an orthogonal direct sum of irreducible representations of $\text{ISpin}(d-1,1)$. In particular, the operator M has continuous spectrum $[2M_1, \infty)$. The continuous spectrum reflects the contributions of multi-particle states.

See the figure for a schematic picture of the spectrum of M.

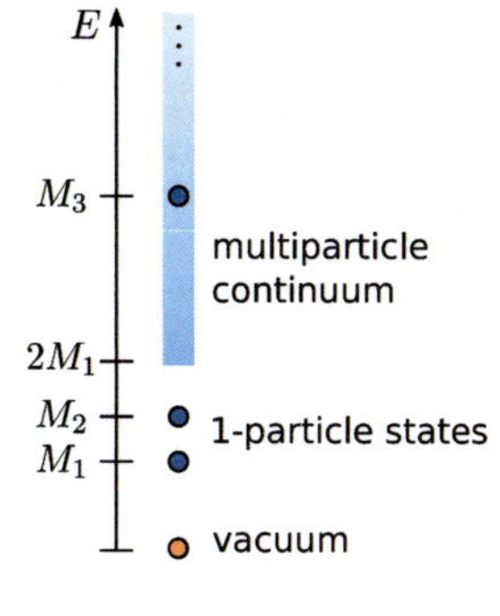

1.6. Supersymmetric field theory Now let's consider the Hilbert spaces of supersymmetric field theories. These will be representations of a super Lie algebra $\mathcal{A}$ extending $\text{Lie}(\text{ISpin}(d-1,1))$, in parallel to Example 1.2.3 above where we met a super extension of $\text{Lie}(\text{ISpin}(0,1))$.

The story has a different flavor depending on the dimension. We will focus on two examples.

Example 1.6.1 ($\mathcal{N}=(2,2)$ supersymmetry in $d=2$)**.** The supersymmetry algebra $\mathcal{A}$ which we will consider in this case has 4 odd generators,

$$Q^\pm, \quad \overline{Q}^\pm, \tag{1.6.2}$$

and 6 even generators,

$$P^0, \quad P^1, \quad B, \quad Z, \quad \overline{Z}, \quad F. \tag{1.6.3}$$

Here B is a generator of $\mathfrak{so}(1,1)$. Z and $\overline{Z}$ are central generators, known as central charges. The nonvanishing commutation relations of the odd generators are

$$[Q^+,\overline{Q}^+] = P^+, \quad [Q^+,Q^-] = Z, \tag{1.6.4}$$

$$[Q^-,\overline{Q}^-] = P^-, \quad [\overline{Q}^+,\overline{Q}^-] = \overline{Z}, \tag{1.6.5}$$

where we defined

$$P^\pm = P^0 \pm P^1 = H \pm P^1. \tag{1.6.6}$$

The generator F, "fermion number," obeys $[F,Q^\pm] = -Q^\pm$ and $[F,\overline{Q}^\pm] = \overline{Q}^\pm$.[5] The only other nonvanishing commutation relations are those between the $\mathfrak{so}(1,1)$ generator B and the translations P (which transform in the fundamental representation) and the odd generators (which are spinors).

Now we want to study unitary $\mathbb{Z}/2\mathbb{Z}$-graded representations of $\mathcal{A}$. As before we start by considering the subalgebra $\mathcal{A}^{\text{rest}}$, generated by $Q^\pm$, $\overline{Q}^\pm$, P^0, P^1, F, Z, $\overline{Z}$. We fix some $M > 0$ and $Z \in \mathbb{C}$, and consider a representation of $\mathcal{A}^{\text{rest}}$ where $P = (P^0, P^1)$ acts by the character $(M, 0)$ for some $M > 0$, and Z, $\overline{Z}$ act by the scalars Z, $\overline{Z}$. Then the brackets of odd generators become

$$[Q^+,\overline{Q}^+] = M, \quad [Q^+,Q^-] = Z, \tag{1.6.7}$$

and

$$[Q^-,\overline{Q}^-] = M, \quad [\overline{Q}^+,\overline{Q}^-] = \overline{Z}. \tag{1.6.8}$$

Now consider the generator

$$Q_\vartheta = \frac{1}{\sqrt{2}}\left(e^{i\vartheta/2}Q^+ + e^{-i\vartheta/2}\overline{Q}^-\right). \tag{1.6.9}$$

This has

$$[Q_\vartheta, \overline{Q}_\vartheta] = M + \frac{1}{2}(e^{i\vartheta}Z + e^{-i\vartheta}\overline{Z}). \tag{1.6.10}$$

We optimize by choosing $\vartheta = \pi - \arg Z$. Then

$$[Q_\vartheta, \overline{Q}_\vartheta] = M - |Z|. \tag{1.6.11}$$

But now

$$(M-|Z|)\|\psi\|^2 = \langle\psi, [Q_\vartheta, \overline{Q}_\vartheta]\psi\rangle = \|Q_\vartheta\psi\|^2 + \|\overline{Q}_\vartheta\psi\|^2 \geqslant 0 \tag{1.6.12}$$

implies the bound (sometimes called BPS bound)

$$M \geqslant |Z|, \tag{1.6.13}$$

[5]Other variants are possible; for what we do here, all that is really important is that F behaves as "fermion number" in the sense that each odd generator Q shifts it by ± 1.

which moreover is saturated if and only if

$$Q_\vartheta \psi = \overline{Q}_\vartheta \psi = 0. \tag{1.6.14}$$

The nature of the representations depends on whether the BPS bound (1.6.13) is saturated. Indeed, the relations (1.6.7)-(1.6.8) say that the four Q's generate a Clifford algebra:

- When $M > |Z|$ this Clifford algebra is nondegenerate, and its irreducible representations have dimension $2^{4/2} = 4$.[6]
- When $M = |Z|$ we've seen that Q_ϑ and $\overline{Q}_\vartheta$ act identically as zero, and the Clifford algebra is degenerate: its irreducible representations have only dimension $2^{2/2} = 2$.

By induction as before, these representations of $\mathcal{A}^{\mathrm{rest}}$ can be extended to representations of the whole algebra $\mathcal{A}$, with the following result:

Fact 1.6.15 (1-particle representations of $\mathcal{N} = (2,2)$ supersymmetry in $d = 2$)**.** Given a complex number $Z \neq 0$, a real scalar $M \geqslant |Z|$, and a real scalar f, there is a unique irreducible unitary representation $V^f_{Z,M}$ of $\mathcal{A}$, in which the Casimir operator ρ acts as M^2, the generators $Z, \overline{Z} \in \mathcal{A}$ act as the complex numbers Z, $\overline{Z}$, and the smallest eigenvalue of F is f.

The representations $V^f_{Z,M}$ with $M = |Z|$ are called *short* or *BPS*, while the $V^f_{Z,M}$ with $M > |Z|$ are called *long*. The short representations are separately rigid, but $V^f_{Z,M} \oplus V^{f+1}_{Z,M}$ can deform to a long representation $V^f_{Z,M}$, in parallel to what we saw earlier in supersymmetric quantum mechanics.

Now let us restrict attention to the case where the eigenvalues of F are all integers.[7] Then we define a new index counting short representations with signs:

Definition 1.6.16 (Index for representations of $\mathcal{N} = (2,2)$ supersymmetry in $d = 2$)**.**

$$\mu(\mathcal{H}) = \sum_{f \in \mathbb{Z}} (-1)^f (\text{\# copies of } V^f_{Z,M=|Z|} \text{ in } \mathcal{H}) \tag{1.6.17}$$

$$= \mathrm{Tr}_{\mathcal{H}^{1,\mathrm{BPS,rest}}} (-1)^F F. \tag{1.6.18}$$

Remark 1.6.19 (The phase of the central charge)**.** Note that the phase of the complex number Z plays a critical role in this story: it determines *which* 2-dimensional subspace of the 4-dimensional $\mathcal{A}^1$ annihilates the BPS states.

Remark 1.6.20 (Charge gradings)**.** We will typically consider situations in which the Hilbert space $\mathcal{H}$ has a further grading by "charges"

$$\mathcal{H} = \bigoplus_{\alpha} \mathcal{H}_\alpha \tag{1.6.21}$$

[6] Recall that, when n is even, the nondegenerate complex Clifford algebra on a vector space of dimension n has a unique irreducible representation, which has dimension $2^{n/2}$.

[7] In examples which occur in nature one sometimes gets a canonically defined F which does not have integer eigenvalues, but then one can define some other operator $F' = F + \delta$, where δ acts as a scalar in each $\mathcal{H}_\alpha$, such that F' does have integer eigenvalues. We will not try to discuss this subtle issue here.

with an additive structure, such that in each sector $\mathcal{H}_a$ the central charge acts by a constant $Z_a \in \mathbb{C}$, and $Z_{a+b} = Z_a + Z_b$. In this situation we can define $\mu(\mathcal{H}_a)$ for each sector separately.

Remark 1.6.22 (Deformation invariance and wall-crossing)**.** Formally it looks like the indices $\mu(\mathcal{H}_a)$ deserve to be invariant under deformations of $\mathcal{H}$, as was the index in Definition 1.2.12, but there is a hitch: as we noted before, in QFT $\mathcal{H}$ always contains continuous spectrum, and this could spoil our arguments. As long as the continuum stays bounded away from $M = |Z|$ we will be safe (we can just restrict to the part of the Hilbert space below the continuum), but if it touches $M = |Z|$ we may be in trouble.

As a rough diagnostic for whether this will happen, we use our earlier remark that the continuous spectrum arises from multiparticle states. Thus let us consider a 2-particle state: take two particles in a common rest frame, and imagine we can neglect their interaction (say by placing them very far apart). This state will have total $M = M_1 + M_2$ and $Z = Z_1 + Z_2$. Now,

$$M = M_1 + M_2 \geqslant |Z_1| + |Z_2| \geqslant |Z_1 + Z_2| = |Z| \tag{1.6.23}$$

with equality iff both particles are BPS *and* $\arg Z_1 = \arg Z_2$. Thus the 2-particle states usually are safely separated from the 1-particle BPS states: the only exception is a state of 2 particles which are themselves BPS and have $\arg Z_1 = \arg Z_2$.

This leads to the expectation that $\mu(\mathcal{H}_a)$ will be invariant under deformations, *except* that it may jump when the central charges Z_b and Z_c of two BPS particles become aligned, where $a = b + c$. This is indeed what one finds in examples (and we will see it in the following lectures). Note that this kind of alignment is a codimension-1 phenomenon in parameter-space; thus what we will find is that the quantities $\mu(\mathcal{H}_a)$ are piecewise constant, but jump at some codimension-1 "walls" in parameter-space.

Example 1.6.24 ($\mathcal{N} = 2$ supersymmetry in $d = 4$)**.** Now we consider 4-dimensional theories with $\mathcal{N} = 2$ supersymmetry.

In this case the supersymmetry algebra $\mathcal{A}$ extending $\mathrm{Lie}(\mathrm{ISpin}(3,1))$ has 8 odd generators, described as follows. $\mathrm{Spin}(3,1)$ has 2 inequivalent spin representations $S^\pm$, both complex and 2-dimensional. Each of $S^\pm$ is equipped with an invariant pairing $\cdot$, and there is an intertwiner $\Gamma : S^+ \otimes S^- \to V$, with $V = \mathbb{R}^{3,1}$ the vector representation. The odd generators of $\mathcal{A}$ are Q^1, Q^2 valued in S^+ and $\overline{Q}^1$, $\overline{Q}^2$ valued in S^-.[8] The odd bracket relations are

$$[Q^I(s), Q^J(s')] = (s \cdot s')\epsilon^{IJ} Z, \quad [\overline{Q}^I(s), \overline{Q}^J(s')] = (s \cdot s')\epsilon^{IJ} \overline{Z}, \tag{1.6.25}$$

$$[Q^I(s), \overline{Q}^J(s')] = \delta^{IJ} \Gamma(s, s') P, \tag{1.6.26}$$

[8] What we mean by "Q is valued in S^+" is that for any $s \in S^+$, there is a corresponding operator $Q(s)$, depending linearly on s. $S^+ \oplus S^-$ admits a conjugate-linear involution exchanging the factors, so given $s \in S^\pm$ we can write $\overline{s} \in S^\mp$; then the adjointness condition means that in unitary representations $Q^I(s)$ is adjoint to $\overline{Q}^I(\overline{s})$, i.e. $\overline{Q(s)} = \overline{Q}(\overline{s})$.

Suppressing the spinor and vector indices, we write these more schematically as

$$[Q^I, Q^J] = \epsilon^{IJ} Z, \qquad [\overline{Q}^I, \overline{Q}^J] = \epsilon^{IJ} \overline{Z}, \qquad [Q^I, \overline{Q}^J] = \delta^{IJ} P. \tag{1.6.27}$$

The analysis of representations runs parallel to what we did in Example 1.6.1, as follows. We consider the "little algebra" $\mathcal{A}^{\text{rest}}$ generated by Q^I, $\overline{Q}^I$, Z, P and generators of $\text{Spin}(3) \subset \text{Spin}(3,1)$, acting on a unitary representation in which P acts by the character $(M, 0, 0, 0)$ for some $M > 0$, and Z acts by some scalar $Z \in \mathbb{C}$.

$\text{Spin}(3) \simeq SU(2)$ has only one spin representation S, which is complex and 2-dimensional. So $S^+ \simeq S^- \simeq S$ when considered as representations of Spin(3). After fixing an isomorphism, we could write the odd brackets in $\mathcal{A}^{\text{rest}}$ as

$$[Q^I, Q^J] = \epsilon^{IJ} Z, \qquad [\overline{Q}^I, \overline{Q}^J] = \epsilon^{IJ} \overline{Z}, \qquad [Q^I, \overline{Q}^J] = \delta^{IJ} M. \tag{1.6.28}$$

Now, we consider the generators (cf. (1.6.9))

$$Q_\vartheta = \frac{1}{\sqrt{2}}(e^{i\vartheta/2} Q^1 + e^{-i\vartheta/2} \overline{Q}^2). \tag{1.6.29}$$

By considering $[Q_\vartheta, \overline{Q_\vartheta}]$, in parallel to what we did in the two-dimensional case, we conclude that $M \geqslant |Z|$ in unitary representations, and if $M = |Z|$, then a 4-dimensional subspace of the 8-dimensional space of odd supercharges acts trivially. Then using the representation theory of Clifford algebras as before, we arrive at the following:

Fact 1.6.30 (1-particle representations of $\mathcal{N} = 2$ supersymmetry in $d = 4$)**.** Given a complex scalar $Z \neq 0$, a real scalar $M \geqslant |Z|$, and an irreducible representation S of $\text{Spin}(3) \simeq SU(2)$, there is a corresponding unitary irreducible representation $V_{Z,M,S}$ of $\mathcal{A}$. The representations with $M = |Z|$ are short or "BPS" while those with $M > |Z|$ are long.

As before, the short representations are individually rigid, but can combine to make long representations. Thus as before we introduce an index which is insensitive to long representations. The only index available is the "second helicity supertrace,"

$$\Omega(\mathcal{H}) = \sum_{n \geqslant 1} (-1)^{n+1} n (\text{\# copies of } V_{Z, M=|Z|, S_n} \text{ in } \mathcal{H}) \tag{1.6.31}$$

$$= -\frac{1}{2} \operatorname{Tr}_{\mathcal{H}^{1,\text{BPS,rest}}} (-1)^{2J_3} (2J_3)^2 \tag{1.6.32}$$

where S_n denotes the (unique up to isomorphism) irreducible n-dimensional representation of Spin(3), and $J_3 \in \mathfrak{so}(3,1)$ denotes the generator of rotations around the x^3-axis. For example,

$$\Omega(V_{Z,|Z|,S_1}) = +1, \tag{1.6.33}$$

$$\Omega(V_{Z,|Z|,S_2}) = -2. \tag{1.6.34}$$

These two representations are referred to as the massive hypermultiplet and the massive vector multiplet, respectively. We will see a geometric realization of states in these representations in the final lecture.

Just as in the 2-dimensional case, we will study examples in which $\mathcal{H}$ is graded, $\mathcal{H} = \bigoplus_{\Gamma} \mathcal{H}_\gamma$; then the indices $\Omega(\gamma)$ are formally deformation invariant, but they can suffer from wall-crossing when the central charges Z_{γ_1} and Z_{γ_2} of two BPS particles become aligned, with $\gamma = \gamma_1 + \gamma_2$.

2. Lectures 2-3: 2d theories, tt^* geometry and Stokes phenomenon

Now we focus on $\mathcal{N} = (2,2)$ supersymmetric theories in $d = 2$. Such theories come in families parameterized by complex manifolds.[9] These parameter spaces carry a rich geometric structure, revealed in a large body of literature: pioneering early work is Witten [52], Dijkgraaf-Verlinde-Verlinde [7], Cecotti-Vafa [5,6], Dubrovin [12].

2.1. Landau-Ginzburg models We begin our discussion by looking at a specific class of $\mathcal{N} = (2,2)$ theories.

Physics Fact 2.1.1. Suppose given a Kähler manifold M and a holomorphic function W on M. There is a 2-dimensional quantum field theory $T_2[M, W]$ called a supersymmetric Landau-Ginzburg model on M, with $\mathcal{N} = (2,2)$ supersymmetry. W is called the superpotential.

In these lectures we will just consider the case where $M = \mathbb{C}^k$ (usually $k = 1$), with its standard metric, and W is a polynomial. We will mainly discuss two examples:

Example 2.1.2 (Superpotentials for the cubic Landau-Ginzburg model)**.** We take $C = \mathbb{C}$ and for any $z \in C$:

$$W_z(x) = \frac{1}{3}x^3 - zx. \tag{2.1.3}$$

Example 2.1.4 (Superpotentials for the quartic Landau-Ginzburg model)**.** We take $C = \mathbb{C}^2$ and for any $z = (z_1, z_2) \in C$:

$$W_z(x) = \frac{1}{4}x^4 - \frac{1}{2}z_1x^2 - z_2x. \tag{2.1.5}$$

We will be interested in how the Landau-Ginzburg model varies as we vary the couplings $z \in C$.

2.2. BPS solitons We now discuss the theory of BPS solitons in the supersymmetric Landau-Ginzburg model, as developed in [6].

Definition 2.2.1 (Classical BPS solitons in Landau-Ginzburg models)**.** Suppose given a superpotential $W(x^1, \dots, x^k)$ which is a holomorphic Morse function (ie all critical points of W are nondegenerate). A *classical BPS soliton with phase* ϑ is a map $x : \mathbb{R} \to \mathbb{C}^k$ obeying

$$\frac{dx^j}{ds} = e^{i\vartheta}\overline{\partial_j W(x(s))}. \tag{2.2.2}$$

[9] Actually the structure can be more general, but we'll simplify by considering only the "chiral" deformations and not the "twisted chiral" ones.

Given two critical points x_i, x_j of W, an ij-*soliton* is a BPS soliton with

(2.2.3) $$\lim_{s\to-\infty} x(s) = x_i, \quad \lim_{s\to+\infty} x(s) = x_j.$$

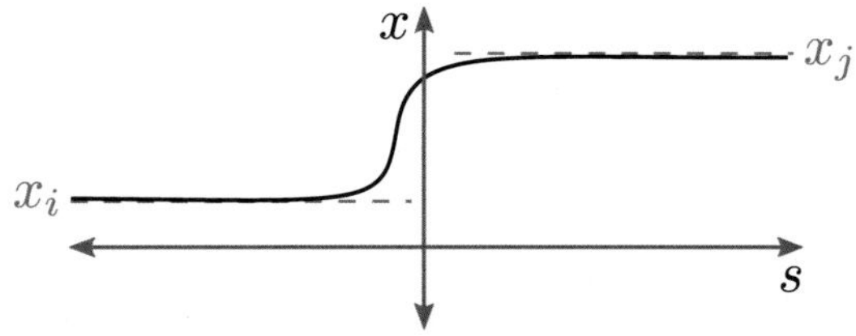

Remark 2.2.4 (Motivation for BPS soliton equations). Equation (2.2.2) arises from the classical field theory counterpart of the Landau-Ginzburg model $T_2[\mathbb{C}^k, W]$. When spacetime is taken to be $\mathbb{R}^{1,1}$, the bosonic fields in this classical field theory are maps $x : \mathbb{R}^{1,1} \to \mathbb{C}^k$. Considering time-independent configurations reduces us to maps $x : \mathbb{R} \to \mathbb{C}^k$, now with s interpreted as the spatial coordinate. The full space of fields carries an action of the superalgebra $\mathcal{A}$ from Example 1.6.1, and requiring that the configuration is invariant under the action of half of the odd generators of $\mathcal{A}$ leads to (2.2.2). This is the classical counterpart of the condition that a state in the Hilbert space of the theory is annihilated by half of the odd generators of $\mathcal{A}$, which we recall defines the BPS states. Indeed, classical BPS solitons give rise to BPS states after quantization.

If (2.2.2) is satisfied we have $\frac{dW(x(s))}{ds} = e^{i\vartheta}|\nabla W(x(s))|^2$. Thus composing with the map W projects a classical BPS soliton with phase ϑ to a line segment in $\mathbb{C}$, inclined by an angle ϑ from the positive real axis. An ij-soliton projects to a segment running from $W(x_i)$ to $W(x_j)$. In particular, for an ij-soliton the angle ϑ is determined:

$$e^{i\vartheta} = \frac{W(x_j) - W(x_i)}{|W(x_j) - W(x_i)|}.$$

To construct solutions of (2.2.2) explicitly is difficult, but if we don't keep track of the parameterization it becomes easier, as follows. Fix some w on the segment connecting $W(x_i)$ to $W(x_j)$, and consider the set of "half-solitons," i.e. solutions of (2.2.2) on the half-line $(-\infty, 0]$ which have

(2.2.5) $$\lim_{s\to-\infty} x(s) = x_i, \quad W(x(0)) = w.$$

Now we consider the possible values of $x(0)$ for such half-solitons. One can show that these values sweep out a cycle $V_i(w) \subset \mathbb{C}^k$ homeomorphic to S^{k-1}. As $w \to w_i$, $V_i(w)$ collapses to the point x_i; thus $V_i(w)$ is called a *vanishing cycle.* A parallel construction for solutions on $[0,\infty)$ with $\lim_{s\to+\infty} x(s) = x_j$ gives another vanishing cycle $V_j(w)$.

This description of the ij-solitons motivates a topological approach to counting them, as follows. Both vanishing cycles have real dimension $k-1$ and lie in the fiber $W^{-1}(w)$, which has complex dimension $k-1$. Thus we are in the correct dimension for the topological intersection number of $V_i(w)$ and $V_j(w)$ to be defined. We choose orientations of $V_i(w)$ and $V_j(w)$, and make the following definition:

Definition 2.2.6 (Soliton degeneracies in Landau-Ginzburg models)**.** Given two critical points x_i, x_j of W, we define $\mu(i,j)$ to be the topological intersection number between $V_i(w)$ and $V_j(w)$ in $W^{-1}(w)$, for any w in the interval connecting $W(x_i)$ to $W(x_j)$.

The ij-solitons are in 1-1 correspondence with points of the setwise intersection $V_i(w) \cap V_j(w)$. The topological intersection number $\mu(i,j)$ gives a signed count of these points. Moreover, as was argued in [6], this signed count turns out to agree with the signed count one would obtain by quantizing the classical solutions and taking the BPS index:

Physics Fact 2.2.7. The $\mu(i,j)$ are BPS indices in the $\mathcal{N} = (2,2)$ supersymmetric Landau-Ginzburg model $T_2[\mathbb{C}^k, W]$. (More precisely: the Hilbert space in this theory is divided up into sectors $\mathcal{H} = \bigoplus_{i,j} \mathcal{H}_{ij}$, and $\mu(i,j) = \mu(\mathcal{H}_{ij})$ in the sense of (1.6.17) above.)

So far we have neglected a subtlety: how to choose the orientations of the vanishing cycles $V_i(w)$? In general there is no completely natural way to do it, and thus there is no completely natural definition of an integer $\mu(i,j)$; only $|\mu(i,j)|$ is independent of the choice of orientations and thus canonical. This difficulty in defining the overall sign of $\mu(i,j)$ is related to the subtlety we mentioned in the first lecture, that in general there is no canonical definition of the fermion number F; indeed, shifting $F \to F+1$ reverses the sign of the BPS index. I will not delve into this here, nor will I try to construct the actual Hilbert spaces of solitons (as opposed to counts of their multiplicities.) A good reference for that material, with vastly more detail than I can give here, is [25].[10]

Example 2.2.8 (Soliton degeneracies in the cubic Landau-Ginzburg model)**.** In this case the structure is very simple: V_i and V_j are distinct 2-element subsets of a 3-element set, so $|V_i \cap V_j|$ is always 1, i.e. for any $z \neq 0$ we have $\mu(1,2;z) = \pm 1$ and $\mu(2,1;z) = \mp 1$.

Example 2.2.9 (Soliton degeneracies in the quartic Landau-Ginzburg model)**.** This case is already more interesting: as we change z_1 and z_2 the $\mu(i,j;z_1,z_2)$ can change. The change occurs when three critical values of W become collinear. In one region there are two solitons, in another region there are three; e.g. in the slice $z_1 = -1$, the picture in the z_2-plane has the shape shown in the figure below.

[10]To be exact, for the construction of the vector spaces, see Section 12.3, page 266. The basic idea: realize solutions of (2.2.2) as critical points of a certain Morse function in infinite dimensions, then promote them to generators of a Morse complex with differentials given by gradient flows. cf. a simpler example: critical points of a Morse function on a finite-dimensional manifold can be promoted to cohomology vector spaces by the same construction.

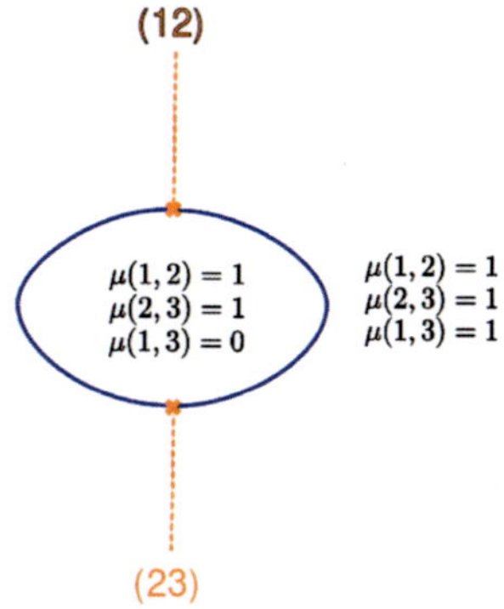

The boundary between the two regions, indicated in blue, is called a *wall of marginal stability*. In this figure we have chosen a specific labeling of the critical points as x_1, x_2, x_3 on the complement of two branch cuts, indicated by orange dashed lines; each cut is labeled by a transposition of the labels.

This is our first concrete example of the wall-crossing phenomenon we mentioned in the first lecture.

What happens to the 13-soliton that disappears at the wall? As we approach the wall from the outside, it starts to look more and more like a combination of a 12-soliton and a 23-soliton separated by a large distance Δs, as indicated in the figure below. This large distance $\Delta s \to \infty$ as z approaches the wall.

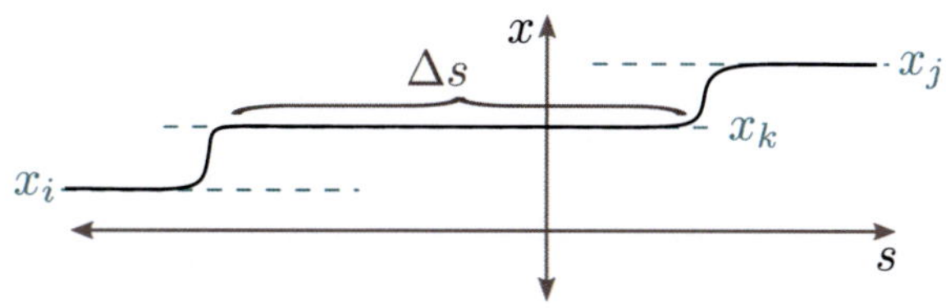

2.3. Wall-crossing formula and spectral networks for 2d $\mathcal{N} = (2,2)$ theories The heuristic picture above suggests that the wall-crossing behavior should be governed by a universal law, as follows. Let

$$Z_{ij} = W(x_i) - W(x_j). \tag{2.3.1}$$

When we cross a locus in parameter-space where Z_{ij} and Z_{jk} become collinear, $\mu(i,k)$ should change by a jump of the form

$$\mu(i,k) \to \mu(i,k) \pm \mu(i,j)\mu(j,k). \tag{2.3.2}$$

This law turns out to be true, not only in supersymmetric Landau-Ginzburg models but in $\mathcal{N} = (2,2)$ theories much more generally; it is the Cecotti-Vafa wall-crossing formula, first derived in [6]. We are going to sketch a route to proving it, but first let's explore how it lets us determine μ in practice.

Definition 2.3.3 (Spectral network for 2d $\mathcal{N} = (2,2)$ theory)**.** For any ϑ, the spectral network $\mathrm{SN}(\vartheta) \subset C$ is the set of all $z \in C$ such that there exist some critical points x_i, x_j with $\mu(i,j;z) \neq 0$ and $\arg Z_{ij} = \vartheta$.

Example 2.3.4 (Spectral network for the cubic Landau-Ginzburg model)**.** In this case the critical values are $W(x_i) = \pm\frac{2}{3}z^{\frac{3}{2}}$. There will be a soliton with phase $\vartheta = 0$ just if the two critical values have the same imaginary part, i.e. just when $z^{\frac{3}{2}}$ is real. This gives the picture of $\mathrm{SN}(\vartheta = 0)$:

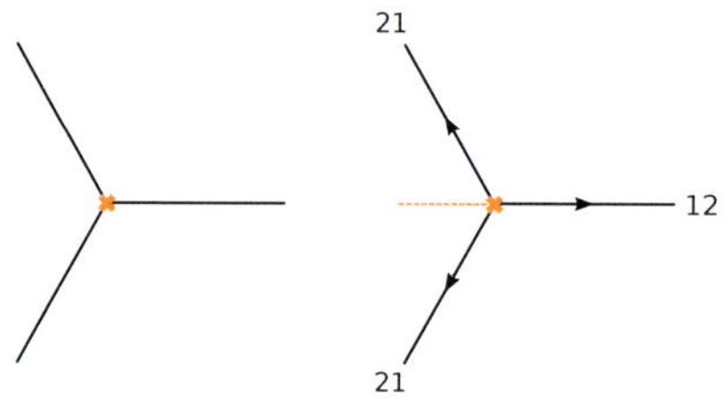

In the second picture we chose a branch cut for the function $z^{\frac{3}{2}}$ and labeled the two critical values as x_1, x_2. The paths are labeled by the ij type of the soliton with phase $\vartheta = 0$. The orientation of each wall shows the direction in which the mass $M = |W(x_i) - W(x_j)| = \frac{4}{3}|z|^{\frac{3}{2}}$ of the soliton increases.

Example 2.3.5 (Spectral network for the quartic Landau-Ginzburg model)**.** In this case the full spectral network $\mathrm{SN}(\vartheta)$ would be drawn on $C = \mathbb{C}^2$. The figure below shows a slice of $\mathrm{SN}(\vartheta = \frac{\pi}{2})$ as the z_2-plane, with $z_1 = -1$ held fixed:

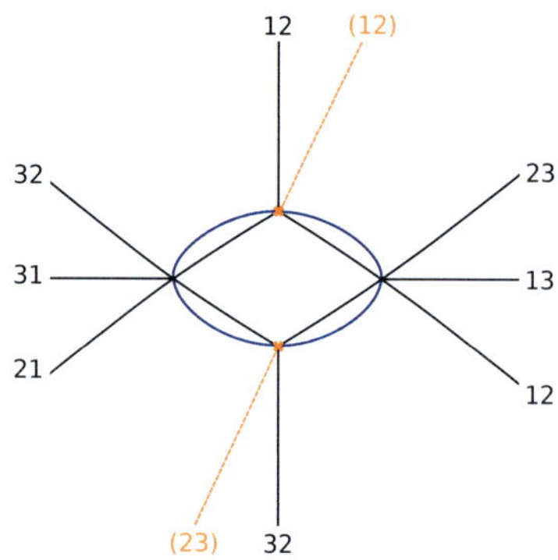

This figure is drawn in the same plane as the figure in Example 2.2.9. (This spectral network and many similar ones were first drawn in [33].)

In the figure above there is an interesting new feature: some trajectories are "born" at intersection points of other trajectories. This is a manifestation of wall-crossing; indeed the intersection point lies on the wall of marginal stability. If we vary the parameter ϑ, the intersection point moves, sweeping out the whole wall of marginal stability.

How was this picture made? In principle it could have been done by directly analyzing the BPS solitons. In fact, it was done in a simpler way, as follows. After restriction to the one-dimensional slice $z_1 = -1$, one can see directly from Definition 2.3.3 that the paths making up $\mathrm{SN}(\vartheta)$ are solutions of a first-order ODE:

$$dZ_{ij}/dt = e^{i\vartheta}. \tag{2.3.6}$$

We consider the locus where some pair of critical points, say $W(x_i) = W(x_j)$, collide. Near this locus, we will have $\mu(i, j; z) = \pm 1$ as in (2.3.4) above. Thus the local structure of $SN(\vartheta)$ around each such point is 3-pronged as in that example. So first we draw these 3 paths coming out of each such point, and then we use the ODE (2.3.6) to continue the paths from there. When an ij-path and a jk-path in $SN(\vartheta)$ cross one another, the point of intersection becomes the initial point of an ik-path which is also included in $SN(\vartheta)$, in the manner indicated in the figure above. On a generic one-dimensional slice of C, this algorithm can be used to determine $SN(\vartheta)$ for any ϑ, and thus determine all the $\mu(i, j; z)$.

This algorithm is a slight variant of one given in [22] in the context of surface defects in 4-dimensional theories, which we will discuss in the last lecture.

2.4. Chiral rings and vacua

Physics Fact 2.4.1 (Chiral ring). Suppose C is the parameter space of chiral deformations of an $\mathcal{N} = (2, 2)$ supersymmetric field theory. Then C carries:

- a holomorphic vector bundle of commutative algebras E over C (chiral rings),
- a holomorphic map of vector bundles $q : TC \to E$.

(The origin of this structure is as follows: E is a certain subspace of the space of local operators in the theory; the map q arises because deformations of the theory are constructed by perturbing the action by local operators.)

Remark 2.4.2 (Vacua). Fix some $z \in C$. Dually, it may be convenient to consider the spectrum Σ_z of the commutative algebra E_z instead of the algebra itself. The points of Σ_z are then in 1-1 correspondence with the *vacua* of the theory; as z varies, they sweep out a space $\Sigma \to C$ (in our examples it will be a branched covering).

Remark 2.4.3 (Chiral ring gives a Higgs bundle). If we define $\varphi : TC \to \mathrm{End}(E)$ by $(\varphi(v))(w) = q(v) \cdot w$ then the pair (E, φ) define a Higgs bundle over C[11]. The space Σ is then also known as the *spectral curve* of this Higgs bundle.

Example 2.4.4 (Chiral rings in Landau-Ginzburg models). Suppose given a family of complex polynomials $W_z(x^1, \dots, x^k)$ parameterized by a space C.[12]

Then the Landau-Ginzburg model gives the following structures:

- E_z is the Jacobian ring,

$$E_z = \mathbb{C}[x^1, \dots, x^k]/\langle \partial_{x^1} W_z, \dots, \partial_{x^k} W_z \rangle. \tag{2.4.5}$$

- The map q takes a vector $\partial_z \in TC$ to $\partial_z W_z \in E$.

The space Σ_z consists of the critical points of W_z, ie solutions of $\partial_x W_z = 0$.

[11]This is a Higgs bundle in the same sense as in Laura Schaposnik's lectures, except that here we allow C to be higher-dimensional; then the definition of Higgs bundle requires the extra condition $[\varphi, \varphi] = 0$, which is satisfied here because the algebra structure on E is commutative.

[12]I'm not sure of the precise conditions one should put on this family of polynomials.

Example 2.4.6 (Chiral rings in the cubic Landau-Ginzburg model)**.** We consider the Landau-Ginzburg model with a family of potentials parameterized by $C = \mathbb{C}$:

$$W(x) = \frac{1}{3}x^3 - zx. \tag{2.4.7}$$

Then E has a basis $\{1, x\}$ and the algebra structure is given by the relation $x^2 = z$, from which we obtain

$$\varphi(\partial_z) = (-x\cdot) = -\begin{pmatrix} 0 & z \\ 1 & 0 \end{pmatrix}. \tag{2.4.8}$$

The vacua make a double cover $\Sigma \to \mathbb{C}$, branched over $z = 0$.

Example 2.4.9 (Chiral rings in the quartic Landau-Ginzburg model)**.** We consider the Landau-Ginzburg model with the potentials parameterized by $C = \mathbb{C}^2$:

$$W(x) = \frac{1}{4}x^4 - \frac{1}{2}z_1x^2 - z_2x. \tag{2.4.10}$$

Now E has a basis $\{1, x, x^2\}$ and the relation $x^3 = z_1x + z_2$, so

$$\varphi(\partial_{z_1}) = \left(-\frac{1}{2}x^2\cdot\right) = -\frac{1}{2}\begin{pmatrix} 0 & z_2 & 0 \\ 0 & z_1 & z_2 \\ 1 & 0 & z_1 \end{pmatrix},$$

and

$$\varphi(\partial_{z_2}) = (-x\cdot) = -\begin{pmatrix} 0 & 0 & z_2 \\ 1 & 0 & z_1 \\ 0 & 1 & 0 \end{pmatrix}.$$

The vacua make a triple cover $\Sigma \to \mathbb{C}^2$, branched over the curve $4z_1^3 = 27z_2^2$.

2.5. The topological connection

Physics Fact 2.5.1 (Topological connection)**.** In addition to the chiral ring structure just discussed, a family of $\mathcal{N} = (2,2)$ theories also has more (ultimately derived from the possibility of making a family of 2d TFTs by topological twisting): there are

- a symmetric bilinear pairing $\langle , \rangle$ on E such that $\langle a, bc\rangle = \langle ab, c\rangle$,
- a pencil of flat connections $\nabla^\hbar$ in E, of the form

$$\nabla^\hbar = \nabla^\infty + \hbar^{-1}\varphi, \tag{2.5.2}$$

such that $\langle , \rangle$ is compatible with ∇^∞.

This structure is closely related to the notion of Frobenius manifold, but slightly different since we are not identifying E with the tangent bundle of C: φ need not be an isomorphism.

Example 2.5.3 (Topological connection in the cubic Landau-Ginzburg model)**.** In the cubic Landau-Ginzburg model, in the basis $\{1, x\}$, the pairing and connections are given by

$$\langle , \rangle = \begin{pmatrix} 0 & 1 \\ 1 & 0 \end{pmatrix}, \qquad \nabla^{\hbar}_{z} = \partial_z - \hbar^{-1}\begin{pmatrix} 0 & z \\ 1 & 0 \end{pmatrix}. \tag{2.5.4}$$

The covariantly constant sections can be written in terms of Airy functions; indeed $\nabla^{\hbar}$ can be obtained from the Airy equation $(\hbar^2\partial_z^2 + z)\psi(z) = 0$ by the usual device of writing a second-order equation as a matrix first-order equation.

Example 2.5.5 (Topological connection in the quartic Landau-Ginzburg model). In the quartic Landau-Ginzburg model, in the basis $\{1, x, x^2\}$, the pairing and connections are respectively

$$\langle , \rangle = \begin{pmatrix} 0 & 0 & 1 \\ 0 & 1 & 0 \\ 1 & 0 & z_1 \end{pmatrix},$$

and

$$\nabla^{\hbar}_{z_1} = \partial_{z_1} + \frac{1}{2}\begin{pmatrix} 0 & 0 & 1 \\ 0 & 0 & 0 \\ 0 & 0 & 0 \end{pmatrix} - \frac{\hbar^{-1}}{2}\begin{pmatrix} 0 & z_2 & 0 \\ 0 & z_1 & z_2 \\ 1 & 0 & z_1 \end{pmatrix},$$

$$\nabla^{\hbar}_{z_2} = \partial_{z_2} - \hbar^{-1}\begin{pmatrix} 0 & 0 & z_2 \\ 1 & 0 & z_1 \\ 0 & 1 & 0 \end{pmatrix}.$$

Note that in this example $\nabla^{\infty}_{z_1} \neq \partial_{z_1}$, i.e., the basis $\{1, x, x^2\}$ is not covariantly constant for ∇^{∞}.

2.6. Contour integrals A general construction of the desired pairings $\langle \cdot, \cdot \rangle$ and connections $\nabla^{\hbar}$ in Landau-Ginzburg models exists, going back to work of Saito; see [5, 12, 28, 45] for accounts of aspects of this construction. Here we cannot describe the construction completely, but just touch on a few key features.

The flat connection $\nabla^{\hbar}$ can be produced by first writing a basis of sections $\psi_i(z)$ of E, then defining $\nabla^{\hbar}$ to be the connection for which each $\psi_i(z)$ is covariantly constant. The $\psi_i(z)$ in turn can be characterized by their pairing with a general $\alpha \in E_z$. In our examples, this pairing is given by an explicit contour integral in the x-plane,

$$\langle \psi_i(z), \alpha \rangle = \int_{C_i} dx\, e^{\hbar^{-1} W_z(x)}\, \alpha(x). \tag{2.6.1}$$

Here $\alpha(x)$ denotes the unique representative for the class α which is a polynomial of degree $\leqslant (\deg W) - 2$ (this choice really matters: the integral (2.6.1) is not invariant under shifting α by a multiple of W_z.) The resulting $\nabla^{\hbar}$ is indeed of the form (2.5.2); verifying this requires some investigation of the asymptotics of the integrals (2.6.1) as $\hbar \to 0$ and $\hbar \to \infty$.

In this construction we need C_i to have no boundary, but we also need the integral (2.6.1) to be convergent. A systematic way to get a contour C_i is to start from a critical point x_i of W and follow the paths of steepest descent for the integrand $e^{\hbar^{-1}W(x)}$, i.e. paths along which $\mathrm{Im}(\hbar^{-1}W)$ is constant while $\mathrm{Re}(\hbar^{-1}W) \to -\infty$. These are simple examples of Lefschetz thimbles.[13] This procedure yields a well defined contour C_i unless the path of steepest descent runs into another critical point x_j. The latter can happen only if there is a soliton connecting x_i to x_j, with phase $\vartheta = \arg \hbar$. As we vary z across a value where such a soliton exists, the contour C_i jumps.

For example, the figure below shows the contours C_i obtained by steepest descent in the case of the cubic Landau-Ginzburg model with $z = 1 - i$ (left) and $z = 1 + i$ (right), when $\arg \hbar = 0$.

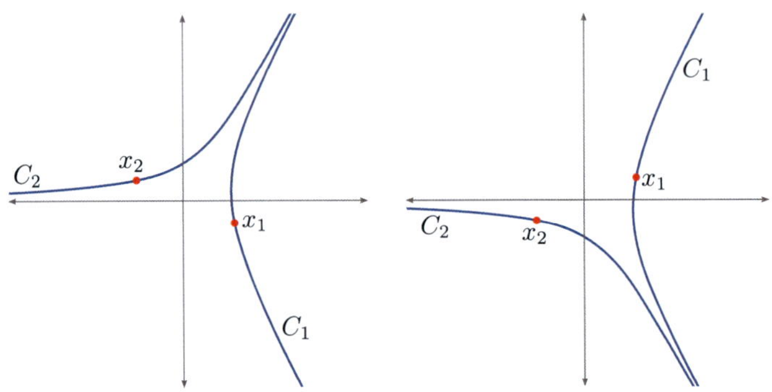

As we vary z along a straight segment from $1-i$ to $1+i$, the contours C_i deform continuously, except at the moment when z crosses the real axis at $z = 1$; at that moment there is a soliton connecting x_1 to x_2 with phase $\vartheta = 0$, and the contour C_2 jumps by the addition of $\pm C_1$ (with the sign depending on how we orient the contours), while C_1 varies continuously. Thus, when z crosses the real axis, the local solutions ψ_i are transformed as $\psi_1 \to \psi_1, \psi_2 \to \psi_2 \pm \psi_1$. This transformation can alternatively be expressed as a unipotent matrix:

$$\begin{pmatrix} \psi_1 \\ \psi_2 \end{pmatrix} \to \begin{pmatrix} 1 & \pm 1 \\ 0 & 1 \end{pmatrix} \begin{pmatrix} \psi_1 \\ \psi_2 \end{pmatrix}. \tag{2.6.2}$$

The upper-right entry of this matrix should be interpreted as the soliton count $\pm\mu(1,2; z = 1)$.

2.7. Spectral network as jumping locus for covariantly constant sections We have arrived at the following picture. In each domain on the complement of the

[13] cf. Pavel Putrov's lectures at this school, which involved much more sophisticated examples of Lefschetz thimbles: there the superpotential W was defined on the infinite-dimensional space of connections on a 3-manifold M, and the critical points x_i were flat connections. To connect the two stories directly one should try to find a natural construction of a 2-dimensional $\mathcal{N} = (2,2)$ supersymmetric theory with this W; it could likely be done by starting with 5-dimensional supersymmetric Yang-Mills theory and compactifying it appropriately on M [55].

spectral network $\mathrm{SN}(\vartheta = \arg\hbar)$, we have a distinguished basis of covariantly constant sections $(\psi_i(z))_{i=1}^N$ for $\nabla^\hbar$ that can change when z crosses $\mathrm{SN}(\vartheta = \arg\hbar)$.

More precisely, as we move from one domain to another across a wall with label ij, the basis $(\psi_i(z))_{i=1}^N$ is transformed by a unipotent change-of-basis matrix S of the form

$$S = \mathbf{Id} \pm \mu(i,j;z)e_{ij}, \tag{2.7.1}$$

where e_{ij} is the elementary matrix. One can establish (2.7.1) by studying how the contours (Lefschetz thimbles) change as z varies, as we did above in the example of the cubic Landau-Ginzburg model.

2.8. Wall-crossing formula via the spectral network Assuming (2.7.1), we now revisit the wall-crossing formula (2.3.2) for the soliton counts $\mu(i,j;z)$. We consider the changes of basis induced by traveling two different paths in C, as shown in the figure below.

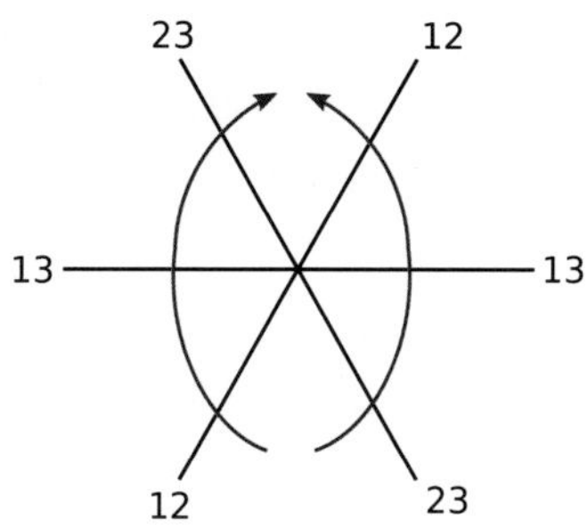

Requiring that these two changes of basis are equal leads to the following relation among the change-of-basis matrices:

$$\begin{pmatrix} 1 & \mu(1,2) & 0 \\ 0 & 1 & 0 \\ 0 & 0 & 1 \end{pmatrix} \begin{pmatrix} 1 & 0 & \mu(1,3) \\ 0 & 1 & 0 \\ 0 & 0 & 1 \end{pmatrix} \begin{pmatrix} 1 & 0 & 0 \\ 0 & 1 & \mu(2,3) \\ 0 & 0 & 1 \end{pmatrix}$$
$$= \begin{pmatrix} 1 & 0 & 0 \\ 0 & 1 & \mu'(2,3) \\ 0 & 0 & 1 \end{pmatrix} \begin{pmatrix} 1 & 0 & \mu'(1,3) \\ 0 & 1 & 0 \\ 0 & 0 & 1 \end{pmatrix} \begin{pmatrix} 1 & \mu'(1,2) & 0 \\ 0 & 1 & 0 \\ 0 & 0 & 1 \end{pmatrix},$$

where $\mu(i,j)$ and $\mu'(i,j)$ refer respectively to the values of $\mu(i,j;z)$ for z on the left and right sides of the figure. This equation holds just if

$$\mu'(1,2) = \mu(1,2), \quad \mu'(2,3) = \mu(2,3), \quad \mu'(1,3) = \mu(1,3) + \mu(1,2)\mu(2,3), \tag{2.8.1}$$

which is the Cecotti-Vafa wall-crossing formula (2.3.2), now interpreted as a consequence of the existence of the sections $\psi_i(z)$.

Remark 2.8.2 (Stokes phenomenon). The construction of the $\psi_i(z)$ by contour integration allows one to determine their complete asymptotic series expansion as $\hbar \to 0$ with $\arg\hbar = \vartheta$, by the method of steepest descent (eg the leading behavior is like $\exp(W(x_i)/\hbar)$.) This asymptotic series only depends on what happens very near the critical point; in particular, it does not change when the contours C_i jump. (Another way of saying this is that when the jump occurs, it shifts the contour by adding the contribution of a contour through a subdominant critical point, which thus doesn't change the leading asymptotic.)

Thus the local solutions $\psi_i(z)$ have the nice property that they have a simple uniform asymptotic expansion as $\hbar \to 0$ with $\arg \hbar = \vartheta$. This is very valuable if one wants to use them to study the parallel transport of $\nabla^{\hbar}$. On the other hand, it turns out that this property alone is already enough to imply that the $\psi_i(z)$ cannot be global analytic objects: they have to jump somewhere. As we saw, they do jump, at the locus $\mathrm{SN}(\vartheta)$. This is (a manifestation of) the Stokes phenomenon, which might loosely be described as a conflict between analyticity and uniform asymptotics. For this reason, $\mathrm{SN}(\vartheta)$ is sometimes called the *Stokes graph*.

2.9. tt^* geometry Another crucial discovery of Cecotti-Vafa [5] is that $\mathcal{N} = (2,2)$ theories have a canonical metric structure, obeying a highly constrained system of PDEs. This structure has been extensively studied in both the physics and mathematics literature, under various names; see e.g. [6,11,29,44]. Our treatment here will be very brief.

Physics Fact 2.9.1 (tt^* structure)**.** Suppose C is the parameter space of chiral deformations of an $\mathcal{N} = (2,2)$ supersymmetric field theory. Then the bundle E carries a Hermitian metric, obeying the tt^* *equations* [5],

$$F_{D_h} + [\varphi, \varphi^{\dagger_h}] = 0, \qquad \partial_{D_h} \varphi = 0. \tag{2.9.2}$$

Here D_h is the Chern connection uniquely determined by compatibility with both the holomorphic structure and the Hermitian metric in E.

Remark 2.9.3 (tt^* equations and Hitchin's equation)**.** The second equation in (2.9.2) expresses the vanishing of a $(2,0)$-form on C. When C has complex dimension 1, this equation holds automatically, and so (2.9.2) reduces to

$$F_{D_h} + [\varphi, \varphi^{\dagger_h}] = 0, \tag{2.9.4}$$

which is also called *Hitchin's equation* (the same equation that appeared in Laura Schaposnik's lectures at this school). More generally, we can view the tt^* equations (2.9.2) as a natural higher-dimensional extension of Hitchin's equation.

Example 2.9.5 (The tt^* structure for the cubic Landau-Ginzburg model)**.** In the cubic Landau-Ginzburg model of Example 2.1.2 there is a tt^* structure for which the metric h is diagonal: making the ansatz

$$h = \begin{pmatrix} e^{-u} & 0 \\ 0 & e^{u} \end{pmatrix} \tag{2.9.6}$$

the tt^* equations reduce to

$$\partial_z \partial_{\bar{z}} u - (e^{2u} - e^{-2u}|z|^2) = 0. \tag{2.9.7}$$

With an appropriate boundary condition as $|z| \to \infty$, this equation has a unique smooth solution $u(|z|)$, expressible in terms of Painleve transcendents [5]. This turns out to be a very useful "model solution" for studying the limiting behavior of solutions of Hitchin's equation when one goes to infinity in moduli space (see e.g. [23, 38].)

Definition 2.9.8 (The improved connections). We define a family of connections ∇^ζ in the bundle E over C by

$$\nabla^\zeta = D_h + \zeta^{-1}\varphi + \zeta\varphi^{\dagger_h}. \tag{2.9.9}$$

The tt^* equations imply that ∇^ζ is *flat* for every $\zeta \in \mathbb{C}^\times$.

The picture we described above for the topological connections $\nabla^\hbar$ holds equally for the connections ∇^ζ: there are covariantly constant sections in each domain, given by a formula like (2.6.1),

$$\langle \psi_i(z), \alpha \rangle = \int_{C_i} dx\, e^{\zeta^{-1} W_z(x) + \zeta \overline{W_z(x)}} \alpha(x), \tag{2.9.10}$$

but in this case the representatives $\alpha(x)$ we choose don't have a simple explicit expression. Nevertheless the formal structure is the same; in particular, the basis sections $\psi_i(z)$ also transform by (2.7.1) at walls.

Although the $\psi_i(z)$ do not admit a simple explicit formula, they can be determined from the data of $W_z(x)$ together with the soliton counts $\mu(i, j; z)$. The idea is that these data are sufficient to determine the analytic properties of $\psi_i(z)$ in the ζ-plane, and these properties in turn identify $\psi_i(z)$ as the solution of a certain integral equation. Once the $\psi_i(z)$ are determined, they determine the connections ∇^ζ, and finally these connections determine the metric h; thus one obtains a solution of the tt^* equations. This strategy is developed in [6, 11].

3. Lecture 4: 4d theories and spectral networks

We have discussed some rich structure which appears on C when C is a moduli space of $\mathcal{N} = (2, 2)$ supersymmetric field theories: in particular C carries

- a Higgs bundle (E, φ),
- two distinguished families of flat connections $\nabla^\hbar$ and ∇^ζ in E,
- BPS state indices and a spectral network, which control the Stokes phenomena of these connections.

The kind of C we met so far were very simple: just vector spaces. However, it turns out that this kind of structure exists far more generally: for example, we can take C to be any compact Riemann surface. Thus we will obtain a physical construction which leads naturally to Higgs bundles and flat connections over Riemann surfaces.

One of the payoffs of this kind of construction is that it gives new tools to study the moduli space of Higgs bundles on C; still, for this lecture we mainly focus on one Higgs bundle at a time.

3.1. Class $\mathcal{S}$ and surface defects The physical underpinnings of the construction depend on facts we mentioned in Example 1.4.2, also mentioned in Pavel Putrov's last lecture at this school:

Physics Fact 3.1.1. There is a 6-dimensional supersymmetric QFT $T_6[\mathfrak{g}]$, depending only on the data of an ADE Lie algebra $\mathfrak{g}$ (see e.g. [46, 53, 54] for some discussion of this theory). This theory supports supersymmetric 2-dimensional defects, labeled by irreducible finite-dimensional representations R of $\mathfrak{g}$.

Theory $T_6[\mathfrak{g}]$ is sometimes referred to as "the six-dimensional $(2,0)$ superconformal field theory."

Now suppose given a Riemann surface C. We consider the theory $T_6[\mathfrak{g}]$ on the spacetime $C \times \mathbb{R}^{3,1}$, in the limit where the metric on C is rescaled to be extremely small. In this way we obtain a 4-dimensional QFT $T_4[\mathfrak{g}, C]$, described in [17, 23]. These theories are known as "theories of class $\mathcal{S}$."

From now on we will specialize to $\mathfrak{g} = \mathfrak{sl}_N$ or $\mathfrak{g} = \mathfrak{gl}_N$. Next we place a surface defect of the theory $T_6[\mathfrak{g}]$ on the subspace $\{z\} \times \mathbb{R}^{1,1} \subset C \times \mathbb{R}^{3,1}$, taking R to be the fundamental (N-dimensional) representation of $\mathfrak{g}$. In this way we obtain a 2-dimensional surface defect in the theory $T_4[\mathfrak{g}, C]$, depending on the parameter $z \in C$. We call the combined 2d-4d system $T_{4+2}[\mathfrak{g}, C, z]$.

As far as the symmetry algebra goes, the 2d-4d system $T_{4+2}[\mathfrak{g}, C, z]$ behaves just like an ordinary 2d $\mathcal{N} = (2,2)$ supersymmetric theory; the introduction of the surface defect reduces the $ISO(3,1)$ symmetry to $ISO(1,1)$, and also reduces the 8 supercharges to 4. Thus we have achieved our goal of realizing C as a moduli space of 2d $\mathcal{N} = (2,2)$ supersymmetric QFTs.

3.2. Chiral rings The chiral ring of $T_{4+2}[\mathfrak{g}, C, z]$ is infinite-dimensional, essentially because it is a module over the infinite-dimensional chiral ring of the bulk theory $T_4[\mathfrak{g}, C]$. The latter is identified with the space of functions on the Coulomb branch of $T_4[\mathfrak{g}, C]$. This Coulomb branch can be identified with the base of Hitchin's integrable system (see e.g. [23]):

Definition 3.2.1 (Hitchin base)**.** Fix a Riemann surface C and a Lie algebra $\mathfrak{g} = \mathfrak{gl}_N$ or $\mathfrak{g} = \mathfrak{sl}_N$. The *Hitchin base* $\mathcal{B}(C, \mathfrak{g})$ is

$$\mathcal{B}(C, \mathfrak{g}) = \begin{cases} \bigoplus_{n=1}^{N} H^0(C, K_C^n) & \text{for } \mathfrak{g} = \mathfrak{gl}_N, \\ \bigoplus_{n=2}^{N} H^0(C, K_C^n) & \text{for } \mathfrak{g} = \mathfrak{sl}_N. \end{cases} \tag{3.2.2}$$

Thus, a point of $\mathcal{B}(C, \mathfrak{g})$ is a tuple $(\phi_1, \phi_2, \dots, \phi_N)$ where ϕ_n is a holomorphic n-differential on C (locally $\phi_n = f(z)dz^n$), and $\phi_1 = 0$ when $\mathfrak{g} = \mathfrak{sl}_N$.

Instead of considering the full chiral ring we consider a quotient, obtained by specializing to a point $u \in \mathcal{B}(C, \mathfrak{g})$. In physical terms this corresponds to fixing a vacuum of the theory $T_4[\mathfrak{g}, C]$ on its Coulomb branch. This reduced chiral ring is finite-dimensional and will be a close analog of the chiral rings we had in the purely 2-dimensional case.

Physics Fact 3.2.3 (Reduced chiral ring for surface defect in class S theory). Fix C, $\mathfrak{g}$, $z \in C$, and $u \in \mathcal{B}(C, \mathfrak{g})$. The reduced chiral ring R_z is (see e.g. [18, 20])

$$R_z = \mathrm{Sym}^*(T_z C)/I \tag{3.2.4}$$

where I is an ideal depending on z and u, generated by $\sigma^N(1 + \phi_1 + \cdots + \phi_N)$, where σ is any element of degree 1 and thus $\sigma^N \phi_i$ is an element of degree $N - i$. In local coordinates, we would represent this as follows: write each $\phi_i = P_i(z)dz^i$, trivialize $T_z C$ by the generator $\sigma = \partial_z$, and then

$$R_z = \mathbb{C}[\sigma]/\langle \sigma^N + P_1 \sigma^{N-1} + \cdots + P_N \rangle. \tag{3.2.5}$$

The map $q : T_z C \to R_z$ is then the obvious embedding, $\partial_z \mapsto \sigma$.

The spectrum of R_z consists generically of N points, which are identified as the vacua of the theory living on the surface defect. As z varies over C, these vacua give a branched N-fold covering of C, explicitly

$$\Sigma_u = \{\lambda^N + \phi_1 \lambda^{N-1} + \cdots + \phi_N = 0\} \subset T^* C. \tag{3.2.6}$$

Also as in the 2d case, q induces a Higgs field φ valued in $\mathrm{End}(R) \otimes K_C$, so that R is naturally a Higgs bundle; then Σ_u is the spectral curve.

As a holomorphic bundle, R is $\mathcal{O} \oplus TC \oplus TC^2 \oplus \cdots \oplus TC^{N-1}$. There is a closely related bundle E over C, consisting of the ground states of the surface defect. That one differs from R by a global twist:

$$E = R \otimes K_C^{\frac{N-1}{2}} = K_C^{-\frac{N-1}{2}} \oplus \cdots \oplus K_C^{\frac{N-1}{2}}. \tag{3.2.7}$$

Then E is an $\mathfrak{sl}_N$-Higgs bundle, lying in the "Hitchin section" described in Laura Schaposnik's lectures.

3.3. BPS solitons As in the pure 2d case, we are interested in understanding the physics of BPS particles living on the surface defect. Unlike the Landau-Ginzburg model, here we won't give a direct description of the BPS particles in terms of solving some equation along the string. Nevertheless we continue to use the term "soliton" for these BPS particles.

3.4. Spectral networks One can compute the spectrum of BPS solitons in the surface defect system using a spectral network, in parallel to what we did in the Landau-Ginzburg model. Here we just summarize what the spectral network looks like, formulating it as a definition; for the physical background, the precise interpretation in terms of solitons, and the computation of BPS soliton counts $\mu(i, j, a)$ generalizing the $\mu(i, j)$ from above, see [22].[14]

Definition 3.4.1 (Spectral network for class S theory). Fix C, $\mathfrak{g} = \mathfrak{gl}_N$ or $\mathfrak{sl}_N$, and $u \in \mathcal{B}(C, \mathfrak{g})$. Label the sheets of the spectral cover $\Sigma_u \to C$ by a (locally defined)

[14] The charges of solitons in the 2d-4d setting are a bit more elaborate than they were in the pure 2d case. In the pure 2d case, the only data specifying the soliton charge was the pair (i, j) of vacua. In the surface defect case, the data is (i, j, a), where $i, j \in \Sigma_u$ are two vacua of the surface defect i.e. preimages of $z \in C$, and a is (roughly) a path from i to j on Σ_u.

index $i = 1, \dots, N$; they correspond to holomorphic 1-forms λ_i on C. Define an ij-*trajectory with phase* ϑ on C to be a path along which

$$e^{-i\vartheta} \int \lambda_i - \lambda_j \in \mathbb{R}. \tag{3.4.2}$$

Locally the ij-trajectories give a foliation of C.

For generic enough u, the spectral curve Σ_u has only simple branch points. In this case the spectral network $SN(\vartheta)$ on C can be constructed by the same algorithm we used in the pure 2d case, as follows. Around each simple branch point, the foliation by ij-trajectories has a 3-pronged singularity, just like the 3-pronged singularities we encountered in the spectral networks for Landau-Ginzburg models; see the figure below.

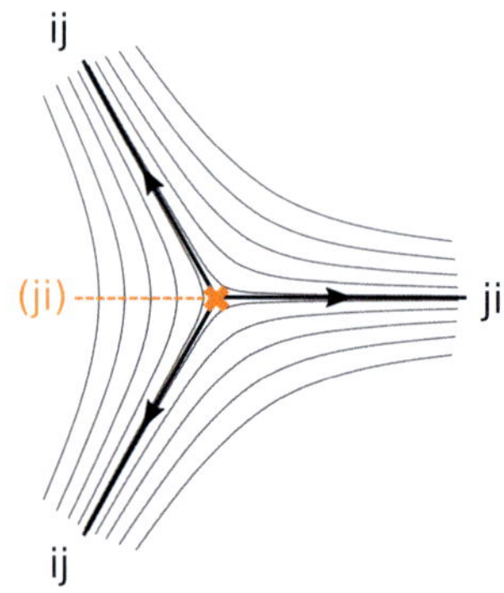

We shoot three ij-trajectories from each ij-branch point, and evolve them following the differential equation (3.4.2). When an ij-trajectory crosses a jk-trajectory we create a new ik-trajectory born from the intersection point, and include it in $SN(\vartheta)$ as well, as indicated in the figure below (again following the same pattern we saw in the 2d case, eg in Example 2.3.5.)

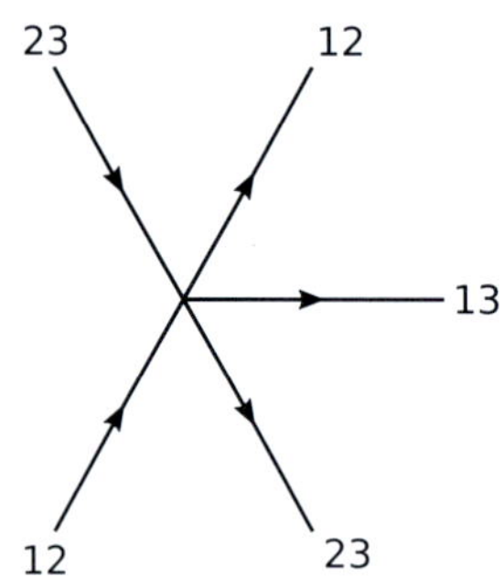

If Σ_u has non-simple branch points, then in order to construct $SN(\vartheta)$ we need to somehow determine the local structure around each branch point. Various examples have been studied, e.g. in [22, 27, 31, 33], but as far as I know, a completely systematic picture has not yet been developed. At any rate, once we have somehow determined the local structure around the branch points, the rest of the construction can proceed as described above.

Remark 3.4.3 (Spectral networks in the $\mathfrak{sl}_2$ case)**.** When $\mathfrak{g} = \mathfrak{sl}_2$, all this becomes simpler:

- The data of u reduces to a holomorphic quadratic differential ϕ_2.
- The branch points of Σ_u are the zeroes of ϕ_2.
- The ij-trajectories are paths along which $e^{-i\vartheta}\sqrt{\phi_2}$ is real. These are also called ϑ-trajectories of ϕ_2. They make up a single, global, singular foliation of the curve C. They are well-studied objects; e.g. they are the main subject of the book [49].

In this case $SN(\vartheta)$ is also known as the critical graph of $e^{-2i\vartheta}\phi_2$.

3.5. Adding punctures If you try to apply the above construction of $SN(\vartheta)$ in an example, say taking C to be a compact genus 2 surface, you run into a problem immediately: a typical ij-trajectory will wind around the surface with dense image, and thus $SN(\vartheta)$ will be dense on C. In principle one expects that all our analysis can be made to work even in this setting, but it is not yet fully understood (see [14] for some work on spectral networks in this setting).

To avoid this problem we now add one extra ingredient: we allow C to have punctures, and make the differentials ϕ_n meromorphic instead of holomorphic, with poles at the punctures. The definition of the theory then includes additional parameters controlling the singular behavior of the ϕ_n at the punctures. For us, the main virtue of including these punctures is that if the poles have sufficiently high order, then the punctures will generically attract the ij-trajectories, making the pictures and the analysis much simpler.

In the physical context, adding a puncture at $p \in C$ corresponds to adding a 4-dimensional defect of theory $T_6[\mathfrak{g}]$ along the locus $\{p\} \times \mathbb{R}^{3,1} \subset C \times \mathbb{R}^{3,1}$.

Example 3.5.1 (Spectral networks in Argyres-Douglas theory)**.** We consider the case[15] where $\mathfrak{g} = \mathfrak{sl}_2$, $C = \mathbb{CP}^1$ with a puncture at $p = \infty$, and

$$\phi_2 = (z^3 + \Lambda z + u)\, dz^2. \tag{3.5.2}$$

The figure below shows $SN(\vartheta)$ when we fix $\Lambda = 1$ and $u = 1$, and $\vartheta = \frac{\pi}{10}$ (left), $\vartheta = \frac{\pi}{5}$ (middle), $\vartheta = \frac{\pi}{2}$ (right).

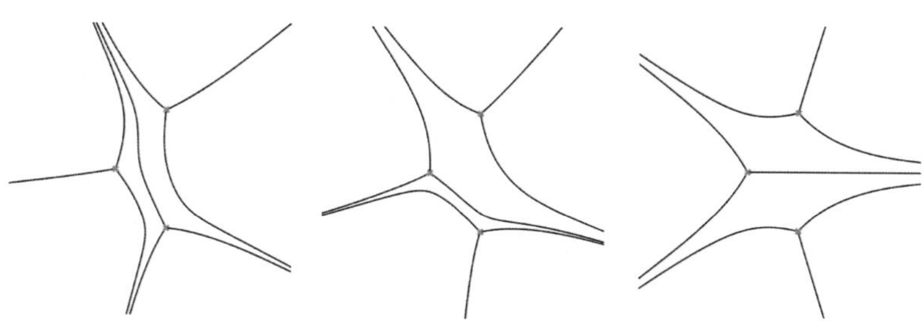

[15]More generally, a very nice class of examples arises from spectral curves obtained by perturbing the equation $y^n = z^m$ by lower-degree terms; physically these are associated with a QFT called the Argyres-Douglas theory of type (A_{n-1}, A_{m-1}). The original Argyres-Douglas theory, studied in [1], is the (A_1, A_2) example. The $(\mathfrak{g}, \mathfrak{g}')$ taxonomy was introduced in [4].

Remark 3.5.3 (Role of parameters). The parameters Λ and u in Example 3.5.1 above play different roles. Λ is a parameter which enters the definition of the theory; it should be thought of as like a complex structure modulus associated to the puncture at $z = \infty$, despite the fact that naively a once-punctured $\mathbb{CP}^1$ has no moduli. (This point of view has been emphasized by Boalch, e.g. [2].) The Hitchin base in this case depends on Λ; we write it as $\mathcal{B}(C, \Lambda, \mathfrak{sl}_2)$. The parameter u is then a (global) coordinate on $\mathcal{B}(C, \Lambda, \mathfrak{sl}_2)$.

3.6. BPS indices in the $\mathfrak{sl}_2$ case Comparing the pictures of spectral networks $\mathrm{SN}(\vartheta)$ in the figure above, one notices that the topology of $\mathrm{SN}(\vartheta)$ abruptly changes at certain critical phases — e.g. by comparing the left and middle panes one sees that there must be such a phase between $\vartheta = \frac{\pi}{10}$ and $\vartheta = \frac{\pi}{5}$. Looking closer one finds that the critical phase is $\vartheta_c \approx \frac{4\pi}{25}$. $\mathrm{SN}(\vartheta)$ for $\vartheta \approx \vartheta_c$ is shown in the figure below:

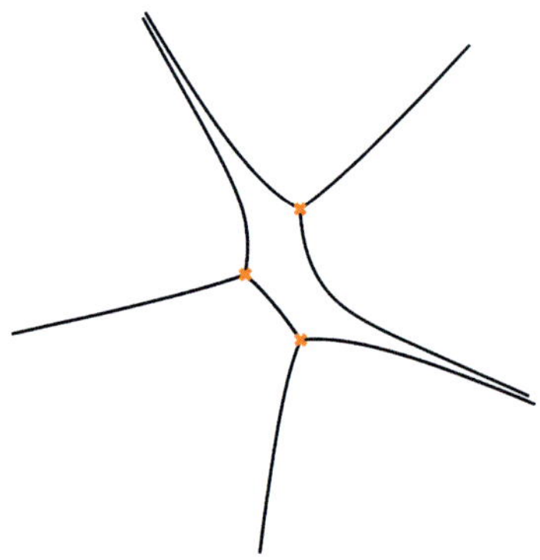

At $\vartheta = \vartheta_c$, $\mathrm{SN}(\vartheta)$ contains a saddle connection:

Definition 3.6.1 (Saddle connection). Given a holomorphic quadratic differential ϕ_2, a *saddle connection of* ϕ_2 *with phase* ϑ is a ϑ-trajectory of ϕ_2, with its two ends on two distinct zeroes of ϕ_2.

Another kind of ϑ-trajectory which could occur in $\mathrm{SN}(\vartheta)$ at special phases ϑ is a loop, which begins and ends at the same branch point. See the figure below, which shows $\mathrm{SN}(\vartheta = 0)$ for the quadratic differential $\phi_2 = \left(z^{-3} - 3z^{-2} + z^{-1}\right) dz^2$.

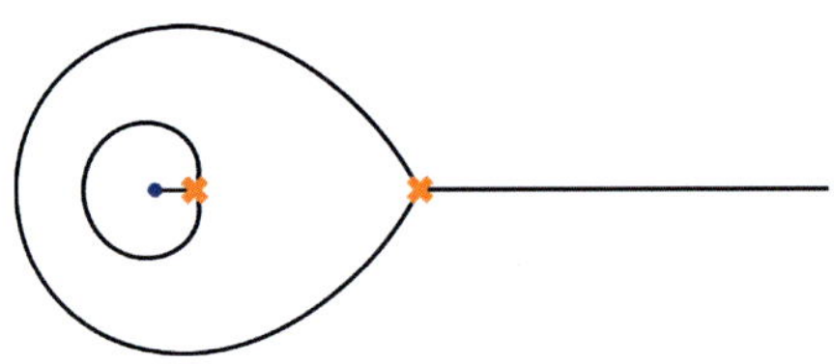

In this figure we see two loops, which are the boundaries of a ring domain:

Definition 3.6.2 (Ring domain). Given a holomorphic quadratic differential ϕ_2, a *ring domain of* ϕ_2 *with phase* ϑ is an annulus or punctured disc on C foliated by ϑ-trajectories of ϕ_2.

This phenomenon occurs generally: whenever we have a closed loop it is the boundary of a ring domain. As with saddle connections, the existence of a ring domain with phase ϑ implies a topology change for $SN(\vartheta)$ at that phase, but one of a more complicated kind; see [23] for a discussion of what happens near such a phase, and some examples.

Now, how should we interpret these topology changes? Recall that $SN(\vartheta)$ captures counts of BPS soliton states. It follows that discontinuous changes in $SN(\vartheta)$ are related to processes where counts of BPS soliton states change. These processes are hard to interpret in purely 2-dimensional terms: unlike the wall-crossing we discussed in pure 2d theories above, they are not related to decay of one soliton into other solitons, or the reverse. Rather, it turns out that they are processes where the soliton decays into another soliton plus a BPS particle in the ambient four-dimensional theory $T_4[\mathfrak{g}, C]$, or the reverse.[16] Indeed, saddle connections and ring domains are associated with BPS particles in the 4d theory $T_4[\mathfrak{sl}_2, C]$:

Physics Fact 3.6.3 (BPS particles in theories $T_4[\mathfrak{sl}_2, C]$)**.** Consider theory $T_4[\mathfrak{sl}_2, C]$, at a point u of its Coulomb branch corresponding to the quadratic differential ϕ_2.

- Saddle connections of ϕ_2 correspond to massive hypermultiplets of this theory, which thus contribute $+1$ to the BPS index, as in (1.6.33).
- Ring domains of ϕ_2 correspond to massive vectormultiplets of this theory, which thus contribute -2 to the BPS index, as in (1.6.34).

This observation was first made in [35], and further developed in various works, e.g. [23, 40, 41, 47].

Each of these 4d BPS particles carries an electromagnetic (and perhaps flavor) charge, which we now describe:

Physics Fact 3.6.4 (Electromagnetic/flavor charge lattice)**.** Let

$$\Gamma = H_1(\Sigma_u, \mathbb{Z}). \tag{3.6.5}$$

Γ is the lattice of electromagnetic and flavor charges in the theory $T_4[\mathfrak{g} = \mathfrak{gl}_N]$ at the point u of its Coulomb branch.

Example 3.6.6 (Charge lattice in the Argyres-Douglas (A_1, A_2) theory)**.** In the example of Example 3.5.1, at $u = 1$ and $\Lambda = 1$, the charge lattice is generated by cycles γ_1 and γ_2 shown in the figure below.

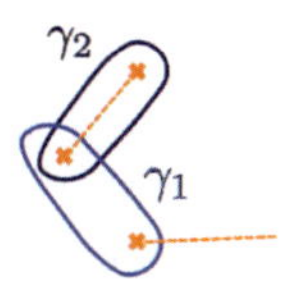

[16] These processes are captured by a more elaborate wall-crossing formula, the "2d-4d wall-crossing formula" described in [20], which is a hybrid between the Cecotti-Vafa [6] and Kontsevich-Soibelman [36] formulas.

Definition 3.6.7 (Charge of a saddle connection or ring domain). For each saddle connection with phase ϑ one can define a class $\gamma \in \Gamma$, represented by a lift to Σ_u of a loop around the saddle connection. The orientation of γ is fixed by the condition that $e^{-i\vartheta} \oint_\gamma \lambda$ is a negative real number. We call γ the *charge* of the saddle connection. Likewise for each ring domain of ϕ_2 with phase ϑ there is a charge $\gamma \in \Gamma$ given by the sum of the two lifts of a generic curve in the ring, with opposite orientations, chosen such that $e^{-i\vartheta} \oint_\gamma \lambda$ is negative.

Physics Fact 3.6.8 (Central charges of 4d particles). A particle with electromagnetic charge $\gamma \in \Gamma$ has central charge given by

$$Z_\gamma = \oint_\gamma \lambda, \tag{3.6.9}$$

where λ denotes the standard Liouville 1-form "$p\,dq$" on T^*C.

Note that a saddle connection with charge γ always has phase $\vartheta = \arg(-Z_\gamma)$. Physics Fact 3.6.3 above motivates the following definition:

Definition 3.6.10 (BPS indices in theories $T_4[\mathfrak{sl}_2, C]$). For $\mathfrak{g} = \mathfrak{sl}_2$, fixed C and ϕ_2 we define

$$\Omega(\gamma) = (\#\text{ saddle conn. of charge }\gamma) - 2(\#\text{ ring domains of charge }\gamma). \tag{3.6.11}$$

Example 3.6.12 (BPS indices in Argyres-Douglas theory). In Example 3.5.1, the BPS indices at $u = 1$, $\Lambda = 1$ are [23, 47]

$$\Omega(\gamma) = \begin{cases} 1 & \text{for } \gamma \in \{\gamma_1, \gamma_2, \gamma_1 + \gamma_2, -\gamma_1, -\gamma_2, -\gamma_1 - \gamma_2\}, \\ 0 & \text{otherwise,} \end{cases} \tag{3.6.13}$$

while at $u = 0$, $\Lambda = 1$ they are

$$\Omega(\gamma) = \begin{cases} 1 & \text{for } \gamma \in \{\gamma_1, \gamma_2, -\gamma_1, -\gamma_2\}, \\ 0 & \text{otherwise.} \end{cases} \tag{3.6.14}$$

This is another example of a wall-crossing phenomenon, where the index that jumps is the 4d index $\Omega(\gamma)$ as a function of u. These jumps are governed by the Kontsevich-Soibelman formula [36] (in this case and also in general for 4d $\mathcal{N} = 2$ theories; see e.g. [8, 21, 23, 37] for more about this).

Remark 3.6.15 (Charge conjugation symmetry). Every saddle connection with phase ϑ is also a saddle connection with phase $\vartheta + \pi$, and similarly for ring domains. This fact implies a symmetry

$$\Omega(\gamma) = \Omega(-\gamma). \tag{3.6.16}$$

Remark 3.6.17 (BPS indices as Donaldson-Thomas invariants). BPS indices in $\mathcal{N} = 2$ supersymmetric QFTs have various structural properties in common with generalized Donaldson-Thomas invariants associated to 3-Calabi-Yau categories — in particular, they obey the same wall-crossing formula. It is natural to conjecture that for each $\mathcal{N} = 2$ theory there is some associated 3-Calabi-Yau category,

[12] B. Dubrovin, *Geometry of 2D topological field theories*, Integrable systems and quantum groups (Montecatini Terme, 1993), Lecture Notes in Math., vol. 1620, Springer, Berlin, 1996, pp. 120–348, DOI 10.1007/BFb0094793. MR1397274 ↑452, 459

[13] D. Dumas and A. Neitzke, *Asymptotics of Hitchin's metric on the Hitchin section*, Comm. Math. Phys. **367** (2019), no. 1, 127–150, DOI 10.1007/s00220-018-3216-7. MR3933406 ↑474

[14] A. Fenyes, *A dynamical perspective on shear-bend coordinates*, available at `1510.05757v2`. ↑467

[15] V. Fock and A. Goncharov, *Moduli spaces of local systems and higher Teichmüller theory*, Publ. Math. Inst. Hautes Études Sci. **103** (2006), 1–211, DOI 10.1007/s10240-006-0039-4. MR2233852 ↑473

[16] L. Fredrickson, *Exponential decay for the asymptotic geometry of the Hitchin metric*, Comm. Math. Phys. **375** (2020), no. 2, 1393–1426, DOI 10.1007/s00220-019-03547-9. MR4083884 ↑474

[17] D. Gaiotto, $\mathrm{N} = 2$ *dualities*, J. High Energy Phys. **8** (2012), 034, front matter + 57, DOI 10.1007/JHEP08(2012)034. MR3006961 ↑464

[18] D. Gaiotto, *Surface operators in* $\mathcal{N} = 2$ *4d gauge theories*, J. High Energy Phys. **11** (2012), 090, front matter + 24, DOI 10.1007/JHEP11(2012)090. MR3036474 ↑465

[19] D. Gaiotto, *Opers and TBA* (2014), available at `1403.6137`. ↑473

[20] D. Gaiotto, G. W. Moore, and A. Neitzke, *Wall-crossing in coupled 2d-4d systems*, J. High Energy Phys. **12** (2012), 082, front matter + 166, DOI 10.1007/JHEP12(2012)082. MR3045271 ↑465, 469

[21] D. Gaiotto, G. W. Moore, and A. Neitzke, *Framed BPS states*, Adv. Theor. Math. Phys. **17** (2013), no. 2, 241–397. MR3250763 ↑470, 473

[22] D. Gaiotto, G. W. Moore, and A. Neitzke, *Spectral networks*, Ann. Henri Poincaré **14** (2013), no. 7, 1643–1731, DOI 10.1007/s00023-013-0239-7. MR3115984 ↑457, 465, 466, 471, 472, 473

[23] D. Gaiotto, G. W. Moore, and A. Neitzke, *Wall-crossing, Hitchin systems, and the WKB approximation*, Adv. Math. **234** (2013), 239–403, DOI 10.1016/j.aim.2012.09.027. MR3003931 ↑463, 464, 469, 470, 473

[24] D. Gaiotto, G. W. Moore, and A. Neitzke, *Spectral networks and snakes*, Ann. Henri Poincaré **15** (2014), no. 1, 61–141, DOI 10.1007/s00023-013-0238-8. MR3147409 ↑473

[25] D. Gaiotto, G. W. Moore, and E. Witten, *Algebra of the infrared: string field theoretic structures in massive* $\mathcal{N} = (2,2)$ *field theory in two dimensions* (2015), available at `1506.04087v1`. ↑454

[26] S. Gukov, A. Schwarz, and C. Vafa, *Khovanov-Rozansky homology and topological strings*, Lett. Math. Phys. **74** (2005), no. 1, 53–74, DOI 10.1007/s11005-005-0008-8. MR2193547 ↑446

[27] Q. Hao, L. Hollands, and A. Neitzke, *BPS states in the Minahan-Nemeschansky* E_7 *theory*, J. High Energy Phys. **4** (2020), 039, 19, DOI 10.1007/jhep04(2020)039. MR4097010 ↑466

[28] C. Hertling, *Frobenius manifolds and moduli spaces for singularities*, Cambridge Tracts in Mathematics, vol. 151, Cambridge University Press, Cambridge, 2002, DOI 10.1017/CBO9780511543104. MR1924259 ↑459

[29] C. Hertling, tt^* *geometry, Frobenius manifolds, their connections, and the construction for singularities*, J. Reine Angew. Math. **555** (2003), 77–161, DOI 10.1515/crll.2003.015. MR1956595 ↑462

[30] L. Hollands and A. Neitzke, *Spectral networks and Fenchel-Nielsen coordinates*, Lett. Math. Phys. **106** (2016), no. 6, 811–877, DOI 10.1007/s11005-016-0842-x. MR3500424 ↑472

[31] L. Hollands and A. Neitzke, *BPS states in the Minahan-Nemeschansky* E_6 *theory*, Comm. Math. Phys. **353** (2017), no. 1, 317–351, DOI 10.1007/s00220-016-2798-1. MR3638316 ↑466

[32] L. Hollands and A. Neitzke, *Exact WKB and abelianization for the* T_3 *equation*, Comm. Math. Phys. **380** (2020), no. 1, 131–186, DOI 10.1007/s00220-020-03875-1. MR4166626 ↑472, 473

[33] K. Hori, C. Y. Park, and Y. Tachikawa, *2d SCFTs from M2-branes*, JHEP **2013** (September 12, 2013), no. 11, available at `1309.3036v2`. ↑456, 466

[34] K. Iwaki and T. Nakanishi, *Exact WKB analysis and cluster algebras*, J. Phys. A **47** (2014), no. 47, 474009, 98, DOI 10.1088/1751-8113/47/47/474009. MR3280000 ↑473

[35] A. Klemm, W. Lerche, P. Mayr, C. Vafa, and N. Warner, *Self-dual strings and* $\mathrm{N} = 2$ *supersymmetric field theory*, Nuclear Phys. B **477** (1996), no. 3, 746–764, DOI 10.1016/0550-3213(96)00353-7. MR1413475 ↑469

[36] M. Kontsevich and Y. Soibelman, *Motivic Donaldson-Thomas invariants: summary of results*, Mirror symmetry and tropical geometry, Contemp. Math., vol. 527, Amer. Math. Soc., Providence, RI, 2010, pp. 55–89, DOI 10.1090/conm/527/10400. MR2681792 ↑469, 470

[37] J. Manschot, B. Pioline, and A. Sen, *Wall crossing from Boltzmann black hole halos*, J. High Energy Phys. **7** (2011), 059, 73, DOI 10.1007/JHEP07(2011)059. MR2875965 ↑470

[38] R. Mazzeo, J. Swoboda, H. Weiss, and F. Witt, *Ends of the moduli space of Higgs bundles*, Duke Math. J. **165** (2016), no. 12, 2227–2271, DOI 10.1215/00127094-3476914. MR3544281 ↑463, 474

[39] R. Mazzeo, J. Swoboda, H. Weiss, and F. Witt, *Asymptotic geometry of the Hitchin metric*, Comm. Math. Phys. **367** (2019), no. 1, 151–191, DOI 10.1007/s00220-019-03358-y. MR3933407 ↑474

[40] A. Mikhailov, *BPS states and minimal surfaces*, Nuclear Phys. B **533** (1998), no. 1-3, 243–274, DOI 10.1016/S0550-3213(98)00524-0. MR1660127 ↑469

[41] A. Mikhailov, N. Nekrasov, and S. Sethi, *Geometric realizations of BPS states in* N = 2 *theories*, Nuclear Phys. B **531** (1998), no. 1-3, 345–362, DOI 10.1016/S0550-3213(98)80001-1. MR1657861 ↑469

[42] A. Neitzke, *Notes on a new construction of hyperkahler metrics*, Homological mirror symmetry and tropical geometry, Lect. Notes Unione Mat. Ital., vol. 15, Springer, Cham, 2014, pp. 351–375, DOI 10.1007/978-3-319-06514-4_8. MR3330790 ↑474

[43] A. Neitzke, *Integral iterations for harmonic maps* (2017), available at 1704.01522. ↑474

[44] C. Sabbah, *Fourier-Laplace transform of a variation of polarized complex Hodge structure*, J. Reine Angew. Math. **621** (2008), 123–158, DOI 10.1515/CRELLE.2008.060. MR2431252 ↑462

[45] K. Saito, *On a linear structure of the quotient variety by a finite reflexion group*, Publ. Res. Inst. Math. Sci. **29** (1993), no. 4, 535–579, DOI 10.2977/prims/1195166742. MR1245441 ↑459

[46] N. Seiberg, *Notes on theories with* 16 *supercharges*, Nuclear Phys. B Proc. Suppl. **67** (1998), 158–171, DOI 10.1016/S0920-5632(98)00128-5. String theory, gauge theory and quantum gravity (Trieste, 1997). MR1628375 ↑464

[47] A. D. Shapere and C. Vafa, *BPS structure of Argyres-Douglas superconformal theories* (1999), available at hep-th/9910182. ↑469, 470

[48] I. Smith, *Quiver algebras as Fukaya categories*, Geom. Topol. **19** (2015), no. 5, 2557–2617, DOI 10.2140/gt.2015.19.2557. MR3416110 ↑471

[49] K. Strebel, *Quadratic differentials*, Ergebnisse der Mathematik und ihrer Grenzgebiete (3) [Results in Mathematics and Related Areas (3)], vol. 5, Springer-Verlag, Berlin, 1984, DOI 10.1007/978-3-662-02414-0. MR743423 ↑467

[50] E. Witten, *Constraints on supersymmetry breaking*, Nuclear Phys. B **202** (1982), no. 2, 253–316, DOI 10.1016/0550-3213(82)90071-2. MR668987 ↑442

[51] E. Witten, *Supersymmetry and Morse theory*, J. Differential Geometry **17** (1982), no. 4, 661–692 (1983). MR683171 ↑442, 444

[52] E. Witten, *On the structure of the topological phase of two-dimensional gravity*, Nuclear Phys. B **340** (1990), no. 2-3, 281–332, DOI 10.1016/0550-3213(90)90449-N. MR1068086 ↑452

[53] E. Witten, *Some comments on string dynamics*, Strings '95 (Los Angeles, CA, 1995), World Sci. Publ., River Edge, NJ, 1996, pp. 501–523. MR1660736 ↑464

[54] E. Witten, *Geometric Langlands from six dimensions*, A celebration of the mathematical legacy of Raoul Bott, CRM Proc. Lecture Notes, vol. 50, Amer. Math. Soc., Providence, RI, 2010, pp. 281–310, DOI 10.1090/crmp/050/23. MR2648898 ↑464

[55] E. Witten, *Analytic continuation of Chern-Simons theory*, Chern-Simons gauge theory: 20 years after, AMS/IP Stud. Adv. Math., vol. 50, Amer. Math. Soc., Providence, RI, 2011, pp. 347–446. MR2809462 ↑460

[56] E. Witten, *Fivebranes and knots*, Quantum Topol. **3** (2012), no. 1, 1–137, DOI 10.4171/QT/26. MR2852941 ↑446

[57] K. Zarembo, *Supersymmetric Wilson loops*, Nuclear Phys. B **643** (2002), no. 1-3, 157–171, DOI 10.1016/S0550-3213(02)00693-4. MR1936917 ↑445

Department of Mathematics, PO Box 208283, New Haven, CT 06520-8283
Email address: andrew.neitzke@yale.edu

Published Titles in This Series

28 **Daniel S. Freed, Sergei Gukov, Ciprian Manolescu, Constantin Teleman, and Ulrike Tillmann, Editors,** Quantum Field Theory and Manifold Invariants, 2021

27 **Carlos E. Kenig, Fang Hua Lin, Svitlana Mayboroda, and Tatiana Toro, Editors,** Harmonic Analysis and Applications, 2020

26 **Alexei Borodin, Ivan Corwin, and Alice Guionnet, Editors,** Random Matrices, 2019

25 **Michael W. Mahoney, John C. Duchi, and Anna C. Gilbert, Editors,** The Mathematics of Data, 2018

24 **Roman Bezrukavnikov, Alexander Braverman, and Zhiwei Yun, Editors,** Geometry of Moduli Spaces and Representation Theory, 2017

23 **Mark J. Bowick, David Kinderlehrer, Govind Menon, and Charles Radin, Editors,** Mathematics and Materials, 2017

22 **Hubert L. Bray, Greg Galloway, Rafe Mazzeo, and Natasa Sesum, Editors,** Geometric Analysis, 2016

21 **Mladen Bestvina, Michah Sageev, and Karen Vogtmann, Editors,** Geometric Group Theory, 2014

20 **Benson Farb, Richard Hain, and Eduard Looijenga, Editors,** Moduli Spaces of Riemann Surfaces, 2013

19 **Hongkai Zhao, Editor,** Mathematics in Image Processing, 2013

18 **Cristian Popescu, Karl Rubin, and Alice Silverberg, Editors,** Arithmetic of L-functions, 2011

17 **Jeffery McNeal and Mircea Mustaţă, Editors,** Analytic and Algebraic Geometry, 2010

16 **Scott Sheffield and Thomas Spencer, Editors,** Statistical Mechanics, 2009

15 **Tomasz S. Mrowka and Peter S. Ozsváth, Editors,** Low Dimensional Topology, 2009

14 **Mark A. Lewis, Mark A. J. Chaplain, James P. Keener, and Philip K. Maini, Editors,** Mathematical Biology, 2009

13 **Ezra Miller, Victor Reiner, and Bernd Sturmfels, Editors,** Geometric Combinatorics, 2007

12 **Peter Sarnak and Freydoon Shahidi, Editors,** Automorphic Forms and Applications, 2007

11 **Daniel S. Freed, David R. Morrison, and Isadore Singer, Editors,** Quantum Field Theory, Supersymmetry, and Enumerative Geometry, 2006

10 **Steven Rudich and Avi Wigderson, Editors,** Computational Complexity Theory, 2004

9 **Brian Conrad and Karl Rubin, Editors,** Arithmetic Algebraic Geometry, 2001

8 **Jeffrey Adams and David Vogan, Editors,** Representation Theory of Lie Groups, 2000

7 **Yakov Eliashberg and Lisa Traynor, Editors,** Symplectic Geometry and Topology, 1999

6 **Elton P. Hsu and S. R. S. Varadhan, Editors,** Probability Theory and Applications, 1999

5 **Luis Caffarelli and Weinan E, Editors,** Hyperbolic Equations and Frequency Interactions, 1999

4 **Robert Friedman and John W. Morgan, Editors,** Gauge Theory and the Topology of Four-Manifolds, 1998

3 **János Kollár, Editor,** Complex Algebraic Geometry, 1997

2 **Robert Hardt and Michael Wolf, Editors,** Nonlinear partial differential equations in differential geometry, 1996

1 **Daniel S. Freed and Karen K. Uhlenbeck, Editors,** Geometry and Quantum Field Theory, 1995